VOLUME FIVE HUNDRED AND EIGHTY

METHODS IN ENZYMOLOGY

Peptide, Protein and Enzyme Design

METHODS IN ENZYMOLOGY

Editors-in-Chief

ANNA MARIE PYLE
Departments of Molecular, Cellular and Developmental Biology and Department of Chemistry Investigator, Howard Hughes Medical Institute Yale University

DAVID W. CHRISTIANSON
Roy and Diana Vagelos Laboratories Department of Chemistry University of Pennsylvania Philadelphia, PA

Founding Editors

SIDNEY P. COLOWICK and NATHAN O. KAPLAN

VOLUME FIVE HUNDRED AND EIGHTY

METHODS IN
ENZYMOLOGY

Peptide, Protein and Enzyme Design

Edited by

VINCENT L. PECORARO
Department of Chemistry
University of Michigan, Ann Arbor
MI, United States

AMSTERDAM • BOSTON • HEIDELBERG • LONDON
NEW YORK • OXFORD • PARIS • SAN DIEGO
SAN FRANCISCO • SINGAPORE • SYDNEY • TOKYO
Academic Press is an imprint of Elsevier

Academic Press is an imprint of Elsevier
50 Hampshire Street, 5th Floor, Cambridge, MA 02139, United States
525 B Street, Suite 1800, San Diego, CA 92101-4495, United States
The Boulevard, Langford Lane, Kidlington, Oxford OX5 1GB, United Kingdom
125 London Wall, London, EC2Y 5AS, United Kingdom

First edition 2016

Copyright © 2016 Elsevier Inc. All rights reserved.

No part of this publication may be reproduced or transmitted in any form or by any means, electronic or mechanical, including photocopying, recording, or any information storage and retrieval system, without permission in writing from the publisher. Details on how to seek permission, further information about the Publisher's permissions policies and our arrangements with organizations such as the Copyright Clearance Center and the Copyright Licensing Agency, can be found at our website: www.elsevier.com/permissions.

This book and the individual contributions contained in it are protected under copyright by the Publisher (other than as may be noted herein).

Notices

Knowledge and best practice in this field are constantly changing. As new research and experience broaden our understanding, changes in research methods, professional practices, or medical treatment may become necessary.

Practitioners and researchers must always rely on their own experience and knowledge in evaluating and using any information, methods, compounds, or experiments described herein. In using such information or methods they should be mindful of their own safety and the safety of others, including parties for whom they have a professional responsibility.

To the fullest extent of the law, neither the Publisher nor the authors, contributors, or editors, assume any liability for any injury and/or damage to persons or property as a matter of products liability, negligence or otherwise, or from any use or operation of any methods, products, instructions, or ideas contained in the material herein.

ISBN: 978-0-12-805380-5
ISSN: 0076-6879

For information on all Academic Press publications
visit our website at https://www.elsevier.com/

Publisher: Zoe Kruze
Acquisition Editor: Zoe Kruze
Editorial Project Manager: Helene Kabes
Production Project Manager: Magesh Kumar Mahalingam
Cover Designer: Maria Ines Cruz

Typeset by SPi Global, India

CONTENTS

Contributors — xiii
Preface — xvii

1. Chemical Posttranslational Modification with Designed Rhodium(II) Catalysts — 1
S.C. Martin, M.B. Minus, and Z.T. Ball

1. Introduction — 2
2. Synthesis of Rhodium(II) Conjugates as Protein Modification Catalysts — 4
3. Modification of an SH3 Domain and Gel-Based Visualization Thereof — 11

References — 18

2. Cell-Binding Assays for Determining the Affinity of Protein–Protein Interactions: Technologies and Considerations — 21
S.A. Hunter and J.R. Cochran

1. Introduction — 22
2. General Binding Theory and Relevance of K_d — 23
3. General Pitfalls in Cell-Based Binding Assays — 25
4. Measuring Binding on the Surface of Yeast — 30
5. Measuring Binding on the Surface of Mammalian Cells — 35
6. Other Methods of Measuring Binding: Kinetic Exclusion Assay and Surface Plasmon Resonance — 37
7. Summary — 42

Acknowledgments — 42
References — 42

3. Protein and Antibody Engineering by Phage Display — 45
J.C. Frei and J.R. Lai

1. Introduction — 46
2. Equipment — 48
3. Materials — 50
4. Phage Display and Library Design — 57
5. Library Production — 66
6. Library Selections and Screening — 73

References — 85

4. Incorporation of Unnatural Amino Acids into Proteins Expressed in Mammalian Cells   89
R. Serfling and I. Coin

1. Introduction   90
2. Fluorescence Assay for Straightforward Evaluation of Uaa-Incorporating Systems   92
3. Incorporation of Uaas into Membrane Proteins   99
4. Conclusion   104
References   104

5. Method for Enzyme Design with Genetically Encoded Unnatural Amino Acids   109
C. Hu and J. Wang

1. Introduction   110
2. The General Procedure for Enzyme Design with Unnatural Amino Acids   112
3. Unnatural Amino Acid Design   112
4. Synthetic Chemistry-Guided Unnatural Amino Acid Design   116
5. Synthetic Methods for Unnatural Amino Acids   117
6. Aminoacyl-tRNA Synthetase Screening   125
7. Protein Design   127
8. Protein Expression   128
9. Conclusion and Perspective   129
References   130

6. Methods for Solving Highly Symmetric De Novo Designed Metalloproteins: Crystallographic Examination of a Novel Three-Stranded Coiled-Coil Structure Containing D-Amino Acids   135
L. Ruckthong, J.A. Stuckey, and V.L. Pecoraro

1. Introduction   136
2. Materials   140
3. Methods   141
Acknowledgments   146
References   146

7. SpyRings Declassified: A Blueprint for Using Isopeptide-Mediated Cyclization to Enhance Enzyme Thermal Resilience   149
C. Schoene, S.P. Bennett, and M. Howarth

1. Introduction   150
2. Isopeptide-Mediated Enzyme Cyclization   151
3. Characterization of Enzyme Resilience After SpyRing Cyclization   159

4. Concluding Remarks	164
Acknowledgments	164
References	164

8. Engineering and Application of LOV2-Based Photoswitches — 169
S.P. Zimmerman, B. Kuhlman, and H. Yumerefendi

1. Introduction	170
2. Engineering LOV2-Based Photoswitches	171
3. Using LOV2-Based Photoswitches	181
4. Summary and Perspectives	188
References	188

9. Minimalist Design of Allosterically Regulated Protein Catalysts — 191
O.V. Makhlynets and I.V. Korendovych

1. Introduction	192
2. Methods	193
Acknowledgments	201
References	201

10. Combining Design and Selection to Create Novel Protein–Peptide Interactions — 203
E.B. Speltz, N. Sawyer, and L. Regan

1. Introduction	204
2. Structure-Guided Rational Design	206
3. Split Fluorescent Protein Assays for Identifying PPIs	207
4. Validation	216
5. Future Perspective: Modular Design of PPI	218
6. Conclusion	218
References	219

11. Metal-Directed Design of Supramolecular Protein Assemblies — 223
J.B. Bailey, R.H. Subramanian, L.A. Churchfield, and F.A. Tezcan

1. Introduction	224
2. Metal-Directed Protein Self-Assembly	227
3. Metal-Templated Interface Redesign	231
4. Methods	237
5. Conclusion	247
Acknowledgments	247
References	247

12. Designing Fluorinated Proteins — 251
E.N.G. Marsh

1. Introduction — 252
2. Methods — 265
Acknowledgments — 272
References — 272

13. Solid Phase Synthesis of Helically Folded Aromatic Oligoamides — 279
S.J. Dawson, X. Hu, S. Claerhout, and I. Huc

1. Introduction — 280
2. Materials — 285
3. Synthetic Protocols — 286
4. Characterization of Foldamers — 297
References — 299

14. Conformational Restriction of Peptides Using Dithiol Bis-Alkylation — 303
L. Peraro, T.R. Siegert, and J.A. Kritzer

1. Introduction — 304
2. Using Thiol Alkylation to Constrain Peptides — 309
3. Protocols for Peptide Synthesis and Cross-Linking — 312
4. Applications, Tips, and Troubleshooting — 316
5. Future Directions — 327
References — 327

15. Engineering Short Preorganized Peptide Sequences for Metal Ion Coordination: Copper(II) a Case Study — 333
L.M.P. Lima and O. Iranzo

1. Introduction — 334
2. Design of Preorganized Peptidic Scaffolds for Metal Ion Coordination — 340
3. Synthesis and Characterization of the Peptidic Scaffolds — 341
4. Preparation of Analytical Stock Solutions — 345
5. Study of the Metal Ion Coordination Properties — 349
6. Concluding Remarks — 360
Acknowledgments — 360
References — 361

16. De Novo Construction of Redox Active Proteins — 365
C.C. Moser, M.M. Sheehan, N.M. Ennist, G. Kodali, C. Bialas, M.T. Englander, B.M. Discher, and P.L. Dutton

1. Introduction — 366
2. Sequence Selection: Binary Patterning — 367
3. Sequence Selection: Cofactor Self-Assembly — 370
4. Binding Heme Redox Cofactors — 371
5. Electron-Transfer Reactions of Heme Maquettes — 374
6. Binding Light-Activated Zn Tetrapyrrole Cofactors — 377
7. Iron–Sulfur Cluster Cofactors — 378
8. Redox Metal Binding Sites — 379
9. Flavin Cofactor Binding and Electron Transfer — 380
10. Translating Aqueous Redox Maquette Designs into Membranes — 382

References — 384

17. Design Strategies for Redox Active Metalloenzymes: Applications in Hydrogen Production — 389
R. Alcala-Torano, D.J. Sommer, Z. Bahrami Dizicheh, and G. Ghirlanda

1. Introduction — 390
2. Design of FeS Clusters — 392
3. Design of [FeFe]-Hydrogenase Mimics — 399
4. Design of Porphyrin Redox Sites — 404
5. Conclusion and Future Outlook — 409

References — 411

18. Equilibrium Studies of Designed Metalloproteins — 417
B.R. Gibney

1. Introduction — 418
2. Metalloprotein Thermodynamics — 419
3. Essential Equilibrium Measurements — 426
4. Thermodynamic Analysis of a Heme Binding Four Helix Bundle, $[\Delta 7\text{-His}]_2$ — 430
5. Experimental Procedures — 435
6. Summary — 436

Acknowledgment — 437

References — 437

19. Reconstitution of Heme Enzymes with Artificial Metalloporphyrinoids — 439
K. Oohora and T. Hayashi

1. Introduction — 440
2. Design and Synthesis of Artificial Metalloporphyrinoids — 442
3. Reconstitution of Hemoproteins — 446
4. Representative Characteristics of Reconstituted Hemoproteins — 451
Acknowledgments — 453
References — 453

20. Creation of a Thermally Tolerant Peroxidase — 455
Y. Watanabe and H. Nakajima

1. Introduction — 456
2. Molecular Design for Cyt c_{552} to Acquire Peroxidase Activity — 458
3. Gene Constructs of Recombinant Wild Type and Mutant Cyt c_{552} — 459
4. Expression and Purification of the Recombinant Proteins — 460
5. Thermal Stability of the Cyt c_{552} Mutants — 461
6. Analysis of a Key Intermediate in the Peroxidase Reaction of Cyt c_{552} V49D/M69A Mutant — 462
7. Effect of Trp45 on the Active Intermediate of Mutants — 465
References — 468

21. Designing Covalently Linked Heterodimeric Four-Helix Bundles — 471
M. Chino, L. Leone, O. Maglio, and A. Lombardi

1. Introduction — 472
2. Selection of the Best Docking Hotspot Given a Predefined Anchor Bolt — 477
3. Broadening the Hotspot and Linker Selections — 486
4. Concluding Remarks — 493
Acknowledgments — 494
References — 495

22. Design of Heteronuclear Metalloenzymes — 501
A. Bhagi-Damodaran, P. Hosseinzadeh, E. Mirts, J. Reed, I.D. Petrik, and Y. Lu

1. Step 1: Computational Design of Heteronuclear Metal-Binding Sites in Mb or CcP — 505
2. Step 2: Purification and Structural Characterization of Rationally Designed Proteins — 513
3. Step 3: Functional Characterization of Designed Heterobinuclear Metalloenzymes — 518

4.	Step 4: Further Improvement of Designed Heteronuclear Metalloenzymes: Case Studies	527
5.	Conclusions	529
	Acknowledgments	529
	References	529

23. Periplasmic Screening for Artificial Metalloenzymes — 539
M. Jeschek, S. Panke, and T.R. Ward

1.	Introduction	540
2.	Anchoring Strategies	541
3.	Increasing the Throughput	543
4.	Advantages of Periplasmic Screening	544
5.	Periplasmic Screening for ArMs	546
6.	Summary	553
	References	554

24. De Novo Designed Imaging Agents Based on Lanthanide Peptides Complexes — 557
A.F.A. Peacock

1.	Introduction	558
2.	Lanthanide Coordination Chemistry	558
3.	Lanthanide Imaging Agents	558
4.	Lanthanide Peptides and Proteins	560
5.	Strategies for Designing CCs for Lanthanide Coordination	562
6.	Conclusions and Perspectives	576
	Acknowledgments	576
	References	577

25. Peptide Binding for Bio-Based Nanomaterials — 581
N.M. Bedford, C.J. Munro, and M.R. Knecht

1.	Introduction	582
2.	Materials	585
3.	Methods	587
4.	Conclusions	596
	Acknowledgments	596
	References	596

Author Index — *599*
Subject Index — *639*

CONTRIBUTORS

R. Alcala-Torano
School of Molecular Sciences, Arizona State University, Tempe, AZ, United States

Z. Bahrami Dizicheh
School of Molecular Sciences, Arizona State University, Tempe, AZ, United States

J.B. Bailey
University of California, San Diego, La Jolla, CA, United States

Z.T. Ball
Rice University, Houston, TX, United States

N.M. Bedford
National Institute of Standards and Technology, Boulder, CO, United States

S.P. Bennett
Sekisui Diagnostics UK Ltd., Maidstone, Kent, United Kingdom

A. Bhagi-Damodaran
University of Illinois at Urbana-Champaign, Urbana, IL, United States

C. Bialas
University of Pennsylvania, Philadelphia, PA, United States

M. Chino
University of Napoli Federico II, Napoli, Italy

L.A. Churchfield
University of California, San Diego, La Jolla, CA, United States

S. Claerhout
Université de Bordeaux, CBMN (UMR5248); CNRS, CBMN (UMR5248), Institut Européen de Chimie et Biologie, Pessac, France

J.R. Cochran
Stanford University, Stanford, CA, United States

I. Coin
Institute of Biochemistry, Leipzig University, Leipzig, Germany

S.J. Dawson
Université de Bordeaux, CBMN (UMR5248); CNRS, CBMN (UMR5248), Institut Européen de Chimie et Biologie, Pessac, France

B.M. Discher
University of Pennsylvania, Philadelphia, PA, United States

P.L. Dutton
University of Pennsylvania, Philadelphia, PA, United States

M.T. Englander
University of Pennsylvania, Philadelphia, PA, United States

N.M. Ennist
University of Pennsylvania, Philadelphia, PA, United States

J.C. Frei
Albert Einstein College of Medicine, Bronx, NY, United States

G. Ghirlanda
School of Molecular Sciences, Arizona State University, Tempe, AZ, United States

B.R. Gibney
Brooklyn College, Brooklyn; Ph.D. Programs in Chemistry and Biochemistry, The Graduate Center of the City University of New York, New York, NY, United States

T. Hayashi
Department of Applied Chemistry, Graduate School of Engineering, Osaka University, Suita, Japan

P. Hosseinzadeh
University of Illinois at Urbana-Champaign, Urbana, IL, United States

M. Howarth
University of Oxford, Oxford, United Kingdom

C. Hu
Laboratory of RNA Biology, Institute of Biophysics, Chinese Academy of Sciences, Beijing, China

X. Hu
Université de Bordeaux, CBMN (UMR5248); CNRS, CBMN (UMR5248), Institut Européen de Chimie et Biologie, Pessac, France

I. Huc
Université de Bordeaux, CBMN (UMR5248); CNRS, CBMN (UMR5248), Institut Européen de Chimie et Biologie, Pessac, France

S.A. Hunter
Stanford University, Stanford, CA, United States

O. Iranzo
Aix Marseille Université, Centrale Marseille, CNRS, iSm2 UMR 7313, Marseille, France

M. Jeschek
ETH Zurich, Basel, Switzerland

M.R. Knecht
University of Miami, Coral Gables, FL, United States

G. Kodali
University of Pennsylvania, Philadelphia, PA, United States

I.V. Korendovych
Syracuse University, Syracuse, NY, United States

J.A. Kritzer
Tufts University, Medford, MA, United States

B. Kuhlman
University of North Carolina Chapel Hill, Chapel Hill, NC, United States

J.R. Lai
Albert Einstein College of Medicine, Bronx, NY, United States

L. Leone
University of Napoli Federico II, Napoli, Italy

L.M.P. Lima
Instituto de Tecnologia Química e Biológica António Xavier, Universidade Nova de Lisboa, Lisbon, Portugal

A. Lombardi
University of Napoli Federico II, Napoli, Italy

Y. Lu
University of Illinois at Urbana-Champaign, Urbana, IL, United States

O. Maglio
University of Napoli Federico II; Institute of Biostructures and Bioimages-IBB, CNR, Napoli, Italy

O.V. Makhlynets
Syracuse University, Syracuse, NY, United States

E.N.G. Marsh
University of Michigan, Ann Arbor, MI, United States

S.C. Martin
Rice University, Houston, TX, United States

M.B. Minus
Rice University, Houston, TX, United States

E. Mirts
University of Illinois at Urbana-Champaign, Urbana, IL, United States

C.C. Moser
University of Pennsylvania, Philadelphia, PA, United States

C.J. Munro
University of Miami, Coral Gables, FL, United States

H. Nakajima
Graduate School of Science, Osaka City University, Osaka, Japan

K. Oohora
Department of Applied Chemistry; Frontier Research Base for Global Young Researchers, Graduate School of Engineering, Osaka University, Suita; PRESTO, Japan Science and Technology Agency, Kawaguchi, Japan

S. Panke
ETH Zurich, Basel, Switzerland

A.F.A. Peacock
School of Chemistry, University of Birmingham, Edgbaston, United Kingdom

V.L. Pecoraro
Department of Chemistry, University of Michigan, Ann Arbor, MI, United States

L. Peraro
Tufts University, Medford, MA, United States

I.D. Petrik
University of Illinois at Urbana-Champaign, Urbana, IL, United States

J. Reed
University of Illinois at Urbana-Champaign, Urbana, IL, United States

L. Regan
Department of Chemistry, Yale University, New Haven, CT, United States

L. Ruckthong
Department of Chemistry, University of Michigan, Ann Arbor, MI, United States

N. Sawyer
Department of Chemistry, Yale University, New Haven, CT, United States

C. Schoene
University of Oxford, Oxford, United Kingdom

R. Serfling
Institute of Biochemistry, Leipzig University, Leipzig, Germany

M.M. Sheehan
University of Pennsylvania, Philadelphia, PA, United States

T.R. Siegert
Tufts University, Medford, MA, United States

D.J. Sommer
School of Molecular Sciences, Arizona State University, Tempe, AZ, United States

E.B. Speltz
Department of Chemistry, Yale University, New Haven, CT, United States

J.A. Stuckey
Life Sciences Institute, University of Michigan, Ann Arbor, MI, United States

R.H. Subramanian
University of California, San Diego, La Jolla, CA, United States

F.A. Tezcan
University of California, San Diego, La Jolla, CA, United States

J. Wang
Laboratory of RNA Biology, Institute of Biophysics, Chinese Academy of Sciences, Beijing, China

T.R. Ward
University of Basel, Basel, Switzerland

Y. Watanabe
Research Center of Materials Science, Nagoya University, Nagoya, Japan

H. Yumerefendi
University of North Carolina Chapel Hill, Chapel Hill, NC, United States

S.P. Zimmerman
University of North Carolina Chapel Hill, Chapel Hill, NC, United States

PREFACE

The diversity of structures found for peptides and proteins in the biosphere is overwhelming with folds as simple as α-helical coiled coils to far more complex membrane conformations. In order to understand biological function at the most basic level, one needs a coherent description of the factors that lead to stable protein structures and, preferably, the ability to construct systems using the basic principles of chemistry and biophysics. It is this latter objective that drives much of the research in the field of protein design. The field benefited greatly from advances in chemistry, biophysics, and molecular biology that occurred in the 1980s. Automated peptide, protein, and nucleic acid synthesis combined with the development of cloning, mutagenesis, and expression of genes made preparation of new sequences routine. At the same time, expanded computer capability, combined with access to synchrotron radiation sources and high-field NMR and mass spectrometers, provided a previously unavailable level of determining sequence, structure, and, most recently, a computational basis for understanding protein structure and dynamics. These advances have led to a broad range of systems that define a significant element of modern synthetic biology.

Peptide, protein, and enzyme design may take on many different flavors. One branch of the field focuses on selective pressure to evolve new activities for proteins, enhancing existing catalysis or stabilizing already efficient systems. This field of molecular evolution has had remarkable impact for arriving rapidly at desired chemical transformations (McIntosh, Heel, Buller, Chio, & Arnold, 2015). A drawback, however, is that "directed evolution" (even if in a test tube) drives the modifications so that intermediate steps that provide insight on function are often missed. Nonetheless, combining computational efforts in conjunction with this directed evolution can now be used to identify key components of catalysis (Dodani et al., 2016). While directed evolution is an important component of the protein reengineering toolkit, there are other treatises that cover this area extensively, so it is not generally reflected in this compendium.

This does not mean that reengineering proteins are not contained herein as numerous methods are presented on this topic. In fact, this can be done by making mutations to existing proteins (Hu, Chan, Sawyer, Yu, & Wang, 2014; Nakajima et al., 2008), adding cofactors to systems that previously did not include such catalytic functionality (Dürrenberger & Ward, 2014; Miner et al., 2012), making protein fusions that provide photoaddressable proteins that localize in membranes (Hallett, Zimmerman, Yumerefendi, Bear, & Kuhlman, 2016), altering cofactors to serve a different purpose, or attaching side chains (Hayashi, Sano, & Onoda, 2015) or metal binding sites to confer new protein–protein interactions (PPIs) (Speltz, Nathan, & Regan, 2015) or added stability (Salgado, Faraone-Mennella, & Tezcan, 2007; Schoene, Fierer, Bennett, & Howarth, 2014). These PPIs are critical for molecular recognition, but often are difficult to measure in their native environments so that new methodologies have been developed to quantify these interactions in vivo (Cherf & Cochran, 2015). Often, proteins are reengineered to enhance stability and so we examine this general strategy as well. A major limitation for the development of novel designed systems has been the limitation of amino acids that can be incorporated into proteins by bacterial expression. For this reason, advances for employing noncoded or even nonnatural amino acids are discussed (Coin, Perrin, Vale, & Wang, 2011; Hu et al., 2014). Additionally, libraries of antibodies or other proteins are often made as libraries so methods are needed to display these products (Frei, Kielian, & Lai, 2015).

While protein engineering relies on a previously evolved protein scaffold, de novo protein design approaches this topic from the most rudimentary levels by attempting to prepare existing or unknown protein folds using first principles of biophysics. DeGrado described the approach well (Bryson et al., 1995; DeGrado, Wasserman, & Lear, 1989) as a peptidic or protein system that is not directly related to any sequence found in nature yet folds into a predicted structure and/or carries out desired reactions. Because of its more straightforward rules for folding (Hartmann et al., 2016), α-helical coiled coils have been one of the major starting points for designed proteins with systems as simple as models for the two-stranded coiled coil GCN4 to elaborate multicomponent structures (Woolfson, Bartlett, Bruning, & Thomson, 2012). Computational methods, the most famous of these being Rosetta (Rohl, Strauss, Misura, & Baker, 2004), both for reengineering proteins and de novo design have appeared (Mitra, Shultis, & Zhang, 2013). While the present issue does not describe the computational aspects in detail, many workers have used these methodologies to develop functional designed scaffolds. There has been effort expended on using nonnatural amino acids to stabilize these systems (Buer, Meagher, Stuckey, & Marsh, 2012). A significant portion of the text is devoted to these de novo systems with additions of related cyclic peptides (Bock, Gavenonis, & Kritzer, 2013) and more recently prepared amide foldamers that mimic protein structure (Kudo, Maurizot, Kauffmann, Tanatani, & Huc, 2013).

It is believed that roughly a third to one-half of native proteins bind metals for their function, their structure, or to transport these ions to or within cells (Thomson & Gray, 1998; Waldron & Robinson, 2009). Thus, it is appropriate that in this volume roughly the same proportion of articles address methods to design metalloproteins (Yu, Penner-Hahn, & Pecoraro, 2013; Yu et al., 2014). In studies such as these one must not only consider the factors that determine a specific secondary structure or PPI, but must also address the thermodynamics, kinetics, and dynamics of metal ion interactions with the peptides and proteins (Farrer & Pecoraro, 2003; Ghosh & Pecoraro, 2004). Thus, one must determine how to place metal ligands in desired constructs in a way that will satisfy the bonding preferences of the metal (ligand type and number and coordination geometry) without destabilizing the existing scaffold. Further complicating these studies is the need to encapsulate metals in more than one oxidation state, a requirement that is essential for electron transfer proteins and redox active catalysts. In essence such multiple redox systems present the problem of simultaneously building two metalloproteins representing each metal oxidation level that must then be merged into one. In some instances, workers have expended effort to develop de novo designed scaffolds as simple coiled coils, sometimes called maquettes (Zhang et al., 2011), and others have taken a redesign strategy. Both approaches are represented in this collection. In terms of systems investigated, many design studies attempt to mimic or reproduce electron transfer (Plegaria, Herrero, Quaranta, & Pecoraro, 2015; Roy et al., 2014; Zhang et al., 2011) and structural (Fragoso et al., 2015; Lombardi et al., 2000; Mocny & Pecoraro, 2015; Plegaria, Duca, et al., 2015) or catalytic function (Cangelosi, Deb, Penner-Hahn, & Pecoraro, 2014; Dürrenberger & Ward, 2014; Miner et al., 2014; Nakajima et al., 2008; Reig et al., 2012; Tegoni, Yu, Berseloni, Penner-Hahn, & Pecoraro, 2012; Zastrow, Peacock, Stuckey, & Pecoraro, 2012). However, this field is not restricted to biomimics as new catalysis is explored by building unique cofactors (Dürrenberger & Ward, 2014; Hayashi et al., 2015; Popp & Ball, 2010). Another area of great interest is the ability of a protein to recognize another protein or an inanimate surface. New methods for linking proteins together using metal-directed protein self-assembly (Salgado et al., 2007) or exploring protein interactions with inorganic surfaces (Bedford et al., 2015) are burgeoning fields. There are even new strategies for obtaining clinically relevant MRI or luminescent imaging agents using designed proteins containing rare-earth ions (Berwick et al., 2014). In all these areas, one will find chapters herein that provide detailed insight on the methods necessary to achieve these types of exciting and heretofore unknown metalloproteins.

As one can see from this Preface, the field of protein design is vast; therefore, it is impossible to present the broad range of methods in a single, limited volume. Fortunately, other volumes of *Methods in Enzymology* and related series provide additional information that the reader may find helpful. It is my hope that the sampling of different techniques represented here will assist individuals presently looking for alternative strategies to solve problems associated with their protein design projects. In particular, I will consider this project a success if new researchers are inspired and assisted by these chapters to develop their own programs in this field.

V.L. PECORARO
University of Michigan, Ann Arbor, MI, United States

REFERENCES

Bedford, N. M., Ramezani-Dakhel, H., Slocik, J. M., Briggs, B. D., Ren, Y., Frenkel, A. I., et al. (2015). Elucidation of peptide-directed palladium surface structure for biologically tunable nanocatalysts. *ACS Nano*, 9, 5082–5092.

Berwick, M. R., Lewis, D. J., Jones, A. W., Parslow, R. A., Dafforn, T. R., Cooper, H. J., et al. (2014). De novo design of Ln(III) coiled coils for imaging applications. *Journal of the American Chemical Society*, 136, 1166–1169.

Bock, J. E., Gavenonis, J., & Kritzer, J. A. (2013). Getting in shape: Controlling peptide bioactivity and bioavailability using conformational constraints. *ACS Chemical Biology*, 8(3), 488–499.

Bryson, J. W., Betz, S. F., Lu, H. S., Suich, D. J., Zhou, H. X., O'Neil, K. T., et al. (1995). Protein design: A hierarchic approach. *Science*, 270, 935–941.

Buer, B. C., Meagher, J. L., Stuckey, J. A., & Marsh, E. N. G. (2012). Structural basis for the enhanced stability of highly fluorinated proteins. *Proceedings of the National Academy of Sciences of the United States of America*, 109(13), 4810–4815.

Cangelosi, V., Deb, A., Penner-Hahn, J. E., & Pecoraro, V. L. (2014). A de novo designed metalloenzyme for the hydration of CO_2. *Angewandte Chemie*, 53, 7900–7903.

Cherf, G. M., & Cochran, J. R. (2015). Applications of yeast surface display for protein engineering. *Methods in Molecular Biology (Clifton, N.J.)*, 1319, 155–175.

Coin, I., Perrin, M. H., Vale, W. W., & Wang, L. (2011). Photo-cross-linkers incorporated into G-protein-coupled receptors in mammalian cells: A ligand comparison. *Angewandte Chemie International Edition in English*, 50, 8077–8081.

DeGrado, W. F., Wasserman, Z. R., & Lear, J. D. (1989). Protein design, a minimalist approach. *Science*, 243, 622–628.

Dodani, S. C., Kiss, G., Cahn, J. K. B., Su, Y., Pande, V. S., & Arnold, F. H. (2016). Discovery of a regioselectivity switch in nitrating P450s guided by molecular dynamics simulations and Markov models. *Nature Chemistry*, 8, 419–425.

Dürrenberger, M., & Ward, T. R. (2014). Recent achievements in the design and engineering of artificial metalloenzymes. *Current Opinions in Chemical Biology*, 19, 99–106.

Farrer, B. T., & Pecoraro, V. L. (2003). Kinetics of Hg(II) encapsulation by a weakly associated three-stranded coiled coil. *Proceedings of the National Academy of Sciences of the United States of America*, 100, 3760–3765.

Fragoso, A., Carvalho, T., Rousselot-Pailley, P., Correia dos Santos, M. M., Delgado, R., & Iranzo, O. (2015). Effect of the peptidic scaffold in copper(II) coordination and the redox properties of short histidine-containing peptides. *Chemistry—A European Journal, 21*, 13100–13111.

Frei, J. C., Kielian, M., & Lai, J. R. (2015). Comprehensive mapping of functional epitopes on dengue virus glycoprotein E DIII for binding to broadly neutralizing antibodies 4E11 and 4E5A by phage display. *Virology, 485*, 371–382.

Ghosh, D., & Pecoraro, V. L. (2004). Studying metalloprotein folding using de novo peptide design. *Inorganic Chemistry, 43*(25), 7902–7915.

Hallett, R. A., Zimmerman, S. P., Yumerefendi, H., Bear, J. E., & Kuhlman, B. (2016). Correlating in vitro and in vivo activities of light-inducible dimers: A cellular optogenetics guide. *ACS Synthetic Biology, 5*(1), 53–64.

Hartmann, M. D., Mendler, C. T., Bassler, J., Karamichali, I., Ridderbusch, O., Lupas, A. N., et al. (2016). α/β Coiled coils. *eLife, 5*, e11861.

Hayashi, T., Sano, Y., & Onoda, A. (2015). Generation of new artificial metalloproteins by cofactor modification of native hemoproteins. *Israel Journal of Chemistry, 55*, 76–84.

Hu, C., Chan, S. I., Sawyer, E. B., Yu, Y., & Wang, J. (2014). Metalloprotein design using genetic code expansion. *Chemical Society Reviews, 43*, 6498–6510.

Kudo, M., Maurizot, V., Kauffmann, B., Tanatani, A., & Huc, I. (2013). Folding of a linear array of α-amino acids within a helical aromatic oligoamide frame. *Journal of the American Chemical Society, 135*, 9628–9631.

Lombardi, A., Summa, C. M., Geremia, S., Randaccio, L., Pavone, V., & DeGrado, W. F. (2000). Retrostructural analysis of metalloproteins: Application to the design of a minimal model for diiron proteins. *Proceedings of the National Academy of Sciences of the United States of America, 97*, 6298–6305.

McIntosh, J. A., Heel, T., Buller, A. R., Chio, L., & Arnold, F. H. (2015). Structural adaptability facilitates histidine heme ligation in a cytochrome P450. *Journal of the American Chemical Society, 137*, 13861–13865.

Miner, K. D., Mukherjee, A., Gao, Y.-G., Null, E. L., Petrik, I. D., Zhao, X., et al. (2012). A designed functional metalloenzyme that reduces O_2 to H_2O with over one thousand turnovers. *Angewandte Chemie, 51*, 5589–5592.

Mitra, P., Shultis, D., & Zhang, Y. (2013). EvoDesign: De novo protein design based on structural and evolutionary profiles. *Nucleic Acids Research, 41*, W273–W280.

Mocny, C. S., & Pecoraro, V. L. (2015). De novo protein design as a methodology for synthetic bioinorganic chemistry. *Accounts of Chemical Research, 48*, 2388–2396.

Nakajima, H., Ichikawa, Y., Satake, Y., Takatani, N., Manna, S. K., Rajbongshi, J., et al. (2008). Engineering of Thermus thermophilus cytochrome c552: Thermally tolerant artificial peroxidase. *ChemBioChem, 9*, 2954–2957.

Plegaria, J. S., Duca, M., Tard, C., Deb, A., Penner-Hahn, J. E., & Pecoraro, V. L. (2015). De novo design of copper metallopeptides inspired by native cupredoxins. *Inorganic Chemistry, 54*, 9470–9482.

Plegaria, J. S., Herrero, C., Quaranta, A., & Pecoraro, V. L. (2015). Electron transfer activity of a de novo designed copper center in a three-helix bundle fold. *Biochimica et Biophysica Acta, 1857*, 522–530.

Popp, B. V., & Ball, Z. T. (2010). Structure-selective modification of aromatic side chains with dirhodium metallopeptide catalysts. *Journal of the American Chemical Society, 132*, 6660–6662.

Reig, A. J., Pires, M. M., Snyder, R. A., Wu, Y., Jo, H., Kulp, D. W., et al. (2012). Alteration of the oxygen-dependent reactivity of de novo Due Ferri proteins. *Nature Chemistry, 4*, 900–906.

Rohl, C. A., Strauss, C. E. M., Misura, K. M. S., & Baker, D. (2004). Protein structure prediction using Rosetta. *Methods in Enzymology, 383*, 66–93.

Roy, A., Sommer, D. J., Schmidt, R., Brown, C., Gust, D., Astashkine, A., et al. (2014). A de novo designed 2[4Fe-4S] ferredoxin mimic mediates electron transfer. *Journal of the American Chemical Society, 136*, 17343–17349.

Salgado, E. N., Faraone-Mennella, J., & Tezcan, F. A. (2007). Controlling protein-protein interactions through metal coordination: Assembly of a 16-helix bundle protein. *Journal of the American Chemical Society, 129*, 13374–13375.

Schoene, C., Fierer, J. O., Bennett, S. P., & Howarth, M. (2014). SpyTag/SpyCatcher cyclization confers resilience to boiling on a mesophilic enzyme. *Angewandte Chemie, 53*, 6101–6104.

Speltz, E. B., Nathan, A., & Regan, L. (2015). Design of protein–peptide interaction modules for assembling supramolecular structures in vivo and in vitro. *ACS Chemical Biology, 10*(9), 2108–2115.

Tegoni, M., Yu, F., Berseloni, M., Penner-Hahn, J. E., & Pecoraro, V. L. (2012). Designing functional type 2 copper centers within alpha-helical coiled coils that have nitrite reductase activity. *Proceedings of the National Academy of Sciences of the United States of America, 109*, 21234–21239.

Thomson, A. J., & Gray, H. B. (1998). Bio-inorganic chemistry. *Current Opinion in Chemical Biology, 2*, 155–158.

Waldron, K. J., & Robinson, N. J. (2009). How do bacterial cells ensure that metalloproteins get the correct metal? *Nature Reviews. Microbiology, 7*(1), 25–35.

Woolfson, D. N., Bartlett, G. J., Bruning, M., & Thomson, A. R. (2012). New currency for old rope: From coiled-coil assemblies to α-helical barrels. *Current Opinion in Structural Biology, 22*, 432–441. Elsevier: New York.

Yu, F., Cangelosi, V., Zastrow, M. L., Tegoni, M., Plegaria, J., Tebo, A., et al. (2014). Protein design: Towards functional metalloenzymes. *Chemical Reviews, 114*, 3495–3578.

Yu, F., Penner-Hahn, J. E., & Pecoraro, V. L. (2013). De novo designed metallopeptides with type 2 copper centers: Modulation of reduction potentials and nitrite reductase activities. *Journal of the American Chemical Society, 135*, 18096–18107.

Zastrow, M. L., Peacock, A. F. A., Stuckey, J. A., & Pecoraro, V. L. (2012). Hydrolytic catalysis and structural stabilization in a designed metalloprotein. *Nature Chemistry, 4*, 118–123.

Zhang, L., Anderson, J. L. R., Ahmed, I., Norman, J. A., Negron, C., Mutter, A. C., et al. (2011). Manipulating cofactor binding thermodynamics in an artificial oxygen transport protein. *Biochemistry, 50*, 10254–10261.

CHAPTER ONE

Chemical Posttranslational Modification with Designed Rhodium(II) Catalysts

S.C. Martin, M.B. Minus, Z.T. Ball[1]
Rice University, Houston, TX, United States
[1]Corresponding author: e-mail address: zb1@rice.edu

Contents

1. Introduction — 2
2. Synthesis of Rhodium(II) Conjugates as Protein Modification Catalysts — 4
   2.1 Discussion — 4
   2.2 Materials and General Considerations — 5
   2.3 Protocol 1: Preparation of $Rh_2(OAc)_3(tfa)_1$ — 7
   2.4 Protocol 2: Preparation of Rhodium(II) Conjugates in Organic Solution — 8
   2.5 Protocol 3: Preparation of Rhodium(II) Conjugates in Aqueous Solution — 10
3. Modification of an SH3 Domain and Gel-Based Visualization Thereof — 11
   3.1 Discussion — 11
   3.2 Materials and General Considerations — 12
   3.3 Protocol 4: Rhodium(II)–Catalyzed Protein Modification — 14
   3.4 Protocol 5: Visualization of an Alkyne-Tagged Protein by Chemical Blotting — 15
References — 18

Abstract

Natural enzymes use molecular recognition to perform exquisitely selective transformations on nucleic acids, proteins, and natural products. Rhodium(II) catalysts mimic this selectivity, using molecular recognition to allow selective modification of proteins with a variety of functionalized diazo reagents. The rhodium catalysts and the diazo reactivity have been successfully applied to a variety of protein folds, the chemistry succeeds in complex environments such as cell lysate, and a simple protein blot method accurately assesses modification efficiency. The studies with rhodium catalysts provide a new tool to study and probe protein-binding events, as well as a new synthetic approach to protein conjugates for medical, biochemical, or materials applications.

1. INTRODUCTION

Site-specific enzymatic modification of complex protein structures is a routine and essential process in living systems. However, designed systems that achieve selective posttranslational modification are a daunting challenge. The design of reagents or catalysts that achieve site-selective protein modification represents an important fundamental challenge for chemists, necessitating robust and predictable reactions, broad functional-group tolerance in aqueous media, and subtle selectivity in a sea of identical functional groups. Selective protein modification is also an enabling tool for protein therapeutics, biomaterials engineering, and other fields. With a few exceptions, it has not been possible to alter or evolve natural enzymes—eg, by methylation, phosphorylation, acylation—to permit modification with new molecules of interest, such as drugs, fluorescent dyes, or reactive handles. Most successful demonstrations of designed protein modification, both chemical and enzymatic, rely on recombinant technology to install uniquely reactive tags of varying sizes or bioorthogonal unnatural amino acids.

Our approach to selective modification of natural proteins is based on rhodium(II)–ligand conjugates, in which a ligand for the desired protein target ensures transient localization of a rhodium(II) catalyst at the target protein (Fig. 1). Catalytic modification of amino-acid side chains by diazo reagents is possible at a diverse range of amino acids, which ensures that modification selectivity is driven by molecular recognition rather than the inherent chemoselectivity of the reagent or catalyst (Vohidov, Coughlin, & Ball, 2015). The benefits of this approach include a modest affinity requirement (20 μM ligands have been routinely successful in our

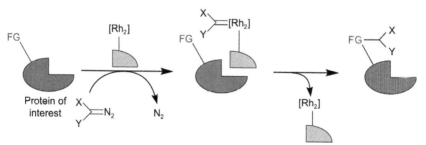

Fig. 1 A schematic depiction of diazo modification with rhodium(II) conjugates. Ligand binding localizes the catalytic rhodium(II) complex, allowing selective modification via a reactive metallocarbene intermediate.

hands), the ability to modify individual natural proteins in lysate, and success at low concentration of the molecular recognition motif (the rhodium catalyst), which minimizes nonselective reactivity and allows studies at biologically relevant concentrations.

From a conceptual standpoint, this wholly synthetic, designed system replicates key features of natural enzymes. Reactivity driven by molecular recognition makes it possible to modify relatively unreactive side chains (eg, Gln, Phe) in the presence of inherently far more reactive functionality (eg, Trp, Tyr). The functional-group tolerance rivals that of natural enzymes. We have also discovered design strategies for uniquely potent binding between a protein and a rhodium conjugate (Coughlin, Kundu, Cooper, & Ball, 2014; Vohidov, Knudsen, et al., 2015). This potent coordination to the rhodium complex blocks modification reactivity, providing a preliminary example of inhibitory regulation that is common in biological systems.

The method described here has been, in our hands, a robust and reliable one. Rhodium(II) conjugates catalyze modification of amino-acid side chains comprising about 40% of protein space, allowing for the potential modification of a large variety of protein domains (Popp & Ball, 2011). Successful side-chain substrates include aromatic (tryptophan, tyrosine, phenylalanine, and histidine), acidic (glutamate, aspartate), and polar side chains (glutamine, asparagine, serine, cysteine). We have demonstrated examples of modification driven by coiled–coil interactions, SH3–peptide assembly, and small molecule interactions with signal transducer and activator of transcription (STAT) proteins (Chen, Popp, Bovet, & Ball, 2011; Popp & Ball, 2010, 2011; Vohidov, Knudsen, et al., 2015). Modifications in lysate, even at natural abundance, are feasible. Recently, this method was used for affinity labeling of STAT3, enabling the discovery of a new small molecule binding site with functional consequences for human disease. A variety of diazo reagents are suitable, allowing introduction of diverse functional-group handles (Chen et al., 2011; Vohidov, Coughlin, & Ball, 2015).

This protocol also includes methods for the preparation of the complex rhodium(II) conjugates needed for this work. Traditional equilibration methods for the synthesis of rhodium(II) carboxylates are not suitable for the complex, functional-group-rich peptides and small molecules required for these studies (Bonge, Kaboli, & Hansen, 2010; Callot & Metz, 1985; Lifsey et al., 1987; Lou, Horikawa, Kloster, Hawryluk, & Corey, 2004; Pirrung & Zhang, 1992). Significant efforts were needed to develop a new synthesis—discussed later—building on the significant prior art of rhodium(II) tetracarboxylate preparation.

2. SYNTHESIS OF RHODIUM(II) CONJUGATES AS PROTEIN MODIFICATION CATALYSTS

2.1 Discussion

Rhodium(II) complexes commonly have four bridging, bidentate ligands, and a typical "paddlewheel" structure (Fig. 2). Tetracarboxylate complexes are the most common and the most widely studied in catalytic applications. Although some paddlewheel complexes can be accessed directly from rhodium(III) salts such as $RhCl_3$ (Kataoka, Yano, Kawamoto, & Handa, 2015; Legzdins, Mitchell, Rempel, Ruddick, & Wilkinson, 1970), tetracarboxylate complexes are traditionally prepared by ligand exchange from the commercially available parent complex, $Rh_2(OAc)_4$ (Kitchens & Bear, 1970; Lou et al., 2004). Thermal equilibration under forcing conditions with excess ligand allows formation of other homoleptic carboxylate complexes (Rh_2L_4), and many related complexes, such as tetraamidate complexes, are prepared similarly (Bonge et al., 2010; Callot & Metz, 1985; Kitchens & Bear, 1970; Lou et al., 2004; Pirrung & Zhang, 1992). The synthesis of heteroleptic complexes—those with nonidentical ligands—often requires separation of statistical mixtures from equilibration reactions (Callot & Metz, 1985; Lifsey et al., 1987; Lou, Remarchuk, & Corey, 2005).

Our protocol provides a uniquely effective solution to the many challenges presented by heteroleptic rhodium(II) complexes, especially for the functional-group-rich ligands, such as peptides, required for these studies. Displacement of trifluoroacetate (tfa) groups occurs selectively over that of acetate and other carboxylate and carboxamidate ligands under a variety of conditions (Lou et al.,

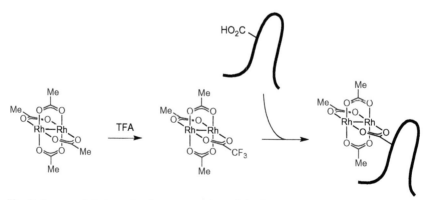

Fig. 2 A general strategy for the construction of rhodium(II) conjugates with bioactive peptides.

2005). The target complexes are thus easily prepared from appropriate mixed complexes, such as $Rh_2(OAc)_3(tfa)_1$ (Fig. 2). Depending on the individual case, two successful conditions for rhodium conjugate synthesis have been identified: mildly basic conditions in organic solvent (Section 2.4) and mildly acidic (pH 4–5) buffered aqueous conditions (Section 2.5).

The preparation of the requisite mixed complexes, $Rh_2(OAc)_x(tfa)_{(4-x)}$, is straightforward, with interesting reaction kinetics (see Fig. 3). It is most instructive to imagine ligand exchange for these neutral carboxylate complexes as a pseudodissociative process. Electron-poor ligands (such as trifluoroacetate) slow ligand exchange. As a result, it is possible to optimize for the formation of $Rh_2(OAc)_3(tfa)_1$ or cis-$Rh_2(OAc)_2(tfa)_2$ complexes, as each subsequent addition of trifluoroacetic acid (TFA) is slower than the previous. Formation of the cis isomer of $Rh_2(OAc)_2(tfa)_2$ is preferred due to the *trans* effect, which disfavors loss of the acetate ligand *trans* to the first displacement (Fig. 3). Accessing other useful precursor complexes, $Rh_2(OAc)_1(tfa)_3$ or trans-$Rh_2(OAc)_2(tfa)_2$, is possible from $Rh_2(tfa)_4$, though product selectivity can be more challenging, as each ligand exchange is now faster than the previous (Bear, Kitchens, & Willcott, 1971; Lou et al., 2005).

The method described here allows preparation of 3:1 *tris*-acetate conjugates via $Rh_2(OAc)_3(tfa)_1$. The final metalation method is general enough to allow preparation of other stoichiometries, including complexes $Rh_2(OAc)_2(L)_1$ or $Rh_2(L)_2$, where L is a chelating *bis*-carboxylic acid (Kundu et al., 2012; Sambasivan & Ball, 2010).

2.2 Materials and General Considerations
Materials

Rhodium(II) tetraacetate, $Rh_2(OAc)_4$ (Pressure Chemical, Pittsburgh, PA, #3745)

TFA (EMD Millipore, Billerica, MA, TX1275)

Diisopropylethylamine, DIEA (Acros, Geel, Belgium, #11522)

2-(N-morpholino)ethanesulfonic acid, MES (Sigma, St. Louis, MO, M3671)

18.2 MΩ water

TLC plates (EMD, Morris Plains, NJ, #1058050001)

Silica gel (SiliCycle, Quebec, Canada, R12030B)

Solvents (THF, MeCN, EtOAc, hexanes, 2,2,2-trifluoroethanol) were purchased from Fisher (Hampton, NH) and used as received.

Analytical reversed-phase high-performance liquid chromatography (RP-HPLC). Analytical column: Phenomenex Jupiter 4 μ Proteo 90 Å (250 mm × 4.6 μm); flow rate = 1.0 mL/min; semipreparative

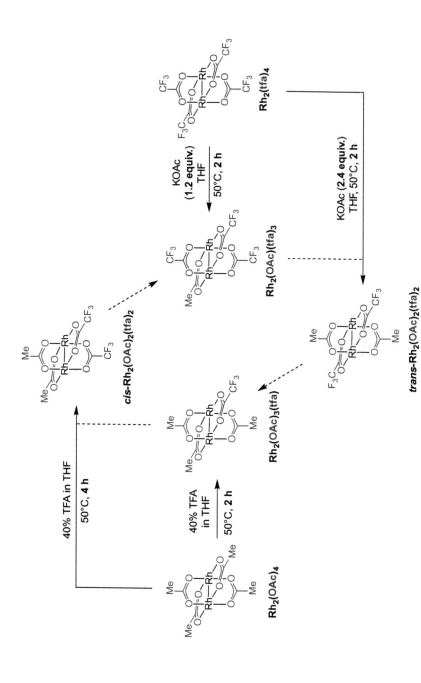

Fig. 3 Synthetic routes to the various $Rh_2(OAc)_x(tfa)_{(4-x)}$ complexes.

column: Phenomenex Jupiter 4 μ Proteo (250 mm × 15 μm); flow rate = 8.0 mL/min

HPLC Method A: 25-Min gradient of 25–90% MeCN/H$_2$O with 0.1% (v/v) TFA

HPLC Method B: 17-Min gradient of 20–80% MeCN/H$_2$O with 0.1% (v/v) TFA

HPLC Method C: 25-Min gradient of 10–60% MeCN/H$_2$O with 0.1% (v/v) TFA

HPLC Method D: 22-Min gradient of 20–50% MeCN/H$_2$O with 0.1% (v/v) TFA

2.2.1 Preparation of Metalation Buffer

MES (190 mg, 1.0 mmol) was dissolved in H$_2$O (8 mL) by sonication. The pH of the solution was adjusted to 4.7 with dilute NaOH and HCl. The solution was then diluted to 10.0 mL to bring the final MES concentration to 100 mM.

2.3 Protocol 1: Preparation of Rh$_2$(OAc)$_3$(tfa)$_1$

An oven-dried round-bottom flask was charged with Rh$_2$(OAc)$_4$ (1.0 g, 2.3 mmol) and THF (52 mL). The flask was swirled until the solid dissolved completely before TFA (34 mL, 440 mmol) and a PTFE (polytetrafluoroethylene)-coated magnetic stir bar were added. The reaction mixture was capped loosely and stirred in an oil bath at 55°C. The reaction was monitored by RP–HPLC (HPLC Method A, see Fig. 4 for details). After 2 h, HPLC showed that the amount of cis-Rh$_2$(OAc)$_2$(tfa)$_2$

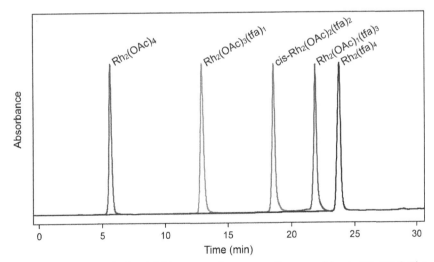

Fig. 4 The various Rh$_2$(OAc)$_x$(tfa)$_{(4-x)}$ species are readily separable via RP–HPLC. The analytical traces shown were acquired using HPLC Method A.

formed was approximately equal to that of $Rh_2(OAc)_4$ remaining. At that point, the reaction was quenched with MeCN (30 mL) and concentrated by rotary evaporation. The crude product was purified by silica chromatography, eluting with 1:2 EtOAc/hexanes. As expected, the products elute in reverse order to that observed by RP–HPLC. Fractions containing the desired product were combined, and the product was dried in vacuo to yield $Rh_2(OAc)_3(tfa)_1$ as a green powder (593 mg, 53%). ^1H NMR (600 MHz, CD_3CN) δ 1.84 (s, 6H), 1.83 (s, 3H). ^{13}C NMR (151 MHz, CD_3CN) δ 207.8, 194.0, 174.1 (q, J_{C-F} = 39 Hz), 111.9 (q, J_{C-F} = 285 Hz), 23.7, 23.6.

Notes

1. This reaction has been successfully run on 0.1–1 gram scale.
2. THF is the optimal solvent because the high solubility of $Rh_2(OAc)_4$ (and many other dirhodium complexes) therein. Ligand exchange in strongly coordinating solvents (eg, MeCN, DMSO, pyridine) is generally not successful.
3. Extending the reaction time to approximately 4 h allows efficient preparation of cis-$Rh_2(OAc)_2(tfa)_2$.
4. Most dirhodium conjugates, including the various $Rh_2(OAc)_x(tfa)_{(4-x)}$ products, are separable by RP–HPLC (see Fig. 4). Silica gel chromatography is an effective purification technique for preparative-scale reactions with simple, relatively nonpolar carboxylate ligands.
5. $Rh_2(tfa)_4$ is commercially available. We have found that commercial material can be contaminated with $Rh_2(OAc)_1(tfa)_3$. $Rh_2(tfa)_4$ can be prepared by heating a dirhodium precursor in neat TFA for 24–48 h.
6. Rhodium byproducts can be recovered and treated with neat acetic acid at elevated temperature to regenerate $Rh_2(OAc)_4$. Thus, all rhodium-containing waste in our lab is collected and recycled when several hundred milligrams have been collected.

2.4 Protocol 2: Preparation of Rhodium(II) Conjugates in Organic Solution

Rhodium(II) conjugates have been successfully synthesized under both organic and aqueous conditions. Because metalation reactions release acid, neutralization is necessary. The addition of 1 equiv. of base relative to the carboxylic acid is generally preferred in organic-phase reactions. In the presence of excess base, the dirhodium core can undergo base-catalyzed decomposition, giving the reaction mixture a yellow hue.

2.4.1 Preparation of a Metalated FKBP Inhibitor (Coughlin et al., 2014)

A 20-mL vial was charged with the carboxylic acid (12.0 mg, 26.1 µmol), $Rh_2(OAc)_3(tfa)_1$ (12.8 mg, 25.8 µmol), 2,2,2-trifluoroethanol (TFE, 15 mL), and a PTFE-coated magnetic stir bar. DIEA (12 µL, 65 µmol) was subsequently added, and the vessel was heated at 50°C. The reaction was monitored by analytical RP–HPLC (HPLC Method B). After 16 h HPLC showed consumption of the starting materials. The reaction mixture was reduced in volume in vacuo and purified by injection onto preparative RP–HPLC. The product fractions were combined and dried by lyophilization to afford the product as a blue solid (12.6 mg, 57%). ^{1}H NMR (400 MHz, CDCl$_3$) δ 7.61–7.20 (m, 10H), δ 5.75 (m, 1H), δ 4.68 (m, 1H), δ 4.36 (s, 2H), δ 3.51 (br s, 6H), δ 3.41 (m, 1H), δ 3.14 (m, 1H), δ 3.10 (m, 1H), δ 2.50 (s, 3H), δ 1.96 (s, 3H), δ 1.91 (s, 6H), δ 1.82–0.75 (m, 10H). ^{13}C NMR (125 MHz, CDCl$_3$) δ 194.2, δ 192.7, δ 192.5, δ 171.0, δ 140.2, δ 131.3, δ 129.5, δ 128.7, δ 128.3, δ 126.5, δ 115.8, δ 59.0, δ 56.0, δ 43.6, δ 36.9, δ 35.7, δ 28.1, δ 25.2, δ 24.0, δ 23.9, δ 21.8, δ 20.3, δ 3.58. IR (thin film): 3249, 2943, 2360, 2341, 1734, 1695, 1636, 1585, 1558, 1417, 1336, 1200, 1151, 1129, 968, 734, 699, 668, 539 cm^{-1}. ESI–MS, m/z: calculated for [M−Na]$^+$: 864.0, found: 864.0.

Notes
1. The first choice of solvent when running a metalation in organic media is TFE. TFE possesses high polarity, low nucleophilicity, and dissolves many rhodium(II) complexes and carboxylate ligands. Other solvents that have met some success include THF, MeOH, and acetone. As in the synthesis of $Rh_2(OAc)_x(tfa)_{(4-x)}$ complexes, strongly coordinating solvents should be avoided.
2. DIEA is the preferred base for metalations because it is highly soluble and weakly coordinating. Nonnucleophilic inorganic bases such as carbonate and phosphate have also proven suitable in some cases.

3. The molar ratio of the carboxylic acid to the dirhodium precursor can be changed to suit the needs of a particular reaction. A 10–20% excess of one reactant is usually sufficient to push the reaction closer to completion. Unique circumstances—such as the substitution of only one site of cis-Rh(OAc)$_2$(tfa)$_2$ or the metalation of a precious carboxylate—may benefit from a larger excess (2–3 equiv.) of one reagent.
4. Rhodium(II) ligation reactions are not moisture sensitive and can be performed open to the air.

2.5 Protocol 3: Preparation of Rhodium(II) Conjugates in Aqueous Solution

Rhodium conjugates can also be synthesized in buffered aqueous media. This method is required where (1) peptides or other polar molecules require aqueous (co)solvent for solubility or (2) the ligand contains primary or secondary amines, or other basic functional groups, which can lead to decomposition of the rhodium center under basic conditions. These syntheses are best performed at pH 4–5, a window which ensures that basic functional groups stay protonated while allowing deprotonation of carboxylic acid groups.

2.5.1 Preparation of the Metallopeptide S2E^{Rh}

Ac—V—E—L—A—R—R—P—L—P—P—L—P—N—NH$_2$
S2E
→ MES buffer, pH 4.7 →
Ac—V—E—L—A—R—R—P—L—P—P—L—P—N—NH$_2$
S2E^{Rh}

As described previously (Vohidov, Knudsen, et al., 2015), a 4-mL vial was charged with purified **S2E** peptide (4.63 mg, 3.06 μmol), Rh$_2$(OAc)$_3$(tfa)$_1$ (1.52 mg, 3.07 μmol), and MES buffer (1.0 mL) and equipped with a PTFE-coated magnetic stir bar. The reaction was stirred at 50°C overnight. After 16 h, analytical RP–HPLC (HPLC Method C) showed that the reactants were consumed. The mixture was then purified by preparative RP–HPLC (HPLC Method D). Fractions were combined and dried by lyophilization, affording the product as light blue solid (4.27 mg, 73%). ESI-MS, m/z: calculated for [M+2H]$^{2+}$: 948.4; found: 948.1. Purity was assessed by analytical HPLC. Solution concentrations were quantified by UV-visible spectroscopy ($\varepsilon^{586\,nm} = 207$ L mol^{-1} cm^{-1}).

Notes
1. These conditions were developed to permit basic peptide side chains (lysine, arginine). Depending on the functional groups in a particular compound, different pH ranges may be tolerated.
2. We have consistently employed MES as the buffering agent due to its buffer range that extends below pH 5 and its lack of metal-chelating capabilities or interfering functional groups (primary or secondary amines or carboxylic acids).
3. Typical concentrations for aqueous reactions are 2–20 μM. The limited aqueous solubility of rhodium(II) complexes will not typically permit higher concentrations.
4. Solubility is a common issue in aqueous metalations; reaction mixtures often require sonication to become homogenous. Up to 50% (v/v) MeOH, THF, acetone, or TFE have all been used successfully as cosolvents. At high proportions of organic cosolvents, care should be taken to ensure that buffer concentrations remain well above reagent concentrations. When full solubility is achieved, the solution pH should be checked and readjusted by addition of dilute HCl or NaOH.
5. As in organic solvent, a wide range of carboxylic acid/dirhodium complex ratios is tolerated. We have performed reactions from 1:3 to 2:1, depending on the cost and accessibility of the substrates.

3. MODIFICATION OF AN SH3 DOMAIN AND GEL-BASED VISUALIZATION THEREOF

3.1 Discussion

Rhodium conjugates catalyze the modification of target proteins at a specific residue near the ligand-binding site. The general approach requires a protein-binding molecule that can be conjugated to a rhodium(II) center via a carboxylate group.

The system tolerates a range of affinities for the protein target. Most of our studies have involved rhodium conjugates in the range of 50 nM to 50 μM. Successful modification based on catalysts with weak binding is a key advantage of our approach, and we have generally focused on the study of such weak, transient interactions. Conjugating a rhodium core to a known ligand for a given protein could, in theory, disrupt or block binding. In practice, we have found it straightforward to design suitable rhodium-attachment points, at least in cases where some SAR information is known. Subtle variation in attachment point (Vohidov, Knudsen, et al., 2015) can alter reaction efficiency, sometimes meaningfully (Vohidov, Coughlin, & Ball, 2015). We have generally verified that rhodium conjugates maintain binding similar to the parent ligand. Typical biophysical methods are suitable for this task, including fluorescence polarization and isothermal titration calorimetry. Care must be taken to ensure that the fluorescence-quenching and absorption properties of rhodium(II) complexes do not affect measurements.

Purified protein modification typically reaches completion at RT within 5 h. When protein modification is conducted in lysate, the complex environment requires some additional considerations. Moderately higher catalyst loading aids detection of the more dilute target protein. In-lysate reactions are usually best conducted at lower temperature (4°C) for a longer period of time (16 h). We have observed modifications at natural abundance, though overexpressed protein targets are easier to analyze.

Imaging modifications, especially within lysate or other protein mixtures, can be challenging. We initially developed a biotin–diazo reagent for avidin-based detection (Chen et al., 2011; Vohidov, Coughlin, & Ball, 2015). While rhodium conjugates catalyzing the modification can be used at low concentration, diazo concentrations of 100–1000 μM are commonplace. At these levels, biotin itself can have effects on the modification chemistry and leads to nonselective labeling. Fluorogenic probe development and the "chemical blotting" method fix these problems, providing a fast, cheap imaging method that does not require any bioreagents (Ohata, Vohidov, Aliyan, et al., 2015; Ohata, Vohidov, & Ball, 2015). The imaging technique described here has proven remarkably robust and valuable in our hands. Relative to enzyme and antibody methods, the cost savings are significant, and the reagents have indefinite shelf life.

3.2 Materials and General Considerations
Protein modification

Protein solution, purified or lysate
Rhodium(II) catalyst: prepared as described earlier (Section 2)
Diazo reagent: prepared according to the reported procedures (Antos & Francis, 2004; Vohidov, Coughlin, & Ball, 2015). Diazo compounds require special handling; see "Notes" later.
TBHA (*N*-(*tert*-butyl)hydroxylamine, TCI, Tokyo, Japan, B1700)
Tris (Thermo Scientific, Waltham, MA, #17926)
Sinapinic acid (Sigma, St. Louis, MO, #85429)
    Acetonitrile (Fisher, A998)
    18 MΩ water

SDS-PAGE
4X LDS sample buffer (Life Technologies, Carlsbad, CA, NP0007)
0.5 mL boilproof microtubes (Phenix Research Products, Chandler, NC, MAX-805)
12% *bis–tris* gel (Life Technologies, Carlsbad, CA, NP0343BOX)
MOPS buffer (3-(*N*-morpholino)propanesulfonic acid, Life Technologies, Carlsbad, CA, NP0001)
Electrophoresis chamber (Life Technologies, Carlsbad, CA, EI0001)
Power supply (Bio-Rad, Hercules, CA, Model 200/2.0)

Protein Transfer
Bicine (Fisher BioReagents, Hampton, NH, BP2646)
Bis–tris (Chem-Impex Int'l Inc., Wood Dale, IL, #00034)
EDTA (Fisher BioReagents, Hampton, NH, BP121)
Chlorobutanol (Sigma-Aldrich, St. Louis, MO, #112054)
NuPAGE Antioxidant (Life Technologies, Carlsbad, CA, NP0005)
Extra thick blot filtrate paper (Bio-Rad, Hercules, CA, #1703966)
Methanol (Fisher Chemical, Hampton, NH, A412)
Low-fluorescence PVDF (polyvinylidene fluoride) membrane (Azure Biosystems, Dublin, CA, AC2108)
Transfer cell (Bio-Rad, Hercules, CA, Trans-Blot SD Semi-Dry)
Power supply (Bio-Rad, Hercules, CA, model 200/2.0)

Chemical blotting
Methanol (Fisher, Hampton, NH, A412)
Petri dish (VWR, Radnor, PA, 25384-302 or 25384-326)
0.5-mL boilproof microtubes (Phenix Research, Chandler, NC, MAX-805)
1.5-mL microcentrifuge tubes (VWR, Radnor, PA, #20170)
Sodium ascorbate (Sigma-Aldrich, St. Louis, MO, A7631)
Tris[(1-benzyl-1*H*-1,2,3-triazol-4-yl)methyl]amine (TBTA, Click Chemistry Tools, Scottsdale, AZ, #1061)
Copper(II) sulfate (ACROS organics, Morris Plains, NJ, AC42361)
3-Azido-7-hydroxycoumarin (prepared by literature procedures, Sivakumar et al., 2004)
Dimethyl sulfoxide (Sigma-Aldrich, St. Louis, MO, #472301)
Parafilm M® (Parafilm, Oshkosh, WI, PM996)
Aluminum foil
200-Proof ethanol (VWR, Radnor, PA, V1001)
50-mL centrifuge tubes (Thermo Scientific, Waltham, MA, #339652)
Shaker (New Brunswick Scientific, Edison, NJ, C25KC)
Vortex (Scientific Industries, Bohemia, NY, S8223)

Imaging

Imager (Fujifilm, Tokyo, Japan, ImageQuant LAS-4000)
Forceps

3.2.1 Preparation of 1X Transfer Buffer

20X Transfer buffer was prepared by dissolving bicine (10.2 g, 25 mM), bis–tris (13.1 g, 25 mM), EDTA (0.75 g, 1 mM), and chlorobutanol (0.025 g, 0.05 mM) in H_2O (final volume 125 mL). Next, 1X transfer buffer was prepared by diluting 20X buffer (50 mL) with methanol (100 mL), water (849 mL), and NuPAGE antioxidant (1 mL).

3.2.2 Preparation of Protein Modification Buffer

The buffering agent TBHA (126 mg, 1.0 mmol) or tris (121 mg, 1.0 mmol) was dissolved in H_2O (8 mL). The solution pH was adjusted to 6.2 or 7.4, respectively, with dilute NaOH or HCl. The solution was then diluted to 10.0 mL to bring the buffer concentration to 100 mM.

3.2.3 Preparation of MALDI–MS Matrix

A 1:1 $MeCN/H_2O$ solution containing 0.1% TFA was saturated with sinapinic acid and centrifuged. The supernatant was spotted 1:1 with the protein sample for MALDI–MS analysis.

3.3 Protocol 4: Rhodium(II)–Catalyzed Protein Modification

3.3.1 Modification of SH3 Domain of Yes Kinase in Mammalian Cell Lysate (Vohidov, Coughlin, & Ball, 2015)

A 600-μL Eppendorf tube was charged with 25 μL of PC-3 lysate (1 mg/mL total protein), 23 μL of TBHA buffer, 1.5 μL of 500 μM **R5D^{Rh}**

metallopeptide (Vohidov, Coughlin, & Ball, 2015), and 0.5 µL of 50-mM alkyne–diazo solution (total reaction volume = 50.0 µL. Final concentrations: protein, 0.5 mg/mL; catalyst, 15 µM; diazo, 0.5 mM). The reaction tube was vortexed for 30 s and gently rocked at 4°C for 16 h. Reaction progress is conveniently monitored by removing 5.0-µL aliquots, quenching with 2.0 µL MeCN, and subjecting to MALDI and/or gel blot analysis.

Notes
1. Diazo compounds must be handled with care. They are moderately unstable at room temperature and when exposed to light. Exposure to light is minimized during synthesis, purification, and handling by covering flasks with aluminum foil and dimming overhead lights. Rotary evaporation is conducted without heating. For storage and convenient use, the diazo reagents are divided into 200-µL solutions in t-BuOH (25–50 mM), which are individually frozen for storage to minimize freeze/thaw cycles.
2. Freshly prepared diazo solutions exhibit a bright orange color. Over time the color of solution becomes pale yellow, indicating decomposition (Clark et al., 2002; Robertson & Sharp, 1984). We generally prepare diazo compounds on a modest scale (5–10 mg) which is sufficient for use over many weeks.
3. Higher concentrations of diazo reagent can improve sluggish reactions.
4. The nature of the amino acid targeted will affect the optimal conditions. In our hands optimal tryptophan reactivity occurs at pH 6.2. Histidine modification is optimal at pH 7.4 due to the basicity of the imidazole side-chain functionality.
5. Temperature effects are small, and ambient temperature is usually suitable.
6. Reaction parameters (eg, catalyst, catalyst loading, diazo concentration, pH, buffer, temperature) can be screened effectively by MALDI–MS, as shown in Fig. 5.
7. This protocol describes modification of an individual protein in lysate. Modifications of purified proteins are typically conducted with protein concentrations of 5–10 µM, catalyst concentrations of 1–5 µM, and diazo concentrations of 0.5–1.0 mM.

3.4 Protocol 5: Visualization of an Alkyne-Tagged Protein by Chemical Blotting

Analysis of modification products of large (>10 kDa) pure SH3 domain fusion proteins and in-lysate reactions is accomplished using the chemical blotting protocol outlined later (Fig. 6). This protocol does not require purification or enrichment steps. After SDS-PAGE, the resolved protein bands

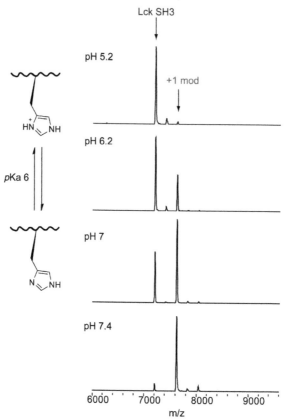

Fig. 5 MALDI–MS analysis of modification reactions of Lck SH3 with 11LRh metallopeptide. Reactions run parallel under identical conditions varying pH. Any of the various reaction parameters can be optimized in this way.

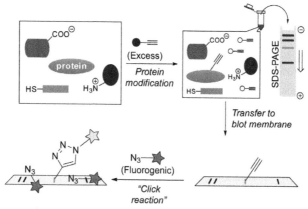

Fig. 6 Schematic description of chemical blotting.

are transferred onto the surface of a PVDF membrane. A copper-catalyzed alkyne–azide cycloaddition reaction is performed directly on the surface of the membrane using a fluorogenic azide that reports on alkyne-modified proteins. Additional discussion and analysis of this chemical blotting method are available elsewhere (Ohata, Vohidov, et al., 2015).

3.4.1 Fluorescent "Chemical Blot" Analysis of a Modified SH3 Domain
SDS-PAGE
1. The protein modification reaction mixture was loaded onto an SDS-PAGE gel and run for 45 min, following manufacturer's protocol.
2. Electrophoresis was conducted using MOPS buffer at 200 V for 50 min.
    Note: Exposure to light during electrophoresis is minimized due to the photosensitivity of fluorophores and azides.
3. The top (comb) and bottom (below the dye) of the gel were cut out.

Protein transfer

4. A PVDF membrane was preactivated in MeOH for 5 min then rinsed with water twice.
    Notes: PVDF membranes are used because they withstand the organic solvents needed for the click reaction and subsequent rinsing step. The membrane must be kept clean; dirt or other contamination will affect the fluorescence imaging results.
5. The membrane, gel, and filtrate papers were soaked in cold 1X transfer buffer for 5–10 min.
6. The membrane and the gel were sandwiched with the filtrate paper in the transfer cell, and bubbles were rolled out.
7. The sandwich was wetted by pipetting cold transfer buffer thereon. Excess buffer was wiped away.
8. The electrotransfer was conducted at 24 V for 30 min.
    Note: Voltage and time for the transfer vary depending on the protein of interest.

Chemical blotting

9. The membrane was soaked in fresh MeOH for two 5-min periods and was then rinsed twice with water.
10. The membrane was immersed in a Petri dish containing approximately 20 mL of the click reaction mixture: a 50% DMSO/H_2O solution containing 3-azido-7-hydroxycoumarin (500 μM), TBTA (20 μM), sodium ascorbate (5 mM), and $CuSO_4$ (500 μM).
11. The Petri dish was sealed with Parafilm, and whole dish was covered with aluminum foil.
12. The mixture was shaken at 70–100 rpm for 1 h at room temperature.

Wash and imaging

13. After the reaction, the membrane was rinsed thrice with 70% aq. EtOH to remove any DMSO, causing it to become opaque.
14. The membrane was imaged with a Fujifilm LAS-4000 instrument using the epi-UV light source (370 nm LED), L41 filter set, and a 4-s exposure time.

    Notes: Because the PVDF membrane is opaque, it is necessary to use epi-mode irradiation. The wavelength settings should complement the fluorophore being used, and exposure time can be optimized for individual membranes. If the fluorescence of unreacted alkyne molecules at the bottom of the membrane hampers acquisition of clear images, the problematic section can be cut out of the membrane.
15. The membrane was then air dried for 30 min and stained with a Ponceau S solution to obtain a total protein image.

    Notes: Incomplete drying will cause the membrane itself to stain. No fluorescence measurements can be made after Ponceau S staining. Membranes can be stored in a 50% EtOH solution for many days before imaging.
16. In order to assess the transfer efficiency and extent of protein modification, an alkyne-modified protein of known concentration can be loaded as an internal standard on the SDS-PAGE gel and quantified by fluorescence intensity.

REFERENCES

Antos, J. M., & Francis, M. B. (2004). Selective tryptophan modification with rhodium carbenoids in aqueous solution. *Journal of the American Chemical Society, 126*, 10256–10257.

Bear, J. L., Kitchens, J., & Willcott, M. R., III. (1971). A kinetic study of the reaction of rhodium(II) acetate with trifluoroacetic acid. *Journal of Inorganic and Nuclear Chemistry, 33*, 3479–3486.

Bonge, H. T., Kaboli, M., & Hansen, T. (2010). Rh(II) catalysts with 4-hydroxyproline-derived ligands. *Tetrahedron Letters, 51*, 5375–5377.

Callot, H. J., & Metz, F. (1985). Rhodium(II) 2,4,6-triarylbenzoates: Improved catalysts for the syn cyclopropanation of Z-olefins. *Tetrahedron, 41*, 4495–4501.

Chen, Z., Popp, B. V., Bovet, C. L., & Ball, Z. T. (2011). Site-specific protein modification with a dirhodium metallopeptide catalyst. *ACS Chemical Biology, 6*, 920–925.

Clark, J. D., Shah, A. S., Peterson, J. C., Patelis, L., Kersten, R. J. A., Heemskerk, A. H., et al. (2002). The thermal stability of ethyl diazoacetate. *Thermochimica Acta, 386*, 65–72.

Coughlin, J. M., Kundu, R., Cooper, J. C., & Ball, Z. T. (2014). Inhibiting prolyl isomerase activity by hybrid organic–inorganic molecules containing rhodium(II) fragments. *Bioorganic & Medicinal Chemistry Letters, 24*, 5203–5206.

Kataoka, Y., Yano, N., Kawamoto, T., & Handa, M. (2015). Isolation of a tetranuclear intermediate complex in the synthesis of paddlewheel-type dirhodium tetraacetate. *European Journal of Inorganic Chemistry, 2015*, 5650–5655.

Kitchens, J., & Bear, J. L. (1970). The thermal decomposition of some rhodium(II) carboxylate complexes. *Thermochimica Acta, 1*, 537–544.

Kundu, R., Cushing, P. R., Popp, B. V., Zhao, Y., Madden, D. R., & Ball, Z. T. (2012). Hybrid organic–inorganic inhibitors of a PDZ–peptide interaction that regulates CFTR endocytic fate. *Angewandte Chemie, International Edition, 51*, 7217–7220.

Legzdins, P., Mitchell, R. W., Rempel, G. L., Ruddick, J. D., & Wilkinson, G. (1970). The protonation of ruthenium- and rhodium-bridged carboxylates and their use as homogeneous hydrogenation catalysts for unsaturated substances. *Journal of the Chemical Society A: Inorganic, Physical, Theoretical*, 3322–3326.

Lifsey, R. S., Lin, X. Q., Chavan, M. Y., Ahsan, M. Q., Kadish, K. M., & Bear, J. L. (1987). Reaction of rhodium(II) acetate with N-phenylacetamide: Substitution products and geometric isomers. *Inorganic Chemistry, 26*, 830–836.

Lou, Y., Horikawa, M., Kloster, R. A., Hawryluk, N. A., & Corey, E. J. (2004). A new chiral Rh(II) catalyst for enantioselective [2+1]-cycloaddition. Mechanistic implications and applications. *Journal of the American Chemical Society, 126*, 8916–8918.

Lou, Y., Remarchuk, T. P., & Corey, E. J. (2005). Catalysis of enantioselective [2+1]-cycloaddition reactions of ethyl diazoacetate and terminal acetylenes using mixed-ligand-complexes of the series Rh2(RCO2)n (L*4-n). Stereochemical heuristics for ligand exchange and catalyst synthesis. *Journal of the American Chemical Society, 127*, 14223–14230.

Ohata, J., Vohidov, F., Aliyan, A., Huang, K., Martí, A. A., & Ball, Z. T. (2015a). Luminogenic iridium azide complexes. *Chemical Communications, 51*, 15192–15195.

Ohata, J., Vohidov, F., & Ball, Z. T. (2015b). Convenient analysis of protein modification by chemical blotting with fluorogenic "click" reagents. *Molecular BioSystems, 11*, 2846–2849.

Pirrung, M. C., & Zhang, J. (1992). Asymmetric dipolar cycloaddition reactions of diazocompounds mediated by a binaphtholphosphate rhodium catalyst. *Tetrahedron Letters, 33*, 5987–5990.

Popp, B. V., & Ball, Z. T. (2010). Structure-selective modification of aromatic side chains with dirhodium metallopeptide catalysts. *Journal of the American Chemical Society, 132*, 6660–6662.

Popp, B. V., & Ball, Z. T. (2011). Proximity-driven metallopeptide catalysis: Remarkable side-chain scope enables modification of the Fos bZip domain. *Chemical Science, 2*, 690–695.

Robertson, I. R., & Sharp, J. T. (1984). A study of periselectivity in the thermal cyclisation reactions of diene-conjugated diazo compounds: 1,7-cyclisation as a route to 3H-1,2-diazepines and 1,5-cyclisation leading to new rearrangement reactions of 3H-pyrazoles. *Tetrahedron, 40*, 3095–3112.

Sambasivan, R., & Ball, Z. T. (2010). Metallopeptides for asymmetric dirhodium catalysis. *Journal of the American Chemical Society, 132*, 9289–9291.

Sivakumar, K., Xie, F., Cash, B. M., Long, S., Barnhill, H. N., & Wang, Q. (2004). A fluorogenic 1,3-dipolar cycloaddition reaction of 3-azidocoumarins and acetylenes. *Organic Letters, 6*, 4603–4606.

Vohidov, F., Coughlin, J. M., & Ball, Z. T. (2015). Rhodium(II) metallopeptide catalyst design enables fine control in selective functionalization of natural SH3 domains. *Angewandte Chemie, International Edition, 54*, 4587–4591.

Vohidov, F., Knudsen, S. E., Leonard, P. G., Ohata, J., Wheadon, M. J., Popp, B. V., et al. (2015). Potent and selective inhibition of SH3 domains with dirhodium metalloinhibitors. *Chemical Science, 6*, 4778–4783.

CHAPTER TWO

Cell-Binding Assays for Determining the Affinity of Protein–Protein Interactions: Technologies and Considerations

S.A. Hunter, J.R. Cochran[1]
Stanford University, Stanford, CA, United States
[1]Corresponding author: e-mail address: jennifer.cochran@stanford.edu

Contents

1. Introduction	22
2. General Binding Theory and Relevance of K_d	23
3. General Pitfalls in Cell-Based Binding Assays	25
3.1 Time to Equilibrium	26
3.2 Ligand Depletion	27
4. Measuring Binding on the Surface of Yeast	30
4.1 Materials	32
4.2 Method	33
5. Measuring Binding on the Surface of Mammalian Cells	35
5.1 Direct Binding	35
5.2 Competition Binding	36
6. Other Methods of Measuring Binding: Kinetic Exclusion Assay and Surface Plasmon Resonance	37
6.1 Kinetic Exclusion Assay	37
6.2 Surface Plasmon Resonance	38
6.3 Comparison	41
7. Summary	42
Acknowledgments	42
References	42

Abstract

Determining the equilibrium-binding affinity (K_d) of two interacting proteins is essential not only for the biochemical study of protein signaling and function but also for the engineering of improved protein and enzyme variants. One common technique for measuring protein-binding affinities uses flow cytometry to analyze ligand binding to proteins presented on the surface of a cell. However, cell-binding assays require specific considerations to accurately quantify the binding affinity of a protein–protein interaction. Here we will cover the basic assumptions in designing a cell-based binding assay, including the relevant equations and theory behind determining binding affinities. Further, two major considerations in measuring binding affinities—time to equilibrium and ligand depletion—will be discussed. As these conditions have the potential to greatly alter the K_d, methods through which to avoid or minimize them will be provided. We then outline detailed protocols for performing direct- and competitive-binding assays against proteins displayed on the surface of yeast or mammalian cells that can be used to derive accurate K_d values. Finally, a comparison of cell-based binding assays to other types of binding assays will be presented.

1. INTRODUCTION

Proteins, as ligands, signaling partners, or messengers, form vast networks that require specific interactions to carry out their functions. Generally, these binding events can be thought of as the propensity of two protein-binding partners to either associate, or conversely dissociate at different concentrations (Kuriyan, Boyana, & Wemmer, 2013). Protein partners that come together rapidly in solution at low relative concentrations, and stay bound together for an extended period of time are thought to have a low propensity for dissociation, and are thus thought of as stronger binders (Kuriyan et al., 2013).

While protein interactions are easy to conceptualize in theory, in practice they can be challenging to measure accurately. Protein interactions are measured using binding assays, each of which follows a general pattern. Methods require a soluble or cell surface displayed protein (eg, receptor) of interest, varying concentrations of soluble ligand, and a mechanism through which bound ligand can be measured (de Jong, Uges, Franke, & Bischoff, 2005; Hulme & Trevethick, 2010; Kuriyan et al., 2013). Bound ligand is measured over a variety of starting concentrations, and the resulting dose response curve can be fit to determine specific binding values (Berson & Yalow, 1959; Hulme & Trevethick, 2010; Kuriyan et al., 2013). Historically, ligand/receptor-binding assays were performed by using radioactively labeled ligands, allowing for quantification of high-affinity binding interactions with great specificity (Crevat-Pisano, Hariton, Rolland, & Cano, 1986; Maguire, Kuc, & Davenport, 2012; McKinney & Raddatz, 2006). Safety concerns and limitations in radioligand labeling, along with recent advances in imaging technology have given way to fluorescent labeling and detection of ligands, which is commonly used in cell-binding assays today (de Jong et al., 2005). A number of label-free methods have also been developed to measure protein-protein binding interactions in solution.

Here we present a practical guide for measuring 1:1 binding events between soluble ligands and binding partners expressed on the surface of yeast and mammalian cells. Further, we describe several important considerations: (1) how to properly set up a cellular-binding assay, (2) how to avoid experimental errors that will lead to inaccurate values, and (3) how cell-based binding assays compare to other assays used in the field.

2. GENERAL BINDING THEORY AND RELEVANCE OF K_d

Analysis of the binding of proteins and small molecules was initially proposed by Scatchard (1949), and later refined for protein–protein interactions by others (Berson & Yalow, 1959; Feldman, 1972; Rosenthal, 1967; Stephenson, 1956). The collective model states that the binding interaction of two proteins (here a ligand and receptor) can be described by the following equation, where L represents free ligand, R represents unbound receptor, and LR represents the bound ligand–receptor complex (Berson & Yalow, 1959; Kuriyan et al., 2013):

$$L + R \leftrightarrow LR \tag{1}$$

A binding reaction consists of a dynamic exchange between bound and unbound states, described by two rate constants of the reaction, the k_{on} and k_{off} (units of $M^{-1}s^{-1}$ and s^{-1}, respectively). Over time, reactions proceed to equilibrium as more and more ligand binds to receptor, until the concentrations of [L], [R], and [LR] are held in steady state.

Under these equilibrium conditions, binding can then be modeled further, using the law of mass action (Guldberg & Waage, 1864; Hulme & Trevethick, 2010; Kuriyan et al., 2013; Pollard, 2010). At equilibrium, the rates of the forward ($k_{on}[L][R]$) and reverse ($k_{off}[LR]$) reactions are equal. This relationship, shown in Eq. (2), can be rearranged to derive a ratio known as the equilibrium-binding association constant, K_a with units of inverse molarity (M^{-1}) (Berson & Yalow, 1959; Hulme & Trevethick, 2010; Pollard, 2010). This relationship can be modeled by the following equations:

$$k_{on}[L][R] = k_{off}[LR] \tag{2}$$

$$\frac{[LR]}{[L][R]} = \frac{k_{on}}{k_{off}} = K_a \tag{3}$$

However, binding is usually not thought of in terms of the association reaction, but instead the dissociation reaction. This is given by the inverse of the association constant and is known as K_d and has the units of molarity (M) (Berson & Yalow, 1959; Hulme & Trevethick, 2010; Kuriyan et al., 2013; Pollard, 2010). The equation for K_d can be written as follows:

$$\frac{1}{K_a} = \frac{[L][R]}{[LR]} = \frac{k_{off}}{k_{on}} = K_d \tag{4}$$

It can be useful to think of the K_d in terms of the total fraction of ligand bound (Kenakin, 2016; Klotz, 1982). To determine this relationship, an equation for the total receptor concentration must first be established. This equation is based on the assumption that the total concentration of receptor $[R]_T$ present is simply the sum of the concentration of free receptor $[R]$ and that of bound receptor $[LR]$. Thus:

$$[R]_T = [R] + [LR] \tag{5}$$

Rearrangement of Eq. (4) gives:

$$[R] = \frac{[LR]K_d}{[L]} \tag{6}$$

Substituting into Eq. (5) for $[R]$, the following equation is derived:

$$[R]_T = \frac{[LR]K_d}{[L]} + [LR] \tag{7}$$

The fraction of receptor bound (f) can be thought of as the ratio of the concentration of bound receptor $[LR]$ to the total receptor $[R]_T$: $[LR]/[R]_T$. Thus, by rearranging Eq. (7), the following equation is derived, which gives the fraction of receptor bound in terms of the concentration of free ligand and K_d:

$$\frac{[LR]}{[R]_T} = f = \frac{[L]}{K_d + [L]} \tag{8}$$

Eq. (8) can be fitted to either a hyperbolic shape (on a normal axis) or a sigmoidal shape (on a logarithmic axis) (Hulme & Trevethick, 2010; Klotz, 1982; Pollard, 2010). Most typically, this equation is depicted as a sigmoidal curve on a semilog plot, with the ligand concentration as the abscissa (on a log-scale axis) and the fraction bound as the ordinate (Fig. 1) (Klotz, 1982). As K_d is thought of in units of molarity, it is easiest to derive the K_d from this equation when the value is equal to that of the ligand concentration. In other words: $K_d = [L]$.

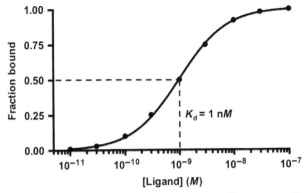

Fig. 1 Sigmoidal binding curve of varying concentrations of ligand bound to cell surface receptor. *Dashed lines* demarcate the K_d (1 nM) as the ligand concentration at 0.5 fraction bound.

Substituting [L] for K_d under these circumstances in Eq. (8), the following is derived for the fraction bound (f):

$$f = \frac{K_d}{K_d + K_d} = \frac{K_d}{2K_d} = \frac{1}{2}$$

Thus, when 50% of the receptor is bound by ligand, the K_d will be equal to the ligand concentration, meaning that its value can be extrapolated from the fit sigmoidal curve (Fig. 1).

For any given binding reaction, the lower the concentration of ligand required for half of the maximal binding events possible to occur, the more tightly the two binding partners must interact. Moreover, as the K_d is defined as k_{off}/k_{on}, there are two mechanisms through which a K_d can take on a low value, a slow off-rate (k_{off}), or a fast on-rate (k_{on}). Thus, when altering a binding interaction to have a lower K_d, both of these parameters may be modulated to achieve a tighter binding interaction (Chen, de Picciotto, Hackel, & Wittrup, 2013).

3. GENERAL PITFALLS IN CELL-BASED BINDING ASSAYS

The K_d derivation equation (Eq. 8) relies on a number of assumptions that, if unaccounted for, may introduce significant error into the assay. Two major considerations are the time to equilibrium of the binding reaction and ligand depletion that occurs during the binding assay.

3.1 Time to Equilibrium

The K_d equation is based on a system at equilibrium, which requires that a binding reaction must come to equilibrium for measured K_d values to be accurate (Berson & Yalow, 1959; Bylund & Toews, 1993; Hulme & Trevethick, 2010; Kuriyan et al., 2013; Pollard, 2010). Allowing binding reactions to come to equilibrium is highly dependent on the concentration of ligand used and the strength of the binding interaction, and thus may require the reaction to be incubated for several hours or days. Mathematically, equilibrium time can be calculated from the k_{off}, [L], and K_d (Chen et al., 2013; Hulme & Trevethick, 2010). A useful parameter is the half-time of the equilibrium equation ($t_{1/2}$):

$$t_{1/2} = \frac{\ln(2)}{k_{off}\left(1 + \frac{[L]}{K_d}\right)} \tag{9}$$

The reaction will reach one half of the remaining concentration ratio toward equilibrium after each half-time. Thus, it takes $5 \times t_{1/2}$ to achieve 97% of the final equilibrium value. As additional time after this point will only lead to minimal increases towards complete equilibrium, $5 \times t_{1/2}$ can be thought of as the time required to reach equilibrium in a binding reaction (Chen et al., 2013; Hulme & Trevethick, 2010).

Establishing the time to equilibrium is critical for assuring that a binding assay has been performed correctly, and requires knowledge of k_{off}, [L], and K_d (Eq. 9). However, for a novel binding interaction, the k_{off} and K_d will be unknown, making determination of the half-time parameter at the outset difficult. The k_{off} can be determined from an off-rate assay (Boder & Wittrup, 1998). This value, in conjunction with an estimated K_d (derived from literature or pilot experiments) can then be used to determine the time to equilibrium at the lowest concentration of ligand tested to determine the appropriate incubation time. In addition, it is often useful to perform the binding assay using a longer incubation time to determine if a similar K_d is obtained. If so, this provides additional confidence that a binding reaction has come to equilibrium.

Failure to allow binding reactions to come to equilibrium results in a right-ward shift in the binding curve (Fig. 2), meaning that the K_d is determined as a higher value than it actually is and the tightness of binding is underestimated (Hulme & Trevethick, 2010; Pollard, 2010). When the reaction has not come to equilibrium, the ratio of free ligand and receptor to bound receptor ([L][R]/[LR]) is higher, as not all of the free ligand has yet bound receptor in contrast to when the system is at equilibrium.

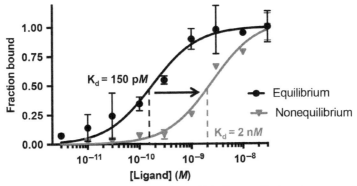

Fig. 2 The effects of equilibrium time on ligand binding to displayed receptor. *Dashed lines* represent the K_d of each binding curve. The equilibrium-binding curve (represented by *black circles*) shows a binding reaction that was allowed to go to equilibrium and reveals the true K_d of 150 pM. The nonequilibrium-binding curve (represented by *red* (*gray* in the print version) *triangles*) shows the results of a binding reaction that was not allowed enough time to come to equilibrium. The measured K_d of this curve is now 2 nM, a roughly tenfold difference above the actual K_d.

3.2 Ligand Depletion

It is essential to be able to determine or assume an accurate value for the concentration of free ligand [L] when performing a binding assay (Hulme & Trevethick, 2010; Kuriyan et al., 2013). While it is theoretically possible to measure the concentration of ligand that remains free at equilibrium, it is much easier to assume that the initial concentration of ligand used is the same as the free ligand concentration at equilibrium. This assumption is only possible if the initial concentration of ligand introduced to the system is far greater than the concentration of receptor present (Kuriyan et al., 2013; Limbird, 1995). Ligand depletion can be modeled by the fraction of ligand bound by receptor, described using the following equation, with δ as ligand depletion, [LR] as the concentration of bound receptor, and $[L]_T$ as the total concentration of ligand in the system (Hulme & Trevethick, 2010):

$$\delta = \frac{[LR]}{[L]_T} \quad (10)$$

In this case, it is assumed that all of the receptor present is bound by ligand; therefore [LR] becomes equivalent to $[R]_T$, or the total concentration of receptor. The ratio for ligand depletion then becomes:

$$\delta = \frac{[R]_T}{[L]_T} \quad (11)$$

If the total concentration of receptor $[R]_T$ is 10% of the concentration of total ligand $[L]_T$ ($\delta = 0.1$), then the initial concentration of ligand introduced will be roughly the same as the final concentration of free ligand [L] when the system reaches equilibrium (Colby et al., 2004). Thus, the $[L]_T$ added at the beginning of the assay can be used in place of [L] when calculating K_d. A more in depth description of the mathematical rationale for ligand depletion is discussed in review elsewhere (Colby et al., 2004; Hulme & Trevethick, 2010).

Based on Eq. (11), ligand depletion can be modulated by controlling two factors of the binding assay: the total concentration of receptor present $[R]_T$ or the total concentration of ligand present $[L]_T$. To decrease the value of δ to 0.1 or below, $[R]_T$ can be reduced or $[L]_T$ can be increased. The main technique used to control for $[R]_T$ in cell-binding assays is to control the number of cells present. However, in most experiments there is a minimum number of cells that must be used to measure a significant binding signal to derive reproducible data ($\sim 10^5$ cells) (Colby et al., 2004). Alternatively, to increase $[L]_T$, the total moles of ligand added to the reaction and the total volume of the reaction can be increased. As a result of increasing the volume of the reaction, the overall concentration of receptor $[R]_T$ will decrease, as the total number of receptors (based on cell number) is held constant in each condition. Thus, volumetric increase is critical to avoid ligand depletion at low ligand concentrations.

Table 1 shows the minimum volume necessary for varying concentrations of ligand, given a receptor number of 10^{10} (10^5 cells $\times$ 10^5 receptors/cell), a standard estimate when testing proteins expressed on the

Table 1 Minimum Volumes Needed to Avoid Ligand Depletion in a Cell-Binding Assay with 10^{10} Receptors Present

Ligand Conc. $[L]_T$ (nM)	Ligand Depletion Volume (μL)	Ligand Conc. $[L]_T$ (pM)	Ligand Depletion Volume (mL)
500	0.33	500	0.33
100	1.66	100	1.66
50	3.32	50	3.32
10	16.6	10	16.6
5	33.2	5	33.2
1	166	1	166
0.5	330	0.5	330

Large volumes are required to avoid ligand depletion for low ligand concentrations.

yeast cell surface (Boder & Wittrup, 2000) or mammalian cell surface (discussed in Sections 4 and 5, respectively). As is clearly evident, the effects of ligand depletion become the most apparent when starting concentrations of ligand become very low (<1 nM). For this reason, it is often challenging to quantify high-affinity binders using cell-based assays because of the difficulty in recovering small numbers of cells from large (>15 mL) assay volumes.

If compensation for ligand depletion is not taken into account, the effect will create a right-ward shift of the binding curve and an overestimation of the K_d (Hulme & Trevethick, 2010; Limbird, 1995; Moore & Cochran, 2012). This is due to the mechanism of analysis of K_d from binding curves, reliant on the assumption that the free ligand concentration [L] used in fitting the curve is equal to the total starting ligand concentration [L]$_T$ (Kuriyan et al., 2013). If ligand depletion exists, then a significant portion of the free ligand will be bound by receptor, and as the assay progresses, the effective free ligand concentration [L] will no longer be equal to the initial [L]$_T$. The binding signal measured will thus be reflective of a concentration of free ligand that is significantly lower than the initial ligand concentration. This phenomenon is demonstrated in Fig. 3 and Table 2.

Given the number of considerations required in order to set up an accurate binding assay, it is perhaps unsurprising that there are often a variety of K_d values (often over-estimations) provided in the literature for the same protein–protein interaction (Kastritis & Bonvin, 2013; Kastritis et al., 2011).

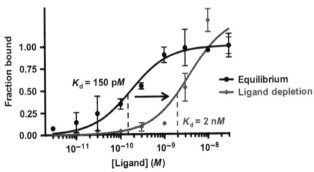

Fig. 3 The effects of ligand depletion on cell-binding assay measurements. *Dashed lines* represent the K_d of each binding curve. The equilibrium-binding reaction curve (represented by *black circles*) was performed using appropriate volumes at each concentration to avoid ligand depletion, while the ligand depletion reaction curve (represented by *blue* (*dark gray* in the print version) *diamonds*) did not. The measured K_d of this curve is 2 nM, a 10-fold difference above the actual K_d of 150 pM. These data are tabulated in Table 2.

Table 2 Binding Assay Conditions Used in Nonligand Depleting and Ligand Depleting Experiments

Perceived [Ligand] (nM)	Equilibrium-Binding Conditions		Ligand-Depleting Conditions	
	Volume (μL)	Fraction Bound	Volume (μL)	Fraction Bound
30.00	20	1.00	20	1.00
10.00	20	0.95	20	1.28
3.00	50	0.98	20	*0.52
1.00	200	0.90	20	*0.12
0.30	500	0.54	20	*0.08
0.10	2000	0.34	60	*0.03
0.03	5500	0.24	200	*0.04
0.01	20,000	0.14	600	*0.05
0.003	55,000	0.07	1500	*0.07

The perceived ligand concentration is the concentration of ligand that was used as a starting point at each listed volume. The resultant fraction bound values are derived from analyzing the final reactions. Values that are under ligand depletion conditions ($\delta > 0.1$) are italicized and asterisked. Data are plotted in Fig. 3. While this table provides useful guidelines, ligand depletion (or lack thereof) should be determined empirically by measuring binding affinities under different volumes to confirm similar K_d values are achieved.

This is an especially important consideration in engineering proteins for tighter binding, as "improved" variants are benchmarked against wild-type counterparts, often based on literature K_d values.

4. MEASURING BINDING ON THE SURFACE OF YEAST

A powerful method for analyzing and engineering novel protein variants and interactions is yeast surface display (YSD), pioneered by Boder and Wittrup. YSD relies on the fusion of a protein of interest to a yeast surface protein (Aga2p), along with epitope tags (c-myc or hemagglutinin (HA)) to monitor protein expression after induction (Fig. 4) (Boder & Wittrup, 1997). YSD as a technology to engineer proteins has been reviewed extensively (Chen et al., 2013; Cherf & Cochran, 2015; Colby et al., 2004; Moore & Cochran, 2012). YSD is also a useful platform for performing characterization of binding interactions with a soluble target, obviating the need for large-scale expression and purification of a protein of interest

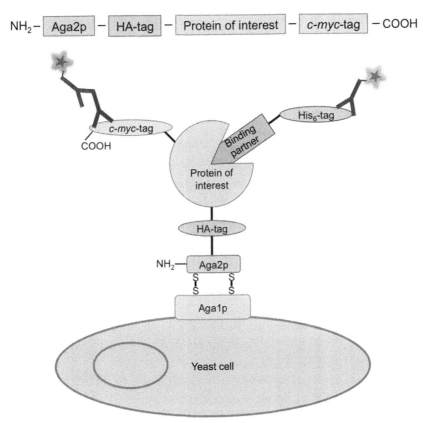

Fig. 4 A schematic of yeast surface display. In this case, the binding partner protein is detected using a fluorescently labeled antibody against a His_6-tag, although other epitope tags or a fluorescently labeled binding partner can be used. The pCTCON2 vector layout and display system is shown. The N-terminus of the protein of interest is fused to the C-terminus of Aga2p; a C-terminal c-myc tag allows the expression of full-length protein to be measured with a fluorescent secondary antibody binding to an anti-c-myc antibody. Alternatively, the pTMY vector layout and display system (not shown) results in a C-terminal fusion of the protein of interest to N-terminus of Aga2p, with an exposed HA expression tag. Expression is then detected with an anti-HA antibody.

(Gai & Wittrup, 2007). Importantly, affinities of proteins that are displayed on the surface of yeast have been demonstrated to be comparable to the same proteins in a soluble form, as measured by other techniques (Gai & Wittrup, 2007). However, binding values can be confounded if the binding interaction being measured on yeast is more complicated in the endogenous setting, for example, due to binding interactions that are not 1:1, or if the displayed protein adopts an unnatural conformation.

Here we will describe the general procedures for binding assays using yeast that have been transformed with the appropriate vector to display the protein of interest (Boder & Wittrup, 1998; Chen et al., 2013; Colby et al., 2004; Moore & Cochran, 2012).

4.1 Materials

4.1.1 Yeast Cells
Saccharomyces cerevisiae yeast strain EBY100 transformed with the pCTCON2 or pTMY vector containing a protein of interest. pCTCON2 displays the protein of interest as an N-terminal fusion to Aga2p (with a C-terminal c-myc expression tag) (Boder & Wittrup, 1998), while pTMY displays the proteins as a C-terminal fusion to Aga2p (with an N-terminal HA expression tag) (Jones et al., 2011) (Fig. 4).

4.1.2 Solutions and Media
SD-CAA Media (Growth media): 20 g/L dextrose, 6.7 g/L yeast nitrogen base (Becton Dickinson, catalog no. 291940), 5 g/L casamino acids (Becton Dickinson, catalog no. BP1424), 5.4 g/L Na_2HPO_4 (anhydrous), 8.6 g/L $NaH_2PO_4 \cdot H_2O$ in deionized H_2O, filter sterilize with a 0.2-μm filter and store at 4°C.

SG-CAA Media (Induction media): identical to SD-CAA, except replace dextrose with galactose (20 g/L).

0.1% BPBS (also known as PBS-BSA, PBSA—"Binding Assay Buffer"): Dissolve 1 g of BSA in 1 L of 1 × PBS. Filter sterilize using a 0.2 μm filter and store at 4°C. In cases of specialized binding buffers, a 1:1 mixture of the selective binding buffer and 0.1% BPBS can be used. Inclusion of a low concentration of BSA in the binding assay buffer helps to prevent nonspecific binding.

4.1.3 Proteins/Antibodies
Binding protein of interest (ligand): Should be tagged for binding recognition (biotinylation, His_6-tag, fluorescent dye, etc.).

Primary expression antibody (to be used against expression tag)—Example: Chicken-Anti *c-myc* (Life Technologies A21281, 1:500 dilution, for pCTCON2).

Fluorophore-conjugated secondary expression antibody (to be used against Primary Expression Antibody, typically R-phycoerythrin [PE] conjugated)—Example: Goat-Anti Chicken–R-phycoerythrin (Santa Cruz Biotechnologies D1715, 1:100 dilution, for pCTCON2).

Fluorophore-conjugated secondary binding antibody (to be used against tag on ligand-binding partner, typically FITC conjugated)—Example: Rabbit-Anti His$_6$–FITC (Bethyl A190-114F, 1:100 dilution).

4.2 Method

Day 1: Inoculation
1. Over flame, inoculate a single pCTCON2- or pTMY-transformed yeast colony into 5 mL SD-CAA. Place at 30°C for 24 h, or until the OD$_{600nm}$ = 3–5.

Day 2: Induction
1. Determine the OD$_{600nm}$ of all samples using a spectrophotometer.
2. Seed induction samples at an OD$_{600nm}$ of 1. Transfer the appropriate volume to a 1.5-mL Eppendorf tube. Spin down all samples at 3600 rpm for 4 min. Discard the supernatant.
3. Resuspend in 500 µL of SG-CAA, add to 5 mL SG-CAA in new culture tube.
4. Place at the optimal induction temperature for 24 h.
   a. *Note*: Yeast are typically induced to express protein at either 20°C or 30°C. The optimal expression condition for the protein of interest should be tested and determined prior to starting a binding assay (Fig. 5).

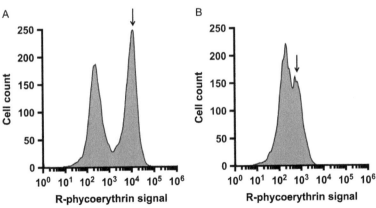

Fig. 5 Expression histograms of example protein displayed on the surface of yeast. The y-axis is the number of cells and the x-axis is the PE signal. (A) Expression at 20°C, with a tightly defined expressing population (*arrow*). (B) Expression at 30°C, with a poorly defined expressing population (*arrow*). Differences between A and B are due to yeast growth and overall protein folding and expression levels.

Day 3

1. Determine the range of ligand concentrations (two orders of magnitude above and below the K_d) (Hulme & Trevethick, 2010; Moore & Cochran, 2012) and volumes needed to avoid ligand depletion. Prepare dilutions of ligand. Include expression, secondary antibody, and cells only controls (Table 3).
2. Measure the OD_{600nm} of each induced yeast culture.
3. Add 10^5 yeast to labeled Eppendorf tubes (OD_{600nm} of 1 is equal to 1×10^7 cells/mL). Wash with 50 μL binding assay buffer. Spin at $14,000 \times g$ for 30 s and discard supernatant.
4. Resuspend in binding assay buffer for the final binding incubation volume (step 1). Transfer yeast to larger tubes for larger volume incubations, if needed.
5. Add the ligand to each sample. Do not add ligand to the control tubes for expression, nonspecific secondary binding, and cells only. Incubate tubes on a rotator at room temperature or 4°C until reactions reach equilibrium (Section 3).

Postincubation

1. Keep samples at 4°C. Spin down samples at $14,000 \times g$ for 30 s at 4°C, and remove the supernatant. For larger tubes (>2 mL), spin at 3600 rpm for 4 min.
2. Resuspend each tube with 50 μL of cold binding assay buffer. Add primary expression antibody at the appropriate dilution (Anti-HA for pTMY or Anti-c-myc for pCTCON2). Do not add primary antibody to the nonspecific binding secondary only and cells only control tubes. Incubate on a rotator at 4°C for 30 min.

Table 3 Preparation for a Binding Assay Set Up

Sample No.	Purpose	Primary Condition	Secondary Condition
1	Binding/expression (100 nM ligand)	0.4 μL of 5 μM ligand stock + primary expression Ab (20 μL total volume)	Secondary binding Ab + secondary expression Ab
2	Expression only	Primary expression Ab	Secondary expression Ab
3	Secondary only	N/A	Secondary binding Ab + secondary expression Ab
4	Cells only	N/A	N/A

An example is provided for binding measured at 100 nM of ligand, stored at a stock concentration of 5 μM (Sample 1). The appropriate detection antibodies to be used are listed. Controls are also shown (Samples 2–4).

3. Wash each sample with 0.5 mL of cold binding assay buffer. Spin at 14,000 × g for 30 s at 4°C. Remove supernatant, leaving samples on ice.
4. Resuspend each sample in 50 μL of secondary antibody containing binding assay buffer (1:100 dilution). Keep tubes with secondary antibody in the dark. Incubate at 4°C on a rotator for 15 min.
5. Wash as in step 3.
6. Analyze the samples on a flow cytometer. Binding values can be determined from the average FITC value of each sample corrected for autofluorescence (expression only). Plot fraction bound vs ligand concentration (log scale). Fit a sigmoidal curve using nonlinear regression analysis. The K_d value can be derived from the ligand concentration at half the fraction bound (Fig. 1).

5. MEASURING BINDING ON THE SURFACE OF MAMMALIAN CELLS

Measuring binding of receptors or membrane-bound proteins on the surface of mammalian cells offers a unique ability to assay for binding values in endogenous settings, instead of using displayed or soluble protein (Bylund & Toews, 1993). Mammalian systems also give the advantage of being able to assay against receptor complexes and can present proteins that may not fold or be displayed properly on the surface of yeast. There are two major types of mammalian cell-binding assays: direct and competition (Moore & Cochran, 2012). Both protocols follow similar steps to the yeast cell-binding assay, however, mammalian cells are much more fragile than yeast and should be treated with care. The potential pitfalls and logic behind the experimental design are the same for both types of cell-based assays.

5.1 Direct Binding

1. Aliquot 5×10^4 mammalian cells expressing the target protein of interest to labeled 1.5 mL Eppendorf tubes. Make sure to include tubes for cells only and cells with antibodies only controls, if needed.
2. Spin down cells at 800 × g for 5 min, ensuring cells are still viable by trypan blue staining. Remove supernatant, and resuspend in binding assay buffer, using appropriate volumes to avoid ligand depletion (Section 3).
   a. *Note*: Binding assay buffer can be anything from media to BPBS. All components necessary for the binding interaction of interest should be included (salts, cofactors, etc.).

3. Add soluble ligand to each tube at varying concentrations, spanning two orders of magnitude above and below the anticipated K_d.
4. Allow the reaction to incubate at 4°C until it has come to equilibrium, generally a number of hours.
5. If the ligand is fluorescently labeled, proceed to step 6. If the ligand has an epitope tag, first wash the tubes with 0.5 mL of cold BPBS, and spin them down at 800 × g for 5 min at 4°C. Remove the supernatant and then resuspend the cells in 50 μL with a 1:100 dilution of the appropriate fluorescently labeled antibody (eg, Rabbit Anti-His$_6$–FITC). Allow the antibody to incubate for 20–30 min at 4°C.
6. Wash cells with 0.5 mL of cold BPBS. Spin down at 800 × g for 5 min at 4°C. Remove the supernatant.
7. Analyze the cells using flow cytometry. Analyze data as described in the yeast binding assay (Postincubation, step 6).

5.2 Competition Binding

The competition assay follows very similar steps to those listed in the direct-binding assay. Differences are noted below.
1. Follow steps 1 and 2 as in the direct-binding assay.
2. For step 3, incubate with the same range of ligand concentrations as described but also include a constant concentration of competitor in each tube. Competitor should bind to the target receptor and should be fluorescently labeled or contain an epitope tag for detection. Competitor should be used at a value lower than its K_d, such that it can still be detected but is able to be competed off. Allow to incubate at 4°C for a number of hours.
3. If competitor is fluorescently labeled, follow steps 6–7 of the direct-binding assay. If not, wash and incubate with the appropriate fluorescently labeled antibody, as described in step 5 of the direct-binding assay. Then follow steps 6–7.
4. When the curve is analyzed it will give the IC$_{50}$ of the binding reaction, not the K_d. The K_d can be calculated from the Cheng–Prusoff equation (Cheng & Prusoff, 1973):

$$K_d = \frac{IC_{50}}{1 + [\text{competitor}]/K_d \text{competitor}} \quad (12)$$

The assumptions of Eq. (12) are very similar to those of the general-binding Eq. (8) described earlier (Krohn & Link, 2003). Notably, the

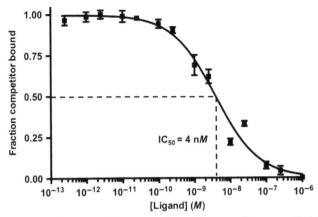

Fig. 6 Mammalian cell competition assay. Competitor was fluorescently labeled with Alexa Fluor 488 prior to assay. Reactions were carried out with varying concentrations of ligand. *Dotted lines* mark the IC$_{50}$ (4 nM).

reaction must be at equilibrium, represent a 1:1 binding interaction, and not contain depleted ligand (Ehlert, Roeske, & Yamamura, 1981; McKinney & Raddatz, 2006). Further, the K_d of the competitor must be known under the binding conditions being used. With these parameters in place, the K_d can be calculated using the competition assay (Fig. 6).

6. OTHER METHODS OF MEASURING BINDING: KINETIC EXCLUSION ASSAY AND SURFACE PLASMON RESONANCE

While cell-binding assays are advantageous in the speed and ease with which they can be carried out, they are limited in their ability to measure high-affinity binding interactions and kinetic parameters (Bylund & Toews, 1993). In these cases, the kinetic exclusion assay (KinExA) or surface plasmon resonance (SPR) can be used (Blake, Pavlov, & Blake, 1999; Darling & Brault, 2004; Myszka, 2000; Patching, 2014). Here we will give a brief introduction to each and list the major advantages and disadvantages of each technology. Note that the designations of ligand and receptor in the examples below are arbitrary and can be reversed.

6.1 Kinetic Exclusion Assay

Binding interactions measured by KinExA (Blake et al., 1999) make use of a column of packed beads to which an immobilized ligand is affixed (Darling & Brault, 2004). Binding reactions are set up with varying concentrations of ligand and constant concentrations of receptor, allowed to come to equilibrium, and are then run over the column to capture unbound receptor by the affixed ligand (Darling & Brault, 2004). The amount of bead-bound receptor is then detected and quantified using a fluorescent antibody (Fig. 7A–C). Thus, KinExA allows direct measurement of free receptor remaining at equilibrium in each binding reaction.

Kinetic exclusion refers to the rapid speed at which the binding reaction is flowed over the column (<0.5 s), such that the bound ligand–receptor complex does not have time to dissociate and increase the free receptor concentration in solution (Darling & Brault, 2004). Thus LR dissociation is "kinetically excluded" from occurring. This critical feature means that the amount of receptor captured by the immobilized ligand is proportional to $[R]_{free}$, providing a method through which K_d can be determined without having to consider factors underlying cell-binding assays, such as ligand depletion (Blake et al., 1999; Darling & Brault, 2004; Drake, Myszka, & Klakamp, 2004).

Since binding is measured directly and extrapolated from a small amount of measured $[R]_{free}$, KinExA allows for the determination of the K_d of ultra high affinity binding interactions, and can also measure kinetic binding parameters (Darling & Brault, 2004). This is done by mixing predetermined concentrations of $[L]_o$ and $[R]_o$ together, and then measuring $[R]_{free}$ over time (Darling & Brault, 2004; Drake et al., 2004). These data can be plotted and fit to the standard biomolecular rate equation and extrapolated to determine the k_{on} (Darling & Brault, 2004). For more details, KinExA as a technology for determining binding affinities has been reviewed extensively elsewhere (Darling & Brault, 2004; Drake et al., 2004).

6.2 Surface Plasmon Resonance

SPR relies on the measurement of refractive index change from receptor attached to a gold surface after soluble ligand, or analyte, is flowed over the chip (Fig. 7D) (Jönsson et al., 1991, 1993; Myszka, 2000; Patching, 2014). These measurements are carried out in real time, and the magnitude of refractive index change is directly proportional to the molecular weight of the analyte (Myszka, 2000; Patching, 2014).

To measure kinetic parameters, the SPR protocol makes use of varying concentrations of analyte flowed over chips of immobilized receptor. Real-time measurements are made of the association phase, as analyte is flowed, and of the dissociation phase, as buffer is flowed (Myszka, 2000; Patching, 2014). These data generate a set of binding curves from

Cell-Binding Assays to Determine Affinity

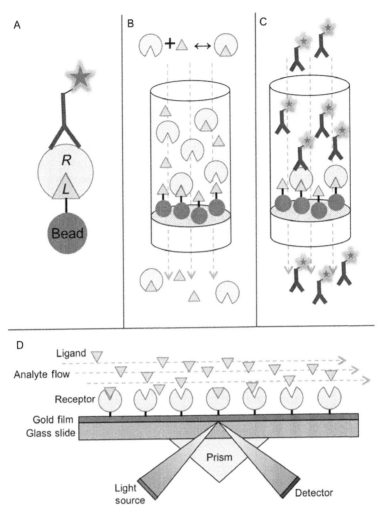

Fig. 7 (A–C) KinExA. (A) Close up of the KinExA bead-based detection system. Ligand affixed to beads in a column bind soluble, free receptor, which is detected via fluorescent antibody binding to receptor. (B and C) Schematic of the KinExA assay. (B) Soluble ligand and receptor are incubated until equilibrium is reached. This reaction is flowed over the column, allowing free receptor at equilibrium to bind bead-affixed ligand. (C) After washing, bound receptor is detected using specific fluorescent antibodies and quantified. (D) Surface plasmon resonance. Receptor is affixed to a gold film surface. Ligand, or "analyte," is flowed over the surface. Alterations in resonance due to binding are measured via a detector from light shone through a prism and analyzed. Note that for KinExA and SPR, the designation of ligand and receptor is arbitrary and can be reversed.

Table 4 Advantages and Disadvantages of Cell-Binding Assays, KinExA, and SPR

Assay	Advantages	Disadvantages
Yeast surface display	• No need for large-scale expression and purification of protein of interest • Easy to measure binding of multiple protein variants • Relatively high throughput • Low technical and time requirements • Protein expression can be quantified and normalized with binding • Compatible with flow cytometry	• Unable to accurately measure high-affinity interactions (<10 pM) • Protein immobilization on the surface of yeast might not mimic natural environment • Measurement of high-affinity binders requires cognizance of ligand depletion and time to equilibrium • Requires separate assays to measure equilibrium and kinetic parameters
Mammalian cell	• Replicates natural binding conditions (for membrane-bound proteins) • No need to express or purify the receptor of interest • Compatible with flow cytometry • Relatively high throughput • Can perform direct- or competition-binding assays	• Unable to accurately measure high-affinity interactions (<10 pM) • Mammalian cells are fragile, dead cells can nonspecifically bind to ligand or reagents • Measurement of high-affinity binders requires cognizance of ligand depletion and time to equilibrium • Requires separate assays to measure equilibrium and kinetic parameters
KinExA	• True solution-based measurements of unmodified proteins • Accurately measures high-affinity interactions (<10 pM), can measure fM • Avoids nonspecific cell binding • Measures amount of $[R]_{free}$ at equilibrium removing concerns of ligand depletion • Measurements can be made with unpurified mixtures	• Relatively low throughput, although autosampler is available • Nonspecific bead binding may occur • Weaker binders can be measured but may require extra care • Solution binding might not mimic natural environment • Requires separate assays to measure equilibrium and kinetic parameters

Table 4 Advantages and Disadvantages of Cell-Binding Assays, KinExA, and SPR—cont'd

Assay	Advantages	Disadvantages
Surface plasmon resonance	• Advanced automation • Low quantities of samples needed • Temperature control • Able to measure k_{on} and k_{off} simultaneously • Reactions do not need to reach equilibrium to be evaluated	• Binding partner is immobilized to a surface • Requires purified proteins • Limited by mass transport effects in measuring very fast association rates or very slow dissociation rates • Can be relatively low throughput, depending on the instrument

which the k_{on} (association), k_{off} (dissociation), and K_d can be extrapolated (Myszka, 2000; Patching, 2014).

Unlike KinExA and cell-based binding assays, SPR does not require reactions to come to equilibrium, advantageous for binding reactions that have very long equilibrium times. However, because it requires measuring k_{on} and k_{off}, if either of these values is too fast or too slow, respectively, SPR can become no longer reliable in measuring these parameters and another type of binding assay should be used. For more details, the use of SPR to determine binding affinities has been extensively reviewed elsewhere (Myszka, 2000; Patching, 2014).

6.3 Comparison

Each binding assay covered in this review has a specific set of circumstances under which its use is optimal. The advantages and disadvantages of each type of binding assay are summarized in Table 4. All four methods of measuring binding interactions can be used to characterize wild-type and engineered proteins. For example, YSD can be used to rapidly express multiple protein variants and test them for equilibrium binding or kinetic off-rate of binding to a protein of interest, without having to carry out large scale purification and expression of individual proteins. Ligands can also be tested using direct or competition mammalian cell-binding assays to probe for receptor or membrane-bound protein interactions in an endogenous setting. Further characterization, including in-depth kinetic analysis and a more rigorous testing of the K_d, especially if the engineered affinity is very tight, can be measured using either KinExA or SPR, depending on the properties of the binding interaction (see Table 4). These four methods can be also used in concert to obtain confidence in the accuracy of binding measurements.

7. SUMMARY

In general, whenever they are possible cell-based binding assays provide facile and robust methods to measure affinities of protein–protein interactions. In this review we cover basic principles underlying the design of cell-based binding assays, discuss potential pitfalls that can occur in determining the K_d of protein–protein interactions, and provide protocols to determine the binding affinities of protein interactions using cell-based assays. We discuss steps to establish the proper time to equilibrium and incubation volumes to avoid or minimize ligand depletion, and demonstrate how these factors can lead to errors in determining K_d values. Additionally, we compare cell-based binding assays to KinExA and SPR, and offer rationale for when each assay might best be used.

ACKNOWLEDGMENTS

S.A.H. is supported by a Stanford Graduate Fellowship, a National Science Foundation Graduate Fellowship, and PHS Grant Number CA09302, awarded by the National Cancer Institute, DHHS. The authors thank Sandra DePorter and Shizuka Yamada for helpful comments and feedback.

REFERENCES

Berson, S. A., & Yalow, R. S. (1959). Quantitative aspects of the reaction between insulin and insulin-binding antibody. *Journal of Clinical Investigation, 38*(11), 1996–2016.
Blake, R. C., Pavlov, A. R., & Blake, D. A. (1999). Automated kinetic exclusion assays to quantify protein binding interactions in homogeneous solution. *Analytical Biochemistry, 272*(2), 123–134.
Boder, E. T., & Wittrup, K. D. (1997). Yeast surface display for screening combinatorial polypeptide libraries. *Nature Biotechnology, 15*(6), 553–557. http://doi.org/10.1038/nbt0697-553.
Boder, E. T., & Wittrup, K. D. (1998). Optimal screening of surface-displayed polypeptide libraries. *Biotechnology Progress, 14*(1), 55–62.
Boder, E. T., & Wittrup, K. D. (2000). Yeast surface display for directed evolution of protein expression, affinity, and stability. *Methods in Enzymology, 328*, 430–444.
Bylund, D. B., & Toews, M. L. (1993). Radioligand binding methods: Practical guide and tips. *The American Journal of Physiology, 265*(5 Pt. 1), L421–L429.
Chen, T. F., de Picciotto, S., Hackel, B. J., & Wittrup, K. D. (2013). Engineering fibronectin-based binding proteins by yeast surface display. *Methods in Enzymology, 523*, 303–326.

Cheng, Y., & Prusoff, W. H. (1973). Relationship between the inhibition constant (K1) and the concentration of inhibitor which causes 50 per cent inhibition (I50) of an enzymatic reaction. *Biochemical Pharmacology, 22*(23), 3099–3108.

Cherf, G. M., & Cochran, J. R. (2015). Applications of yeast surface display for protein engineering. *Methods in Molecular Biology (Clifton, NJ), 1319*, 155–175.

Colby, D. W., Kellogg, B. A., Graff, C. P., Yeung, Y. A., Swers, J. S., & Wittrup, K. D. (2004). Engineering antibody affinity by yeast surface display. *Methods in Enzymology, 388*, 348–358.

Crevat-Pisano, P., Hariton, C., Rolland, P. H., & Cano, J. P. (1986). Fundamental aspects of radioreceptor assays. *Journal of Pharmaceutical and Biomedical Analysis, 4*(6), 697–716.

Darling, R. J., & Brault, P. A. (2004). Kinetic exclusion assay technology: Characterization of molecular interactions. *Assay and Drug Development Technologies, 2*(6), 647–657.

de Jong, L. A. A., Uges, D. R. A., Franke, J. P., & Bischoff, R. (2005). Receptor–ligand binding assays: Technologies and Applications. *Journal of Chromatography B, 829*(1–2), 1–25.

Drake, A. W., Myszka, D. G., & Klakamp, S. L. (2004). Characterizing high-affinity antigen/antibody complexes by kinetic- and equilibrium-based methods. *Analytical Biochemistry, 328*(1), 35–43.

Ehlert, F. J., Roeske, W. R., & Yamamura, H. I. (1981). Mathematical analysis of the kinetics of competitive inhibition in neurotransmitter receptor binding assays. *Molecular Pharmacology, 19*(3), 367–371.

Feldman, H. A. (1972). Mathematical theory of complex ligand-binding systems at equilibrium: Some methods for parameter fitting. *Analytical Biochemistry, 48*(2), 317–338.

Gai, S. A., & Wittrup, K. D. (2007). Yeast surface display for protein engineering and characterization. *Current Opinion in Structural Biology, 17*(4), 467–473.

Guldberg, C. M., & Waage, P. (1864). Studies concerning affinity. *Forhandlinger: Videnskabs-Selskabet i Christiana, 2*(1), 35–45.

Hulme, E. C., & Trevethick, M. A. (2010). Ligand binding assays at equilibrium: Validation and interpretation. *British Journal of Pharmacology, 161*(6), 1219–1237.

Jones, D. S., Tsai, P. C., & Cochran, J. R. (2011). Engineering hepatocyte growth factor fragments with high stability and activity as Met receptor agonists and antagonists. *Proceedings of the National Academy of Sciences of the United States of America, 108*(32), 13035–13040.

Jönsson, U., Fägerstam, L., Ivarsson, B., Johnsson, B., Karlsson, R., Lundh, K., et al. (1991). Real-time biospecific interaction analysis using surface plasmon resonance and a sensor chip technology. *BioTechniques, 11*(5), 620–627.

Jönsson, U., Fägerstam, L., Löfas, S., Stenberg, E., Karlsson, R., Frostell, A., et al. (1993). Introducing a biosensor based technology for real-time biospecific interaction analysis. *Annales De Biologie Clinique, 51*(1), 19–26.

Kastritis, P. L., & Bonvin, A. M. J. J. (2013). On the binding affinity of macromolecular interactions: Daring to ask why proteins interact. *Journal of the Royal Society Interface, 10*(79), 20120835.

Kastritis, P. L., Moal, I. H., Hwang, H., Weng, Z., Bates, P. A., Bonvin, A. M. J. J., et al. (2011). A structure-based benchmark for protein–protein binding affinity. *Protein Science, 20*(3), 482–491.

Kenakin, T. (2016). The mass action equation in pharmacology. *British Journal of Clinical Pharmacology, 81*(1), 41–51.

Klotz, I. M. (1982). Numbers of receptor sites from Scatchard graphs: Facts and fantasies. *Science, 217*(4566), 1247–1249.

Krohn, K. A., & Link, J. M. (2003). Interpreting enzyme and receptor kinetics: Keeping it simple, but not too simple. *Nuclear Medicine and Biology, 30*(8), 819–826.

Kuriyan, J., Boyana, K., & Wemmer, D. (2013). Molecular recognition: The thermodynamics of binding. *The Molecules of Life: Physical and Chemical Principles, 1,* 531–548.

Limbird, L. E. (1995). *Cell surface receptors: A short course on theory and methods.* Norwell, MA: Kluwer Academic Publishers.

Maguire, J. J., Kuc, R. E., & Davenport, A. P. (2012). Radioligand binding assays and their analysis. *Methods in Molecular Biology (Clifton, N.J.), 897,* 31–77.

McKinney, M., & Raddatz, R. (2006). Practical aspects of radioligand binding. *Current Protocols in Pharmacology, 33,* 1.3.1–1.3.42, Unit 1.3. Hoboken, NJ: John Wiley & Sons, Inc.

Moore, S. J., & Cochran, J. R. (2012). Engineering knottins as novel binding agents. *Methods in Enzymology, 503,* 223–251.

Myszka, D. G. (2000). Kinetic, equilibrium, and thermodynamic analysis of macromolecular interactions with BIACORE. *Methods in Enzymology, 323,* 325–340.

Patching, S. G. (2014). Surface plasmon resonance spectroscopy for characterisation of membrane protein–ligand interactions and its potential for drug discovery. *Biochimica et Biophysica Acta (BBA)—Biomembranes, 1838*(1 Pt. A), 43–55.

Pollard, T. D. (2010). A guide to simple and informative binding assays. *Molecular Biology of the Cell, 21*(23), 4061–4067.

Rosenthal, H. E. (1967). A graphic method for the determination and presentation of binding parameters in a complex system. *Analytical Biochemistry, 20*(3), 525–532.

Scatchard, G. (1949). The attractions of proteins for small molecules and ions. *Annals of the New York Academy of Sciences, 51*(4), 660–672.

Stephenson, R. P. (1956). A modification of receptor theory. *British Journal of Pharmacology and Chemotherapy, 11*(4), 379–393.

CHAPTER THREE

Protein and Antibody Engineering by Phage Display

J.C. Frei, J.R. Lai[1]
Albert Einstein College of Medicine, Bronx, NY, United States
[1]Corresponding author: e-mail address: jon.lai@einstein.yu.edu

Contents

1. Introduction 46
2. Equipment 48
3. Materials 50
   3.1 Cell Lines 50
   3.2 M13KO7 Helper Phage 52
   3.3 Phagemid Considerations 53
   3.4 Reagents 54
4. Phage Display and Library Design 57
   4.1 Humanization of a Murine SUDV Antibody 60
   4.2 Mapping Hotspot Residues of Protein–Protein Interfaces 61
5. Library Production 66
   5.1 Kunkel Mutagenesis 66
   5.2 Primer Design 72
   5.3 Library Quality Control 72
6. Library Selections and Screening 73
   6.1 Selection 73
   6.2 Screening 75
   6.3 Characterization of Selected Clones 78
   6.4 Troubleshooting 84
References 85

Abstract

Phage display is an in vitro selection technique that allows for the rapid isolation of proteins with desired properties including increased affinity, specificity, stability, and new enzymatic activity. The power of phage display relies on the phenotype-to-genotype linkage of the protein of interest displayed on the phage surface with the encoding DNA packaged within the phage particle, which allows for selective enrichment of library pools and high-throughput screening of resulting clones. As an in vitro method, the conditions of the binding selection can be tightly controlled. Due to the high-throughput nature, rapidity, and ease of use, phage display is an excellent technological platform for engineering antibody or proteins with enhanced properties. Here, we describe methods for synthesis, selection, and screening of phage libraries with particular emphasis on designing humanizing antibody libraries and combinatorial scanning mutagenesis libraries. We conclude with a brief section on troubleshooting for all stages of the phage display process.

1. INTRODUCTION

Phage display is an in vitro selection technique for the isolation of binding peptides or proteins from diverse mutagenic libraries. The method was first described in Smith (1985) using peptides as fusions to the pIII minor coat protein. Since then, a range of different peptide/protein sizes, scaffolds, and expression formats have been described (elaborated later); we will focus our discussion here on phage display of proteins or antibodies. Displayed elements are expressed on the surface of filamentous M13 bacteriophage typically as a fusion to either the major (pVIII) or minor (pIII) phage coat protein; the DNA of each library member is packaged within the phage particle, thus linking its phenotype with its genotype (Fig. 1). This critical phenotype-to-genotype linkage allows for the high-throughput selection of variants and rapid identification of selectants with desirable characteristics such as high affinity and increased specificity. Proteins up to 100 kDa, such as bivalent Fabs, have been successfully displayed on the surface of phage, and therefore this technique is amenable to a diverse size range of protein scaffolds (Lee, Sidhu, & Fuh, 2004).

Diversity can be incorporated into phage libraries via both random and targeted mutagenesis methods. In this chapter, we focus on the use of degenerate oligonucleotides and Kunkel mutagenesis for the targeted incorporation of diversity during library production. After a phage library is produced, biopanning can be used to selectively enrich for clones with desired properties such as affinity, specificity, stability, or even new or improved enzymatic properties (Forrer, Jung, & Plückthun, 1999). For binding selections, the target is immobilized onto a solid support, such as the surface of a 96-well plate or magnetic beads, then the library phage added, nonbinders washed away, and binders eluted and amplified for subsequent rounds of selection (Fig. 1). Various strategies can be employed to enrich for binders with very specific preferences such as negative selection, and/or competitive selection to drive recognition of one target or binding region over another (Clackson & Lowman, 2004).

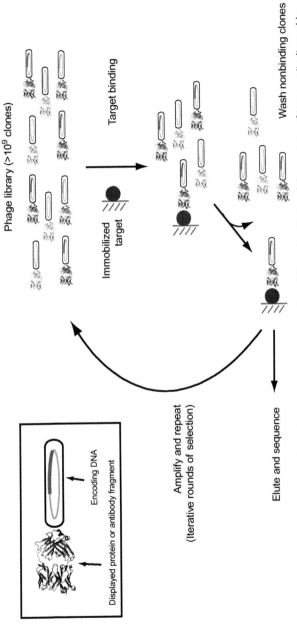

Fig. 1 Phage display selection. An initial population of phage, expressing different variants of the protein of interest (indicated by *different colors* (*different gray shades* in the print version)), are selected against the target. The target is first immobilized, the phage library is added with nonbinders washed away, followed by the elution of bound phage. The output phage population can then be amplified by infecting *E. coli* cells. After several rounds of selection, clones demonstrating desirable properties can be sequenced.

The utility of phage display is illustrated by its extensive use for the engineering of widely different protein scaffolds for a variety of applications. Some examples include: antibodies that target specific fusion-relevant intermediates of virus glycoproteins (Koellhoffer et al., 2012), humanized antibodies that retain high affinity for their original antigens (Baca, Presta, O'Connor, & Wells, 1997; Chen et al., 2014), human growth hormone variants with increased affinity for the growth hormone receptor (Lowman & Wells, 1993), cytochrome b562 variants that selectively bind BSA conjugated to N-methyl-p-nitrobenzylamine derivative 1 (Ku & Schultz, 1995), fibronectin type III domain variants that bind ubiquitin (Koide, Bailey, Huang, & Koide, 1998), Zinc finger domains that bind novel DNA sequences (Choo & Klug, 1994; Jamieson, Kim, & Wells, 1994; Rebar & Pabo, 1994), knottins that bind glycosylated alkaline phosphatase with micromolar affinity (Smith et al., 1998), cytotoxic T lymphocyte-associated antigen 4 variants that bind specifically to human $av\beta3$ integrin (Hufton et al., 2000), tendamistat, a β-sheet protein, that binds to the monoclonal antibody A8, which recognizes endothelin (McConnell & Hoess, 1995), and bilin binding protein, a lipocalin, that specifically binds to fluorescein (Beste, Schmidt, Stibora, & Skerra, 1999).

Phage display can also be used to rapidly dissect the energetics of protein interaction surfaces using combinatorial scanning mutagenesis, a technique in which libraries are designed to allow interface residues of one protein to vary between wild type and alanine (combinatorial alanine scanning) or other variations such as homolog scanning. This technique readily identifies "hotspot residues," or residues that contribute substantial binding energy to the interaction. Combinatorial alanine scanning with phage display has been used to probe the energetics of the growth hormone and growth hormone receptor interface, as well as, to characterize residues that contribute substantially to the binding of a Dengue virus (DENV) broadly neutralizing antibody to its antigens (Frei, Kielian, & Lai, 2015; Pal, Kossiakoff, & Sidhu, 2003; Weiss, Watanabe, Zhong, Goddard, & Sidhu, 2000). We describe here how to design, synthesize, and screen phage libraries, including a list of troubleshooting steps for common problems that arise during the phage display process.

2. EQUIPMENT

- Pipettes
- Pipette tips with filters (phage manipulations should be done with filter tips)
- Multichannel pipettes—10 and 300 μL volumes

- Reagent reservoirs (USA Scientific, Ocala, FL, catalog #: 1346-2510)
- Pipette gun
- Serological pipettes
- pH indicator strips (Sigma-Aldrich; Whatman, St. Louis, MO, catalog #: Z134147)
- 96-Well plates (USA Scientific, Ocala, FL; CytoOne, catalog #: CC7682-7596)
- 96-Well high-binding plates (Corning Incorporated; Costar, Corning, NY, catalog #: 3590)
- Plastic or foil plate covers (USA Scientific, Ocala, FL, catalog #: 2921-0000 or 2923-0100)
- 96-Well 2 mL deep-well plate (USA Scientific, Ocala, FL; PlateOne, catalog #: 1896-2000)
- Gas-permeable plate covers (Thermo Scientific, Waltham, MA, catalog #: AB-0718)
- Repeater pipette and tips
- 250 mL, 500 mL, 1 L, 2 L, and 4 L baffled glass flasks
- Electroporator (BioRad; GenePulser Xcell™, Hercules, CA, catalog #: 1652660)
- 2 mm- and 1 mm-gap cuvettes (Fisher Scientific, Hampton, NH; Fisherbrand™ Electroporation Cuvettes Plus™, catalog #: FB102 and FB101)
- Shaking incubators
- Incubator
- 1 L Bottle Top Filters (Corning, Corning, NY, catalog #: 431174)
- Plate shaker (USA Scientific, Ocala, FL, catalog #: 7400-4000)
- Plate washer
- Plate reader (Molecular Devices, Sunnyvale, CA, SpectraMax M5 Multi-Mode plate reader)

Other disposables
- Plates for agar plates
- ColiRollers Plating Beads (Novagen, Darmstadt, Germany, catalog #: 71013)
- Amicon Ultra 4 Centrifugal Filters (Millipore, Darmstadt, Germany, catalog #s: UFC800396 (3 k MWCO), UFC801096 (10 k MWCO), and UFC903096 (30 k MWCO))
- SnakeSkin Dialysis Tubing (Thermo Scientific, Waltham, MA, catalog #s: 68035 (3500 MWCO) and 88245 (10 k MWCO))
- PD-10 Desalting Columns (GE Healthcare, Piscataway, NJ, catalog #: 17-0851-01)

3. MATERIALS

3.1 Cell Lines

SS320 cells. The success of any phage display project relies on the ability to electroporate a large amount of library DNA into *Escherichia coli* with high efficiency. The *E. coli* strain SS320 was developed by crossing XL1-Blue, which contains the F' pilus that is required for phage entry on a tet-selectable episome, and MC1061, which tolerates high cell densities and thus can be prepared in high electroporation competency (Sidhu, Lowman, Cunningham, & Wells, 2000). The SS320 strain cell line can be propagated in-house as detailed in Protocol 1, or can be purchased from Lucigen (Middleton, WI, catalog #s: 60512-1 or 60512-2).

If using cells from Lucigen, a critical step of library production is the addition of M13KO7 helper phage after electroporation of library DNA into SS320 cells. If produced in-house following Protocol 1, SS320 cells will already be infected with M13KO7 phage, so that addition of exogenous helper phage is unnecessary.

3.1.1 Protocol 1: Production of SS320 Cells (Modified from Tonikian, Zhang, Boone, & Sidhu, 2007)

1. Day 1:
   a. Streak a fresh LB/Tet plate with SS320 cells
   b. Incubate overnight at 37°C
2. Day 2:
   a. Inoculate SS320 colony into 25 mL 2xYT/25 µL Tet and grow to $OD_{600}=0.6$ at 37°C, 220 RPM
   b. Serially dilute M13KO7 phage 1:10 in PBS, pH 7.4 (add 30 µL phage to 270 µL PBS)
   c. Mix 500 µL SS320 cells with 200 µL of each M13KO7 dilution
   d. Add to 4 mL Top Agar and pour onto LB/Tet plates
   e. Incubate overnight at 37°C
3. Day 3:
   a. Pick a plaque and inoculate into 1 mL 2xYT media/0.5 µL kanamycin (final kanamycin concentration = 25 µg/mL)
   b. Incubate at 37°C, 220 RPM, for 6–8 h
   c. Transfer into 100 mL 2xYT/50 µL kanamycin (final kanamycin concentration = 25 µg/mL)
   d. Incubate overnight at 37°C and 220 RPM

4. Day 4:
   a. To twelve 2 L baffled flasks add the following: 500 mL SB media with 250 μL kanamycin (final kanamycin concentration = 25 μg/mL)
   b. Add 5 mL of the overnight culture to each flask
   c. Incubate at 37°C, 220 RPM until culture reaches $OD_{600} = 0.5$–0.8
      i. *Note*: Monitor the OD_{600} closely
   d. Immediately chill cells on ice and transfer to preautoclaved and pre-chilled 1 L centrifuge bottles containing stir bars
   e. Spin down cells at 5000 RPM for 7 min at 4°C—discard supernatant
   f. Wash cells with 4 L of cold HEPES buffer (gently swirl centrifuge bottles to detach cells from bottle)—consolidate cells to end with highly concentrated cells
   g. Spin down as in step 4e—discard supernatant
   h. Wash cells with 2 L of cold HEPES buffer and consolidate cells as in step 4f
   i. Spin as in step 4e—discard supernatant
   j. Wash cells with cold 10% glycerol (in water, filter sterilized) and consolidate into 1 spin bottle
   k. Spin as in step 4e—discard supernatant
   l. Aliquot cells into preautoclaved ependorf tubes (350 μL aliquots) and flash freeze with liquid nitrogen—store at −80°C

XL1-Blue cells. *E. coli* XL1-Blue cells harbor an episome that contains the F′ (pilus) that is required for M13 bacteriophage infection, which allows for amplification of phage libraries. This episome also contains the tetracycline resistance gene, so cells should be grown in the presence of tetracycline to maintain the F′ pilus. An original stock can be purchased from Agilent Technologies (Santa Clara, CA, catalog #: 200249) and propogated in-house following Protocol 2. A fresh XL1-Blue plate should be streaked out from a −80°C glycerol stock each week for all phage-based experiments.

3.1.2 Protocol 2: Production of XL1-Blue Cells
1. Day 1:
   a. Streak out XL1-Blue cells from glycerol stock onto LB/Tet plate
   b. Incubate overnight at 37°C
2. Day 2:
   a. Pick single colony into 10 mL LB media/10 μL tetracycline
   b. Incubate overnight at 37°C and 220 RPM

3. Day 3:
   a. Inoculate 1 L LB/1 mL tetracycline with 5 mL of overnight culture (dilution of 1:2000)
   b. Incubate at 37°C and 220 RPM until cell density reaches $OD_{600} = 0.3–0.6$
      i. *Note*: $OD_{600} = 0.4$ seems to work the best
   c. During cell incubation: prechill centrifuge bottles, autoclaved Eppendorf tubes, 10% glycerol, and sterile water (the rest of the protocol should be carried out either on ice or in a cold room)
   d. Once cells reach OD_{600} of 0.3–0.6, immediately chill on ice for 10–15 min
   e. Transfer cells to centrifuge bottles and spin at 4000 RPM, 4°C for 20 min—discard supernatant
   f. Rinse cells with 1/10th original volume of ice cold sterile water (do not resuspend the cells at this time)
   g. Gently pour off the water
   h. Resuspend the cells in 1/10th original volume of ice cold sterile water
   i. Centrifuge at 4000 RPM, 4°C for 20 min—gently discard the supernatant
   j. Resuspend the cells with 1/50th volume of ice cold 10% glycerol and aliquot (200–300 µL) into autoclaved, prechilled Eppendorf tubes
   k. Flash freeze in liquid nitrogen and store at −80°C

Other cell lines needed
1. CJ236 (Lucigen, Middleton, WI, catalog #: 60701-1 or 60701-2)
2. One Shot TOP10 Chemically Competent cells (Invitrogen, Carlsbad, CA, catalog #: C4040-03)
3. One Shot BL21(DE3) Chemically Competent cells (Invitrogen, Carlsbad, CA, catalog #: C6000-03)
4. HEK239F (Invitrogen, Carlsbad, CA, catalog #: R790-07)

3.2 M13KO7 Helper Phage

Our group utilizes M13KO7 (NEB, Ipswich, MA, catalog #: N0315), although other helper phage strains are available (for example, R408 and VCSM13). Helper phage provide all of the necessary wild-type phage proteins required for the production of progeny phage particles; the phagemid (discussed later) contains only the engineered protein-pIII or protein-pVIII fusion and thus does not result in a replicative particle in the absence of the helper phage (Bass, Greene, & Wells, 1990). The helper phage genome contains mutations that reduce its packaging ability, so that the genome of the phagemid is preferentially packaged into progeny phage, preserving the phenotype-to-genotype linkage. The M13KO7 phage contain a kanamycin resistance gene; therefore, *E. coli* that are infected with M13KO7 can readily be selected by the addition of kanamycin to the media. For the production of M13KO7 phage from a glycerol stock, see Protocol 3.

3.2.1 Protocol 3: Production of M13KO7 Helper Phage (Modified from Tonikian et al., 2007)

1. Day 1:
   a. Streak out a new XL1-Blue plate from glycerol stock
   b. Incubate overnight at 37°C
2. Day 2:
   a. Pick a single colony into 2 mL 2xYT/2 µL Tet media
   b. Incubate at 37°C and 220 RPM for 6–8 h
   c. Add M13KO7 helper phage from glycerol stock to a final concentration of 10^{9-10} phage per mL
   d. Incubate at 37°C and 220 RPM for 30 min
   e. Transfer to 1 L of 2xYT/1 mL Kan media in a 4 L baffled flask
   f. Incubate overnight at 37°C and 220 RPM
3. Day 3:
   a. Centrifuge cells at 4000 RPM for 15 min at 4°C—phage are in supernatant
   b. Transfer supernatant to a flask containing 1/5th volume of 5 × PEG/NaCl and incubate with stirring for 1 h at 4°C
   c. Centrifuge at 12,000 RPM for 20 min at 4°C—discard supernatant
   d. Resuspend phage in 10 mL PBS, pH 7.4
   e. Spin down in microcentrifuge at 7600 RPM for 10 min at 4°C
   f. Transfer supernatant to fresh tubes and either store at 4°C or add 10% glycerol and store at −80°C
      i. *Note*: Helper phage stored at 4°C is stable for 6 months

3.3 Phagemid Considerations

Phagemids are plasmids that contain only one of the phage structural proteins (pIII or pVIII), which is fused with the desired display protein. During phage production, the remaining phage proteins are supplied by coinfection of helper phage (see Section 3.2). Phagemids contain a bacterial origin of replication for propagation of the plasmid in *E. coli* and an F1 phage origin for replication and packaging of single-stranded DNA (ssDNA) (Qi, Lu, Qiu, Petrenko, & Liu, 2012).

Several phagemids are available for the production of phage libraries; the reader is directed to tables 1 and 2 of Qi et al., 2012 for a comprehensive list (Qi et al., 2012). Our lab has had the most success with phagemid HP153 (Persson et al., 2013). HP153 drives expression of the protein of interest fused to the C-terminal domain of the phage minor coat protein-pIII via the weak phosphatase A (phoA) promoter. This promoter is activated under low phosphate conditions. Thus, in phosphate-rich media the promoter is only minimally active, allowing for slow and steady expression of the pIII- or pVIII-fused protein (Lowman, Bass, Simpson, & Wells, 1991). HP153 contains a StII signal sequence N-terminal to the displayed protein, which delivers the protein to the periplasmic space, where folding and formation of disulfide bonds occur before it is incorporated into nascent phage particles. Additionally, this phagemid contains a dimerization domain allowing for the bivalent expression of the protein at the surface of phage. Cloning into HP153 using NsiI and FseI restriction sites results in bivalent display. Selection for *E. coli* containing HP153 plasmid is achieved by the addition of carbenicillin, as HP153 contains a β-lactamase gene to confer ampicillin resistance.

3.4 Reagents
3.4.1 Media
1. 2xYT media (Fisher Scientific, Hampton, NH, catalog #: BP97432)
2. Super Broth (SB) media
    a. Per liter: 10 g MOPS, 30 g tryptone, 20 g yeast extract, pH 7.0
3. SOC media:
    a. Per liter: 20 g tryptone, 5.0 g yeast extract, 0.5 g NaCl, 0.186 g KCl
    b. Optional: add 10 mL 1 M $MgCl_2$ (filtered) and 20 mL 1 M Glucose
4. Low phosphate media:
    a. Per liter: 3.57 g ammonium sulfate, 0.517 g sodium citrate monobasic anhydrous, 1.07 g potassium chloride, 5.36 g Hy-Case SF casein acid hydrolysate from bovine milk, 5.36 g yeast extract, pH 7.3 with KOH
    b. After autoclaving, add 7 mL 1 M $MgSO_4$ and 14 mL 1 M Glucose (filtered) before use
    c. Optional: add 1 M MOPS (filtered) to buffer media for optimal cell growth
5. LB Agar, Miller (Fisher Scientific, Hampton, NH, catalog #: BP9724-2):
    a. Add either carbenicillin, tetracycline, kanamycin, carbenicillin and tetracycline, or carbenicillin and chloramphenicol at 1:1000 dilution

6. 2xYT Top Agar:
   a. Per liter: 16 g tryptone, 10 g yeast extract, 5 g NaCl, 7.5 g granulated agar, pH 7.0 with NaOH

3.4.2 Buffers (Filter Sterilize)

1. Phosphate-buffered saline (PBS): 20 mM Na_3PO_4, 150 mM NaCl, pH 7.4 and 8.0 (pH 8.0 used as coating buffer)
2. PB-T: 20 mM Na_3PO_4, 150 mM NaCl, 1% BSA (w/v), 0.5% Tween-20 (v/v)
3. PBS-T: 20 mM Na_3PO_4, 150 mM NaCl, 0.5% Tween-20 (v/v)
4. Phage elution buffer: 100 mM Glycine, pH 2.0
5. Phage neutralization buffer: 2 M Tris, pH 7.5
6. Tris buffered saline (TBS): 20 mM Tris–HCl, 150–500 mM NaCl, pH 7.8
7. MOPS (1 M): 209.27 g in 1 L H_2O, pH 7.3
8. HEPES: 1 mM, pH 7
9. 8 M urea in TBS
10. 0.2% SDS, 8 M urea in TBS
11. MLB buffer: 1 M Sodium Perchlorate, 30% Isopropanol in water
12. MP buffer: 3.3 g citric acid monohydrate in 3 mL sterile H_2O; incubate with stirring for 5 min at room temperature, filter sterilize through 0.2 μm filter
13. Gentle Ag/Ab Binding Buffer, pH 8.0 (Pierce, Thermo Fisher Scientific, Waltham, MA, catalog #: 21020)
14. Gentle Ag/Ab Elution Buffer, pH 6.6 (Pierce, Thermo Fisher Scientific, Waltham, MA, catalog #: 21027)
15. 2% SDS (w/v) in H_2O

3.4.3 Solutions

1. 5× PEG/NaCl—filter sterilize
   a. Per liter: 200 g PEG-8000 (20% w/v), 150 g NaCl (15% w/v)
2. 1–3% BSA (w/v) in PBS, pH 7.4
3. 2–5% Nonfat dry milk (NFDM) (w/v) in PBS, pH 7.4

3.4.4 Antibiotics (Filter Sterilize)

1. Carbenicillin: stock concentration of 100 mg/mL
2. Tetracycline: stock concentration of 5 mg/mL (made up in 70% ethanol)
3. Kanamycin: stock concentration of 50 mg/mL
4. Chloramphenicol: stock concentration of 34 mg/mL (made up in 100% ethanol)

3.4.5 Other Reagents

1. 3,3′,5,5′-Tetramethylbenzidine (TMB) (Sigma, St. Louis, MO, catalog #: T0440)
2. 0.5 M H_2SO_4
3. Anti-M13-HRP conjugated antibody (GE Healthcare, Marlborough, MA, catalog #: GE27-9421-01)
4. T7 DNA Polymerase (NEB, Ipswich, MA, catalog #: M0274L)
5. T4 DNA Ligase (Invitrogen, Carlsbad, CA, catalog #: 15224-017)
6. 25 mM dNTPs (Fisher, Hampton, NH, catalog #: FERR1121)
7. 25 mg/mL Uridine (Sigma, St. Louis, MO, catalog #: U3750-100G)
8. BugBuster 10X Protein Extraction Reagent (Millipore, Billerica, MA, catalog #: 70921)
9. QIAquick PCR Purification Kit (Qiagen, Hilden, Germany, catalog #: 28104)
10. QIAquick Gel Extraction Kit (Qiagen, Hilden, Germany, catalog #: 28704)
11. QIAprep Spin Miniprep Kit (Qiagen, Hilden, Germany, catalog #: 27104)
12. NucleoBond Xtra Maxi Plus (Macherey-Nagel, Düren, Germany, catalog #: 740416.50)
13. Ni-NTA Agarose (Qiagen, Hilden, Germany, catalog #: 30210)
14. DNaseI (Invitrogen, Carlsbad, CA, catalog #: 18047-019)
15. SIGMAFast EDTA-free protease cocktail inhibitor (Sigma, St. Louis, MO, catalog #: S8830)
16. Precision Plus Protein Dual Color Standards (BioRad, Hercules, CA, catalog #: 161-03741)
17. Anti-His (C-term)-HRP antibody (Invitrogen, Carlsbad, CA, catalog #: 46-0707)
18. Protein A Agarose (Pierce, Thermo Fisher Scientific, Waltham, MA, catalog #: 15918-014)
19. Phusion High Fidelity PCR Master Mix with HF Buffer (NEB, Ipswich, MA, catalog #: M0531S)
20. Restriction enzymes: BssHII, BsiWI, NheI, NsiI, FseI (NEB, Ipswich, MA, catalog #s: R0199S, R0553S, R0131S, R0127S, R0558S, respectively)
21. Alkaline Phosphatase, Calf Intestinal (NEB, Ipswich, MA, catalog #: M0290S)
22. Quick Ligation Kit (NEB, Ipswich, MA, catalog #: M2200S)
23. Polyethylenimine (PEI) Linear, 25,000 MW (Polysciences, Warrington, PA, catalog #: 23966)

24. FreeStyle™ 293 Expression Medium (ThermoFisher; Invitrogen, Carlsbad, CA, catalog #: 12338-018)
25. M2 antibody (Sigma-Aldrich, St. Louis, MO, catalog #: F1804)
26. Protein A-HRP conjugate (Invitrogen, Carlsbad, CA, catalog #: 101023)

4. PHAGE DISPLAY AND LIBRARY DESIGN

The first step in any phage display project is to clone the protein of interest into a suitable phagemid and verify that it is functionally expressed on the surface of phage. In our lab, phagemid HP153 has been shown to readily display many different proteins including domain III (DIII) of the glycoprotein E of DENV (Frei et al., 2015), Sudan virus (SUDV) Fabs (Chen et al., 2014), Dengue Fab E106 (unpublished results), HIV-1 Fab 2G12 (Stewart, Liu, & Lai, 2012), and protein G domain B1 from *Streptococcus* (unpublished results). To detect the protein of interest on phage and to monitor expression levels, we include a FLAG epitope (DYKDDDDK) either at the N-terminus (Dengue DIII, protein G) or C-terminus (light chain of Fabs), and probe for it using the anti-FLAG antibody M2. Protocol 4 details cloning into HP153. The protein insert should be designed with 5′-NsiI and 3′-FseI cut sites, 5′-"cc" (two cytosines) to keep the protein sequence in frame with the StII secretion sequence, and an N- or C-terminal FLAG-tag, or any tag of choice for which there is a well-characterized detection antibody.

Once cloned into HP153, expression of the protein at the surface of phage is confirmed by phage ELISA, in which the anti-FLAG antibody M2 is immobilized on a well, phage expressing the fusion protein are added, and bound phage are detected by the addition of an anti-M13-HRP conjugated antibody. ELISAs are developed using TMB and quenched with sulfuric acid, with absorbance measured at 450 nm (Abs. 450 nm) serving as a readout for phage binding. Although a positive ELISA result with M2 indicates expression of the protein, it does not necessarily confirm proper folding/function; function of the displayed protein can be determined by performing an ELISA with a known binding agent or antibody, preferably one that requires a conformational epitope for recognition. See Protocol 5 for phage amplification and titering steps and Protocol 6 for phage ELISA. We typically observe Abs. 450 nm values of 2.5 or higher for well-expressing constructs with 1–20 min developing time (Fig. 2).

After confirming functional expression of a protein of interest on phage, library design can then ensue. The type of randomization scheme employed and the library format is highly dependent on the end goal of the project. If

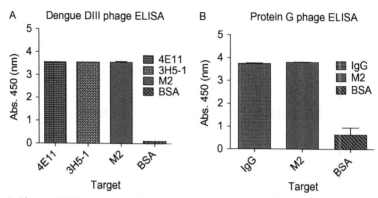

Fig. 2 Phage ELISA at single high-point concentration of phage for Dengue DIII-displaying phage (A) and protein G-displaying phage (B). (A) DIII displayed on phage binds specifically to target antibodies 4E11 and 3H5-1 but not negative control BSA. The N-terminal FLAG-tag can also be probed by M2 binding. (B) Protein G displayed as a fusion to the pIII coat protein also shows specific binding to its target IgG over negative control and displays the N-terminal FLAG-tag as probed by M2 binding.

the goal is to optimize or alter selectivity of a given protein, then soft randomization libraries (discussed below) may be most appropriate. However, if the goal is to impart a protein scaffold with new function, then either hard randomization or limited diversity mutagenesis may be required. Furthermore, designing a library is easiest when high-resolution structures of the displayed protein or related homologues are available, however, this is not a requirement. In projects where the objective is to improve or broaden the specificity of an existing scaffold, the known contact residues can be targeted for mutagenesis. However, in other cases where one seeks to engineer new function into a scaffold or engineer an alternative binding interface, several sublibraries or a semiempirical step-wise approach may be required to identify which residues should be varied (Koide, Wojcik, Gilbreth, Hoey, & Koide, 2012; Mandal et al., 2012). Our lab has successfully used phage display to humanize an anti-SUDV mouse antibody (Chen et al., 2014), isolate antibodies specific to different proteolytic intermediates of the Ebola glycoprotein (Koellhoffer et al., 2012), isolate Fabs against proapoptosis protein BAX (Uchime et al., 2016), and map the hotspot residues in antibody–antigen interactions on both the antigen (DENV DIII) or the antibody (Da Silva, Harrison, & Lai, 2010; Frei et al., 2015; Liu, Regula, Stewart, & Lai, 2011; Stewart, Harrison, Regula, & Lai, 2013; Stewart et al., 2012).

The targeted introduction of diversity into any protein is accomplished through the use of degenerate oligonucleotides. Many randomization schemes have been developed including hard randomization (allowing all 20 amino acids), soft randomization (high bias toward WT residue identity), and tailored/restricted diversity (Clackson & Lowman, 2004). It is important to keep in mind that the actual size of the library must surpass the theoretical diversity (all possible variants) of a phage library by 100-fold in order for all variants to be plausibly surveyed. This consideration will thus limit the number of residues that are randomized based on the diversification scheme selected. However, in cases where the library randomization scheme is highly biased, as in soft randomization or tailored/restricted diversity, there is often not a strict requirement that library size surpass the theoretical complexity of the library. Our lab typically achieves electroporation efficiencies of 10^8–10^9 transformants per reaction, and thus library sizes of 10^9–10^{10} can be achieved by pooling cultures from 10 to 20 electroporations.

In hard randomization, a position is allowed to vary among all 20 amino acids using the degenerate codon "NNS," in which N codes for any of the four nucleotides and S codes for guanosine or cytosine. This randomization scheme encodes 32 of the 64 possible codons, including all 20 amino acids (some in duplicate) and one stop codon. When using a hard randomization strategy the number of residues that can be varied is limited to approximately 6, as the theoretical diversity becomes much too large past this (theoretical diversity $= 32^6$ at the genetic level $= 1.1 \times 10^9$). In soft randomization, a position is allowed to vary among all 20 amino acids, but is biased toward the wild-type sequence by doping the four bases at each codon position at nonequivalent levels. To achieve approximately 50% of the wild-type sequence, each base of each codon is 70% wild type and 10% each of the other three nucleotides. With soft randomization, additional positions can be varied because, although the theoretical diversity is still 32^N (where N is the number of randomized residues), the library contains strong biases toward wild-type residue identity at any particular position. Soft randomization is best instituted in cases where one seeks to optimize, broaden, or even switch the specificity of a binding protein using the same contact residues. Lastly in tailored diversity strategies, degenerate codons are chosen that code for amino acids with desirable physicochemical properties, or in cases where only a subset of the 20 amino acids is to be considered. The reader is directed to table 1 of chapter 2 in Clackson and Lowman (2004) for a list of such degenerate codons.

In this section, we will focus the discussion of library design on the humanization of antibodies and mapping of hotspot residues in protein–protein interfaces.

4.1 Humanization of a Murine SUDV Antibody (Chen et al., 2014)

Full humanization of murine antibodies can be achieved rapidly by phage display if there is significant sequence and structural homology between the murine antibody and a human or humanized antibody. For the humanization of SUDV murine antibody 16F6, the structural alignment with a humanized template (YADS1, based on Herceptin®) was 1.7 Å in the non–complementarity determining region (CDR) portions of the variable domains. The CDR loops of 16F6 and YADS1 began and ended at approximately the same locations within the Fv scaffolds, but with diverse structural configurations in some cases. A "grafted" chimeric clone, in which the CDR segments of 16F6 were grafted on YADS1 was found to be inactive. In our experience, a simple CDR graft between a murine antibody and a closely related human scaffold often results in significant diminishment or complete loss of binding activity.

For humanization of 16F6, the variable heavy (VH) and variable light (VL) domains of the grafted antibody were cloned into HP153 for display on phage protein-pIII using the restriction sites NsiI and FseI. The synthetic gene was ordered to contain the following 5′ to 3′: NsiI cut site, two cytosines, VL sequence, CGGS (amino acids), FLAG-tag, TAA stop codon, DNA region containing an internal ribosome binding site and StII sequence (5′-ttaactcgaggctgagcaaagcagactactaataacataaagtctacgccggacgcatcgtggccctagtacgcaagttcacgtaaaaagggtaactagaggttgaggtgattttatgaaaaagaatatcgcatttcttcttgcatctatgttcgttttttctattgctacaaacgcgtacgctgagatctcc-3′), VH sequence, cysteine, gacaaaactcacacatgc (DNA sequence 5′ to 3′), and FseI cut site. This format drives the bicistronic expression of both VH and VL and will result in bivalent display of a LC C-terminally FLAG-tagged Fab. Expression can be checked via phage ELISA against the anti-FLAG antibody M2 (see Protocol 6).

If a structure of the antibody–antigen complex is available, all paratope residues (residues of the antibody that contact the antigen) can be determined by visual inspection or calculation of buried surface area. In the case of 16F6, the humanization library maintained most of these contact residues as the native murine residue identity (ie, the residue present in the starting antibody). Noncontact residues that differed between 16F6 and YADS1 were varied among the human antibody identity and the starting antibody identity, as well as, residues with similar physicochemical properties (Chen et al., 2014). Two positions that differed between 16F6 and YADS1, whose contributions to direct binding or productive loop conformation were unclear, were varied among all 20 amino acids using the hard randomization "NNS" codon.

This randomization scheme minimizes the diversity of the library so that a phage library large enough to cover the theoretical diversity can be synthesized. Maintaining only the contacting residues as the starting antibody identity minimizes the nonhuman regions of the antibody and the potential for a human–antimouse–antibody response. Other methods for humanization of antibodies using phage display have been previously described and can serve as complementary or alternative approaches (Baca et al., 1997; Rader, Cheresh, & Barbas, 1998). In the case of 16F6 humanization, the strong sequence and structural agreement between the murine antibody 16F6 and a humanized scaffold YADS1 allowed precise prediction of sites that would require randomization. However, other cases for which the alignment of murine and human/humanized scaffolds is not as clear may necessitate randomization in positions of the framework that are distal to the CDRs.

4.2 Mapping Hotspot Residues of Protein–Protein Interfaces (Frei et al., 2015; Stewart et al., 2013)

While surfaces involved in protein–protein interfaces can be quite large and encompass many residues from both proteins, it has been well established that only a small subset of the contact residues contribute significantly to the energetics of binding and form the "functional" epitope (Bogan & Thorn, 1998; Clackson & Wells, 1995). Combinatorial alanine scanning mutagenesis by phage display is a method for rapidly mapping protein–protein interfaces to statistically determine hotspot residues, or residues that contribute significantly (>1 kcal/mole) to the energetics of binding. We have successfully applied this strategy to mapping the hotspot residues of the DENV domain III (DIII) protein binding to the broadly neutralizing antibody 4E11, and others have applied it to the human growth hormone binding to its receptor (Frei et al., 2015; Pal et al., 2003; Weiss et al., 2000). Here we describe the library design for mapping the functional epitopes of DENV DIII binding to antibody 4E11; however, this strategy can be generalized to other protein–protein interactions.

N-terminally FLAG-tagged DENV DIIIs from all four serotypes (DENV1–4) were cloned into phagemid HP153 using NsiI and FseI restriction sites for bivalent display (see Protocol 4). For each serotype, the antibody-contacting residues (structural epitope) were determined from the crystal structure of the antibody–antigen complex (Cockburn et al., 2012). These residues were allowed to vary among the wild-type identity and alanine using the degenerate codons listed in table 1 of Weiss et al. (2000). Due to the degenerate nature of the genetic code, some residues varied among wild type, alanine and two other residues. For example if the wild-type residue is lysine, then at that position the library encodes for lysine, alanine, glutamate, or threonine. This strategy was applied to all four DIIIs and resulted in the production of four phage libraries (Frei et al., 2015).

This approach can be generalized to any protein–protein interaction. A solved crystal structure of the complex, while useful, is not required for this analysis as multiple alanine scanning libraries can be synthesized and screened in parallel to map the complete binding interface. The reader is directed to Weiss et al. for a detailed explanation of the combinatorial shotgun scanning analysis that results in estimates of the $\Delta\Delta G$ of binding upon mutation to alanine (Weiss et al., 2000).

4.2.1 Protocol 4: Cloning into HP153
1. Day 1: Digestions (dilute vector and insert to 100 ng/mL in sterile H_2O)
   a. Add the following:

HP153		Insert	
15 µL	DNA	15 µL	DNA
5 µL	NEBuffer2.1	5 µL	NEBuffer2.1
1 µL	NsiI	1 µL	NsiI
1 µL	FseI	1 µL	FseI
28 µL	H_2O	28 µL	H_2O
50 µL	Total vol.	50 µL	Total vol.
Incubate at 37°C for 2.5 h			
Add 1 µL CIP to vector			
Incubate at 37°C for 0.5 h			
Run a 1% agarose gel on digestion products			

   b. Gel purify the appropriately sized bands (digested HP153 runs between 3000 and 4000 bp)
      i. Gel purify using Qiagen QIAquick Gel Purification Kit as per manufacturer's instructions
      ii. Elute with 30 µL H_2O
      iii. Measure concentrations

2. Day 1: Ligation (NEB Quick Ligation Kit—*Note*: Modified manufacturer's protocol)
   a. Add the following:

Ligation	
7.5 μL	2× Quick Ligase Buffer
2 μL	Digested vector
5.5 μL	Digested insert
1 μL	Quick ligase
Incubate at room temperature for 5 min	
Incubate on ice for 5 min	

      i. *Note*: May need to change the molar ratio of vector:insert for best results

3. Day 1: Transformation
   a. Add 5 μL ligation product to 25 μL One Shot TOP10 Chemically Competent cells
   b. Incubate on ice 10 min
   c. Heat shock at 42°C for 45 s
   d. Incubate on ice 2 min
   e. Add 200 μL SOC media and recover 45–60 min at 37°C, 220 RPM
   f. Plate out 10 and 200 μL onto LB/Carb plates
   g. Incubate overnight at 37°C
4. Day 2: Verify sequences
   a. *Note*: We send several colonies to Genewiz for sequencing by making grid plates
   b. Day 3: Send for sequencing
      i. Send 1 plate along with 400 μL of 5 μM primer for sequencing; store second plate at 4°C
      ii. Primer sequences for sequencing (5' to 3'):
         - VH-For (VH domain): TGTAAAACGACGGCCAGTG GACGCATCGTGGCCCTA
         - VH-Rev (anneals in CH domain): CAGGAAACAGCT ATGACCCCTTGGTG

- VL-For (VL domain): TGTAAAACGACGGCCAGTCTGTCATAAAGTTGTCACGG
- VL-Rev: CAGGAAACAGCTATGACCCCTTGGTACCCTGTCCG
- HP153-For (other proteins): CTGTCATAAAGTTGTCACGG

4.2.2 Protocol 5: Amplifying and Titering Phage

1. Day 1: Electroporate phagemid into XL1-Blue cells
   a. Thaw XL1-Blue cells on ice and place 2 mm-cuvette on ice
   b. To 1 tube of XL1-Blue cells add 2–10 µL of phagemid DNA
   c. Transfer cells and DNA to cuvette on ice—tap to get rid of air bubbles
   d. Electroporate using the following protocol:
      i. *E. coli*—2 mm, 2.5 kV
      ii. Voltage: 2500 V
      iii. Capacitance: 25 µF
      iv. Resistance: 200 Ω
      v. Cuvette: 2 mm
   e. Resuspend cells in 2 × 200 µL SOC media
   f. Recover at 37°C, 220 RPM, 45–60 min
   g. Plate various dilutions onto LB/Carb/Tet plates
   h. Incubate overnight at 37°C
2. Day 2: Amplify phage from single colony
   a. Add 1 colony to 1 mL 2xYT media/0.5 µL Carb/1 µL Tet
   b. Incubate at 37°C, 220 RPM until slightly dense
   c. Add a final concentration of 10^{9-10} M13KO7 phage/mL
   d. Incubate for 30–60 min at 37°C, 220 RPM
   e. Transfer 1 mL culture to 25–50 mL 2xYT/25–50 µL Carb/25–50 µL Kan
   f. Incubate overnight at 37°C, 220 RPM
3. Day 3: Precipitate and titer phage
   a. Precipitate phage:
      i. Spin down cells: 4000 RPM, 4°C, 15 min—phage are in supernatant
      ii. Transfer supernatant to 5.5 mL 5× PEG/NaCl
      iii. Incubate on ice 30 min
      iv. Spin down (SS-34 rotor): 12,000 RPM, 4°C, 20 min—discard supernatant

v. Resuspend phage in 1–2 mL PB-T
vi. Spin down in microcentrifuge: 7600 RPM, 4°C, 5 min
vii. Transfer supernatant to new tube
  b. Titering phage (use multichannel pipettes):
     i. Grow XL1-Blue cells in 2xYT/Tet media to $OD_{600}=0.6$
     ii. Make serial 1:10 dilutions of phage in 2xYT media—switch tips between dilutions
     iii. Transfer 10 μL of each dilution to 90 μL XL1-Blue cells
     iv. Incubate at 37°C for 45–60 min
     v. Using a multichannel pipette, plate out 5 μL of each dilution onto LB/Carb plate
     vi. Incubate overnight at 37°C
4. Day 4: Calculating phage titer (infectious units/mL (IU/mL) or colony forming units/mL (CFU/mL)):
   a. Count number of colonies at a dilution containing between 5 and 30 colonies
   b.

$$\text{Titer} = (\# \text{ colonies})\left(\frac{1}{5}\right)(1000)\left(10^{\text{dilution}+1}\right)$$

   i. *Note*: The dilution factor contains "+1" to account for the 1:10 dilution from media to cells

4.2.3 Protocol 6: Phage ELISA
1. Day 1: Amplify phage (Protocol 5)
2. Day 2: Titer phage (Protocol 5) and coat plates
   a. Coat antigen or antibody at 0.1–2 μg per well in coating buffer (100 μL total volume per well)
      i. *Note*: Empirically determine which amount is necessary for each protein pair
   b. Coating buffer: PBS, pH 8.0, serves as negative control
   c. If coating with M2 (anti-FLAG antibody): dilute 1:500 in PBS, pH 8.0
      i. *Note*: Can either coat overnight at 4°C or for 2 h at room temperature
3. Day 3: Phage ELISA
   a. Make serial 1:10 dilutions of amplified phage stock in PB-T—switch tips between dilutions

b. Block plate with 1–3% BSA (w/v) or 2–5% NFDM (w/v) for 1.5–2 h at room temperature with shaking
   c. Wash plate 5× with PBS-T
   d. Add 100 µL of phage dilutions per well—incubate for 1 h at room temperature with shaking
   e. Repeat step c
   f. Add 100 µL per well of 1:1000 diluted anti-M13-HRP (dilute in PB-T)—incubate for 45–60 min at room temperature with shaking
   g. Repeat step c
   h. Develop with 100 µL per well TMB and quench with 100 µL per well 0.5 M H_2SO_4
   i. Read absorbance at 450 nm

5. LIBRARY PRODUCTION

There are several methods to produce phage libraries; we favor the use of degenerate oligonucleotides and Kunkel mutagenesis to introduce targeted diversity as originally described by Kunkel, Roberts, and Zakourts, and Zakour (1987). Therefore, we chose to focus on this strategy for this section. The reader is encouraged to review "Phage display: A practical approach" for other methods of library synthesis (Clackson & Lowman, 2004).

5.1 Kunkel Mutagenesis

The process of Kunkel mutagenesis is summarized in Fig. 3. Kunkel mutagenesis relies on the synthesis of a ssDNA template enriched in uracil (dU-ssDNA) by isolating the ssDNA from phage produced in an *E. coli* strain that lacks dUTPase and uracil-N-glycosylase (dut^-/ung^-) such as CJ236 cells. Deficiencies in these two enzymes result in production of DNA that has a high incorporation rate of uracil (dU-ssDNA). CJ236 cells are chloramphenicol (cmp) resistant and should be plated on LB/cmp plates, or on LB/cmp/carb plates after electroporating in the phagemid.

Once dU-ssDNA has been produced, 5′-phosphorylated degenerate oligonucleotides are annealed to the dU-ssDNA. These primers are designed so that the 5′ and 3′ ends are exactly complementary to the 5′ and 3′ DNA surrounding varied positions to allow for annealing. The mismatched region between the complementary segments contains the degenerate codons that encode for diversity according to the library design. Degenerate primers are

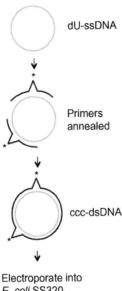

Fig. 3 Kunkel mutagenesis. Single-stranded DNA with uracil incorporated (dU-ssDNA; *green* (*gray* in the print version) above) is isolated from phage produced in CJ236 cells. Degenerate primers (*red* (*black* in the print version)) are annealed (mutagenesis regions indicated by *asterices*) to the dU-ssDNA. T7 DNA polymerase and T4 DNA ligase and dNTPs are added to synthesize the complementary strand based on the dU-ssDNA template. This covalently closed circular double-stranded DNA (CCC-dsDNA) is then electroporated into *E. coli* SS320 cells, in which the uracil-containing template is deactivated and the new, mutagenic strand is preferentially replicated.

specifically incorporated during the annealing step of library synthesis—thus allowing for the targeted incorporation of diversity. After the primers are annealed, the complementary strand is synthesized based on the dU-ssDNA template by the addition of all four dNTPs (dATP, dCTP, dGTP, and dTTP), T7 polymerase, and T4 ligase, which results in a hybrid double-stranded DNA (dsDNA) molecule. One strand contains the uracil-containing template while the newly synthesized strand contains thymine. At this point, there is a heterogeneous mixture of hybrid covalently closed circular double-stranded DNA (CCC-dsDNA). Hybrid CCC-dsDNA is electroporated into a high efficiency electroporatable cell line that is dut^+/ung^+, such as SS320. In SS320 cells the uracil-containing DNA strand is inactivated by cleavage of the uracil, and the newly synthesized (mutated) strand is preferentially replicated. During replication, the mutations encoded by the degenerate oligonucleotides are incorporated. Addition of M13KO7 helper phage after electroporation results in the production of library phage; each phage displays a different version of the mutated protein.

While Kunkel mutagenesis is a highly efficient mutagenesis technique, with an overall efficiency approaching 80% when a single primer is used, this still results in 20% of clones maintaining the wild-type sequence. If three to four primers are used, corresponding to three or four distal regions of sequence space that are randomized, then mutagenesis efficiency decreases to approximately 50% (Clackson & Lowman, 2004). One must consider that the wild-type population will dominate the selection, as 20% or more of the sequences will be wild type depending on the efficiency of primer incorporation. To circumvent the wild-type phenotype dominating selections, an inactive template may be synthesized in which residues that are to be varied in the library are replaced with stop codons (TAA) or rare arginine codons (AGA or AGG). Incorporation of stop codons or rare arginine codons results in the decreased expression of the protein at the surface of phage (Sidhu & Geyer, 2015). Therefore even though 20% of the clones will maintain the wild-type genotype, they are not represented at the phenotypic level, as they are not expressed.

An inactive template is synthesized using Kunkel mutagenesis as described above; however, because only one clone containing the inactivating mutations is required, the hybrid DNA can be electroporated into XL1-Blue cells instead of SS320. The inactivated template will match the wild-type protein sequence but will contain either stop or rare arginine codons at positions of diversity. Once the inactive dU-ssDNA template is synthesized, library synthesis continues with annealing of library primers to the inactive template as described earlier.

After library synthesis is completed, library phage should be titered and a number of clones should be sequenced to determine the incorporation rate of library primers and library diversity. While the efficiency for incorporation of mutations with one primer can reach 80%, this efficiency realistically decreases to 20–30% upon increasing the number of primers to be incorporated. Sequencing data from the naïve library allow one to determine whether or not to begin the next step of phage display: selection.

5.1.1 Protocol 7: dU-ssDNA Synthesis (Modified from Tonikian et al., 2007)

1. Electroporate into CJ236 cells
   a. Electroporate 1 μL phagemid DNA into 25 μL CJ236 cells using 1 mm-gap cuvette
   b. Recover in 3 × 200 μL recovery medium (provided with cells)

c. Incubate at 37°C, 220 RPM for 1 h
   d. Plate out 0.5, 1, 10, and 200 µL onto LB/Carb/cmp plates—want individual colonies
   e. Incubate overnight at 37°C
2. Grow CJ236 cells
   a. Add 1 colony to 1 mL 2xYT/2 µL cmp/5 µL Carb
      i. *Note*: Cells will grow slowly
   b. Incubate at 37°C, 220 RPM until slightly dense
      i. *Note*: Can require overnight incubation, therefore include a negative control
      ii. Negative control: media + antibiotics but no colony
   c. Add M13KO7 helper phage to a final concentration of 10^{9-10} CFU/mL
   d. Incubate at 37°C, 220 RPM for 2 h
   e. Add 1 mL culture to 25 mL 2xYT/12.5 µL Carb/12.5 µL Kan
   f. Incubate at 37°C, 220 RPM for 3 h
   g. Expand to 1 L 2xYT/500 µL Carb/500 µL Kan/10 µL of 25 mg/mL Uridine
   h. Incubate at 37°C, 220 RPM until dense—can require 24–36 h
3. Precipitate phage
   a. Centrifuge at 4000 RPM, 4°C for 15 min to pellet cells
   b. Pour supernatant into 250 mL 5× PEG/NaCl
   c. Incubate at 4°C for 1 h with stirring
   d. Centrifuge at 12,000 RPM and 4°C for 20 min to precipitate phage
   e. Resuspend phage in 10 mL PBS, pH 7.4
   f. Microcentrifuge at 7600 RPM and 4°C for 10 min
   g. Transfer supernatant to clean 15 mL Falcon tube
4. Isolate dU-ssDNA (modified from Qiagen's QIAprep Spin M13 Kit protocol)
   a. Add 7 µL Buffer MP (see Section 3.4.2) per 500 µL phage, vortex, and incubate at room temperature for 10 min
   b. Apply 0.7 mL of phage to spin columns
      i. *Note*: For 10 mL of phage, use four spin columns
   c. Microcentrifuge at 8000 RPM for 30 s—discard the flow through
      i. Continue loading and centrifuging spin columns until all phage has been loaded
   d. Add 0.7 mL Buffer MLB (see Section 3.4.2) and centrifuge at 8000 RPM for 30 s
   e. Add 0.7 mL Buffer MLB and incubate for 1 min at room temperature

f. Centrifuge at 8000 RPM for 1 min
   g. Add 0.7 mL Buffer PE and centrifuge at 8000 RPM for 1 min—discard flow through
   h. Repeat step g
   i. Centrifuge at 8000 RPM for an additional 3 min
   j. Add 50–100 μL sterile H_2O directly to membrane and incubate at room temperature for 10 min
   k. Centrifuge at 8000 RPM for 1 min; combine into 1 tube

5.1.2 Protocol 8: Kunkel Mutagenesis (Modified from Kunkel et al., 1987; Tonikian et al., 2007)

1. Synthesis of hybrid covalently closed circular DNA (CCC-DNA)
   a. Determine concentration of dU-ssDNA
      i. Per reaction, use 10 μg dU-ssDNA
      ii. For library synthesis start with four reactions
      iii. For template synthesis use one reaction with 5 μg dU-ssDNA
   b. Resuspend primers to 1 mg/mL in sterile H_2O and dilute 1:10 in sterile H_2O
   c. Determine number of base pairs (bp) in each primer
   d. Calculate dU-ssDNA:Primer ratio
      i. Start with a dU-ssDNA:Primer ratio of 1:3
      ii. Sample calculation: $\dfrac{1}{3} = \dfrac{\frac{10\ \mu g(DNA)}{\#bp(phagemid)}}{\frac{x}{\#bp(primer)}}$
      iii. May need to change the DNA:Primer ratio for best incorporation results
   e. Set up reactions as follows:

Per Kunkel reaction	
10 μg	dU-ssDNA
X_1 μL	Primer 1
X_2 μL	Primer 2
X_3 μL	Primer 3
25 μL	5X T4 DNA Ligase Buffer (Invitrogen)
12.5 μL	10X T7 DNA Polymerase Buffer (NEB)
X_4 μL	Sterile H_2O
250 μL	Total vol.

f. For each reaction: mix and distribute to three PCR tubes (83 µL per tube)
g. Run the following PCR cycle:
    i. 90°C (5 min), 75°C (45 s), 70°C (1 min), 65°C (1 min), 60°C (1 min), 55°C (1 min), 52°C (3 min), 45°C (30 s), 35°C (30 s), 30°C (45 s), 25°C (45 s), 22°C (1.5 min), 20°C (indefinitely)
h. Combine each starting reaction back into one Eppendorf tube (four tubes total)
i. Add 10 µL 25 mM dNTPs, 2 µL T4 DNA Ligase (Invitrogen), and 3 µL T7 DNA polymerase (NEB) to each
j. Incubate at room temperature for 3 h
k. PCR purify using four spin columns using Qiagen QIAquick PCR purification kit, but wash 2× with buffer PE
    i. Elute with 50 µL sterile H$_2$O and combine into one tube
l. Measure DNA concentration

2. Electroporate into SS320 cells
    a. If using Lucigen SS320 cells, note that each tube contains two reactions worth of cells (50 µL cells per tube)
    b. Per 25 µL SS320 cells add 15–20 µL of DNA and electroporate using a 1 mm-gap cuvette (*E. coli*—1 mm, 1.8 kV)
        i. *Note*: Time constants of 5.0 ms are desirable
        ii. We start with 2–6 electroporations
    c. Resuspend immediately in 2× 500 mL SOC media for each electroporation
    d. Recover in a total volume of 100 mL SOC media with 10^{9-10} CFU/mL M13KO7 helper phage at 37°C, 220 RPM for 1 h
    e. Titer recovered transformants:
        i. Serially dilute 10 µL of the recovered phage in 90 µL 2xYT (1:10 dilutions)
        ii. Plate out 80 µL of each dilution (10^{-1} through 10^{-8}) onto LB/Carb plates
        iii. Incubate overnight at 37°C
        iv. Transformants = (#colonies)$\left(\dfrac{1}{80}\right)$(1000)(100)($10^{\text{dilution}}$)
        - *Note*: Library size should exceed theoretical diversity by 100-fold
    f. Add 100 mL recovery culture to 500 mL 2xYT/250 µL Carb/250 µL Kan
    g. Incubate overnight at 37°C and 220 RPM

3. Precipitating and storing library phage
   a. Centrifuge at 4500 RPM, 4°C for 15 min
   b. Add supernatant to 125 mL 5× PEG/NaCl for 1 h at 4°C with stirring
   c. Centrifuge at 12,000 RPM, 4°C for 20 min
   d. Resuspend phage in 10 mL PBS, pH 7.4
   e. Microcentrifuge at 7600 RPM, 4°C, for 10 min
   f. Collect supernatant in clean 15 mL Falcon tube
      i. *Note*: Use this library phage immediately in the first round of selection
   g. Add final 50% (v/v) of pure glycerol to phage and thoroughly combine
   h. Aliquot and store phage library at −80°C

5.2 Primer Design

Degenerate oligonucleotides should be designed so that there are at least 15 complementary bases upstream and downstream of the targeted mutated sequence to ensure targeted and proper annealing during Kunkel mutagenesis. Primer length is limited to approximately 100 bp due to decreasing oligonucleotide synthesis efficiency past this size. The regions of randomization can be longer or shorter than the corresponding region of the native template protein, to either add or remove residues. For CDR loop segments of antibodies, the size range can be especially beneficial. All nonmutated bases covered by the primer are exactly complementary to the wild-type DNA sequence. Up to four primers can be incorporated in a library design; however, the sequences they cover cannot overlap, as annealing will be mutually exclusive in those cases.

In order for the ligation reaction during complementary strand synthesis to occur, the degenerate primers must be 5′-phosphorylated. We order our degenerate primers with a 5′-phosphorylation modification from commercial suppliers. However, primers can be ordered without the modification and then 5′-phosphorylated following the protocol found in Tonikian et al. (2007). We dilute our primers to 1 mg/mL in sterile H_2O and store them at −20°C. Before library synthesis, dilute primers 1:10 in sterile H_2O.

5.3 Library Quality Control

To determine the quality of the library, it is recommended that the titer plates (naïve library) be sequenced in large scale ($n > 100$). The primer incorporation rate should be determined for each of the primers individually and the total incorporation rate if more than one primer was used for mutagenesis. Although initial incorporation rates can be low (20–30%) with many clones containing the template sequence, the first round of selection against target often results in enrichment of library-encoded sequences. However, in cases where the selection is difficult, such as when there is little or no initial affinity for the target of interest, enrichment for library members may not be observed until later rounds. In this case, it is often best to optimize the synthesis of the library by empirically determining the best primer to dU-ssDNA ratio and annealing conditions that provide the highest incorporation levels of the primers. From the naïve library sequences, one should also verify the diversity of the sequences. At this stage no selective pressure has been applied, so a single sequence should not dominate. Additionally, the entire sequence of the protein should be checked to ensure no deletions or truncations have occurred.

6. LIBRARY SELECTIONS AND SCREENING

Libraries that are produced in sufficient quality can be subjected to selection to enrich for phage clones that demonstrate increased affinity or specificity for the target. Typically, two to four rounds of library selections are performed; however, it is important to avoid overselection since the "fittest" clones are frequently those that harbor some activity toward the target but also a distinct growth advantage and are not necessarily the highest affinity binders. Protocols for phage selection and screening can be found below (Protocols 9 and 10).

6.1 Selection

Library selection is accomplished through the immobilization of target in wells, followed by blocking of nonspecific sites. The phage library is added in a binding buffer, which contains an irrelevant protein (BSA, NFDM, etc.) and/or a nonionic detergent (Tween-20), the wells are washed to remove phage that do not bind to the target, and bound phage are eluted by addition of an acidic solution. The phage solution is neutralized and amplified by infection of an XL1-Blue culture for subsequent rounds of selection. During each round of selection, a negative control well coated with an irrelevant protein, such as BSA or NFDM, should be included to determine enrichment for binding to target (see later).

Over the course of selection, the stringency should be increased to select for phage clones that demonstrate increased affinity or specificity for the target. This can be accomplished by decreasing the amount of target coated (ie, decrease the number of wells), increasing the number of washes between phage binding and elution, increasing the incubation time of each wash step, or increasing the phage binding time to select for clones with larger k_{off} values and thus smaller K_Ds. If increased binding to the negative control becomes evident in later rounds of selection, it may be useful to switch binding and blocking buffers and repeat that round of selection.

During selection, it is important to monitor enrichment. This is done by titering the input, output, and control phage onto LB/carb and LB/kan plates. The carb plate provides information on the enrichment, the ratio of phage recovered from wells coated with target vs phage recovered from negative control wells. Often very little enrichment is seen in the first round of selection. Ideally enrichment ratios of 10-fold or more will be observed, but often this is not observed until later rounds. Once background enrichment is no longer observed, continuing with selection may result in overselection. The kan plate provides information on the specific enrichment of displaying phage vs nondisplaying helper phage. Over rounds of selection, the output population should decrease on the kan plate, as the control (nondisplaying) helper phage should not be binding to the target of interest.

After each round of selection, we send clones for sequencing to monitor the enrichment process. If a selection is proceeding efficiently, we see higher percentages of library members emerge as rounds of selection progress. After several rounds of selection, the sequences may converge to a consensus sequence. In this case, screening should begin from an earlier round, as convergence to a single sequence may indicate overselection. It is best to screen populations that have been selectively enriched for functional binders but that still retain some degree of sequence variation. It is also important to monitor if the selection is selecting clones with truncation mutations.

6.1.1 Protocol 9: Selection

1. Day 1: Coat plate with antigen
   a. Coat seven wells of a high-binding plate with 0.2–10 µg per well of antigen (or antibody if antigen is on phage) in coating buffer (PBS, pH 8.0)
      i. Add 100 µL per well
   b. Coat one well with 100 µL PBS, pH 8.0 as negative control
   c. Incubate overnight at 4°C; alternatively, incubate for 2 h at room temperature with shaking

2. Day 2: Selection
   a. Discard coating solution and block with either 1–3% BSA (w/v) or 2–5% NFDM (w/v)
   b. Incubate for 1.5–2 h at room temperature with shaking
   c. Wash wells 3–5 × with PBS-T
   d. Add 100 µL phage per well
   e. Incubate for 1 h at room temperature with shaking
   f. Repeat step c
   g. Elute phage with 100 mM glycine, pH 2.0
      i. Add 100 µL per well including the negative control
      ii. Incubate for 5 min with shaking
      iii. Remove phage from each well by vigorously pipetting—keep the negative control phage separate
      iv. Per 100 µL phage add 30 µL 2 M Tris, pH 7.5—invert to mix
      v. Confirm neutral pH with pH indicator strips
3. Day 2: Amplification of output phage population
   a. Grow XL1-Blue cells in 2xYT/Tet media in a baffled flask at 37°C, 220 RPM until $OD_{600} = 0.6$
   b. To 5 mL XL1-Blue cells add half of the volume of the output phage from the previous round of selection
   c. Incubate at 37°C, 220 RPM for 30–60 min
   d. Add a final concentration of 10^{9-10} CFU/mL M13KO7 helper phage
   e. Incubate at 37°C, 220 RPM for 1 h
   f. Expand to 25–50 mL 2xYT/25–50 µL Carb/25–50 µL Kan
   g. Incubate at 37°C, 220 RPM overnight
4. Day 2: Phage titering
   a. See Protocol 5 step 3b—titer the input, output, and control phage
   b. Using a multichannel pipette, plate out 5 µL of each dilution of the input, output, and control phage onto LB/Carb, LB/Tet, and LB/Kan plates, respectively
      i. *Note*: LB/Tet is a control for XL1-Blue cells
   c. Incubate at 37°C overnight
   d. Calculate phage titer as in Protocol 5 step 4b

6.2 Screening

A reliable and rapid way to interrogate a selected population of phage for binding to the target of interest is to use monoclonal phage ELISAs. In monoclonal phage ELISA, individual phage are amplified in small scale in a 96-well format. These phage are then bound at a single high-point concentration to the target, which is immobilized on a well of a 96-well high-binding plate. The phage are incubated separately in parallel with a negative control protein (BSA or NFDM) and a positive control (anti-M2 antibody for detection of the FLAG-tag). Unbound phage are removed by washing, and binding to all three targets (target protein, negative control, and positive control) is probed by the addition of an antibody that recognizes the major coat protein-pVIII of the M13 bacteriophage and is conjugated to horse radish peroxidase (HRP). Substrate (TMB) is added and the reaction is quenched with sulfuric acid. The readout for binding is the absorbance measured at 450 nm (Abs(450 nm)) for each well for each plate.

Monoclonal phage ELISA data is analyzed by comparing the absorbance of the target protein wells to the negative control wells. Generally, we prefer to carry forward only those clones that have a strong Abs(450 nm) (2.5 or higher after 1–20 min development) and with a high preference over background. The most robust clones show 10- or more fold preference for the target over negative control wells, but we have in the past characterized clones with more moderate preferences (2- or 3-fold) on phage and found them in some cases to be well behaved as soluble proteins. Furthermore, one should consider the absorbance values for binding to both the target and the positive expression control, M2. If the values for binding to target and to M2 are low, it is likely that the clone does not have high affinity for the target and is not expressed well at the surface of the phage. If the absorbance value for M2 is high and the absorbance value for target is low, it is likely that this phage clone expresses very well but does not have high affinity for the target. If both absorbance values are high it is likely that this clone is a true hit, and one should proceed with validating it (see Section 6.3 characterization of selected clones).

The protocol for monoclonal phage ELISAs can be found below in Protocol 10 and protocols for characterizing hits can be found in Section 6.3.

6.2.1 Protocol 10: Monoclonal ELISA

1. Day 1: Titer and plate out phage
    a. Grow XL1-Blue cells in 2xYT/Tet media until $OD_{600} = 0.6$
    b. Titer output phage (for different rounds of selection) as described in Protocol 5 step 3b
    c. Plate out 50 µL of phage on LB/Carb plates from different dilutions—want individual colonies
    d. Incubate at 37°C overnight
2. Day 2: Coat plates
    a. For each screen, you will need a 96-well high-binding plate for the target(s), negative control (NFDM or BSA), and positive control (anti-FLAG M2)

b. Target plate: per well add 100 µL of 0.2–2 µg of target in PBS, pH 8.0
  c. Negative control plate: per well add 100 µL of PBS, pH 8.0
  d. Positive control plate: per well add 100 µL of 1:500 diluted M2 antibody in PBS, pH 8.0
  e. Incubate overnight at 4°C, or alternatively incubate at room temperature for 2 h with shaking
3. Day 2: Make 96-deep-well plate and grid plates
  a. To each well of a 96-deep-well plate add 1 mL 2xYT, 1 µL Carb, and final concentration of 10^{9-10} CFU/mL M13KO7 helper phage
     i. *Note*: Make up 100 mL 2xYT/Carb/M13KO7 and distribute to deep-well plate using repeater pipette
  b. Prewarm two LB/Carb plates and draw a 96-spot grid on each
  c. Using sterile pipette tips, pick colonies from phage plates (made on Day 1) and spot onto each grid plate and then place in the deep-well plate
     i. *Note*: Match up the grid plate numbers with the deep-well plate
  d. Place a gas-permeable membrane cover over the deep-well plate
  e. Incubate deep-well plate at 37°C, 220 RPM overnight
  f. Incubate grid plates at 37°C overnight
4. Day 3: Monoclonal ELISA
  a. Discard coating solutions and block with 2–5% NFDM (w/v) or 1–3% BSA (w/v) at room temperature with shaking for 1.5–2 h
  b. Wash plates 5× with PBS-T; pat out excess solution onto paper towels
  c. Spin down the deep-well plate at 4000 RPM, 4°C for 20 min
  d. Using a multichannel pipette, add 100 µL phage from deep-well plate to each high-binding plate
     i. *Note*: Match up the rows of the deep-well plate with those of the high-binding plates
     ii. *Note*: If high background binding is encountered, try adding binding buffer to wells before adding the phage
  e. Incubate at room temperature with shaking for 1 h
  f. Repeat step b
  g. Add 100 µL of 1:1000 α-M13-HRP in PB-T to each well
  h. Incubate at room temperature with shaking for 1 h
  i. Repeat step b
  j. Develop by adding 100 µL TMB to each well and shaking at room temperature for 1–10 min, or until a dark blue color appears
  k. Quench the reaction by adding 100 µL of 0.5 M H_2SO_4 to each well
  l. Read the absorbance at a wavelength of 450 nm (A_{450}) for each plate

m. Phage clones are considered "hits" if the (A_{450} target)/(A_{450} negative control) > 2
      i. *Note*: In some cases the cutoff value needs to be increased; this should be determined empirically for each selection
   n. Send colonies corresponding to hits for sequencing using appropriate primers (see Protocol 4)

6.3 Characterization of Selected Clones

Monoclonal phage ELISA measures binding to target only at a single high concentration of phage. It is therefore important to determine the binding reactivity over a range of phage concentrations. This can be accomplished by electroporating the phagemid DNA into XL1-Blue cells, amplifying phage from a single colony as described in Protocol 5, and probing binding to target at multiple serial dilutions of phage. Plotting absorbance vs log (phage titer) determines the half-maximal binding titer, which often provides a relative readout for the relative affinity of the interaction between the protein expressed on phage and the target (Fig. 4).

While the first stages of phage display are relatively rapid, the downstream steps often require much more time and optimization and are therefore the bottleneck to many phage-based projects. Even if a displayed

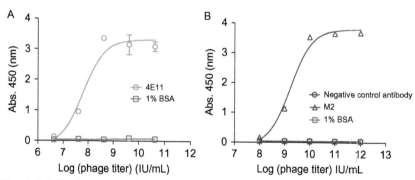

Fig. 4 Full binding curve phage ELISA. (A) Serial dilutions of DENV-2 DIII-expressing phage were incubated with DIII-binding antibody 4E11 (*orange* (*gray* in the print version) *circles*) or negative control, 1% BSA (*green* (*dark gray* in the print version) *squares*). The absorbance measured at 450 nm is a readout for phage binding. (B) As a negative control, dilutions of DENV-2 DIII-expressing phage were incubated with nonbinding antibody (*purple* (*dark gray* in the print version) *circles*), M2 (*blue* (*dark gray* in the print version) *triangles*) or 1% BSA (*green* (*gray* in the print version) *squares*). No binding to the negative control antibody was observed (cf. panel A), but DIII is expressed on the surface of phage as indicated by the M2 signal.

protein demonstrates desirable properties on phage, it may not behave the same expressed as a soluble protein. Therefore, we include protocols for expressing phage-displayed proteins as soluble proteins (Protocols 11 and 12). Once the protein has been expressed off phage, biochemical characterization of the soluble protein can be done.

6.3.1 Protocol 11: Expressing Soluble Fabs (Or Other Proteins) in E. coli

If the phagemid was designed to include an amber stop codon (TAG) between the end of the VH domain (or protein of interest) and the beginning of the pIII phage protein, then expressing soluble Fab is achieved by transforming the phagemid into a nonsuppressor strain of *E. coli* such as BL21(DE3). In XL1-Blue cells, an amber suppressor cell line, the amber stop codon will be read through and the protein will be expressed as a fusion to pIII. In the nonsuppressor cell line, a release factor will be recruited to the ribosome at the amber stop codon and the soluble protein will be expressed. If an amber stop codon was included in the construct, proceed to step 2 of Protocol 11. If no stop codon was included between the protein of interest and the phage pIII coat protein, see below.

1. Insertion of stop codon between protein of interest and phage pIII
   a. Prepare dU-ssDNA as described in Protocol 7 for the phagemid
   b. Order 5′-phosphorylated primer that is complementary to the last five C-terminal residues of the protein of interest and to the first five residues of the pIII gene with a histidine tag and stop codon (TAA) placed between the complementary bases.
      i. *Note*: The primer can include a protease cut site for removal of His-tag
      ii. *Note*: If the protein of interest is cloned into HP153 with NsiI and FseI, then annealing the primer to the last five C-terminal residues of the protein will result in the expression of monovalent soluble protein
      iii. *Note*: If the protein of interest has an N-terminal FLAG-tag, this can be removed by annealing a primer complementary to the last five residues of the StII sequence followed by the first 5 N-terminal residues of the protein
   c. Perform Kunkel mutagenesis as described in Protocol 8 and electroporate into XL1-Blue cells
      i. *Note*: Only one clone containing the mutation needs to be isolated; therefore decrease the amount of dU-ssDNA to 5 μg and use one reaction

d. Screen for clones containing the correct mutations by sequencing
   e. Isolate DNA using a Qiagen QIAquick Miniprep Spin Kit
   f. Transform into BL21(DE3) cells for protein expression
2. Expression
   a. Pick a single colony from a freshly transformed plate into 25–50 mL 2xYT/25–50 μL Carb.
   b. Incubate at 37°C, 220 RPM overnight
   c. Expand culture to Low Phosphate Media/Carb
   d. Grow at 30°C and 220 RPM for 24 h
   e. Harvest cells by centrifugation and determine weight of cell pellet—freeze cells
3. Purification
   a. Per gram of cell pellet, add 5 mL of 1X BugBuster
   b. Resuspend cell pellet in PBS, pH 7.4/EDTA-free protease cocktail inhibitor/DNaseI
   c. Add enough 10X BugBuster to dilute to 1X
   d. Incubate at room temperature with gentle rocking for 20 min
   e. Centrifuge at 12,000 RPM (SS-34 rotor), 4°C for 30 min
   f. Keep the supernatant and the pellet
      i. *Note*: Fab should express periplasmically
      ii. Western blot the supernatant (periplasm/cytosol) and pellet (inclusion body), dissolved in 0.2% SDS/8 M Urea/PBS buffer, with an anti-His (C-term)-HRP antibody to confirm expression location
   g. Wash Ni-NTA agarose beads with TBS, pH 7.4
   h. Apply clarified periplasmic fraction to beads and collect flow through
   i. Wash with 10 column volumes (CV) of TBS + 20–25 mM Imidazole and collect wash fractions
   j. Elute with 5 × 1 mL TBS + 250 mM Imidazole—collect individual fractions
   k. Elute with 5 mL TBS + 500 mM Imidazole—collect
   l. Run SDS-PAGE on all fractions, pool relevant fractions, and concentrate using a 10 k MWCO Amicon Ultra 4
   m. Optional secondary purification step with Protein A agarose:
      i. Dialyze into TBS, pH 8.0
      ii. Wash 1 mL Protein A agarose beads with Gentle Ag/Ab Binding Buffer, pH 8.0

iii. Batch bind Fab with beads at 4°C for 1–2 h
iv. Collect flow through
v. Wash with 2 × 10 CV of Gentle Ag/Ab Binding Buffer, pH 8.0—collect wash fractions
vi. Elute with 5 × 2.5 mL Gentle Ag/Ab Elution Buffer, pH 6.6—collect elution fractions
vii. Desalt using disposable PD-10 desalting columns into TBS, pH 7.4
viii. Run SDS-PAGE, pool relevant fractions, and concentrate using 10 k MWCO Amicon Ultra 4

6.3.2 Protocol 12: Expressing Fab as IgG in HEK239F Cells

1. PCR out VH and VL domains
   a. PCR primer sequences:
      i. IgG-VL-For (Bold=BssHII cut site, underlined=annealing site): CAG**GCGCGC**ACTCCGATATCCAGATGACCCAG
      ii. IgG-VL-Rev (Bold=BsiWI cut site, underlined=annealing site): CAC**CGTACG**TTTGATCTCCACCTTGGTAC
      iii. IgG-VH-For (Bold=BssHII cut site, underlined=annealing site): CAG**GCGCGC**ACTCCGAGGTTCAGCTGGTGGAGTC
      iv. IgG-VH-Rev (Bold=NheI cut site, underlined=annealing site): GGT**GCTAGC**CGAGGAGACGGTGACC
   b. Setup PCR reactions for VH and VL by adding the following:

	Vol. (µL):
Phusion Master Mix	10
For Primer (diluted to 32 µM in sterile H$_2$O)	1
Rev Primer (diluted to 32 µM in sterile H$_2$O)	1
Phagemid DNA (want 10–30 ng DNA)	1
Sterile H$_2$O	7
Total vol.	20

c. Run the following PCR cycle:

98°C	0.5 min	
98°C	0.5 min	
53°C	0.5 min	×30
72°C	1.0 min	
72°C	10–15 min	
4°C	Indefinitely	

d. PCR purify using QIAquick PCR purification kit following manufacturer's protocol—elute with 30 μL sterile H_2O
e. Measure concentration

2. Digestion and ligation into pMAZ vectors (Mazor, Barnea, Keydar, & Benhar, 2007)
   a. The reader is referred to Mazor et al. for a detailed description of the pMAZ vectors. In brief, pMAZ-IgH contains human IgG1 constant domain and pMAZ-IgL contains human kappa constant domain. To clone VH and VL into their respective plasmids, digest both insert and vector with BssHII and BsiWI for VL or with BssHII and NheI for VH.
   b. Digestion (dilute vector DNA to 100 ng/μL in sterile H_2O):

pMAZ-IgH		VH (Insert)		pMAZ-IgL		VL (Insert)	
15 μL	DNA	23 μL	DNA	15 μL	DNA	23 μL	DNA
5 μL	Buffer2.1	5 μL	Buffer2.1	5 μL	Buffer3.1	5 μL	Buffer 3.1
1 μL	BssHII	1 μL	BssHII	1 μL	BssHII	1 μL	BssHII
1 μL	NheI	1 μL	NheI	1 μL	BsiWI	1 μL	BsiWI
28 μL	H_2O	20 μL	H_2O	28 μL	H_2O	20 μL	H_2O
50 μL	Total	50 μL	Total	50 μL	Total	50 μL	Total
Digest for 2.5 h at 37°C							
Add 1 μL CIP to vectors only							
Incubate at 37°C for 0.5 h							

PCR purify VH and VL inserts following Qiagen's QIAquick PCR purification kit—elute with 30 μL sterile H_2O

Run a 1% agarose gel on the vector digestion products and excise the upper band

Gel extract the DNA following Qiagen's QIAquick Gel Extraction Kit protocol—elute with 30 μL H_2O

Measure the concentrations of digested inserts and vectors

    c. Ligations:
       i. Ligate VH into pMAZ-IgH and VL into pMAZ-IgL as described in Protocol 7
    d. Transformations:
       i. Transform 5 μL ligation products into 25 μL One Shot TOP10 Chemically Competent cells as described in Protocol 7

3. Maxiprep IgH and IgL DNA
   a. Transform purified DNA into One Shot TOP10 Chemically Competent cells and plate onto LB/Carb
   b. Pick colony into 500 mL 2xYT/500 μL Carb for IgH and IgL
   c. Incubate at 37°C, 220 RPM, overnight
   d. Centrifuge cells at 4000 RPM, 4°C for 15 min
   e. Isolate DNA following manufacturer's protocol using Macherey–Nagel maxiprep kit
   f. Measure concentrations
4. Transfection: 600 mL scale; using sterile disposable 2 L baffled flask
   a. Seed 3×10^7 HEK293F cells in 600 mL FreeStyle™ Media
   b. To 50 mL sterile PBS add 201 μg IgH, 201 μg IgL DNA, and 1.2 mL PEI at a stock concentration of 1 mg/mL (3:1 ratio of PEI:DNA)
   c. Vortex 15 s
   d. Incubate at room temperature 15 min
   e. Exchange media
   f. Add PBS/DNA/PEI mixture to cells while swirling flask
   g. Incubate at 37°C, 120 RPM, 8% CO_2, and 85% humidity for 5 days
5. Antibody purification
   a. Centrifuge culture at 4000 RPM, 4°C for 15 min
   b. pH supernatant to 8.0 with diluted NaOH
   c. Wash 1 mL of Protein A agarose beads with 10 CV of Gentle Ag/Ab Binding Buffer
   d. Incubate beads with supernatant for 1–2 h at 4°C with stirring
   e. Apply to column—collect flow through

f. Wash with 2× 10 CV Gentle Ag/Ab Binding Buffer—collect washes
  g. Elute with 5 × 2.5 mL Gentle Ag/Ab Elution Buffer—collect elutes
  h. Desalt into TBS, pH 7.4 using PD-10 Desalting columns
  i. Run SDS-PAGE, pool relevant fractions, concentrate using a 30 k MWCO Amicon Ultra 4
  j. Measure concentration by measuring the absorbance at wavelength 280 nm

6.4 Troubleshooting

1. If expression of the protein of interest on phage results in no appreciable affinity for the target, increase the linker length between the protein of interest and the pIII minor coat protein.
2. dU-ssDNA synthesis:
   a. Decrease the amount of antibiotics used by approximately half.
   b. If large-scale production of dU-ssDNA results in low yields, try multiple small-scale preps (30 mL and using 1 spin column) and pooling preps together.
   c. Run 1% agarose gel and verify that dU-ssDNA runs as predominantly 1 band—heat the DNA at 65°C prior to running the gel to reduce secondary structure.
3. Library synthesis:
   a. Allow synthesis reaction to run overnight at 20°C.
   b. Add additional ATP (10 mM) to assist ligation reaction.
   c. Phosphorylate primers enzymatically instead of chemically modifying them—chemical phosphorylation is often incomplete.
   d. Try concentrating ccc-dsDNA to greater than 100 ng/mL for more efficient electroporation.
   e. Run 1% agarose gel on CCC-dsDNA next to dU-ssDNA. The majority of the CCC-dsDNA should run at the expected molecular weight, however, higher molecular bands corresponding to strand displaced DNA and nicked DNA may be visible. These higher molecular weight bands transform *E. coli* less efficiently and do not have a high rate of mutation incorporation.
   f. If there is a low incorporation rate of primers during library synthesis, try different DNA:Primer ratios during synthesis and verify incorporation by sequencing the naïve library.

4. Library selection:
   a. Change binding and blocking buffers to decrease background phage binding.
   b. Change concentration of coated target to enrich for specifically bound phage vs nonspecific binding.
   c. After elution of phage, check for complete elution by blocking the wells and running an ELISA. Make sure to include a blocked well that has no phage added. If needed, increase elution time to 10–30 min.
   d. If output population of phage does not amplify, check that the output phage titer is greater than 10^4. If so, streak out a fresh XL1-Blue plate and grow a fresh batch of XL1-Blue cells for infection. Check infectivity of M13KO7 helper phage by infecting XL1-Blue cells and titering.
5. Library screening:
   a. If 300–400 clones have been screened with no hits observed, check the titer of the phage in the deep-well plate. If titers are low, try growing phage at higher shaker speed and with a lower concentration of carbenicillin to increase titers.
   b. If high background (binding to negative control) is seen, before adding the phage, add binding buffer to the high-binding plates to decrease background binding.
   c. Check sequences of clones from different rounds of selection to verify that there are diverse clones available, and that frameshift or deletion mutants did not dominate the selection. Sequencing should also be done in parallel with selections to confirm this.

REFERENCES

Baca, M., Presta, L. G., O'Connor, S. J., & Wells, J. A. (1997). Antibody humanization using monovalent phage display. *Journal of Biological Chemistry*, *272*(16), 10678–10684.

Bass, S., Greene, R., & Wells, J. A. (1990). Hormone phage: An enrichment method for variant proteins with altered binding properties. *Proteins: Structure, Function, and Bioinformatics*, *8*(4), 309–314.

Beste, G., Schmidt, F. S., Stibora, T., & Skerra, A. (1999). Small antibody-like proteins with prescribed ligand specificities derived from the lipocalin fold. *Proceedings of the National Academy of Sciences*, *96*(5), 1898–1903.

Bogan, A. A., & Thorn, K. S. (1998). Anatomy of hot spots in protein interfaces. *Journal of Molecular Biology*, *280*(1), 1–9.

Chen, G., Koellhoffer, J. F., Zak, S. E., Frei, J. C., Liu, N., Long, H., ... Chandran, K. (2014). Synthetic antibodies with a human framework that protect mice from lethal Sudan ebolavirus challenge. *ACS Chemical Biology*, *9*(10), 2263–2273.

Choo, Y., & Klug, A. (1994). Selection of DNA binding sites for zinc fingers using rationally randomized DNA reveals coded interactions. *Proceedings of the National Academy of Sciences, 91*(23), 11168–11172.

Clackson, T., & Lowman, H. B. (2004). Phage display: A practical approach. *Vol. 266*. Oxford: OUP.

Clackson, T., & Wells, J. A. (1995). A hot spot of binding energy in a hormone–receptor interface. *Science, 267*(5196), 383–386.

Cockburn, J. J. B., Sanchez, M. E. N., Fretes, N., Urvoas, A., Staropoli, I., Kikuti, C. M., ... Rey, F. A. (2012). Mechanism of dengue virus broad cross-neutralization by a monoclonal antibody. *Structure, 20*(2), 303–314.

Da Silva, G. F., Harrison, J. S., & Lai, J. R. (2010). Contribution of light chain residues to high affinity binding in an HIV-1 antibody explored by combinatorial scanning mutagenesis. *Biochemistry, 49*(26), 5464–5472.

Forrer, P., Jung, S., & Plückthun, A. (1999). Beyond binding: Using phage display to select for structure, folding and enzymatic activity in proteins. *Current Opinion in Structural Biology, 9*(4), 514–520.

Frei, J. C., Kielian, M., & Lai, J. R. (2015). Comprehensive mapping of functional epitopes on dengue virus glycoprotein E DIII for binding to broadly neutralizing antibodies 4E11 and 4E5A by phage display. *Virology, 485*, 371–382.

Hufton, S. E., van Neer, N., van den Beuken, T., Desmet, J., Sablon, E., & Hoogenboom, H. R. (2000). Development and application of cytotoxic T lymphocyte-associated antigen 4 as a protein scaffold for the generation of novel binding ligands. *FEBS Letters, 475*(3), 225–231.

Jamieson, A. C., Kim, S.-H., & Wells, J. A. (1994). In vitro selection of zinc fingers with altered DNA-binding specificity. *Biochemistry, 33*(19), 5689–5695.

Koellhoffer, J. F., Chen, G., Sandesara, R. G., Bale, S., Saphire, E. O., Chandran, K., ... Lai, J. R. (2012). Two synthetic antibodies that recognize and neutralize distinct proteolytic forms of the Ebola virus envelope glycoprotein. *ChemBioChem, 13*(17), 2549–2557.

Koide, A., Bailey, C. W., Huang, X., & Koide, S. (1998). The fibronectin type III domain as a scaffold for novel binding proteins. *Journal of Molecular Biology, 284*(4), 1141–1151.

Koide, A., Wojcik, J., Gilbreth, R. N., Hoey, R. J., & Koide, S. (2012). Teaching an old scaffold new tricks: Monobodies constructed using alternative surfaces of the FN3 scaffold. *Journal of Molecular Biology, 415*(2), 393–405.

Ku, J., & Schultz, P. G. (1995). Alternate protein frameworks for molecular recognition. *Proceedings of the National Academy of Sciences, 92*(14), 6552–6556.

Kunkel, T. A., Roberts, J. D., & Zakour, R. A. (1987). Rapid and efficient site-specific mutagenesis without phenotypic selection. *Methods in Enzymology, 154*, 367–382.

Lee, C. V., Sidhu, S. S., & Fuh, G. (2004). Bivalent antibody phage display mimics natural immunoglobulin. *Journal of Immunological Methods, 284*(1), 119–132.

Liu, Y., Regula, L. K., Stewart, A., & Lai, J. R. (2011). Synthetic Fab fragments that bind the HIV-1 gp41 heptad repeat regions. *Biochemical and Biophysical Research Communications, 413*(4), 611–615.

Lowman, H. B., Bass, S. H., Simpson, N., & Wells, J. A. (1991). Selecting high-affinity binding proteins by monovalent phage display. *Biochemistry, 30*(45), 10832–10838.

Lowman, H. B., & Wells, J. A. (1993). Affinity maturation of human growth hormone by monovalent phage display. *Journal of Molecular Biology, 234*(3), 564–578.

Mandal, K., Uppalapati, M., Ault-Riché, D., Kenney, J., Lowitz, J., Sidhu, S. S., et al. (2012). Chemical synthesis and X-ray structure of a heterochiral {D-protein antagonist plus vascular endothelial growth factor} protein complex by racemic crystallography. *Proceedings of the National Academy of Sciences, 109*(37), 14779–14784.

Mazor, Y., Barnea, I., Keydar, I., & Benhar, I. (2007). Antibody internalization studied using a novel IgG binding toxin fusion. *Journal of Immunological Methods, 321*(1), 41–59.

McConnell, S. J., & Hoess, R. H. (1995). Tendamistat as a scaffold for conformationally constrained phage peptide libraries. *Journal of Molecular Biology, 250*(4), 460–470.

Pal, G., Kossiakoff, A. A., & Sidhu, S. S. (2003). The functional binding epitope of a high affinity variant of human growth hormone mapped by shotgun alanine-scanning mutagenesis: Insights into the mechanisms responsible for improved affinity. *Journal of Molecular Biology, 332*(1), 195–204.

Persson, H., Ye, W., Wernimont, A., Adams, J. J., Koide, A., Koide, S., et al. (2013). CDR-H3 diversity is not required for antigen recognition by synthetic antibodies. *Journal of Molecular Biology, 425*(4), 803–811.

Qi, H., Lu, H., Qiu, H.-J., Petrenko, V., & Liu, A. (2012). Phagemid vectors for phage display: Properties, characteristics and construction. *Journal of Molecular Biology, 417*(3), 129–143.

Rader, C., Cheresh, D. A., & Barbas, C. F. (1998). A phage display approach for rapid antibody humanization: Designed combinatorial V gene libraries. *Proceedings of the National Academy of Sciences, 95*(15), 8910–8915.

Rebar, E. J., & Pabo, C. O. (1994). Zinc finger phage: Affinity selection of fingers with new DNA-binding specificities. *Science, 263*(5147), 671–673.

Sidhu, S. S., & Geyer, C. R. (2015). *Phage display in biotechnology and drug discovery*. Boca Raton, FL: CRC Press.

Sidhu, S. S., Lowman, H. B., Cunningham, B. C., & Wells, J. A. (2000). Phage display for selection of novel binding peptides. *Methods in Enzymology, 328*, 333.

Smith, G. P. (1985). Filamentous fusion phage: Novel expression vectors that display cloned antigens on the virion surface. *Science, 228*(4705), 1315–1317.

Smith, G. P., Patel, S. U., Windass, J. D., Thornton, J. M., Winter, G., & Griffiths, A. D. (1998). Small binding proteins selected from a combinatorial repertoire of knottins displayed on phage. *Journal of Molecular Biology, 277*(2), 317–332.

Stewart, A., Harrison, J. S., Regula, L. K., & Lai, J. R. (2013). Side chain requirements for affinity and specificity in D5, an HIV-1 antibody derived from the VH1-69 germline segment. *BMC Biochemistry, 14*(1), 9.

Stewart, A., Liu, Y., & Lai, J. R. (2012). A strategy for phage display selection of functional domain-exchanged immunoglobulin scaffolds with high affinity for glycan targets. *Journal of Immunological Methods, 376*(1), 150–155.

Tonikian, R., Zhang, Y., Boone, C., & Sidhu, S. S. (2007). Identifying specificity profiles for peptide recognition modules from phage-displayed peptide libraries. *Nature Protocols, 2*(6), 1368–1386.

Uchime, O., Dai, Z., Biris, N., Lee, D., Sidhu, S. S., Li, S., ... Gavathiotis, E. (2016). Synthetic antibodies inhibit Bcl-2-associated X protein (BAX) through blockade of the N-terminal activation site. *Journal of Biological Chemistry, 291*(1), 89–102.

Weiss, G. A., Watanabe, C. K., Zhong, A., Goddard, A., & Sidhu, S. S. (2000). Rapid mapping of protein functional epitopes by combinatorial alanine scanning. *Proceedings of the National Academy of Sciences, 97*(16), 8950–8954.

CHAPTER FOUR

Incorporation of Unnatural Amino Acids into Proteins Expressed in Mammalian Cells

R. Serfling, I. Coin[1]
Institute of Biochemistry, Leipzig University, Leipzig, Germany
[1]Corresponding author: e-mail address: irene.coin@uni-leipzig.de

Contents

1. Introduction   90
2. Fluorescence Assay for Straightforward Evaluation of Uaa-Incorporating Systems   92
   2.1 Materials   95
   2.2 Transfection of HEK293 Cells and Fluorescence Measurement   96
   2.3 Evaluation of Different UaaRS–tRNA Pairs   97
3. Incorporation of Uaas into Membrane Proteins   99
   3.1 Materials   100
   3.2 Transfection of HEK293T Cells and Cross-Linking Procedure   102
   3.3 Confirming Uaa Incorporation and Cross-Linking Events   103
4. Conclusion   104
References   104

Abstract

The site-specific incorporation of unnatural amino acids (Uaas) via genetic code expansion provides a powerful method to introduce synthetic moieties into specific positions of a protein directly in the live cell. The technique, first developed in bacteria, is nowadays widely applicable in mammalian cells. In general, different Uaas are incorporated with different efficiency. By comparing the incorporation efficiency of several Uaas recently designed for bioorthogonal chemistry, we present here a facile dual-fluorescence assay to evaluate relative yields of Uaa incorporation. Several biological questions can be addressed using Uaas tools. In recent years, photo-cross-linking Uaas have been extensively applied to map ligand-binding sites on G protein-coupled receptors (GPCRs). We describe a simple and efficient two-plasmid system to incorporate a photoactivatable Uaa into a class B GPCR, and demonstrate cross-linking to its non-modified natural ligand.

1. INTRODUCTION

Biophysical probes, chemical anchors, posttranslational modification mimics, and other artificial moieties can be introduced site specifically into proteins in a straightforward manner by the biosynthetic incorporation of unnatural amino acids (Uaas) (Lang & Chin, 2014; Liu & Schultz, 2010; Neumann, 2012; Wang, Parrish, & Wang, 2009). Synthetic groups installed on the side chain of tailored Uaas are incorporated into the protein of interest (POI) by the ribosomal machinery, directly in the live cell and without the need of any chemical step. Uaas have been incorporated into a number of proteins in live mammalian cells, which range from soluble proteins to membrane proteins including challenging G protein-coupled receptor (GPCR) targets (Coin et al., 2013; Coin, Perrin, Vale, & Wang, 2011; Ye et al., 2010). The Uaa is usually incorporated in response to the amber stop codon UAG, which is decoded by the complementary suppressor tRNA (CUA anticodon) with higher efficiency and specificity respect to opal and ochre codons (O'Donoghue et al., 2012; Xiao et al., 2013). The suppressor tRNA is charged with the Uaa by an ad hoc engineered aminoacyl tRNA synthetase (aaRS). Overall, the system needs to be orthogonal, ie, there must be no cross talk with endogenous tRNAs and aaRSs. Such condition is best satisfied by importing into the host cell heterologous aaRS/tRNA pairs derived from a different organism, whereas the Uaa must be sterically divergent from the 20 canonical amino acids.

Each Uaa is recognized by a dedicated UaaRS. In general, UaaRSs are derived from a naturally orthogonal aaRS, which undergoes a series of randomization and selection rounds until it recognizes the noncanonical moiety specifically. UaaRS selection is most efficient in bacteria, it is possible but less efficient in yeast, and it is currently not feasible in mammalian cells. Three main orthogonal systems have been developed for use in eukaryotic hosts. Two are derived from the pairs, respectively, dedicated to the incorporation of Tyr (Chin et al., 2003) and Leu (Brustad, Bushey, Brock, Chittuluru, & Schultz, 2008; Chatterjee, Guo, Lee, & Schultz, 2013; Lemke, Summerer, Geierstanger, Brittain, & Schultz, 2007; Summerer et al., 2006; Wu, Deiters, Cropp, King, & Schultz, 2004) in *Escherichia coli* (*Ec*-TyrRS/tRNA$_{CUA}^{Tyr}$; *Ec*-LeuRS/tRNA$_{CUA}^{Leu5}$), while the third naturally incorporates the unusual amino acid pyrrolysine (Pyl) in response to the amber codon in some methanogenic archaea of the genus *Methanosarcina* (*Mb*-PylRS/tRNA^{Pyl},

Mm-PylRS/tRNA^{Pyl}) (Chen et al., 2009). Between them, the Pyl system is the most versatile. In contrast to the other two pairs, the PylRS/tRNA^{Pyl} pair is orthogonal both in bacteria and in eukaryotes, so that UaaRS derived from PylRS can be selected using robust protocols in bacteria and directly shuttled to eukaryotic hosts, whereas the UaaRSs imported from *E. coli* must be evolved in yeast.

More than 100 amino acids have been encoded using either PylRS itself or UaaRSs derived from it, including anchors for bioorthogonal chemistry, fluorophores, photocaged amino acids, photoswitchable, and photo-activatable moieties, among many others (Chen et al., 2009; Groff, Chen, Peters, & Schultz, 2010; Hoppmann et al., 2014; Lang & Chin, 2014; Li et al., 2013; Nikic, Kang, Girona, Aramburu, & Lemke, 2015; Schmidt, Borbas, Drescher, & Summerer, 2014). Most of them are structurally composed of a Lys bearing the synthetic moiety on the N$^{\varepsilon}$ through either amide or urethane bonds (see Fig. 2A), but most recently PylRS-based UaaRSs that incorporate Tyr-like Uaas have also been evolved (Takimoto, Dellas, Noel, & Wang, 2011; Wang et al., 2013, 2011; Xiao et al., 2014). Only a few Uaas, all structurally similar to Tyr, are incorporated by UaaRSs evolved from Ec-TyrRS, including the popular cross-linking Uaas p-azido-Phe (Azi) and p-benzoyl-Phe (Bpa) (Chin et al., 2003). Synthetases derived from Ec-LeuRS have been evolved to recognize a few fluorescent and photocaged amino acids based on Ser, Cys, and 2,3-diaminopropionic acid, besides some Uaas bearing aliphatic chains (Brustad et al., 2008; Chatterjee, Guo, et al., 2013; Lemke et al., 2007; Summerer et al., 2006; Wu et al., 2004).

Three genes need to be expressed in the cell for Uaa incorporation: (1) the target gene, bearing an amber stop codon TAG in the site designated for mutation; (2) the UaaRS; (3) the suppressor tRNA. In most common systems, both the POI and the synthetase gene are controlled by a strong constitutive promoter, such as the CMV promoter. The use of the elongation factor promoter (EF1α) has also been demonstrated (Schmied, Elsasser, Uttamapinant, & Chin, 2014). Expression of the suppressor tRNA in mammalian cells is instead more challenging, as transcription of eukaryotic tRNAs by the RNA polymerase III (Pol III) is regulated by intragenic promoter elements (A box and B box) that are usually missing in prokaryotic tRNAs. tRNA genes are therefore put under control of external Pol III promoters, such as the H1 and the U6 promoter (Wang et al., 2007). To guarantee the correct processing of the acceptor stem the by the eukaryotic machinery, the universal 3′-CCA is replaced by a poly T tail in the expression cassette. Highest yield of tRNA transcripts is achieved by combining more of such tRNA cassettes in tandem, usually in a series of four cassettes.

We have initially designed an all-in-one (AIO) construct that combines cassettes for the expression of the three genes in a single plasmid (Coin et al., 2011). Transient transfection of 293T cells with this AIO-plasmid allowed incorporation of Azi into the CRF class B GPCR for photo-cross-linking studies. All three components are combined in a single plasmid also in a viral system developed by the Schultz group (Chatterjee, Xiao, Bollong, Ai, & Schultz, 2013). However, for most applications in which several positions of the POI shall be tested with a Uaa, a more convenient option is to separate the gene of the POI from the translational elements into two plasmids. Crucial is that the plasmid coding the POI should not contain tRNA tandem cassettes, as these repeated sequences are not always stable during cloning, especially when single-step PCR-mutagenesis protocols are applied. We routinely express cytosolic and membrane proteins containing different Uaas by cotransfecting cells with two plasmids, one exclusively dedicated to the POI (usually based on the pcDNA3 vector) and the other bearing the synthetase gene under control of either the PGK or the CMV promoter and three or four copies of a U6-tRNA cassette.

The enzymatic activity varies significantly between different UaaRSs and Uaa substrates. While the most efficient systems are likely to give high yield of Uaa incorporation into many kinds of proteins, weaker systems may give low expression of challenging POIs, as it is often the case with membrane proteins. We describe here a simple assay to evaluate the efficiency of a Uaa-incorporation system, based on the use of an EGFP reporter. Furthermore, we present a reliable protocol for Uaa incorporation into a membrane protein and an example of application for a cross-linking Uaa.

2. FLUORESCENCE ASSAY FOR STRAIGHTFORWARD EVALUATION OF UAA-INCORPORATING SYSTEMS

Nowadays, Uaas can be smoothly incorporated into POIs also by non-specialized labs. In fact, once a new orthogonal pair is described in the literature, a suitable plasmid encoding for the desired UaaRS and its cognate suppressor tRNA can be easily built up using standard cloning methods, especially given that the prices for custom gene synthesis have dropped tremendously over the last years. Many plasmids encoding for UaaRS/tRNA pairs are shared by Uaa specialists, and some are available from Addgene. However, yields of Uaa incorporation may vary significantly when using different constructs, so that it is important to have available a facile assay to evaluate the efficiency of orthogonal pairs and to compare plasmids for UaaRS/tRNA$_{CUA}$ expression.

A fluorescence-based functional assay in mammalian cells has been developed by Wang and coworkers (2007), who used a mutated *EGFP* gene bearing a TAG stop codon at a permissive site as a fluorescent reporter. When the *EGFP_TAG* gene is coexpressed with a UaaRS/tRNA$_{CUA}$ pair, the efficiency of amber suppression and Uaa incorporation determines the yield of full-length EGFP, which linearly correlates to the green fluorescence emitted by the sample. This allows comparing different Uaa-incorporation systems by simply comparing the total intensity of cell fluorescence, measured on intact cells by flow cytometry (FACS) (Chatterjee, Xiao, et al., 2013; Wang et al., 2007). We have now modified the assay to reduce the time required for measuring the samples and analyzing the data, thus increasing the throughput. In addition to the expression cassette for the *EGFP_TAG* gene, we have included an independent cassette for expression of mCherry in the reporter plasmid (Fig. 1). We measure both green and red fluorescence on the crude cell lysate using a plate reader and normalize EGFP fluorescence to mCherry fluorescence. In this way, we obviate the need of performing an additional assay (eg, Bradford assay) to normalize fluorescence to the total amount of protein and at the same time include a control for the transfection efficiency.

A higher sensitivity can be achieved by substituting the fluorescent reporter with a luciferase reporter for luminescence readings (Schmied et al., 2014; Ye et al., 2008). A dual-luminescence assay to evaluate amber codon suppression efficiency has also been reported (Schmied et al., 2014). However, the fluorescence assay gave in our hands more stable and reproducible data respect to the luciferase assay. We describe here the procedure for our dual-fluorescence assay. For a preliminary screening, we usually cotransfect cells with equal amounts of the plasmid encoding for the UaaRS/tRNA pair and the reporter plasmid. Later on, the assay can also be used to identify optimal ratios of the two plasmids, which can vary from one system to another.

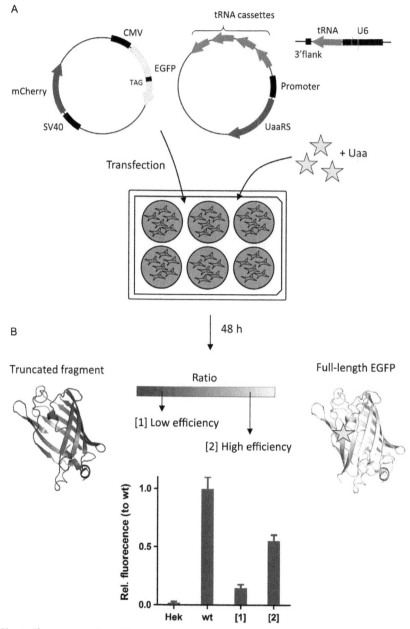

Fig. 1 Fluorescence-based functional assay to evaluate a UaaRS/tRNA pair. (A) Cotransfect HEK293 cells with two plasmids in the presence of the Uaa. One plasmid encodes for the desired UaaRS and suppressor tRNA. The second plasmid encodes for the EGFP gene bearing a TAG stop-codon at a tolerant position together with a mCherry control. (B) Two days after transfection, harvest cells and prepare crude lysates. Determine green fluorescence using a plate reader. As the EGFP N-segment upstream the stop codon is not fluorescent, the yield of full-length EGFP directly correlates to the efficiency of Uaa incorporation.

2.1 Materials
Reagents for culture and transfection of HEK293 cells
- Plasmids:
  pcDNA3.0_EGFP_mCherry, EGFP wild type or variant with amber codon at a tolerant position (Y183) under control of the CMV promoter and mCherry controlled by SV40 promoter; *pNEU_UaaRS_4xtRNA$_{CUA}$*, specific UaaRS under control of CMV promoter and four copies of U6-tRNA expression cassettes
- Uaa stock solutions:
  500 mM N^{ε}-(*tert*-Butyloxycarbonyl)-L-lysine (Lys(Boc); Bachem) in 0.5 M NaOH
  100 mM Cyclopropene-L-lysine (CPK; Sichem) in 0.2 M NaOH with 15% (v/v) DMSO
  100 mM Propargyl-L-lysine (PrK; Sichem) in 0.2 M NaOH with 15% (v/v) DMSO
  500 mM N-Benzyloxycarbonyl-L-lysine (Lys(Z); NovaBiochem) in 0.5 M NaOH
  100 mM *endo*-Bicyclo[6.1.0]nonyne-L-lysine (BCNK; Sichem) in 0.2 M NaOH with 15% (v/v) DMSO
  100 mM *trans*-Cyclooctene-L-lysine (TCO*K; Sichem) in 0.2 M NaOH
  100 mM Cyclooctyne-L-lysine (SCOK; Sichem) in 0.2 M NaOH with 15% (v/v) DMSO
- HEK293 cell line
- 1 M HEPES in distilled H_2O
- Lactate-buffered saline (LBS): 20 mM sodium lactate, 150 mM NaCl, pH 4.0
- Polyethylenimine 25kD linear (PEI; Polysciences); stock solution 10 mg/mL in 2 N HCl; dilute stock solution 1/10 with LBS to obtain working solution 1 µg/µL in LBS. The stock solution is stable at least 1 year at $-80\,°C$; we do not use the working solution for more than 1 week.
- Complete growth medium:
  Dulbecco's modified eagle's medium (DMEM; high glucose, 4 mM glutamine, pyruvate; Gibco) added 10% (v/v) of fetal bovine serum (FBS superior; Biochrom) and 100 units/mL of Penicillin and 100 µg/mL Streptomycin (Gibco)

Reagents for harvesting cells and fluorescence assay
- Phosphate-buffered saline (PBS): 10 mM Na$_2$HPO$_4$, 154 mM NaCl, pH 7.4
- PBS/EDTA (0.5 mM EDTA)
- PBS/MgCl$_2$ (5 mM MgCl$_2$)
- Triton lysis buffer: 50 mM Tris–HCl, pH 8.0, 150 mM NaCl, 1% Triton, 1 mM EDTA
- 100 mM Phenylmethylsulfonylfluoride (PMSF) in isopropanol

Equipment for fluorescence measurement
- Plate reader: FLUOstar Omega (BMG Labtech)
- Plates: Cellstar® 96-well, black (Greiner)

2.2 Transfection of HEK293 Cells and Fluorescence Measurement

Cultivate the cells in complete growth medium at 37°C in a 5% CO$_2$ atmosphere. The day before transfection, seed HEK293 cells in six-well plates at a density of 5.0–6.5×10^5 cells per well in 2 mL complete growth medium and let them reach 60–80% confluency. One hour prior to transfection, add Uaas to the corresponding wells. For Lys(Boc) and LysZ, transfer 200 μL medium from each well into Eppendorf tubes and add 2 μL of the 0.5 M Uaa stock. Mix and transfer the solution back to each well (final Uaa concentration = 0.5 mM). For BCNK, TCO*K, SCOK, CPK, and PrK, transfer the 100 mM stock solution (10 μL per well) to a 1.5-mL tube and dilute in four volumes of 1 M HEPES. Mix thoroughly and add 50 μL of the diluted stock dropwise to each well leaving a final Uaa concentration of 0.5 mM. Prepare the transfection reagents at room temperature as follows. Mix 1 μg of *pNEU_UaaRS_4xtRNA$_{CUA}$* and 1 μg of *pcDNA3.0_EGFP_mCherry* in 50 μL LBS in an Eppendorf tube. Dilute 6 μg PEI working solution (ratio: 3 μg PEI per 1 μg of DNA) in LBS to a total volume of 50 μL. Add the PEI solution to the DNA solution, mix well, and incubate 15 min. Add four volumes (400 μL) of complete growth medium, mix, and apply to the cells (Fukumoto et al., 2010).

Forty-eight hours after transfection aspirate medium and rinse cells once with prewarmed PBS (37°C). Detach cells by incubating 20 min at 37°C in 800 μL of PBS/EDTA. Resuspend cells and transfer to 1.5 mL tubes containing 200 μL of PBS/MgCl$_2$. Pellet cells for 2 min at $800 \times g$ and discard supernatant. Prepare cold Triton lysis buffer by adding PMSF from stock at a final concentration of 0.5 mM. Resuspend pellet in 100 μL Triton lysis buffer and place cells on ice for 30 min. Facilitate lysis by vortexing every 5 min. Spin down cell debris for 10 min at 4°C and 14,000 × g. Transfer 90 μL of the supernatant containing mCherry and EGFP to black 96-well plates. Include three wells containing only Triton lysis buffer for measuring blank values. Quantify the amount of EGFP and mCherry using a plate reader with fluorescence module. In our laboratory, we use a FLUOstar Omega equipped with a filter set for green fluorescence (Ex485-12/Em520) and a filter set for red fluorescence (Ex584/Em620-10). Blank the instrument using the wells containing lysis buffer. If the reader does not support this function, fluorescence readings of the buffer should be subtracted from all absolute fluorescence values. Normalize EGFP fluorescence to fluorescence of mCherry. We define as efficiency of a translational system the ratio between the fluorescence of wild-type EGFP (no TAG codon) and the green fluorescence of the sample, which is calculated as follows:

$$\text{Translational efficiency} = \frac{\text{EGFP}\,x}{\text{mCherry}\,x} \div \frac{\text{EGFP wt}}{\text{mCherry wt}} \times 100\%$$

The denotations EGFP and mCherry refer to the measured fluorescence units of green and red fluorescence after automated subtraction of background.

2.3 Evaluation of Different UaaRS–tRNA Pairs

Synthetic Uaas bearing functional handles for bioorthogonal chemistry enable installing many kinds of probes regardless of their size on target proteins and have therefore an enormous application potential. Using Uaa-mediated two-step labeling protocols, proteins can be site specifically equipped with last generation fluorophores suitable for single-molecule and super-resolution imaging (Nikic et al., 2015). Particularly attractive are Uaas bearing strained alkenes or alkynes (Fig. 2A) that under physiological conditions react selectively with tetrazines over a few minutes and enable rapid labeling of proteins on live cells (Nikic et al., 2015).

All recently developed Uaas for protein labeling are incorporated by the Pyl system. Several synthetases have been derived either from the archaea *Methanosarcina mazei* (*Mm*-PylRS) or from *Methanosarcina barkeri* (*Mb*-PylRS), which share a much conserved catalytic site and esterify Pyl to the same tRNA^{Pyl}. The wt PylRS itself has a very tolerant binding pocket and accepts Uaas with relatively slim yet divergent acyl groups on the N$^\varepsilon$, such as *tert*-butyloxycarbonyl in Lys(Boc), (2-methylcycloprop-2-*en*-1-yl)

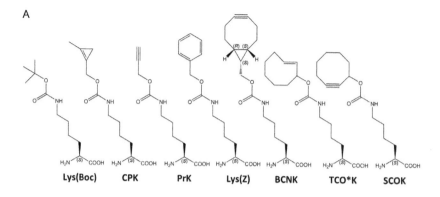

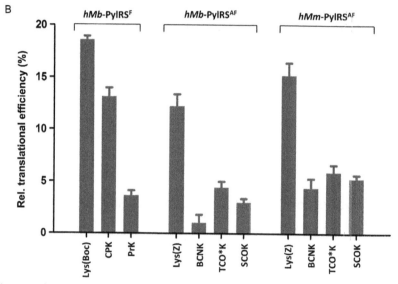

Fig. 2 Evaluating the incorporation of Uaas anchors for bioorthogonal chemistry. The letter "h" before the UaaRS name indicates that the synthetases have been expressed using optimized genes with human codon usage. (A) Structures of Uaas. (B) Incorporation of different Uaas mediated by different variants of PylRS as determined via the dual-fluorescence assay described above. The data present the means (±SEM) of three biological replicates ($n=3$).

methoxycarbonyl in CPK (Elliott et al., 2014), propynyloxycarbonyl in PrK (Nguyen et al., 2009), and a norbornene moiety (Lang, Davis, Torres-Kolbus, et al., 2012). Most labs employ a PylRS^F variant bearing a Tyr to Phe mutation close to the amino acid binding pocket (Y384F for *Mm*-PylRS, homologue to Y349F in *Mb*-PylRS), which has been shown to enhance incorporation of all amino acids accepted by *Mm*-PylRS

(Yanagisawa et al., 2008). Substrate selectivity can be relaxed by enlarging the pocket with a single rationally designed mutation: Y306A for *Mm*-PylRS and Y271A for *Mb*-PylRS (Yanagisawa et al., 2008). PylRS^{AF} variants accept Uaas bearing bulkier groups in the side chain such as benzyloxycarbonyl in Lys(Z) (Yanagisawa et al., 2008), strained cyclooctyne in SCOK (Plass, Milles, Koehler, Schultz, & Lemke, 2011), bicyclo[6.1.0] nonyne in BCNK (Borrmann et al., 2012), and *trans*-cyclooctenes in TCO*K (Nikic et al., 2014; Plass et al., 2012). Other synthetases evolved from *Mb*-PylRS for similar amino acids have also been published (Lang, Davis, Wallace, et al., 2012).

Using our dual-fluorescence assay, we compare the incorporation efficiency of Uaas designed for bioorthogonal chemistry by the PylRS^F and PylRS^{AF} variants. Similar to the observation, that using a codon optimized version of the gene coding for *Mb*-PylRS gives better outcome of mutant protein in *E. coli* (Chatterjee, Sun, Furman, Xiao, & Schultz, 2013), we have improved Uaa incorporation in HEK293 cells by using humanized genes for the synthetases. The overall system is further improved by mutating an unfavorable U:G wobble pair in the anticodon stem to a C:G pair in tRNA^{Pyl} (Chatterjee, Sun, et al., 2013), a stabilizing mutation which has been shown to rise tRNA^{Pyl} levels in mammalian cells (Schmied et al., 2014). Cells were cotransfected with equimolar amounts of the EGFP/mCherry dual-reporter plasmid and a plasmid bearing both the gene for the indicated UaaRS controlled by the CMV promoter and four copies of the expression cassette for the stabilized tRNA^{Pyl} (Fig. 1A). Results of triplicate experiments are reported in Fig. 2B. It is evident that the incorporation efficiency varies significantly from one Uaa to another. The assay can also serve to determine optimal ratios of the two plasmids to maximize protein yield.

3. INCORPORATION OF UAAS INTO MEMBRANE PROTEINS

We routinely use biosynthetically incorporated cross-linkers to investigate protein–peptide and protein–protein interactions. In particular, we use Uaa photochemical probes to map ligand-binding pockets on GPCRs. This allows determining how ligands bind to their receptor on very small samples even in the absence of crystallographic data. By incorporating the photo-cross-linking amino acid Azi throughout the juxtamembrane domain of the class B GPCR corticotropin-releasing factor receptor type 1

(CRF1R), we have revealed the binding mode of its natural neuropeptide agonist Urocortin 1 (Ucn1) at a single amino acid resolution (Coin et al., 2013, 2011).

We describe here a general procedure to incorporate the cross-linking Uaa Bpa into a GPCR using a facile two-plasmid system that in our hands has guaranteed robust Uaa incorporation into several proteins. We use an enhanced version of the *Ec*-BpaRS (EBpaRS), which bears the Asp265Arg mutation to improve recognition of the suppressor tRNA by the synthetase in the anticodon-binding domain (Takimoto, Adams, Xiang, & Wang, 2009). Cross-linking is triggered by 365 nm UV light upon incubating live cells expressing the receptor with the ligand. Whole cell lysates are then analyzed by SDS-PAGE and immunoblotting, using both an antibody targeted to the receptor and an antibody targeted to the ligand. If cross-linking occurs, the anti-ligand antibody will give high-molecular weight signals corresponding to the covalent ligand–receptor complex (Fig. 3A).

3.1 Materials
Reagents for transfection of HEK293T cells
- Plasmids:
    pcDNA3.1_CRF1R$_x$-FLAG, CRF1R wild-type or amber codon variant at position V172 under control of the CMV promoter. The stop codon was TAG was introduced into the *CRF1R* gene at the triplet corresponding to Val172 via one-step mutagenesis based on the "QuikChange®" strategy. *pNEU_EBpaRS_4xBstYam*, CMV promoter controlled enhanced BpaRS and four copies of U6-BstYam expression cassettes
- Uaa:
    500 mM *p*-Benzoylphenylalanine (Bpa; Bachem) in 0.5 M NaOH
- HEK293T cell line
- Polyethylenimine (PEI; 1 µg/µL; Polysciences), see above
- LBS: 20 mM sodium lactate, 150 mM NaCl, pH 4.0
- Growth medium: DMEM (high glucose, 4 mM glutamine, pyruvate; Gibco) added 10% (v/v) of fetal bovine serum (FBS superior; Biochrom) and 100 units/mL of Penicillin and 100 µg/mL Streptomycin (Gibco)

Reagents and equipment for ligand binding, UV-cross-linking, and sample preparation
- HEPES dissociation buffer (HDB): 12.5 mM HEPES, 140 mM NaCl, 5 mM KCl

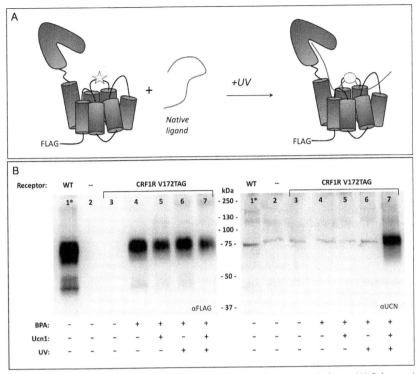

Fig. 3 Investigating ligand–receptor interactions via photo-cross-linking. (A) Schematic illustration of the photo-cross-linking experiment. A receptor mutant containing a photo-cross-linking Uaa (*yellow* (*white* in the print version) *star*) is incubated with its native ligand. If the ligand lies in close proximity of the Uaa, it will be covalently captured by the photoactivated moiety upon photoactivation with UV light. (B) Specific photo-cross-linking of Ucn1 to CRF1R_Val172Bpa. Western blots of cell lysates resolved by SDS-PAGE were probed with αFLAG antibody (*left*) to detect the receptor bearing a C-terminal FLAG tag and with αUcn1 antibody (*right*) to detect the ligand. The mature glycosylated CRF1R runs at an apparent molecular weight of 65–75 kDa (Coin et al., 2013). The *asterisk* indicates one-fourth of total protein loaded.

- Ligand-binding buffer: HDB supplemented with 5 mM MgCl$_2$, 0.1% (w/v) BSA, and 0.01% (v/v) Triton X-100
- Ucn1 peptide synthesized on the solid phase using standard Fmoc-protocols (Coin, Beyermann, & Bienert, 2007) on a Syro parallel peptide synthesizer and purified on reverse phase HPLC to >95% purity. The peptide is also commercially available. Stock solution: 1 mM in DMSO (store at −20°C).
- HDB/EDTA (0.5 mM EDTA)
- HDB/MgCl$_2$ (5 mM MgCl$_2$)

- HEPES lysis buffer (HLB): 50 mM HEPES, pH 7.5, 150 mM NaCl, 10% glycerol, 1% Triton X-100, 1.5 mM MgCl$_2$, 1 mM EGTA, 1 mM dithiothreitol (DTT), 1× protease inhibitors w/o EDTA
- Cross-linker: Bio-Link (Lamps: 5 × 8 W, 365 nm; BioBudget)

Reagents and equipment for SDS-PAGE and Western blotting
- Sample solubilization buffer 4× (SSB): 250 mM Tris–HCl, pH 6.8, 8% SDS, 40% glycerol, bromophenol blue 0.2 mg/mL
- 1 M DTT in dH$_2$O
- SDS-PAGE running buffer: 25 mM Tris–HCl, 190 mM glycine, 0.1% (w/v) SDS, pH 8.8
- Western transfer buffer: 25 mM Tris–HCl, 192 mM glycine, 20% (v/v) methanol
- Tris-buffered saline (TBS): 50 mM Tris–HCl, 150 mM NaCl, pH 7.4
- TBS-T: TBS supplemented with 0.1% Tween 20
- Blocking solution: TBS-T with 5% (w/v) nonfat dry milk
- Immobilon®-P transfer membrane (pore size 0.45 µm, hydrophobic PVDF; Millipore)
- Antibodies:

    Monoclonal mouse M2 αFLAG lag-HRP conjugate (Sigma)
    Homemade polyclonal rabbit αUcn1 (kindly provided by Paul Sawchenko, Peptide Biology Laboratory, The Salk Institute, La Jolla, CA)
    Secondary α-rabbit-HRP conjugate (Santa Cruz Biotechnology)

3.2 Transfection of HEK293T Cells and Cross-Linking Procedure

The day before transfection, seed HEK293T cells on six-well culture plates at a density of $4.0–5.0 \times 10^5$ cells per well in 2 mL complete growth medium. Maintain cells until they reach about 70% of confluence. One hour before transfection, add Bpa to the growth medium to a final concentration of 0.5 mM. Cotransfect cells with the plasmid encoding for CRF1R variants (wt or V172TAG) and the plasmid harboring the EBpaRS and four copies of BstYam expression cassette in a 1:2 ratio. Dilute 0.7 µg of *pcDNA3.1_CRF1R$_x$-FLAG* and 1.4 µg of *pNEU_EBpaRS_4xBstYam* in 50 µL LBS and incubate the DNA solution 15 min with 6.3 µg of PEI previously diluted in 50 µL LBS.

Forty-eight hours after transfection, aspirate growth medium from the corresponding wells and incubate cells with 2 mL binding buffer containing 100 nM Ucn1 for 10 min at 37°C. Irradiate cells 15 min with UV light

(365 nm). Aspirate binding buffer or growth medium and detach cells from the culture plates with 800 μL HDB/EDTA for 20 min at 37 °C. Transfer cells to 1.5 mL tubes containing 200 μL of PBS/MgCl$_2$, pellet at $800 \times g$ for 2 min, and discard supernatant. Add 40 μL HDB (added protease inhibitors), vortex briefly to resuspend samples, and flash freeze in liquid nitrogen. Thaw cells at 37 °C for ~30 s, vortex quickly, chill tubes on ice, and then centrifuge at $2500 \times g$ at 4 °C for 10 min. Discard supernatant (cytosolic proteins). Add 40 μL HLB and incubate on ice for 30 min, vortexing every 5 min. Clear cell debris by centrifugation at $14,000 \times g$ at 4 °C for 10 min. Discard pellet and proceed immediately to sample analysis. Membrane extracts can be stored at -20 °C and analyzed later, but quality of Western blots will decrease at every freeze–thaw cycle.

3.3 Confirming Uaa Incorporation and Cross-Linking Events

Prepare samples for SDS-PAGE as follows: dilute 3 μL of in 6 μL distilled H$_2$O and add 3 μL of SSB ($4\times$) supplemented with DTT from stock to a final concentration of 200 mM. Solubilize for 20–30 min at 40 °C, load on polyacrylamide gel (4% stacking gel and a 10% resolving gel), and run 20 min at 80 V and 50 min at 150 V. To achieve sharper bands, take care that the temperature of the tank remains low. Samples are then transferred to PVDF membrane. To analyze membrane proteins, we prefer using classic wet transfer (2 h at 80 V) instead of semidry system. It is recommendable to run gels in duplicate so that immunoblotting can be performed in parallel on two replicate membranes with the two antibodies. Block membranes for 1 h to overnight (4 °C) with blocking solution. Rinse one membrane at least once with TBS-T (5 min) and probe with αFLAG-HRP conjugate (1:10,000 in TBS-T) for 1 h to detect the receptor via the C-terminal FLAG tag. Wash 3×10 min in TBS-T and proceed with detection of specific bands via horseradish peroxidase catalyzed enhanced chemiluminescence reaction (ECL). Incubate the second membrane for 1 h at room temperature and later overnight at 4 °C with the anti-ligand antibody rabbit-αUcn1 (1:5000 in blocking solution). Wash 3×10 min in TBS-T and incubate for 1 h at room temperature with secondary α-rabbit-HRP conjugate (1:10,000 in blocking solution). Proceed with another three wash steps followed by ECL detection.

In Fig. 3B, we demonstrate the example of Ucn1 cross-linked to a CRF1R variant bearing Bpa in position 172, which is known to lie close to the bound ligand (Coin et al., 2013). The αFLAG-antibody detects the receptor only when Bpa is added to the culture medium, thus enabling stop codon suppression and synthesis of the full-length protein. A ligand-specific band is detected only when the Bpa-receptor is irradiated in the presence of Ucn1. The receptor–ligand complex runs at an apparent molecular weight similar to that of the receptor alone (MW of Ucn1 $\sim$ 4 kDa). Screening several positions in a binding pocket allows determining the ligand-binding site (Coin et al., 2013; Grunbeck, Huber, & Sakmar, 2013).

4. CONCLUSION

The site-specific incorporation of unnatural moieties via the biosynthetic machinery has provided a wide palette of chemical tools for investigating the most disparate biological processes in a natural cellular environment. Although the technique is very robust for general application in bacteria, there is still room for optimizing Uaa-incorporation systems and improving the overall Uaa-incorporation rate in mammalian cells. This aspect is crucial especially when incorporating Uaas into membrane proteins, which are synthesized at the translocon and are notoriously more difficult objects for Uaa mutagenesis than cytosolic proteins. We have presented here a facile dual-fluorescence assay to evaluate and optimize the efficiency of UaaRS–tRNA pair for usage in mammalian cells. We have then showed an example for a cross-linking experiment aimed at determining a ligand-binding site on a GPCR, directly on the intact fully glycosylated receptor at the cell membrane. Importantly, Uaa tools allow investigating single proteins in the natural setting of the live cell.

REFERENCES

Borrmann, A., Milles, S., Plass, T., Dommerholt, J., Verkade, J. M., Wiessler, M., et al. (2012). Genetic encoding of a bicyclo[6.1.0]nonyne-charged amino acid enables fast cellular protein imaging by metal-free ligation. *Chembiochem, 13*, 2094–2099.

Brustad, E., Bushey, M. L., Brock, A., Chittuluru, J., & Schultz, P. G. (2008). A promiscuous aminoacyl-tRNA synthetase that incorporates cysteine, methionine, and alanine homologs into proteins. *Bioorganic & Medicinal Chemistry Letters, 18*, 6004–6006.

Chatterjee, A., Guo, J., Lee, H. S., & Schultz, P. G. (2013). A genetically encoded fluorescent probe in mammalian cells. *Journal of the American Chemical Society, 135*, 12540–12543.

Chatterjee, A., Sun, S. B., Furman, J. L., Xiao, H., & Schultz, P. G. (2013). A versatile platform for single- and multiple-unnatural amino acid mutagenesis in *Escherichia coli*. *Biochemistry, 52*, 1828–1837.

Chatterjee, A., Xiao, H., Bollong, M., Ai, H. W., & Schultz, P. G. (2013). Efficient viral delivery system for unnatural amino acid mutagenesis in mammalian cells. *Proceedings of the National Academy of Sciences of the United States of America, 110*, 11803–11808.

Chen, P. R., Groff, D., Guo, J., Ou, W., Cellitti, S., Geierstanger, B. H., et al. (2009). A facile system for encoding unnatural amino acids in mammalian cells. *Angewandte Chemie (International Ed. in English), 48*, 4052–4055.

Chin, J. W., Cropp, T. A., Anderson, J. C., Mukherji, M., Zhang, Z., & Schultz, P. G. (2003). An expanded eukaryotic genetic code. *Science, 301*, 964–967.

Coin, I., Beyermann, M., & Bienert, M. (2007). Solid-phase peptide synthesis: From standard procedures to the synthesis of difficult sequences. *Nature Protocols, 2*, 3247–3256.

Coin, I., Katritch, V., Sun, T., Xiang, Z., Siu, F. Y., Beyermann, M., et al. (2013). Genetically encoded chemical probes in cells reveal the binding path of urocortin-I to CRF class B GPCR. *Cell, 155*, 1258–1269.

Coin, I., Perrin, M. H., Vale, W. W., & Wang, L. (2011). Photo-cross-linkers incorporated into G-protein-coupled receptors in mammalian cells: A ligand comparison. *Angewandte Chemie (International Ed. in English), 50*, 8077–8081.

Elliott, T. S., Townsley, F. M., Bianco, A., Ernst, R. J., Sachdeva, A., Elsasser, S. J., et al. (2014). Proteome labeling and protein identification in specific tissues and at specific developmental stages in an animal. *Nature Biotechnology, 32*, 465–472.

Fukumoto, Y., Obata, Y., Ishibashi, K., Tamura, N., Kikuchi, I., Aoyama, K., et al. (2010). Cost-effective gene transfection by DNA compaction at pH 4.0 using acidified, long shelf-life polyethylenimine. *Cytotechnology, 62*, 73–82.

Groff, D., Chen, P. R., Peters, F. B., & Schultz, P. G. (2010). A genetically encoded epsilon-N-methyl lysine in mammalian cells. *Chembiochem, 11*, 1066–1068.

Grunbeck, A., Huber, T., & Sakmar, T. P. (2013). Mapping a ligand binding site using genetically encoded photoactivatable crosslinkers. *Methods in Enzymology, 520*, 307–322.

Hoppmann, C., Lacey, V. K., Louie, G. V., Wei, J., Noel, J. P., & Wang, L. (2014). Genetically encoding photoswitchable click amino acids in *Escherichia coli* and mammalian cells. *Angewandte Chemie (International Ed. in English), 53*, 3932–3936.

Lang, K., & Chin, J. W. (2014). Cellular incorporation of unnatural amino acids and bioorthogonal labeling of proteins. *Chemical Reviews, 114*, 4764–4806.

Lang, K., Davis, L., Torres-Kolbus, J., Chou, C., Deiters, A., & Chin, J. W. (2012a). Genetically encoded norbornene directs site-specific cellular protein labelling via a rapid bioorthogonal reaction. *Nature Chemistry, 4*, 298–304.

Lang, K., Davis, L., Wallace, S., Mahesh, M., Cox, D. J., Blackman, M. L., et al. (2012b). Genetic encoding of bicyclononynes and trans-cyclooctenes for site-specific protein labeling in vitro and in live mammalian cells via rapid fluorogenic Diels-Alder reactions. *Journal of the American Chemical Society, 134*, 10317–10320.

Lemke, E. A., Summerer, D., Geierstanger, B. H., Brittain, S. M., & Schultz, P. G. (2007). Control of protein phosphorylation with a genetically encoded photocaged amino acid. *Nature Chemical Biology, 3*, 769–772.

Li, F., Zhang, H., Sun, Y., Pan, Y., Zhou, J., & Wang, J. (2013). Expanding the genetic code for photoclick chemistry in *E. coli*, mammalian cells, and *A. thaliana*. *Angewandte Chemie (International Ed in English), 52*, 9700–9704.

Liu, C. C., & Schultz, P. G. (2010). Adding new chemistries to the genetic code. *Annual Review of Biochemistry, 79*, 413–444.

Neumann, H. (2012). Rewiring translation—Genetic code expansion and its applications. *FEBS Letters, 586*, 2057–2064.

Nguyen, D. P., Lusic, H., Neumann, H., Kapadnis, P. B., Deiters, A., & Chin, J. W. (2009). Genetic encoding and labeling of aliphatic azides and alkynes in recombinant proteins via a pyrrolysyl-tRNA Synthetase/tRNA(CUA) pair and click chemistry. *Journal of the American Chemical Society, 131*, 8720–8721.

Nikic, I., Kang, J. H., Girona, G. E., Aramburu, I. V., & Lemke, E. A. (2015). Labeling proteins on live mammalian cells using click chemistry. *Nature Protocols, 10*, 780–791.

Nikic, I., Plass, T., Schraidt, O., Szymanski, J., Briggs, J. A., Schultz, C., et al. (2014). Minimal tags for rapid dual-color live-cell labeling and super-resolution microscopy. *Angewandte Chemie (International Ed. in English), 53*, 2245–2249.

O'Donoghue, P., Prat, L., Heinemann, I. U., Ling, J., Odoi, K., Liu, W. R., et al. (2012). Near-cognate suppression of amber, opal and quadruplet codons competes with aminoacyl-tRNAPyl for genetic code expansion. *FEBS Letters, 586*, 3931–3937.

Plass, T., Milles, S., Koehler, C., Schultz, C., & Lemke, E. A. (2011). Genetically encoded copper-free click chemistry. *Angewandte Chemie (International Ed. in English), 50*, 3878–3881.

Plass, T., Milles, S., Koehler, C., Szymanski, J., Mueller, R., Wiessler, M., et al. (2012). Amino acids for Diels-Alder reactions in living cells. *Angewandte Chemie (International Ed. in English), 51*, 4166–4170.

Schmidt, M. J., Borbas, J., Drescher, M., & Summerer, D. (2014). A genetically encoded spin label for electron paramagnetic resonance distance measurements. *Journal of the American Chemical Society, 136*, 1238–1241.

Schmied, W. H., Elsasser, S. J., Uttamapinant, C., & Chin, J. W. (2014). Efficient multisite unnatural amino acid incorporation in mammalian cells via optimized pyrrolysyl tRNA synthetase/tRNA expression and engineered eRF1. *Journal of the American Chemical Society, 136*, 15577–15583.

Summerer, D., Chen, S., Wu, N., Deiters, A., Chin, J. W., & Schultz, P. G. (2006). A genetically encoded fluorescent amino acid. *Proceedings of the National Academy of Sciences of the United States of America, 103*, 9785–9789.

Takimoto, J. K., Adams, K. L., Xiang, Z., & Wang, L. (2009). Improving orthogonal tRNA-synthetase recognition for efficient unnatural amino acid incorporation and application in mammalian cells. *Molecular BioSystems, 5*, 931–934.

Takimoto, J. K., Dellas, N., Noel, J. P., & Wang, L. (2011). Stereochemical basis for engineered pyrrolysyl-tRNA synthetase and the efficient in vivo incorporation of structurally divergent non-native amino acids. *ACS Chemical Biology, 6*, 733–743.

Wang, Y. S., Fang, X., Chen, H. Y., Wu, B., Wang, Z. U., Hilty, C., et al. (2013). Genetic incorporation of twelve meta-substituted phenylalanine derivatives using a single pyrrolysyl-tRNA synthetase mutant. *ACS Chemical Biology, 8*, 405–415.

Wang, Q., Parrish, A. R., & Wang, L. (2009). Expanding the genetic code for biological studies. *Chemistry & Biology, 16*, 323–336.

Wang, Y. S., Russell, W. K., Wang, Z., Wan, W., Dodd, L. E., Pai, P. J., et al. (2011). The de novo engineering of pyrrolysyl-tRNA synthetase for genetic incorporation of L-phenylalanine and its derivatives. *Molecular BioSystems, 7*, 714–717.

Wang, W. Y., Takimoto, J. K., Louie, G. V., Baiga, T. J., Noel, J. P., Lee, K. F., et al. (2007). Genetically encoding unnatural amino acids for cellular and neuronal studies. *Nature Neuroscience, 10*, 1063–1072.

Wu, N., Deiters, A., Cropp, T. A., King, D., & Schultz, P. G. (2004). A genetically encoded photocaged amino acid. *Journal of the American Chemical Society, 126*, 14306–14307.

Xiao, H., Chatterjee, A., Choi, S. H., Bajjuri, K. M., Sinha, S. C., & Schultz, P. G. (2013). Genetic incorporation of multiple unnatural amino acids into proteins in mammalian cells. *Angewandte Chemie (International Ed. in English), 52*, 14080–14083.

Xiao, H., Peters, F. B., Yang, P. Y., Reed, S., Chittuluru, J. R., & Schultz, P. G. (2014). Genetic incorporation of histidine derivatives using an engineered pyrrolysyl-tRNA synthetase. *ACS Chemical Biology*, *9*, 1092–1096.

Yanagisawa, T., Ishii, R., Fukunaga, R., Kobayashi, T., Sakamoto, K., & Yokoyama, S. (2008). Multistep engineering of pyrrolysyl-tRNA synthetase to genetically encode N(epsilon)-(o-azidobenzyloxycarbonyl) lysine for site-specific protein modification. *Chemistry & Biology*, *15*, 1187–1197.

Ye, S. X., Kohrer, C., Huber, T., Kazmi, M., Sachdev, P., Yan, E. C. Y., et al. (2008). Site-specific incorporation of keto amino acids into functional G protein-coupled receptors using unnatural amino acid mutagenesis. *The Journal of Biological Chemistry*, *283*, 1525–1533.

Ye, S., Zaitseva, E., Caltabiano, G., Schertler, G. F., Sakmar, T. P., Deupi, X., et al. (2010). Tracking G-protein-coupled receptor activation using genetically encoded infrared probes. *Nature*, *464*, 1386–1389.

CHAPTER FIVE

Method for Enzyme Design with Genetically Encoded Unnatural Amino Acids

C. Hu, J. Wang[1]

Laboratory of RNA Biology, Institute of Biophysics, Chinese Academy of Sciences, Beijing, China
[1]Corresponding author: e-mail address: jwang@ibp.ac.cn

Contents

1. Introduction	110
2. The General Procedure for Enzyme Design with Unnatural Amino Acids	112
3. Unnatural Amino Acid Design	112
4. Synthetic Chemistry-Guided Unnatural Amino Acid Design	116
5. Synthetic Methods for Unnatural Amino Acids	117
5.1 Unnatural Amino Acid Toolkit	119
5.2 Metal-Chelating Amino Acid	120
5.3 Redox Mediators	122
5.4 Click Chemistry Reaction Reagents	124
6. Aminoacyl-tRNA Synthetase Screening	125
7. Protein Design	127
8. Protein Expression	128
8.1 Plasmid Preparation	128
8.2 Protein Expression	129
9. Conclusion and Perspective	129
References	130

Abstract

We describe the methodologies for the design of artificial enzymes with genetically encoded unnatural amino acids. Genetically encoded unnatural amino acids offer great promise for constructing artificial enzymes with novel activities. In our studies, the designs of artificial enzyme were divided into two steps. First, we considered the unnatural amino acids and the protein scaffold separately. The scaffold is designed by traditional protein design methods. The unnatural amino acids are inspired by natural structure and organic chemistry methods, and synthesized by either organic chemistry methods or enzymatic conversion. With the increasing number of published unnatural amino acids with various functions, we described an unnatural amino acids toolkit containing metal chelators, redox mediators, and click chemistry reagents. These efforts enable a researcher to search the toolkit for appropriate unnatural amino acids for the study, rather than design and synthesize the unnatural amino acids from the beginning. After the first step, the model enzyme was optimized by computational methods and directed evolution. Lastly, we describe a general method for evolving aminoacyl-tRNA synthetase and expressing unnatural amino acids incorporated into a protein.

1. INTRODUCTION

All natural proteins are composed of 20 canonical amino acids with only a few exceptions. Tested by millions of years evolution, it seems that encoding only these 20 amino acids is sufficient for an organisms' survival. However, encoding unnatural amino acids have been proven to be helpful for the design of artificial enzymes (Hu, Chan, Sawyer, Yu, & Wang, 2014). The significance of using unnatural amino acids can be explained by the following three reasons:

1. Incorporating an unnatural amino acid significantly simplifies the mimicking of natural structures. In nature, the catalytic structures are formed by collaboration of several residues, cofactors, protein subunits, and even chaperones. The natural synthesis routes are often too difficult for traditional molecule biology methods. However, the genetic code expansion method provides an opportunity to synthesize and manipulate the biological structure by powerful chemical methods and strategies (Martin & Schultz, 1999). Assisted by the unnatural amino acid that already contains the essence of the natural structure, researchers can conveniently construct the natural or designed structure by directly incorporating an appropriate unnatural amino acid into the protein scaffold.

2. Unnatural amino acid incorporation may enhance natural enzymes' performance. The evolution of a natural enzyme is restricted by the genetic code and amino acids that are naturally accessibility. Some unnatural amino acids may be more appropriate for a certain chemical reaction, but they are unavailable under natural environment. Thus, organisms have chosen a natural amino acid as an inferior replacement. Examples have proven that some natural enzyme could be improved by inserting an appropriate unnatural amino acid (Jackson, Duffy, Hess, & Mehl, 2006).

3. Incorporating unnatural amino acids with unnatural chemical groups can enhance a biomolecules' ability to catalyze chemical reactions not normally undertaken by the enzyme. Natural enzymes with 20 encoded canonical amino acids are evolved under natural environment by natural

selection. However, synthetic biologists are also interested in unnatural synthetic chemical reactions, such as the preparation of macromolecule material, inorganic nanoparticle, and computationally designed drugs (Keasling, 2008). Through incorporating unnatural amino acid with unnatural organic chemical groups, researchers can "teach" proteins to catalyze those chemical conversions and synthesize organic molecule by highly developed chemical synthetic methods (Liu & Schultz, 2010). Thus, the application of synthetic biology would be greatly expanded. Genetic code expansion has become a comprehensive method for site-specific unnatural amino acid incorporation. The mechanism and process of genetic code expansion is shown in Fig. 1. The method reassigns the nonsense codon TAG, termed as amber codon, or a quadruplet codon to an unnatural amino acid by introducing an orthogonal exogenous amber suppressor tRNA/aminoacyl-tRNA synthetase pair (Anderson et al., 2004; Wang, Brock, Herberich, & Schultz, 2001). The evolved aminoacyl-tRNA synthetase only recognizes the exogenous amber suppressor tRNA together with the unnatural amino acid, while it does not respond to any natural tRNA or natural amino acids. The amber suppressor tRNA is only charged by the aminoacyl-tRNA synthetase. Under these conditions, the evolved aminoacyl-tRNA synthetase could charge the unnatural amino acid

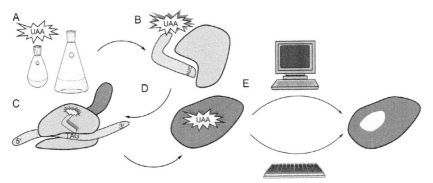

Fig. 1 Mechanism and general process of enzyme design with an expanded genetic code. (A) An unnatural amino acid was synthesized by organic chemistry methods or enzymatic conversion. (B) The unnatural amino acid was charged by an exogenous bio-orthogonal amber codon suppressor tRNA/aminoacyl-tRNA synthetase. (C) The ribosome read amber codon, the activated unnatural amino acid could be incorporated into a nascent peptide, and the translation process would continue. (D) An enzyme model was achieved, by incorporating the unnatural amino acids into a designed protein scaffold. (E) The enzyme mode with an unnatural amino acid was evolved by computational methods or high-throughput screening.

specifically. Thus, when the ribosome reaches the TAG codon, instead of recruiting release factor I (and stopping the translation), the peptide synthesis would continue and would integrate the unnatural amino acid as one residue. Today, more than 100 unnatural amino acids have been encoded in response to an amber codon. They have been widely used as spectrum probes, posttranslation modification mimics, bioorthogonal reaction reagents, etc. (Fekner, Li, Lee, & Chan, 2009; Guo, Wang, Lee, & Schultz, 2008; Wang, Xie, & Schultz, 2006).

2. THE GENERAL PROCEDURE FOR ENZYME DESIGN WITH UNNATURAL AMINO ACIDS

In most of our studies, the designs of artificial enzyme were divided into two steps (Fig. 1). First, we considered the unnatural amino acids and the protein scaffold separately. Then, the model enzymes were optimized by computational methods and directed evolution. We believe this process is both practical and efficient. In many studies, such as the cytochrome c oxidase model, only completing the first stage is already sufficient for getting an effective enzyme model with comparable activity. The reason for that is listed as below.

1. Unnatural amino acid incorporation can be treated as a single normal site-mutation process, which usually does not cause much disturbance of the protein structure.
2. The unnatural amino acid was usually designed to function alone. They can function well even if the surrounding protein structure is not optimized. For example, a metal-binding amino acid, such as a metal-chelating unnatural amino acid, can coordinate metal ions with high affinity without the assistance from surrounding residues.

3. UNNATURAL AMINO ACID DESIGN

Designing the unnatural amino acid is the first and crucial step in artificial enzyme design with an expanded genetic code. The amino acid should facilitate the construction of a certain structure; therefore, it largely determines the activity of the model enzyme. Different amino acids are specifically designed for different enzyme studies. However, in order to ensure that the amino acid can be incorporated properly, the following two conditions must be followed.

1. The amino acid should not be too close to the natural amino acids (usually tyrosine and lysine), nor too deviate from them.

   The incorporation of unnatural amino acids depends on the evolved aminoacyl-tRNA synthetase mutants, which specifically recognize and charge the unnatural amino acid molecule. Currently most commonly used aminoacyl-tRNA synthetases are evolved from tyrosyl-tRNA synthetase from *Methanocaldococcus jannaschii* (Kobayashi et al., 2003) and pyrrol-lysine aminoacyl-tRNA synthetase from *Methanosarcina barkeri* (Polycarpo et al., 2004). Their substrates are tyrosine and pyrrolysine under native conditions, and can be expanded to various unnatural amino acids by directly evolving the residues, which participates in the amino acid recognition process. However, if the properties of UAA were too different from those of tyrosine or lysine, including molecular size, shape, polarity, and other factors, it might be beyond the aminoacyl-tRNA synthetase's ability and cannot be activated. On the other hand, if the unnatural amino acids were too close to natural amino acids, the aminoacyl-tRNA synthetase mutants would not be able to distinguish them from natural amino acids. Therefore, the natural amino acid would be incorporated at the amber codon, instead of the unnatural amino acid.

   Several strategies are helpful for those unqualified unnatural amino acids. One is discarding unnecessary elements of the unnatural amino acid, while keeping the essential chemical groups intact. The other is adding extra chemical elements or removable protection groups to make the unnatural amino acid unique. An example of this approach is the incorporation of methyllysine (Ai, Lee, & Schultz, 2010). Detail of these methods will be discussed in the following examples.

2. The amino acid should be able to penetrate the cell's membrane and be stable in the cytoplasm.

   The unnatural amino acid has to reach the aminoacyl-tRNA synthetase as well as ribosome in the cytoplasm to be charged. Most unnatural amino acids do not have problems in penetrating the cell membrane. We assumed that they are transported by natural tyrosine and lysine permeases (Brown, 1971; Steffes, Ellis, Wu, & Rosen, 1992). But some exceptions should be noticed. Most unnatural amino acids we have used are stable in the cell, and their toxicity can be tolerated. However, if the unnatural amino acid were too sensitive to the environment, special treatment should be considered. For example, reductive amino acids such as dopa-alanine should be incorporated together with the extra addition of sodium dithionite (Alfonta, Zhang, Uryu, Loo, & Schultz, 2003).

Another approach is using a chemically protected unnatural amino acid instead, and disassociating the protective groups by light illuminating or pH changes after protein expression. This approach is also suitable for toxic unnatural amino acids.

Two ideas are commonly used in unnatural amino acid designs. The first one is learn-from-nature. The nature has already provided numerous inspirations in various scientific investigations, as well as unnatural amino acid designs. Genetically encoded unnatural amino acids provided good opportunities for simulating natural catalytic structures, especially residues with post-translation modifications. Some modifications are crucial for the enzymes' catalytic performance, and some others play important roles in regulating the activities (Uy & Wold, 1977). However, lots of important modifications only occur in eukaryote cells, and some others are also difficult to control in engineering bacteria. That brought forth great obstacles in constructing them. Fortunately, through genetic code expansion, the modifications can be synthesized in vitro and be incorporated in vivo. Some residues with modifications can be treated as an unnatural amino acids directly, such as acetyllysine and sulfated tyrosine (Guo et al., 2008; Liu, Cellitti, Geierstanger, & Schultz, 2009). The others need modifications to make them "looks like" a common amino acid. The criteria are the two rules mentioned earlier. Some particular examples are discussed as following.

The studies on mimicking cytochrome c oxidase are a good example to elucidate how nature inspired the design of unnatural amino acids. Cytochrome c oxidase is a huge membrane protein complex located on the inner membrane of mitochondrion, which catalyzes the oxidation of cytochrome c and the reduction of oxygen. Coupled with one molecule oxygen reduced to water, four protons are transported from the matrix to the inner membrane space for ATP generation (Chan, 2009; Yoshikawa & Shimada, 2015). A remarkable feature of the oxygen reduction is its high turnover number and high selectivity. Nearly all oxygen is completely reduced by four electrons and only less than 5% toxic reactive oxygen species are released. The catalytic center of cytochrome c oxidase was resolved by X-ray diffraction, which is composed of a central heme molecule, a copper ion coordinated by three histidine residues, and a histidine cross-linked with tyrosine (Yoshikawa et al., 1998). Theoretic calculations and small-molecule models suggested that the crossing-linking between Tyr244 and His240 is crucial for the high-efficient four-electron reduction (Collman et al., 2007). This posttranslational modification decreased the pK_a and the reduction potential of the tyrosine, making the proton and electron transfer more easy (Wikstrom, 1998). In the contrast, the small-molecule model and small-protein model which lack the cross-link between tyrosine and histidine produce large amount of reactive oxygen species and are quickly damaged (Fig. 2) (Sigman, Kwok, & Lu, 2000).

Inspired by the nature, we hoped to improve previously reported myoglobin heme-copper oxidase models by incorporating a histidine–tyrosine cross-linked unnatural amino acid (Liu, Yu, et al., 2012). We designed the unnatural amino acid 2-amino-3-(4-hydroxy-3-(1H-imidazol-1-yl) phenyl) propanoic acid (imiTyr). The unnatural amino acid inherited the *ortho*-imidazole–phenol structure from the natural enzyme. And other irrelevant parts, such as the backbone of the histidine, are discarded. Though direct evolution, we encoded it in *Escherichia coli* and incorporated it in the myoglobin model Cu$_B$Mb at the 33rd position. We tested the new model and discovered that it can perform more than 1000 cycles of oxygen reduction, while only less than 6% of reactive oxygen species was produced. The model would drastically lose the oxygen reduction activity, if the unnatural amino acid imiTyr was replaced by a tyrosine, which demonstrated that the unnatural amino acid

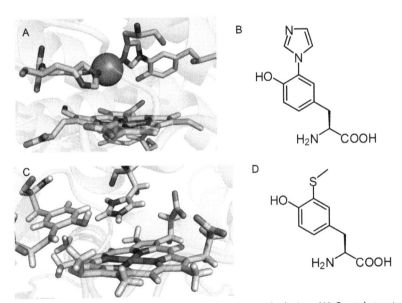

Fig. 2 A scheme for natural inspired unnatural amino acids design. (A) Crystal structure of the Cu$_B$ site of bovine cytochrome *c* oxidase. (B) The unnatural amino acid imiTyr, with a 3-imidazole phenol group inspired by the natural tyrosine–histidine cross-link. (C) Structure of the catalytic center of *T. nitratireducens* cytochrome *c* nitrite reductase. (D) The unnatural amino acid 3-methylthio tyrosine inspired by the natural cross-link between tyrosine and cysteine. (See the color plate.)

imiTyr is crucial for the CcO model and successfully mimicked the function of the histidine–tyrosine cross-link modification in natural cytochrome *c* oxidase. This study is helpful for developing artificial cytochrome *c* models or fuel cell electrodes with higher activity and selectivity as well as taking a deeper look at the mechanism of cytochrome *c* oxidase.

The design of thioether-bonded tyrosine–cysteine cross-links mimicking unnatural amino acid is another example. The thioether-bond tyrosine–cysteine cross-link widely exists in natural metalloenzymes, such as galactose oxidase, glyoxal oxidase, cysteine dioxygenase, and sulfite reductase (Verma, Pratt, Storr, Wasinger, & Stack, 2011). Studies on tyrosine–cysteine model compounds stated that the cross-link between tyrosine and cysteine significantly decreases the pK_a and the reduction potential of the phenol group and facilitating proton and electron transfer.

In order to mimicking the natural cofactor and design superior artificial nitrite reductase, we designed the unnatural amino acid MtTyr, which contains a 3-methylthio substitution on the aromatic ring of native tyrosine (Zhou et al., 2013). We left a methyl substitution on the sulfur molecule in order to keep its redox potential similar with the natural cysteine cross-linked tyrosine. And other cysteine parts were discarded to make it a tyrosine derivative. To test whether the unnatural amino acid has the similar chemical properties with natural Tyr–Cys cross-link, we incorporated the unnatural amino acid at site 33rd position of a myoglobin scaffold with a histidine 29 mutation. The myoglobin model then formed a *Thioalkalivibrio nitratireducens* cytochrome *c* nitrite reductase (TvNiR) structural mimic (Polyakov et al., 2009). Compared to the TvNiR myoglobin model with normal tyrosine residue, the hydroxylamine conversion activity of MtTyrMb was fourfold higher, although their subtract affinity is nearly the same. The result demonstrated that the unnatural amino acid inspired by natural tyrosine–cysteine cross-link significantly enhanced the artificial enzyme's activity. It also exhibited its potential in designed artificial galactose oxidase and other enzyme containing a tyrosine–cysteine cross-link modification.

4. SYNTHETIC CHEMISTRY-GUIDED UNNATURAL AMINO ACID DESIGN

Genetic code expansion enables the usage of unnatural chemical groups, which are widely used in organic chemistry but is rare in organisms. At least three advantages can be achieved by doing that. First, the protein scaffold provides a secondary coordination sphere for the organic catalyst, which may enhance their performance, including turnover numbers and enantioselectivity (Durrenberger & Ward, 2014). Second, the structure containing unnatural organic amino acid is genetically encoded. As a consequence, its self-assembly can be easily amplified or improved by directed evolution. Finally, adding unnatural organic molecules enables researchers to solve biological chemistry problems by organic chemical methods, which may be helpful in green chemistry and synthetic biology.

Some unnatural amino acids were inspired by organic chemistry studies. Thus, some unnatural organic molecules and powerful and highly developed synthetic chemistry methods can be introduced to molecular biology (Mann, 1989). In order to be properly incorporated, those organic molecules are required to be converted into an unnatural amino acid first. Based on their structure features, they can be converted into either a "tyrosine" or a "lysine". For instance, if the molecule contained an aromatic ring, it would be suitable for mimicking tyrosine. The tyrosine type unnatural amino acid often contains an aromatic ring that bears the unnatural chemical groups, and a covalently linked aliphatic amino acid part (usually alanine) as the amino acid back bone. On the other hand, a lysine host is more suitable for flexible aliphatic chemical groups. The lysine mimic is usually composed of the unnatural aliphatic chemical groups and a lysine molecule. Usually they are covalently connected by an amide bond or carbamate.

5. SYNTHETIC METHODS FOR UNNATURAL AMINO ACIDS

Due to the structural similarities among artificial tyrosine mimicking unnatural amino acids or lysine derived unnatural amino acids, some universal chemical synthetic protocol are used for the synthesis. For instance, most tyrosine derivatives can be synthesized by the nucleophile substitution reaction between acetamidomalonate and corresponding benzyl chloride. The reaction is initiated by adding 1 equiv. solid sodium to water-free ethanol. After the sodium was totally dissolved, 2 equiv. of diethyl acetylaminomalonate, 1 equiv. of benzyl chloride, and DMAP (cat.) are added. The mixture needs to be stirred for 1 h at room temperature, and then be refluxed for additional 12 h. After cooling, the solution was extracted by EtOAc and then evaporated to get the crude product. After being purified by a silicon column, the product was then reflux in 12 N HCl for 12 h. Then the racemic unnatural amino acid containing the unnatural organic chemical groups is achieved, which can be further purified by HPLC with a C18 column (Fig. 3).

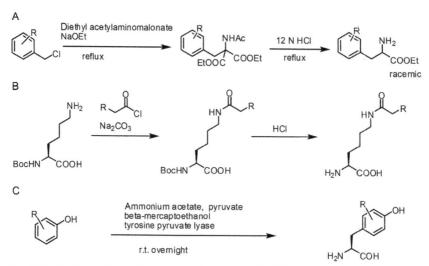

Fig. 3 Synthetic methods for unnatural amino acids. (A) Tyrosine-derived unnatural amino acids can be chemical synthesized by the substitution reaction between acetylaminomalonate and benzyl chloride. (B) Lysine-derived unnatural amino acids can be chemically synthesized by the substitution reaction between BOC protected lysine and acyl chloride. (C) Some phenol-derived unnatural amino acids can be prepared by TPL-catalyzed enzymatic conversion.

Some aromatic molecules with *para*-hydroxyl group can be converted by enzymatic conversion. The enzymatic route is cleaner, more environment friendly, and enantio specific. The enzyme tyrosine pyruvate lyase (TPL) catalyzes reversible degeneration of tyrosine to phenol and pyruvate. TPL from *Citrobacter freundii* has a relative board substrate scope. Phenol derivatives such as halogenated phenol can be converted by the wild-type enzyme (Seyedsayamdost, Yee, & Stubbe, 2007). Using these phenol derivatives as starting material, the unnatural amino acid can be obtained as the following procedure:

1. The plasmid-encoding TPL gene was transformed into a protein expression strain. The cell was amplified in LB media with ampicillin. Then 1 mL starter culture was used to inoculate 10 mL of LB media with appropriate antibiotic. Cells were grown at 37°C to an OD_{600} of 0.8 and induced by addition of 1 mM isopropyl thiogalactoside (IPTG). The cell was cultured at 18°C for another 12 h. The cell was then collected for purification.
2. The TPL enzyme was purified form the soluble fraction by using Ni-NTA affinity column.

3. The 1-L TPL conversion solution contains 10 mg TPL enzyme, 30 mM ammonium acetate, 60 mM sodium pyruvate, 5 mM beta-mercaptoethanol, 20 mM phenol derivatives, and 0.05 mM pyridoxal phosphate.
4. The solution was kept in the dark at room temperature for 2 days. The solvent was removed by vacuum evaporation. The product was dissolved in 1 N HCl and was purified by HPLC with C18 column.

The substrate scope of TPL can be further expanded by directed evolution. The evolution process can be carried out as below (Liu et al., 2013):

1. A TPL library with more than 4096 members was constructed by randomizing residues Phe448, Phe36, and Met288 using an overlapping-extension PCR method.
2. Clones expressing different TPL mutants were picked from the plate. They are transferred to liquid minimal medium in microtiter plates and allowed to grow to saturation.
3. Cells were then lysed with lysozyme. The corresponding phenol derivative, ammonia chloride, and pyruvate were added. The plates were then incubated at room temperature for 4 h. Each well of the microtiter plate was then analyzed by ninhydrin thin-layer chromatography. The clones with significant chromogenic reaction was selected and sequenced.

Lysine-derived amino acids are generally chemically synthesized, though the nucleophile substitution reaction between acyl chlorides contains the unnatural chemical group and BOC protected lysine. Here is a universal method for lysine derivatives synthesis to a solution of N-α-Boc-lysine (2.46 g, 10.0 mmol) and Na_2CO_3 (2.12 g, 20.0 mmol) in ethylacetate/H_2O (1:1, 200 mL total) at 0°C was added acryl chloride 11.0 mmol) dropwise over 10 min. The solution was then stirred for 12 h at room temperature. Acetic acid was then added till the pH of the aqueous phase reaches 3. The mixture was extracted with ethyl acetate (200 mL, three times). HCl gas (prepared by adding H_2SO_4 to NH_4Cl) was then passed into the combined organic phase, affording the unnatural amino acid.

5.1 Unnatural Amino Acid Toolkit

Generally, a different unnatural amino acid is designed for each different study. Over the past few years, a large number of unnatural amino acids have been encoded and reported (Liu & Schultz, 2010). Due to the increasing unnatural amino acid toolkit, various challenges in enzyme design can be solved by previously published unnatural amino acids. That provides an opportunity for researchers to search the unnatural amino acid toolkit to get a appropriate unnatural amino acid, rather than design and discover a synthetic route all form the beginning (Xie & Schultz, 2006). Here is a brief summary on published functional unnatural amino acids. Based on their functions, they were grouped as metal-chelating amino acid, redox mediators, and click chemistry reagents.

5.2 Metal-Chelating Amino Acid

Over one-third of natural enzymes are metalloprotein. They catalyze some of the most important chemical reactions on the earth, including photosynthesis, oxygen reduction, and nitrogen fixation (Lu, Yeung, Sieracki, & Marshall, 2009). In nature, a metal ion is usually anchored by the collaboration of several metal-binding residues, such as histidine, aspartic acid, glutamic acid, and arginine. Compared to natural metal ion coordination amino acids, metal-chelating unnatural amino acid has much stronger binding affinity. Thus, the metal-chelating amino acid can anchor a metal ion at any site of a protein by its own affinity. Consequently, the design of metalloenzyme and metal-binding sites can be significantly facilitated.

Currently published metal-binding amino acids are derived from organic metal chelators bipyridine, pyrazole derivatives and 8-hydroxyl quinoline. Bipyridine has a strong binding affinity to transition-metal ions, such as $Fe^{2+/3+}$, Cu^{2+}, $Co^{2+/3+}$, and $Ru^{2+/3+}$. This ability is inherited by the bipyridine unnatural amino acid (Xie, Liu, & Schultz, 2007). The application of bipyridine was exampled by the artificial DNA-cleaving enzyme (Lee & Schultz, 2008). Researchers site specifically incorporated this metal-chelating amino acid to a DNA-binding protein, thus enabled it to coordinate an metal ion near the binded DNA, and functions its catalytic ability. Upon adding copper or ferrous ion, the bound double-stranded DNA would be cleaved, due to the Fenton reaction catalyzed by the metal ion coordinated by the bipyridine unnatural amino acid (Fig. 4).

The unnatural amino acids pyrazole tyrosine was derived from pyrazole derivatives (Liu, Li, et al., 2012). We synthesized this unnatural amino acid by coupling the *ortho*-carbon of tyrosine and the nitrogen of pyrazole. The amino acid exhibited unique metal selectivity to copper ion, and the affinity was measured in a green fluorescent protein system. Another metal-chelating amino acid was derived from 8-hydroxyl quinoline. 8-Hydroxyl quinoline is widely used metal chelator in industry. Compared to bipyridine, 8-hydroxyl quinoline has stronger metal-binding affinity and a wider metal ion scope. We used the metal-binding ability of 8-hydroxyl quinoline

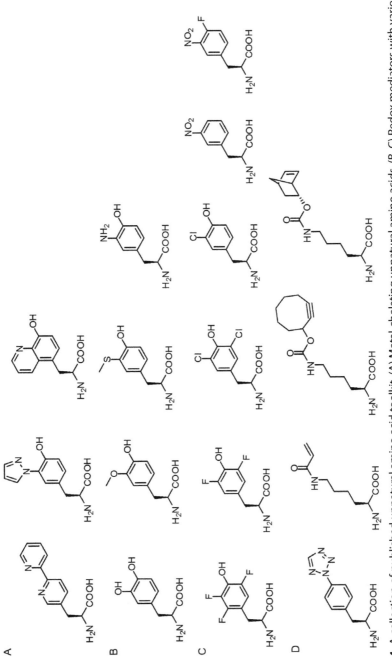

Fig. 4 A collection of published unnatural amino acids toolkit. (A) Metal-chelating unnatural amino acids. (B, C) Redox mediators with various redox potential and pK_a. (D) Click chemistry reagents.

unnatural amino acid to convert green fluorescent protein into a metal responsive fluorescent protein. When incorporated the 8-hydroxyl quinoline unnatural amino acid into the site 66th of super folded green fluorescent protein, the 8-hydroxyl quinoline was integrated into the chromophore, and red shifted the fluorescence by 30 nm. Besides that, upon adding zinc ion the fluorescence is strongly enhanced in cell. As the 8-hydroxyl quinoline unnatural amino acid can be synthesized in vivo by TPL-catalyzed conversion; we were looking forward to perform directed evolution for screening novel metalloenzymes with 8-hydroxyl quinoline as metal chelator.

5.3 Redox Mediators

The electron transfer in proteins can be described by Marcus equation, which states that the electron transfer rate is mainly determined by driving force, recombination energy, and distance (Marcus & Sutin, 1985). In natural enzymes, electrons are tunneling from one-electron mediator to another to cover a long distance (Gray & Winkler, 2005; Winkler & Gray, 2014). With the help of genetic code expansion methods and a series of redox active unnatural amino acids, a researcher would be able to put a redox mediator with a certain redox potential at any site of the protein scaffold; therefore, simplifying and directly constructing a highly efficient electron transfer pathway which may match those of nature.

We have reported various unnatural amino acids with redox activity. The first type is metal chelators. Their redox activity and redox potential depends on the valance change of the coordinated transition-metal ion, such as $Fe^{2+/3+}$, Cu^{2+}, $Co^{2+/3+}$, $Ru^{2+/3+}$, and rare earth. The second type is tyrosine derivatives. The redox potential changes when then aromatic hydrogen was substituted by other chemical groups. A series of tyrosine derivatives have been genetically encoded, including redox potential enhanced halogenated tyrosine, and redox potential decreased methoxyl/methylthio dopa and 3-amino substituted tyrosine. Additionally, an unnatural amino acid derived from nitrobenzene with a remarkable low-redox potential was also reported (Fig. 5).

We used fluorescent proteins to evaluate the electron transfer performance of unnatural amino acids. It is reported that the chromophore of green fluorescent protein is both oxidative and reductive when excited (Bogdanov et al., 2009). We incorporated the unnatural amino acid at various sites of the green fluorescence protein, and evaluated the electron

Method for Enzyme Design

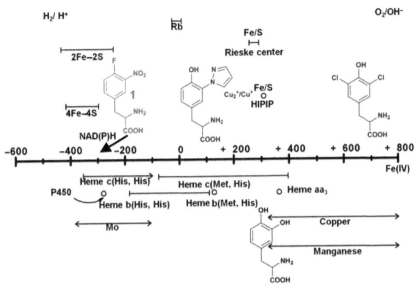

Fig. 5 The redox potential scope of some published redox active unnatural amino acids (Lv et al., 2015).

transfer between protein chromophore and the unnatural amino acids by measuring the fluorescence intensity and lifetime. We discovered drastic fluorescence quenching when metal-binding unnatural amino acid pyrazole tyrosine and 8-hydroxyl quinoline alanine was incorporated on the outside of the beta-barrel upon adding copper ion. The lifetime decreased when the unnatural amino acid was placed closer to the chromophore, which fit the Marcus theory. When the unnatural amino acid is incorporated at site 149th, the quenching efficiency reaches maxima. And the fastest observed electron transfer rate was calculated to be comparable to the primary charge separation in photosynthesis. The experiment demonstrated that metal-binding unnatural amino acids can be used as highly efficient electron acceptors.

With its low remarkable redox potential, the unnatural amino acid derived from nitrophenol can be inserted as an electron donor, which is also demonstrated in green fluorescence protein test (Lv et al., 2015). The nitrobenzene unnatural amino acid has peak potential at −470 mV, which is comparable to reductive cofactors in biosystem such as NADH and NADPH. When the nitrobenzene unnatural amino acid was incorporated close to GFP's chromophore, the fluorescence quenched drastically. Moreover, we observed one-electron oxidized intermediate of the nitrobenzene and reduced GFP chromophore in transient absorption spectrum, suggesting that the fluorescence quenching was caused by one-electron transfer from nitrobenzene to the excited GFP chromophore. Thus, this observation demonstrated that nitrobenzene-based unnatural amino acids could be used as electron donors.

The redox potential of tyrosine derivatives varies with its substitution. Together with previously reported 3-amino- and dopa-substituted tyrosine (Alfonta et al., 2003), we achieved a series of tyrosine-derived unnatural amino acids with their redox potentials ranging from 300 to 800 mV. Considering their potential redox diversity, they can optimize artificial enzymes by incorporating a redox potential-tuned unnatural amino acid to replace natural redox cofactors. The idea has been demonstrated on the study on cytochrome c oxidase models (Yu et al., 2015). The tyrosine at site 33th of a cytochrome c oxidase model Cu_bMb was replaced by a serial of tyrosine derivatives. Finally, it was discovered that the oxygen reduction increased along with the redox potential of tyrosine.

When tyrosine acts as an electron donor in an organism, its redox potential is always connected with its pK_a value. For example, a separated tyrosine is difficult to be oxidized under neutral conditions because of the formation of a high-energy tyrosyl radical intermediate. In photosystem II, this obstacle was solved by proton-coupled electron transfer facilitated by a hydrogen bond connected histidine residue (Huynh & Meyer, 2007). We used another route to pass by the proton transfer problem by genetically encoding halogenated tyrosines with lower pK_a values (Liu et al., 2014). These unnatural amino acids were deprotonated under neutral conditioned; thus, they can be easily oxidized despite their high oxidative potential. We incorporated them into a flavin-dependent fluorescent protein iLOV. The fluorescence quenched under neutral conditions, and can be recovered by acidifying the solution, which suggests that highly efficient electron transfer between flavin and deprotonated tyrosine derivatives under neutral conditions can be achieved. The experiment suggests the application of halogenated tyrosine as electron donors and probes to study and design proton-coupled electron transfer processes in enzymes.

5.4 Click Chemistry Reaction Reagents

Click chemistry, referring to the chemical reactions with high selectivity, high yield, and fast reaction rate, has been widely used in biological large molecule labeling and super-molecule structure construction (Kolb, Finn, & Sharpless, 2001). Genetically encoded click chemistry reaction reagents enables researchers to covalently link an unnatural chemical groups, including inorganic catalysts or photosensitizers, to any site of a target protein with high site-selectivity. The click chemistry reagent can also be transformed into another unnatural amino acid such as glycosylated amino acids, which might be too difficult to be incorporated (Kaya et al., 2009). The click chemistry-functionalized unnatural amino acids were derived from bioorthogonal reaction reagents. Considering the toxicity of Cu(I), ring strain assists cycloaddition between azide and alkyne, and photoassisted cycloaddition reaction between tetrazole and alkyne are more suitable for biological system (Jewett & Bertozzi, 2010; Lim & Lin, 2011). Several unnatural amino acids derived from them have been reported and they have been tested in protein site-specific labeling (Plass, Milles, Koehler, Schultz, & Lemke, 2011).

6. AMINOACYL-tRNA SYNTHETASE SCREENING

The bioorthogonal acyl-tRNA synthetase was screened through positive screening and negative screening (Wang et al., 2001). In positive screening, the aminoacyl-tRNA synthetase gene was cotransformed with TAG-mutated chloramphenicol resistance gene. Then the cell is cultured on a glycerol minimal media with leucine (GMML) plate with antibiotics and the unnatural amino acid. If the aminoacyl-tRNA synthetase could charge exogenous added unnatural amino acid or natural amino acids, the cell would survive. Thus, the aminoacyl-tRNA synthetase that recognizes the unnatural amino acid or natural amino acids is selected. In negative screening, the aminoacyl-tRNA synthetase gene was cotransformed with TAG-mutated toxic protein. The cell was grown on a LB plate without the unnatural amino acid. And the cell would die if natural amino acid were encoded. Thus, though negative screening, synthetases that charge natural amino acids are discarded. Combined with three rounds of positive screening and two rounds of negative screening, the selected synthetase mutants only charge the certain unnatural amino acid and do not response to any natural amino acids. The detail procedure is list as following:

Screening protocol
1. The gene of aminoacyl-tRNA synthetase was coded on a PBK plasmid. A library with more than 109 diversity was construct by randomize residue Tyr32, Leu65, Phe108, Gln109, Asp158, and Leu162. And any one of the six residues (Ile63, Ala67, His70, Tyr114, Ile159, and Val164) was either mutated to Gly or kept unchanged. The plasmid pREP (2)/YC encodes the tRNA, and TAG-mutated chloramphenicol acetyltransferase

gene. Plasmid pLWJ17B3 encodes the tRNA and a toxic barnase gene with three amber codons at residues 2, 44, and 65.

2. *Positive screening.* E. coli DH10B harboring the pREP(2)/YC plasmid was used as the host strain for the positive screening. Cells are transformed with the pBK-aminoacyl-tRNA synthetase, washed twice with glycerol minimal media with leucine and then plate on GMML-agar plates supplemented with kanamycin, chloramphenicol, tetracycline, and the unnatural 1 mM the unnatural amino acid. The plate was incubated at 37°C for over 60 h. The surviving clones were collected and their plasmid DNA was extracted. By gel electrophoresis, the PBK plasmid was purified and was prepared for the negative screening.

3. *Negative screening.* E. coli DH10B harboring the pREP(2)/YC plasmid was used as the host strain for the positive screening. Cells are transformed with the pBK-acyl-tRNA synthetase, washed twice with glycerol minimal media with leucine and then plate on GMML-agar plates supplemented with kanamycin, chloramphenicol, tetracycline, and the unnatural 1 mM the unnatural amino acid. The plate was incubated at 37°C for over 60 h. The surviving clones were collected and their plasmid DNA was extracted. By gel electrophoresis, the PBK plasmid was purified and was prepared for the negative screening.

4. *Chloramphenicol resistance assay.* In the third round of positive screening, instead of scraping the plate, 96 individual clones were selected and suspended in 50 μL of LB media in a 96-well plate, and printed on two sets of LB plates. One set was supplemented with tetracycline, kanamycin, and chloramphenicol at concentrations of 60, 80, 100, and 120 μg/mL with 1 mM of the unnatural amino acid. The other sets were identical but completed in the absence of the unnatural amino acid, and the chloramphenicol concentration were 0, 20, and 40 μg/mL. After incubation 37°C for 60 h, the clones those grown on the unnatural amino acid plate and do not grow on the unnatural amino acid absent plate were selected, and sent for sequencing.

5. *Bioorthogonal verification.* Newly evolved aminoacyl-tRNA synthetase needs extra tests on their bioorthogonal properties. The first test compares the protein expression yield with or without an extra added unnatural amino acid, in order to ensure that the unnatural amino acid rather than natural amino acid was incorporated at the amber codon. Usually the plasmid containing the evolved aminoacyl-tRNA synthetase and amber codon suppressor tRNA is cotransformed with the plasmid, encoding a model protein with the TAG mutation (for example,

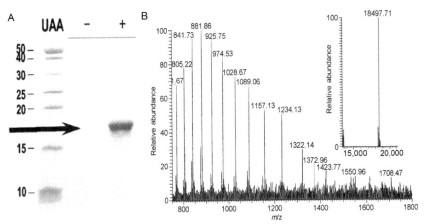

Fig. 6 An example of bioorthogonal verification for newly evolved aminoacyl-tRNA synthetase. (A) A TAG-mutated protein (myoglobin 4TAG) was expressed with and without the unnatural amino acid (imiTyr). The aminoacyl-tRNA synthetase is bioorthogonal if only the unnatural amino acid group achieved full-length protein. (B) The expressed protein was tested by mass spectrum. The measured molecule weight should exactly fit theoretical calculation (Liu, Yu, et al., 2012).

myoglobin fourth TAG). The protein is expressed with and without the unnatural amino acids. The group with the unnatural amino acid in the media should be able to achieve the expressed target protein and the control group with no unnatural amino acid should not be able to detect any target protein production. The second test calculates the molecule weight of the incorporated amino acid by mass spectrum, in order to ensure that it is the designed unnatural amino acid that is to be encoded. The reliability of the aminoacyl-tRNA synthetase can be trusted only when both tests are qualified (Fig. 6).

7. PROTEIN DESIGN

In the first stage of enzyme design, the unnatural amino acid and the scaffold protein are designed separately. The host protein can be designed by traditional protein design methods, which will not be discussed here. The main problem of the first stage is determining which site the unnatural amino acid should be incorporated. Currently, we cannot precisely predict which site is more suitable for the unnatural amino acid incorporation. However, the following empirical rules can be used as criteria.

1. The unnatural amino acids favor that site whose initial amino acids are structural similar to the unnatural amino acid. For example, aromatic unnatural amino acids are best to take the place of aromatic residues, such as tyrosine or tryptophan. And the sites with original long-chain aliphatic amino acids are suitable for lysine derivatives incorporation.
2. An unnatural amino acid favors the sites close to the N-terminal, rather than those close to the C-terminal. According to an empirical law, but if the TAG codon is close to the C-terminal, it would be more likely to be translated as a terminal signal.
3. Currently only one unnatural amino acid can be incorporated in one single peptide. The protein yield decreases drastically if more than one amber codon exists in the peptide.

The second stage treats the protein scaffold and the unnatural amino acid as an entirety and improvement its catalytic performance. Directed evolution and high-throughput screen are two strategies for such refinement. As directed evolution methods for normal enzymes, first a library was constructed by replacing some residues around the catalytic center by NNK mutations. Then the library was transformed into a protein expression strain together with the corresponding aminoacyl-tRNA synthetase and amber codon suppressor tRNA. The protein was expressed under normal conditions with extra addition of the unnatural amino acid. And the improved artificial enzyme was identified by in vivo activity assay and high-throughput screening. Computational methods for refining unnatural amino acid containing amino acid were also reported. Baker and coworkers reported precisely designed metalloprotein and an intermediate trapped protein by their computation software, which showed great promise in artificial metalloprotein design (Mills et al., 2013; Pearson et al., 2015).

8. PROTEIN EXPRESSION

8.1 Plasmid Preparation

Currently the plasmid pEVOL reported by Schultz is commonly used to support the evolved aminoacyl-tRNA synthetase (Young, Ahmad, Yin, & Schultz, 2010). pEVOL is chloramphenicol-resistant plasmids with amber codon suppressor tRNA and two copies of aminoacyl-tRNA synthetase which were promoted by AraB and Glsn. The acyl-tRNA synthetase on the vector can be replaced by another aminoacyl-tRNA synthetase mutants through digest and ligation (*Bgl*II-*Sal*I and *Nde*I-*Pst*I). Take the first open-reading frame as an example. First, the plasmid was digested by *Bgl*II and *Sal*I in order to remove the original aminoacyl-tRNA synthetase. Then the evolved aminoacyl-tRNA synthetase was amplified by primers containing *Bgl*II and *Sal*I sites. The PCR product was digested with *Bgl*II and *Sal*I, and then ligated into the previous digested pEVOL vector. The second reading frame was operated in a same method to afford the plasmid encoding necessary elements for unnatural amino acid incorporation.

The protein gene can be encoded on another plasmid that is comparable with pEVOL. The arranged unnatural amino acid incorporation site is mutated to a TAG amber codon by site-specific mutation method. It should be noticed that the TAG codon can be translated as either the unnatural amino acid or termination signal. Due to several essential TAG codons in *E. coli*, the later possibility cannot be totally avoided in normal *E. coli* strains. Thus, in order to remove the truncate translation products, the affinity tag, such as histidine affinity tag, should be attached to the C-terminal of the protein.

8.2 Protein Expression

The following is a universal protocol for unnatural amino acid containing protein expression.

1. The plasmid encodes the target protein with TAG mutants were cotransformed with the pEVOL expressing corresponding unnatural amino acid acyl-tRNA synthetase into BL21(DE3) *E. coli* or other strain used for protein expression.
2. After being grown overnight, a single colony was selected and then grown overnight at 37°C in 4 mL LB media with antibiotic (25 μg/mL chloramphenicol and another antibiotic which depends on the plasmid coding the target protein).
3. 2 mL culture is transferred to 100 mL fresh LB media with appropriate antibiotics. When the OD_{600} reach 1.1, 1 mM the unnatural amino acid, 1 mM IPTG and 0.02% arabinose is added to induce the protein expression.
4. After being grown another 4–12 h at 30°C or 37°C, the cells are harvested and frozen at −80°C for further purification.

9. CONCLUSION AND PERSPECTIVE

We have described the methods for enzyme design with genetically encoded unnatural amino acids. Unnatural amino acids can simply the simulating of natural structures, as well as provide opportunities to introduce powerful synthetic chemistry methods into molecular biology and synthetic biology. Yet some limitations remain. The type and the number of inserted unnatural amino acids in one peptide are strongly restricted. For instance, usually only one unnatural amino acid incorporation is permitted in one peptide. And, with only a few exceptions, currently most available unnatural amino acids are tyrosine derivatives and lysine derivatives.

However, recent progress in genomically recorded organisms exhibited great promise in breaking such limitations (Lajoie, Soll, & Church, 2015). The authors replaced all TAG codons in *E. coli* with nonsense codons TAA, which permitted the deletion of release factor 1 and reassignment of the TAG translation function (Lajoie et al., 2013). With the genomically recorded *E. coli* and evolved bioorthogonal aminoacyl-tRNA synthetase, they were able to incorporate multiple unnatural amino acids in one peptide with high yield (Rovner et al., 2015). We believe this progress in genetic code expansion will bring new opportunities in constructing novel functional protein structure.

REFERENCES

Ai, H. W., Lee, J. W., & Schultz, P. G. (2010). A method to site-specifically introduce methyllysine into proteins in E. coli. *Chemical Communications (Cambridge, England), 46*, 5506–5508.

Alfonta, L., Zhang, Z. W., Uryu, S., Loo, J. A., & Schultz, P. G. (2003). Site-specific incorporation of a redox-active amino acid into proteins. *Journal of the American Chemical Society, 125*, 14662–14663.

Anderson, J. C., Wu, N., Santoro, S. W., Lakshman, V., King, D. S., & Schultz, P. G. (2004). An expanded genetic code with a functional quadruplet codon. *Proceedings of the National Academy of Sciences of the United States of America, 101*, 7566–7571.

Bogdanov, A. M., Mishin, A. S., Yampolsky, I. V., Belousov, V. V., Chudakov, D. M., Subach, F. V., et al. (2009). Green fluorescent proteins are light-induced electron donors. *Nature Chemical Biology, 5*, 459–461.

Brown, K. D. (1971). Maintenance and exchange of the aromatic amino acid pool in Escherichia coli. *Journal of Bacteriology, 106*, 70–81.

Chan, S. I. (2009). A physical chemist's expedition to explore the world of membrane proteins. *Annual Review of Biophysics, 38*, 1–27.

Collman, J. P., Devaraj, N. K., Decreau, R. A., Yang, Y., Yan, Y. L., Ebina, W., et al. (2007). A cytochrome c oxidase model catalyzes oxygen to water reduction under rate-limiting electron flux. *Science, 315*, 1565–1568.

Durrenberger, M., & Ward, T. R. (2014). Recent achievements in the design and engineering of artificial metalloenzymes. *Current Opinion in Chemical Biology, 19*, 99–106.

Fekner, T., Li, X., Lee, M. M., & Chan, M. K. (2009). A pyrrolysine analogue for protein click chemistry. *Angewandte Chemie International Edition, 48*, 1633–1635.

Gray, H. B., & Winkler, J. R. (2005). Long-range electron transfer. *Proceedings of the National Academy of Sciences of the United States of America, 102*, 3534–3539.

Guo, J. T., Wang, J. Y., Lee, J. S., & Schultz, P. G. (2008). Site-specific incorporation of methyl- and acetyl-lysine analogues into recombinant proteins. *Angewandte Chemie International Edition, 47*, 6399–6401.

Hu, C., Chan, S. I., Sawyer, E. B., Yu, Y., & Wang, J. (2014). Metalloprotein design using genetic code expansion. *Chemical Society Reviews, 43*, 6498–6510.

Huynh, M. H. V., & Meyer, T. J. (2007). Proton-coupled electron transfer. *Chemical Reviews, 107*, 5004–5064.

Jackson, J. C., Duffy, S. P., Hess, K. R., & Mehl, R. A. (2006). Improving nature's enzyme active site with genetically encoded unnatural amino acids. *Journal of the American Chemical Society, 128*, 11124–11127.

Jewett, J. C., & Bertozzi, C. R. (2010). Cu-free click cycloaddition reactions in chemical biology. *Chemical Society Reviews, 39*, 1272–1279.

Kaya, E., Gutsmiedl, K., Vrabel, M., Muller, M., Thumbs, P., & Carell, T. (2009). Synthesis of threefold glycosylated proteins using click chemistry and genetically encoded unnatural amino acids. *Chembiochem, 10*, 2858–2861.

Keasling, J. D. (2008). Synthetic biology for synthetic chemistry. *ACS Chemical Biology, 3*, 64–76.

Kobayashi, T., Nureki, O., Ishitani, R., Yaremchuk, A., Tukalo, M., Cusack, S., et al. (2003). Structural basis for orthogonal tRNA specificities of tyrosyl-tRNA synthetases for genetic code expansion. *Nature Structural Biology, 10*, 425–432.

Kolb, H. C., Finn, M. G., & Sharpless, K. B. (2001). Click chemistry: Diverse chemical function from a few good reactions. *Angewandte Chemie International Edition, 40*, 2004–2021.

Lajoie, M. J., Rovner, A. J., Goodman, D. B., Aerni, H. R., Haimovich, A. D., Kuznetsov, G., et al. (2013). Genomically recoded organisms expand biological functions. *Science, 342*, 357–360.

Lajoie, M. J., Soll, D., & Church, G. M. (2015). Overcoming challenges in engineering the genetic code. *Journal of Molecular Biology, 428*, 1004–1021.

Lee, H. S., & Schultz, P. G. (2008). Biosynthesis of a site-specific DNA cleaving protein. *Journal of the American Chemical Society, 130*, 13194–13195.

Lim, R. K. V., & Lin, Q. (2011). Photoinducible bioorthogonal chemistry: A spatiotemporally controllable tool to visualize and perturb proteins in live cells. *Accounts of Chemical Research, 44*, 828–839.

Liu, C. C., Cellitti, S. E., Geierstanger, B. H., & Schultz, P. G. (2009). Efficient expression of tyrosine-sulfated proteins in E-coli using an expanded genetic code. *Nature Protocols, 4*, 1784–1789.

Liu, X. H., Jiang, L., Li, J. S., Wang, L., Yu, Y., Zhou, Q., et al. (2014). Significant expansion of fluorescent protein sensing ability through the genetic incorporation of superior photo-induced electron-transfer quenchers. *Journal of the American Chemical Society, 136*, 13094–13097.

Liu, X. H., Li, J. S., Dong, J. S., Hu, C., Gong, W. M., & Wang, J. Y. (2012). Genetic incorporation of a metal-chelating amino acid as a probe for protein electron transfer. *Angewandte Chemie International Edition, 51*, 10261–10265.

Liu, X. H., Li, J. S., Hu, C., Zhou, Q., Zhang, W., Hu, M. R., et al. (2013). Significant expansion of the fluorescent protein chromophore through the genetic incorporation of a metal-chelating unnatural amino acid. *Angewandte Chemie International Edition, 52*, 4805–4809.

Liu, C. C., & Schultz, P. G. (2010). Adding new chemistries to the genetic code. *Annual Review of Biochemistry, 79*, 413–444.

Liu, X. H., Yu, Y., Hu, C., Zhang, W., Lu, Y., & Wang, J. Y. (2012). Significant increase of oxidase activity through the genetic incorporation of a tyrosine-histidine cross-link in a myoglobin model of heme-copper oxidase. *Angewandte Chemie International Edition, 51*, 4312–4316.

Lu, Y., Yeung, N., Sieracki, N., & Marshall, N. M. (2009). Design of functional metalloproteins. *Nature, 460*, 855–862.

Lv, X. X., Yu, Y., Zhou, M., Hu, C., Gao, F., Li, J. S., et al. (2015). Ultrafast photoinduced electron transfer in green fluorescent protein bearing a genetically encoded electron acceptor. *Journal of the American Chemical Society, 137,* 7270–7273.

Mann, J. (1989). The logic of chemical synthesis—E. J. Corey, X. M. Cheng, *Nature, 341,* 118.

Marcus, R. A., & Sutin, N. (1985). Electron transfers in chemistry and biology. *Biochimica et Biophysica Acta, 811,* 265–322.

Martin, A. B., & Schultz, P. G. (1999). Opportunities at the interface of chemistry and biology. *Trends in Biochemical Sciences, 24,* M24–M28.

Mills, J. H., Khare, S. D., Bolduc, J. M., Forouhar, F., Mulligan, V. K., Lew, S., et al. (2013). Computational design of an unnatural amino acid dependent metalloprotein with atomic level accuracy. *Journal of the American Chemical Society, 135,* 13393–13399.

Pearson, A. D., Mills, J. H., Song, Y. F., Nasertorabi, F., Han, G. W., Baker, D., et al. (2015). Trapping a transition state in a computationally designed protein bottle. *Science, 347,* 863–867.

Plass, T., Milles, S., Koehler, C., Schultz, C., & Lemke, E. A. (2011). Genetically encoded copper-free click chemistry. *Angewandte Chemie International Edition, 50,* 3878–3881.

Polyakov, K. M., Boyko, K. M., Tikhonova, T. V., Slutsky, A., Antipov, A. N., Zvyagilskaya, R. A., et al. (2009). High-resolution structural analysis of a novel octaheme cytochrome c nitrite reductase from the haloalkaliphilic bacterium Thioalkalivibrio nitratireducens. *Journal of Molecular Biology, 389,* 846–862.

Polycarpo, C., Ambrogelly, A., Berube, A., Winbush, S. A. M., McCloskey, J. A., Crain, P. F., et al. (2004). An aminoacyl-tRNA synthetase that specifically activates pyrrolysine. *Proceedings of the National Academy of Sciences of the United States of America, 101,* 12450–12454.

Rovner, A. J., Haimovich, A. D., Katz, S. R., Li, Z., Grome, M. W., Gassaway, B. M., et al. (2015). Recoded organisms engineered to depend on synthetic amino acids. *Nature, 518,* 89–93.

Seyedsayamdost, M. R., Yee, C. S., & Stubbe, J. (2007). Site-specific incorporation of fluorotyrosines into the R2 subunit of E. coli ribonucleotide reductase by expressed protein ligation. *Nature Protocols, 2,* 1225–1235.

Sigman, J. A., Kwok, B. C., & Lu, Y. (2000). From myoglobin to heme-copper oxidase: Design and engineering of a Cu-B center into sperm whale myoglobin. *Journal of the American Chemical Society, 122,* 8192–8196.

Steffes, C., Ellis, J., Wu, J., & Rosen, B. P. (1992). The lysP gene encodes the lysine-specific permease. *Journal of Bacteriology, 174,* 3242–3249.

Uy, R., & Wold, F. (1977). Posttranslational covalent modification of proteins. *Science, 198,* 890–896.

Verma, P., Pratt, R. C., Storr, T., Wasinger, E. C., & Stack, T. D. P. (2011). Sulfanyl stabilization of copper-bonded phenoxyls in model complexes and galactose oxidase. *Proceedings of the National Academy of Sciences of the United States of America, 108,* 18600–18605.

Wang, L., Brock, A., Herberich, B., & Schultz, P. G. (2001). Expanding the genetic code of Escherichia coli. *Science, 292,* 498–500.

Wang, J. Y., Xie, J. M., & Schultz, P. G. (2006). A genetically encoded fluorescent amino acid. *Journal of the American Chemical Society, 128,* 8738–8739.

Wikstrom, M. (1998). Proton-coupled electron transfer in cytochrome c oxidase. *Abstracts of Papers of the American Chemical Society, 216,* U95.

Winkler, J. R., & Gray, H. B. (2014). Electron flow through metalloproteins. *Chemical Reviews, 114,* 3369–3380.

Xie, J. M., Liu, W. S., & Schultz, P. G. (2007). A genetically encoded bidentate, metal-binding amino acid. *Angewandte Chemie International Edition, 46,* 9239–9242.

Xie, J. M., & Schultz, P. G. (2006). Innovation: A chemical toolkit for proteins—An expanded genetic code. *Nature Reviews. Molecular Cell Biology, 7,* 775–782.

Yoshikawa, S., & Shimada, A. (2015). Reaction mechanism of cytochrome c oxidase. *Chemical Reviews, 115,* 1936–1989.

Yoshikawa, S., Shinzawa-Itoh, K., Nakashima, R., Yaono, R., Yamashita, E., Inoue, N., et al. (1998). Redox-coupled crystal structural changes in bovine heart cytochrome c oxidase. *Science, 280,* 1723–1729.

Young, T. S., Ahmad, I., Yin, J. A., & Schultz, P. G. (2010). An enhanced system for unnatural amino acid mutagenesis in E. coli. *Journal of Molecular Biology, 395,* 361–374.

Yu, Y., Lv, X. X., Li, J. S., Zhou, Q., Cui, C., Hosseinzadeh, P., et al. (2015). Defining the role of tyrosine and rational tuning of oxidase activity by genetic incorporation of unnatural tyrosine analogs. *Journal of the American Chemical Society, 137,* 4594–4597.

Zhou, Q., Hu, M. R., Zhang, W., Jiang, L., Perrett, S., Zhou, J. Z., et al. (2013). Probing the function of the Tyr-Cys cross-link in metalloenzymes by the genetic incorporation of 3-methylthiotyrosine. *Angewandte Chemie International Edition, 52,* 1203–1207.

CHAPTER SIX

Methods for Solving Highly Symmetric De Novo Designed Metalloproteins: Crystallographic Examination of a Novel Three-Stranded Coiled-Coil Structure Containing D-Amino Acids

L. Ruckthong*, J.A. Stuckey[†], V.L. Pecoraro*,[1]
*Department of Chemistry, University of Michigan, Ann Arbor, MI, United States
[†]Life Sciences Institute, University of Michigan, Ann Arbor, MI, United States
[1]Corresponding author: e-mail address: vlpec@umich.edu

Contents

1. Introduction	136
2. Materials	140
3. Methods	141
3.1 Synthesis and Purification	141
3.2 Peptide Purification	141
3.3 Protein Crystallization	141
3.4 Data Collection and Processing	142
3.5 Structure Determination and Refinements	142
Acknowledgments	146
References	146

Abstract

The core objective of de novo metalloprotein design is to define metal–protein relationships that control the structure and function of metal centers by using simplified proteins. An essential requirement to achieve this goal is to obtain high resolution structural data using either NMR or crystallographic studies in order to evaluate successful design. X-ray crystal structures have proven that a four heptad repeat scaffold contained in the three-stranded coiled coil (3SCC), called CoilSer (CS), provides an excellent motif for modeling a three Cys binding environment capable of chelating metals into geometries that resemble heavy metal sites in metalloregulatory systems. However, new generations of more complicated designs that feature, for example, a D-amino acid or multiple metal ligand sites in the helical sequence require a more stable construct. In doing so, an extra heptad was introduced into the original CS sequence, yielding a GRAND-CoilSer (GRAND-CS) to retain the 3SCC folding. An apo-(GRAND-CSL12$_D$LL16C)$_3$ crystal structure, designed for Cd(II)S$_3$ complexation, proved to be a well-folded parallel 3SCC. Because this structure is novel, protocols for crystallization, structural determination, and refinements of the apo-(GRAND-CSL12$_D$LL16C)$_3$ are described. This report should be generally useful for future crystallographic studies of related coiled-coil designs.

1. INTRODUCTION

De novo designed proteins provide a simplified scaffold that allows us to construct a metal site to learn the basic principles of metal–protein interactions that are, often times, challenging to address in a natural protein system due to its high level of complexity (Beasley & Hecht, 1997; DeGrado, Summa, Pavone, Nastri, & Lombardi, 1999; Kaplan & DeGrado, 2004; Kohn & Hodges, 1998; Lu, Yeung, Sieracki, & Marshall, 2009; Peacock, Iranzo, & Pecoraro, 2009; Zastrow & Pecoraro, 2013). For decades, the Pecoraro group has been interested in studying heavy metal chemistry in proteins by introducing a *tris*-thiolate rich site into the hydrophobic core of a three-stranded coiled-coil (3SCC) scaffold based on the TRI-family peptide sequence (sequence shown in Table 1). The helix is designed based on a heptad repeat approach. Hydrophobic residues are placed at the *a* and *d* positions which undergo hydrophobic collapse when helices associate into a tertiary coiled-coil structure in solution. Hydrophilic residues at *e* and *g* facilitate salt bridge interactions between strands to stabilize the 3SCC orientation at an appropriate pH. The replacement of a core Leu residue with a Cys yields a three Cys environment for heavy metal binding studies (Fig. 1). Spectroscopic studies have shown that the engineered peptides are capable of direct binding to a variety of metals (Hg(II), Cd(II), Pb(II), As(III), Zn(II), and Bi(III)) in which the knowledge gained can be applied to investigate heavy metal interactions in metalloregulatory systems (Chakraborty, Touw, Peacock, Stuckey, & Pecoraro, 2010; Dieckmann, 1995; Dieckmann et al., 1998, 1997; Ghosh, Lee, Demeler, & Pecoraro, 2005; Ghosh & Pecoraro, 2005; Iranzo, Chakraborty, Hemmingsen, & Pecoraro, 2011; Iranzo, Ghosh, & Pecoraro, 2006; Iranzo, Jakusch, Lee, Hemmingsen, & Pecoraro, 2009; Iranzo, Thulstrup, Ryu, Hemmingsen, &

Table 1 Peptide Sequences

Peptides		abcdefg 2	abcdefg 9 12	abcdefg 16 19	abcdefg	abcdefg 30
TRI	Ac-G	LKALEEK	LKALEEK	LKALEEK	LKALEEK	G-NH$_2$
TRIL12$_D$LL16C	Ac-G	LKALEEK	LKA**D**L**EEK	**C**KALEEK	LKALEEK	G-NH$_2$
CoilSer (CS)	Ac-E	WEALEKK	LAALESK	LQALEKK	LEALEHG	–NH$_2$
CSL16C	Ac-E	WEALEKK	LAALESK	**C**QALEKK	LEALEHG	–NH$_2$
GRAND-CoilSer	Ac-E	WEALEKK	LAALESK	LQALEKK	LQALEKK	LEALEHG –NH$_2$
GRANDCSL12$_D$LL16C	Ac-E	WEALEKK	LAA**D**L**ESK	**C**QALEKK	LQALEKK	LEALEHG –NH$_2$

Bold and underlined residues indicate substitutions. C– and N-termini are capped by Ac and NH$_2$ groups, respectively.

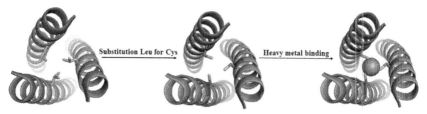

Fig. 1 PyMOL visualization demonstrating how to engineer a *tris*-thiolate binding site into the interior of 3SCC. The substitution of Leu residues (shown as *pink* (gray in the print version) *sticks*) with Cys yields three Cys environment that is capable of binding heavy metal atoms. *The figure was made based on the As(III)(CSL9C)$_3$ (PDB code: 2JGO) (Touw, Nordman, Stuckey, & Pecoraro, 2007).*

Pecoraro, 2007; Lee, Matzapetakis, Mitra, Marsh, & Pecoraro, 2004; Łuczkowski et al., 2008; Matzapetakis et al., 2002; Matzapetakis, Ghosh, Weng, Penner-Hahn, & Pecoraro, 2006; Matzapetakis & Pecoraro, 2005; Neupane & Pecoraro, 2010; Peacock, Hemmingsen, & Pecoraro, 2008; Peacock, Iranzo, et al., 2009; Pecoraro, Peacock, Iranzo, & Luczkowski, 2009; Touw, 2007; Touw et al., 2007; Zampella, Neupane, De Gioia, & Pecoraro, 2012; Zastrow, Peacock, Stuckey, & Pecoraro, 2012).

Despite the success of spectroscopic experiments from solution studies, structural details of the metal structures are needed in order to provide a thorough understanding of how these *tris*-thiolate environments lead to each specific metal coordination mode based on their geometric preferences. Unfortunately, the TRI-family peptides were not capable of generating high-quality protein crystals, therefore, a closely related CoilSer (CS) peptide has been used to serve as a crystallographic analogue. The CS sequence was also designed based on a heptad repeat approach (Lovejoy et al., 1993) and self-assemble into a parallel 3SCC in a similar manner to TRI when the Cys-rich site is incorporated (Touw et al., 2007). Spectroscopic studies demonstrated that TRI and CS derivatives bind to metals at the sulfur site in an analogous manner (Iranzo et al., 2006); however, the presence of a His residue at the 28th position allows CS to create a 3D packing of peptides by forming external Zn(II) sites between symmetry-related trimers. X-ray crystal structures have been obtained for CS peptides with thiol containing ligand substitution (Chakraborty et al., 2010; Peacock, Stuckey, & Pecoraro, 2009; Ruckthong, 2016; Touw et al., 2007; Zastrow et al., 2012); however, these few structures do not allow one to assess the consequences of new features incorporated in more advanced protein designs where, for example, multiple metal sites are present or a destabilizing residue is introduced into the parent sequence. Because the stability of these amphipathic 3SCCs is dependent on exclusion of residues such as leucine from the aqueous environment, the number of unperturbed *a* and *d* hydrophobic layers determines the aggregate stability. Designs including multiple modifications in the hydrophobic interior are no longer stable with the relative short TRI scaffold so GRAND peptides, containing an additional heptad, needed to be developed (Ghosh et al., 2005). Herein, we report a GRAND-CoilSer (GRAND-CS) in which an additional heptad is incorporated into the original CS sequence to enhance the stability of the aggregates for crystallographic studies. The His residue, now placed at the 35th position (last heptad) of the elongating peptide, allows for high-quality crystal formation. The designed sequence reported in this article is GRAND-CSL12$_D$LL16C where a D-Leu residue (DLE) at the 12th position was designed to generate steric encumbrance above the Cys site to enforce an unusual Cd(II)S$_3$ center in a 3SCC environment (Peacock et al., 2008; Ruckthong, 2016). The GRAND-CSL12$_D$LL16C serves as a crystallographic analogue for TRIL12$_D$LL16C. Original structural studies were attempted using the CSL12$_D$LL16C derivative; however, crystals were not obtained due to the decreased stability associated with the packing of a D-Leu into the sequence (Ruckthong, 2016). By adding an extra heptad, the resulting apo-(GRAND-CSL12-$_D$LL16C)$_3$ crystal structure reveals a parallel 3SCC peptide resembling the previous Cys-rich CS helices (Fig. 2). The two types of peptides, however, were crystallized in different space groups in which the GRAND-CS helices are packed in the high symmetry space group R32 (#155) that produces a crystallographically imposed threefold axis along the three helices, while the shorter CS derivatives pack in space group C2 (#5) having one 3SCC in the asymmetric unit (ASU). A methodology for crystallization and structure determination is provided since this is the first GRAND-CS derivative that has been solved. This structure emphasizes the benefits of the expansion of the heptad repeats to achieve sufficient overall stability in the coiled-coil system when drastic mutations are included and demonstrates the potential impact of D-amino acid chirality on controlling a metal center coordination. This observation may open up new vistas for metalloprotein engineering that could greatly expand the structural understanding of new designed scaffolds for further biophysical, catalytic, and therapeutic applications.

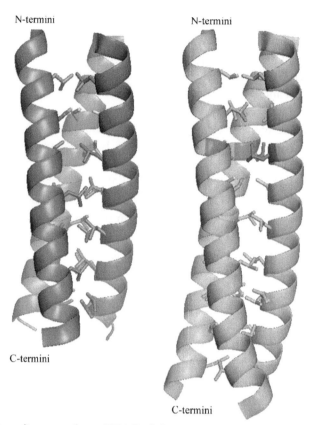

Fig. 2 Ribbon diagrams of apo-(CSL9C)$_3$ (*left*) and apo-(GRAND-CSL12$_D$LL16C)$_3$ (*right*) representing the difference in length of the two peptides. The apo-(CSL9C)$_3$ contains four heptad repeats while an additional heptad is added in the apo-(GRAND-CSL12$_D$LL16C)$_3$ subsequently resulting in a longer trimer. Helical core residues are shown as *sticks*. D-Leu in the 12th position of apo-(GRAND-CSL12$_D$LL16C)$_3$ is *red*. Thiols of Cys residues are *yellow*. (See the color plate.)

2. MATERIALS

Synthesis and purification: Fmoc-protected amino acids and the MBHA rink amide resin were purchased from Novabiochem; *N*-hydroxybenzotriazole (HOBt) and 2-(1H-benzotriazol-1-yl)-1,1,3,3-tetramethyluronium hexafluorophosphate (HBTU) from Anaspec Inc.; diisopropylethylamine (DIEA), acetic anhydride, piperidine, and pyridine from Sigma-Aldrich; *N*-methylpyrrolidinone, *N,N*-dimethylformamide (DMF), anisole, thioanisole, ethanedithiol, trifluoroacetic acid (TFA), and acetronitile (high performance liquid chromatography (HPLC) grade) from Fisher Scientific.

Crystallization: Tris(hydroxymethyl)aminomethane and zinc acetate dehydrate (colorless solid) were purchased from Sigma-Aldrich. Nextal PEG screen was obtained from Qiagen.

3. METHODS
3.1 Synthesis and Purification

The GRAND-CSL12$_D$LL16C peptide was synthesized on an Applied Biosystems 433A automated peptide synthesizer with Fmoc-protected amino acids using the standard Fmoc protocol (Applied Biosystems) (Chan & White, 2000). The C-terminus of the peptides was amidated on the solid support MBHA rink amide resin (0.25 mmol scale) with HBTU/HOBt/DIEA coupling methods. The N-terminus was acetylated with a solution of 4% (v/v) acetic anhydride, 4.3% (v/v) pyridine, and 91.7% DMF. The peptide was cleaved from the resin using a cleavage mixture of 90% TFA, 5% anisole, 3% thioanisole, and 2% ethanedithiol for 3.5 h. The cleaved peptide solution was filtered and evaporated under a dry N_2-flow until a glassy film appeared on the surface. Cold diethyl ether was then added to the thin film to obtain a precipitated white crude peptide. This crude was redissolved in ddH_2O and lyophilized to get a fluffy white powder which was subsequently dissolved in 10% acetic acid.

3.2 Peptide Purification

The peptide was purified by reversed-phase HPLC on a Waters 600 Semi-prep HPLC peptide C-18 using a linear gradient of 0.1% TFA in water to 0.1% TFA in 9:1 CH_3CN/H_2O program over 30 min (flow rate 10 mL/min). The purified peptides were identified by electrospray mass spectrometry. The actual mass of the GRAND-CSL12$_D$LL16C was determined to be 4147.55 g/mol corresponding to the theoretical calculation (http://rna.rega.kuleuven.ac.be/masspec/pepcalc.htm). The purified solution was frozen under liquid N_2 and lyophilized until the peptide was completely dry.

3.3 Protein Crystallization

About 25 mg of white power GRAND-CSL12$_D$LL16C peptide was dissolved in degassed 250 µL ddH_2O as a peptide stock. The sample was prepared to yield the final concentration of 20 mg/mL peptide with a total volume of 420 µL. Tris buffer at pH 8.5 was added to make 5 mM concentration followed by an addition of Zn(OAc)$_2$ solution to reach a final concentration of 15 mM Zn(OAc)$_2$. After every addition of a new reagent, the sample was vortexed and centrifuged at 14,000 rpm for a minute to ensure that the sample was well dissolved.

Crystallization was performed robotically with a Crystal Grypton machine from Art Robbins Instruments using the Nextal PEG screen from Qiagen. The screen was setup on a 96-well sitting drop plate with drops containing equal volumes of peptide (0.75 μL) and precipitant (0.75 μL) solutions. The crystals were grown by vapor diffusion at 20°C. The appropriate precipitant condition for GRAND-CSL12$_D$LL16C crystallization contains 25% (v/v) PEG-2000 MME and 0.1 M MES buffer pH 6.5. Crystals were cryoprotected in a mother liquor containing 20% glycerol prior to supercooling in liquid N_2 for data collection.

3.4 Data Collection and Processing

Data collection was performed at the Advanced Photon Source of the Argonne National Laboratory on the LS-CAT Beamline 21-ID-F equipped with a Mar 225 CCD detector at a wavelength of 0.97872 Å. The data were collected with a 1 degree oscillation and 1 s exposure time. The data were processed with HKL2000 (Otwinowski & Minor, 1997). The GRAND-CSL12$_D$LL16C crystal diffracted to 1.42 Å with the R32 (#155) space group. The unit cell parameters are shown in Table 2.

3.5 Structure Determination and Refinements

The structure determination process began with the estimation of the numbers of molecules in an ASU using a method developed by Matthews (1968). The GRAND-CSL12$_D$LL16C, crystallized in the R32 space group, was determined to contain one single strand of the 3SCC peptide per ASU. The Matthew's coefficient of 2.38 corresponds to 47.67% solvent content. (The 3SCC is obtained by the combination of three adjacent ASUs that are crystallographic imposed by the threefold axis.) This is in marked contrast with the previous CS crystal structures where most were crystallized in the C2 space group, consequently containing a single trimer per ASU (Chakraborty et al., 2010; Peacock, Stuckey, et al., 2009; Ruckthong, 2016; Touw et al., 2007; Zastrow et al., 2012). Due to the high sequence identity, the GRAND-CSL12-$_D$LL16C structure was solved using a four heptad chain A of the published 3SCC apo-(CSL9C)$_3$ (PDB code: 3LJM) (Chakraborty et al., 2010) to serve as a search model in auto MolRep (Vagin & Teplyakov, 2010) from the CCP4 suite of programs (McCoy et al., 2007; Potterton, Briggs,

Table 2 Crystallography Data Collection and Refinement Statistics

Peptide	GRAND-CSL12$_D$LL16C (1SCC per ASU)
Data collection	
Space group	R32
Unit cell	
a, b, c (Å)	38.213, 38.213, 140.655
α, β, γ (degree)	90.00, 90.00, 120.00
Wavelength (Å)	0.97872
Resolution (Å)[a]	1.42 (1.42–1.40)
Rsym (%)[b]	5.6 (43.4)
$\langle I/\sigma I \rangle$[c]	>50 (2)
Completeness (%)[d]	99.3 (100)
Redundancy	5.6 (5.5)
Refinement	
Resolution (Å)	1.42
R-factor (%)[e]	19.6
R_{free} (%)[f]	20.3
Protein atoms	302
Metal ions	Zn(II)
Water molecules	52
Unique reflections	8093
RMSD[g]	
Bonds	0.010
Angles	1.15
MolProbity score[h]	1.11
Clash score[h]	3.17

[a]Statistics for the highest resolution bin of reflections in parentheses.
[b]$R_{sym} = \Sigma_h \Sigma_j |I_{hj} - \langle I_h \rangle| / \Sigma_h \Sigma_j I_{hj}$, where I_{hj} is the intensity of observation j of reflection h and $\langle I_h \rangle$ is the mean intensity for multiply recorded reflections.
[c]Intensity signal-to-noise ratio.
[d]Completeness of the unique diffraction data.
[e]R-factor $= \Sigma_h |IF_o I - IF_c I| / \Sigma_h |F_o|$, where F_o and F_c are the observed and calculated structure factor amplitudes for reflection h.
[f]R_{free} is calculated against a 10% random sampling of the reflections that were removed before structure refinement.
[g]Root mean square deviation of bond lengths and bond angles.
[h]Chen et al. (2010).

Turkenburg, & Dodson, 2003; Winn et al., 2011), in which the Cys9 and Leu16 were mutated to Leu and Cys, respectively. The phases calculated from the molecular replacement solution were then used in autobuilding Arp/warp software (Langer, Cohen, Lamzin, & Perrakis, 2008) to generate a model that has five heptads in length corresponding to the GRAND-CS structure. The percentage ratio of R/R_{free} that resulted from the autobuilding Arp/warp was 35.3/42.7. Then, the model underwent iterative rounds of electron density fitting and refining in Coot (Emsley & Cowtan, 2004) and Buster 2.11.2 program (Bricogne et al., 2016), respectively. After the first round of rigid body refinement, the DLE was built at the 12th position using the difference density present in $F_o - F_c$ map. Based on the $F_o - F_c$ map, a Zn(II) ion was constructed near the C-terminus of the structure. The refined metal site displays a tetrahedral geometry with 31Glu, 34Glu, and 35His from the corresponding ASU and the fourth ligand from 3Glu of a symmetry-related molecule. The site is located at the crystallographic interface to link the helices into the lattice form. Acylated and amidated groups were built to cap the N- and C-termini, respectively. A PEG chain and waters were added upon refinements. The validity of the model was verified using the MolProbity software (Chen et al., 2010). All nonglycine residues of this structure fall in the preferred right handed α-helical region of the Ramanchandran plot. Every side chain is present in the preferred rotameric conformations. The final model was refined to 1.42 Å ($R_{working} = 19.6\%$, $R_{free} = 20.3\%$). The 3SCC visualization was achieved through threefold symmetry-related helices from three adjacent ASUs. The lack of $F_o - F_c$ electron difference for a metal at the Cys site (16th position) reveals the (GRAND-CSL12$_D$LL16C)$_3$ to be an apo-protein. The apo-(GRAND-CSL12$_D$LL16C)$_3$ is unperturbed by D-Leu as clearly demonstrated by the similarity in secondary structure of the helical backbones between the longer apo-(GRAND-CSL12$_D$LL16C)$_3$ and shorter apo-(CSL9C)$_3$ helices (Fig. 3). Moreover, the major Cys conformations of apo-(GRAND-CSL12$_D$LL16C)$_3$ are similar to the apo-(CSL9C)$_3$ (Chakraborty et al., 2010) and the apo-(CSL16C)$_3$ (Ruckthong, 2016) indicating that the thiols rotamers in the *a* site of the 3SCCs, regardless of the length of the helices, are identical in the absence of metal (Fig. 4). Therefore, the apo-(GRAND-CSL12$_D$LL16C)$_3$ crystal structure provides structural insight of elongating Cys-rich CS derivative and represents a useful scaffold for future active site metalloprotein modelling.

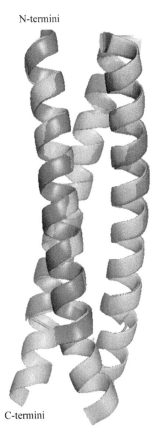

Fig. 3 Side view of an overlay between the trimeric apo-(GRAND-CSL12$_D$LL16C)$_3$ and apo-(CSL9C)$_3$ structures indicating that there is no change in the overall secondary structures of GRAND-CS and CS derivatives. Moreover, the alignment demonstrates that the incorporation of a nonnatural amino acid, D-Leu, does not perturb the secondary structure of the three-stranded coiled coil. Shown in *green* is the apo-(GRAND-CSL12-$_D$LL16C)$_3$ and in *pink* is apo-(CSL9C)$_3$. Side chain residues are omitted for clarity. (See the color plate.)

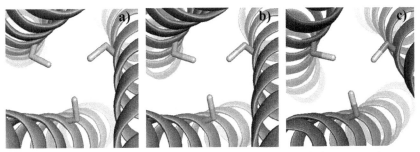

Fig. 4 Top down view of 3SCC represents the similarity of the major Cys conformer orientations between (A) apo-(GRAND-CSL12$_D$LL16C)$_3$, (B) apo-(CSL16C)$_3$, and (C) apo-(CSL9C)$_3$ structures. The Cys side chains are shown as *sticks* with sulfurs labelled as *yellow* (light gray in the print version).

ACKNOWLEDGMENTS
V.L.P. and L.R. thank the National Institute of Health for support of this research (Grant R01 ES012236). J.A.S. thanks the University of Michigan, Center for Structural Biology. L.R. thanks Dr. Jennifer Meagher and David Smith for crystal screening. This research used resources of the Advanced Photon Source, a U.S. Department of Energy (DOE) Office of Science User Facility operated for the DOE Office of Science by Argonne National Laboratory under Contract No. DE-AC02-06CH11357. Use of the LS-CAT Sector 21 was supported by the Michigan Economic Development Corporation and the Michigan Technology Tri-Corridor (Grant 085P1000817).

REFERENCES
Beasley, J. A. R., & Hecht, M. H. (1997). Protein design: The choice of de novo sequences. *The Journal of Biological Chemistry*, (272), 2031–2034.

Bricogne, G., Blanc, E., Brandl, M., Flensburg, C., Keller, P., Paciorek, W., et al. (2016). *BUSTER version X.Y.Z.* Cambridge, UK: Global Phasing Ltd.

Chakraborty, S., Touw, D. S., Peacock, A. F. A., Stuckey, J. A., & Pecoraro, V. L. (2010). Structural comparisons of apo- and metalated three-stranded coiled coils clarify metal binding determinants in thiolate containing designed peptides. *Journal of the American Chemical Society*, (132), 13240–13250.

Chan, W. C., & White, P. D. (2000). *Fmoc solid phase peptide synthesis: A practical approach.* New York: Oxford University Press.

Chen, V. B., Arendall, W. B., Headd, J. J., Keedy, D. A., Immormino, R. M., Kapral, G. J., et al. (2010). MolProbity: All-atom structure validation for macromolecular crystallography. *Acta Crystallographica, Section D: Biological Crystallography*, (66), 12–21.

DeGrado, W. F., Summa, C. M., Pavone, V., Nastri, F., & Lombardi, A. (1999). De novo design and structural characterization of proteins and metalloproteins. *Annual Review of Biochemistry*, (68), 779–819.

Dieckmann, G. R. (1995). *Use of metal-binding de novo-designed alpha-helical peptides in the study of metalloprotein structures.* PhD dissertation, University of Michigan.

Dieckmann, G. R., McRorie, D. K., Lear, J. D., Sharp, K. A., DeGrado, W. F., & Pecoraro, V. L. (1998). The role of protonation and metal chelation preferences in defining the properties of mercury-binding coiled coils. *Journal of Molecular Biology*, (280), 897–912.

Dieckmann, G. R., Mcrorie, D. K., Tierney, D. L., Utschig, L. M., Singer, C. P., O'Halloran, T. V., et al. (1997). De novo design of mercury-binding two- and three-helical bundles. *Journal of the American Chemical Society*, (119), 6195–6196.

Emsley, P., & Cowtan, K. (2004). Coot: Model-building tools for molecular graphics. *Acta Crystallographica Section D: Biological Crystallography*, 2126–2132.

Ghosh, D., Lee, K.-H., Demeler, B., & Pecoraro, V. L. (2005). Linear free-energy analysis of mercury(II) and cadmium(II) binding to three-stranded coiled coils. *Biochemistry*, (44), 10732–10740.

Ghosh, D., & Pecoraro, V. L. (2005). Probing metal–protein interactions using a de novo design approach. *Current Opinion in Chemical Biology*, (9), 97–103.

Iranzo, O., Chakraborty, S., Hemmingsen, L., & Pecoraro, V. L. (2011). Controlling and fine tuning the physical properties of two identical metal coordination sites in de novo designed three stranded coiled coil peptides. *Journal of the American Chemical Society*, (133), 239–251.

Iranzo, O., Ghosh, D., & Pecoraro, V. L. (2006). Coiled coils in the presence of metal ions. *Inorganic Chemistry*, (45), 9959–9973.

Iranzo, O., Jakusch, T., Lee, K.-H., Hemmingsen, L., & Pecoraro, V. L. (2009). The correlation of ^{113}Cd NMR and ^{111m}Cd PAC spectroscopies provides a powerful approach for the characterization of the structure of Cd(II)-substituted Zn(II) proteins. *Chemistry*, (15), 3761–3772.

Iranzo, O., Thulstrup, P. W., Ryu, S.-B., Hemmingsen, L., & Pecoraro, V. L. (2007). The application of ^{199}Hg NMR and ^{199m}Hg perturbed angular correlation (PAC) spectroscopy to define the biological chemistry of Hg(II): A case study with designed two- and three-stranded coiled coils. *Chemistry*, (13), 9178–9190.

Kaplan, J., & DeGrado, W. F. (2004). De novo design of catalytic proteins. *Proceedings of the National Academy of Sciences of the United States of America*, (101), 11566–11570.

Kohn, W. D., & Hodges, R. S. (1998). De novo design of α-helical coiled coils and bundles: Models for the development of protein-design principles. *Trends in Biotechnology*, (16), 379–389.

Langer, G. G., Cohen, S. X., Lamzin, V. S., & Perrakis, A. (2008). Automated macromolecular model building for X-ray crystallography using ARPwARP version 7 Gerrit. *Nature Protocols*, 31171–31179.

Lee, K.-H., Matzapetakis, M., Mitra, S., Marsh, E. N. G., & Pecoraro, V. L. (2004). Control of metal coordination number in de novo designed peptides through subtle sequence modifications. *Journal of the American Chemical Society*, (126), 9178–9179.

Lovejoy, B., Choe, S., Cascio, D., Mcrorie, D. K., William, F., Eisenberg, D., et al. (1993). Crystal structure of a synthetic triple-stranded alpha-helical bundle. *Science*, (259), 1288–1293.

Lu, Y., Yeung, N., Sieracki, N., & Marshall, N. M. (2009). Design of functional metalloproteins. *Nature*, (460), 855–862.

Łuczkowski, M., Stachura, M., Schirf, V., Demeler, B., Hemmingsen, L., & Pecoraro, V. L. (2008). Design of thiolate rich metal binding sites within a peptidic framework. *Inorganic Chemistry*, (47), 10875–10888.

Matthews, B. W. (1968). Solvent content of protein crystals. *Journal of Molecular Biology*, (33), 491–497.

Matzapetakis, M., Farrer, B. T., Weng, T.-C., Hemmingsen, L., Penner-Hahn, J. E., & Pecoraro, V. L. (2002). Comparison of the binding of cadmium(II), mercury(II), and arsenic(III) to the *de novo* designed peptides TRI L12C and TRI L16C. *Journal of the American Chemical Society*, (124), 8042–8054.

Matzapetakis, M., Ghosh, D., Weng, T.-C., Penner-Hahn, J. E., & Pecoraro, V. L. (2006). Peptidic models for the binding of Pb(II), Bi(III) and Cd(II) to mononuclear thiolate binding sites. *The Journal of Biological Chemistry*, (11), 876–890.

Matzapetakis, M., & Pecoraro, V. L. (2005). Site-selective metal binding by designed alpha-helical peptides. *Journal of the American Chemical Society*, (127), 18229–18233.

McCoy, A. J., Grosse-Kunstleve, R. W., Adams, P. D., Winn, M. D., Storoni, L. C., & Read, R. J. (2007). Phaser crystallographic software. *Journal of Applied Crystallography*, (40), 658–674.

Neupane, K. P., & Pecoraro, V. L. (2010). Probing a homoleptic PbS$_3$ coordination environment in a designed peptide using ^{207}Pb NMR spectroscopy: Implications for understanding the molecular basis of lead toxicity. *Angewandte Chemie International Edition in English*, (49), 8177–8180.

Otwinowski, W., & Minor, Z. (1997). Processing of X-ray diffraction data collected in oscillation mode. *Methods in Enzymology*, (276), 307–326.

Peacock, A. F. A., Hemmingsen, L., & Pecoraro, V. L. (2008). Using diastereopeptides to control metal ion coordination in proteins. *Proceedings of the National Academy of Sciences of the United States of America*, (105), 16566–16571.

Peacock, A. F. A., Iranzo, O., & Pecoraro, V. L. (2009). Harnessing natures ability to control metal ion coordination geometry using de novo designed peptides. *Dalton Transactions*, (13), 2271–2280.

Peacock, A. F. A., Stuckey, J. A., & Pecoraro, V. L. (2009). Switching the chirality of the metal environment alters the coordination mode in designed peptides. *Angewandte Chemie International Edition in English*, (48), 7371–7374.

Pecoraro, V. L., Peacock, A. F. A., Iranzo, O., & Luczkowski, M. (2009). Understanding the biological chemistry of mercury using a *de novo* protein design strategy. *Bioinorganic Chemistry ACS Symposium Series*, 183–197.

Potterton, E., Briggs, P., Turkenburg, M., & Dodson, E. (2003). A graphical user interface to the CCP4 program suite. *Acta Crystallographica, Section D: Biological Crystallography*, (59), 1131–1137.

Ruckthong, L. (2016). *Crystallographic comparison of tris-thiolate sites in designed proteins to control metal geometries*. PhD dissertation, Ann Arbor: University of Michigan.

Theoretical mass of the peptide was determined from http://rna.rega.kuleuven.ac.be/masspec/pepcalc.htm (accessed 09.09.09).

Touw, D. (2007). *Structural and spectroscopic studies of heavy metal binding to de novo designed coiled coil peptides*. PhD dissertation, Ann Arbor, MI: University of Michigan.

Touw, D. S., Nordman, C. E., Stuckey, J. A., & Pecoraro, V. L. (2007). Identifying important structural characteristics of arsenic resistance proteins by using designed three-stranded coiled coils. *Proceedings of the National Academy of Sciences of the United States of America*, (104), 11969–11974.

Vagin, A., & Teplyakov, A. (2010). Molecular replacement with MOLREP. *Acta Crystallographica Section D: Biological Crystallography*, 22–27.

Winn, M. D., Ballard, C. C., Cowtan, K. D., Dodson, E. J., Emsley, P., Evans, P. R., et al. (2011). Overview of the CCP 4 suite and current developments. *Acta Crystallographica, Section D: Biological Crystallography*, (67), 235–242.

Zampella, G., Neupane, K. P., De Gioia, L., & Pecoraro, V. L. (2012). The importance of stereochemically active lone pairs for influencing Pb(II) and As(III) protein binding. *European Journal of Inorganic Chemistry*, (18), 2040–2050.

Zastrow, M. L., Peacock, A. F. A., Stuckey, J. A., & Pecoraro, V. L. (2012). Hydrolytic catalysis and structural stabilization in a designed metalloprotein. *Nature Chemistry*, (4), 118–123.

Zastrow, M., & Pecoraro, V. L. (2013). Designing functional metalloproteins: From structural to catalytic metal sites. *Coordination Chemistry Reviews*, 257(17–18), 2565–2588.

CHAPTER SEVEN

SpyRings Declassified: A Blueprint for Using Isopeptide-Mediated Cyclization to Enhance Enzyme Thermal Resilience

C. Schoene*, S.P. Bennett[†], M. Howarth*,[1]
*University of Oxford, Oxford, United Kingdom
[†]Sekisui Diagnostics UK Ltd., Maidstone, Kent, United Kingdom
[1]Corresponding author: e-mail address: mark.howarth@bioch.ox.ac.uk

Contents

1. Introduction	150
2. Isopeptide-Mediated Enzyme Cyclization	151
2.1 Partners for Spontaneous Isopeptide Bond Formation	151
2.2 Cyclization with Isopeptide Bonds	153
2.3 Selection of Enzymes for Cyclization	153
2.4 Cloning Genes of Interest into the SpyRing Cassette	154
2.5 Confirming Successful Cyclization	155
3. Characterization of Enzyme Resilience After SpyRing Cyclization	159
3.1 Industrial Relevance of Phytase	159
3.2 Measuring Aggregation Can Provide Evidence for an Increase in Thermal Resilience	159
3.3 Recovered Activity Is the Most Important Measurement When Testing Thermal Resilience	160
3.4 FLuc Is an Example of a Poor Target for SpyRing Cyclization	162
3.5 Dynamic Scanning Calorimetry Is a Useful Biophysical Tool to Study SpyRing Cyclization	163
4. Concluding Remarks	164
Acknowledgments	164
References	164

Abstract

Enzymes often have marginal stability, with unfolding typically leading to irreversible denaturation. This sensitivity is a major barrier, both for de novo enzyme development and for expanding enzyme impact beyond the laboratory. Seeking an approach to enhance resilience to denaturation that could be applied to a range of different enzymes, we developed SpyRing cyclization. SpyRings contain genetically encoded SpyTag (13 amino acids) on the N-terminus and SpyCatcher (12 kDa) on the

C-terminus of the enzyme, so that the Spy partners spontaneously react together through an irreversible isopeptide bond. SpyRing cyclization gave major increases in thermal resilience, including on a model for enzyme evolution, β-lactamase, and an industrially important enzyme in agriculture and nutrition, phytase. We outline the SpyRing rationale, including comparison of SpyRing cyclization to other cyclization strategies. The cloning strategy is presented for the simple insertion of enzyme genes for recombinant expression. We discuss structure-based approaches to select suitable enzyme cyclization targets. Approaches to evaluate the cyclization reaction and its effect on enzyme resilience are described. We also highlight the use of differential scanning calorimetry to understand how SpyRing cyclization promotes enzyme refolding. Efficiently searching sequence space will continue to be important for enzyme improvement, but the SpyRing platform may be a valuable rational adjunct for conferring resilience.

1. INTRODUCTION

Understanding how to extend the scope of enzyme use is a central challenge in protein engineering. The potential of enzymes is illustrated by the fact that there are more than 100 enzymes already in large-scale industrial use. These applications range from laundry detergents and food processing to the cosmetics industry (Li, Yang, Yang, Zhu, & Wang, 2012). Major challenges for taking promising enzymes from the laboratory into large-scale application include the high cost of production (Howard, Abotsi, Jansen van Rensburg, & Howard, 2003) and poor stability at high temperatures (Lei, Weaver, Mullaney, Ullah, & Azain, 2013). The most common strategies to enhance the thermal resilience of enzymes are structure based or library based, but these approaches are usually performed on a case-by-case basis (Packer & Liu, 2015; Yang, Liu, Li, Chen, & Du, 2015).

To identify generic approaches to enhance stability, protein cyclization has been explored. Protein cyclization consists of covalently tethering the N- and C-termini of a protein, thereby reducing flexibility in two of the most mobile parts of the protein, while reducing the entropy gain on unfolding (Schumann et al., 2015). Previous work on protein cyclization has utilized carbodiimide cross-linking (Goldenberg & Creighton, 1983), split inteins (Camarero & Muir, 1999; Evans, Benner, & Xu, 1999; Iwai & Plückthun, 1999; Scott, Abel-Santos, Wall, Wahnon, & Benkovic, 1999), and sortase (Parthasarathy, Subramanian, & Boder, 2007). In some cases, an increase in thermal resilience has been reported

(Antos et al., 2009; Iwai & Plückthun, 1999; Popp, Dougan, Chuang, Spooner, & Ploegh, 2011; van Lieshout et al., 2012). We have developed an approach to cyclize proteins based on spontaneous isopeptide bond formation, which has been able to achieve >60°C increases in thermal resilience (Reddington & Howarth, 2015; Schoene, Bennett, & Howarth, 2016; Schoene, Fierer, Bennett, & Howarth, 2014).

2. ISOPEPTIDE-MEDIATED ENZYME CYCLIZATION
2.1 Partners for Spontaneous Isopeptide Bond Formation

We have developed a range of different peptide/protein pairs able to react irreversibly. The story starts when Ted Baker's group solved the crystal structure of Spy0128, a pilin subunit originating from *Streptococcus pyogenes*, discovering the formation of an isopeptide bond in each of two domains between lysine and asparagine side chains (Kang, Coulibaly, Clow, Proft, & Baker, 2007). Mass spectrometry of Spy0128 validated isopeptide bond formation, while mutational analysis confirmed the three residues necessary for the formation of this bond (Kang et al., 2007). We were able to split the C-terminal β-strand of Spy0128 into a 16 amino acid tag, termed Isopeptag. Residues 18–299 of Spy0128 were separately expressed and called Pilin-C. By genetically fusing Isopeptag to maltose-binding protein (MBP), we found that Isopeptag-MBP could reconstitute and form an isopeptide bond with Pilin-C (Zakeri & Howarth, 2010).

Further work was then carried out to find a faster and smaller reactive pair than Pilin-C/Isopeptag. The CnaB2 domain of the fibronectin-binding protein (FbaB) of *S. pyogenes* was found to form a spontaneous isopeptide bond between aspartic acid and lysine (Hagan et al., 2010; Oke et al., 2010). SpyTag, a 13 amino acid peptide tag containing the reactive aspartic acid residue, consisted of the C-terminal β-strand of the CnaB2 domain as well as the succeeding five residues of FbaB not included in the crystal structure. SpyCatcher was generated from the remainder of CnaB2, which was further optimized by mutating two exposed hydrophobic residues (Fig. 1A). SpyTag-MBP was able to react efficiently with SpyCatcher (second-order rate constant 1.4×10^3 $M^{-1}s^{-1}$). Reaction occurred in a wide range of buffers, including reducing or oxidizing conditions (Zakeri et al., 2012).

Next, we developed an orthogonal pair to SpyTag/SpyCatcher. The D4 domain of RrgA adhesin from *Streptococcus pneumoniae* was found to have

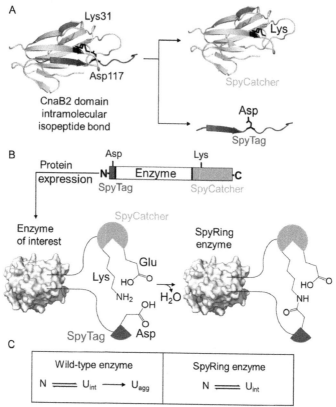

Fig. 1 SpyTag and SpyCatcher for enzyme cyclization. (A) The CnaB2 domain of FbaB was split into SpyTag and SpyCatcher. Cartoon based on PDB 2X5P (Oke et al., 2010). (B) Expression of the SpyRing genetic construct leads to in vivo cyclization. The formation of the isopeptide bond is dependent on the three illustrated residues: reaction between Lys and Asp promoted by an apposed Glu. (C) Proposed principle of SpyRing-mediated thermal resilience, by favoring reversible transition between the native (N) and unfolded intermediate states (U_{int}), avoiding irreversible transition to the unfolded aggregate state (U_{agg}).

an isopeptide bond between lysine and asparagine side chains (Izoré et al., 2010). The N-terminal β-strand containing the reactive lysine residue (residues 734–745) was split and named SnoopTag. Residues 749–860 provided SnoopCatcher, which was further optimized by mutating two residues to stabilize secondary structure. SnoopTag/SnoopCatcher and SpyTag/SpyCatcher were found to be mutually unreactive. The combined use of both Tag/Catcher systems made it possible to build sequence-programmed protein chains (Veggiani et al., 2016).

2.2 Cyclization with Isopeptide Bonds

Our initial studies with these peptide/protein ligation tools all focused on opportunities for locking together different protein units (Fairhead et al., 2014; Fierer, Veggiani, & Howarth, 2014; Veggiani et al., 2016). However, the efficient reaction and flexible tag location suggested to us the opportunity to apply these pairs to drive intramolecular reaction. SpyTag/SpyCatcher (SpyRing, Fig. 1B), Pilin-C/Isopeptag (PilinRing), and SnoopTag/SnoopCatcher (SnoopRing) were all shown to cyclize β-lactamase (BLA) efficiently (Schoene et al., 2016). All these cyclized constructs exhibited majorly enhanced thermal resilience, in comparison to the untagged control. We placed Pilin-C, SpyTag, and SnoopTag on the N-terminus and Isopeptag, SpyCatcher, and SnoopCatcher on the C-terminus of BLA but we have not verified whether flipping the components around might work better. Therefore, further work could be carried out on testing whether the Tag should be on the N or C-terminus and vice versa for the Catcher. However, we found SpyRing cyclization to give the greatest enhancement in resilience (Schoene et al., 2016), so this approach will be the focus of this chapter. Such comparisons, along with testing linear controls, show how SpyRing resilience exceeds that simply from cyclization or fusion to a thermostable domain (Schoene et al., 2016, 2014). We proposed a mechanism whereby SpyRing cyclization confers resilience through favoring enzyme refolding after denaturation, blocking irreversible transitions to an unfolded aggregate (Fig. 1C) (Schoene et al., 2016, 2014; Schumann et al., 2015).

We describe the selection of appropriate enzyme targets for cyclization, their insertion into the SpyRing construct, and the methods used to verify successful cyclization. Moreover, we focus on the cyclization of a β-propeller phytase (PhyC) and the enzymatic, solubility, and biophysical characterization approaches one might use to study SpyRing-cyclized proteins.

2.3 Selection of Enzymes for Cyclization

SpyRing cyclization was envisioned to provide a simple method to stabilize different enzymes. In 2 weeks, one can go from obtaining the DNA for an enzyme of interest to testing the SpyRing enzyme's thermal resilience. However, we postulate that the following criteria are helpful when prioritizing enzyme candidates for SpyRing cyclization.

If a structure of the enzyme (or a close ortholog) is available, it is helpful to consider:
- the distance between the N- and C-termini,
- whether the active site is on the same face as the termini.

We suggest that enzymes with a short distance (ie, <15 Å) between the N- and C-termini and where the active site is not closely apposed to the termini would be ideal candidates for SpyRing cyclization. In such cases, our experience is that SpyRing cyclization gave little or no change in enzyme activity. For BLA (7 Å from N- to C-terminus), we showed minimal difference in k_{cat} or K_m (Schoene et al., 2016) between wild-type and SpyRing forms. The distribution of termini in the Protein Data Bank has been surveyed (Krishna & Englander, 2005) and a large fraction of single-domain proteins should be amenable based on this distance criterion. However, even where the termini are 29 Å apart as in PhyC, we obtained efficient SpyRing cyclization and increased thermal resilience (Schoene et al., 2016).

Previous successes with other cyclization approaches, as well as with circular permutations or disulfide bond engineering, might also give confidence that SpyRing cyclization would work. Intein-mediated cyclization of BLA led to an increase in aggregation resistance of ~5°C (Iwai & Plückthun, 1999). Dihydrofolate reductase (DHFR) had previously been stabilized via cyanocysteine-mediated cyclization (Takahashi et al., 2007). We were able to show that both DHFR and BLA were rendered resilient by SpyRing cyclization (Schoene et al., 2014). However, if no data on cyclization are available, it is worth searching the literature for whether the enzyme remains active when fused to another protein, or whether a fusion to a thermostable protein increases the thermal resilience (Pierre, Xiong, Hayles, Guntaka, & Kim, 2011).

SpyRing cyclization is unlikely to substitute for the activity of chaperones. If a protein requires chaperones to fold or refold, like RuBisCO (Liu et al., 2010) or firefly luciferase (FLuc) (Svetlov, Kommer, Kolb, & Spirin, 2006), SpyRing cyclization might not be the right approach. Enzyme cofactors might also cause complications, especially if an enzyme cannot refold correctly following the loss of a cofactor, eg, glucose oxidase (Zoldák, Zubrik, Musatov, Stupák, & Sedlák, 2004).

So far we have worked primarily with monomeric enzymes. Generic strategies for stabilizing multimeric proteins will require further study.

2.4 Cloning Genes of Interest into the SpyRing Cassette

SpyRing cyclization is achieved by genetically encoding the SpyTag on the N-terminus and the SpyCatcher on the C-terminus of an enzyme of interest. The expressed construct is then able to spontaneously form an isopeptide bond inside the expressing cell (Fig. 1B). Therefore, SpyRing cyclization requires the insertion of the open-reading frame of the enzyme within the SpyRing expression cassette.

To achieve a quick, reliable, and modular approach to insert different enzymes into the SpyRing cassette, we designed a ligation-independent cloning strategy (Fig. 2). The cloning method requires two custom primers which can anneal to the coding sequence of the enzyme. The custom primers have tail-regions which can hybridize to the vector. The 5′ tail-region is 27 bp long and the 3′ tail-region is 23 bp long and both regions overlap with the vector sequence. This approach produces a glycine/serine spacer (GSGGSG) on both sides of the enzyme, to facilitate efficient cyclization by SpyTag/SpyCatcher and to minimize interference with enzyme folding. The template from which to amplify the vector sequence is readily available from the Addgene plasmid repository (ID: 52656). Once the insert and vector have been amplified and purified, the fragments can be linked together by circular polymerase extension cloning (CPEC) (Quan & Tian, 2009). We use KOD Hot Start DNA Polymerase (Roche) in the CPEC reaction. 500 ng of vector are mixed with an equimolar amount of insert in a 50 μL reaction. Cycling conditions are: 95°C for 3 min followed by 10 cycles of 95°C for 30 s, 64°C for 30 s, and 68°C for 4 min. 1 μL of the CPEC reaction can be used to transform competent *Escherichia coli* cells (a RecA⁻ strain such as XL-1 or DH5α).

We recommend CPEC because the method does not require any additional materials not already used to amplify the vector and insert. However, methods such as Gibson assembly (Gibson et al., 2009) should also be suitable.

2.5 Confirming Successful Cyclization

The cloned construct can then be expressed in *E. coli* protein production strains. We have obtained good yields using the BL21 DE3 RIPL strains from Agilent, which enables efficient translation even if the gene contains rare codons (Schoene et al., 2016, 2014). Two important parameters to consider are the temperature at which the constructs are expressed and the concentration of isopropyl β-D-1-thiogalactopyranoside (IPTG). There are two types of covalent interactions that can occur once the SpyRing protein has been expressed: intramolecular covalent interactions leading to enzyme cyclization and intermolecular covalent interactions leading to enzyme oligomerization (Fig. 3A). By decreasing the induction temperature and IPTG concentration, one can increase the proportion of monomeric-cyclized

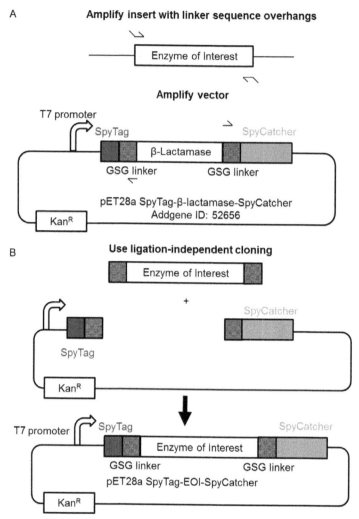

Fig. 2 Cloning strategy to insert an enzyme into the SpyRing expression cassette. (A) The enzyme of interest is amplified by PCR using primers containing adaptor sequences, to generate an overlap at each end with the vector. The vector is amplified using pET28a SpyTag-BLA-SpyCatcher as a template, with the primers hybridizing to linker regions of the construct. (B) The vector and insert are then combined through a ligation-independent cloning method, such as CPEC, to generate the desired SpyRing expression plasmid.

protein (Zhang, Sun, Tirrell, & Arnold, 2013). We found that growing the culture to an OD_{600} of ~0.5 and then inducing protein expression at 18°C for 16 h using 0.4 mM IPTG worked well to obtain a good yield of monomeric-cyclized product.

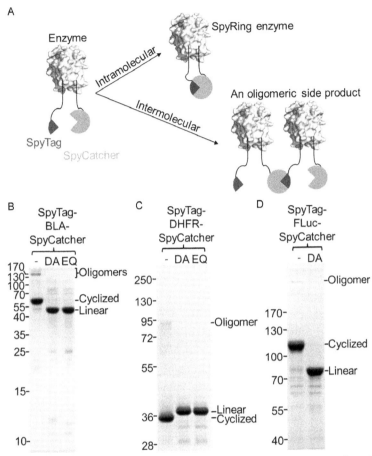

Fig. 3 Gel analysis of SpyRing enzyme cyclization. (A) Cartoon illustrating how fusing SpyTag on the N-terminus and SpyCatcher at the C-terminus can lead to intermolecular or intramolecular covalent bond formation. (B) SpyRing cyclization decreased the mobility of BLA. Reducing SDS–PAGE with Coomassie staining, showing SpyTag-BLA-SpyCatcher compared to the noncyclized controls of SpyTag DA-BLA-SpyCatcher and SpyTag-BLA-SpyCatcher EQ. "Oligomers" indicates the side products from intermolecular SpyTag/SpyCatcher reaction. (C) SpyRing cyclization increased the mobility of DHFR. Reducing SDS–PAGE with Coomassie staining for SpyTag-DHFR-SpyCatcher and DA or EQ linear controls. (D) SpyRing cyclization decreased the mobility of FLuc. Reducing SDS–PAGE with Coomassie staining for SpyTag-FLuc-SpyCatcher and the DA control. *Panels (B) and (C): Data previously published in Schoene, C., Fierer, J. O., Bennett, S. P., & Howarth, M. (2014). SpyTag/SpyCatcher cyclization confers resilience to boiling on a mesophilic enzyme.* Angewandte Chemie, 53, 6101–6104.

The SpyRing cassette has an N-terminal His_6 tag, which can be used to purify the cyclized proteins via immobilized metal ion affinity chromatography (IMAC). To avoid eluting most of the oligomeric protein, we recommend using low concentrations of imidazole (ie, 75 mM) to favor elution of proteins that have only a single His_6 tag (Howarth & Ting, 2008). If desired, complete removal of oligomeric side products can be achieved by a subsequent size-exclusion chromatography step (Schoene et al., 2016).

An alternative method for the purification of the SpyRing enzyme is to use heat to precipitate *E. coli*'s native proteins, while the SpyRing enzyme largely remains soluble. The SpyRing enzyme in the soluble fraction can be separated from the precipitate using centrifugation (Schoene et al., 2016). Heat-mediated purification is an attractive option because it is rapid and cheap. An important consideration is the total concentration of protein in the cell lysate. If the concentration is too high, the cyclized protein may be lost to the precipitate as well, through interaction with other unfolded proteins. A cell lysate OD_{280} of ~ 6 is recommended. Furthermore, we found that the presence of NaCl can help improve the solubility of the cyclized protein (Schoene et al., 2016). Heat-mediated purification has also been achieved previously by using thermostable proteins as fusion partners (de Marco, Casatta, Savaresi, & Geerlof, 2004).

A first quick test for cyclization is to perform SDS–PAGE with Coomassie staining on the SpyRing enzyme, looking for a mobility shift compared to a linear point mutant unable to form an isopeptide bond (SpyTag DA or SpyCatcher EQ) (Zakeri et al., 2012). Complicated effects on mobility are seen from cyclization, depending on enzyme size and charge (Fig. 3B–D). Moreover, the acrylamide percentage of the SDS–PAGE can alter the difference in mobility between the cyclized enzyme and the non-cyclized point mutant. The change in mobility resulting from cyclization is much larger than expected simply from the effect of the mutation on protein charge, reflecting the complex interaction between the conformation of the unfolded protein coated in SDS and the polyacrylamide network (Rath, Cunningham, & Deber, 2013).

For a more definitive method to confirm isopeptide bond formation, electrospray ionization mass spectrometry (ESI-MS) can be employed to determine the exact mass of the protein constructs (Schoene et al., 2016, 2014). Since the formation of the isopeptide bond between SpyTag and SpyCatcher leads to the loss of a water molecule (Fig. 1B), the mass determined by ESI-MS should be 18 Da lighter than the predicted molecular weight of the linear protein from the ProtParam online tool

(Gasteiger et al., 2005). A possible complication in the spectrum from *E. coli* expression is gluconoylation of the His$_6$ tag (Geoghegan et al., 1999). For ESI-MS, samples should be expressed in *E. coli* B834 DE3, grown in the absence of any supplementary glucose in the LB media to an OD$_{600}$ of ~0.5, and induced at 30°C for 3 h using 0.4 m*M* IPTG. Samples should then be purified using the previously described IMAC approach, before dialyzing into 10 m*M* ammonium acetate buffer (a low ionic strength buffer with volatile components is important to obtain clean spectra).

3. CHARACTERIZATION OF ENZYME RESILIENCE AFTER SpyRing CYCLIZATION

3.1 Industrial Relevance of Phytase

We illustrate the process of characterizing enzyme resilience with particular reference to SpyRing cyclization of phytase. Most plants store phosphates as phytic acid (inositol hexakisphosphate). Nonruminant animals such as chicken, pigs, and salmon are unable to degrade phytic acid, and therefore find it difficult to absorb the necessary dietary phosphate for a healthy diet. Moreover, phytic acid is an antinutrient, which is able to chelate and thereby block absorption of micronutrients like zinc and iron (Bohlke, Thaler, & Stein, 2005; Selle, Ravindran, Bryden, & Scott, 2006). Therefore, the enzyme phytase is widely used to release phosphate from phytic acid in animal feed. However, during the pelleting process, feedstock is heated to 75–90°C to kill pathogenic bacteria, notably *Salmonella* (Lei et al., 2013). A phytase variant able to regain full activity after this pelleting process would be a significant advance.

Humans are also unable to degrade phytic acid. This antinutrient may contribute to the global health problem of micronutrient deficiency (Bailey, West, & Black, 2015), the "silent hunger" with a range of short-term and long-term health consequences (Allen, de Benoist, Dary, & Hurrell, 2006). Food security is a major global issue; increasing the nutritional value of food that is already available will be an important part of the solution.

3.2 Measuring Aggregation Can Provide Evidence for an Increase in Thermal Resilience

The first simple and quick test we use to evaluate a SpyRing enzyme is to measure aggregation at elevated temperature. Heat-induced aggregation correlates with a loss of activity and irreversible inactivation of enzymes.

To test whether SpyRing cyclization has improved an enzyme's thermal resilience, we recommend that a SpyTag D → A linear point mutant as well as a construct lacking both SpyTag and SpyCatcher are used as controls. The test should be run in the wild-type enzyme's preferred buffer. Enzyme concentration can be critical to the extent of aggregation. Therefore, we recommend using a concentration relevant to the application one has in mind for the enzyme of interest. 10 μM enzyme is a good starting point.

We recommend the use of a PCR machine to heat and then cool the protein for a precise amount of time with a reproducible ramp rate. For example, heat the proteins at 25°C, 37°C, 55°C, 75°C, 90°C, and 100°C for 10 min and then cool to 10°C for 1 h. After cooling for 1 h, centrifuge the samples for 30 min at 17,000 × g, before analyzing the soluble fraction by SDS–PAGE. Coomassie-stained bands can then be quantified using densitometry.

In the case of PhyC, SpyRing cyclization improved the solubility of the enzyme at temperatures at and above 75°C (Fig. 4A). However, an aggregation assay alone should not be used as conclusive evidence on whether SpyRing cyclization is helpful.

3.3 Recovered Activity Is the Most Important Measurement When Testing Thermal Resilience

Aggregation can be used as an initial indicator to determine whether SpyRing cyclization increases the thermal resilience of an enzyme. However, the aggregation assay does not take into account the formation of soluble aggregates and the possibility that an enzyme does not aggregate but misfolds and becomes irreversibly inactivated. For example, the noncyclized control SpyTag DA-DHFR-SpyCatcher did not aggregate at 100°C but the activity assay showed that the enzyme was irreversibly inactivated at this temperature (Schoene et al., 2014). Therefore, the best assay to use is one that measures the residual enzyme activity after heating.

The enzyme's temperature treatment should be the same as described earlier, so a correlation can be drawn between the enzyme's recovered activity and solubility. However, since the enzyme will be at a high concentration (~10 μM) for the solubility assay, one might need to dilute the enzyme down to nM concentrations for the activity assay. Finding the right concentration of enzyme for your assay is important, so an enzyme titration to where the initial activity can be measured using at least five time points is a good starting place. When diluting the enzyme, reagent stability will need to be taken into account. Enzymes at low concentrations will readily stick to

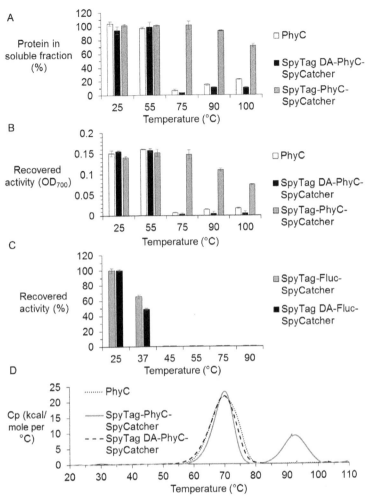

Fig. 4 SpyRing cyclization effect on thermal resilience. (A) SpyRing cyclization improved the solubility of PhyC at high temperatures. Samples were heated to the indicated temperature for 10 min, cooled to 10°C, and the supernatant was analyzed by SDS–PAGE and Coomassie staining (mean of triplicates ± 1 s.d.). (B) Cyclization improved the recovered activity of PhyC. Samples were heated to the indicated temperature for 10 min and cooled to 10°C. Phosphate release from phytase was quantified colorimetrically (mean of triplicate ± 1 s.d.). (C) Cyclization had a minor effect on FLuc activity recovered after heating. Samples at 5 μM in PBS, pH 7.4, containing 100 mM DTT were heated to the indicated temperature for 10 min and cooled to 10°C. 12.5 nM enzyme was reacted with 300 μM luciferin in 50 mM Tris-acetate, pH 7.8, 10 mM MgSO$_4$, 4 mM ATP, and 2 mM EDTA. Luminescence was measured after a 15 s delay (mean of triplicates ± 1 s.d.). (D) DSC of SpyTag-PhyC-SpyCatcher overlaid with the untagged PhyC construct and the DA point mutant. Scanning from 20°C to 110°C at 1°C/min. *Panels (A) and (D): Data previously published in Schoene, C., Bennett, S. P., & Howarth, M. (2016). SpyRing interrogation: Analyzing how enzyme resilience can be achieved with phytase and distinct cyclization chemistries. Scientific Reports, 6, 21151.*

plastic surfaces and be removed from solution. Therefore, plastic containers should be preblocked with 1–3% bovine serum albumin (BSA) at 37°C for 2 h and the dilution buffer should include BSA at a concentration of 0.1–1%.

We recommend that the activity recovery assay is set up in 96-well plate format to allow for high throughput. Ideally, an enzyme substrate should be used with an absorbance or fluorescence readout. For plate-reader measurement, the delay time between each read should be considered and how that might affect the reading of the initial catalytic activity.

In the case of PhyC, there are no fluorescent or colorimetric analogs of phytic acid currently available, so reactions were quenched with trichloroacetic acid and the released phosphate was measured colorimetrically using the ammonium molybdate method (Schoene et al., 2016). Cyclization provided a large improvement in recovered activity of PhyC after heating to the industrially relevant temperatures of 75°C and 90°C (Fig. 4B).

3.4 FLuc Is an Example of a Poor Target for SpyRing Cyclization

FLuc is an important reporter enzyme used to study gene expression, monitor protein–protein interactions, and sequence DNA. However, FLuc's instability over 30°C is a problem for various applications (Ebrahimi, Hosseinkhani, Heydari, Khavari-Nejad, & Akbari, 2012). SpyRing cyclization of FLuc decreased the mobility on SDS–PAGE, compared to the linear DA control (Fig. 3D), and therefore we postulated that the cyclization reaction had been successful. However, when we tested the recovered activity of the cyclized FLuc, improvements were only marginal when compared to the linear point mutant (Fig. 4C).

The following tips might help to avoid testing cyclization on poor targets. A circular permutation of FLuc in combination with intein-mediated cyclization has been shown to inactivate the enzyme (Kanno, Yamanaka, Hirano, Umezawa, & Ozawa, 2007). Cyclization of a circularly permuted FLuc was used to create a reporter for caspase-3, by placing a caspase cleavage site between the N- and C-termini. Upon cleavage the termini are able to move and the enzyme becomes active (Kanno et al., 2007). While our cyclized FLuc was still active (Fig. 4C), enzymes which have been shown to be less active when the termini are restrained will probably be hard to work with.

FLuc has also been shown to require the help of chaperones when refolding from a heat-denatured state (Schröder, Langer, Hartl, & Bukau, 1993). Moreover, misfolding involves the formation of a stable nonnative conformation where the N- and C-terminal residues are interacting (Scholl, Yang, & Marszalek, 2014). Therefore, FLuc SpyRing cyclization might not be able to stop this interaction from occurring unless chaperones are present.

3.5 Dynamic Scanning Calorimetry Is a Useful Biophysical Tool to Study SpyRing Cyclization

Dynamic scanning calorimetry (DSC) has provided an effective approach to follow the thermal unfolding and refolding of SpyRing-cyclized proteins (Schoene et al., 2014). DSC measures the difference in heat required to increase the temperature of a protein sample in comparison to a reference cell. When proteins unfold, the hydration of the core causes an increase in heat capacity (Privalov & Makhatadze, 1992). Preparation of samples for DSC is simple: no dyes are necessary since the signal monitored is the change in specific heat capacity. We have used the VP Cap DSC system from GE Healthcare, which requires 400 µL of 20 µM protein and ~4 mL of buffer for a single run.

The ramp rate for a single scan from 20°C to 110°C was set to 1°C/min, which produced high-resolution data (Fig. 4D). DSC data indicated that the SpyRing did not increase the transition temperature of PhyC (PhyC: 69.8°C, SpyTag DA-PhyC-SpyCatcher: 69.7°C, SpyTag-PhyC-SpyCatcher: 70.0°C). SpyTag-PhyC-SpyCatcher unfolded as two transitions. The second transition had a T_m of 91.5°C, which is close to the T_m we measured for SpyCatcher + SpyTag peptide melting (Schoene et al., 2016).

Such data are consistent with the mechanism proposed for the increase in thermal resilience observed in PhyC, DHFR, and BLA, where SpyRing cyclization does not block unfolding but promotes efficient refolding (Fig. 1C). The increase in thermal resilience of the cyclized enzyme also extended higher than the T_m of the SpyTag/SpyCatcher domain (Schoene et al., 2016, 2014).

A downside of using DSC is that it is low throughput. Even automated DSC systems are unable to process more than 50 samples/day (Plotnikov et al., 2002). Dynamic scanning fluorimetry (DSF) is able to provide a much higher throughput (Niesen, Berglund, & Vedadi, 2007). However, DSF requires the use of a dye which can interfere with the transition temperature of a protein (Shi, Semple, Cheung, & Shameem, 2013). Alternatively, microscale thermophoresis is a recently developed high-throughput method, which can monitor unfolding from the intrinsic protein fluorescence (Alexander et al., 2014).

4. CONCLUDING REMARKS

We have developed three distinct peptide–protein pairs (Veggiani et al., 2016; Zakeri et al., 2012; Zakeri & Howarth, 2010) which are able to cyclize inside cells and increase enzyme resilience (Schoene et al., 2016). SpyRing cyclization requires the genetic encoding of a 13 amino acid tag, SpyTag, at the N-terminus of a protein and a 12 kDa protein, SpyCatcher, on the C-terminus. Cloning of an enzyme of interest into the SpyRing cassette is simple and modular. SpyRing enzymes can easily be purified through IMAC or heat purification. SpyRing cyclization may provide a quick and generic route toward enzyme stabilization. We have provided guidelines for the selection of appropriate targets for SpyRing cyclization but exploring this strategy on a wider diversity of enzymes (expression hosts, protein folds, multimerization state, and enzyme classes) will help to advance the scope of this approach.

ACKNOWLEDGMENTS

Funding was provided by BBSRC (C.S.), Sekisui Diagnostics (C.S., S.P.B.), and Oxford University Department of Biochemistry (M.H.).

REFERENCES

Alexander, C. G., Wanner, R., Johnson, C. M., Breitsprecher, D., Winter, G., Duhr, S., et al. (2014). Novel microscale approaches for easy, rapid determination of protein stability in academic and commercial settings. *Biochimica et Biophysica Acta (BBA)—Proteins and Proteomics, 1844*, 2241–2250.

Allen, L., de Benoist, B., Dary, O., & Hurrell, R. (2006). *Guidelines on food fortification with micronutrients*. Geneva, Switzerland: World Health Organization.

Antos, J. M., Popp, M. W.-L., Ernst, R., Chew, G.-L., Spooner, E., & Ploegh, H. L. (2009). A straight path to circular proteins. *The Journal of Biological Chemistry, 284*, 16028–16036.

Bailey, R. L., West, K. P., & Black, R. E. (2015). The epidemiology of global micronutrient deficiencies. *Annals of Nutrition & Metabolism, 66*(Suppl. 2), 22–33.

Bohlke, R. A., Thaler, R. C., & Stein, H. H. (2005). Calcium, phosphorus, and amino acid digestibility in low-phytate corn, normal corn, and soybean meal by growing pigs. *Journal of Animal Science, 83*, 2396–2403.

Camarero, J. A., & Muir, T. W. (1999). Biosynthesis of a head-to-tail cyclized protein with improved biological activity. *Journal of the American Chemical Society, 121*, 5597–5598.

de Marco, A., Casatta, E., Savaresi, S., & Geerlof, A. (2004). Recombinant proteins fused to thermostable partners can be purified by heat incubation. *Journal of Biotechnology, 107*, 125–133.

Ebrahimi, M., Hosseinkhani, S., Heydari, A., Khavari-Nejad, R. A., & Akbari, J. (2012). Improvement of thermostability and activity of firefly luciferase through [TMG][Ac] ionic liquid mediator. *Applied Biochemistry and Biotechnology, 168*, 604–615.

Evans, T. C., Jr., Benner, J., & Xu, M. Q. (1999). The in vitro ligation of bacterially expressed proteins using an intein from *Methanobacterium thermoautotrophicum*. *The Journal of Biological Chemistry, 274*, 3923–3926.

Fairhead, M., Veggiani, G., Lever, M., Yan, J., Mesner, D., Robinson, C. V., et al. (2014). SpyAvidin hubs enable precise and ultrastable orthogonal nanoassembly. *Journal of the American Chemical Society, 136*, 12355–12363.

Fierer, J. O., Veggiani, G., & Howarth, M. (2014). SpyLigase peptide-peptide ligation polymerizes affibodies to enhance magnetic cancer cell capture. *Proceedings of the National Academy of Sciences, 111*, E1176–E1181.

Gasteiger, E., Hoogland, C., Gattiker, A., Duvaud, S., Wilkins, M. R., Appel, R. D., et al. (2005). Protein identification and analysis tools on the ExPASy server. *The proteomics protocols handbook* (pp. 571–607). New York City, NY: Humana Press.

Geoghegan, K. F., Dixon, H. B., Rosner, P. J., Hoth, L. R., Lanzetti, A. J., Borzilleri, K. A., et al. (1999). Spontaneous alpha-N-6-phosphogluconoylation of a 'His tag' in *Escherichia coli*: The cause of extra mass of 258 or 178 Da in fusion proteins. *Analytical Biochemistry, 267*, 169–184.

Gibson, D. G., Young, L., Chuang, R.-Y., Venter, J. C., Hutchison, C. A., & Smith, H. O. (2009). Enzymatic assembly of DNA molecules up to several hundred kilobases. *Nature Methods, 6*, 343–345.

Goldenberg, D. P., & Creighton, T. E. (1983). Circular and circularly permuted forms of bovine pancreatic trypsin inhibitor. *Journal of Molecular Biology, 165*, 407–413.

Hagan, R. M., Bjornsson, R., McMahon, S. A., Schomburg, B., Braithwaite, V., Buhl, M., et al. (2010). NMR spectroscopic and theoretical analysis of a spontaneously formed Lys-Asp isopeptide bond. *Angewandte Chemie, 49*, 8421–8425.

Howard, R. I., Abotsi, E., Jansen van Rensburg, E. I., & Howard, S. (2003). Lignocellulose biotechnology: Issues of bioconversion and enzyme production. *African Journal of Biotechnology, 2*, 602–619.

Howarth, M., & Ting, A. Y. (2008). Imaging proteins in live mammalian cells with biotin ligase and monovalent streptavidin. *Nature Protocols, 3*, 534–545.

Iwai, H., & Plückthun, A. (1999). Circular β-lactamase: Stability enhancement by cyclizing the backbone. *FEBS Letters, 459*, 166–172.

Izoré, T., Contreras-Martel, C., El Mortaji, L., Manzano, C., Terrasse, R., Vernet, T., et al. (2010). Structural basis of host cell recognition by the pilus adhesin from *Streptococcus pneumoniae*. *Structure, 18*, 106–115.

Kang, H. J., Coulibaly, F., Clow, F., Proft, T., & Baker, E. N. (2007). Stabilizing isopeptide bonds revealed in gram-positive bacterial pilus structure. *Science, 318*, 1625–1628.

Kanno, A., Yamanaka, Y., Hirano, H., Umezawa, Y., & Ozawa, T. (2007). Cyclic luciferase for real-time sensing of caspase-3 activities in living mammals. *Angewandte Chemie, 46*, 7595–7599.

Krishna, M. M. G., & Englander, S. W. (2005). The N-terminal to C-terminal motif in protein folding and function. *Proceedings of the National Academy of Sciences, 102*, 1053–1058.

Lei, X. G., Weaver, J. D., Mullaney, E., Ullah, A. H., & Azain, M. J. (2013). Phytase, a new life for an 'old' enzyme. *Annual Review of Animal Biosciences, 1*, 283–309.

Li, S., Yang, X., Yang, S., Zhu, M., & Wang, X. (2012). Technology prospecting on enzymes: Application, marketing and engineering. *Computational and Structural Biotechnology Journal, 2*, 1–11.

Liu, C., Young, A. L., Starling-Windhof, A., Bracher, A., Saschenbrecker, S., Rao, B. V., et al. (2010). Coupled chaperone action in folding and assembly of hexadecameric Rubisco. *Nature, 463*, 197–202.

Niesen, F. H., Berglund, H., & Vedadi, M. (2007). The use of differential scanning fluorimetry to detect ligand interactions that promote protein stability. *Nature Protocols, 2*, 2212–2221.

Oke, M., Carter, L. G., Johnson, K. A., Liu, H., McMahon, S. A., Yan, X., et al. (2010). The Scottish structural proteomics facility: Targets, methods and outputs. *Journal of Structural and Functional Genomics, 11*, 167–180.

Packer, M. S., & Liu, D. R. (2015). Methods for the directed evolution of proteins. *Nature Reviews. Genetics, 16*, 379–394.

Parthasarathy, R., Subramanian, S., & Boder, E. T. (2007). Sortase A as a novel molecular 'stapler' for sequence-specific protein conjugation. *Bioconjugate Chemistry, 18*, 469–476.

Pierre, B., Xiong, T., Hayles, L., Guntaka, V. R., & Kim, J. R. (2011). Stability of a guest protein depends on stability of a host protein in insertional fusion. *Biotechnology and Bioengineering, 108*, 1011–1020.

Plotnikov, V., Rochalski, A., Brandts, M., Brandts, J. F., Williston, S., Frasca, V., et al. (2002). An autosampling differential scanning calorimeter instrument for studying molecular interactions. *Assay and Drug Development Technologies, 1*, 83–90.

Popp, M. W., Dougan, S. K., Chuang, T.-Y., Spooner, E., & Ploegh, H. L. (2011). Sortase-catalyzed transformations that improve the properties of cytokines. *Proceedings of the National Academy of Sciences, 108*, 3169–3174.

Privalov, P. L., & Makhatadze, G. I. (1992). Contribution of hydration and non-covalent interactions to the heat capacity effect on protein unfolding. *Journal of Molecular Biology, 224*, 715–723.

Quan, J., & Tian, J. (2009). Circular polymerase extension cloning of complex gene libraries and pathways. *PloS One, 4*, e6441.

Rath, A., Cunningham, F., & Deber, C. M. (2013). Acrylamide concentration determines the direction and magnitude of helical membrane protein gel shifts. *Proceedings of the National Academy of Sciences, 110*, 15668–15673.

Reddington, S. C., & Howarth, M. (2015). Secrets of a covalent interaction for biomaterials and biotechnology: SpyTag and SpyCatcher. *Current Opinion in Chemical Biology, 29*, 94–99.

Schoene, C., Bennett, S. P., & Howarth, M. (2016). SpyRing interrogation: Analyzing how enzyme resilience can be achieved with phytase and distinct cyclization chemistries. *Scientific Reports, 6*, 21151.

Schoene, C., Fierer, J. O., Bennett, S. P., & Howarth, M. (2014). SpyTag/SpyCatcher cyclization confers resilience to boiling on a mesophilic enzyme. *Angewandte Chemie, 53*, 6101–6104.

Scholl, Z. N., Yang, W., & Marszalek, P. E. (2014). Chaperones rescue luciferase folding by separating its domains. *Journal of Biological Chemistry, 289*, 28607–28618.

Schröder, H., Langer, T., Hartl, F. U., & Bukau, B. (1993). DnaK, DnaJ and GrpE form a cellular chaperone machinery capable of repairing heat-induced protein damage. *The EMBO Journal, 12*, 4137–4144.

Schumann, F. H., Varadan, R., Tayakuniyil, P. P., Grossman, J. H., Camarero, J. A., & Fushman, D. (2015). Changing the topology of protein backbone: The effect of backbone cyclization on the structure and dynamics of a SH3 domain. *Frontiers in Chemistry, 3*, 26.

Scott, C. P., Abel-Santos, E., Wall, M., Wahnon, D. C., & Benkovic, S. J. (1999). Production of cyclic peptides and proteins in vivo. *Proceedings of the National Academy of Sciences, 96*, 13638–13643.

Selle, P. H., Ravindran, V., Bryden, W. L., & Scott, T. (2006). Influence of dietary phytate and exogenous phytase on amino acid digestibility in poultry: A review. *The Journal of Poultry Science, 43*, 89–103.

Shi, S., Semple, A., Cheung, J., & Shameem, M. (2013). DSF method optimization and its application in predicting protein thermal aggregation kinetics. *Journal of Pharmaceutical Sciences, 102*, 2471–2483.

Svetlov, M. S., Kommer, A., Kolb, V. A., & Spirin, A. S. (2006). Effective cotranslational folding of firefly luciferase without chaperones of the Hsp70 family. *Protein Science, 15*, 242–247.

Takahashi, H., Arai, M., Takenawa, T., Sota, H., Xie, Q. H., & Iwakura, M. (2007). Stabilization of hyperactive dihydrofolate reductase by cyanocysteine-mediated backbone cyclization. *Journal of Biological Chemistry, 282*, 9420–9429.

van Lieshout, J. F. T., Pérez Gutiérrez, O. N., Vroom, W., Planas, A., de Vos, W. M., van der Oost, J., et al. (2012). Thermal stabilization of an endoglucanase by cyclization. *Applied Biochemistry and Biotechnology, 167*, 2039–2053.

Veggiani, G., Nakamura, T., Brenner, M. D., Gayet, R. V., Yan, J., Robinson, C. V., et al. (2016). Programmable polyproteams built using twin peptide superglues. *Proceedings of the National Academy of Sciences, 113*, 1202–1207.

Yang, H., Liu, L., Li, J., Chen, J., & Du, G. (2015). Rational design to improve protein thermostability: Recent advances and prospects. *ChemBioEng Reviews, 2*, 87–94.

Zakeri, B., Fierer, J. O., Celik, E., Chittock, E. C., Schwarz-Linek, U., Moy, V. T., et al. (2012). Peptide tag forming a rapid covalent bond to a protein, through engineering a bacterial adhesin. *Proceedings of the National Academy of Sciences, 109*, E690–E697.

Zakeri, B., & Howarth, M. (2010). Spontaneous intermolecular amide bond formation between side chains for irreversible peptide targeting. *Journal of the American Chemical Society, 132*, 4526–4527.

Zhang, W.-B., Sun, F., Tirrell, D. A., & Arnold, F. H. (2013). Controlling macromolecular topology with genetically encoded SpyTag–SpyCatcher chemistry. *Journal of the American Chemical Society, 135*, 13988–13997.

Zoldák, G., Zubrik, A., Musatov, A., Stupák, M., & Sedlák, E. (2004). Irreversible thermal denaturation of glucose oxidase from *Aspergillus niger* is the transition to the denatured state with residual structure. *Journal of Biological Chemistry, 279*, 47601–47609.

CHAPTER EIGHT

Engineering and Application of LOV2-Based Photoswitches

S.P. Zimmerman, B. Kuhlman[1], H. Yumerefendi[1]
University of North Carolina Chapel Hill, Chapel Hill, NC, United States
[1]Corresponding authors: e-mail address: bkuhlman@email.unc.edu; yumer@unc.edu

Contents

1. Introduction 170
2. Engineering LOV2-Based Photoswitches 171
   - 2.1 Rational Engineering of a LOV2-Based Photoswitch 172
   - 2.2 Validating LOV2-Based Photoswitches 174
   - 2.3 Improving Initial Switches by Known Mutations 177
   - 2.4 Library-Based Screening and Selection for Functional LOV2-Based Photoswitches 177
   - 2.5 Tuning the Activity Timescales of the LOV2-Based Photoswitch 180
3. Using LOV2-Based Photoswitches 181
   - 3.1 Single-Cell Microscopy 181
   - 3.2 Image Analysis 185
   - 3.3 Functional Assays 187
4. Summary and Perspectives 188
References 188

Abstract

Cellular optogenetic switches, a novel class of biological tools, have improved our understanding of biological phenomena that were previously intractable. While the design and engineering of these proteins has historically varied, they are all based on borrowed elements from plant and bacterial photoreceptors. In general terms, each of the optogenetic switches designed to date exploits the endogenous light-induced change in photoreceptor conformation while repurposing its effect to target a different biological phenomenon. We focus on the well-characterized light–oxygen–voltage 2 (LOV2) domain from *Avena sativa* phototropin 1 as our cornerstone for design. While the function of the LOV2 domain in the context of the phototropin protein is not fully elucidated, its thorough biophysical characterization as an isolated domain has created a strong foundation for engineering of photoswitches. In this chapter, we examine the biophysical characteristics of the LOV2 domain that may be exploited to produce an optogenetic switch and summarize previous design efforts to provide guidelines for an effective design. Furthermore, we provide protocols for assays including fluorescence polarization, phage display, and microscopy that are optimized for validating, improving, and using newly designed photoswitches.

1. INTRODUCTION

The ability to control biological processes with light is revolutionizing how we study biological phenomena. Via the use of lasers and modern microscopes, light-activatable systems allow researchers to stimulate and monitor signaling events with amazing spatiotemporal resolution in living cells and organisms. In general, light-activatable systems can be classified into two types: those that involve chemical modification of macromolecules and those that are fully genetically encoded (optogenetic). This chapter focuses on genetically encoded biological photoswitches that are based on light-sensing plant proteins. Typically, upon light stimulation the proteins undergo conformational changes that are used to modulate signaling events. These photoreceptors can be placed into two groups—red-light and blue-light photoreceptors.

The red-light photoreceptor that has been used most extensively to control signaling is phytochrome B (PhyB). PhyB exists in two distinct states P_R and P_{FR} that reversibly interconvert when stimulated with red light (660 nm) or far-red light (720 nm). When phyB is in the P_{FR} state it binds to phytochrome-interacting family proteins (PIF), which mediate its biological activity (Ni, Tepperman, & Quail, 1999). This provides a natural red light inducible heterodimerization system that can readily be used for recruiting a protein of interest to a specific location in a cell (Levskaya, Weiner, Lim, & Voigt, 2009). However, this photoswitch has a few drawbacks that can restrict engineering efforts: (1) there is a lack of structural understanding on how this heterodimerizable pair works; (2) PhyB requires the chromophore phytochromobilin, which is not present in organisms other than some plants and thus necessitates external supplementation; and (3) many optical systems are not fitted with lasers in the far-red light range.

There are two main photoreceptors used for tool engineering that are excited by blue light: cryptochromes and light–oxygen–voltage (LOV) domains from phototropins. The cryptochrome used extensively is the *Arabidopsis thaliana* Cry2, which requires an FAD chromophore, commonly found throughout various organisms (Liu et al., 2008). Similarly to PhyB–PIF interaction heterodimer pair, Cry2 interacts naturally with the protein CIB1 in its photoexcited state (Kennedy et al., 2010). This represents a natural blue-light inducible heterodimerizable pair, which has been used for the recruitment of proteins of interest to the plasma membrane for the study of cell motility as well as in a two-hybrid fashion for the control of transcription and others (Kennedy et al., 2010). An intriguing property of Cry2 discovered recently is that it also homo-oligomerizes upon blue-light illumination, which can be used to control cellular processes (Bugaj et al., 2015; Taslimi et al., 2014). Thus far, the most characterized photoreceptor is the LOV2 domain from phototropin 1 of *Avena sativa*. This protein contains an FMN chromophore, which like FAD is found in most organisms. Structural studies have established that upon blue-light stimulation the C-terminal helix of the domain, the Jα-helix, undocks from the PAS core domain and unfolds (Halavaty & Moffat, 2007; Harper, Neil, & Gardner, 2003). This dramatic conformational change has served as the basis for most engineering efforts involving blue-light inducible photoswitches (Baarlink, Wang, & Grosse, 2013; Lungu et al., 2012; Niopek et al., 2014; Renicke, Schuster, Usherenko, Essen, & Taxis, 2013; Spiltoir, Strickland, Glotzer, & Tucker, 2015; Strickland et al., 2012; Wu et al., 2009; Yi, Wang, Vilela, Danuser, & Hahn, 2014; Yumerefendi et al., 2015; Yumerefendi et al., 2016). LOV2-based switches can be grouped into two categories: single- or two-component systems. Single component LOV2-based photoswitches generally exploit endogenous cellular mechanisms. Examples of such are the photoactivatable GTPases and recently described systems that exploit endogenous trafficking mechanisms (Wu et al., 2009; Yumerefendi et al., 2015). Two-component systems usually couple two functional elements into a synthetic heterodimerizable system (Guntas et al., 2015; Lungu et al., 2012; Strickland et al., 2012). Here we describe how LOV2-based photoswitches can be created, validated, and optimized. Additionally, we provide some guidelines for using them in yeast and mammalian tissue culture systems.

2. ENGINEERING LOV2-BASED PHOTOSWITCHES

All LOV2-based photoswitches to date are efforts of rational engineering. The engineering process is centered on testing a small number of variants, which requires a method and criteria for validation. While there is not a single best test for validation, most LOV2-based switches function by regulating binding affinity for another protein, and therefore we describe a relatively universal approach based on fluorescence polarization measurements preillumination and postillumination to measure the binding affinities of the photoswitch for binding partners. The criteria for a successful design should involve achieving a change in binding affinities that is relevant for the biological process being manipulated. The design process is iterative in nature, and we have shown that an initial hit can be optimized using directed evolution. Finally, LOV2-based switches can also be tuned to suit the timescale of biological process under investigation.

2.1 Rational Engineering of a LOV2-Based Photoswitch

Structural studies of LOV2 have yielded atomic resolution information for the protein in absence of light, which revealed a PAS-fold with a tightly embedded flavin mononucleotide cofactor (Crosson & Moffat, 2001). Upon blue-light irradiation the C(4a) atom in the FMN ring forms a covalent adduct with a proximal cysteine C450 (Halavaty & Moffat, 2007). The transient thiol bond formation leads to small conformational changes in the domain which propagate to the Jα-helix and cause it to become unstructured and undocked from the rest of the PAS-fold (Fig. 1A). This, Jα-helix conformational change, allows for allosteric block of a peptide of interest in the dark and its release in the light. There are successful examples for both the control of short peptides and entire functional protein domains. In all cases, the goal is to sterically block key amino acids from binding to an interaction partner.

2.1.1 Embedding Peptides

There are a few photoswitches that have been designed by embedding a peptide in the Jα-helix. In most successful cases the peptide was embedded near the C-terminus of the Jα-helix (Guntas et al., 2015; Lungu et al., 2012; Mart et al., 2016; Niopek et al., 2014, 2016; Spiltoir et al., 2015; Strickland et al., 2012; Yi et al., 2014; Yumerefendi et al., 2015; Yumerefendi et al., 2016) (Fig. 1B). We have tested constructs that place a peptide sequence near the N-terminal or middle of the Jα-helix, and in these cases observed only weak binding affinity for the target protein in the lit state (Lungu et al., 2012). When we aligned the Jα-helix sequence of the best-reported successful photoswitches, we observed that they make use of five primary truncation points (E537, D540, E541, A542, and K544) (Fig. 1B). The total length for caged peptides counting from E537 and extending to the longest extension of Jα-helix is 29 amino acids. Generally, significant sequence homology is followed up to residue I539 with exceptions mostly to mutations described to stabilize the LOV2 dark state (Guntas et al., 2015; Strickland et al., 2010). After residue I539, significant sequence divergence can be seen but a pronounced preference exists for alanine at residue 542, small hydrophobic amino acid at residue 543 and leucine at residue 551 (Fig. 1C). Notably, when the conserved hydrophobic amino acid at position 543 was mutated

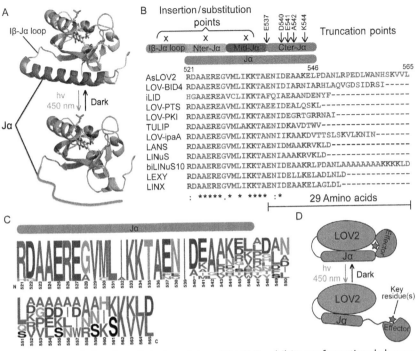

Fig. 1 Design of LOV2-based photoswitches. (A) LOV2 exhibits conformational change upon blue-light illumination, which manifests by the partial undocking followed by unfolding of its C-terminal helix, the Jα-helix. *Note*: while the structure for the dark state of the protein is known, thorough characterization of the lit state structure is missing and thus is illustrated in the *bottom* part of this figure. (B) Sequence alignment of the Jα-helix of LOV2-based photoswitches to date demonstrates emerging patterns. All of the current successful photoswitches are created by embedding a peptide epitope after truncation of the Jα-helix at its C-terminal region (Guntas et al., 2015; Lungu et al., 2012; Mart, Meah, & Allemann, 2016; Niopek et al., 2014; Niopek, Wehler, Roensch, Eils, & Di Ventura, 2016; Spiltoir et al., 2015; Strickland et al., 2012; Yi et al., 2014; Yumerefendi et al., 2015; Yumerefendi et al., 2016). (C) Consensus amino acids within the Jα-helix of the photoswitches aligned in (B) (Crooks, Hon, Chandonia, & Brenner, 2004). (D) Schematic of how LOV2 allows the control of a protein domain. Key residues (illustrated with a *red star* (*dark gray star* in the print version)) allosteric blockage in the dark is essential for this strategy.

to arginine, the switch (PA-PKI) exhibited low dynamic range with only modest control (twofold) of its target PKA kinase.

2.1.2 Fusing Protein Domains

There are two prominent successful examples for LOV2-based switches that allosterically block a functional protein domain, PA-Rac and LOV-DAD (Baarlink et al., 2013; Wu et al., 2009). PA-Rac is a fusion of LOV2

(404–546) and the small GTPase Rac1 (4–192). In the dark state the constitutively active Rac1 mutant Q61L has weaker affinity for its effector, the endogenous CRIB domain of PAK1 kinase. LOV-DAD is a fusion of LOV2 (403–543) and the autoinhibitory DAD domain (1038–1171) of the actin nucleation and elongation factor, the formin mDia2. Upon light illumination of LOV-DAD, the DAD domain is released to bind the N-terminal part of the endogenous mDia2 and thus locks it in an active conformation. In both photoswitches, successful obstruction of key residues in the dark state of the protein has been achieved. In the case of PA-Rac, its crystal structure reveals occlusion of a beta-strand required for its interaction to its effector CRIB in the dark state of the protein. This was achieved with no truncation further into the LOV2 domain but with the identification of a suitable truncation position for Rac1. In the case of LOV-DAD, a single truncation point within the Jα-helix was tested (A543) and a successful truncation of mDia2 DAD was identified such that the critical residue M1041 is immediately at the C-terminus of the Jα-helix and thus is likely to pack against the PAS domain of the LOV2 domain. These examples imply that packing a single key residue against the PAS domain may be sufficient for caging (Fig. 1D).

2.2 Validating LOV2-Based Photoswitches

The validation of initial designs depends largely on the targeted effector protein. In the majority of cases the goal is to regulate binding affinity to an effector protein, and therefore in vitro binding measurements can be used to validate the switch. Previously, we have demonstrated a good correlation between the biophysical behavior of a photoswitch and its functional activity (Hallett, Zimmerman, Yumerefendi, Bear, & Kuhlman, 2016). While there are numerous techniques for the measurement of protein–protein affinities few of the necessary instruments can be adapted for light illumination during the measurement process. Most systems are in fact enclosed and thus permit measurement of affinities only in the dark state of a photoswitch. One solution to this issue is the application of mutations that mimic its lit (open) conformation (Table 1). Such mutations have been proven useful but unsatisfactory for the recapitulation of the behavior of the switch in their absence, ie, the light-stimulated photoswitch. To address this issue and measure the affinities of light sensitive switches, we have modified a fluorometer with a 465 nm LED lamp to successfully employ a variation of a fluorescence polarization competition assay (Fig. 2A and B). This setup enables us to illuminate a sample and subsequently make a measurement. To perform the assay we used a fluorescently labeled peptide with the same sequence as the peptide

Table 1 List of Mutations That Affect the Caging of LOV2 Domain

Mutation	Effect	References
V529N	Lit mimetic	Strickland et al. (2012)
I539E	Lit mimetic	Strickland et al. (2012)
I532E/A536E	Lit mimetic	Nash, Ko, Harper, and Gardner (2008)
C450A	Dark mimetic	Nash et al. (2008)
C450S	Dark mimetic	Nash et al. (2008)
T406A/T407A	Improve caging	Strickland et al. (2012)
G528A	Improve caging	Strickland et al. (2010)
N538E	Improve caging	Strickland et al. (2010)
I532A	Improve caging	Strickland et al. (2010)
L514K/L531E	Improve caging	Lungu et al. (2012)

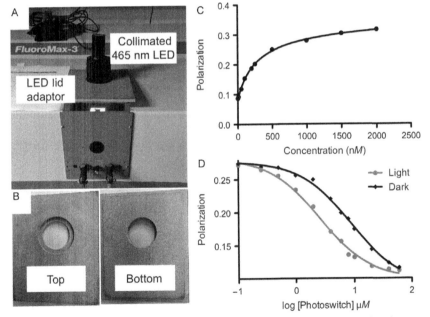

Fig. 2 Validation and screening of initial LOV2-based photoswitch designs. (A) A FluoroMax-3 setup used for measuring the affinities of photoswitches using blue LED light. (B) LED lid adaptor made to fit the fluorometer and the LED light. (C) An example of a titration of a TAMRA-labeled peptide against a target. (D) An example of competitive binding of a photoswitch upon light illumination or absence of light.

embedded in the Jα-helix of LOV2. In the dark the labeled peptide binds to the target protein (realized as a high polarization state) but upon illumination it is competed off by the photoswitch (realized by a low polarization state) (Fig. 2C and D).

2.2.1 Fluorescence Polarization Competition Assay to Measure Binding Affinities in the Lit and Dark States

Equipment and materials
1. Fluorimeter equipped with polarizers (Jobin Yvon Horiba FluoroMax-3)
2. 5-TAMRA-labeled peptide
3. LED light (Thorlabs M455L3-C1—Royal Blue (455 nm))
4. LED lid adaptor
5. Quartz fluorometer open top or microcell cuvettes (Starna Cells, Inc. #3-3.45-Q-3)
6. Purified proteins
7. Buffers

Protocol
1. Establish the binding affinity of the fluorescent peptide to the target protein by titrating the target protein into 150 μL of buffer containing 20 nM of the fluorescent peptide (see Fig. 2C). NOTE: *The concentration of the peptide may need to be higher to obtain good signal-to-noise ratio for the assay.*
2. Establish concentration of the target peptide at which it binds the fluorescent protein at about 60%.
3. Prepare 150 μL of buffer containing 20 nM of the fluorescent peptide and the target protein bound at 60% (Fig. 2C).
4. Titrate the photoswitch into the cuvette in a dark room using a blue LED lamp to cover the sample chamber. Illuminate the sample for 2 min with 6 mW/cm^2. Quickly turn off the lamp and take a reading (light binding). NOTE: *The faster the lamp can be turned off and the measurement made the more accurate the reading will be as the process will begin reversion immediately after the lamp has been turned off. This makes measurement for switches with <30 s half-lives difficult to measure.* Take a second reading after 5 more minutes keeping the sample in the dark (dark binding). Repeat the process over 12 titration points.
5. Fit the fluorescence polarization values with Prism 5 to identify IC50. Use this value, the starting concentration of the target protein and peptide as well as their affinity to calculate Ki (Nikolovska-Coleska et al., 2004).

2.3 Improving Initial Switches by Known Mutations

The efficiency of a photoswitch depends on how well its inherent dynamic range aligns to that of the biologically relevant process it controls. In our hands, photoswitches with a twofold change in binding affinity within the low nanomolar and low micromolar range have not proven useful for the control of transcription or protein colocalization. While such switches may be useful in some cases, most protein interactions occur between the low nanomolar and the low micromolar affinity range, which suggests that within this affinity range a photoswitch that exhibits such a low dynamic range may not be effective. If an initial photoswitch does not exhibit dynamic range larger than twofold we would recommend to test a set of mutations previously described to enhance the dynamic range of a LOV2-based photoswitches (Strickland et al., 2012, 2010) (Table 1). These generally have been demonstrated to stabilize the Jα-helix in the dark state of the LOV2 domain by increasing its helical propensity or its packing against the PAS domain thus increasing the stability of the dark conformation. A common side effect, we have observed, is that the use of such mutations may be accompanied by a decrease of activity upon light illumination. Although the number of LOV2-based photoswitches is not large, it appears that the aforementioned mutations may be particularly helpful if the peptide embedded does not have high helical propensity (the case of TULIPS and iLID).

2.4 Library-Based Screening and Selection for Functional LOV2-Based Photoswitches

Ideally, a photoswitch would be a binary system with a defined on-state that activates a biological process and an off-state that is completely devoid of function. While this may never be achieved due to the stochastic nature of biological systems, a large dynamic range between the active and inactive state of the protein is sufficient. One approach to achieve a large dynamic range is to create synthetic libraries of the photoswitch of interest and isolate the most efficient ones through screening or selection strategies. A possible solution we have successfully explored is to use computationally predicted mutations that stabilize the dark state of the LOV2 domain thus aiming to decrease the affinity of the targeted interaction in the dark while maintaining the affinity of the interaction under blue light. The key to successful strategies for evolving dynamic range is to maintain selective pressure for both states of the designed photoswitch and perform rounds of dual—positive and negative selections. The first aims to increase or maintain the activity in the

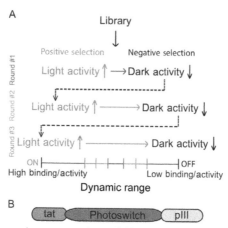

Fig. 3 Dynamic range enhancement through library-based methods. (A) A schematic of a selection strategy for the enhancement of the dynamic range of a photoswitch. Starting from a library of mutants one can perform a positive or negative selection at first and then rounds of dual positive and negative selections. (B) A schematic of the pIII construct used to functionally display photoswitches using phage display.

light immediately followed by the second, which selects for variants from the first that have turned off and exhibit low levels of activity in the dark (Fig. 3A).

Phage display is a common selection methodology used to evolve protein–protein interactions (Packer & Liu, 2015). Typically, a coat protein of the M13 filamentous phage such as pIII is fused to a target peptide or protein for its display (Packer & Liu, 2015). As part of the phage life cycle the coat phage proteins are secreted and fold in the periplasm (Steiner, Forrer, Stumpp, & Plückthun, 2006). Our initial attempts to use phage display on LOV2 switches using the typical Sec secretion pathway proved unsuccessful resulting in nonfunctional display of the LOV2 domain probably due to misfolding. To overcome this hurdle we used the twin-arginine translocation (tat) pathway, which allows the folding of a target protein to occur in the cytoplasm and its translocation to the periplasm in the folded state (DeLisa, Tullman, & Georgiou, 2003) (Fig. 3B).

Finally, we have incorporated the aforementioned dual selection concepts into a phage display strategy with which we were able to improve a photoswitch from an 8-fold dynamic range to 58-fold (Guntas et al., 2015).

2.4.1 Phage Display Panning

There are numerous protocols for performing phage display panning. Here, we describe the protocol employed for the development of the iLID photoswitch (Guntas et al., 2015). In this case, after the fourth round of dual selection, the gene pool was cloned in pET21b introducing a FLAG tag and soluble protein ELISA was performed on 96 clones. The ELISA allows for ranking of hits from the selection experiment as well as prioritizing clones for further validation and testing.

Equipment and materials
1. 96-well Maxisorp immunoplates (Nunc, cat. no. 430341)
2. M13K07 helper phage (Invitrogen)
3. SS320 *Escherichia coli* cells (Invitrogen)
4. BSA (Sigma, cat. no. A3059)
5. Sodium bicarbonate buffer pH 9.6
6. Purified target protein
7. 2YT media
8. PBS-Tween 20 0.1% (PBS-T)

Protocol

Day 1:
   1. Coat a MaxiSorp 96-well plate (Nunc, cat. no. 430341) with 100 μL/well of 5 μg/mL His-MBP-target protein in 50 mM NaHCO$_3$ buffer at pH 9.6 over night at 4°C

Day 2:
   2. Wash three times the plate with PBS-T
   3. Block with 200 μL of PBS-T and 5 mg/mL BSA for 2 h at room temperature.
   4. Wash three times the plate with PBS-T
   5. Incubate 1×10^{11} to 1×10^{12} phage library for 1 h under blue light at about 1 mW/cm^3
   6. Wells are washed 10 times with PBS-T
   7. The washed plate is moved to a dark room and incubated for 10 min at room temperature
   8. The supernatant is carefully collected (this represents the unbound portion of the phage library in the dark) and early log SS320 cells (OD$_{600}$ 0.2–0.5) are infected and grown in 25 mL of 2YT media supplemented with 100 μg/mL ampicillin
   9. Incubate at 37°C shaking for 20–30 min
   10. Add M13K07 helper phage to the culture with 20:1 multiplicity of infection
   11. Incubate for 20 min at 37°C without shaking
   12. Add 0.1 mM IPTG and 5 μM FMN and incubate at 30°C for 1.5 h in a shaking incubator covered with aluminum foil to ensure darkness over the sample

13. Add kanamycin to 25 μg/mL final to the culture and continue growing it at 30°C over night
14. Precipitate the enriched phage library and repeat the selection three times

2.5 Tuning the Activity Timescales of the LOV2-Based Photoswitch

Biological processes happen at various timescales. Furthermore, long-term exposure to intense light can affect biological samples (Cheng, Kiernan, Eliceiri, Williams, & Watters, 2016). Therefore, it is important to consider the timescale of photoswitch activation. The timescales at which a photoswitch is effective depends on its photocycle. In the case of the LOV2 domain the thiol bond formation is reversible with a half-life of ∼30 s, leading to complete dark state reversion in absence of illumination. A large number of mutations that affect the half-life of the thiol bond between C450 and FMN in the light state of the photoswitch have been reported (Nash et al., 2008; Zayner, Antoniou, & Sosnick, 2012; Zayner & Sosnick, 2014). Half-lives from 2 s to 4300 s have been described. Depending on the application of interest one may consider tuning the half-life of the photoswitch using such mutations (Table 2).

Table 2 List of Mutations That Affect the Photocycle Half-Life of the LOV2 Domain

Mutation	Half-Life(s)	References
N414A/Q513H	2	Zayner and Sosnick (2014)
V416T	2.6	Kawano, Aono, Suzuki, and Sato (2013)
I427V	4	Christie et al. (2007)
Q513D	5	Zayner and Sosnick (2014)
Q513N	27.3	Nash et al. (2008)
Q513H	30	Zayner and Sosnick (2014)
N414D	69	Zayner and Sosnick (2014)
WT	80	Zayner and Sosnick (2014)
L453V	160	Zayner and Sosnick (2014)
Q513A	261	Zayner and Sosnick (2014)
N414Q	280	Zayner and Sosnick (2014)

Table 2 List of Mutations That Affect the Photocycle Half-Life of the LOV2 Domain—cont'd

Mutation	Half-Life(s)	References
N416S	685	Zayner and Sosnick (2014)
N414T	892	Zayner and Sosnick (2014)
V416I/L496I	1000	Zoltowski, Vaccaro, and Crane (2009)
Q513L	1080	Nash et al. (2008)
N414A	1427	Zayner and Sosnick (2014)
Q513L	1793	Zayner and Sosnick (2014)
N414A/Q513A	2081	Zayner and Sosnick (2014)
V416L	4300	Kawano et al. (2013)

3. USING LOV2-BASED PHOTOSWITCHES

LOV2-based photoswitches can be used to modulate proteins at various size and timescales, from subcellular to whole organism and seconds to days. Population-based assays can be performed using a variety of illumination methods. One difficult aspect is maintaining proper growth conditions while performing the exposure. We have designed ways to outfit incubators with light-emitting apparatuses that do not interfere with growth and describe their use in this section (Fig. 4A). At the level of single cell and subcellular assays, a microscope outfitted with a laser that can be directed to a region of interest (ROI) for activation is necessary. In most cases a laser scanning confocal microscope contains the required hardware and software modules out of the box. These instruments are commonly used for photobleaching assays, which are performed in a similar manner to photoactivation and can therefore be easily modified to do so.

3.1 Single-Cell Microscopy

Using a laser scanning confocal microscope, a small group or an individual cell can be imaged during photoactivation. Many out of the box microscopes have the ability to perform photoactivation/photobleaching experiments. The microscope described later is meant as a reference to what sorts of features are necessary. When designing your experiment, it is important to use fluorescent proteins or other molecules that are spectrally distinct from

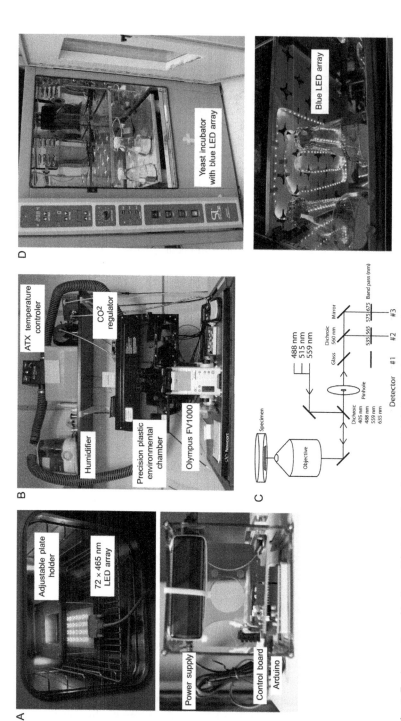

Fig. 4 Equipment used in the validation and application of LOV2-based photoswitches. (A) Adjustable and programmable tissue culture dish illuminator. *Top*—image of 72 × 465 nm LED array placed in incubator for cell culture experiments. The plate holder can be adjusted to control the distance from the LED array. *Bottom*—image of Arduino-based controller, which allows for programming of LED array brightness and timing. (B) Image of Olympus FV1000 with equipment necessary for imaging and activation of LOV2-based photoswitches in live cells. (C) Diagram of the light path used to image and activate LOV2-based photoswitches on a confocal microscope. (D) Image of yeast incubator outfitted with a blue LED array for growth under activating conditions. (See the color plate.)

LOV2 so that activation and imaging are separate processes. Any visible light below 500 nm can activate LOV2, while optimal activation occurs at 450 nm. Therefore we use fluorophores with excitation and emission spectra greater than 500 nm (eg, mVenus, tagRFPt, mCherry, Cy5). While the filter sets described do not perfectly match the spectra of LOV2 and the fluorescent proteins used, they provide a usable signal-to-noise ratio that can be easily quantified.

Equipment and reagents
1. Olympus FV1000 laser scanning confocal microscope equipped with (Fig. 4B):
   a. 40× oil immersion objective lens, NA 1.3
   b. 100× oil immersion objective lens, NA 1.4
   c. Multiline argon laser, 30 mW
   d. 559 nm diode laser, 15 mW
   e. 405/488/559/635 nm dichroic excitation mirror
   f. 560 nm dichroic emission mirror
   g. 505–540, 535–565, and 575–675 nm bandpass filters
   h. Precision plastics black environmental chamber
   i. Air-Therm ATX-H heater
   j. Vix Humidifier
2. 35 mm, No. 1.5 coverglass MatTek cell culture dish
3. Cell culture grade phosphate buffered saline (PBS)
4. Dulbecco's modified eagle's medium (DMEM)
5. DMEM, supplemented with 10% FBS, 100 U/mL penicillin, 100 μg/mL streptomycin, 292 μg/mL L-glutamine
6. Trypsin–EDTA (TE)
7. Mammalian expression vector containing the sequence for your photoswitch fused to a fluorescent protein spectrally compatible with LOV2 (mVenus, TagRFPt, mCherry)
8. Cell line of interest (mouse embryonic fibroblasts are recommended as they spread out flat over a large area and survive at sparse concentrations which make them easy to work with and analyze image data from. One caveat is that they do not have high transfection efficiency)
9. 10 μg/mL fibronectin
10. NanoJuice transfection reagent (for improved MEF transfection efficiency)

Protocol
Day 1:
1. Using supplemented DMEM and TE, lift and plate cells in a 35 mm cell culture dish so that they are 50–80% confluent the following day

Day 2:
2. Using DMEM and NanoJuice transfect cells with the vector(s) containing your photoswitch according to the NanoJuice protocol
3. Four to six hour later wash cells with PBS and change supplemented DMEM

Day 3:
4. Coat glass area of MatTek dish with 1 mL of 10 μg/mL fibronectin
5. Incubate cover slip for 1 h at 37°C
6. Wash dish with PBS
7. Using TE, lift transfected cells and triturate gently with pipette to separate cells
8. Plate cells on coated MatTek dish with 3 mL of supplemented DMEM at ~10% confluency so that individual cells can be imaged the next day

Day 4:
Using the FV1000 and accompanying Fluoview software image and activate the cells using the following suggested settings as a good starting point. Fig. 4C represents the light path used in most experiments. While the excitation filter is not specific for 515 nm light it allows enough to pass in order to excite the mVenus fluorophore. Activate the cells using the 488 nm laser line and image them with the 515 and 559 nm laser as to not activate the LOV2 domain while imaging (*Caution*: while low power excitation with the 515 nm laser does not activate the LOV2 domain we have noticed that use at higher power produces low levels of activation). There is always a tradeoff between speed and image quality while imaging. Collecting an 800 × 800 pixel image with a 2 μs/pixel dwell time without averaging provides an image with enough signal to quantify while being sufficiently fast to measure the kinetics of your switch. To increase the signal while maintaining speed, the pinhole can be opened wider. However, this will decrease confocality of your image. Using the Time Controller module in the Fluoview software we have designed a timeline of imaging and activation. Table 3 represents a timeline in which a 60 × 60 ROI is activated followed by an 800 × 800 image acquisition. This is repeated in order to measure activation over time. The sample is then repeatedly imaged without activation in order to measure the kinetics of inactivation. For membrane localization, a 60 × 60 pixel ROI is chosen near the edge of a cell where the thickness is relatively uniform so that volume artifacts are avoided during quantification and the cytoplasmic background signal is lower. For whole-cell activation the ROI can be expanded but keep in mind this will lengthen

Table 3 Imaging Timeline

	Imaging Preactivation	Repeat 50× (Total Time = 10 s)		Imaging Postactivation
		Activation	Imaging Activation	
Timing	5 scans, 1 every 10 s	10 scans, as fast as possible	1 image	50 images, 1 every 10 s
Size (pixels)	800 × 800	60 × 60	800 × 800	800 × 800
Lasers and power	5%—515 and 5%—559 nm	1%—488 nm	5%—515 and 5%—559 nm	5%—515 and 5%—559 nm
Pixel dwell time	2 μs/pixel	8 μs/pixel	2 μs/pixel	2 μs/pixel

the time between images, as the activation time will be extended. At the end of an experiment, images collected can be appended together using the append module in the Fluoview software.

3.2 Image Analysis

Many of the switches previously engineered function through translocation. In these cases microscopy and image analysis can be a useful method to verify and compare the function of the photoswitches. To quantify the translocation within a cell the FIJI software package (Schindelin et al., 2012) provides many useful tools and allows you to string them together using its macro function so that analysis is automated. In Hallett et al. the translocation of the iLID switch from the cytoplasm to the mitochondria of a cell is quantified in this way. The analysis consists of four steps (Fig. 5).

For each cell imaged collect two channels; Venus-iLID-Mito (acts as a mitochondria label) tagRFPt-SspB

1. Identify the cell and mitochondria—threshold tagRFPT and Venus channels to create ROIs that demarcate the entire cell and the mitochondria, respectively.
2. Deduce a cytoplasmic ROI—create a cytoplasmic ROI by extending the mitochondrial ROI by 10 pixels. Then remove the pixels that represent the mitochondria and any pixels that extend outside of the cell body.
3. Measure the average tagRFPt fluorescence intensity in the mitochondria and cytoplasm ROIs.
4. Repeat steps 1–3 for every image in the series.

The supplement (http://dx.doi.org/10.1016/bs.mie.2016.05.058) contains a macro script that was used to perform this analysis.

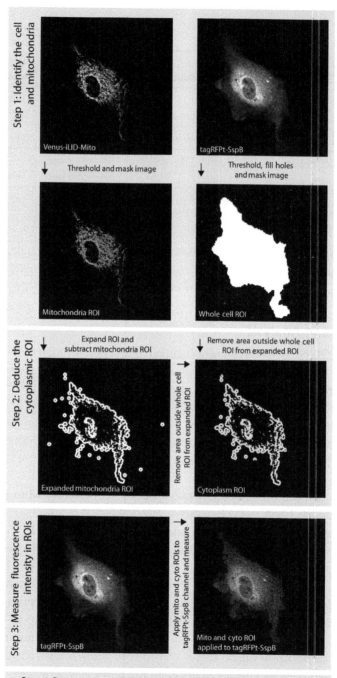

Fig. 5 Image analysis schematic for quantification of mitochondrial/cytoplasmic fluorescence intensity. (See the color plate.)

3.3 Functional Assays

LOV2-based photoswitches could be used in a number of applications for the study of the kinetics of a given process or simply for the rapid activation of a desired signaling pathway. One of the most common biological processes rendered under control of light is transcription. Here, we describe the setups and experimental procedure for the blue-light activation of transcription (Fig. 4D).

3.3.1 Yeast Transcription Using Photoswitch for Light-Induced Dimerization

Equipment and materials
1. Matchmaker Gold Yeast Two-Hybrid Kit (Clontech)

Protocol
1. Clone and sequence verify the two parts of the light inducible dimerizer into pGBKT7 and pGADT7 plasmids
2. Transform the plasmids in Y2HGold and Y187 yeast strains, respectively, selecting for their auxotrophic selection
3. Combine a single colony of each in 0.5 mL YPD culture overnight for mating
4. Pellet and plate on double dropout plates
5. Select single colony and inoculate 5 mL double dropout media culture overnight
6. Next morning use the overnight for setting a 5 mL culture at $OD_{600} = 0.2$ which is then split into two 2.5 mL cultures for dark and blue-light treatment (465 nm LED of 0.5–1 mW/cm^3) in a shaking incubator at 30°C for 4 h. The dark state tube is wrapped in aluminum foil to ensure dark condition.
7. Take three aliquots of 600 µL from each culture and pellet them. Cultures could be frozen at −20°C at this point for testing at a later time.
8. Perform β-galactosidase assay to test the functionality of the light inducible dimerization.

3.3.2 Tissue Culture Illumination Setup

For many experiments it is necessary to illuminate an entire culture dish or multiwell dish in order to obtain enough material to properly perform the experiment. In such cases it is important to evenly illuminate your samples while maintaining the necessary growth conditions for your cultures. For these experiments we have designed a 72×465 nm LED array with an adjustable rack that holds a standard-sized tissue culture plate (Fig. 4A).

The LEDs are powered through a 9 V and 2 A power supply with timing and intensity controlled through an Arduino board. The LEDs are arranged under the rack to match the pattern of a 24-well plate. However, the rack height is fully adjustable by thumbscrew so that the array can evenly illuminate any dish. This apparatus easily fits within a standard tissue culture incubator so that cells can be grown under blue light with optimal temperature, CO_2, and humidity (Fig. 4D).

4. SUMMARY AND PERSPECTIVES

Here we described the process of engineering a LOV2-based photoswitch. There are several steps in the rational design of such a switch, which all ultimately fall into an iterative process of trial and error. We have summarized successful strategies and provided protocols for validation and high-throughput library-based screening for evolved variants of such photoswitches. We have provided protocols for the test of switches via microscopy for single cells as well as entire organisms. Finally, details are provided for example applications in yeast and tissue culture. We expect that the number of cellular optogenetic tools based on *A. sativa* LOV2 will continue to grow in the upcoming years.

REFERENCES

Baarlink, C., Wang, H., & Grosse, R. (2013). Nuclear actin network assembly by formins regulates the SRF coactivator MAL. *Science, 340*(6134), 864–867. http://dx.doi.org/10.1126/science.1235038.

Bugaj, L. J., Spelke, D. P., Mesuda, C. K., Varedi, M., Kane, R. S., & Schaffer, D. V. (2015). Regulation of endogenous transmembrane receptors through optogenetic Cry2 clustering. *Nature Communications, 6*, 6898. http://dx.doi.org/10.1038/ncomms7898.

Cheng, K. P., Kiernan, E. A., Eliceiri, K. W., Williams, J. C., & Watters, J. J. (2016). Blue light modulates murine microglial gene expression in the absence of optogenetic protein expression. *Scientific Reports, 6*, 21172. http://dx.doi.org/10.1038/srep21172.

Christie, J. M., Corchnoy, S. B., Swartz, T. E., Hokenson, M., Han, I. S., Briggs, W. R., & Bogomolni, R. A. (2007). Steric interactions stabilize the signaling state of the LOV2 domain of phototropin 1. *Biochemistry, 46*(32), 9310–9319. http://dx.doi.org/10.1021/bi700852w.

Crooks, G. E., Hon, G., Chandonia, J. M., & Brenner, S. E. (2004). WebLogo: A sequence logo generator. *Genome Research, 14*(6), 1188–1190. http://dx.doi.org/10.1101/gr.849004.

Crosson, S., & Moffat, K. (2001). Structure of a flavin-binding plant photoreceptor domain: Insights into light-mediated signal transduction. *Proceedings of the National Academy of Sciences of the United States of America, 98*(6), 2995–3000. http://dx.doi.org/10.1073/pnas.051520298.

DeLisa, M. P., Tullman, D., & Georgiou, G. (2003). Folding quality control in the export of proteins by the bacterial twin-arginine translocation pathway. *Proceedings of the National*

Academy of Sciences of the United States of America, 100(10), 6115–6120. http://dx.doi.org/10.1073/pnas.0937838100.

Guntas, G., Hallett, R. A., Zimmerman, S. P., Williams, T., Yumerefendi, H., Bear, J. E., & Kuhlman, B. (2015). Engineering an improved light-induced dimer (iLID) for controlling the localization and activity of signaling proteins. *Proceedings of the National Academy of Sciences of the United States of America, 112*(1), 112–117. http://dx.doi.org/10.1073/pnas.1417910112.

Halavaty, A. S., & Moffat, K. (2007). N- and C-terminal flanking regions modulate light-induced signal transduction in the LOV2 domain of the blue light sensor phototropin 1 from *Avena sativa*. *Biochemistry, 46*(49), 14001–14009. http://dx.doi.org/10.1021/bi701543e.

Hallett, R. A., Zimmerman, S. P., Yumerefendi, H., Bear, J. E., & Kuhlman, B. (2016). Correlating in vitro and in vivo activities of light-inducible dimers: A cellular optogenetics guide. *ACS Synthetic Biology, 5*(1), 53–64. http://dx.doi.org/10.1021/acssynbio.5b00119.

Harper, S. M., Neil, L. C., & Gardner, K. H. (2003). Structural basis of a phototropin light switch. *Science, 301*(5639), 1541–1544. http://dx.doi.org/10.1126/science.1086810.

Kawano, F., Aono, Y., Suzuki, H., & Sato, M. (2013). Fluorescence imaging-based high-throughput screening of fast- and slow-cycling LOV proteins. *PloS One, 8*(12), e82693. http://dx.doi.org/10.1371/journal.pone.0082693.

Kennedy, M. J., Hughes, R. M., Peteya, L. A., Schwartz, J. W., Ehlers, M. D., & Tucker, C. L. (2010). Rapid blue-light-mediated induction of protein interactions in living cells. *Nature Methods, 7*(12), 973–975. http://dx.doi.org/10.1038/nmeth.1524.

Levskaya, A., Weiner, O. D., Lim, W. A., & Voigt, C. A. (2009). Spatiotemporal control of cell signalling using a light-switchable protein interaction. *Nature, 461*(7266), 997–1001. http://dx.doi.org/10.1038/nature08446.

Liu, H., Yu, X., Li, K., Klejnot, J., Yang, H., Lisiero, D., & Lin, C. (2008). Photoexcited CRY2 interacts with CIB1 to regulate transcription and floral initiation in *Arabidopsis*. *Science, 322*(5907), 1535–1539. http://dx.doi.org/10.1126/science.1163927.

Lungu, O. I., Hallett, R. A., Choi, E. J., Aiken, M. J., Hahn, K. M., & Kuhlman, B. (2012). Designing photoswitchable peptides using the AsLOV2 domain. *Chemistry & Biology, 19*(4), 507–517. http://dx.doi.org/10.1016/j.chembiol.2012.02.006.

Mart, R. J., Meah, D., & Allemann, R. K. (2016). Photocontrolled exposure of pro-apoptotic peptide sequences in LOV proteins modulates Bcl-2 family interactions. *Chembiochem, 17*(8), 698–701. http://dx.doi.org/10.1002/cbic.201500469.

Nash, A. I., Ko, W. H., Harper, S. M., & Gardner, K. H. (2008). A conserved glutamine plays a central role in LOV domain signal transmission and its duration. *Biochemistry, 47*(52), 13842–13849. http://dx.doi.org/10.1021/bi801430e.

Ni, M., Tepperman, J. M., & Quail, P. H. (1999). Binding of phytochrome B to its nuclear signalling partner PIF3 is reversibly induced by light. *Nature, 400*(6746), 781–784. http://dx.doi.org/10.1038/23500.

Nikolovska-Coleska, Z., Wang, R., Fang, X., Pan, H., Tomita, Y., Li, P., … Wang, S. (2004). Development and optimization of a binding assay for the XIAP BIR3 domain using fluorescence polarization. *Analytical Biochemistry, 332*(2), 261–273. http://dx.doi.org/10.1016/j.ab.2004.05.055.

Niopek, D., Benzinger, D., Roensch, J., Draebing, T., Wehler, P., Eils, R., & Di Ventura, B. (2014). Engineering light-inducible nuclear localization signals for precise spatiotemporal control of protein dynamics in living cells. *Nature Communications, 5*, 4404. http://dx.doi.org/10.1038/ncomms5404.

Niopek, D., Wehler, P., Roensch, J., Eils, R., & Di Ventura, B. (2016). Optogenetic control of nuclear protein export. *Nature Communications, 7*, 10624. http://dx.doi.org/10.1038/ncomms10624.

Packer, M. S., & Liu, D. R. (2015). Methods for the directed evolution of proteins. *Nature Reviews. Genetics*, *16*(7), 379–394. http://dx.doi.org/10.1038/nrg3927.

Renicke, C., Schuster, D., Usherenko, S., Essen, L. O., & Taxis, C. (2013). A LOV2 domain-based optogenetic tool to control protein degradation and cellular function. *Chemistry & Biology*, *20*(4), 619–626. http://dx.doi.org/10.1016/j.chembiol.2013.03.005.

Schindelin, J., Arganda-Carreras, I., Frise, E., Kaynig, V., Longair, M., Pietzsch, T., … Cardona, A. (2012). Fiji: An open-source platform for biological-image analysis. *Nature Methods*, *9*(7), 676–682. http://dx.doi.org/10.1038/nmeth.2019.

Spiltoir, J. I., Strickland, D., Glotzer, M., & Tucker, C. L. (2015). Optical control of peroxisomal trafficking. *ACS Synthetic Biology*. http://dx.doi.org/10.1021/acssynbio.5b00144.

Steiner, D., Forrer, P., Stumpp, M. T., & Plückthun, A. (2006). Signal sequences directing cotranslational translocation expand the range of proteins amenable to phage display. *Nature Biotechnology*, *24*(7), 823–831. http://dx.doi.org/10.1038/nbt1218.

Strickland, D., Lin, Y., Wagner, E., Hope, C. M., Zayner, J., Antoniou, C., … Glotzer, M. (2012). TULIPs: Tunable, light-controlled interacting protein tags for cell biology. *Nature Methods*, *9*(4), 379–384. http://dx.doi.org/10.1038/nmeth.1904.

Strickland, D., Yao, X., Gawlak, G., Rosen, M. K., Gardner, K. H., & Sosnick, T. R. (2010). Rationally improving LOV domain-based photoswitches. *Nature Methods*, *7*(8), 623–626. http://dx.doi.org/10.1038/nmeth.1473.

Taslimi, A., Vrana, J. D., Chen, D., Borinskaya, S., Mayer, B. J., Kennedy, M. J., & Tucker, C. L. (2014). An optimized optogenetic clustering tool for probing protein interaction and function. *Nature Communications*, *5*, 4925. http://dx.doi.org/10.1038/ncomms5925.

Wu, Y. I., Frey, D., Lungu, O. I., Jaehrig, A., Schlichting, I., Kuhlman, B., & Hahn, K. M. (2009). A genetically encoded photoactivatable Rac controls the motility of living cells. *Nature*, *461*(7260), 104–108. http://dx.doi.org/10.1038/nature08241.

Yi, J. J., Wang, H., Vilela, M., Danuser, G., & Hahn, K. M. (2014). Manipulation of endogenous kinase activity in living cells using photoswitchable inhibitory peptides. *ACS Synthetic Biology*, *3*(11), 788–795. http://dx.doi.org/10.1021/sb5001356.

Yumerefendi, H., Dickinson, D. J., Wang, H., Zimmerman, S. P., Bear, J. E., Goldstein, B., … Kuhlman, B. (2015). Control of protein activity and cell fate specification via light-mediated nuclear translocation. *PloS One*, *10*(6), e0128443. http://dx.doi.org/10.1371/journal.pone.0128443.

Yumerefendi, H., Lerner, A. M., Zimmerman, S. P., Hahn, K., Bear, J. E., Strahl, B. D., & Kuhlman, B. (2016). Light-induced nuclear export reveals rapid dynamics of epigenetic modifications. *Nature Chemical Biology*, *12*(6), 399–401. http://dx.doi.org/10.1038/nchembio.2068.

Zayner, J. P., Antoniou, C., & Sosnick, T. R. (2012). The amino-terminal helix modulates light-activated conformational changes in AsLOV2. *Journal of Molecular Biology*, *419*(1–2), 61–74. http://dx.doi.org/10.1016/j.jmb.2012.02.037.

Zayner, J. P., & Sosnick, T. R. (2014). Factors that control the chemistry of the LOV domain photocycle. *PloS One*, *9*(1), e87074. http://dx.doi.org/10.1371/journal.pone.0087074.

Zoltowski, B. D., Vaccaro, B., & Crane, B. R. (2009). Mechanism-based tuning of a LOV domain photoreceptor. *Nature Chemical Biology*, *5*(11), 827–834. http://dx.doi.org/10.1038/nchembio.210.

CHAPTER NINE

Minimalist Design of Allosterically Regulated Protein Catalysts

O.V. Makhlynets, I.V. Korendovych[1]

Syracuse University, Syracuse, NY, United States
[1]Corresponding author: e-mail address: ikorendo@syr.edu

Contents

1. Introduction — 192
2. Methods — 193
   2.1 Required Tools — 193
   2.2 Overall Approach — 194
   2.3 Starting Points for the Design — 195
   2.4 Testing the Possibility of the Substrate Associating with the Protein Chosen as a Starting Point for the Design — 197
   2.5 Identification of the Positions to Mutate — 198
   2.6 Establishing of the Impact of the Mutations on the Overall Fold — 198
   2.7 Predicting the Feasibility of Catalysis — 200
Acknowledgments — 201
References — 201

Abstract

Nature facilitates chemical transformations with exceptional selectivity and efficiency. Despite a tremendous progress in understanding and predicting protein function, the overall problem of designing a protein catalyst for a given chemical transformation is far from solved. Over the years, many design techniques with various degrees of complexity and rational input have been developed. Minimalist approach to protein design that focuses on the bare minimum requirements to achieve activity presents several important advantages. By focusing on basic physicochemical properties and strategic placing of only few highly active residues one can feasibly evaluate in silico a very large variety of possible catalysts. In more general terms minimalist approach looks for the mere *possibility of catalysis*, rather than trying to identify the *most active catalyst possible*. Even very basic designs that utilize a single residue introduced into nonenzymatic proteins or peptide bundles are surprisingly active. Because of the inherent simplicity of the minimalist approach computational tools greatly enhance its efficiency. No complex calculations need to be set up and even a beginner can master this technique in a very short time. Here, we present a step-by-step protocol for minimalist design of functional proteins using basic, easily available, and free computational tools.

1. INTRODUCTION

Nature facilitates chemical transformations with exceptional selectivity and efficiency. Precise positioning of functional groups and finely tuned dynamics enable proteins to catalyze a large variety of reactions. Nonetheless, as the repertoire of practically applicable chemical reactions is constantly expanding, the need for efficient catalysts is ever growing (Bornscheuer et al., 2012). Many such reactions have no obvious natural analogs and thus development of tools to create catalysts for new chemical transformations is of great importance (Nanda & Koder, 2010). Proteins provide an obvious choice for a starting point as they represent many natural enzymes, a great deal of structural information is available for them, and fairly sophisticated (semi)empirical computational tools provide reasonably accurate predictions of protein structure and stability. Despite a tremendous progress in understanding and predicting protein function in the past several decades, the overall problem of designing a protein catalyst for a given chemical transformation "a la carte" is far from solved (Korendovych & DeGrado, 2014). Over the years, many design techniques with various degrees of complexity and rational input have been developed (Kries, Blomberg, & Hilvert, 2013; Maeda, Makhlynets, Matsui, & Korendovych, 2016; Wijma & Janssen, 2013). Nonetheless, the overwhelming majority of the reported artificial protein catalysts owe their success to the (semi)rational design only in part; most of the improvement in catalytic efficiency comes from subsequent directed evolution. While there seems to be a correlation between the catalytic efficiency of the starting point of the design and the activity of the evolved enzyme, proteins with even modest initial activity can be improved drastically (Brustad & Arnold, 2011). These considerations underscore both the difficulty of the task and the opportunity presented by the fact that the designed protein catalyst can be far from perfect. Indeed, if most of the catalytic efficiency does come from improvement via directed evolution, why do we need to waste resources trying to *accurately* predict the sequence of the *most active* protein catalyst? Importantly, even the most efficient computational algorithms do not even try to model all states of the enzyme's catalytic cycle—they focus only on the transition state geometry. The enormous combinatorial cost of the task simply makes any other path not feasible.

Minimalist approach to protein design that focuses on the bare minimum requirements to achieve activity presents several important advantages (DeGrado, Wasserman, & Lear, 1989). By focusing on basic physicochemical properties and strategic placing of only few highly active residues one can easily sample a very large variety of proteins to identify starting points for design of catalysts. In more general terms minimalist approach looks for the mere *possibility of catalysis*, rather than trying to identify the *most active catalyst possible*. Even very basic designs that utilize a single residue introduced into nonenzymatic proteins or peptide bundles are surprisingly active (Muller, Windsor, Pomerantz, Gellman, & Hilvert, 2009). Because of the inherent simplicity of the minimalist approach, computational tools greatly enhance its efficiency. No complex calculations need to be set up and even a beginner can master this technique in a very short time. Here we present a step-by-step protocol for minimalist design of functional proteins using basic, easily available, and free computational tools. We will show the design process of AlleyCatE, an allosterically regulated catalyst of ester hydrolysis. AlleyCatE was generated by introducing a single histidine residue into a nonenzymatic protein calmodulin (CaM) (Moroz et al., 2015).

2. METHODS
2.1 Required Tools

Computational tools for design are easily available online; most of them are free for academic use.
1. Autodock Vina (available from http://vina.scripps.edu).
2. Autodock Tools (available from http://mgltools.scripps.edu).
3. ChemDraw or comparable software package (available from http://www.cambridgesoft.com/software/overview.aspx).
4. Rosetta (available from https://www.rosettacommons.org/software).
5. PyMOL or comparable software package for protein structure viewing. PyMOL is available from https://www.pymol.org.

Note: While different operating systems can be successfully used, Mac OS has provided us with the best overall usability. Installation of the Rosetta package might take up to a day, but the manual provided on the RosettaCommons is very detailed and a person with no extensive computational skills can successfully install it. Familiarity with basic Unix commands is necessary.

Minimalist design is inherently rational and it relies on the knowledge of the mechanism of the reaction to be catalyzed. Simplicity of the minimalist approach allows for a stepwise consideration of all of the states along the Michaelis path, such as binding of the substrate by the protein, transition state, and product dissociation. The overall approach is summarized in

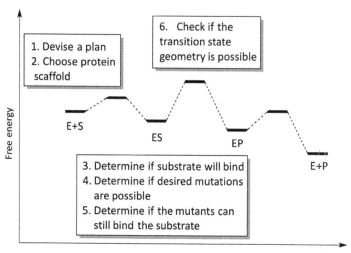

Fig. 1 Steps of the minimalist protein design aligned with the reaction coordinate in enzymatic catalysis.

Fig. 1. First, an overall idea for potential catalysis is rationally devised, then an appropriate protein scaffold is selected, next docking is performed to determine whether the desired substrate can associate with the protein to any degree, potential placement of the functional groups that are absolutely critical for catalysis is then evaluated, and finally feasibility of the transition state geometry is established.

2.2 Overall Approach

This is the step that requires the most input from protein designer. We will present design protocol for creating a catalyst for ester hydrolysis, a highly practically important chemical transformation. Specifically, we chose to look at hydrolysis of *p*-nitrophenyl ester **(pNPP)** for the ease of experimental characterization and straightforward comparison to previously developed catalysts (Richter et al., 2012). **pNPP** also possesses a stereo center, which allows one to evaluate stereoselectivity of minimalist protein catalysts. We decided to introduce a single amino acid residue that is capable of nucleophilic attack on the carbonyl group of the ester (Scheme 1). Oxygen nucleophiles (Ser and Thr) require additional activity modulating interactions, care should be exercised with sulfur nucleophiles (Cys) to avoid oxidation and cross-linking. This leaves histidine (His) as a viable general option for nucleophile incorporation into a desired protein scaffold.

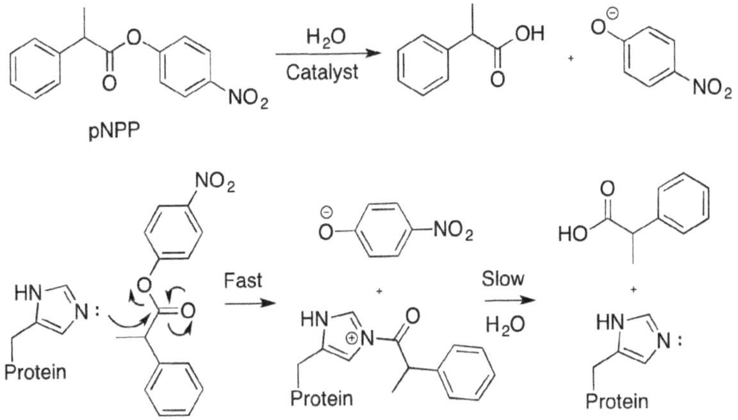

Scheme 1 Hydrolysis of **pNPP** by a single histidine residue in a protein.

$$E + S \underset{}{\overset{K_s}{\rightleftharpoons}} ES \xrightarrow{k_2} ES' + P_1 \xrightarrow{k_3} E + P_2$$

To serve as a good starting point for design of an allosterically regulated catalyst a protein has to satisfy a number of conditions: at a minimum it has to possess a "reasonable" substrate binding site, be stable (ideally thermophilic) and evolvable to preserve its fold during productive incorporation of multiple potentially very disruptive mutations (Tokuriki & Tawfik, 2009), and finally have an allosteric regulation site. Allosteric regulation in protein catalysts provides a number of opportunities for modulation of in vivo function (Makhlynets, Raymond, & Korendovych, 2015) and design of catalytically amplified sensors for metal ions (Mack et al., 2013). Despite having no known enzymatic activity, calmodulin satisfies all of the above requirements and thus we chose it as a scaffold for the design. A step-by-step application of the minimalist approach (Fig. 1) to create an allosterically regulated esterase is shown in Fig. 2.

2.3 Starting Points for the Design

Availability of reliable structural information is critical for the success of the project. Crystal structures obtained at the conditions relevant for catalysis tend to provide the best starting points. Care should be exercised when selecting an NMR structure, one should use regions within models with tightest backbone RMSD. Critical evaluation of both X-ray and NMR structures can be very informative for the subsequent design. Model 1 (the one with lowest energy by default) is normally used as a starting point

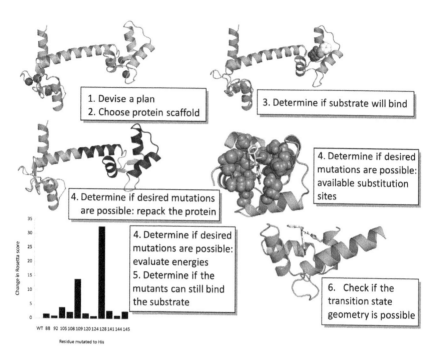

Fig. 2 Minimalist design of AlleyCatE.

in the design based on NMR structures, however several models might have to be examined to ensure complete coverage of all possibilities. In order to split models in NMR structures use PyMOL **split_states** command followed by the corresponding pdb code. Flexible disordered regions are normally not considered in the design process unless they are very close to the active site. Disordered residues can be removed in PyMOL by selecting the sequence of interest and performing Extract Object command followed by saving the resulting obj01 as a pdb file (menu command File/Save Molecule, …).

Substrate models are generated in ChemDraw as a chemical structure and then saved in the MDL Molfile format. The resulting file should be converted into the pdb format (save as Ligand.pdb using a National Cancer Institute online converter found at https://cactus.nci.nih.gov). The pdb file of the model should be opened in AutoDockTools-1.5.6 using the menu Ligand/Input/Open Ligand.pdb. During this process, AutoDockTools-1.5.6 program automatically merges nonpolar hydrogen atoms. The **pNPP** model was then saved in the pdbqt format using the following: Ligand/Output/Save as PDBQT, …

2.4 Testing the Possibility of the Substrate Associating with the Protein Chosen as a Starting Point for the Design

Docking experiment needs to be performed to evaluate whether the overall design is feasible and the substrate can be associated with the protein chosen as initial point for the design. A video tutorial on how to use AutoDockTools is available from http://vina.scripps.edu. The key steps are outlined below. The pdb file with the structure of the protein should be opened in AutoDockTools-1.5.6 using the menu items File/Read molecule/Open pdb file. Note that the program will merge nonpolar hydrogen atoms. Most structures do not have hydrogen atoms; if this is the case then polar hydrogens need to be added to the structure using AutoDockTools-1.5.6 (Edit/Hydrogens/Add/Polar only). The file should be then saved in the pdbqt format by going to Grid/Macromolecule/Choose/select pdb file. Search space can be visualized using Grid/Grid box and by changing the center of the grid box ($x=0.106$, $y=1.764$, $z=1.454$ in the example presented here) and a number of points in x, y, z directions (change spacing dial to 1 Å, adjust dimensions of the box to $x=18$, $y=16$, $z=18$). All input parameters were saved in a configuration file (conf.txt) as outlined below, which contains names of the input files (receptor and ligand) and output file (out), geometry of the search space, number of saved output configurations (num_modes), and how much time the program searches for global minimum (# exhaustiveness). At least 20 docking models for each mutant should be generated.

```
receptor = 4BYA.pdbqt
ligand = pNPP.pdbqt

out = all.pdbqt

center_x = 0.106
center_y = 1.764
center_z = 1.454

size_x = 18
size_y = 16
size_z = 18

num_modes = 20
# exhaustiveness = 20
```

pNPP was then docked into AlleyCatE using AutoDock Vina. To start Autodock Vina, open Terminal (use C shell, **csh** command), go to the directory containing Vina executable and run it with the following arguments:

~/vina --config conf.txt --log log.txt

After this step, Vina will run docking and write an output file with the name as specified in conf.txt file. The resulting output file (all.pdbqt) that contains docked substrate poses can be opened in PyMOL together with the pdb file of the structure used to generate the pdbqt file in the same session. Output parameters including the affinity score will be recorded in the log.txt file.

2.5 Identification of the Positions to Mutate

In the first round of computational screening we will assume that every single residue in van der Waals contact with the docked substrate can serve as a place to introduce a catalytic residue. In our experience this provides a reasonably extensive, but not too large, set of possibilities to explore. Importantly, multiple docking poses have to be considered to ensure that all positions are properly identified. While appropriate scripts can be devised, we found that simple examination of the docking structures in PyMOL provides a quick and easy way to determine positions in direct contact with the docked substrate. It is the easiest strategy for inexperienced users.

2.6 Establishing of the Impact of the Mutations on the Overall Fold

While association of the substrate with the protein provides a necessary condition for activity, it is by no means sufficient for catalysis. Next we perform computational examination of the overall impact of the mutations on the protein fold and stability. What if the introduced mutations disrupt the fold so much to render the computational prediction meaningless? Overall protein fold stability plays a major role in applicability of the protein for catalysis. To determine the impact of the introduction of the histidine on stability of CaM, we employed the fixbb algorithm in Rosetta developed by Kuhlman et al. (Leaver-Fay et al., 2011). Fixbb computes a score (thought to be proportional to free energy) for the protein and optimizes rotamers according to the Rosetta score, a procedure that is sometimes referred to as repacking. In our experience, it is necessary to repack initial crystal structure to obtain a reliable baseline. Assuming that Rosetta was successfully installed on a computer with Mac OS operating system under admin_account in the folder

Rosetta the following command will run the fixbb program from the Terminal window (we normally utilize C shell, **csh** command).

/Users/admin_account/rosetta/rosetta_source/bin/fixbb.macosgccreleases 1CLL.pdb -database/Users/admin_account/rosetta/rosetta_database -resfile CaM_repack.txt -nstruct 5 -minimize_sidechains -ex1 -ex2-overwrite

This command will utilize a parameter file CaM_repack.txt and an initial structure from the file 1CLL.pdb. It will generate five pdb structures with minimized side chain orientations and a corresponding score.sc file with output parameters. Detailed description of all optional commands is given on the corresponding tutorial web page:

https://www.rosettacommons.org/manuals/archive/rosetta3.5_user_guide/d4/d68/fixbb.html

The input file set up is critical. A fragment of **CaM_repack.txt** is presented below (the columns are departed by tabs):

start			
#1	A	NATRO	
#2	A	NATRO	
#3	A	NATRO	
4	A	NATRO	
....			
74	A	NATRO	
75	A	NATRO	
76	A	PIKAA	M
77	A	PIKAA	K
...			
94	A	NATRO	

The first column shows amino acid number, it can be commented out using # to be removed from consideration, the second column represents chain identifier followed by a command. **NATRO** command preserves the rotameric state (and, of course, the identity of the original amino acid) for the calculation and it is used for the regions of the protein not involved in catalysis. In this case, EF-hands that bind calcium ions and a disordered

N-terminal region were left intact. **PIKAA** command followed by a one letter code signifying the corresponding amino acid to allow all rotamers for a corresponding residue to be sampled with or without changing its identity was used for the rest of the protein. We keep the identities of the original amino acids intact for repacking of the starting structure. The output of **fixbb** is five pdb structures (the number of output structures can be adjusted by changing **–nstruct**) and a score.sc file. In the score file a number of various parameters is present. We used the overall Rosetta score to evaluate the impact of the mutations on the overall stability. We normally find a very tight distribution of scores in the repacked structures if starting from a high quality structure. The lowest score structure is used as a reference and a starting point for computational mutagenesis.

Mutations in the positions identified during Section 2.5 are then performed using the procedure outlined earlier. Histidine is introduced using the **PICKAA H** command and the score is then evaluated relative to the score of the repacked starting point protein. Mutants with the scores that are lower or approximately the same as the score of the starting protein are selected for further consideration. The definition of "approximately the same" is vague and subjective. When in doubt all possible mutants should be considered. Note that repacked structures produce only best rotamers. Multiple rotameric states need to be considered in certain cases, especially for flexible residues; we evaluate them on a case-by-case basis.

Finally, it is a good idea to perform docking experiments on the obtained mutant to establish whether the newly introduced residue precludes binding.

2.7 Predicting the Feasibility of Catalysis

While protein fold stability is critical for proper positioning of the active residue, for the efficient catalysis to occur the active nucleophile must be present sufficiently close to the carbonyl; moreover the geometry of the nascent coordination site has to be favorable. This presents a very simple docking computational problem. Using the mutants that are determined to be suitable in Section 2.6, we perform docking of the substrate into the computationally created mutant structures in a manner identical to that described in Section 2.4. The distances between nitrogen atom in the histidine ring and the carbonyl carbon in the substrate were measured using the Wizard/Measurement tool in PyMOL. The docking poses of **pNPP** were considered capable of supporting catalysis only if (a) the distance between carbonyl carbon of the ester and the nitrogen atom of imidazole ring (either Nε or Nδ) was less than 3.5 Å and (b) the angle between carbonyl carbon of **pNPP** and the nitrogen atoms of the histidine imidazole was in the range of 90–140 degrees. If the protein conformation with the lowest energy selected from pdb file does not give satisfactory docking models, other conformations need to be tested. This step requires critical evaluation of the possible rotamers. In the case of histidine, only a handful of allowed rotameric states is present, however lysine residues have a very large number of possible rotamers. Then docking should be performed for all feasible rotameric states. In our experience, a large number of poses (50+) is required to properly evaluate feasibility of catalysis. In the case of low complementarity of the docked substrate to protein's cavity this number can be even larger. Nonetheless, even very crude constraints applied in a sequential manner greatly limit the number of mutants for experimental characterization. Importantly, the predictive power of this approach is very high—every single mutant that positioned the functional group in close proximity of the substrate ended up showing above the baseline activity (Moroz et al., 2015, 2013; Raymond et al., 2015).

ACKNOWLEDGMENTS

This work was supported in part by a grant number 1332349 from NSF-EFRI, ORAU Ralph E. Powe Junior Faculty Enhancement award and a Humboldt Research Fellowship to I.V.K.

REFERENCES

Bornscheuer, U. T., Huisman, G., Kazlauskas, R. J., Lutz, S., Moore, J., & Robins, K. (2012). Engineering the third wave of biocatalysis. *Nature*, 485, 185–194.
Brustad, E. M., & Arnold, F. H. (2011). Optimizing non-natural protein function with directed evolution. *Current Opinion in Chemical Biology*, 15, 201–210.
DeGrado, W. F., Wasserman, Z. R., & Lear, J. D. (1989). Protein design, a minimalist approach. *Science*, 243, 622–628.
Korendovych, I. V., & DeGrado, W. F. (2014). Catalytic efficiency of designed catalytic proteins. *Current Opinion in Structural Biology*, 27, 113–121.
Kries, H., Blomberg, R., & Hilvert, D. (2013). De novo enzymes by computational design. *Current Opinion in Chemical Biology*, 17, 221–228.
Leaver-Fay, A., Tyka, M., Lewis, S. M., Lange, O. F., Thompson, J., Jacak, R., et al. (2011). ROSETTA3: An object-oriented software suite for the simulation and design of macromolecules. *Methods in Enzymology*, 487, 545–574.
Mack, K. L., Moroz, O. V., Moroz, Y. S., Olsen, A. B., McLaughlin, J. M., & Korendovych, I. V. (2013). Reprogramming EF-hands for design of catalytically amplified lanthanide sensors. *Journal of Biological Inorganic Chemistry*, 18, 411–418.
Maeda, Y., Makhlynets, O. V., Matsui, H., & Korendovych, I. V. (2016). Non-computational design of catalytic peptides and proteins through rational and combinatorial approaches. *Annual Review of Biomedical Engineering*, 18, 311–328.

Makhlynets, O. V., Raymond, E. A., & Korendovych, I. V. (2015). Design of allosterically regulated protein catalysts. *Biochemistry*, *54*, 1444–1456.

Moroz, Y. S., Dunston, T. T., Makhlynets, O. V., Moroz, O. V., Wu, Y., Yoon, J. H., et al. (2015). New tricks for old proteins: Single mutations in a nonenzymatic protein give rise to various enzymatic activities. *Journal of the American Chemical Society*, *137*, 14905–14911.

Moroz, O. V., Moroz, Y. S., Wu, Y., Olsen, A. B., Cheng, H., Mack, K. L., et al. (2013). A single mutation in a regulatory protein produces evolvable allosterically regulated catalyst of unnatural reaction. *Angewandte Chemie International Edition*, *52*, 6246–6249.

Muller, M. M., Windsor, M. A., Pomerantz, W. C., Gellman, S. H., & Hilvert, D. (2009). A rationally designed aldolase foldamer. *Angewandte Chemie International Edition*, *48*, 922–925.

Nanda, V., & Koder, R. L. (2010). Designing artificial enzymes by intuition and computation. *Nature Chemistry*, *2*, 15–24.

Raymond, E. A., Mack, K. L., Yoon, J. H., Moroz, O. V., Moroz, Y. S., & Korendovych, I. V. (2015). Design of an allosterically regulated retroaldolase. *Protein Science*, *24*, 561–570.

Richter, F., Blomberg, R., Khare, S. D., Kiss, G., Kuzin, A. P., Smith, A. J. T., et al. (2012). Computational design of catalytic dyads and oxyanion holes for ester hydrolysis. *Journal of the American Chemical Society*, *134*, 16197–16206.

Tokuriki, N., & Tawfik, D. S. (2009). Stability effects of mutations and protein evolvability. *Current Opinion in Structural Biology*, *19*, 569–604.

Wijma, H. J., & Janssen, D. B. (2013). Computational design gains momentum in enzyme catalysis engineering. *FEBS Journal*, *280*, 2948–2960.

CHAPTER TEN

Combining Design and Selection to Create Novel Protein–Peptide Interactions

E.B. Speltz[1], N. Sawyer[1,2], L. Regan[3]
Department of Chemistry, Yale University, New Haven, CT, United States
[3]Corresponding author: e-mail address: lynne.regan@yale.edu

Contents

1. Introduction	204
2. Structure-Guided Rational Design	206
3. Split Fluorescent Protein Assays for Identifying PPIs	207
3.1 Screening Protein Libraries Using Split FPs	209
3.2 Quantifying PPIs Using Split FPs	212
3.3 Selecting for PPIs Using Split FPs	213
4. Validation	216
4.1 Cloning, Expression, and Purification of Hits	216
4.2 Qualitative Measurements of Protein–Peptide Interactions	216
4.3 Quantifying PPIs In Vitro	217
4.4 Incorporating Redesign	218
5. Future Perspective: Modular Design of PPI	218
6. Conclusion	218
References	219

Abstract

The ability to design new protein–protein interactions (PPIs) has many applications in biotechnology and medicine. The goal of designed PPIs is to achieve both high affinity and specificity for the target protein. A great challenge in protein design is to identify such proteins from an enormous number of potential sequences. Many computational and experimental methods have been developed to contend with this challenge. Here we describe one particularly powerful approach—semirational design—that combines design and selection. This approach has been applied to generate new PPIs for many applications, including novel affinity reagents for protein detection/purification and bioorthogonal modules for synthetic biology (Jackrel, Valverde, & Regan, 2009; Sawyer et al., 2014; Speltz, Brown, Hajare, Schlieker, & Regan, 2015; Speltz, Nathan, & Regan, 2015).

---

[1] These authors contributed equally to this work.
[2] Current address: Department of Chemistry, New York University, 100 Washington Square East, New York, NY 10003, United States.

1. INTRODUCTION

Protein–protein interactions (PPIs) are vital for all cellular processes. For instance, there are an estimated 30,000–70,000 PPI in *Saccharomyces cerevisiae* that contribute to cell division, cell migration, signal transduction, and metabolism (Gavin et al., 2002; Hart, Ramani, & Marcotte, 2006; Ho et al., 2002; Krogan et al., 2006). There is great interest in engineering PPIs for applications in biotechnology and medicine. Such designs test our fundamental knowledge about the principles of molecular recognition and can also be used to probe PPI networks in living cells.

In the design of novel PPIs, it is critical to delineate the design goals based on their potential applications. Some applications, such as affinity purification, may require high-affinity binding to withstand repeated washing steps. Specificity is also an important consideration, particularly for intracellular applications. In this case, it is necessary to ensure that designs do not exhibit unwanted interactions with cellular components. Finally, if more than one designed PPI is required for an application, it is critical to limit cross-talk between protein pairs.

Many approaches have been developed to contend with the challenge of identifying functional protein sequences from the vast realm of protein sequence space. One can consider PPI design approaches as falling on a spectrum from 100% rational design to 100% selection from large, randomized libraries (Fig. 1). Rational protein design involves mutation of a subset of protein residues to predefined amino acids. This approach requires an intimate knowledge of the complexes' structure and a thorough understanding of the chemical forces that contribute to molecular recognition. While rational protein design has a small number of notable successes, it is not robust with all proteins of interest (Lee et al., 2012). To contend with these limitations, directed evolution uses random mutagenesis followed by stringent selection to identify variants with the desired function from a large pool of nonfunctional variants (Boder & Wittrup, 1997; Bonsor & Sundberg, 2011; Levin et al., 2009). Although this approach circumvents the need for structural information, it requires a very efficient screen/selection and typically requires many intensive rounds of mutagenesis. Furthermore, the randomization protocols involved do not typically target a specific protein interface.

Here, we describe the synergistic combination of rational design and selection. In semirational design, libraries are tailored to mutate specific sequence positions, often focusing on a particular amino acid side chain

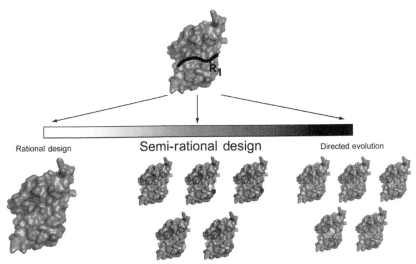

Fig. 1 Spectrum of PPI design strategies. Given a PPI of interest (*top*, with protein shown in surface representation and peptide shown as a *black squiggle*), there are many available design strategies for complementing changes in one of the partners (in this case, a peptide mutation at the R_1 position). At one extreme, rational design (*left*) can be used with detailed structural information to introduce specific amino acid mutations in the protein to interact favorably with the peptide mutation without destabilizing the protein. At the other extreme, directed evolution (*right*) can be used without structural information to produce a randomly mutated protein library. Rare mutants that complement the peptide mutation can be identified by rigorous screening or selection. In between, semirational design (*middle*) uses site-directed randomization to produce a focused protein library with mutations near the peptide mutation site. Screening and/or selection is then performed to identify library members that complement the peptide mutation.

chemistry (eg, hydrophobic, charged, polar). By creating smaller and more functionally relevant libraries, semirational design reduces both the time and resources required for screening and/or selection. The relatively small library size also allows complete library sampling. As with rational design, structural information is required to decide which residues to mutate to achieve the desired behavior (Koide, Gilbreth, Esaki, Tereshko, & Koide, 2007; Koide & Huang, 2013; Koide, Wojcik, Gilbreth, Hoey, & Koide, 2012; Sawyer et al., 2014; Sidhu & Kossiakoff, 2007; Speltz, Nathan, & Regan, 2015). Here, we outline the steps involved in the semirational design of PPIs, including the design and generation of protein libraries, screening and selection, and hit validation (Fig. 2). Each of these steps is described in the following sections.

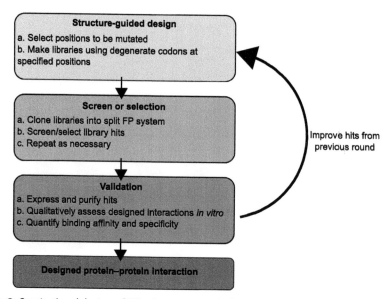

Fig. 2 Semirational design of PPIs. Structure-guided rational design is used to produce a protein library modified at specific positions and a predefined set of amino acids. These libraries are subjected to screening and/or selection for binding to a target protein or peptide using split FPs. Finally, the designed PPIs are validated in vitro using a variety of qualitative and quantitative approaches. If necessary, the biophysical properties of a designed PPI can be improved through iterative cycles of this design approach. The end result is a bona fide PPI.

2. STRUCTURE-GUIDED RATIONAL DESIGN

In discussing our approach, we will focus on the design and selection of tetratricopeptide repeat affinity proteins (TRAPs), which are derived from the ubiquitous tetratricopeptide repeat (TPR) protein motif. TRAPs bind to short linear peptide sequences and are found in all living organisms from *Escherichia coli* to humans (D'Andrea, 2003; Sawyer, Speltz, & Regan, 2013).

TRAPs have many appealing features for the development of protein-based tools and therapeutics, including tunable thermal stability and lack of disulfide bonds permitting intracellular applications (Cortajarena & Regan, 2011; Kajander, Cortajarena, Main, Mochrie, & Regan, 2005). From a design perspective, TRAPs are appealing because of the clear distinction between the residues that form the hydrophobic core and the residues that confer diverse binding affinities and specificities

(Magliery & Regan, 2004; Sawyer, Chen, & Regan, 2013). Thus, one can manipulate functional residues with minimal effects on protein stability.

To reduce library complexity, we typically focus our design approach on specific types of interactions (ie, charge or hydrophobic interactions). Charge complementation, based on the electromagnetic attraction of residues with opposing charges, is often successful and can be incorporated through small library screens (Sawyer et al., 2014; Speltz, Nathan, et al., 2015). Charge–charge interactions depend on the separation between opposing charges and the dielectric constant of the intervening *milieu*. Unlike hydrogen bonds, they do not require stringent distance and bond angle requirements to contribute substantially to the binding free energy. Thus, they are relatively easier to design and incorporate into a protein interface.

The application of our design strategy to hydrophobic interactions has also been successful. These interactions present a protein design challenge because high-affinity interactions require the close packing of nonpolar side chains. However, these designs are easy to incorporate through small library screens using a reduced set of amino acids (Jackrel, Valverde, & Regan, 2009; Speltz, Nathan, et al., 2015). In our experience, we find that specificity in hydrophobic interactions is achieved by designed pockets that match a given amino acid side chain size and shape. For instance, a small hydrophobic pocket may bind small nonpolar residues and sterically prohibit binding of bulky nonpolar side chains. Conversely, a larger pocket will favor large nonpolar residues with high affinity and bind smaller residues more weakly because of the creation of energetically unfavorable cavities.

For constructing TRAP libraries, we typically subdivide the ~400 bp TRAP genes into a set of eight 80–100 bp oligonucleotides with overlapping regions (labeled as primers in Fig. 3). Oligonucleotides with degenerate sites are used to make the libraries (*stars* in Fig. 3). We generate TRAP gene libraries by PCR as shown in Table 1 and Fig. 3.

For screening, we clone these libraries into a split fluorescent protein assay for screening and/or selection (vide infra).

3. SPLIT FLUORESCENT PROTEIN ASSAYS FOR IDENTIFYING PPIS

Protein complementation assays build on a serendipitous observation by Richards that ribonuclease A spontaneously refolds and maintains enzymatic activity even when split into two parts (Richards, 1958). A number of

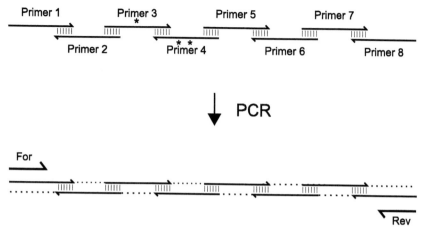

Fig. 3 Generation of semirational protein gene libraries using PCR. (*Top*) The libraries are encoded by eight primers with overlapping ends. Degenerate codons (indicated by *stars*) are used to specify the libraries. (*Bottom*) PCR primers (consisting of a forward and a reverse primer) are used to extend and amplify the inserts.

Table 1 Library Construction Using PCR

Component	Volume (µL)
Insert primers (1 µM stock concentration each)	0.5
Forward primer (25 µM stock concentration)	1
Reverse primer (25 µM stock concentration)	1
5X GC buffer	10
10 mM dNTP mix	5
Phusion DNA polymerase	1
dd H$_2$O	28
Total volume	50

Insert primers refer to the overlapping oligonucleotides with degenerate sites.
5xGC Buffer is 5 × Phusion GC Buffer (NEB #B0519S).
10 mM dNTP mix contains 10 mM each of dATP, dCTP, dGTP, and dTTP in ddH$_2$O.

different protein complementation assays have been developed using enzyme, luminescent protein, and fluorescent protein reporters (Blakeley, Chapman, & McNaughton, 2012; Fan et al., 2008; Filonov & Verkhusha, 2013; Ghosh, Hamilton, & Regan, 2000; Johnsson & Varshavsky, 1994; Magliery et al., 2005; Paulmurugan & Gambhir, 2003; Wilson, Magliery, & Regan, 2004). To avoid false positives, it is critical that the protein fragments do not reassemble unless they are fused to a pair of interacting proteins. Assembly can be monitored in living cells using simple enzymatic assays or spectroscopic methods. We commonly employ split fluorescent protein (FP) assays because fluorescence is easy to detect inside living cells and does not require the addition of exogenous substrates (Fig. 4). An advantage of split FP assays over other in vitro screens or selections is that the assay is performed inside living cells, which favors the isolation of PPIs that are functional in the presence of other cellular components. Furthermore, this assay is available in different colors and is scalable for both low- and high-throughput formats. We have successfully used split green fluorescent protein (GFP) and split mCherry for such assays. Later, we describe methodologies for using split FPs to screen, select, and quantify PPIs in *E. coli*.

3.1 Screening Protein Libraries Using Split FPs

Split FPs have been used widely as screens to monitor PPIs in many cellular systems (Grove, Hands, & Regan, 2010; Jackrel, Cortajarena, Liu, & Regan, 2010; Sarkar & Magliery, 2008; Speltz, Nathan, et al., 2015; Sung & Huh, 2007). Small protein libraries ($<10^3$) can be screened directly on plates of colonies expressing the split FP components (Fig. 4B). An important consideration for screening PPIs using split FPs is that the fluorophore reassembly is irreversible. This feature enables one to screen PPIs with relatively weak binding affinities, which may be beneficial for some applications. It is also critical to choose screening conditions (ie, inducer concentrations and assay time) carefully. Later, we describe protocols for using split GFP to screen protein–peptide interactions with low micromolar binding affinities. For higher affinity interactions, it may be necessary to screen at an earlier time point and/or at lower induction conditions. Conversely, for lower affinity interactions, it may be necessary to screen at a later time point and/or under higher induction conditions.

3.1.1 Cloning

We typically express the two halves of GFP from separate compatible plasmids because it allows for mixing and matching of bait and prey constructs without the need for sequential cloning on a single plasmid (Wilson et al., 2004). The first plasmid expresses the N-terminal half of GFP (NGFP) fused to the N-terminus of a bait protein under the control of a T7 promoter (pET11a-NGFP-link). The second plasmid expresses the C-terminal half of GFP (CGFP) fused to the C-terminus of a prey protein under the control

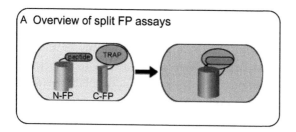

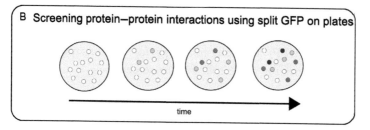

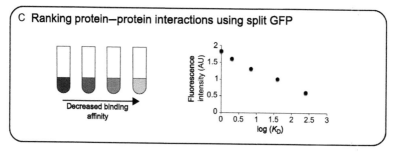

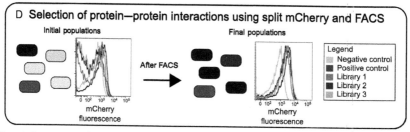

Fig. 4 Overview of screening and selection methods using split fluorescent proteins (FPs). (A) The FP is split into two parts (indicated by complementary cylinder pieces labeled N-FP and C-FP) and fused to a pair of interacting proteins. N-FP is fused to a peptide (shown as an *oblong* linked to N-FP) and C-FP is fused a TRAP (shown as a *yellow circle* linked to C-FP (*gray* in the print version)). Interaction of the TRAP–peptide pair reconstitutes the FP's chromophore, resulting in cellular fluorescence as indicated by the overall *green E. coli* (surrounding *green oblong* (*gray* in the print version)). (B) Split GFP can be used to screen libraries of TRAPs for peptide binding directly on

of an arabinose promoter (pMRBAD-link-CGFP). We illustrate the application of this assay as a screen using TRAP–peptide pairs, as described below.
1. Generate NGFP peptide fusion constructs using PCR. Clone gene inserts into pET11a-NGFP-link, as described previously (Wilson et al., 2004).
2. Generate TRAP gene libraries by PCR as described in Section 2 using degenerate oligonucleotides. Alternatively, one can generate libraries by Klenow extension of overlapping oligonucleotides containing degenerate codons (Jackrel et al., 2009). Clone TRAP gene libraries into pMRBAD-link-CGFP vector using ligation-dependent or ligation-independent cloning techniques (Speltz, Nathan, et al., 2015; Speltz & Regan, 2013; Wilson et al., 2004).

3.1.2 Expression of Split GFP Components
1. Transform pET11a and pMRBAD vectors into BL21-Gold(DE3) *E. coli*. Rescue for 1 h at 30–37°C with shaking. Although any strain lysogenized with DE3 is suitable for expression, we recommend BL21-Gold(DE3) cells because they produce high quality DNA for sequencing.

Note: It is advisable to include both positive and negative controls for screening purposes. As a positive control, we recommend using split GFP plasmids with known interacting proteins with quantified affinities (Addgene plasmids 52734 + 52735, 61966). As negative controls, we recommend transforming plasmids containing NGFP and CGFP without the bait or prey protein, respectively (Addgene plasmids 52732 + 52733, 61965).

---

plates. The earliest colonies to fluoresce (*green colonies* on the second plate (*gray* in the print version)) contain the highest affinity TRAP–peptide pairs. (C) (*Left*) Schematic of using split FPs to quantitatively rank the binding affinity of PPIs. At a particular time point, the most fluorescent samples indicate the highest affinity PPIs. (*Right*) Fluorescence intensity of TRAP–peptide pairs with varying K_D values. Fluorescence intensity correlates linearly with the logarithm of their dissociation constant. (D) Split mCherry can be used to select high affinity TRAP–peptide pairs using FACS. Starting from initial library populations with many TRAP–peptide pairs with widely varying fluorescence intensities, several rounds of FACS yield populations with fluorescence intensity distributions similar to the positive control. The positive control is a known TRAP–peptide interacting pair fused to the split FP components. The negative control has the same TRAP fusion protein but no peptide fused to N-FP.

2. Plate dilutions of the transformations on inducing LB agar plates (LB agar supplemented with 100 μg/mL ampicillin, 50 μg/mL kanamycin, 0.05% arabinose, and 50 μM isopropyl β-D-1-thiogalactopyranoside (IPTG)) and incubate at 25–30°C for 16 h. It is critical to incubate plates below 37°C to improve the folding and solubility of the split GFP components. The target colony density is approximately 100–200 per 15 cm Petri dish.

  Note: To screen a practical number of colonies and to achieve adequate library coverage, we recommend screening 10 × library size. Thus, it will be necessary to plate the libraries on multiple inducing plates. The number of plates required to adequately screen a library is $10S/X$, where X is the average number of colonies per plate and S is the library size.

3.1.3 Screening PPIs in E. coli

For screening purposes, one can either identify the brightest colonies at a given time point or identify the earliest colonies to fluoresce. We find that identifying the first colonies to fluoresce produces fewer false positives (Fig. 4B). We describe this approach in detail below.

1. After ~16 h on plates at 30°C, examine the colonies for fluorescence using a 1 W blue LED flashlight and lab goggles containing a 500 nm LP filter (NightSea Technologies, Lexington, MA). We generally find it useful to mark fluorescent colonies on the bottom of the Petri dish.
2. Repeatedly view and observe colonies for fluorescence every hour until the majority of colonies are fluorescent (~6–8 h). We find it useful to mark colonies on the bottom of a Petri dish with different colors to represent different time points.
3. To identify the protein sequences that gave rise to the first fluorescent colonies, inoculate colonies in LB with 50 μg/mL kanamycin. Isolate the plasmid DNA by miniprep and sequence the TRAP gene insert.

3.2 Quantifying PPIs Using Split FPs

Split FP technologies can also be used as a quantitative tool to estimate the affinity of PPIs in vivo (Fig. 4C). Under appropriate induction conditions, the fluorescence intensity correlates with the binding affinity for a particular class of PPI. This approach has been used to quantify the binding affinity of SH3-peptide, leucine zipper, and TRAP–peptide interactions (Magliery et al., 2005; Morell, Espargaró, Avilés, & Ventura, 2007; Speltz, Nathan, et al., 2015). We describe a protocol to analyze TRAP–peptide interactions with affinities in the low micromolar range using a fluorimeter. This approach can also be adapted to analyze PPIs in a plate reader.

3.2.1 Expression and Quantification of Split GFP Components

1. Cotransform constructs encoding NGFP peptides and CGFP TRAP fusion proteins in an *E. coli* strain lysogenized with DE3. Rescue for 1 h at 30–37°C.
2. Plate on selective LB agar plates (containing 100 μg/mL ampicillin and 50 μg/mL kanamycin).
3. Isolate individual colonies in LB media with appropriate antibiotics. Grow for ~16 h with shaking at 30–37°C.
    Note: Colonies can be grown at 37°C until induction.
4. Dilute the overnight culture 1/100 in LB with appropriate antibiotics at 30°C. Induce the expression of the split GFP components at an OD_{600} of ~0.5–0.8 using 50 μM IPTG and 0.05% arabinose and continue growth for 6 h at 20°C.
5. Harvest the cells by centrifugation at ~5000 × g for 15 min. To remove excess media, wash the cell pellet twice with 1xPBS (phosphate-buffered saline, pH 7.4 contains 137 mM NaCl, 2.7 mM KCl, 4.3 mM Na_2HPO_4, and 1.47 mM KH_2PO_4).
6. Resuspend the cell pellet in 1xPBS to an optical density of 0.3–0.5. Measure the resulting cellular fluorescence using a fluorimeter with excitation at 468 nm and emission detection at 505 nm. Fluorescence should be normalized by optical density.
    Note: To quantify PPIs using split GFP, it is important to ensure that the fluorescence intensity correlates linearly with optical density for a given sample. To do so, measure both optical density and fluorescence intensity for serial dilutions of each sample in PBS. Standard curves of fluorescence vs. optical density can be used to normalize other samples in the same set of experiments.

3.3 Selecting for PPIs Using Split FPs

Split FP technologies can also be coupled to fluorescence-activated cell sorting (FACS) for selection of PPIs (Fig. 4D). This approach is recommended for libraries with >10^3 members. The combination of split FPs and FACS has been used successfully with both split GFP and split mCherry (Cortajarena, Liu, Hochstrasser, & Regan, 2010; Jackrel et al., 2010; Sawyer et al., 2014). Later, we describe a general protocol for FACS selection using split mCherry.

3.3.1 Cloning

This technique is performed using a duet vector (pNAS1B, Addgene plasmid 61965), which contains both the N- and C-terminal halves of mCherry (NmC and CmC, respectively) under the respective control of the P_{LtetO} and P_{BAD} promoters. Cloning of genes encoding protein and peptide interaction pairs is performed sequentially as described below:
1. Generate a gene encoding NmC tagged with the target peptide by PCR amplification (Sawyer et al., 2014). Clone it as desired into the pNAS1B vector using ligation-dependent or ligation-independent cloning techniques.
2. Generate TRAP gene libraries by PCR as described in Section 2.
3. Purify the PCR product. Digest the PCR product and the pNAS1B duet vector containing the NmC-tagged target peptide gene with AatII and NotI-HF. Ligate the PCR product and vector.
4. Transform several microliters of ligation mixture directly into the expression strain. The exact volume of ligation mixture transformed should be optimized based on transformation efficiency. Transformation efficiency can be estimated by plating on selective media and counting colonies. In-frame ligation can be estimated by picking individual colonies and sequencing.
5. Incubate the 500 µL transformation mixture at 30°C with shaking for 1 h. Then, inoculate the transformation mixture into 4.5 mL 2xYT with antibiotics as a starter culture and grow for 24 h.
    Note: These conditions were optimized for a recoded *E. coli* strain with a slow growth phenotype (Heinemann et al., 2012; Lajoie et al., 2013; Sawyer et al., 2014). Growth temperature and time may need to be optimized for normal *E. coli* strains (ie, 12–16 h growth at 37°C).

3.3.2 Selection

In one round of FACS, a stream of suspended cells is passed through a narrow nozzle and fragmented into droplets, with the goal of separating each cell into a different droplet. The fluorescence of each droplet is then analyzed and sorted based on fluorescence intensity. We typically collect the top 1% of droplets in each round, and we perform multiple rounds of sorting to enrich the final population for the most fluorescent cells.
1. Inoculate 1 mL of starter culture into 100 mL 2xYT containing antibiotics and grow to $OD_{600} = 0.8–1.0$. Add 0.1% (w/v) arabinose and 100 ng/mL anhydrotetracycline to induce expression of split mCherry components. Incubate with shaking. Incubation time and temperature

vary depending on the expression strain and fusion proteins. For instance, TRAP–phosphopeptide interactions were studied with shaking at 20°C for 16 h (Sawyer et al., 2014).

2. Harvest cells by centrifugation and wash three times with 1xPBST (1xPBS containing 0.1% Tween-20).
3. Prepare several dilutions (1:1000–1:100,000) of the final cell pellet in 1xPBST to achieve appropriate concentration for flow cytometry (target event rate is 5000–10,000 per second).
4. Inject cells into a flow cytometer. A 561 nm laser and 610/40 band pass filter are recommended for mCherry. Use a small nozzle (eg, 70 μm) to minimize the number of droplets containing multiple cells. Analyze a test population (approximately 10,000 cells), and set a sorting threshold based on the fluorescence profile. Typically, the threshold is set such that the most fluorescent 1% of cells is collected.
5. Sort cells into 5 mL of 2xYT media. Inoculate the collected cells in 45 mL of fresh media to make a new starter culture. Grow for 24 h at 30°C with shaking.
6. Repeat steps 1–5.
7. After the final round of selection, plate an aliquot of the final sorting mixture on antibiotic containing media to obtain isolated colonies. Create glycerol stocks from the remainder.

   Note: The number of sorting rounds required to enrich for PPI interactions with the desired affinity depends on the desired affinity and the fluorescence separation between negative and positive control samples. In the case of TRAP–peptide interactions, the positive control is a known TRAP–peptide pair fused to CmC and NmC, respectively. The negative control is the same TRAP fused to CmC but no peptide fused to NmC. The number of rounds can be determined in one of two ways. The first and more preferred strategy is to assess the overlap of the library event fluorescence histogram with a positive control event fluorescence histogram (ie, a known PPI with the desired affinity expressed in split mCherry format). Once the library histogram matches or surpasses the positive control, the population should be dominated by PPI interactions with the desired affinity. Alternatively, if this goal is not achieved, sorting should be stopped once the library event fluorescence histogram does not change significantly between consecutive rounds.
8. To obtain TRAP gene sequences, pick individual colonies, isolate plasmid DNA, and sequence the TRAP gene insert.

9. Subclone TRAP genes into the original pNAS1B vector from "Cloning," part A. Perform the split mCherry enhanced reassembly and fluorescence assay, as described previously (Sawyer et al., 2014) to confirm that the TRAP gene, and not other mutations acquired elsewhere in the plasmid, is responsible for the increase in fluorescence. This step is particularly important after many sorting rounds.

After each round of FACS selection, one can make glycerol stocks from the combined starter culture (step 6) to assess library diversity by sequencing.

4. VALIDATION

All screens and selections are prone to false positives. For protein complementation assays, false positives can result from nonspecific interactions between components in the assay or result from interactions with other cellular components that improve fragment reassembly. We typically validate all designed PPIs in vitro. Later, we provide a general approach to analyze the fidelity of a series of designed protein–peptide interactions.

4.1 Cloning, Expression, and Purification of Hits

For these applications, we typically purify both the prey (TRAP) and bait (peptide) protein in recombinant form in *E. coli*.

1. We typically subclone TRAP hits into an expression vector with a TEV protease-cleavable hexahistidine tag (such as the pPROEX-HTa vector). The protein can then be expressed in *E. coli* and purified using Ni-NTA resin following the manufacturer's recommendations.
2. We typically fuse peptide sequences via a flexible peptide linker to glutathione-S-transferase (GST) to enable recombinant expression and purification. The GST tag is also used for pull-down assays as described in Section 4.2. As a negative control, we recommend constructing, expressing, and purifying unfused GST proteins or GST fusion proteins with a mutated peptide sequence that was not used during the screen or selection process. GST proteins can be expressed in *E. coli* and purified using Glutathione Sepharose resin according to manufacturer's recommendations.

4.2 Qualitative Measurements of Protein–Peptide Interactions

As a first test, we typically analyze the fidelity of designed interactions using pull-down assays, which allow us to quickly and directly compare the binding affinity and specificity of many hits from a screen or selection. Below we describe a general pull-down assay protocol using GST-tagged peptides and TRAP proteins as specific examples.

1. Incubate 2 nmol GST-peptide with ~50 μL bed volume of glutathione sepharose resin in 200 μL 1xPBS for 1 h with gentle shaking at 4°C.
2. Centrifuge resin at ~6000 × g for 1 min. Pipette off the supernatant, taking care not to disturb the resin. Add 200 μL 1xPBS to the beads. This is one wash.
3. Repeat step 2 four times for a total of five washes.
4. Incubate resin with 2 nmol TRAP protein in 200 μL 1xPBS for 1 h with gentle shaking at 4°C.
    Note: The concentration of the prey protein may need to be optimized for individual systems and depends on the affinity of a given interaction.
5. Repeat step 2 five times.
    Note: The number of washes used in this step will depend on the affinity of the PPI and unexpected, nonspecific interactions with assay components. It will need to be optimized for each system.
6. Elute protein bound to the glutathione beads with 15 μL SDS dye, 10 μL water, and boiling for 10 min. Load 15 μl of eluate onto a SDS-PAGE gel, perform gel electrophoresis, and stain for protein using Coomassie Blue. The amount of TRAP protein bound to a target peptide provides a qualitative measure of the binding affinity of the interaction.

4.3 Quantifying PPIs In Vitro

Pull-down assays provide a convenient and rapid qualitative assessment of library hits. As a complementary approach, we use many biophysical approaches to quantify the binding affinity and specificity of hits. The three most common methods, isothermal titration calorimetry (ITC), surface plasmon resonance (SPR), and fluorescence anisotropy (FA), have been compared in detail previously (Rossi & Taylor, 2011). We choose the method that is most suited to a particular application. A few considerations are whether a label is required and the amount of time and resources required for the measurements. Of these three methods, ITC is the only label-free method; however, it is also the most time- and resource-intensive. SPR and FA require relatively less purified material; however, they also require significant modifications to the bait peptide or protein. Thus, it is necessary to perform additional experiments to ensure that there are no artifacts arising from these modifications. Given these limitations, we typically use one or a combination of these methods to quantify the binding affinity and specificity of designed PPIs.

4.4 Incorporating Redesign

The strategy outlined here typically produces bona fide PPIs. However, the designed interactions may not have the desired affinity or specificity. We incorporate redesign into our approach by iterative cycles of Sections 2–4, which can be used to improve the biophysical properties of a designed PPI (Fig. 2). In this case, we use a hit from a previous round as the starting point for design. Additionally, iterative design cycles can be used to design new PPI as outlined in Section 5 vide infra.

5. FUTURE PERSPECTIVE: MODULAR DESIGN OF PPI

The development of modular protein–peptide interactions is an attractive concept (Reichen, Hansen, & Plückthun, 2014). Here, the term "modular" refers to a scenario in which each peptide segment (eg, amino acid residue) interacts with a distinct protein binding pocket and does not interact with other peptide segments or protein regions. Such 1:1 pairings between peptide segments and protein regions are similar to modular protein–nucleic acid interactions observed in transcription activator-like effector nucleases, pentatricopeptide repeat proteins, Pumilio homology domains, and zinc finger repeats (Bhakta & Segal, 2010; Li et al., 2011; Wang, McLachlan, Zamore, & Hall, 2002; Yin et al., 2013) (Fig. 5).

A major benefit of modular protein–peptide interactions is the opportunity to mix and match designed binding pockets combinatorially to generate proteins that bind to a predefined set of peptides. This strategy dramatically decreases the selection burden on the researcher, saving valuable time and resources. Although the modular design of PPIs has yet to be realized, some recent exciting work demonstrates that this strategy may be a possibility in the future (Sawyer, Chen, & Regan, 2013; Speltz, Nathan, et al., 2015; Varadamsetty, Tremmel, Hansen, Parmeggiani, & Plückthun, 2012).

6. CONCLUSION

Semirational PPI design provides an invaluable tool to expand Nature's PPI repertoire. The design of novel PPIs provides valuable insight into the underlying basis of molecular recognition. Such designs also provide novel tools to study PPIs in vitro and in living cells.

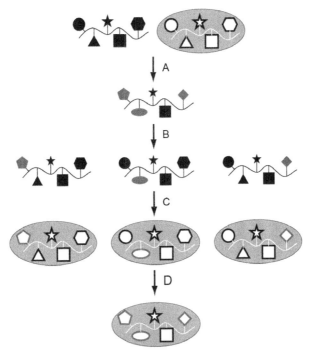

Fig. 5 Modular design of PPIs. The design begins with a characterized TRAP–peptide interaction pair, indicated by the *gray oval* and the *squiggle*, respectively. The peptide binds to the TRAP in an extended conformation such that each of the peptide side chains (indicated by *shapes*) interact with distinct regions of the TRAP surface (indicated by *complementary cut-outs*). In step (A), a new target peptide of interest is identified, in this case with three mutations relative to the initial peptide (indicated as *different red shapes* (*dark gray* in the print version)). In the first modular design step (B), where three intermediate target peptides are generated from the new target peptide. These peptides are identical to the initial peptide except with one mutation each from the new target peptide. In the second modular design step (C), TRAP libraries are generated to complement each intermediate peptide. TRAP binding pockets that complement the peptide mutation are identified by screening and/or selection, as indicated by the *complementary red* (*dark gray* in the print version) *cut-outs*. In the final modular design step (D), TRAP binding pockets are combined on a single TRAP to generate a TRAP that binds the new target peptide.

REFERENCES

Bhakta, M. S., & Segal, D. J. (2010). The generation of zinc finger proteins by modular assembly. In J. P. Mackay & D. J. Segal (Eds.), *Engineered zinc finger proteins: Vol. 649* (pp. 3–30). Totowa, NJ: Humana Press.

Blakeley, B. D., Chapman, A. M., & McNaughton, B. R. (2012). Split-superpositive GFP reassembly is a fast, efficient, and robust method for detecting protein–protein interactions in vivo. *Molecular BioSystems*, 8(8), 2036–2040.

Boder, E. T., & Wittrup, K. D. (1997). Yeast surface display for screening combinatorial polypeptide libraries. *Nature Biotechnology*, 15(6), 553–557.

Bonsor, D. A., & Sundberg, E. J. (2011). Dissecting protein–protein interactions using directed evolution. *Biochemistry, 50*(13), 2394–2402.

Cortajarena, A. L., Liu, T. Y., Hochstrasser, M., & Regan, L. (2010). Designed proteins to modulate cellular networks. *ACS Chemical Biology, 5*(6), 545–552.

Cortajarena, A. L., & Regan, L. (2011). Calorimetric study of a series of designed repeat proteins: Modular structure and modular folding. *Protein Science, 20*(2), 336–340.

D'Andrea, L. (2003). TPR proteins: The versatile helix. *Trends in Biochemical Sciences, 28*(12), 655–662.

Fan, J.-Y., Cui, Z.-Q., Wei, H.-P., Zhang, Z.-P., Zhou, Y.-F., Wang, Y.-P., et al. (2008). Split mCherry as a new red bimolecular fluorescence complementation system for visualizing protein–protein interactions in living cells. *Biochemical and Biophysical Research Communications, 367*(1), 47–53.

Filonov, G. S., & Verkhusha, V. V. (2013). A near-infrared BiFC reporter for in vivo imaging of protein–protein interactions. *Chemistry & Biology, 20*(8), 1078–1086.

Gavin, A.-C., Bösche, M., Krause, R., Grandi, P., Marzioch, M., Bauer, A., et al. (2002). Functional organization of the yeast proteome by systematic analysis of protein complexes. *Nature, 415*(6868), 141–147.

Ghosh, I., Hamilton, A. D., & Regan, L. (2000). Antiparallel leucine zipper-directed protein reassembly: Application to the green fluorescent protein. *Journal of the American Chemical Society, 122*(23), 5658–5659.

Grove, T. Z., Hands, M., & Regan, L. (2010). Creating novel proteins by combining design and selection. *Protein Engineering, Design and Selection, 23*(6), 449–455.

Hart, G. T., Ramani, A. K., & Marcotte, E. M. (2006). How complete are current yeast and human protein interaction networks? *Genome Biology, 7*(11), 120.

Heinemann, I. U., Rovner, A. J., Aerni, H. R., Rogulina, S., Cheng, L., Olds, W., et al. (2012). Enhanced phosphoserine insertion during *Escherichia coli* protein synthesis via partial UAG codon reassignment and release factor 1 deletion. *FEBS Letters, 586*(20), 3716–3722.

Ho, Y., Gruhler, A., Heilbut, A., Bader, G. D., Moore, L., Adams, S.-L., et al. (2002). Systematic identification of protein complexes in *Saccharomyces cerevisiae* by mass spectrometry. *Nature, 415*(6868), 180–183.

Jackrel, M. E., Cortajarena, A. L., Liu, T. Y., & Regan, L. (2010). Screening libraries to identify proteins with desired binding activities using a split-GFP reassembly assay. *ACS Chemical Biology, 5*(6), 553–562.

Jackrel, M. E., Valverde, R., & Regan, L. (2009). Redesign of a protein–peptide interaction: Characterization and applications. *Protein Science, 18*(4), 762–774.

Johnsson, N., & Varshavsky, A. (1994). Split ubiquitin as a sensor of protein interactions in vivo. *Proceedings of the National Academy of Sciences of the United States of America, 91*(22), 10340–10344.

Kajander, T., Cortajarena, A. L., Main, E. R. G., Mochrie, S. G. J., & Regan, L. (2005). A new folding paradigm for repeat proteins. *Journal of the American Chemical Society, 127*(29), 10188–10190.

Koide, A., Gilbreth, R. N., Esaki, K., Tereshko, V., & Koide, S. (2007). High-affinity single-domain binding proteins with a binary-code interface. *Proceedings of the National Academy of Sciences of the United States of America, 104*(16), 6632–6637.

Koide, S., & Huang, J. (2013). Generation of high-performance binding proteins for peptide motifs by affinity clamping. *Methods in Enzymology, 523*, 285–302.

Koide, A., Wojcik, J., Gilbreth, R. N., Hoey, R. J., & Koide, S. (2012). Teaching an old scaffold new tricks: Monobodies constructed using alternative surfaces of the FN3 scaffold. *Journal of Molecular Biology, 415*(2), 393–405.

Krogan, N. J., Cagney, G., Yu, H., Zhong, G., Guo, X., Ignatchenko, A., et al. (2006). Global landscape of protein complexes in the yeast *Saccharomyces cerevisiae*. *Nature*, *440*(7084), 637–643.

Lajoie, M. J., Kosuri, S., Mosberg, J. A., Gregg, C. J., Zhang, D., & Church, G. M. (2013). Probing the limits of genetic recoding in essential genes. *Science*, *342*(6156), 361–363.

Lee, S.-C., Park, K., Han, J., Lee, J.-J., Kim, H. J., Hong, S., et al. (2012). Design of a binding scaffold based on variable lymphocyte receptors of jawless vertebrates by module engineering. *Proceedings of the National Academy of Sciences of the United States of America*, *109*(9), 3299–3304.

Levin, K. B., Dym, O., Albeck, S., Magdassi, S., Keeble, A. H., Kleanthous, C., et al. (2009). Following evolutionary paths to protein–protein interactions with high affinity and selectivity. *Nature Structural and Molecular Biology*, *16*(10), 1049–1055.

Li, T., Huang, S., Zhao, X., Wright, D. A., Carpenter, S., Spalding, M. H., et al. (2011). Modularly assembled designer TAL effector nucleases for targeted gene knockout and gene replacement in eukaryotes. *Nucleic Acids Research*, *39*(14), 6315–6325. http://doi.org/10.1093/nar/gkr188.

Magliery, T. J., & Regan, L. (2004). Beyond consensus: Statistical free energies reveal hidden interactions in the design of a TPR motif. *Journal of Molecular Biology*, *343*(3), 731–745.

Magliery, T. J., Wilson, C. G. M., Pan, W., Mishler, D., Ghosh, I., Hamilton, A. D., et al. (2005). Detecting protein–protein interactions with a green fluorescent protein fragment reassembly trap: Scope and mechanism. *Journal of the American Chemical Society*, *127*(1), 146–157.

Morell, M., Espargaró, A., Avilés, F. X., & Ventura, S. (2007). Detection of transient protein–protein interactions by bimolecular fluorescence complementation: The Abl-SH3 case. *Proteomics*, *7*(7), 1023–1036.

Paulmurugan, R., & Gambhir, S. S. (2003). Monitoring protein–protein interactions using split synthetic renilla luciferase protein-fragment-assisted complementation. *Analytical Chemistry*, *75*(7), 1584–1589.

Reichen, C., Hansen, S., & Plückthun, A. (2014). Modular peptide binding: From a comparison of natural binders to designed armadillo repeat proteins. *Journal of Structural Biology*, *185*(2), 147–162.

Richards, F. M. (1958). On the enzymic activity of subtilisin-modified ribonuclease. *Proceedings of the National Academy of Sciences of the United States of America*, *44*(2), 162–166.

Rossi, A. M., & Taylor, C. W. (2011). Analysis of protein–ligand interactions by fluorescence polarization. *Nature Protocols*, *6*(3), 365–387.

Sarkar, M., & Magliery, T. J. (2008). Re-engineering a split-GFP reassembly screen to examine RING-domain interactions between BARD1 and BRCA1 mutants observed in cancer patients. *Molecular BioSystems*, *4*(6), 599–605.

Sawyer, N., Chen, J., & Regan, L. (2013). All repeats are not equal: A module-based approach to guide repeat protein design. *Journal of Molecular Biology*, *425*(10), 1826–1838.

Sawyer, N., Gassaway, B. M., Haimovich, A. D., Isaacs, F. J., Rinehart, J., & Regan, L. (2014). Designed phosphoprotein recognition in *Escherichia coli*. *ACS Chemical Biology*, *9*(11), 2502–2507.

Sawyer, N., Speltz, E. B., & Regan, L. (2013). NextGen protein design. *Biochemical Society Transactions*, *41*(5), 1131–1136.

Sidhu, S. S., & Kossiakoff, A. A. (2007). Exploring and designing protein function with restricted diversity. *Current Opinion in Chemical Biology*, *11*(3), 347–354.

Speltz, E. B., Brown, R. S. H., Hajare, H. S., Schlieker, C., & Regan, L. (2015). A designed repeat protein as an affinity capture reagent. *Biochemical Society Transactions*, *43*(5), 874–880.

Speltz, E. B., Nathan, A., & Regan, L. (2015). Design of protein–peptide interaction modules for assembling supramolecular structures in vivo and in vitro. *ACS Chemical Biology, 10*(9), 2108–2115.

Speltz, E. B., & Regan, L. (2013). White and green screening with circular polymerase extension cloning for easy and reliable cloning. *Protein Science, 22*(6), 859–864.

Sung, M.-K., & Huh, W.-K. (2007). Bimolecular fluorescence complementation analysis system for in vivo detection of protein–protein interaction in *Saccharomyces cerevisiae*. *Yeast (Chichester, England), 24*(9), 767–775.

Varadamsetty, G., Tremmel, D., Hansen, S., Parmeggiani, F., & Plückthun, A. (2012). Designed armadillo repeat proteins: Library generation, characterization and selection of peptide binders with high specificity. *Journal of Molecular Biology, 424*(1–2), 68–87.

Wang, X., McLachlan, J., Zamore, P. D., & Hall, T. M. T. (2002). Modular recognition of RNA by a human pumilio-homology domain. *Cell, 110*(4), 501–512.

Wilson, C. G. M., Magliery, T. J., & Regan, L. (2004). Detecting protein–protein interactions with GFP-fragment reassembly. *Nature Methods, 1*(3), 255–262.

Yin, P., Li, Q., Yan, C., Liu, Y., Liu, J., Yu, F., et al. (2013). Structural basis for the modular recognition of single-stranded RNA by PPR proteins. *Nature, 504*(7478), 168–171.

CHAPTER ELEVEN

Metal-Directed Design of Supramolecular Protein Assemblies

J.B. Bailey, R.H. Subramanian, L.A. Churchfield, F.A. Tezcan[1]

University of California, San Diego, La Jolla, CA, United States
[1]Corresponding author: e-mail address: tezcan@ucsd.edu

Contents

1. Introduction	224
2. Metal-Directed Protein Self-Assembly	227
2.1 Choosing a Protein Building Block for MDPSA	227
2.2 MDPSA Using Metal Chelating Motifs Composed of Natural Amino Acids	228
2.3 MDPSA Using Synthetic Metal-Coordinating Ligands	230
3. Metal-Templated Interface Redesign	231
3.1 Redesign of Noncovalent Interfaces with MeTIR	231
3.2 Disulfide Cross-Links to Enhance Scaffold Robustness	233
3.3 In Vivo Assembly and Functional Screening of Metal-Templated Protein Assemblies	234
4. Methods	237
4.1 Cytochrome cb_{562} Variant Expression and Purification	237
4.2 Iodoacetamide Ligand Synthesis and Protein Labeling	238
4.3 Determining Oligomeric State by Sedimentation Velocity Experiments	240
4.4 Measuring Protein–Protein Affinities with Sedimentation Equilibrium Experiments	243
4.5 Crystallization of cyt cb_{562} Variants	243
4.6 Periplasmic Extraction of cyt cb_{562} Assemblies from *E. coli*	244
4.7 In Vivo Screening of cyt cb_{562} Assemblies with β-Lactamase Activity	245
5. Conclusion	247
Acknowledgments	247
References	247

Abstract

Owing to their central roles in cellular signaling, construction, and biochemistry, protein–protein interactions (PPIs) and protein self-assembly have become a major focus of molecular design and synthetic biology. In order to circumvent the complexity of constructing extensive noncovalent interfaces, which are typically involved in natural PPIs and protein self-assembly, we have developed two design strategies, metal-directed protein self-assembly (MDPSA) and metal-templated interface redesign

(MeTIR). These strategies, inspired by both the proposed evolutionary roles of metals and their prevalence in natural PPIs, take advantage of the favorable properties of metal coordination (bonding strength, directionality, and reversibility) to guide protein self-assembly with minimal design and engineering. Using a small, monomeric protein (cytochrome cb_{562}) as a model building block, we employed MDPSA and MeTIR to create a diverse array of functional supramolecular architectures which range from structurally tunable oligomers to metalloprotein complexes that can properly self-assemble in living cells into novel metalloenzymes. The design principles and strategies outlined herein should be readily applicable to other protein systems with the goal of creating new PPIs and protein assemblies with structures and functions not yet produced by natural evolution.

ABBREVIATIONS

CA-quin N-(8-hydroxyquinolin-5-yl)-2-chloroacetamide
ccm cytochrome *c* maturation cassette
Cyt cb_{562} cytochrome cb_{562}
DCC N,N'-dicyclohexylcarbodiimide
DTT dithiothreitol
EDTA ethylenediaminetetraacetic acid
FPLC fast protein liquid chromatography
HCM hybrid coordination motif
IA-phen 2-iodo-N-(1,10-phenanthroline-5-yl)acetamide
IA-quin N-(8-hydroxyquinolin-5-yl)-2-iodoacetamide
IA-terpy N-([2,2':6',2''-terpyridin]-4'-yl)-2-iodoacetamide
LB lysogeny broth
MBPC1 metal-binding protein construct 1
MDPSA metal-directed protein self-assembly
MeTIR metal-templated interface redesign
PCR polymerase chain reaction
PDB protein data bank
Phen 1,10-phenanthroline
PPI protein–protein interaction
Quin 8-hydroxyquinoline
RIDC1 Rosetta interface designed cytochrome 1
RZ reinheitzahl
SE-AUC sedimentation equilibrium analytical ultracentrifugation
SV-AUC sedimentation velocity analytical ultracentrifugation
Terpy 2,2':6'2''-terpyridine
TRIS 2-amino-2-hydroxymethyl-propane-1,3-diol

1. INTRODUCTION

Proteins are nature's most versatile building blocks for the construction of supramolecular architectures which fulfill essential structural and catalytic roles in cellular metabolism, signaling, and biomaterial formation (Petsko & Ringe, 2004). The protein–protein interactions (PPIs) that mediate the self-assembly of these protein architectures occur primarily through the formation of numerous noncovalent interfacial contacts (salt bridges, hydrophobic contacts, hydrogen bonds) that extend over large molecular surfaces (Conte, Chothia, & Janin, 1999; Fletcher & Hamilton, 2006; Tsai, Lin, Wolfson, & Nussinov, 1996; Xu, Tsai, & Nussinov, 1997). There has been considerable progress in the de novo design of PPIs and protein assemblies through computationally guided approaches (Der et al., 2012; Gonen, DiMaio, Gonen, & Baker, 2015; Huang, Love, & Mayo, 2007; King et al., 2014; Mandell & Kortemme, 2009; Mou, Huang, Hsu, Huang, & Mayo, 2015; Schreiber & Fleishman, 2013). Yet, it still remains a formidable challenge to accurately capture the scope, specificity, and functionality of natural PPIs and protein assemblies purely through the design of noncovalent interactions.

Transition metal ions are powerful molecular templating/directing agents because of their ability to form strong bonds (relative to noncovalent interactions) which are highly directional and reversible (Harding, Nowicki, & Walkinshaw, 2010; Rulíšek & Vondrášek, 1998). Accordingly, installing chelating motifs on a protein provides a simple route to creating conditionally interactive protein surfaces without extensive design and engineering (Fig. 1). In fact, metal ions have been proposed to serve as templates to guide the formation of PPIs during the structural/functional evolution of some protein superfamilies. In such systems, it has been suggested that metals directed the formation of a simple aggregate species that subsequently evolved a more specialized binding interface around the metal ion (Liu & Xu, 2002). For example, the vicinal oxygen chelate superfamily, a diverse group of metalloproteins (Armstrong, 2000; Bergdoll, Eltis, Cameron, Dumas, & Bolin, 1998), as well as the quintessential oxygen transport proteins (hemocyanin, hemerythrin, and hemoglobin; Volbeda & Hol, 1989) are thought to have originated from a single metal-bridged dimeric precursor, which underwent extensive rearrangement of the overall fold and metal-containing active site to give descendant proteins with diverse structures and functions. It is estimated that at least 4–5% of all structurally characterized protein oligomers contain a mid- to late-first row transition metal ion (Mn, Fe, Co, Ni, Cu, and Zn) or a metallocofactor in an interface that serves an essential structural and/or functional role (Song, Sontz, Ambroggio, & Tezcan, 2014).

Taking this hypothetical evolutionary trajectory as a synthetic blueprint, we have implemented the advantageous properties of metal coordination

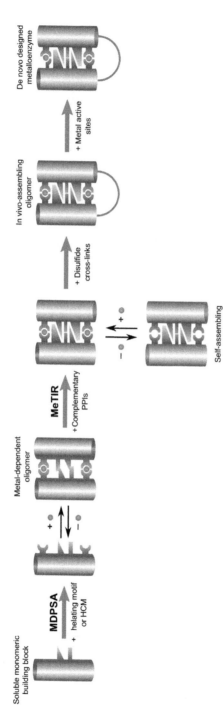

Fig. 1 Scheme showing the iterative engineering of a soluble, nonself-associating protein building block by MDPSA and MeTIR. By MDPSA, chelating motifs (eg, bis-His clamps) can be incorporated into the protein surface, enabling the formation of metal-directed oligomers. By MeTIR, the nascent interfaces in these oligomers can be reinforced with the installation of complementary PPIs to generate self-assembling complexes that can form in the absence of templating metals. The installation of disulfide cross-links increases scaffold robustness and the fidelity of self-assembly. These robust scaffolds can accommodate reactive metal sites to generate de novo designed metalloenzymes.

(bonding strength, directionality, and reversibility) for developing a new strategy to design PPIs and self-assembled protein architectures: metal-directed protein self-assembly (MDPSA). In proof-of-principle studies, MDPSA was applied to a small monomeric protein, cytochrome cb_{562} (cyt cb_{562}), to generate discrete metal-mediated complexes whose oligomerization state and geometry were strictly controlled by metal coordination geometry (Salgado, Faraone-Mennella, & Tezcan, 2007; Salgado, Lewis, Faraone-Mennella, & Tezcan, 2008). The newly generated PPIs in these metal-mediated complexes provided a platform for the subsequent installation of favorable covalent and noncovalent interactions, an approach we have termed metal-templated interface redesign (MeTIR; Salgado et al., 2010). Together, the MDPSA and MeTIR approaches provide a straightforward, yet powerful route for the design of new metalloprotein scaffolds and metalloenzymes.

2. METAL-DIRECTED PROTEIN SELF-ASSEMBLY

2.1 Choosing a Protein Building Block for MDPSA

The choice of the protein building block is critical to successfully developing metal-mediated protein complexes through MDPSA. Ideally, a protein should possess the following characteristics:
- Recombinantly expressible in high yields
- Nonself-associating even at high concentrations (so as to eliminate nonspecific aggregation)
- Thermodynamically stable (so as to enable extensive surface modifications/mutations)
- Uniform overall topology (so as to promote the formation of supramolecular assemblies with well-defined topologies and to facilitate crystallographic investigation)

In proof-of-principle studies, we used a model building block, cyt cb_{562} that fit all these criteria (Section 4.1). Cyt cb_{562} is a highly stable, monomeric electron-transfer protein containing a c-type heme cofactor. The protein adopts a classic 4-helix bundle fold with a rigid, cylindrical shape (Faraone-Mennella, Tezcan, Gray, & Winkler, 2006). Given such a simple uniform architecture, we could readily envision arranging multiple copies of this protein to rationally design larger assemblies mediated by strategically placed intermolecular interactions.

2.2 MDPSA Using Metal Chelating Motifs Composed of Natural Amino Acids

Once a suitable building block like cyt cb_{562} has been selected, a region on its surface can be chosen to implement MDPSA (Fig. 1). MDPSA takes advantage of the chelate effect through the installation of multidentate metal coordination motifs onto the protein surface. Such motifs can outcompete the many monodentate metal-binding functionalities present on the surface of a typical protein. α-Helices are well suited for this purpose as their structural periodicity readily allows the pairing of histidine (His), aspartate (Asp), glutamate (Glu), and cysteine (Cys) residues at i and $i+4$ or i and $i+3$ positions to create bidentate motifs (clamps) with low-micromolar metal-binding affinities (Arnold & Haymore, 1991; Ghadiri & Choi, 1990; Handel, Williams, & Degrado, 1993; Krantz & Sosnick, 2001). In principle, installation of chelating motifs is possible on surface-facing β-strands through i and $i+2$ pairings of His, Asp, Glu, and Cys side chains, or any two appropriately oriented residues on a well-defined surface feature (Harding, 2004).

For installing metal-binding motifs on a building block, protein–protein packing interactions in the crystal lattice can provide a good structural guide to select an appropriate region of the protein surface. In the case of cyt cb_{562}, each protein monomer is paired in an antiparallel fashion with another monomer along the helix–helix face (residues 56–81) formed by Helix3 (Fig. 2). Given these extensive packing interactions (~775 Å^2 of buried surface area), we envisioned incorporating bis-His clamps at both ends of Helix3 to facilitate metal-mediated oligomerization. Since residues 59 and 63 in one monomer oppose residues 73 and 77 of another, a pair of bis-His clamps was installed around the existing H63 with three additional mutations (W59H, D73H, K77H), affording metal-binding protein construct 1 (MBPC1; Fig. 2; Salgado et al., 2007).

A central tenet of protein self-assembly is the reversibility of PPIs such that kinetic traps are avoided and desired thermodynamically favored architectures can be formed under ambient conditions. The fulfillment of this design feature in MDPSA requires the use of metal ions that strongly associate with amino acid residues, yet possess fast ligand exchange rates (ie, are substitution-labile). These requirements are best met by late-first-row transition metal ions (Co^{2+}, Ni^{2+}, Cu^{2+}, Zn^{2+}; Bertini, Gray, Stiefel, & Valentine, 2007; Frausto da Silva & Williams, 2001; Lippard & Berg, 1994). Although earth-alkaline metal ions such as Mg^{2+} and Ca^{2+} are labile,

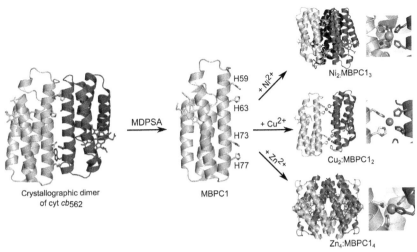

Fig. 2 Application of MDPSA to the natively monomeric cyt cb_{562} (PDB ID: 2BC5), giving the MBPC1 metalloprotein scaffold. In the crystal lattice, cyt cb_{562} forms extensive packing interactions along Helix 3 (residues 56–81), which was chosen as the surface to install two bis-His clamps (H59/H63 and H73/H77). MBPC1 forms three distinct metal-bound oligomers dictated by the stereochemical preferences of the metal ions: a Ni-bound trimer (PDB ID: 3DE9), a Cu-bound dimer (PDB ID: 3DE8), and a Zn-bound tetramer (PDB ID: 2QLA). Metal ligands are shown as *sticks*, and metal ions are shown as *spheres*. Close-up images show the metal coordination environments: Ni^{2+} is coordinated either by three H59/H63 (shown) or three H73/H77 bis-His clamps from three different protomers. Cu^{2+} is coordinated by H59, H63, H73′, and H77′ from two different protomers. Zn^{2+} is coordinated by H63, H73′, H77′, and D74″ from three different protomers.

they have relatively weak affinities for proteinaceous functionalities and their coordination is largely limited to carboxylate and hydroxyl groups. In contrast, second- and third-row transition metal ions can tightly associate with protein surfaces, but possess very slow ligand exchange rates at room temperature (Bertini et al., 2007; Lippard & Berg, 1994).

A further advantage of late-first-row transition metal ions is their distinct preferences for coordination geometry (Ni^{2+}, octahedral; Cu^{2+}, tetragonal/square planar; Zn^{2+}, tetrahedral; Harding et al., 2010; Rulíšek & Vondrášek, 1998). Thus, they can also dictate the oligomerization state and oligomerization geometry of a protein building block. Indeed, MPBC1 self-assembles into a C_3 symmetric trimer upon Ni^{2+} binding (via octahedral coordination by three bis-His clamps), a C_2 symmetric dimer upon Cu^{2+} binding (via square planar coordination by two bis-His clamps), and a D_2 symmetric tetramer upon Zn^{2+} binding (via tetrahedral coordination by a bis-His clamp from one protomer, D74 from a second protomer, and H63 from a third protomer) (Fig. 2; Salgado et al., 2007; Salgado, Lewis, Mossin, Rheingold, & Tezcan, 2009). Notably, the Zn_4:$MBPC1_4$ structure demonstrates the formation of the thermodynamically favored, albeit unforeseen, tetrameric architecture rather than the predicted Zn_2:$MBPC1_2$ dimer mediated by tetrahedral coordination via two bis-His clamps. These studies indicate the following: (1) in terms of designing protein assemblies, metal coordination provides structural versatility and controllability that is difficult and perhaps impossible to match through the design of noncovalent interactions, (2) this versatility is achieved through minimal alterations of the protein building blocks, and (3) protein oligomers formed through MDPSA give rise to nascent protein–protein interfaces and internalized metal centers that provide a rich platform for further design efforts (Section 3).

2.3 MDPSA Using Synthetic Metal-Coordinating Ligands

Precise placement of multiple amino acid side chains to form a chelating motif can be challenging, particularly in the case of proteins with irregular surface features. An alternative route for installing multidentate metal-coordinating motifs on a protein surface is to employ nonnatural metal chelating functionalities, such as the bidentate 8-hydroxyquinoline (quin) and 1,10-phenanthroline (phen), or the tridentate 2,2′:6′,2″-terpyridine (terpy). These ligands have metal-binding affinities many orders of magnitude higher than those of monodentate amino acid residues (Smith, Martell, & Motekaitis, 2004) and can be readily conjugated to surface-exposed cysteine residues through the use of their iodoacetamide or maleimide derivatives (Section 4.2). In the case of cyt cb_{562}, a cysteine (C70) was installed near the middle of the solvent-exposed face of Helix3 to provide an appropriate attachment site for a metal chelating group (Fig. 3A). ^{C70}Cyt cb_{562} was modified with iodoacetamide derivatives of quin, phen, or terpy, and the resulting adducts readily dimerized upon metal coordination (Fig. 3A–C; Radford, Nguyen, & Tezcan, 2010).

These motifs are easily generalizable, since the only requirement is a surface-exposed cysteine residue. Nevertheless, the flexibility of the Cys-attachment site can give rise to unpredictability in the architecture of the resulting protein oligomer. To circumvent this potential problem, we envisioned using the Cys-attached phen or quin functionalities in tandem with a nearby His residue to form a tridentate hybrid coordination motif

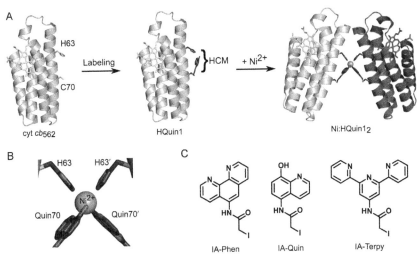

Fig. 3 MDPSA via hybrid coordination motifs (HCMs). (A) ^{C70}Cyt cb562 is labeled with 5-(iodoacetamido)-8-hydroxyquinoline to afford HQuin1. Upon the addition of Ni^{2+}, HQuin1 forms the Ni:HQuin1$_2$ dimer (3L1M). (B) Ni^{2+} (*green* (*light gray* in the print version) *sphere*) is coordinated by the two bidentate quin ligands, and the two H63 residues (*purple* (*dark gray* in the print version) *sticks*). (C) Chemical structures of the iodoacetamide-derivatized metal chelating ligands (IA-phen, IA-quin, and IA-terpy).

(HCM). In exploratory experiments with cyt cb_{562}, this strategy was implemented by pairing C70-attached phen or quin with H63 in an *i* and *i*+7 arrangement on the Helix3 surface (Fig. 3A). The resulting tridentate HCM not only binds Co^{2+}, Ni^{2+}, Cu^{2+}, and Zn^{2+} with nanomolar to femtomolar affinities, but it also allows the formation of a discrete dimeric architecture upon Ni^{2+} binding in an octahedral geometry (Radford, Nguyen, Ditri, Figueroa, & Tezcan, 2010; Radford, Nguyen, & Tezcan, 2010; Fig. 3A). HCMs could, in principle, be applied to any α-helical domain on a protein surface, provided that the helix spans at least two turns.

3. METAL-TEMPLATED INTERFACE REDESIGN

3.1 Redesign of Noncovalent Interfaces with MeTIR

Metals are proposed to have been drivers of evolution in some protein superfamilies through the formation of metal-bridged protein/peptide aggregates (Armstrong, 2000; Bergdoll et al., 1998; Liu & Kuhlman, 2006;

Volbeda & Hol, 1989). These ancestral oligomers subsequently developed stable and specific protein interfaces around the metal-nucleation sites, eventually leading to the emergence of modern proteins with evolved networks of bonding interactions to stabilize their tertiary or quaternary architectures. Whether or not such an evolutionary pathway was indeed operative, one can still use it as a blueprint for protein design in an approach we have termed MeTIR (Fig. 1). Self-assembled protein architectures obtained by MDPSA possess, by default, nascent PPIs between nonself-complementary surfaces surrounding the metal templates. These interfaces can be rendered self-complementary by replacing unfavorable amino acid contacts with favorable ones through rational design. In this way, MeTIR can furnish novel self-assembling proteins that form stable protein architectures with tunable metal-binding sites in their interiors (see Sections 4.3–4.5 for methods).

The Zn-directed MBPC1 tetramer, Zn_4:$MBPC1_4$, was used as a model system for MeTIR (Salgado et al., 2007). This tetramer features an extensive buried surface (>5000 Å^2) between the MBPC1 protomers, distributed among three pairs of interconnected, C_2 symmetric interfaces termed $i1$, $i2$, and $i3$ (protein data bank (PDB) ID: 2QLA; Fig. 4). We chose $i1$ (2 × 1080 Å^2) as a promising candidate for initial redesign because, as the most extensive interface, it could accommodate a greater number of favorable contacts than $i2$ (2 × 870 Å^2) or $i3$ (2 × 490 Å^2). Additionally, $i1$ brings the protomers into closer proximity than $i2$, a prerequisite for forming favorable side-chain packing interactions. The redesign process was carried out by a computationally guided approach using Rosetta (Liu & Kuhlman, 2006). In this process, the residues that coordinate the Zn^{2+} ions and the protein backbone were both held fixed as demanded by the templating strategy. Clusters of $i1$ residues were subjected to concerted mutation and optimization with respect to rotamer geometry and packing energy (Salgado et al., 2010). This procedure led to the selection of six mutations (R34A, L38A, Q41W, K42S, D66W, and V69I), which were installed to convert MBPC1 into the variant RIDC1 (Rosetta interface designed cytochrome 1; Fig. 4).

As expected from the hydrophobicity of the engineered surface residues and determined by molecular docking simulations with RosettaDock (Gray et al., 2003), the geometric specificity of $i1$ interactions is low. Even so, these interactions are strong enough to dimerize RIDC1 monomers in solution ($K_D = 25$ μM) without metal (Salgado et al., 2010). More importantly, in the presence of Zn^{2+}, RIDC1 forms a tetramer (Zn_4:$RIDC1_4$)

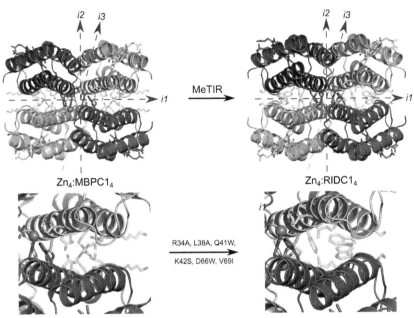

Fig. 4 Application of MeTIR to the metal-directed Zn_4:$MBPC1_4$ tetramer (PDB ID: 2QLA) to give Zn_4:$RIDC1_4$ (PDB ID: 3HNI). *Red* (*dark gray* in the print version) *axes* denote the C_2-symmetric interfaces *i1*, *i2*, and *i3*. The redesigned *i1* residues are shown as *cyan* (*light gray* in the print version) *sticks*. Metal-coordinating ligands are shown as *gray sticks*. Zn^{2+} ions are shown as *gray spheres*. Beneath are close-up views of the *i1* PPIs between two MBPC1 or RIDC1 protomers within the Zn-bound tetramers.

that is isostructural with Zn_4:$MBPC1_4$, but is considerably stabilized by the redesigned *i1* interfaces (Fig. 4).

3.2 Disulfide Cross-Links to Enhance Scaffold Robustness

For an oligomeric protein complex with multiple interaction surfaces, such as the D_2 symmetric Zn_4:$MBPC1_4$ and Zn_4:$RIDC1_4$ tetramers, the redesign of a single set of interfaces is insufficient to maintain the scaffold in the absence of bound metal. A second set of interfaces must be redesigned to hold the tetrameric assembly intact in the absence of the metal templates. However, computational redesign of noncovalent interactions is not only labor-intensive, but can be suboptimal in interfaces that are too wide or too small. A straightforward alternative is to install disulfide bonds across these interfaces that—in combination with computationally redesigned interfaces—can lead to the formation of stable protein complexes in a metal-independent manner. In this regard, the unique properties of disulfide bonds must be emphasized: they are strong covalent bonds that are very specific (ie, they only pair with themselves), yet they are reversible (Fass, 2012). Thus, in terms of protein self-assembly, disulfide bonds share many of the advantageous characteristics of metal coordination mentioned previously.

Cysteine residues can be installed at a pair of symmetry-related sites with the appropriate α-C separation and relative orientation to form disulfide bonds (Petersen, Jonson, & Petersen, 1999). In the case of Zn_4:$RIDC1_4$, the T96/T96' residue pairs across the *i2* interfaces and the E81/E81' across the *i3* interfaces were both suitable for this purpose. Accordingly, we generated a single-cysteine (^{C96}RIDC1; Brodin et al., 2010) and a double-cysteine ($^{C81/C96}$RIDC1; Medina-Morales, Perez, Brodin, & Tezcan, 2013) variant which properly self-assembled into stable metal-free tetramers (^{C96}RIDC1$_4$ and $^{C81/C96}$RIDC1$_4$). Moreover, these tetramers adopted the same structure as the parent tetramer Zn_4:$MBPC1_4$ upon binding Zn^{2+}. Importantly, the quadruply disulfide-linked $^{C81/C96}$RIDC1$_4$ scaffold is sufficiently preorganized to permit altering the inner-coordination sphere of the templating Zn ions without perturbing the quaternary structure (Fig. 5A–C; Medina-Morales et al., 2013).

3.3 In Vivo Assembly and Functional Screening of Metal-Templated Protein Assemblies

A major goal in protein design (and synthetic biology) is to create artificial constructs that function in living systems. We envisioned that, through the combined action of noncovalent interface redesign and disulfide bond formation, self-assembly of artificial protein architectures could also occur in bacterial cells. The ^{C96}RIDC1$_4$ and $^{C81/C96}$RIDC1$_4$ assemblies were indeed found to properly self-assemble in vivo and bind Zn^{2+} ions with high selectivity (Medina-Morales et al., 2013; Song & Tezcan, 2014). To form disulfide bonds in vivo, the protein must be expressed in the oxidizing environment of the periplasm in *Escherichia coli* (Inaba & Ito, 2008). Additionally, the periplasmic space exhibits comparatively relaxed metallostasis (Braymer & Giedroc, 2014; Hu, Gunasekera, Spadafora, Bennett, & Crowder, 2008), which permits the incorporation of metal ions inside the cell (Medina-Morales et al., 2013; Song et al., 2014). Since proteins are synthesized in the cytoplasm, a genetically encoded localization sequence is required for its translocation into the periplasm (Mergulhão,

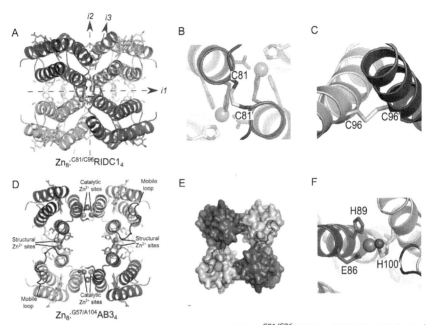

Fig. 5 (A) The quadruply disulfide cross-linked Zn_4:$^{C81/C96}$RIDC1$_4$ (PDB ID: 4JEA). Red (dark gray in the print version) axes denote the C_2-symmetric interfaces i1, i2, and i3. Disulfide cross-links (B) C81/C81' across i3 and (C) C96/C96' across i2. (D) The artificial metallo-β-lactamase Zn_8:$^{G57/A104}$AB3$_4$ (PDB ID: 4U9E) bearing redesigned i1 contacts (cyan sticks (light gray in the print version)), C96/C96' disulfide cross-links (green (light gray in the print version) sticks), and sets of structural Zn^{2+} sites and catalytic Zn^{2+} sites (gray sticks). Zn^{2+} ions are shown as gray spheres, and the catalytic water molecules are shown as red (dark gray in the print version) spheres. Shown in purple (dark gray in the print version) is a natively well-structured loop which became highly mobile (residues 43–57, not observed crystallographically) as a result of directed evolution. (E) Surface representation of the structured regions of Zn_8:$^{G57/A104}$AB3$_4$ showing the proximity of the solvent-exposed catalytic Zn^{2+} site, with ligands colored in blue (dark gray in the print version), to the mobile loop, shown in purple (dark gray in the print version). (F) Close-up view of a catalytic Zn^{2+} site.

Summers, & Monteiro, 2005). In the case of cyt cb_{562}, the signaling sequence was derived from *Rhodopseudomonas palustris* cytochrome c_{556} (Faraone-Mennella et al., 2006; McGuirl et al., 2003). When expressed in *E. coli*, a periplasmic extraction procedure allows for the recovery of intact oligomers (eg, Zn_4:$^{C81/C96}$RIDC1$_4$) to examine in vivo assembly and Zn^{2+} content (Section 4.6).

A robust oligomeric protein scaffold that self-assembles in vivo is an ideal candidate for the de novo design of a functional, metalloenzyme.

A catalytic metal site should be coordinatively unsaturated, such as a hydrolytic Zn center that coordinates and activates water (Alberts, Nadassy, & Wodak, 1998; Holm, Kennepohl, & Solomon, 1996). In ^{C96}RIDC1$_4$ and $^{C81/C96}$RIDC1$_4$, replacing a coordinating ligand (D74) of the core (templating) Zn^{2+} ions with a noncoordinating alanine residue did not yield catalytically competent constructs, presumably because the Zn sites were inaccessible or did not possess a proper secondary sphere environment for substrate binding and activation (Medina-Morales et al., 2013). Therefore, interfacial sites in the solvent-exposed peripheral regions of Zn$_4$:^{C96}RIDC1$_4$ were investigated as potential locations for installing catalytic Zn centers (Song et al., 2014). Examination of i2 in ^{C96}RIDC1$_4$ revealed several three-residue permutations that could accommodate a tripodal Zn-coordination motif. One such permutation was the E86/H89/H100 triad on ^{C96}RIDC1$_4$. The installation of these residues gave the AB3 construct, which upon Zn-coordination yielded the tetrameric Zn$_8$:AB3$_4$ with four core Zn sites and four potentially catalytic, peripheral Zn sites. Removal of an occluding lysine residue (K104) near the peripheral Zn sites resulted in the formation of tetrahedral Zn-H$_2$O/OH$^-$ motifs reminiscent of natural metalloenzyme hydrolytic sites (Crowder, Spencer, & Vila, 2006).

In vitro enzymatic activity assays showed that Zn$_8$:^{A104}AB3$_4$ was indeed catalytically active in the hydrolysis of activated ester bonds (Song et al., 2014). This observation, combined with the ability of ^{A104}AB3$_4$ to properly self-assemble in the periplasm of *E. coli*, raised the possibility of evolving the activity of this nascent, artificial enzyme in vivo (Jäckel, Kast, & Hilvert, 2008). Enzymatic hydrolysis of an antibiotic, such as ampicillin, could be coupled to cellular survival, providing a selection mechanism for directed evolution. The β-lactamase activity of Zn$_8$:^{A104}AB3$_4$ was sufficient to permit *E. coli* growth on media containing ampicillin, which allowed for screening of a library of active site point mutants for improved enzymatic activity (Section 4.7). Library variants were identified and subsequently purified to quantify the enzymatic activities in vitro (Song et al., 2014), which correlated with the in vivo survival frequencies of the variants. X-ray crystallographic studies indicated that best performing variant, Zn$_8$:$^{G57/A104}$AB3$_4$, possessed a highly mobile loop atop the catalytic Zn centers, which is a conserved feature of natural metallo-β-lactamases (Fig. 5D–F; Scrofani et al., 1999; Tomatis et al., 2008). Thus, by iterative protein design, cyt *cb*$_{562}$ was converted into an in vivo active β-lactamase while conserving nearly 90% sequence identity.

4. METHODS
4.1 Cytochrome cb_{562} Variant Expression and Purification

The gene encoding cyt cb_{562} was housed in a modified pET20b+ vector encoding the N-terminal periplasmic localization sequence of R. palustris cytochrome c_{556}. Variants of the cyt cb_{562} gene were obtained by site-directed mutagenesis. Plasmids encoding cyt cb_{562} variants were transformed into BL21(DE3) E. coli cells harboring the cytochrome c maturation cassette (ccm or pEC86; Arslan, Schulz, Zufferey, Kunzler, & Thony-Meyer, 1998). Cells were plated onto selective lysogeny broth (LB)/agar plates (34 μg/mL chloramphenicol and 100 μg/mL ampicillin) and the protein was expressed and purified as follows:

1. Transfer single colonies into three 5 mL of LB medium supplemented with 100 μg/mL ampicillin and 34 μg/mL chloramphenicol.
2. Shake cultures at 250 rpm, 37°C until cultures are visibly turbid ($OD_{600} > 0.6$, 6–8 h)
3. Remove an 1-mL portion of each culture and pellet the cells by centrifugation (1 min, $10,000 \times g$). The color of the cell pellets provides a convenient indicator for the expression and folding of cyt cb_{562} through proper heme incorporation.
4. Transfer 75 μL of the culture which gave the most intensely red-colored pellet into 1 L of LB medium (typically 16 L total) supplemented with 100 μg/mL ampicillin and 34 μg/mL chloramphenicol.
5. Shake inoculated cell cultures at 250 rpm and 37°C for 16–20 h. Protein expression occurs by autoinduction.
6. Harvest the cells by centrifugation ($5000 \times g$, 4°C, 10 min), decant the supernatant, and retain the red cell pellet.
7. Resuspend the cell pellet in 100 mL of a buffered solution (5 mM sodium acetate, pH 5) and freeze the suspension at −80°C.
8. Thaw the cell suspension in a bath of water at room temperature, add ~100 mg lysozyme, and sonicate on ice.
9. Titrate the cell lysate with 40% (w/v) sodium hydroxide to pH 10 and then immediately with 50% (v/v) acetic acid to pH 5. Cream-colored precipitates will form in the red lysate.
10. Subject the lysate to centrifugation ($10,000 \times g$, 4°C, 20 min) and retain the clear, red supernatant. Repeat steps 9–10 to recover additional protein as needed.
11. Dilute the cleared lysate by 150 mL per 1 L cell culture, and apply to a CM Sepharose (GE Healthcare Life Sciences) cation exchange column

equilibrated in a buffered solution (5 mM sodium acetate, pH 5). If the cyt cb_{562} variant of interest in unstable at pH 5, the purification can be carried out with a Q Sepharose (GE Healthcare Life Sciences) column equilibrated in 5 mM sodium phosphate, pH 8.

12. Elute the protein using a gradient of 0–1 M sodium chloride and recover eluate that is visibly red.
13. Concentrate the recovered protein sample using a Diaflow concentrator (Amicon) fitted with a 3-kDa cutoff membrane to a final volume of 50 mL.
14. Dialyze the protein sample overnight at 4°C against 4 L of buffered solution (10 mM sodium phosphate, pH 8) to remove residual salt.
15. Apply the dialyzed sample to a fast protein liquid chromatography (FPLC) workstation equipped with a Macroprep High Q-cartridge (BioRad) anion exchange column equilibrated with a buffered solution (10 mM sodium phosphate, pH 8).
16. Elute protein using a 0–0.5 M sodium chloride gradient.
17. Estimate protein purity spectrophotometrically and combine fractions with a Reinheitzahl (RZ) value (A_{415}/A_{280}) above 3.
18. Concentrate the combined fractions to a volume of <5 mL and apply the sample to an FPLC workstation equipped with a preparative-scale Superdex 75 size-exclusion column equilibrated in a buffered solution (20 mM 2-amino-2-hydroxymethyl-propane-1,3-diol (TRIS), pH 7, and 150 mM sodium chloride).
19. Estimate protein purity spectrophotometrically and combine fractions with an RZ value (A_{415}/A_{280}) above 4.
20. Concentrate the combined fractions to a volume of ~500 μL and add >10-fold excess ethylenediaminetetraacetic acid (EDTA) to remove any residual bound metal ions.
21. Apply the protein sample to a desalting column preequilibrated as desired to remove the EDTA.
22. The protein concentration can be determined using the extinction coefficient of the Soret band of cyt cb_{562} ($\varepsilon_{415} = 148{,}000\ M^{-1}\ cm^{-1}$).
23. Flash freeze aliquots of the protein stock solutions at the desired concentration in liquid nitrogen for storage at −80°C.

4.2 Iodoacetamide Ligand Synthesis and Protein Labeling

Below is a protocol for the synthesis of N-(8-hydroxyquinolin-5-yl)-2-iodoacetamide (IA-quin) for labeling ^{C70}cyt cb_{562} based on previously reported procedures (Radford, Nguyen, Ditri, et al., 2010; Radford, Nguyen, & Tezcan, 2010; Smith, Du, Radford, & Tezcan, 2013). This procedure can be readily adapted for the synthesis of iodoacetamide-derivatized ligands and the Cys-labeling of proteins in general.

1. In a 50-mL round-bottom flask, combine 1.19 g (6.4 mmol) of iodoacetic acid dissolved in 10 mL ethyl acetate with 660 mg (3.2 mmol) of N,N'-dicyclohexylcarbodiimide (DCC) dissolved in 10 mL of ethyl acetate. A white precipitate should form immediately.
2. Stir at room temperature for 2 h protected from light using aluminum foil.
3. Remove and discard the white precipitate (dicyclohexylurea) by filtration.
4. Remove solvent in vacuo protected from light using aluminum foil to recover iodoacetic anhydride.
5. In a 25-mL round-bottom flask, combine 500 mg (2.1 mmol) of 5-amino-8-hydroxyquinolate dihydrochloric acid dissolved in 10 mL of acetonitrile with 1 mL (7 mmol) triethylamine.
6. Reflux the mixture at 80°C for 2 h. This should afford a clear, dark solution. Allow this solution to cool to room temperature.
7. Dissolve the iodoacetic anhydride in 5 mL of acetonitrile and add dropwise to the reaction mixture.
8. Stir overnight at room temperature protected from light using aluminum foil.
9. Isolate the desired product by filtration and wash the retained solid with 5% (w/v) sodium bicarbonate in water.
10. Dry the retained solid in vacuo to afford the desired IA-quin. Store the sample protected from light.

A major side product of this reaction is N-(8-hydroxyquinolin-5-yl)-2-chloroacetamide (CA-quin), which would also react site-specifically with a cysteine residue to append the desired chelating motif, albeit much more slowly than IA-quin. The crude mixture of IA-quin is approximately 50% pure and forms in 80% yield as synthesized. Though not necessary, conversion of the CA-quin to IA-quin can be achieved as follows:

1. Reflux the CA-quin with five molar equivalents of sodium iodide in acetone overnight.
2. Remove solvent in vacuo.
3. Resuspend the dried solid in water to remove excess sodium chloride and recover the solid by filtration.
4. Dry the retained solid in vacuo, affording IA-quin.

Once the iodoacetamide derivatives are prepared, labeling of surface-exposed cysteine residues can be carried out as follows:

1. Add a 10-fold molar excess of dithiothreitol (DTT) to a 5 mL solution of 0.3 mM cyt $^{C70}cb_{562}$ in 0.1 M TRIS, pH 7.75 and incubate for 30 min at room temperature reduce any disulfide bonds.
2. Transfer the cyt $^{C70}cb_{562}$ solution into a glass vial sealed with a gas-tight rubber septum.
3. Degas the cyt $^{C70}cb_{562}$ solution either on a Schlenk line using 3×10 cycles of vacuum/argon-fill or by purging with argon or nitrogen.
4. Transfer the protein sample into an anaerobic glove box (<10 ppm oxygen) under an inert (argon or nitrogen) atmosphere.
5. Dialyze the protein sample twice against 1 L of degassed 0.1 M TRIS, pH 7.75 in the anaerobic glove box to remove the DTT.
6. Recover the protein sample and place in a glass vial.
7. To the protein sample, add a 10-fold molar excess of an IA-quin stock dissolved in degassed dimethylformamide dropwise over the course of 1 min.
8. Incubate the reaction mixture at room temperature overnight, protected from light with aluminum foil.
9. Remove the protein sample from the glove box and dialyze it twice against 1 L of 10 mM sodium acetate, pH 5, supplemented with 1 mM EDTA to remove any trace metals from the sample.
10. Purify the crude-labeled protein by applying it to an FPLC workstation equipped with an Uno-S cation exchange column equilibrated in 10 mM sodium acetate, pH 5.
11. Elute protein using a 0–0.5 M sodium chloride gradient to isolate the quin-functionalized protein sample. Labeling efficiency was >60%.
12. Confirm the identity of the functionalized protein by matrix-assisted laser desorption/ionization or electrospray ionization mass spectrometry.

4.3 Determining Oligomeric State by Sedimentation Velocity Experiments

Protein oligomerization can be characterized in solution by sedimentation velocity analytical ultracentrifugation (SV-AUC). An SV-AUC experiment requires an optical signal (absorbance or fluorescence) which can be used to monitor protein sedimentation. The heme cofactor of cyt cb_{562} provides a convenient optical handle (ε_{max} at 415 nm = 148,000 M^{-1} cm^{-1}) that allows the AUC experiments to be conducted from very low (≥ 5 μM; near the Soret peak) to very high protein concentrations (>600 μM; near the shoulders of the Soret band; Fig. 6A; Faraone-Mennella et al., 2006).

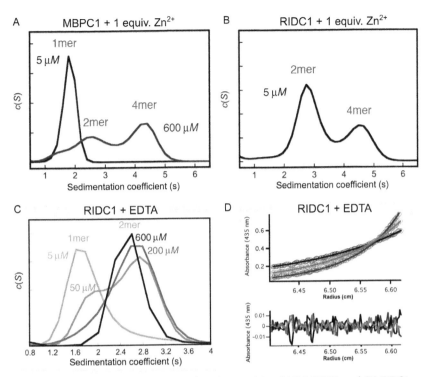

Fig. 6 SV-AUC profiles showing Zn-induced assembly of (A) MBPC1 and (B) RIDC1 as well as (C) RIDC1 in the absence of metal ions. (D) SE-AUC scans of metal-free RIDC1 (*circles*) globally fit to a monomer–dimer equilibrium model for centrifugation at 20,000 (*red* (*dark gray* in the print version)), 25,000 (*green* (*light gray* in the print version)), and 30,000 (*blue* (*gray* in the print version)) rpm along with the residuals of the data fitting. Adapted from Salgado, E. N., Faraone-Mennella, J., & Tezcan, F. A. (2007). Controlling protein-protein interactions through metal coordination: Assembly of a 16-helix bundle protein. Journal of the American Chemical Society, 129, 13374–13375.

Metal-induced protein oligomerization can be assessed under both metal-free and metal-loaded conditions (Fig. 6A–C). To assess metal-induced oligomerization, the protein sample is preincubated with a stoichiometric amount (one molar equivalent) of M^{2+} per coordination site. For substitution-labile metals, brief incubations (<1 h at 4°C) are sufficient, whereas longer incubations (~20 h at 4°C) may be necessary for substitution-inert metals. The metal stock should be added slowly (~0.25 equivalents at a time) to prevent nonspecific protein aggregation. The choice of buffering agent is also an important consideration for examining protein oligomerization. Weakly chelating buffering agents, such as TRIS, will suppress nonspecific metal binding. For weak metal-binding events, a nonchelating buffering agent such as 3-(*N*-morpholino)propanesulfonic acid is more appropriate. Buffering agents that form insoluble metal salts, such as phosphates, should be avoided.

Samples are sedimented by high-speed centrifugation at a fixed velocity. The sedimentation is monitored by changes in absorbance at different radii of rotation. To examine oligomerization by SV-AUC, the hydrodynamic behavior of the protein is accounted for by applying the Lamm equation (Correia & Stafford, 2015). AUC-fitting programs can be used to model the sedimentation behavior of the protein. The range of AUC-fitting programs and procedures available is beyond the scope of this review and is discussed elsewhere (Cole, Lary, Moody, & Laue, 2008; Lebowitz, Lewis, & Schuck, 2002). The procedure used for cyt cb_{562} variants is briefly described:

1. Load the protein sample into a two-sector sample cell with a slight excess of an appropriate buffer blank (eg, 420 μL sample and 450 μL buffer blank).
2. Sediment the protein sample at $135,000 \times g$ at 25°C, monitoring continuously at the wavelength(s) of interest, typically for 500 scans over 20 h.
3. Load the sedimentation velocity scans into HETEROANALYSIS (http://biotech.uconn.edu/auf/) and use the match function to determine when sedimentation has ended.
4. Load the sedimentation velocity scans up to the point where sedimentation has ended into SEDFIT (Schuck, 2000).
5. Manually set the cell limits and data-fitting limits of the overlaid scans. Hold these fixed during the fitting procedure.
6. Estimate the partial specific volume in milliliter per gram by taking the quotient of the protein volume and the molecular weight.
7. Estimate the viscosity and density of the buffer using SEDNTERP (http://sednterp.unh.edu/).
8. Using the appropriate partial specific volume, buffer viscosity, and buffer density, fit the data to a continuous distribution of molecular weight ($c(M)$ model) or sedimentation coefficient ($c(S)$ model).
9. Using the "Run" command in SEDFIT, set the baseline of the overlaid velocity scans.
10. Using the "Run" command in SEDFIT, subtract the time-invariant or radius-invariant noise of the overlaid velocity scans.
11. Fit the weight-averaged frictional coefficient (f/f_0) of the protein using the "Fit" command in SEDFIT with the confidence set to 0. A value of 1.2–1.3 is typical for symmetric proteins.
12. Obtain the final distribution profile by setting the confidence to 0.95 and executing the "Run" command in SEDFIT.

Using this procedure, the peaks of the sedimentation profile can be assigned to protein oligomeric states. With a $c(M)$ profile, this can be done from the reported molecular weights. With a $c(S)$ profile, this can be done by comparing the sedimentation profile with those of related proteins, or by modeling the sedimentation coefficients using a program such as HYDROPRO (Ortega, Amoros, & Garcia de la Torre, 2011).

4.4 Measuring Protein–Protein Affinities with Sedimentation Equilibrium Experiments

Once the oligomerization behavior of a cyt cb_{562} construct has been examined across a range of protein concentrations and metal ions by SV-AUC (Fig. 6A–C), one can measure the dissociation constants (K_d) of the oligomers in the absence of metal, or the apparent dissociation constant ($K_{d,app}$) in the presence of metal by sedimentation equilibrium AUC (SE-AUC; Cole et al., 2008; Fig. 6D). In an SE-AUC experiment, samples are subjected to angular velocities that permit the counteracting forces of sedimentation and diffusion to equilibrate. These velocities are slower than what is used in a typical SV-AUC experiment. An important consideration is the time needed to achieve equilibrium. This can be done by manual inspection of overlaid scans, or through the scan match functionalities in programs like HETEROANALYSIS (http://biotech.uconn.edu/auf/). Once equilibrium is established, the extent of sedimentation correlates with the molecular weight of the protein. A protein sample under a given set of conditions is sedimented at various speeds, and the resulting datasets are analyzed globally using programs such as SEDPHAT (Fig. 6D; Schuck, 2004). Noninteracting (eg, monomer-only) and self-assembly (eg, monomer–dimer) models can be considered to find the best model based on the goodness-of-fit statistics. In more complex cases, SV-AUC data can help to eliminate models that are difficult to rule out. For Zn-induced RIDC1 assembly, SV-AUC experiments suggested the trimer was not formed, and assembly was modeled as a monomer–dimer–tetramer equilibrium.

4.5 Crystallization of cyt cb_{562} Variants

As discussed previously, crystallography is a powerful and necessary tool for assessing the outcome and progress of MDPSA/MeTIR, because both design approaches are intimately dependent on detailed structural knowledge at the atomic level. Crystals of cyt cb_{562} variants can be obtained by sitting drop vapor diffusion. In each well of a 24-well crystal tray, 500 μL of a screening solution containing a precipitant, a buffering agent, and a salt

Table 1 Components of Homemade Crystallography Screening Solutions

Buffering Agent (100 mM)	Salt (200 mM)	Precipitant (%, w/v)	
100 mM BIS-TRIS pH 6.5	CaCl$_2$	45% MPD	25% PEG 3350
100 mM HEPES pH 7.5	NaCl	40% PEG 400	25% PEG 4000
100 mM TRIS pH 8.5	MgCl$_2$	30% PPG 400	25% PEG 5000
	NH$_4$OAc	25% PEG 1500	25% PEG 6000
	(NH$_4$)$_2$SO$_4$	25% PEG 2000	25% PEG 8000

(see Table 1 for details on screening solutions) is added to the reservoir. The sitting drop consists of 2 μL of ~3 mM of a cyt cb_{562} variant and 2 μL of the reservoir solution. Crystal formation and growth can be monitored on a timescale of days to weeks. The initial screens can used to determine fine-screening conditions, in which the precipitant concentration is systematically lowered (typically as low as half the initial concentration). Mature crystals were cryoprotected with either 20% glycerol or perfluoropolyether cryo oil (Hampton) and frozen in liquid nitrogen for X-ray crystallographic analysis. We have typically used synchrotron beam lines with tunable energies as this enables confirmation of metal identity via anomalous scattering. Diffraction data are typically indexed and integrated with Mosflm (Battye, Kontogiannis, Johnson, Powell, & Leslie, 2011) and scaled with SCALA (Winn et al., 2011). Crystallographic phases are readily determined through molecular replacement with either PHASER (McCoy et al., 2007) or MOLREP (Vagin & Teplyakov, 1998) using an appropriate cyt cb_{562} monomer as a search model. Structure determination and refinement are accomplished by rigid-body, thermal, and positional refinement with REFMAC (Murshudov, Vagin, & Dodson, 1997). COOT is used for manual model rebuilding and placement of water/ligands (Emsley & Cowtan, 2004). The resultant models provide detailed structural information and serve as a guide for further rational design efforts.

4.6 Periplasmic Extraction of cyt cb_{562} Assemblies from *E. coli*

A periplasmic extraction procedure is necessary to prevent disulfide bond reduction during purification. This protocol, adapted from a previously reported method (Nossal & Heppel, 1966), allows for the recovery of intact cyt cb_{562} oligomers for further analysis:

1. Pellet liquid cell culture expressing the protein of interest by centrifugation (10 min, 5000 × g, 4°C) in a tared vessel.

2. Decant the supernatant and note the cell pellet mass.
3. Wash cell pellet three times with an ice-cold low-salt buffered solution (10 mM TRIS, pH 7.3 with 30 mM sodium chloride).
4. Gently resuspend the cell pellet in 10 volumes (v/w) of 33 mM TRIS, pH 7.3.
5. Add 10 volumes (v/w) of 33 mM TRIS, pH 7.3 supplemented with 40% (w/v) sucrose and 0.4 mM EDTA.
6. Gently shake the sample for 10 min at room temperature.
7. Pellet cells by centrifugation (10 min, $5000 \times g$, 4°C)
8. Discard the supernatant and wash the cell pellet with ice-cold 33 mM TRIS, pH 7.3 supplemented with 40% sucrose.
9. Resuspend the cells in 20 volumes (v/w) of ice-cold 33 mM TRIS, pH 7.3 supplemented with 0.5 mM magnesium chloride to lyse the outer membrane by osmotic shock.
10. Gently swirl protein sample for 10 min at room temperature.
11. Pellet the cells by centrifugation (20 min, $5000 \times g$, 4°C) and retain the supernatant as the periplasmic extract.
12. Immediately add excess iodoacetamide (5 mM) to prevent postlytic formation of disulfide bonds.
13. Immediately apply the periplasmic extract to a High Q-cartridge column (BioRad) equilibrated in 10 mM sodium phosphate, pH 8 and elute the protein using a 0–0.5 M sodium chloride gradient.

Successful periplasmic extraction can be assessed by using UV–visible spectrophotometry to verify that the heme cofactor of cyt cb_{562} variant has remained oxidized, which can be noted by a shift in the Soret band peak (Fe^{3+}-cyt cb_{562} $\varepsilon_{415}=148{,}000\ M^{-1}\ cm^{-1}$; Fe^{2+}-cyt cb_{562} $\varepsilon_{421}=162{,}000\ M^{-1}\ cm^{-1}$). Protein samples purified by periplasmic extraction can be examined for oligomerization and disulfide bond formation by analytical size-exclusion chromatography and nonreducing sodium dodecyl sulfate polyacrylamide gel electrophoresis, respectively. Samples can be assayed for metal content by inductively coupled plasma optical emission spectrometry.

4.7 In Vivo Screening of cyt cb_{562} Assemblies with β-Lactamase Activity

Screening for β-lactamase activity required the removal of the ampicillin-resistance gene from the pET20b(+) vector housing the ^{C96}RIDC1 gene, and replacement with the kanamycin resistance gene from the pET24 vector by molecular cloning (Song et al., 2014). The resultant construct can be used to generate libraries for in vivo β-lactamase screening.

The following protocol outlines saturation mutagenesis at a single site to achieve 95% library coverage:
1. Use available crystal structures to identify residues in the secondary coordination sphere of the peripheral metal sites.
2. Design a primer pair that contains a single NNK codon in the sense primer and a single MNN codon in the antisense primer to generate variants at the site of interest. N is any nucleotide, K is a guanine or thymine, and M is cytosine or adenosine.
3. Carry out saturation mutagenesis using the above primers via polymerase chain reaction (PCR). This can be accomplished with a commercially available product, such as the QuikChange Site-Directed Mutagenesis Kit (Agilent, Santa Clara, CA).
4. Transform the library PCR into XL-1 Blue cells and spread onto LB/agar plates containing 50 μg/mL kanamycin.
5. Swab >100 colonies to inoculate a single culture of 100 mL of LB medium containing 50 μg/mL kanamycin. The number of colonies needed to achieve the desired library coverage can be estimated using CASTER (http://www.kofo.mpg.de/en/research/biocatalysis).
6. Incubate the culture overnight at 37°C, while shaking at 250 rpm.
7. Extract the plasmid library using a Qiagen Plasmid purification kit.
8. Confirm sequence randomization by DNA sequence analysis.
9. Transform the plasmid library into BL21(DE3) cells housing the ccm plasmid and spread onto LB/agar plates containing 50 μg/mL kanamycin and 30 μg/mL chloramphenicol.
10. Swab >100 colonies to inoculate a single culture of 5 mL LB medium supplemented with 50 μg/mL kanamycin, 30 μg/mL chloramphenicol, and 50 μM zinc chloride (to ensure formation of Zn-bound protein).
11. Incubate cultures at 37°C for 20 h, shaking at 250 rpm.
12. Prepare a triplicate series of LB/agar screening plates that contain 50 μg/mL kanamycin, 30 μg/mL chloramphenicol, 50 μM zinc chloride, 25 μM isopropyl β-D-1-thiogalactopyranoside, and varied ampicillin concentrations (0–1.6 μg/mL). Store plates overnight at 4°C. Storage of the plates for longer periods of time is not recommended.
13. Incubate the LB/agar screening plates at 37°C for 30–60 min to prewarm and dry the media.
14. Spread 20 μL aliquots of the cell culture onto half of each LB/agar screening plate. Library culture should be spread onto half of each plate and ^{C96}RIDC1 culture spread onto the other half to detect background cell growth.

15. Select individual colonies from plates that exhibit no ^{C96}RIDC1 growth and separately inoculate each into 5 mL of LB medium containing 50 μg/mL of kanamycin and 30 μg/mL chloramphenicol.
16. Isolate β-lactamase plasmids individually and subject to DNA sequencing analysis as outlined in steps 6–8.

5. CONCLUSION

Inspired by both the evolutionary roles of metal ions and their utility in nature's proteome, MDPSA and MeTIR provide a straightforward means to design functional protein assemblies from simple building blocks. The utility of these approaches has been primarily demonstrated with a model protein, cyt cb_{562}, although the design strategies and principles described here can be readily extended to other building blocks. While MDPSA and MeTIR were originally motivated by the desire to circumvent the design of extensive noncovalent interactions, these strategies are not meant to bypass computational protein design approaches, but rather to complement them. Indeed, it is only with the prudent combination of rational (chemical and computational) design approaches and laboratory evolution that it will be possible to engineer systems that surpass the structural, functional, and dynamic complexity of natural PPIs and protein complexes.

ACKNOWLEDGMENTS

We thank Drs. Cedric Owens, Jonathan Rittle, and Woon Ju Song for their helpful comments. X-ray crystal structures were collected at the Stanford Synchrotron Radiation Laboratory, which is supported by the U.S. Department of Energy, Offices of Basic Energy Sciences and Biological and Environmental Research, as well as by the NIH. The work reported here was supported in part by the University of California, San Diego. F.A.T. was supported by a Hellman Faculty Scholar Award, a Beckman Young Investigator Award, the Arnold and Mabel Beckman Foundation, the Sloan Foundation, the National Science Foundation (CHE-0908115 and CHE-1306646 for MeTIR), and the Department of Energy (DE-FG02-10ER46677 for Metal-Directed Protein Self-Assembly). L.A.C. was supported by the Molecular Biophysics Training Grant (NIH). R.H.S. was supported by the Chemistry-Biology Interface Training Program (NIH).

REFERENCES

Alberts, I. L., Nadassy, K., & Wodak, S. J. (1998). Analysis of zinc binding sites in protein crystal structures. *Protein Science, 7*, 1700–1716.

Armstrong, R. N. (2000). Mechanistic diversity in a metalloenzyme superfamily. *Biochemistry, 39*, 13625–13632.

Arnold, F. H., & Haymore, B. L. (1991). Engineered metal-binding proteins: Purification to protein folding. *Science, 252*, 1796–1797.

Arslan, E., Schulz, H., Zufferey, R., Kunzler, P., & Thony-Meyer, L. (1998). Overproduction of the *Bradyrhizobium japonicum* c-type cytochrome subunits of the cbb3

oxidase in *Escherichia coli*. *Biochemical and Biophysical Research Communications*, *251*, 744–747.
Battye, T. G., Kontogiannis, L., Johnson, O., Powell, H. R., & Leslie, A. G. (2011). iMOSFLM: A new graphical interface for diffraction-image processing with MOSFLM. *Acta Crystallographica. Section D, Biological Crystallography*, *67*, 271–281.
Bergdoll, M., Eltis, L. D., Cameron, A. D., Dumas, P., & Bolin, J. T. (1998). All in the family: Structural and evolutionary relationships among three modular proteins with diverse functions and variable assembly. *Protein Science*, *7*, 1661–1670.
Bertini, I., Gray, H. B., Stiefel, E. I., & Valentine, J. S. (2007). *Biological inorganic chemistry, structure and reactivity*. Sausalito, CA: University Science Books.
Braymer, J. J., & Giedroc, D. P. (2014). Recent developments in copper and zinc homeostasis in bacterial pathogens. *Current Opinion in Chemical Biology*, *19*, 59–66.
Brodin, J. D., Medina-Morales, A., Ni, T., Salgado, E. N., Ambroggio, X. I., & Tezcan, F. A. (2010). Evolution of metal selectivity in templated protein interfaces. *Journal of the American Chemical Society*, *132*, 8610–8617.
Cole, J. L., Lary, J. W., Moody, T., & Laue, T. M. (2008). Analytical ultracentrifugation: Sedimentation velocity and sedimentation equilibrium. *Methods in Cell Biology*, *84*, 143–179.
Conte, L. L., Chothia, C., & Janin, J. (1999). The atomic structure of protein-protein recognition sites. *Journal of Molecular Biology*, *285*, 2177–2198.
Correia, J. J., & Stafford, W. F. (2015). Sedimentation velocity: A classical perspective. *Methods in Enzymology*, *562*, 49–80.
Crowder, M. W., Spencer, J., & Vila, A. J. (2006). Metallo-beta-lactamases: Novel weaponry for antibiotic resistance in bacteria. *Accounts of Chemical Research*, *39*, 721–728.
Der, B. S., Machius, M., Miley, M. J., Mills, J. L., Szyperski, T., & Kuhlman, B. (2012). Metal-mediated affinity and orientation specificity in a computationally designed protein homodimer. *Journal of the American Chemical Society*, *134*, 375–385.
Emsley, P., & Cowtan, K. (2004). Coot: Model-building tools for molecular graphics. *Acta Crystallographica. Section D, Biological Crystallography*, *D60*, 2126–2132.
Faraone-Mennella, J., Tezcan, F. A., Gray, H. B., & Winkler, J. R. (2006). Stability and folding kinetics of structurally characterized cytochrome cb_{562}. *Biochemistry*, *45*, 10504–10511.
Fass, D. (2012). Disulfide bonding in protein biophysics. *Annual Review of Biophysics*, *41*, 63–79.
Fletcher, S., & Hamilton, A. D. (2006). Targeting protein-protein interactions by rational design: Mimicry of protein surfaces. *Journal of the Royal Society, Interface*, *3*, 215–233.
Frausto da Silva, J. J. R., & Williams, R. J. P. (2001). *The biological chemistry of the elements*. Oxford: Oxford University Press.
Ghadiri, M. R., & Choi, C. (1990). Secondary structure nucleation in peptides. Transition-metal ion stabilized alpha-helices. *Journal of the American Chemical Society*, *112*, 1630–1632.
Gonen, S., DiMaio, F., Gonen, T., & Baker, D. (2015). Design of ordered two-dimensional arrays mediated by noncovalent protein-protein interfaces. *Science*, *348*, 1365–1368.
Gray, J. J., Moughon, S., Wang, C., Schueler-Furman, O., Kuhlman, B., Rohl, C. A., et al. (2003). Protein-protein docking with simultaneous optimization of rigid-body displacement and side-chain conformations. *Journal of Molecular Biology*, *331*, 281–299.
Handel, T. M., Williams, S. A., & Degrado, W. F. (1993). Metal-ion dependent modulation of the dynamics of a designed protein. *Science*, *261*, 879–885.
Harding, M. M. (2004). The architecture of metal coordination groups in proteins. *Acta Crystallographica. Section D, Biological Crystallography*, *60*, 849–859.
Harding, M. M., Nowicki, M. W., & Walkinshaw, M. D. (2010). Metals in protein structures: A review of their principal features. *Crystallography Reviews*, *16*, 247–302.
Holm, R. H., Kennepohl, P., & Solomon, E. I. (1996). Structural and functional aspects of metal sites in biology. *Chemical Reviews*, *96*, 2239–2314.

Hu, Z., Gunasekera, T. S., Spadafora, L., Bennett, B., & Crowder, M. W. (2008). Metal content of metallo-β-lactamase L1 is determined by the bioavailability of metal ions. *Biochemistry, 47*, 7947–7953.

Huang, P.-S., Love, J. J., & Mayo, S. L. (2007). A de novo designed protein–protein interface. *Protein Science, 16*, 2770–2774.

Inaba, K., & Ito, K. (2008). Structure and mechanisms of the DsbB–DsbA disulfide bond generation machine. *Biochimica et Biophysica Acta (BBA)—Molecular Cell Research, 1783*, 520–529.

Jäckel, C., Kast, P., & Hilvert, D. (2008). Protein design by directed evolution. *Annual Review of Biophysics, 37*, 153–173.

King, N. P., Bale, J. B., Sheffler, W., McNamara, D. E., Gonen, S., Gonen, T., et al. (2014). Accurate design of co-assembling multi-component protein nanomaterials. *Nature, 510*, 103–108.

Krantz, B. A., & Sosnick, T. R. (2001). Engineered metal binding sites map the heterogeneous folding landscape of a coiled coil. *Nature Structural Biology, 8*, 1042–1047.

Lebowitz, J., Lewis, M. S., & Schuck, P. (2002). Modern analytical ultracentrifugation in protein science: A tutorial review. *Protein Science, 11*, 2067–2079.

Lippard, S., & Berg, J. (1994). *Principles of bioinorganic chemistry*. Mill Valley, CA: University Science Books.

Liu, Y., & Kuhlman, B. (2006). RosettaDesign server for protein design. *Nucleic Acids Research, 34*, W235–W238.

Liu, C. L., & Xu, H. B. (2002). The metal site as a template for the metalloprotein structure formation. *Journal of Inorganic Biochemistry, 88*, 77–86.

Mandell, D. J., & Kortemme, T. (2009). Computer-aided design of functional protein interactions. *Nature Chemical Biology, 5*, 797–807.

McCoy, A. J., Grosse-Kunstleve, R. W., Adams, P. D., Winn, M. D., Storoni, L. C., & Read, R. J. (2007). Phaser crystallographic software. *Journal of Applied Crystallography, 40*, 658–674.

McGuirl, M. A., Lee, J. C., Lyubovitsky, J. G., Thanyakoop, C., Richards, J. H., Gray, H. B., et al. (2003). Cloning, heterologous expression, and characterization of recombinant class II cytochromes c from *Rhodopseudomonas palustris*. *Biochimica et Biophysica Acta—General Subjects, 1619*, 23–28.

Medina-Morales, A., Perez, A., Brodin, J. D., & Tezcan, F. A. (2013). In vitro and cellular self-assembly of a Zn-binding protein cryptand via templated disulfide bonds. *Journal of the American Chemical Society, 135*, 12013–12022.

Mergulhão, F. J. M., Summers, D. K., & Monteiro, G. A. (2005). Recombinant protein secretion in *Escherichia coli*. *Biotechnology Advances, 23*, 177–202.

Mou, Y., Huang, P. S., Hsu, F. C., Huang, S. J., & Mayo, S. L. (2015). Computational design and experimental verification of a symmetric protein homodimer. *Proceedings of the National Academy of Sciences of the United States of America, 112*, 10714–10719.

Murshudov, G. N., Vagin, A. A., & Dodson, E. J. (1997). Refinement of macromolecular structures by the maximum-likelihood method. *Acta Crystallographica. Section D, Biological Crystallography, 53*, 240–255.

Nossal, N. G., & Heppel, L. A. (1966). The release of enzymes by osmotic shock from *Escherichia coli* in exponential phase. *Journal of Biological Chemistry, 241*, 3055–3062.

Ortega, A., Amoros, D., & Garcia de la Torre, J. (2011). Prediction of hydrodynamic and other solution properties of rigid proteins from atomic- and residue-level models. *Biophysical Journal, 101*, 892–898.

Petersen, M. T. N., Jonson, P. H., & Petersen, S. B. (1999). Amino acid neighbours and detailed conformational analysis of cysteines in proteins. *Protein Engineering, 12*, 535–548.

Petsko, G. A., & Ringe, D. (2004). *Protein structure and function*. London: New Science Press.

Radford, R. J., Nguyen, P. C., Ditri, T. B., Figueroa, J. S., & Tezcan, F. A. (2010a). Controlled protein dimerization through hybrid coordination motifs. *Inorganic Chemistry, 49*, 4362–4369.

Radford, R. J., Nguyen, P. C., & Tezcan, F. A. (2010b). Modular and versatile hybrid coordination motifs on alpha-helical protein surfaces. *Inorganic Chemistry, 2010*, 7106–7115.

Rulíšek, L., & Vondrášek, J. (1998). Coordination geometries of selected transition metal ions ($Co2+$, $Ni2+$, $Cu2+$, $Zn2+$, $Cd2+$, and $Hg2+$) in metalloproteins. *Journal of Inorganic Biochemistry, 71*, 115–127.

Salgado, E. N., Ambroggio, X. I., Brodin, J. D., Lewis, R. A., Kuhlman, B., & Tezcan, F. A. (2010). Metal-templated design of protein interfaces. *Proceedings of the National Academy of Sciences of the United States of America, 107*, 1827–1832.

Salgado, E. N., Faraone-Mennella, J., & Tezcan, F. A. (2007). Controlling protein-protein interactions through metal coordination: Assembly of a 16-helix bundle protein. *Journal of the American Chemical Society, 129*, 13374–13375.

Salgado, E. N., Lewis, R. A., Faraone-Mennella, J., & Tezcan, F. A. (2008). Metal-mediated self-assembly of protein superstructures: Influence of secondary interactions on protein oligomerization and aggregation. *Journal of the American Chemical Society, 130*, 6082–6084.

Salgado, E. N., Lewis, R. A., Mossin, S., Rheingold, A. L., & Tezcan, F. A. (2009). Control of protein oligomerization symmetry by metal coordination: C_2 and C_3 symmetrical assemblies through Cu^{II} and Ni^{II} coordination. *Inorganic Chemistry, 48*, 2726–2728.

Schreiber, G., & Fleishman, S. J. (2013). Computational design of protein–protein interactions. *Current Opinion in Structural Biology, 23*, 903–910.

Schuck, P. (2000). Size-distribution analysis of macromolecules by sedimentation velocity ultracentrifugation and Lamm equation modeling. *Biophysical Journal, 78*, 1606–1619.

Schuck, P. (2004). A model for sedimentation in inhomogeneous media: Dynamic density gradients from sedimenting co-solutes. *Biophysical Chemistry, 108*, 187–200.

Scrofani, S. D., Chung, J., Huntley, J. J., Benkovic, S. J., Wright, P. E., & Dyson, H. J. (1999). NMR characterization of the metallo-beta-lactamase from *Bacteroides fragilis* and its interaction with a tight-binding inhibitor: Role of an active-site loop. *Biochemistry, 38*, 14507–14514.

Smith, S. J., Du, K., Radford, R. J., & Tezcan, F. A. (2013). Functional, metal-based crosslinkers for alpha-helix induction in short peptides. *Chemical Science, 4*, 3740–3747.

Smith, R. M., Martell, A. E., & Motekaitis, R. J. (2004). *NIST critically selected stability constants of metal complexes*. NIST standard reference database 46.6.0.

Song, W. J., Sontz, P. A., Ambroggio, X. I., & Tezcan, F. A. (2014). Metals in protein-protein interfaces. *Annual Review of Biophysics, 43*, 409–431.

Song, W. J., & Tezcan, F. A. (2014). A designed supramolecular protein assembly with *in vivo* enzymatic activity. *Science, 346*, 1525–1528.

Tomatis, P. E., Fabiane, S. M., Simona, F., Carloni, P., Sutton, B. J., & Vila, A. J. (2008). Adaptive protein evolution grants organismal fitness by improving catalysis and flexibility. *Proceedings of the National Academy of Sciences of the United States of America, 105*, 20605–20610.

Tsai, C.-J., Lin, S. L., Wolfson, H. J., & Nussinov, R. (1996). Protein-protein interfaces: Architectures and interactions in protein-protein interfaces and in protein cores—Their similarities and differences. *Critical Reviews in Biochemistry and Molecular Biology, 31*, 127–152.

Vagin, A., & Teplyakov, A. (1998). MOLREP: An automated program for molecular replacement. *Journal of Applied Crystallography, 30*, 1022–1025.

Volbeda, A., & Hol, W. G. J. (1989). Pseudo 2-fold symmetry in the copper-binding domain of arthropodan hemocyanins—Possible implications for the evolution of oxygen-transport proteins. *Journal of Molecular Biology, 206*, 531–546.

Winn, M. D., Ballard, C. C., Cowtan, K. D., Dodson, E. J., Emsley, P., Evans, P. R., et al. (2011). Overview of the CCP4 suite and current developments. *Acta Crystallographica. Section D, Biological Crystallography, 67*, 235–242.

Xu, D., Tsai, C. J., & Nussinov, R. (1997). Hydrogen bonds and salt bridges across protein-protein interfaces. *Protein Engineering, 10*, 999–1012.

CHAPTER TWELVE

Designing Fluorinated Proteins

E.N.G. Marsh[1]
University of Michigan, Ann Arbor, MI, United States
[1]Corresponding author: e-mail address: nmarsh@umich.edu

Contents

1. Introduction — 252
   1.1 Basis of Protein Stabilization by Fluorinated Amino Acid Residues — 254
   1.2 Choice of Fluorinated Amino Acids — 255
   1.3 Synthesis of Fluorinated Proteins — 255
   1.4 Designing Fluorinated Amino Acids into Proteins — 257
   1.5 Evaluating Structural and Thermodynamic Effects of Incorporating Fluorinated Side-Chains — 261
   1.6 Concluding Remarks — 264
2. Methods — 265
   2.1 Measuring the Thermodynamic Stability of Fluorinated Proteins by Circular Dichroism — 265
   2.2 Determining $\Delta G°$ from GuHCl-Induced Protein Unfolding — 266
   2.3 Determining $\Delta H°'$, $\Delta S°'$, and $\Delta C°_p$ from Heat and GuHCl-Induced Protein Unfolding — 267
   2.4 MATLAB Code for Determining $\Delta G°$ Using GuHCl Unfolding — 269
   2.5 MATLAB Code for Determining $\Delta H°$, $\Delta S°$, and $\Delta C°_p$ Using Heat and GuHCl Unfolding — 270
Acknowledgments — 272
References — 272

Abstract

As methods to incorporate noncanonical amino acid residues into proteins have become more powerful, interest in their use to modify the physical and biological properties of proteins and enzymes has increased. This chapter discusses the use of highly fluorinated analogs of hydrophobic amino acids, for example, hexafluoroleucine, in protein design. In particular, fluorinated residues have proven to be generally effective in increasing the thermodynamic stability of proteins. The chapter provides an overview of the different fluorinated amino acids that have been used in protein design and the various methods available for producing fluorinated proteins. It discusses model proteins systems into which highly fluorinated amino acids have been introduced and the reasons why fluorinated residues are generally stabilizing, with particular reference to thermodynamic and structural studies from our laboratory. Lastly, details of the methodology we have developed to measure the thermodynamic stability of oligomeric fluorinated proteins are presented, as this may be generally applicable to many proteins.

1. INTRODUCTION

Fluorine is all but absent from biology; however, this element has proved remarkably useful as a probe to investigate the function of biological molecules. For example, ^{19}F NMR is a valuable tool for studying the, dynamics and interactions of fluorine-labeled biomolecules including proteins, peptides, nucleic acids, and lipids, whereas fluorinated substrates have been used to investigate the mechanisms of numerous enzymes (Buer, Chugh, Al-Hashimi, & Marsh, 2010; Dalvit & Vulpetti, 2011; Danielson & Falke, 1996; Evanics, Kitevski, Bezsonova, Forman-Kay, & Prosser, 2007; Gerig, 1994; Khan, Kuprov, Craggs, Hore, & Jackson, 2006; Luchette, Prosser, & Sanders, 2002; Suzuki, Brender, Hartman, Ramamoorthy, & Marsh, 2012; Suzuki et al., 2013; Suzuki, Buer, Al-Hashimi, & Marsh, 2011; Ye et al., 2013; Yu, Kodibagkar, Cui, & Mason, 2005). The importance of fluorinated molecules in medicine is illustrated by the fact that ~20% of all pharmaceuticals contain fluorine, which is thought to improve their pharmacokinetic properties (Müller, Faeh, & Diederich, 2007).

Various methods now allow a wide variety of unnatural amino acids to be incorporated into proteins. This has greatly expanded our ability to modify protein structure by introducing a diverse range of chemical functionality that is not present in natural proteins (Odar, Winkler, & Wiltschi, 2015; Ravikumar, Nadarajan, Yoo, Lee, & Yun, 2015a, 2015b). Over the last decade, our laboratory conducted a number of studies in which we have incorporated fluorinated amino acids into de novo designed proteins (Buer, de la Salud-Bea, Al Hashimi, & Marsh, 2009; Buer, Levin, & Marsh, 2012; Buer, Meagher, Stuckey, & Marsh, 2012a, 2012b; Gottler, de la Salud-Bea, & Marsh, 2008; Lee, Lee, Al-Hashimi, & Marsh, 2006; Lee, Lee, Slutsky, Anderson, & Marsh, 2004) and bioactive peptides. Our initial objective was to understand how fluorination stabilizes proteins against unfolding. We also realized that fluorination could be a useful tool to modify (Gottler, de la Salud Bea, Shelburne, Ramamoorthy, & Marsh, 2008; Gottler, Lee, Shelburne, Ramamoorthy, & Marsh, 2008) and report on the mechanisms of bioactive peptides such as antimicrobial peptides (Buer et al., 2010; Buer, Levin, & Marsh, 2013; Suzuki et al., 2011). In particular, we have exploited the NMR activity of fluorine to report on peptide–membrane interactions (Buer et al., 2010, 2013; Suzuki et al., 2011) and the pathways by which amyloid-forming peptides aggregate (Suzuki et al., 2012, 2013). Working on a variety of systems and using a wide variety of fluorinated amino acids numerous laboratories have made important contributions to our understanding of how fluorination modulates the properties of peptides and proteins (Bilgiçer, Fichera, & Kumar, 2001; Bilgiçer & Kumar, 2002, 2004; Bilgiçer, Xing, & Kumar, 2001; Campos-Olivas, Aziz, Helms, Evans, & Gronenborn, 2002; Chiu, Kokona, Fairman, & Cheng, 2009; Chiu et al., 2006; Cho et al., 2014; Clark, Baleja, & Kumar, 2012; Cornilescu et al., 2007; Horng & Raleigh, 2003; Huhmann et al., 2015; Jäckel, Salwiczek, & Koksch, 2006; Jäckel, Seufert, Thust, & Koksch, 2004; Kwon et al., 2010; Meng, Krishnaji, Beinborn, & Kumar, 2008; Meng & Kumar, 2007; Montclare, Son, Clark, Kumar, & Tirrell, 2009; Mortenson, Satyshur, Guzei, Forest, & Gellman, 2012; Naarmann, Bilgiçer, Meng, Kumar, & Steinem, 2006; Niemz & Tirrell, 2001; Pace, Zheng, Mylvaganam, Kim, & Gao, 2011; Pendley, Yu, & Cheatham, 2009; Salwiczek & Koksch, 2009; Senguen, Doran, Anderson, & Nilsson, 2011; Son, Tanrikulu, & Tirrell, 2006; Tang, Ghirlanda, Petka, et al., 2001; Tang, Ghirlanda, Vaidehi, et al., 2001; Tang & Tirrell, 2001; Wang, Fichera, Kumar, & Tirrell, 2004; Wang, Tang, & Tirrell, 2003; Woll, Hadley, Mecozzi, & Gellman, 2006; Zheng, Comeforo, & Gao, 2008; Zheng & Gao, 2010). This field continues to expand as evidenced in numerous review articles (Biava & Budisa, 2014; Buer & Marsh, 2012; Jäckel & Koksch, 2005; Krishnamurthy & Kumar, 2013; Marsh, 2014; Marsh, Buer, & Ramamoorthy, 2009; Marsh & Suzuki, 2014; Pace & Gao, 2013; Salwiczek, Nyakatura, Gerling, Ye, & Koksch, 2012; Yoder & Kumar, 2002; Yoder, Yüksel, Dafik, & Kumar, 2006).

In addition to their academic interest, the potential of fluorinated amino acids to increase the thermodynamic stability of proteins lends them to technological applications. Increasing the stability of protein-based therapeutics should decrease their susceptibility to proteases and thereby increase their potency and bioavailability. Similarly industrial processes that use enzymes as catalysts would benefit from methods to increase their resistance toward heat and chemical denaturation (Ravikumar et al., 2015a). In this article I discuss the basis for the stabilizing effects of fluorinated amino acids and some factors that should be considered when attempting to incorporate them into proteins and peptides. I also discuss some of the methods that our laboratory has found most useful in synthesizing and characterizing fluorinated proteins. The current article represents a revised, updated, and abbreviated version of a chapter we published in *Methods in Molecular Biology* in 2014 (Buer & Marsh, 2014). The reader is referred to that article for detailed protocols on peptide synthesis and crystallization that are omitted here.

1.1 Basis of Protein Stabilization by Fluorinated Amino Acid Residues

To utilize fluorinated amino acids effectively in protein design it is important to understand how they stabilize proteins. Extensively fluorinated analogs of hydrophobic amino acids generally stabilize protein structure and minimally perturb protein function. Soluble proteins derive their stability primarily from the sequestering of nonpolar residues into the hydrophobic protein interior, ie, the hydrophobic effect. The hydrophobic effect is entropic and involves the release of the ordered clathrate of water molecules that surround exposed hydrophobic residues in the unfolded state. The favorable free energy of folding due to the hydrophobic effect is thus proportional to the change in buried hydrophobic surface area on folding. Various studies on proteins and hydrophobic small molecules have quantified the hydrophobic effect, with the consensus being that it contributes 25–30 cal mol^{-1} $Å^{-2}$ of buried surface area to the stability of a globular protein (Baldwin, 2013; Chothia, 1974; Eriksson et al., 1992; Richards, 1977).

Early studies suggested that extensively fluorinated amino acids might manifest the hydrophobic effect in a different way from natural (hydrocarbon) amino acids (Bilgiçer, Fichera, et al., 2001; Bilgiçer & Kumar, 2002, 2004; Bilgiçer, Xing, et al., 2001; Jäckel et al., 2006, 2004; Lee et al., 2006, 2004; Marsh, 2000; Naarmann et al., 2006; Salwiczek & Koksch, 2009; Tang & Tirrell, 2001), this phenomenon has been described as the "fluorous" or "polar hydrophobic" effect (Biffinger, Kim, & DiMagno, 2004). For example, perfluorinated solvents exhibit unusual phase-segregating properties that have been effectively exploited in "fluorous" separation methodology to extract small molecules equipped with perfluorocarbon "tags" from reaction mixtures (Luo, Zhang, Oderaotoshi, & Curran, 2001). However, in the context of protein folding, our studies, which are discussed in detail below, indicate that highly fluorinated amino acids behave similarly to natural hydrophobic amino acids.

1.2 Choice of Fluorinated Amino Acids

A large number of fluorinated amino acids analogs are now commercially available and a selection of those most commonly used in designing fluorinated proteins and peptides are listed in Table 1. Unfortunately, hFLeu which has a high fluorine content and has been used extensively in the design of fluorinated peptides is not currently commercially available. However, several syntheses of this amino acid have been developed, including one from our laboratory and the reader is referred to the original literature for the synthetic procedures (Anderson, Toogood, & Marsh, 2002; Chiu & Cheng, 2007; Xing, Fichera, & Kumar, 2001). Often commercial fluorinated amino acids are available only as racemic mixtures that are not useful for peptide synthesis. We have found that enzymatic resolution (Tsushima et al., 1988) racemic of fluorinated amino acids using porcine kidney acylase I (Sigma) works well for resolving the pure enantiomers in quantities sufficient for peptide synthesis. The electronegativity of fluorine possesses problems for the incorporation of amino acids such as trifluoroalanine and hexafluorovaline; they are difficult to couple and easily racemize during peptide synthesis due to the acidity of the α-carbon (Sani et al., 2003; Vine, Hsieh, & Marshall, 1981).

1.3 Synthesis of Fluorinated Proteins

In general, the solid-phase peptide synthesis (SPPS) has been the method of choice for synthesizing fluorinated proteins, as this provides the greatest control in placing noncanonical amino acids within the sequence. For larger proteins, in vivo methods for incorporation of fluorinated amino acids (Merkel & Budisa, 2012) such as tFLeu, tFIle, tFVal, and hFLeu, which can be activated by endogenous tRNA synthetases, have been developed by the Tirrell laboratory (Montclare et al., 2009; Son et al., 2006; Tang, Ghirlanda, Petka, et al., 2001; Tang & Tirrell, 2001; Wang et al., 2004, 2003). One drawback is that the presence of a background of natural amino acids, derived from cellular proteins, reduces the efficiency of fluorinated amino acid incorporation, typically to 70–90%. This method is also limited to global substitution of a particular amino acid at all sites within a protein, which limits some applications (Tang, Ghirlanda, Vaidehi, et al., 2001; Tang & Tirrell, 2001; Wang et al., 2004, 2003). Site-specific incorporation of fluorinated amino acids could be achieved through the use of an orthogonal tRNA synthetase/amber-suppressing tRNA pair (Wang, Xie, & Schultz, 2006), as has been recently demonstrated for a fluorinated tyrosine

Table 1 Highly Fluorinated Amino Acids That Have Been Incorporated into Proteins

Amino Acid	Source	Racemic	Reference
Hexafluoroleucine (hFLeu)	Synthesized	No	Anderson et al. (2002), Chiu and Cheng (2007), Xing et al. (2001)
Trifluoroleucine (tFLeu)	Synthesized	—	Xing, Fichera, and Kumar (2002)
Trifluorovaline (tFVal)	Oakwood Chemical	Yes	Erdbrink et al. (2012), Xing et al. (2002)
Trifluoroisoleucine (tFIle)	Synthesized	—	Wang et al. (2003)
Trifluoromethylmethionine (tFMet)	Synthesized	No	Suzuki et al. (2013)
Trifluoroethylglycine (tFeG)	SynQuest, Oakwood Chemical	Yes	Tsushima et al. (1988)
Perfluoro-t-butylhomoserine (pFtBSer)	Synthesized	No	Buer et al. (2013)
Pentafluorophenylalanine (pFPhe)	SynQuest, Sigma	No	
Trifluoromethylphenylalanine (tFmPhe)	Chem-Impex International	No	Suzuki et al. (2012)

Commercial suppliers are listed, along with protocols to resolve enantiomers of racemic mixtures. Amino acids listed that are not commercially available include references to synthetic protocols.
*A racemic stereocenter.

analog (Li et al., 2013). However, although over 100 nonnatural amino acids have been site-specifically incorporated into numerous proteins, we are not aware of the method having been used to incorporate highly fluorinated amino acid analogs so far. Perhaps the best way to specifically incorporate fluorinated amino acids within a larger protein (longer than ~50 residues) is through expressed protein ligation techniques (Muir, 2003); again, though, there are no specific examples to our knowledge of extensively fluorinated proteins produced by this technique.

Our laboratory has synthesized of various fluorinated peptides using both Fmoc- and Boc-protection strategies (Fmoc, fluorenylmethyloxycarbonyl; Boc, *tert*-butyloxycarbonyl) to incorporate the following fluorinated amino acids (Table 1): tFeG, tFmPhe, tFMet, pFtBSer, and hFLeu. Fmoc-protected peptide synthesis has the advantages of mild basic Fmoc-deprotection using 20% piperidine, easy cleavage and deprotection of peptides at the end of the synthesis using trifluoroacetic acid, and amenability to automation. We have found standard Fmoc SPPS protocols work well for the synthesis of peptides containing limited numbers of tFeG, tFmPhe, tFMet, or pFtBSer. However, we encountered large decreases in coupling efficiency when attempting to synthesize peptides containing multiple hFLeu residues by Fmoc methodology, which occur after hFLeu incorporation. It is unclear why this happens, but manual Boc-SPPS methods have given reliable results with all the hFleu-containing peptides we have attempted to synthesize. Cleavage from the resin and deprotection of Boc-synthesized peptides require HF (Schnölzer, Alewood, Jones, Alewood, & Kent, 1992), which is very toxic and needs specialized equipment to handle, which reduces the popularity of this method. However, various peptide synthesis companies provide commercial HF cleavage, and for most laboratories this provides the easiest route to cleaving peptides. In our experience, Boc-SPPS more efficiently deprotects peptide side-chains, resulting in clean peptides that are easy to purify. This benefit justifies the additional time and expense of HF cleavage.

1.4 Designing Fluorinated Amino Acids into Proteins

Various model protein systems have been used to investigate the effects of fluorinated amino acids on protein structure and function. As will be seen, the successfulness of these designs depends on the context in which fluorine is introduced. Studies in our lab have focused on a model 4-helix bundle protein called α_4H (Lee et al., 2004). α_4H is a 27-residue peptide that is designed to adopt a tetrameric, antiparallel 4-α-helix bundle structure (Fig. 1). This provides a simple system to study the effects of fluorination because it possesses a well-defined and very regular hydrophobic core created by contacts between residues of adjacent helices. The hydrogen bonding pattern of an α-helix is defined by backbone amide bonding between every fourth residue (i to $i+4$). In a helical bundle motif, the α-helices supercoil around each other so that the pattern repeats every seven residues resulting in two turns of the helix. This repeating unit is referred to as the "heptad repeat," designated by the letters a through g. By convention, the a and d positions point into the hydrophobic core and therefore should

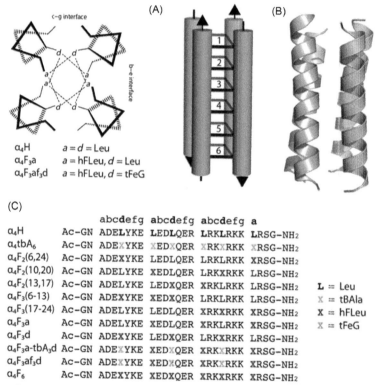

Fig. 1 Sequences and structures of α_4 proteins. (A) *Left* helical wheel diagram for an antiparallel 4-helix bundle represented by the α_4 proteins, illustrating the positions residues within the heptad repeat and *right* a cartoon representation of the six hydrophobic layers formed by two a and two d residues. (B) Ribbon diagram of α_4H derived from its crystal structure. (C) The 27-residue sequence of α_4 proteins designed in our studies, with noncanonical amino acid substitutions at a and d positions denoted by "X."

contain hydrophobic residues, whereas the b, c, e, and g positions form stabilizing salt bridges and hydrogen bonding interactions between helices and generally contain polar and charged residues. For α_4H which was designed to contain antiparallel helices, there are two distinct polar interfaces that are formed by interactions between residues in the b and e positions (b–e interface) and the c and g positions (c–g interface). Because of this, antiparallel helical bundles are more structurally robust than the parallel α-helix bundles in which all interfaces are identical. This topology can switch between oligomeric states in response to even subtle changes to the residues at a and d (Harbury, Zhang, Kim, & Alber, 1993).

We have exploited the regular hydrophobic packing of α_4H to examine the effects of various permutations of Leu and hFLeu residues at the hydrophobic a and d positions on the stability of the protein. For all the various proteins we have synthesized incorporating hFLeu results in a more stable protein that retains its intended structure. Incorporating hFLeu into a and d positions in the central layers of the core results in a larger increase in stability, $\sim$0.25 kcal mol^{-1} hFLeu residue^{-1}, than introducing hFLeu at the terminal a and d positions, $\sim$0.1 kcal mol^{-1} hFLeu residue^{-1}. This is explained by the fact that the terminal positions are partially solvent exposed even in the folded protein. The stability of the protein increases in proportion to the number of hFLeu residues incorporated. This observation suggests that, more generally, it should be possible to fine-tune protein stability by substituting fluorinated analogs of hydrocarbon residues into the hydrophobic cores of proteins. However, even in the very simple, regularly packed core of α_4H the increase in stability depends upon the context. Thus, we found that an alternating pattern of two Leu residues and two hFleu residues per layer of the hydrophobic core that is produced by introducing hFLeu at either only a (α_4F$_3$a), or only d positions (α_4F$_3$d) gives a significantly the greater per-residue stability of 0.80 and 0.72 kcal mol^{-1} hFLeu residue^{-1}, respectively (Buer et al., 2009).

Studies from other groups using different model systems similarly conclude that incorporating fluorinated residues into proteins enhances their stability. In a recent study Koksch and coworkers found that both hFLeu and tFVal increased the stability of a heterotrimeric coiled-coil system. However, the context is clearly important as the following examples from other small protein motifs demonstrate. An early study from the Kumar group found that introducing hFLeu into a two-stranded parallel coiled-coil significantly stabilized the structure but also resulted in a change in oligomerization state to a four-stranded coiled-coil (Bilgiçer & Kumar, 2004; Bilgiçer, Xing, et al., 2001).

In an interesting study, the Cheng group attempted to remove the influence of tertiary packing effects by evaluating the intrinsic α-helix and β-sheet-stabilizing propensities of various fluorinated amino acids in the context of a monomeric α-helix or in a solvent-exposed β-strand of a protein GB1 domain (Chiu et al., 2006, 2009). In the context of a monomeric α-helical (with alanine side-chains flanking in the $i+4$, and $i-4$ positions) fluorinated analogs of Phe, Leu, and ethylglycine were all, surprisingly, slightly *destabilizing* with respect to the nonfluorinated counterparts. In contrast, in the context of a solvent-exposed β-sheet the fluorinated analogs of Phe, Leu, and ethylglycine were stabilizing. However, when a hFLeu:hFleu cross-strand interaction was engineered into a β-hairpin-forming peptide it was found to be destabilizing with respect to the equivalent Leu:Leu interaction (Clark et al., 2012), this is most likely due to unfavorable steric effects and, again, illustrates the importance of context. Koksch and coworkers reached a similar conclusion when they investigated the effects of increasing the fluorine content atom-by-atom using monofluoro-, difluoro-, and trifluoroethylglycine residues, which were introduced into a model coiled-coil protein. In this case, it appears both steric effects and polarity effects, due to the strong electron-withdrawing effect of fluorine, contributed to modulating the stability of the protein (Jäckel et al., 2006).

Two studies have investigated the effects of fluorinated aromatic residues on protein stability. Gellman's group investigated the effect of substituting pFPhe at each of the three Phe residues in the small villin headpiece protein. This is a natural example of a short, folded protein fragment that possesses tertiary structure, and as such its core lacks the regularity of the de novo designed peptides used in most other studies. Interestingly, pFPhe was only stabilizing at one Phe position and destabilizing at the other two (Woll et al., 2006). More recently, the Schepartz group has investigated β-pentafluoro-homo-phenylalanine in the context of a β-peptide bundle (Molski et al., 2013). These are foldamers constructed of β-amino acids in which the side-chain is attached to the β-carbon of the peptide backbone. Designs have been produced that fold into helical bundles that mimic the more familiar α-helical bundle proteins. In this case, it was found that replacing a β-homo-phenylalanine residue in the hydrophobic core with β-pentafluoro-homo-phenylalanine resulted in a β-peptide bundle with a better structured, less molten globule-like as judged by NMR spectroscopy and the protection factors for backbone H/D exchange.

1.5 Evaluating Structural and Thermodynamic Effects of Incorporating Fluorinated Side-Chains

The above discussion illustrates the various contexts in which fluorinated amino acids have been found to stabilize protein structure, and occasionally destabilize it. It is important to consider the origin of this stabilizing effect if one is to exploit it to full effect in protein design. With this in mind, we now discuss some detailed studies conducted in our laboratory on the structures and thermodynamics of a set of fluorinated proteins based on the α_4H design.

1.5.1 Thermodynamic Analysis

Determining such quantities ΔG°_{unfold} and $\Delta H^{\circ\prime}$, $\Delta S^{\circ\prime}$, and ΔC°_p provides valuable information on the physicochemical basis for protein stability. These measurements rely on perturbing the equilibrium between the folded and unfolded state by using denaturants such as guanidinium hydrochloride (GuHCl) and thermal unfolding. For determination of ΔG°_{unfold} we prefer to use GuHCl to generate unfolding curves as their analysis requires fewer assumptions than thermal unfolding. For determining $\Delta H^{\circ\prime}$, $\Delta S^{\circ\prime}$, and ΔC°_p it is necessary to employ a combination of chemical and thermal denaturation to obtain robust fits to the Gibbs–Helmholtz equation. In our research, circular dichroism spectroscopy has proved a very useful technique to monitor protein unfolding curves. In particular, GuHCl-induced unfolding curves can be generated using commercially available autotitration apparatus that interfaces directly with the CD spectrometer and allows unfolding curves to be determined automatically. An important advantage of automation is that this typically yields much more accurate results than can be obtained by hand. Although information on the thermodynamics of protein stability can be obtained by a relatively simple combination of unfolding experiments, fitting such data to extract thermodynamic quantities is a little more involved. Typically for a full thermodynamic analysis we measure thermal denaturation curves at 10–12 concentrations of GuHCl, determining measurements every 2°C between 4°C and 90°C, generating data sets of 400–500 points (Fig. 2A). For this reason, and the fact that this method may be generally useful to fit data for a wide range of proteins we have included a fairly detailed explanation for how to analyze such data in Section 2 of this chapter.

We undertook a thermodynamic analysis of 12 α_4 variants in which fluorinated residues were introduced at different positions within the hydrophobic core. From analysis of the thermal and GuHCl-induced unfolding curves

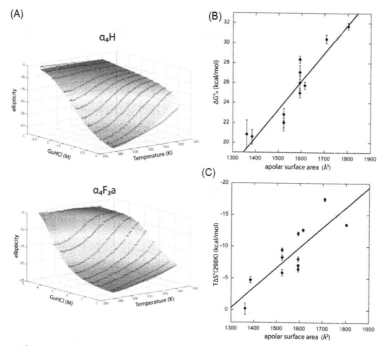

Fig. 2 Thermodynamic analysis of the effects of fluorination on protein folding. (A) Denaturation using heat and GuHCl with experimental data represented by *black circles* and data fitting to determine $\Delta H°$, $\Delta S°$, and $\Delta C_p°$ as *colored (gray shades* in the print version) *surface*. The unfolding of α_4H and α_4F$_3$a was monitored by changes in ellipticity at 222 nm. (B) The correlation between $\Delta G°_{unfold}$ ($T=298$ K) and buried apolar (hydrophobic) surface area for a series of 11 α_4 proteins containing varying numbers of fluorinated residues. (C) The correlation between $T\Delta S°_{unfold}$ ($T=298$ K) and buried apolar (hydrophobic) surface area for the same 11 α_4 proteins. *Data from Buer, B. C., Levin, B. J., & Marsh, E. N. G. (2012). Influence of fluorination on the thermodynamics of protein folding. Journal of the American Chemical Society, 134(31), 13027–13034.*

we determined $\Delta G°$, $\Delta H°$, $\Delta S°$, and $\Delta C_p°$ for unfolding for each of these variants. From these data (Fig. 2B and C) we were able to show that there is a strong correlation between the increase in $\Delta G°_{unfold}$ and the increase in buried hydrophobic surface area associated with replacing a methyl group with a trifluoromethyl group (the volume increase is ~16 Å^3 (Buer, Levin, & Marsh, 2012)). This increase is mainly accounted for by an increase in $T\Delta S°$, as expected for the hydrophobic effect. If one calculates the stabilization energy in terms of $\Delta G°_{unfold}$ per Å^2 of buried surface area one obtains a value of ~28 cal mol^{-1} Å^{-2}; this is very similar to that found for natural hydrophobic amino acids. Therefore, there appear to be no special fluorine–fluorine interactions or "fluorous" effects operating, which would be expected to show up as a change $\Delta H°$, operating to stabilize these proteins.

1.5.2 Crystallographic Analysis

To answer the question of why fluorinated amino acids appear to be so generally effective at stabilizing protein structure, we have determined the structures of several α_4 variants by crystallography. The α_4 variants we studied contained Leu (the parent structure), hFLeu (a larger fluorinated side-chain), β-t-butylalanine (tBAla, a larger nonfluorinated side-chain), or trifluoroethylglycine (tFeG, a smaller fluorinated side-chain) within their hydrophobic cores (Buer et al., 2012a, 2012b). Analysis of these structures demonstrated that the fluorinated residues integrate into the hydrophobic core of the protein with minimal perturbation to its structure. Thus, fluorination appears to be uniquely suited to stabilizing proteins because, while replacing hydrogen atoms with fluorine atoms increases its volume, it very closely preserves the side-chain's shape and it is the shape of the side-chain that is critical to packing.

Two examples illustrate this feature of fluorine. Fig. 3 compares the structures of α_4H protein with its variant containing hFLeu at the a positions, α_4F_3a, and containing hFLeu at the a positions and tFeG at the d positions, α_4F_3a,f_3d. Overall, the structures of the hydrophobic cores α_4H and α_4F_3d are very similar despite the fact that the hFLeu residue is significantly larger and contributes $\sim$0.8 kcal mol^{-1} residue^{-1} to the protein's stability. If, in addition, the leucine at the d position is replaced with tFeG, to give α_4F_3a,f_3d, again the structural changes are minimal. However, this combination of large and small fluorinated residues energetically cancel each other out, so that even though this protein contains more fluorine it is no more stable than the original α_4H protein.

In contrast, the introduction of β-t-butylalanine at both a and d positions, to give the α_4 variant $\alpha_4$tbA$_6$, is much more disruptive (Fig. 3). β-t-Butylalanine is slightly stabilizing relative to Leu, by $\sim$0.18 kcal.mol^{-1} residue^{-1}, due to its increased size and hydrophobicity. However, although the increase in hydrophobic core volume is almost the same as for α_4F_3d, the change in shape imparted by the additional methyl group causes the hydrophobic core to reorganize so that a destabilizing channel now runs through the center of the protein.

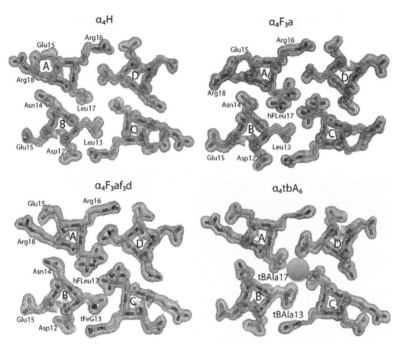

Fig. 3 X-ray structures of four α_4 proteins packed with either fluorinated or non-fluorinated amino acid residues. The figure displays a cross-section of one layer of the hydrophobic core packing (the same layer in each protein). α_4H is the parent protein, packed with Leu at *a* and *d* positions. hFLeu is well accommodated in the structure of α_4F$_3$a which has hFLeu at the *a* position, and in α_4F$_3$a,f$_3$d which has tFeG in the *d* position. However, the introduction of tBAla into the structure of $\alpha_4$tbA$_6$ reorganizes the protein core so that a void runs through the center of the protein, represented by the *circle*. Electron density maps ($2F_o$–F_c) are contoured at 1.0σ.

1.6 Concluding Remarks

Highly fluorinated analogs of hydrophobic amino acids have shown great potential in stabilizing folded proteins. However, it should be evident the above discussion that although fluorinated amino acids can be incorporated into a wide variety of protein structures, increases in stability are not guaranteed. In general, researchers have considered fluorinated amino acids to be closely isosteric with their hydrocarbon counterparts, and have substituted them into protein structures on the assumption that they are interchangeable—ie, they do not perturb the structure significantly. In the context of α-helical structures, this assumption appears to be reasonable, as our investigations have demonstrated. The highly fluorinated amino acids have also been found to stabilize β-sheet structures and more complex structural motifs found in natural proteins, although, as discussed above, the evidence is mixed.

The design of a fluorinated β-peptide foldamer, which utilized the automated protein repacking software suite *Rosetta* developed by Baker and coworkers (Das & Baker, 2008; Leaver-Fay et al., 2011) in the initial design stages, points to the possibility of explicitly designing fluorinated residues into protein structures. This would allow local repacking of natural residues to accommodate the fluorinated residue(s) of interest. We are not aware of an example where this has been done, but there is now sufficient information regarding the properties of commonly used fluorinated side-chains for them to be parameterized for use in protein design programs such as *Rosetta*.

Section 2 has primarily focused on the principles by which fluorination stabilizes proteins with the aim of providing guidance for others who may wish to utilize fluorinated proteins for various applications. Although beyond the scope of this review, the excellent NMR properties of fluorine open up many avenues for the study of protein dynamics and their interaction with other biological macromolecules, which we and others have begun to explore (Buer et al., 2010, 2013; Li et al., 2013; Liu, Horst, Katritch, Stevens, & Wuthrich, 2012; Pomerantz et al., 2012; Smith, Zhou, Gorensek, Senske, & Pielak, 2016; Suzuki et al., 2012, 2013, 2011; Ye et al., 2013; Zigoneanu, Yang, Krois, Haque, & Pielak, 2012). As our ability to design and incorporate noncanonical amino acids into proteins becomes more sophisticated, other applications will open up. For example, the design of enzyme active sites using highly fluorinated amino acids is particularly interesting. Fluorinated residues could be used to stabilize the enzyme's structure in "harsh" environments, eg, organic solvents and/or high temperature. Another more speculative possibility is to design active sites that are resistant to oxidative damage caused by reactive oxidizing species, such as those generated by cytochrome P450 enzymes. This could allow the design of enzymes "fire-proof" that generate highly reactive species capable of oxidizing unreactive organic molecules.

2. METHODS

2.1 Measuring the Thermodynamic Stability of Fluorinated Proteins by Circular Dichroism

Our lab routinely employs GuHCl titrations to measure the changes in $\Delta G^{\circ}_{unfold}$ that occur when fluorinated residues are introduced into synthetic proteins. This method is generally superior to thermal denaturation for the coiled-coil proteins we study as they are characterized by very broad thermal transitions. The free energy of unfolding ($\Delta G°_{unfold}$) can be determined from a single titration at fixed protein concentration and temperature, using circular dichroism spectroscopy to monitor the change in secondary structure associated with the unfolding transition (Kelly & Price, 1997) (Fig. 3A). The analysis of unfolding transitions is straightforward for monomeric and dimeric proteins (assuming it is a two-state process) and the data can be fitted to standard equations (Jackson & Fersht, 1991). For trimeric and tetrameric proteins exact solutions can, in principle, be obtained to model the unfolding curves, but in practice these are too cumbersome to fit to experimental data. Therefore, we use approximate methods to fit such data using programs such as MATLAB (MathWorks Inc.). The routine that we use to fit the data is included at the end of this chapter.

2.2 Determining $\Delta G°$ from GuHCl-Induced Protein Unfolding

It should first be noted that this analysis assumes that the proteins in question unfold reversibly and follow a two-state equilibrium between the folded oligomeric protein (F) and the unfolded monomeric peptide (U), as shown in Eq. (1). This is characterized by an equilibrium constant $K([\text{GuHCl}])$ that is dependent on GuHCl concentration

$$F \Leftrightarrow nU. \tag{1}$$

Eq. (2) relates $K([\text{GuHCl}])$ to [F], [U], and [P], which are the concentrations of folded tetramer, unfolded monomer, and total protein, respectively, so that $[P] = n[F] + [U]$:

$$K([\text{GuHCl}]) = \frac{[U]^n}{[F]} = \frac{n[U]^n}{[P] - [U]}. \tag{2}$$

Rearrangement of Eq. (2) results in the polynomial expression Eq. (3):

$$[U]^n + \frac{K([\text{GuHCl}])}{n}[U] - \frac{K([\text{GuHCl}])}{n}[P] = 0. \tag{3}$$

For a fixed [P] and given any nonnegative value of $K([\text{GuHCl}])$, Eq. (3) has a unique solution for [U] between 0 and [P]. Eq. (3) can be solved numerically, which allows $K([\text{GuHCl}])$ to be calculated at each GuHCl concentration. The protein's stability as a function of GuHCl concentration is modeled by the following relationship:

$$\Delta G°([\text{GuHCl}]) = \Delta G°(0M\,\text{GuHCl}) - m^*[\text{GuHCl}] \tag{4}$$

$$K([\text{GuHCl}]) = \exp\left(\frac{-(\Delta G°(0M\,\text{GuHCl}) - m^*[\text{GuHCl}])}{RT}\right), \quad (5)$$

where $\Delta G°(0\,M\,\text{GuHCl})$ is the stability of a protein in the absence of GuHCl and m is the dependence of stability on GuHCl concentration. $K(\text{GuHCl}])$ is then given by Eq. (5) and global fitting of $K([\text{GuHCl}])$ as a function of [GuHCl] allows the values of $\Delta G°$ and m to be calculated.

2.2.1 Treatment of Baselines

The unfolding curve of a protein as a function of GuHCl concentration is sigmoidal with the pre- and posttransition baselines corresponding to the ellipticity of folded protein (θ_f) and unfolded protein (θ_u). Note, it is important that good measurements of baselines are obtained on either side of the unfolding transition to accurately fit the unfolding curve and reliably determine $\Delta G°_{\text{unfold}}$. The ellipticity of the unfolded and folded proteins is assumed to vary linearly with [GuHCl] and is modeled using Eqs. (6) and (7), where the parameters a and d are the baseline intercept at $0\,M$ GuHCl while c and f describe the baseline slope for unfolded and folded protein, respectively

$$\theta_u([\text{GuHCl}]) = a + c^*[\text{GuHCl}], \quad (6)$$
$$\theta_f([\text{GuHCl}]) = d + f^*[\text{GuHCl}]. \quad (7)$$

The observed ellipticity is the sum of the contributions from the unfolded and folded fractions of protein and is described by Eq. (8)

$$\theta_{O\,\text{bsd}} = \theta_u([\text{GuHCl}])\frac{[U]}{[P]} + \theta_f([\text{GuHCl}])\frac{[P]-[U]}{[P]}. \quad (8)$$

Eqs. (3) and (5)–(7) are substituted implicitly into Eq. (8), which can be fitted to the data using the program MATLAB; see Section 2.4 for MATLAB code to calculate values for a, c, d, f, $\Delta G°$, and m.

2.3 Determining $\Delta H°\prime$, $\Delta S°\prime$, and $\Delta C°_p$ from Heat and GuHCl-Induced Protein Unfolding

The thermodynamic parameters $\Delta H°\prime$, $\Delta S°\prime$, and $\Delta C°\prime_p$ are helpful in diagnosing what types of physical interactions may be contributing to the stability of a protein. Obtaining these parameters requires a more detailed thermodynamic analysis involving fitting denaturation profiles to the Gibbs–Helmholtz equation (9). A single unfolding curve does not contain sufficient information to provide a unique fit to the Gibbs–Helmholtz equation. Instead, a two-dimensional approach is used in which the protein is thermally denatured at different GuHCl concentrations (Fig. 2A). The resulting unfolding surface can then be fitted to the Gibbs–Helmholtz equation by approximate methods as detailed below.

To assess an individual protein, temperature-unfolding data from each GuHCl concentration experiment are assembled into a spreadsheet with columns corresponding to temperature (K), CD output (ellipticity), and GuHCl concentration (M). The thermal unfolding of the protein is modeled by the unfolding of a folded n-mer to n unfolded monomers as in Eq. (1) in which the equilibrium constant $K(T,[\text{GuHCl}])$ is similar to that in Eqs. (2) and (3), but is now dependent on both temperature and denaturant concentration.

The values of $\Delta H°$, $\Delta S°$, and $\Delta C_p°$ associated with protein unfolding are calculated by fitting $K(T,[\text{GuHCl}])$ to the Gibbs–Helmholtz equation (9), which is modified by assuming that the Gibbs free energy, $\Delta G°$, varies linearly with GuHCl concentration, as described by Eq. (4), to give Eq. (10)

$$\Delta G°(T) = \Delta H° - T\Delta S° + \Delta C_p^{\circ*}\left(T - T_0 + T\ln\frac{T_0}{T}\right) \quad (9)$$

$$\Delta G°(T,[\text{GuHCl}]) = \Delta H° - T\Delta S° + \Delta C_p^{\circ*}\left(T - T_0 + T\ln\frac{T_0}{T}\right) - m^*[\text{GuHCl}] \quad (10)$$

In these equations T is temperature, T_0 is the reference temperature of 25°C, $\Delta H°$ is the change in enthalpy, $\Delta S°$ is the change in entropy, and $\Delta C_p°$ is the change in heat capacity, each at the reference temperature T_0. It has been observed that $\Delta C_p°$ and m change little over the measured range of denaturant concentration and temperature and is assumed to be constant (Kuhlman & Raleigh, 1998; Yi, Scalley, Simons, Gladwin, & Baker, 1997). $K(T,[\text{GuHCl}])$ is then given by Eq. (11) and global fitting of $K(T,[\text{GuHCl}])$ as a function of T and $[\text{GuHCl}]$ allows the values of $\Delta H°$, $\Delta S°$, $\Delta C_p°$, $\Delta G°$, and m to be calculated (Kuhlman & Raleigh, 1998).

$$K(T,[\text{GuHCl}]) = \exp\left(\frac{-\left(\Delta H° - T\Delta S° + \Delta C_p^{\circ*}\left(T - T_0 + T\ln\frac{T_0}{T}\right) - m^*[\text{GuHCl}]\right)}{RT}\right). \quad (11)$$

2.3.1 Treatment of Baseplanes

Plotting the ellipticity of α_4 proteins as a function of GuHCl concentration and temperature results in a two-dimensional surface with the pre- and post-transition base planes corresponding to the ellipticity of folded protein (θ_f) and unfolded protein (θ_u). The ellipticity of the unfolded and folded proteins is assumed to vary linearly with T and [GuHCl], and is modeled using Eqs. (12) and (13), where the parameters a, b, c, d, e, and f describe the ellipticity of the folded and unfolded states at various temperatures and GuHCl concentrations

$$\theta_u(T, [\text{GuHCl}]) = a + b*T + c*[\text{GuHCl}], \quad (12)$$
$$\theta_f(T, [\text{GuHCl}]) = d + e*T + f*[\text{GuHCl}]. \quad (13)$$

The observed ellipticity is the sum of the contributions from the unfolded and folded fractions of protein and is described by Eq. (14)

$$\theta_{\text{Obsd}} = \theta_u(T, [\text{GuHCl}])\frac{[U]}{[P]} + \theta_f(T, [\text{GuHCl}])\frac{[P]-[U]}{[P]}. \quad (14)$$

Eqs. (3) and (11)–(13) are substituted implicitly into Eq. (14) which can be fitted to the data using the program MATLAB; the MATLAB code to calculate values for a, b, c, d, e, f, $\Delta H°$, $\Delta S°$, $\Delta C_p°$, and m is included in Section 2.5. In our experiments, we aim to collect between 400 and 500 data points, measured at at least five different GuHCl concentrations, to obtain a robust fit. Values for e and f sometimes cannot be obtained for marginally stable proteins because there are insufficient data points to define the folded base plane. In such cases e and f best set to zero. We have found that this approximation does not usually introduce significant errors into the analysis.

2.4 MATLAB Code for Determining $\Delta G°$ Using GuHCl Unfolding

```
%Change the parameters below
n=4; %This represents monomer to n-mer folding
P=40e-6; %Free monomer concentration in Molarity (SI units only!)
T0=298.15; %Reference Temperature in K (298.15 is room temperature)

%Make sure the data is saved in two columns where each row is a math of
%(Concentration denaturant, Ellipticity)

%Concentration must be in Molarity
%For reference, %1 kcal=4184 J exactly
```

```
%Do not change anything below this point for standard use
R=8.3145; %Ideal Gas Constant

data=uiimport; %Import data; it will be saved as 'data'
data=data.data; %MatLab Glitch; this is the workaround
Den=data(:,1);
Theta_Obsd=data(:,2);

K=@(b,Den) exp(-(b(1).*ones(length(Den),1)-b(2).*Den)./(R*T0));
U=@(b,Den) arrayfun(@(k) fzero(@(x) n*x^n+k*x-k*P, [0 P]), K(b,Den));
f=@(b,Den) (b(3).*ones(length(Den),1) + b(5).*Den).*U(b,Den)./P +
(b(4).*ones(length(Den),1) + b(6).*Den).*(P.*ones(length(Den),1)-U
(b,Den))./P;

%Initial Values Module
beta0=zeros(6,1);
beta0(3)=max(Theta_Obsd)+1;
beta0(4)=min(Theta_Obsd)-1;
Utest=P.*(Theta_Obsd-(beta0(4).*ones(length(Theta_Obsd),1)))./
(beta0(3)-beta0(4));
Ktest=n.*(Utest).^n./(P.*ones(length(Utest),1)-Utest);
DGtest=-R.*T0.*log(Ktest);

TestMat=ones(length(DGtest),2);
for i=1:length(DGtest)
    TestMat(i,1)=1;
    TestMat(i,2)=-Den(i);
end

ParaEst=linsolve(TestMat,DGtest);
beta0(1)=ParaEst(1);
beta0(2)=ParaEst(2);

[beta, r, J, COVB, mse]=nlinfit(Den, Theta_Obsd, f, beta0);

ci=nlparci(beta,r,'covar', COVB);
```

2.5 MATLAB Code for Determining $\Delta H°$, $\Delta S°$, and $\Delta C°_p$ Using Heat and GuHCl Unfolding

```
%Change the parameters below
n=4; %This represents monomer to n-mer folding
P=40e-6; %Free monomer concentration in Molarity (SI units only!)
```

```
T0=298.15; %Reference Temperature in K (298.15 is room temperature)
%Make sure data is saved in three columns:
    %Temp (Celcius), Theta, (Molar)
%For reference, %1 kcal=4184 J exactly

%Do not change anything below this point for standard use
R=8.3145; %Ideal Gas Constant in SI units
data=uiimport; %Import data; it will be saved as 'data'
data=data.data; %MatLab Glitch; this is the workaround
Temp=data(:,1)+273.15; %Converts to Kelvin
Theta_Obsd=data(:,2);
Den=data(:,3);
Cond=[Temp Den];
%Cond is the n x 2 matrix of Temp and Den, and Theta_Obsd is output

%Define additional functions K, U, and f
K=@(b,Cond)exp(-(b(3)-Cond(:,1)*b(4)+b(5)*(Cond(:,1)-T0 ...
    +Cond(:,1).*log(T0./Cond(:,1)))-Cond(:,2)*b(6))./(R*Cond(:,1)));
U=@(b,Cond) arrayfun(@(k) fzero(@(x) n*x^n+k*x-k*P, [0 P]),K(b,Cond));
f=@(b,Cond)1/P*((b(1)*ones(length(Cond(:,1)),1) + b(7).*Cond(:,1) +
b(8).*Cond(:,2)).*U(b,Cond)...
+(b(2)*ones(length(Cond(:,1)),1)+b(9).*Cond(:,1)+b(10).*(Cond(:,2))).
*(P*ones(length(Cond(:,1)),1)-U(b,Cond)));

%Initial Values Module
beta0=zeros(10,1);
beta0(1)=max(Theta_Obsd)+1;
beta0(2)=min(Theta_Obsd)-1;
Uest=P*(Theta_Obsd-beta0(2)*ones(length(Theta_Obsd),1))/(beta0(1)-
beta0(2));
Kest=n*Uest.^n./(P*ones(length(Temp),1)-Uest);
DGest=-R*Temp.*log(Kest);

TestMat=ones(length(Temp),4);
for i=1:length(Temp)
    TestMat(i,1)=1;
    TestMat(i,2)=-Temp(i);
    TestMat(i,3)=Temp(i)-T0+Temp(i)*log(T0/Temp(i));
    TestMat(i,4)=-Den(i);
end
```

```
ParaEst=linsolve(TestMat,DGest);

beta0=[max(Theta_Obsd); min(Theta_Obsd); ParaEst(1); ParaEst(2);
ParaEst(3);ParaEst(4);0;0;0;0];
%Actual Curve Fitting Portion
[beta, r, J, COVB, mse]=nlinfit(Cond, Theta_Obsd, f, beta0);

ci =nlparci(beta,r,'covar', COVB);
```

ACKNOWLEDGMENTS

I would like to acknowledge the contributions of the following scientists who have worked on various aspects of the design of fluorinated proteins and peptides in my laboratory: James Anderson, Morris Slutsky, Kyung-Hoon Lee, Hwang-Yeol Lee, Lindsey Gottler, Roberto de la Salud-Bea, Benjamin Buer, Yuta Suzuki, and Benjamin Levin. These projects have been funded, in part, by the following organizations: National Science Foundation (CHE 0640934), the Army Research Office (W911NF-11-1-0251), and the Defense Threat Reduction Agency (HDTRA1-11-1-0019).

REFERENCES

Anderson, J. T., Toogood, P. L., & Marsh, E. N. G. (2002). A short and efficient synthesis of 1-5, 5, 5, 5', 5', 5'-hexafluoroleucine from N-Cbz-l-serine. *Organic Letters*, *4*(24), 4281–4283.

Baldwin, R. L. (2013). Properties of hydrophobic free energy found by gas–liquid transfer. *Proceedings of the National Academy of Sciences of the United States of America*, *110*(5), 1670–1673.

Biava, H., & Budisa, N. (2014). Evolution of fluorinated enzymes: An emerging trend for biocatalyst stabilization. *Engineering in Life Sciences*, *14*(4), 340–351. http://dx.doi.org/10.1002/elsc.201300049.

Biffinger, J. C., Kim, H. W., & DiMagno, S. G. (2004). The polar hydrophobicity of fluorinated compounds. *ChemBioChem*, *5*(5), 622–627.

Bilgiçer, B., Fichera, A., & Kumar, K. (2001). A coiled coil with a fluorous core. *Journal of the American Chemical Society*, *123*(19), 4393–4399. http://dx.doi.org/10.1021/ja002961j.

Bilgiçer, B., & Kumar, K. (2002). Synthesis and thermodynamic characterization of self-sorting coiled coils. *Tetrahedron*, *58*(20), 4105–4112. http://dx.doi.org/10.1016/s0040-4020(02)00260-0.

Bilgiçer, B., & Kumar, K. (2004). De novo design of defined helical bundles in membrane environments. *Proceedings of the National Academy of Sciences of the United States of America*, *101*(43), 15324–15329. http://dx.doi.org/10.1073/pnas.0403314101.

Bilgiçer, B., Xing, X., & Kumar, K. (2001). Programmed self-sorting of coiled coils with leucine and hexafluoroleucine cores. *Journal of the American Chemical Society*, *123*(47), 11815–11816. http://dx.doi.org/10.1021/ja016767o.

Buer, B. C., Chugh, J., Al-Hashimi, H. M., & Marsh, E. N. G. (2010). Using fluorine nuclear magnetic resonance to probe the interaction of membrane-active peptides with the lipid bilayer. *Biochemistry*, *49*(27), 5760–5765. http://dx.doi.org/10.1021/bi100605e.

Buer, B. C., de la Salud-Bea, R., Al Hashimi, H. M., & Marsh, E. N. G. (2009). Engineering protein stability and specificity using fluorous amino acids: The importance of packing effects. *Biochemistry*, *48*(45), 10810–10817. http://dx.doi.org/10.1021/bi901481k.

Buer, B. C., Levin, B. J., & Marsh, E. N. G. (2012). Influence of fluorination on the thermodynamics of protein folding. *Journal of the American Chemical Society, 134*(31), 13027–13034.

Buer, B. C., Levin, B. J., & Marsh, E. N. G. (2013). Perfluoro-tert-butyl-homoserine as a sensitive 19 F NMR reporter for peptide–membrane interactions in solution. *Journal of Peptide Science: An Official Publication of the European Peptide Society, 19*(5), 308–314.

Buer, B. C., & Marsh, E. N. G. (2012). Fluorine: A new element in protein design. *Protein Science, 21*(4), 453–462.

Buer, B. C., & Marsh, E. N. G. (2014). Design, synthesis, and study of fluorinated proteins. In V. Kohler (Ed.), *Protein Design: Methods and Applications: Vol. 1216* (pp. 89–116) (2nd ed.).

Buer, B. C., Meagher, J. L., Stuckey, J. A., & Marsh, E. N. G. (2012). Comparison of the structures and stabilities of coiled-coil proteins containing hexafluoroleucine and t-butylalanine provides insight into the stabilizing effects of highly fluorinated amino acid side-chains. *Protein Science, 21*(11), 1705–1715.

Buer, B. C., Meagher, J. L., Stuckey, J. A., & Marsh, E. N. G. (2012). Structural basis for the enhanced stability of highly fluorinated proteins. *Proceedings of the National Academy of Sciences of the United States of America, 109*(13), 4810–4815. http://dx.doi.org/10.1073/pnas.1120112109.

Campos-Olivas, R., Aziz, R., Helms, G. L., Evans, J. N. S., & Gronenborn, A. M. (2002). Placement of 19 F into the center of GB1: Effects on structure and stability. *FEBS Letters, 517*(1–3), 55–60. http://dx.doi.org/10.1016/s0014-5793(02)02577-2.

Chiu, H.-P., & Cheng, R. P. (2007). Chemoenzymatic synthesis of (S)-hexafluoroleucine and (S)-tetrafluoroleucine. *Organic Letters, 9*(26), 5517–5520.

Chiu, H.-P., Kokona, B., Fairman, R., & Cheng, R. P. (2009). Effect of highly fluorinated amino acids on protein stability at a solvent-exposed position on an internal strand of protein G B1 domain. *Journal of the American Chemical Society, 131*(37), 13192–13193.

Chiu, H.-P., Suzuki, Y., Gullickson, D., Ahmad, R., Kokona, B., Fairman, R., … Cheng, R. P. (2006). Helix propensity of highly fluorinated amino acids. *Journal of the American Chemical Society, 128*(49), 15556–15557.

Cho, J., Sawaki, K., Hanashima, S., Yamaguchi, Y., Shiro, M., Saigo, K., … Ishida, Y. (2014). Stabilization of beta-peptide helices by direct attachment of trifluoromethyl groups to peptide backbones. *Chemical Communications, 50*(69), 9855–9858. http://dx.doi.org/10.1039/c4cc02136c.

Chothia, C. (1974). Hydrophobic bonding and accessible surface area in proteins. *Nature, 248*(5446), 338–339.

Clark, G. A., Baleja, J. D., & Kumar, K. (2012). Cross-strand interactions of fluorinated amino acids in β-hairpin constructs. *Journal of the American Chemical Society, 134*(43), 17912–17921.

Cornilescu, G., Hadley, E. B., Woll, M. G., Markley, J. L., Gellman, S. H., & Cornilescu, C. C. (2007). Solution structure of a small protein containing a fluorinated side chain in the core. *Protein Science, 16*(1), 14–19.

Dalvit, C., & Vulpetti, A. (2011). Fluorine–protein interactions and 19 F NMR isotropic chemical shifts: An empirical correlation with implications for drug design. *ChemMedChem, 6*(1), 104–114. http://dx.doi.org/10.1002/cmdc.201000412.

Danielson, M. A., & Falke, J. J. (1996). Use of F-19 NMR to probe protein structure and conformational changes. *Annual Review of Biophysics and Biomolecular Structure, 25*, 163–195.

Das, R., & Baker, D. (2008). Macromolecular modeling with Rosetta. *Annual Review of Biochemistry, 77*, 363–382.

Erdbrink, H., Peuser, I., Gerling, U. I., Lentz, D., Koksch, B., & Czekelius, C. (2012). Conjugate hydrotrifluoromethylation of α, β-unsaturated acyl-oxazolidinones: Synthesis of chiral fluorinated amino acids. *Organic & Biomolecular Chemistry, 10*(43), 8583–8586.

Eriksson, A., Baase, W. A., Zhang, X., Heinz, D., Blaber, M., Baldwin, E. P., ... Matthews, B. (1992). Response of a protein structure to cavity-creating mutations and its relation to the hydrophobic effect. *Science, 255*(5041), 178–183.

Evanics, F., Kitevski, J. L., Bezsonova, I., Forman-Kay, J., & Prosser, R. S. (2007). 19 F NMR studies of solvent exposure and peptide binding to an SH3 domain. *BBA-General Subjects, 1770*(2), 221–230. http://dx.doi.org/10.1016/j.bbagen.2006.10.017.

Gerig, J. T. (1994). Fluorine NMR of proteins. *Progress in Nuclear Magnetic Resonance Spectroscopy, 26*, 293–370. http://dx.doi.org/10.1016/0079-6565(94)80009-x.

Gottler, L. M., de la Salud Bea, R., Shelburne, C. E., Ramamoorthy, A., & Marsh, E. N. G. (2008). Using fluorous amino acids to probe the effects of changing hydrophobicity on the physical and biological properties of the β-hairpin antimicrobial peptide protegrin-1. *Biochemistry, 47*(35), 9243–9250. http://dx.doi.org/10.1021/bi801045n.

Gottler, L. M., de la Salud-Bea, R., & Marsh, E. N. G. (2008). The fluorous effect in proteins: Properties of α4F6, a 4-α-helix bundle protein with a fluorocarbon core. *Biochemistry, 47*(15), 4484–4490. http://dx.doi.org/10.1021/bi702476f.

Gottler, L. M., Lee, H.-Y., Shelburne, C. E., Ramamoorthy, A., & Marsh, E. N. G. (2008). Using fluorous amino acids to modulate the biological activity of an antimicrobial peptide. *ChemBioChem, 9*(3), 370–373. http://dx.doi.org/10.1002/cbic.200700643.

Harbury, P. B., Zhang, T., Kim, P. S., & Alber, T. (1993). A switch between two-, three-, and four-stranded coiled coils in GCN4 leucine zipper mutants. *Science, 262*(5138), 1401.

Horng, J.-C., & Raleigh, D. P. (2003). Φ-Values beyond the ribosomally encoded amino acids: Kinetic and thermodynamic consequences of incorporating trifluoromethyl amino acids in a globular protein. *Journal of the American Chemical Society, 125*(31), 9286–9287.

Huhmann, S., Nyakatura, E. K., Erdbrink, H., Gerling, U. I. M., Czekelius, C., & Koksch, B. (2015). Effects of single substitutions with hexafluoroleucine and trifluorovaline on the hydrophobic core formation of a heterodimeric coiled coil. *Journal of Fluorine Chemistry, 175*, 32–35. http://dx.doi.org/10.1016/j.jfluchem.2015.03.003.

Jäckel, C., & Koksch, B. (2005). Fluorine in peptide design and protein engineering. *European Journal of Organic Chemistry, 2005*(21), 4483–4503. http://dx.doi.org/10.1002/ejoc.200500205.

Jäckel, C., Salwiczek, M., & Koksch, B. (2006). Fluorine in a native protein environment—How the spatial demand and polarity of fluoroalkyl groups affect protein folding. *Angewandte Chemie International Edition, 45*(25), 4198–4203. http://dx.doi.org/10.1002/anie.200504387.

Jäckel, C., Seufert, W., Thust, S., & Koksch, B. (2004). Evaluation of the molecular interactions of fluorinated amino acids with native polypeptides. *ChemBioChem, 5*(5), 717–720. http://dx.doi.org/10.1002/cbic.200300840.

Jackson, S. E., & Fersht, A. R. (1991). Folding of chymotrypsin inhibitor 2. 1. Evidence for a two-state transition. *Biochemistry, 30*(43), 10428–10435. http://dx.doi.org/10.1021/bi00107a010.

Kelly, S. M., & Price, N. C. (1997). The application of circular dichroism to studies of protein folding and unfolding. *BBA - Protein Structure and Molecular Enzymology, 1338*(2), 161–185.

Khan, F., Kuprov, I., Craggs, T. D., Hore, P. J., & Jackson, S. E. (2006). 19 F NMR studies of the native and denatured states of green fluorescent protein. *Journal of the American Chemical Society, 128*(33), 10729–10737. http://dx.doi.org/10.1021/ja060618u.

Krishnamurthy, V. M., & Kumar, K. (2013). Fluorination in the design of membrane protein assemblies. In G. Ghirlanda & A. Senes (Eds.), *Membrane Proteins: Folding, Association, and Design: Vol. 1063* (pp. 227–243).

Kuhlman, B., & Raleigh, D. P. (1998). Global analysis of the thermal and chemical denaturation of the N-terminal domain of the ribosomal protein L9 in H2O and D2O. Determination of the thermodynamic parameters, $\Delta H°$, $\Delta S°$, and $\Delta C°$ p, and evaluation of solvent isotope effects. *Protein Science, 7*(11), 2405–2412.

Kwon, O.-H., Yoo, T. H., Othon, C. M., Van Deventer, J. A., Tirrell, D. A., & Zewail, A. H. (2010). Hydration dynamics at fluorinated protein surfaces. *Proceedings of the National Academy of Sciences of the United States of America*, *107*(40), 17101–17106.

Leaver-Fay, A., Tyka, M., Lewis, S. M., Lange, O. F., Thompson, J., Jacak, R., ... Bradley, P. (2011). Rosetta3: An object-oriented software suite for the simulation and design of macromolecules. In M. L. Johnson & L. Brand (Eds.), *Computer Methods*: *Vol. 487. Methods in enzymology* (pp. 545–574). Pt. C.

Lee, H.-Y., Lee, K.-H., Al-Hashimi, H. M., & Marsh, E. N. G. (2006). Modulating protein structure with fluorous amino acids: Increased stability and native-like structure conferred on a 4-helix bundle protein by hexafluoroleucine. *Journal of the American Chemical Society*, *128*(1), 337–343. http://dx.doi.org/10.1021/ja0563410.

Lee, K.-H., Lee, H.-Y., Slutsky, M. M., Anderson, J. T., & Marsh, E. N. G. (2004). Fluorous effect in proteins: De novo design and characterization of a four-α-helix bundle protein containing hexafluoroleucine. *Biochemistry*, *43*(51), 16277–16284. http://dx.doi.org/10.1021/bi049086p.

Li, F., Shi, P., Li, J., Yang, F., Wang, T., Zhang, W., ... Wang, J. (2013). A genetically encoded 19 F NMR probe for tyrosine phosphorylation. *Angewandte Chemie International Edition*, *52*(14), 3958–3962. http://dx.doi.org/10.1002/anie.201300463.

Liu, J. J., Horst, R., Katritch, V., Stevens, R. C., & Wuthrich, K. (2012). Biased signaling pathways in {beta}2-adrenergic receptor characterized by 19 F-NMR. *Science*, *335*(6072), 1106–1110. http://dx.doi.org/10.1126/science.1215802.

Luchette, P. A., Prosser, R. S., & Sanders, C. R. (2002). Oxygen as a paramagnetic probe of membrane protein structure by cysteine mutagenesis and 19 F NMR spectroscopy. *Journal of the American Chemical Society*, *124*(8), 1778–1781. http://dx.doi.org/10.1021/ja016748e.

Luo, Z. Y., Zhang, Q. S., Oderaotoshi, Y., & Curran, D. P. (2001). Fluorous mixture synthesis: A fluorous-tagging strategy for the synthesis and separation of mixtures of organic compounds. *Science*, *291*(5509), 1766–1769.

Marsh, E. N. G. (2000). Towards the nonstick egg: Designing fluorous proteins. *Chemical Biology*, *7*(7), R153–R157. http://dx.doi.org/10.1016/s1074-5521(00)00139-3.

Marsh, E. N. G. (2014). Fluorinated proteins: From design and synthesis to structure and stability. *Accounts of Chemical Research*, *47*(10), 2878–2886. http://dx.doi.org/10.1021/ar500125m.

Marsh, E. N. G., Buer, B. C., & Ramamoorthy, A. (2009). Fluorine—A new element in the design of membrane-active peptides. *Molecular Biosystems*, *5*(10), 1143–1147.

Marsh, E. N. G., & Suzuki, Y. (2014). Using F-19 NMR to probe biological interactions of proteins and peptides. *ACS Chemical Biology*, *9*(6), 1242–1250. http://dx.doi.org/10.1021/cb500111u.

Meng, H., Krishnaji, S. T., Beinborn, M., & Kumar, K. (2008). Influence of selective fluorination on the biological activity and proteolytic stability of glucagon-like peptide-1. *Journal of Medicinal Chemistry*, *51*(22), 7303–7307. http://dx.doi.org/10.1021/jm8008579.

Meng, H., & Kumar, K. (2007). Antimicrobial activity and protease stability of peptides containing fluorinated amino acids. *Journal of the American Chemical Society*, *129*(50), 15615–15622. http://dx.doi.org/10.1021/ja075373f.

Merkel, L., & Budisa, N. (2012). Organic fluorine as a polypeptide building element: In vivo expression of fluorinated peptides, proteins and proteomes. *Organic & Biomolecular Chemistry*, *10*(36), 7241–7261. http://dx.doi.org/10.1039/c2ob06922a.

Molski, M. A., Goodman, J. L., Chou, F.-C., Baker, D., Das, R., & Schepartz, A. (2013). Remodeling a beta-peptide bundle. *Chemical Science*, *4*(1), 319–324. http://dx.doi.org/10.1039/c2sc21117c.

Montclare, J. K., Son, S., Clark, G. A., Kumar, K., & Tirrell, D. A. (2009). Biosynthesis and stability of coiled-coil peptides containing (2S,4R)-5,5,5-trifluoroleucine and (2S,4S)-5,5,5-trifluoroleucine. *ChemBioChem*, *10*(1), 84–86. http://dx.doi.org/10.1002/cbic.200800164.

Mortenson, D. E., Satyshur, K. A., Guzei, I. A., Forest, K. T., & Gellman, S. H. (2012). Quasiracemic crystallization as a tool to assess the accommodation of noncanonical residues in nativelike protein conformations. *Journal of the American Chemical Society*, *134*(5), 2473–2476.

Muir, T. W. (2003). Semisynthesis of proteins by expressed protein ligation. *Annual Review of Biochemistry*, *72*, 249–289. http://dx.doi.org/10.1146/annurev.biochem.72.121801.161900.

Müller, K., Faeh, C., & Diederich, F. (2007). Fluorine in pharmaceuticals: Looking beyond intuition. *Science*, *317*(5846), 1881–1886. http://dx.doi.org/10.1126/science.1131943.

Naarmann, N., Bilgiçer, B., Meng, H., Kumar, K., & Steinem, C. (2006). Fluorinated interfaces drive self-association of transmembrane α helices in lipid bilayers. *Angewandte Chemie International Edition*, *45*(16), 2588–2591.

Niemz, A., & Tirrell, D. A. (2001). Self-association and membrane-binding behavior of melittins containing trifluoroleucine. *Journal of the American Chemical Society*, *123*(30), 7407–7413. http://dx.doi.org/10.1021/ja004351p.

Odar, C., Winkler, M., & Wiltschi, B. (2015). Fluoro amino acids: A rarity in nature, yet a prospect for protein engineering. *Biotechnology Journal*, *10*(3), 427–446. http://dx.doi.org/10.1002/biot.201400587.

Pace, C. J., & Gao, J. (2013). Exploring and exploiting polar-pi interactions with fluorinated aromatic amino acids. *Accounts of Chemical Research*, *46*(4), 907–915. http://dx.doi.org/10.1021/ar300086n.

Pace, C. J., Zheng, H., Mylvaganam, R., Kim, D., & Gao, J. (2011). Stacked fluoroaromatics as supramolecular synthons for programming protein dimerization specificity. *Angewandte Chemie International Edition*, *51*(1), 103–107. http://dx.doi.org/10.1002/ange.201105857. In Press.

Pendley, S. S., Yu, Y. B., & Cheatham, T. E. (2009). Molecular dynamics guided study of salt bridge length dependence in both fluorinated and non-fluorinated parallel dimeric coiled-coils. *Proteins*, *74*(3), 612–629. http://dx.doi.org/10.1002/prot.22177.

Pomerantz, W. C., Wang, N., Lipinski, A. K., Wang, R., Cierpicki, T., & Mapp, A. K. (2012). Profiling the dynamic interfaces of fluorinated transcription complexes for ligand discovery and characterization. *ACS Chemical Biology*, *7*(8), 1345–1350. http://dx.doi.org/10.1021/cb3002733.

Ravikumar, Y., Nadarajan, S. P., Yoo, T. H., Lee, C.-S., & Yun, H. (2015a). Incorporating unnatural amino acids to engineer biocatalysts for industrial bioprocess applications. *Biotechnology Journal*, *10*(12), 1862–1876. http://dx.doi.org/10.1002/biot.201500153.

Ravikumar, Y., Nadarajan, S. P., Yoo, T. H., Lee, C.-s., & Yun, H. (2015b). Unnatural amino acid mutagenesis-based enzyme engineering. *Trends in Biotechnology*, *33*(8), 462–470. http://dx.doi.org/10.1016/j.tibtech.2015.05.002.

Richards, F. M. (1977). Areas, volumes, packing, and protein structure. *Annual Review of Biophysics and Bioengineering*, *6*(1), 151–176.

Salwiczek, M., & Koksch, B. (2009). Effects of fluorination on the folding kinetics of a heterodimeric coiled coil. *ChemBioChem*, *10*(18), 2867–2870. http://dx.doi.org/10.1002/cbic.200900518.

Salwiczek, M., Nyakatura, E. K., Gerling, U. I., Ye, S., & Koksch, B. (2012). Fluorinated amino acids: Compatibility with native protein structures and effects on protein-protein interactions. *Chemical Society Reviews*, *41*(6), 2135–2171.

Sani, M., Bruché, L., Chiva, G., Fustero, S., Piera, J., Volonterio, A., ... Zanda, M. (2003). Highly stereoselective tandem Aza-michael addition–enolate protonation to form partially modified retropeptide mimetics incorporating a trifluoroalanine surrogate.

Angewandte Chemie International Edition, 42(18), 2060–2063. http://dx.doi.org/10.1002/anie.200250711.

Schnölzer, M., Alewood, P., Jones, A., Alewood, D., & Kent, S. B. H. (1992). In situ neutralization in Boc-chemistry solid phase peptide synthesis. *International Journal of Peptide and Protein Research, 40*(3-4), 180–193.

Senguen, F. T., Doran, T. M., Anderson, E. A., & Nilsson, B. L. (2011). Clarifying the influence of core amino acid hydrophobicity, secondary structure propensity, and molecular volume on amyloid-β 16–22 self-assembly. *Molecular Biosystems, 7*(2), 497–510.

Smith, A. E., Zhou, L. Z., Gorensek, A. H., Senske, M., & Pielak, G. J. (2016). In-cell thermodynamics and a new role for protein surfaces. *Proceedings of the National Academy of Sciences of the United States of America, 113*(7), 1725–1730. http://dx.doi.org/10.1073/pnas.1518620113.

Son, S., Tanrikulu, I. C., & Tirrell, D. A. (2006). Stabilization of bzip peptides through incorporation of fluorinated aliphatic residues. *ChemBioChem, 7*(8), 1251–1257. http://dx.doi.org/10.1002/cbic.200500420.

Suzuki, Y., Brender, J. R., Hartman, K., Ramamoorthy, A., & Marsh, E. N. G. (2012). Alternative pathways of human islet amyloid polypeptide aggregation distinguished by 19 F nuclear magnetic resonance-detected kinetics of monomer consumption. *Biochemistry, 51*(41), 8154–8162.

Suzuki, Y., Brender, J. R., Soper, M. T., Krishnamoorthy, J., Zhou, Y., Ruotolo, B. T., ... Marsh, E. N. G. (2013). Resolution of oligomeric species during the aggregation of aβ1-40 using 19 F NMR. *Biochemistry, 52*(11), 1903–1912.

Suzuki, Y., Buer, B. C., Al-Hashimi, H. M., & Marsh, E. N. G. (2011). Using fluorine nuclear magnetic resonance to probe changes in the structure and dynamics of membrane-active peptides interacting with lipid bilayers. *Biochemistry, 50*(27), 5979–5987. http://dx.doi.org/10.1021/bi200639c.

Tang, Y., Ghirlanda, G., Petka, W. A., Nakajima, T., DeGrado, W. F., & Tirrell, D. A. (2001). Fluorinated coiled-coil proteins prepared in vivo display enhanced thermal and chemical stability. *Angewandte Chemie International Edition, 40*(8), 1494–1496. http://dx.doi.org/10.1002/1521-3773(20010417)40:8<1494::aid-anie1494>3.0.co;2-x.

Tang, Y., Ghirlanda, G., Vaidehi, N., Kua, J., Mainz, D. T., Goddard, W. A., ... Tirrell, D. A. (2001). Stabilization of coiled-coil peptide domains by introduction of trifluoroleucine. *Biochemistry, 40*(9), 2790–2796. http://dx.doi.org/10.1021/bi0022588.

Tang, Y., & Tirrell, D. A. (2001). Biosynthesis of a highly stable coiled-coil protein containing hexafluoroleucine in an engineered bacterial host. *Journal of the American Chemical Society, 123*(44), 11089–11090. http://dx.doi.org/10.1021/ja016652k.

Tsushima, T., Kawada, K., Ishihara, S., Uchida, N., Shiratori, O., Higaki, J., ... Hirata, M. (1988). Fluorine containing amino acids and their derivatives. 7. Synthesis and antitumor activity of α-and γ-substituted methotrexate analogs. *Tetrahedron, 44*(17), 5375–5387.

Vine, W. H., Hsieh, K.-H., & Marshall, G. R. (1981). Synthesis of fluorine-containing peptides. Analogs of angiotensin II containing hexafluorovaline. *Journal of Medicinal Chemistry, 24*(9), 1043–1047. http://dx.doi.org/10.1021/jm00141a005.

Wang, P., Fichera, A., Kumar, K., & Tirrell, D. A. (2004). Alternative translations of a single RNA message: An identity switch of (2S,3R)-4,4,4-trifluorovaline between valine and isoleucine codons. *Angewandte Chemie International Edition, 43*(28), 3664–3666. http://dx.doi.org/10.1002/anie.200454036.

Wang, P., Tang, Y., & Tirrell, D. A. (2003). Incorporation of trifluoroisoleucine into proteins in vivo. *Journal of the American Chemical Society, 125*(23), 6900–6906. http://dx.doi.org/10.1021/ja0298287.

Wang, L., Xie, J., & Schultz, P. G. (2006). Expanding the genetic code. *Annual Review of Biophysics and Biomolecular Structure, 35*, 225–249. http://dx.doi.org/10.1146/annurev.biophys.35.101105.121507.

Woll, M. G., Hadley, E. B., Mecozzi, S., & Gellman, S. H. (2006). Stabilizing and destabilizing effects of phenylalanine → F5-phenylalanine mutations on the folding of a small protein. *Journal of the American Chemical Society, 128*(50), 15932–15933. http://dx.doi.org/10.1021/ja0634573.

Xing, X., Fichera, A., & Kumar, K. (2001). A novel synthesis of enantiomerically pure 5, 5, 5, 5', 5', 5'-hexafluoroleucine. *Organic Letters, 3*(9), 1285–1286.

Xing, X., Fichera, A., & Kumar, K. (2002). A simple and efficient method for the resolution of all four diastereomers of 4, 4, 4-trifluorovaline and 5, 5, 5-trifluoroleucine. *The Journal of Organic Chemistry, 67*(5), 1722–1725.

Ye, Y., Liu, X., Zhang, Z., Wu, Q., Jiang, B., Jiang, L., ... Li, C. (2013). F-19 NMR spectroscopy as a probe of cytoplasmic viscosity and weak protein interactions in living cells. *Chemistry–A European Journal, 19*(38), 12705–12710. http://dx.doi.org/10.1002/chem.201301657.

Yi, Q., Scalley, M. L., Simons, K. T., Gladwin, S. T., & Baker, D. (1997). Characterization of the free energy spectrum of peptostreptococcal protein L. *Folding & Design, 2*(5), 271–280.

Yoder, N. C., & Kumar, K. (2002). Fluorinated amino acids in protein design and engineering. *Chemical Society Reviews, 31*(6), 335–341.

Yoder, N. C., Yüksel, D., Dafik, L., & Kumar, K. (2006). Bioorthogonal noncovalent chemistry: Fluorous phases in chemical biology. *Current Opinion in Chemical Biology, 10*(6), 576–583.

Yu, J.-X., Kodibagkar, V. D., Cui, W., & Mason, R. P. (2005). 19 F: A versatile reporter for non-invasive physiology and pharmacology using magnetic resonance. *Current Medicinal Chemistry, 12*(7), 819–848. http://dx.doi.org/10.2174/0929867053507342.

Zheng, H., Comeforo, K., & Gao, J. (2008). Expanding the fluorous arsenal: Tetrafluorinated phenylalanines for protein design. *Journal of the American Chemical Society, 131*(1), 18–19. http://dx.doi.org/10.1021/ja8062309.

Zheng, H., & Gao, J. (2010). Highly specific heterodimerization mediated by quadrupole interactions. *Angewandte Chemie International Edition, 49*(46), 8635–8639. http://dx.doi.org/10.1002/anie.201002860.

Zigoneanu, I. G., Yang, Y. J., Krois, A. S., Haque, M. E., & Pielak, G. J. (2012). Interaction of alpha-synuclein with vesicles that mimic mitochondrial membranes. *Biochimica et Biophysica Acta—Biomembranes, 1818*(3), 512–519. http://dx.doi.org/10.1016/j.bbamem.2011.11.024.

CHAPTER THIRTEEN

Solid Phase Synthesis of Helically Folded Aromatic Oligoamides

S.J. Dawson*,†, X. Hu*,†, S. Claerhout*,†, I. Huc*,†,1

*Université de Bordeaux, CBMN (UMR5248), Institut Européen de Chimie et Biologie, Pessac, France
†CNRS, CBMN (UMR5248), Institut Européen de Chimie et Biologie, Pessac, France
[1]Corresponding author: e-mail address: i.huc@iecb.u-bordeaux.fr

Contents

1. Introduction	280
2. Materials	285
2.1 Reagents	285
2.2 Equipment	285
2.3 Reagent Setup	285
2.4 Equipment Setup	286
3. Synthetic Protocols	286
3.1 8-Amino-2-quinolinecarboxylic Acid-Based Oligoamides	286
3.2 7-Amino-8-fluoro-2-quinolinecarboxylic Acid-Based Oligoamides	293
3.3 Quinoline/α-Amino Acid Hybrid Oligoamides	294
4. Characterization of Foldamers	297
4.1 8-Amino-2-quinolinecarboxylic Acid-Based Oligoamides	297
4.2 7-Amino-8-fluoro-2-quinolinecarboxylic Acid-Based Oligoamides	297
4.3 Quinoline/α-Amino Acid Hybrid Oligoamides	297
References	299

Abstract

Aromatic amide foldamers constitute a growing class of oligomers that adopt remarkably stable folded conformations. The folded structures possess largely predictable shapes and open the way toward the design of synthetic mimics of proteins. Important examples of aromatic amide foldamers include oligomers of 7- or 8-amino-2-quinoline carboxylic acid that have been shown to exist predominantly as well-defined helices, including when they are combined with α-amino acids to which they may impose their folding behavior. To rapidly iterate their synthesis, solid phase synthesis (SPS) protocols have been developed and optimized for overcoming synthetic difficulties inherent to these backbones such as low nucleophilicity of amine groups on electron poor aromatic rings and a strong propensity of even short sequences to fold on the solid phase during synthesis. For example, acid chloride activation and the use of microwaves are required to bring coupling at aromatic amines to completion. Here, we report detailed SPS protocols for the rapid production of: (1) oligomers of 8-amino-2-quinolinecarboxylic acid; (2) oligomers containing 7-amino-8-fluoro-2-quinolinecarboxylic acid; and (3) heteromeric oligomers of 8-amino-2-quinolinecarboxylic acid and α-amino acids. SPS brings the advantage to quickly produce sequences having varied main chain or side chain components without having to purify multiple intermediates as in solution phase synthesis. With these protocols, an octamer could easily be synthesized and purified within one to two weeks from Fmoc protected amino acid monomer precursors.

1. INTRODUCTION

Foldamers, ie, artificial folded molecular architectures, are the object of very active research investigations (Guichard & Huc, 2011). While inspired by the folding behavior seen in natural biopolymers, they possess a fundamental difference, in that they are comprised of either nonnatural building blocks or natural building blocks arranged in a nonnatural sequence. Foldamers closely ressembling their natural counterpart have been termed "biotic" and include peptide nucleic acids (Nielsen, Egholm, Berg, & Buchardt, 1991), peptoids (Simon et al., 1992), β-peptides (Appella, Christianson, Karle, Powell, & Gellman, 1996; Seebach et al., 1996), γ-peptides, and δ-peptides. They also include peptide sequences where amides have been replaced by urea, hydrazide, or hydroxyamide linkages (Li & Yang, 2006; Salaun, Potel, Roisnel, Gall, & Le Grel, 2005; Semetey et al., 2002). In contrast, "abiotic" foldamers consist of entirely unnatural building blocks, giving rise to backbones, and folding modes which are inaccessible to natural motifs. These architectures may have unique physical properties, or may be capable of interacting with biomolecules in unforeseen and interesting ways. Many of these sequences are aromatic rich, examples including oligo-phenylene-ethynylenes (Nelson, Saven, Moore, & Wolynes, 1997), alternating aromatic electron donors and acceptors (Lokey & Iverson, 1995), aryl-oligomers (often based on aza-heterocycles: pyridines, pyrimidines, pyridazines, etc.; Bassani, Lehn, Baum, & Fenske, 1997), aromatic tertiary amide, imide, or urea oligomers, and aromatic oligoamides (Berl, Huc, Khoury, Krische, & Lehn, 2000; Hamuro, Geib, & Hamilton, 1996; Huc, 2004; Jiang, Léger, & Huc, 2003; Zhu et al., 2000). Oligoamide foldamers are particularly attractive, since the amide linkage offers a high level of synthetic feasibility; indeed, this is the motif which has been selected by nature.

Aromatic oligoamide foldamers consisting of 8-amino-2-quinolinecarboxylic acid (Fig. 1) form single helical architectures made up of 2.5 monomer units per turn, with a pitch of 3.4 Å that corresponds

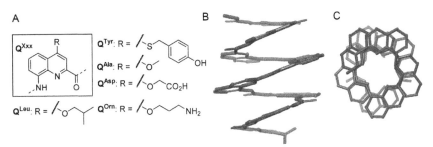

Fig. 1 (A) Structure of quinoline-based aromatic oligoamide foldamers with examples of available monomer side chains. Side (B) and top (C) views of the crystal structures of O_2N-$(Q^{Leu})_8$-OMe. Side chains and hydrogen atoms are omitted for clarity.

to the thickness of one aromatic ring (Fig. 1B and C) (Dolain et al., 2005; Jiang, Léger, Dolain, Guionneau, & Huc, 2003; Jiang, Léger, & Huc, 2003). They feature key properties which make them ideal candidates for the recognition of sizeable surface areas of biomolecules: they are medium sized (0.5–5.0 kDa), resistant to proteolytic degradation, and conformationally stable in a wide range of solvents and in particular in water, even at high temperatures (Gillies, Deiss, Staedel, Schmitter, & Huc, 2007; Qi et al., 2012). Indeed, to date no conditions have been found under which they do not adopt a helically folded conformation. Another advantage is that their helical shape is extremely predictable, and since monomer side chains may be positioned away from the backbone amide motifs, folding is essentially independent of R-group functionality. This allows the display of arrays of side chains to be tuned in order to optimize interactions with a specific biomolecule, affording foldamers with cell-penetrating properties (Gillies et al., 2007; Iriondo-Alberdi, Laxmi-Reddy, Bouguerne, Staedel, & Huc, 2010), high affinity for G-quadruplex DNA (Delaurière et al., 2012; Müller et al., 2014), or the potential to interact with protein surfaces (Buratto et al., 2014).

This high stability of helix shape also brings benefits in terms of synthetic availability. Once a trimer has been reached, addition of further monomer units only results in elongation of the helix under identical conditions regardless of helix length. This is in contrast to biopolymers such as peptides, where each synthetic intermediate may have a different conformation, potentially affecting reactivity in subsequent steps (eg, by aggregation).

A logical extension from homogenous backbones (ie, those consisting of exclusively one monomer type) is to combine a variety of different monomer types to produce a hybrid sequence. This concept highlights one of the advantages of synthetic foldamers, in that specific monomers can be included at any desired point to globally or locally direct sequence architecture in a predictable manner, and thus potentially access unsuspected areas of structural and functional space. For quinoline-based oligoamide foldamers, considerable inroads have been made into understanding the effect on helical structure when a diverse array of aromatic building blocks are included (Fig. 2). Monomers such as those based on fluoroquinoline ("Q_F," **1**) or anthracene ("A," **2**) can be included to code for a wider helix diameter, which, combined in the centre of a quinoline sequence, can lead to the formation of a capsule shape. An example would be $Q_3(Q_F)_3A(Q_F)_3Q_3$. Through iterative design, cavity size, and functionality can be reliably tuned to encapsulate a specific guest, examples ranging from simple short-chain alkanes, to monosaccharides (Bao et al., 2008; Chandramouli et al., 2015; Garric, Léger, & Huc, 2005; Singleton, Pirotte, Kauffmann, Ferrand, & Huc, 2014). Short Q_3 quinoline sequences at each terminus of the capsule function as a type of "cap" effectively insulating the cavity from the exterior (Ferrand et al., 2012). If these quinoline caps are removed, the wider diameter of, for example, a homomeric fluoroquinoline sequence results in hybridization into double (Fig. 2B and C) and even quadruple helices (Gan et al., 2008). This behavior has even been shown to occur in aqueous conditions, representing one of the very few examples of water-soluble synthetic double helices reported in current literature (Shang et al., 2014).

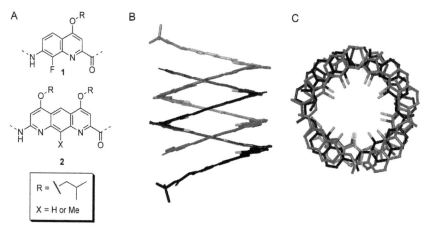

Fig. 2 (A) Fluoroquinoline (**1**) and anthracene (**2**)-based monomers. Side (B) and top (C) views of the crystal structure of double helical $\left(\text{Boc} - (Q_F^{Leu})_8 - \text{OMe}\right)_2$. The backbone of each strand is colored in *green* (*dark gray* in the print version) or *red* (*black* in the print version). Fluorine atoms are colored *blue* (*light gray* in the print version). Side chains and hydrogen atoms are omitted for clarity.

Recent work in the Huc group has also focused on the inclusion of aliphatic units into aromatic oligoamide foldamer sequences, in particular, α-amino acids. This is a natural progression from exclusively abiotic peptidomimetics, in that the exact functional groups which mediate biomolecule recognition (ie, α-amino acid side chains) are included. This also brings synthetic benefits, since a vast range of orthogonally protected α-amino acids are commercially available, avoiding the need for (a perhaps laborious) bespoke monomer synthesis.

In the foldamer world, a single amino acid may for example be added at the extremity of an α-helix mimetic (Barnard et al., 2015). When multiple α-amino acids are incorporated in an abiotic sequence to form a hybrid scaffold, the completely different folding principles of biotic and abiotic units may offer access to secondary structures distinct from those of biopolymers or synthetic homo-oligomers (Nair, Vijayadas, Roy, & Sanjayan, 2014). In other cases, the folding of abiotic units may be so effective that it forces α-amino acids to adopt conformations distinct from those found in peptides. We recently reported examples of this kind using helically folded quinoline oligoamides. Sequences combining α-amino acid (**X**) and quinoline (**Q**) units together in an **XQ₂** trimer repeat motif (Fig. 3) were found to adopt a single well-defined canonical aromatic helical conformation in both organic and aqueous conditions (Hu et al., 2016; Kudo, Maurizot, Kauffmann, Tanatani, & Huc, 2013). In contrast, hybrids based on an

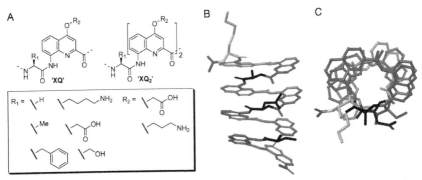

Fig. 3 (A) **XQ** and **XQ₂** motifs with α-amino acids currently validated for incorporation. Side (B) and top (C) views of NMR structure of Ac-K(Q^{Orn})₂A(Q^{Orn})₂D(Q^{Orn})₂S(Q^{Orn})₂-OH (Hu et al., 2016). The aromatic backbone is colored in *green* (*dark gray* in the print version) and the α-amino acid residues are in *pink* (*gray* in the print version) (Lys), *blue* (*black* in the print version) (Ala), *light blue* (*black* in the print version) (Asp), and *red* (*black* in the print version) (Ser). Quinoline side chains and hydrogen atoms are omitted for clarity.

XQ dimer repeat motif were found to adopt a partially folded zig–zag tape conformation with local conformational variability precluding long range order. This behavior was also evident in protic solvents, where the increased solvent accessibility of the hydrophobic aromatic surfaces appeared to drive nonspecific aggregation at lower temperatures (Hu et al., 2016; Kudo, Maurizot, Masu, Tanatani, & Huc, 2014). These types of hybrid foldamer sequences offer access to currently untapped areas of chemical space and may provide promising candidates for the recognition of "difficult" biological targets such as protein–protein interactions.

With foldamers becoming increasingly complex, organic chemists require synthetic methodologies capable of meeting the requirements of their designs. The production of oligomers by solid phase synthesis (SPS) is particularly attractive in that it offers a method for rapidly generating sequence analogues where any monomer unit can be substituted for another, without the laborious resynthesis of intermediates required by a more convergent solution phase approach. In addition, the pseudodilution effect of solid supported synthesis reduces intermolecular reactions between individual oligomer chains, and this "site–site isolation" phenomenon (Shi, Wang, & Yan, 2007) can help minimize reactivity issues which might occur otherwise in solution due to aggregation, or hybridization behavior (eg, as seen in the synthesis of fluoroquinoline double helical foldamers). While solid phase peptide synthesis methods are now widely standardized, the use of an SPS strategy for the production of aromatic foldamers based on 8-amino-2-quinolinecarboxylic acid is not without its challenges. The aromatic amine is a relatively poor nucleophile and thus coupling requires activation of monomers as acid chlorides and microwave assistance in order to be both rapid and essentially quantitative. Over the last number of years we have reported the SPS of increasingly elaborate oligomers, from homomeric quinoline sequences (Baptiste, Douat-Casassus, Laxmi-Reddy, Godde, & Huc, 2010) to heteromeric sequences containing fluoroquinoline monomers and α-amino acids (Hu et al., 2016; Shang et al., 2014). We are now in a position to report here optimized protocols for the microwave-assisted SPS of these foldamers, including variations and improvements from previously published work. These protocols are specific to the requirements of these classes of oligomers and are to be compared to SPS methods developed for other aromatic amide oligomers (König, Abbel, Schollmeyer, & Kilbinger, 2006; Murphy et al., 2013; Puckett, Green, & Dervan, 2012; Wurtz, Turner, Baird, & Dervan, 2001).

2. MATERIALS

2.1 Reagents

Acetyl chloride, acetic anhydride, CaH_2, CBr_4, 1-chloro-N,N,2-trimethyl-1-propenylamine (Ghosez reagent), CsI, 1-methyl-2-(4′-nitrophenyl)-imidazo[1,2-a]pyrimidinium perchlorate (DESC), dimethylformamide (DMF), dichloromethane, N,N-diisopropylethylamine (DIEA), isopropanol, Fmoc α-amino acids, quinoline and fluoroquinoline monomers, methanol, nitrogen gas, piperidine, Sieber amide resin ("low loading" ~0.6–0.7 mmol g^{-1}), tetrahydrofuran (THF), trichloroacetonitrile (TCAN), trifluoroacetic acid (TFA), triisopropylsilane (TIS), triphenylphosphine, 2,4,6-collidine, Wang resin ("low loading" ~0.3–0.4 mmol g^{-1}), and water.

2.2 Equipment

Balloons, CEM discover SPS microwave oven (CEM Corporation, Matthews, USA) equipped with an infrared optical fiber probe internal to the reaction mixture linked to an IR detector for temperature control, CEM SPS vacuum station with membrane vacuum pump (Vacuubrand model ME1, VACUUBRAND GMBH, Wertheim, Germany), CEM SPS reactor vessel (25 mL, polypropylene) with end cap, vial heating block, micropipettes (Gilson, 20 and 100 μL, Gilson, Middleton, USA), high vacuum pump (Vacuubrand model RC6) with glass vacuum manifold and cold finger, microsyringe (Hamilton, 100 μL, Hamilton, Reno, USA), needles (disposable, 2.5 cm and 10 cm), Pasteur pipettes, plastic syringes (disposable, 2, 5, and 10 mL), round bottomed flasks (10 mL) with corresponding rubber septa, sample vials (approx. 1–2 mL), sintered glass filter funnel (approx. 2 mL, porosity grade 3), squeeze solvent wash bottles, stoppered vacuum adapters, rotary evaporator (Büchi, model R-3000, Büchi, Flawil, Switzerland) with vacuum pump (Büchi, model V-500), and vacuum controller (Büchi, model V-800).

2.3 Reagent Setup

Resins: "low loading" resins are to be preferred, in order to avoid potential steric crowding when synthesizing longer oligoamide sequences.

Dichloromethane: dry dichloromethane is obtained by filtration through activated alumina using a dedicated purification system (MBRAUN SPS-800, M. Braun Inertgas-Systeme GmbH, Garching, Germany) and should be used immediately.

THF: dry THF is obtained by filtration through activated alumina using a dedicated purification system (MBRAUN SPS-800) and should be used immediately.

N,N-diisopropylethylamine: should be freshly distilled over CaH_2 to remove traces of water and used immediately.

20% (v/v) piperidine in DMF: solution should be freshly prepared.

0.1 M DESC in anhydrous DMF: solution should be freshly prepared.

Where not stated as anhydrous, all solvents used for rinsing the resin and reaction vessel can be delivered using plastic squeeze bottles.

The synthesis of various Fmoc protected 8-amino-2-quinolinecarboxylic acid and 7-amino-8-fluoro-2-quinolinecarboxylic acid has been reported before (Baptiste et al., 2010; Buratto et al., 2014; Shang et al., 2014).

2.4 Equipment Setup

Dry syringes and needles: Plastic syringes and needles directly used from the packaging are considered dry. When they are reused, they should be cleaned with acetone and dried under vacuum (at least for 1 h).

Microwave: As of publication, these protocols have not been validated on microwave systems other than the CEM discover. Follow manufacturer's instructions for setup of microwave cavity and vacuum station for SPS. The microwave should be run in "open vessel" mode, with temperature control via an internal fiber optic probe. A standard temperature program should be used, with a 5 min ramp time and medium stirring speed. Hold times vary (see specific protocol).

3. SYNTHETIC PROTOCOLS

3.1 8-Amino-2-quinolinecarboxylic Acid-Based Oligoamides

Notes: We show here as an example the synthesis of a water-soluble octameric quinoline oligoamide (Fig. 4) on a 19 μmol scale. In our experience, this protocol can be reliably scaled to approx. 80 μmol with no decrease in efficiency. We have successfully synthesized oligomers up to 24 units in length with these protocols. Protocols 3–6 and 8 have also been validated using low-loading Sieber amide resin.

3.1.1 Protocol 1: Bromination of Low-Loading Wang Resin

Notes: Method based on the procedure of Morales and coworkers (Morales, Corbett, & DeGrado, 1998) where it is reported at larger scales. We

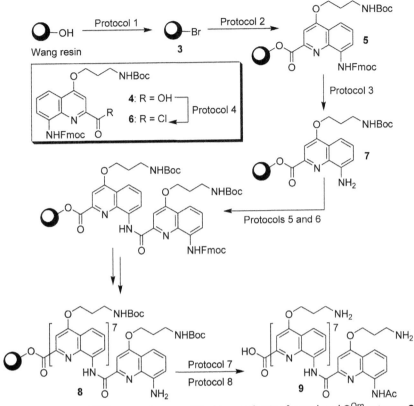

Fig. 4 Summary of SPS protocols exemplified by synthesis of acetylated Q^{Orn} octamer **9**.

exemplify here the procedure on a 76 μmol scale, which is routinely carried out in our laboratory.

Time required: 17 h

(1) Prepare a clean, dry 10 mL round bottomed flask equipped with small magnetic stirring bar, septum, and nitrogen balloon.

(2) Add Wang resin (200 mg, 76 μmol for 0.38 mmol g^{-1} loading) to the reaction flask, followed by 2 mL anhydrous DMF via dry syringe through the septum. Ensure all resin is rinsed from the sides of the flask, then flush with N_2. Note that success of this reaction is highly dependent on the quality of the anhydrous DMF.

(3) Leave resin to swell for 1 h under N_2 atmosphere.

(4) Remove the septum. While stirring the resin, add PPh$_3$ (99.7 mg, 0.38 mmol, 5 equiv.), and CBr$_4$ (126.1 mg, 0.38 mmol, 5 equiv.) rapidly in that order. Then equip with the septum and nitrogen balloon.

Note that flask should warm noticeably, and the solution should develop a lasting yellow–orange coloration.
(5) Stir for 15 h under N_2 atmosphere.
(6) Collect the resin by filtration and wash sequentially with anhydrous DMF (5 mL), anhydrous CH_2Cl_2 (5 mL), and isopropanol (5 mL). Repeat the washing cycle three times in total.
(7) Dry the resin under vacuum in the desiccator.

3.1.2 Protocol 2: Loading of Wang-Bromide Resin
Time required: 1.5 h
(1) Prepare a clean, dry 10 mL round bottomed flask, and equip with septum and nitrogen balloon.
(2) Add Wang-bromide resin (3, 50 mg, 19 µmol for 0.38 mmol g^{-1} loading) to the flask, followed by approximately 1 mL of anhydrous DMF via dry syringe through the septum. Ensure all resin is rinsed from the sides of the flask, then flush with N_2.
(3) Leave resin to swell for 1 h under N_2 atmosphere.
(4) Equip a 25 mL polypropylene SPS microwave reaction vessel with the appropriate size magnetic stirring bar.
(5) Transfer the swollen resin to the microwave reaction vessel as a slurry in anhydrous DMF. Rinse the round bottomed flask with small quantities of anhydrous DMF to allow transfer of residual resin. Remove residual anhydrous DMF from the reaction vessel using the vacuum station.
(6) Remove reaction vessel from the vacuum station and fit end cap to avoid leakage.
(7) To the reaction vessel, add 0.45 mL anhydrous DMF, followed by Fmoc-$Q^{Orn-Boc}$-OH (4, 33.2 mg, 0.057 mmol, 3 equiv.), and CsI (14.8 mg, 0.057 mmol, 3 equiv.). Add DIEA (9.9 µL, 0.057 mmol, 3 equiv.) using a micropipette.
(8) Rinse the inside surface of the reaction vessel with a further 0.45 mL anhydrous DMF, to ensure all reagents are at the resin bed.
(9) Insert optical fiber probe into the reactor vessel ensuring the tip is immersed in the reaction mixture and then place the vessel into the microwave cavity with the appropriate insert. Treat with microwaves: 50 W, ramp to 50°C, with a 5 min hold time.
(10) Remove reaction vessel from microwave cavity, remove end cap, and remove reagent solution on vacuum station. Rinse the inside surfaces of the reaction vessel with anhydrous DMF (5 mL) to remove reagents

and ensure all displaced resin is moved down to the resin bed. Remove DMF on vacuum station and fit end cap to the vessel.
(11) Repeat steps (7) to (9).
(12) Remove reaction vessel from microwave cavity, remove fiber optic probe and end cap, then remove reagent solution using the vacuum station. Using DMF thoroughly rinse the inside surfaces of the reaction vessel (approx. 10 mL DMF) and the resin itself (approx. 10 mL DMF) to ensure all reagents have been removed.
(13) Loading efficiency can be calculated using the same UV spectroscopic methods (based on liberated dibenzofulvene adducts from Fmoc deprotection) reported widely in the literature for solid phase peptide synthesis (White & Chan, 2000).

3.1.3 Protocol 3: Fmoc Deprotection

Time required: allow 40 min.
(1) To the loaded resin **5** in reaction vessel, add approximately 2–3 mL of a 20% (v/v) solution of piperidine in DMF. Stir at medium speed at room temperature for 10 min (can be carried out on a normal stirring plate if desired).
(2) Remove end cap from reaction vessel and remove reagent solution on vacuum station. Rinse the inside surfaces of the reaction vessel with anhydrous DMF (5 mL) to remove reagents and ensure all displaced resin is moved down to the resin bed. Remove DMF on vacuum station and fit end cap to the vessel.
(3) Repeat steps (1) and (2) twice more.
(4) Using DMF thoroughly rinse the inside surfaces of the reaction vessel (approx. 10 mL DMF) and the resin itself (approx. 10 mL DMF) to ensure all reagents have been removed.

3.1.4 Protocol 4: Conversion of N-Fmoc Quinoline Carboxylic Acid Monomer *4* to the Corresponding Acid Chloride, *6*

Time required: allow 3.5 h.
(1) Prepare a clean dry 10 mL round bottomed flask and equip with septum and nitrogen balloon.
(2) To the reaction flask, add monomer **4** (66.4 mg, 0.114 mmol, 6 equiv. relative to resin loading) followed by 1.7 mL anhydrous CH_2Cl_2 via dry syringe through the septum. Note that full dissolution is not required for reaction success. For the cases where solubility is low, sonication is required to obtain a homogeneous slurry before starting next step.

(3) While stirring, add 1-chloro-*N*,*N*,2-trimethyl-1-propenylamine (30.2 µL, 0.228 mmol, 12 equiv. relative to resin loading) via dry microsyringe through the septum. Stir for 1 h at room temperature, under N_2.

(4) Fit flask with vacuum adaptor and, while stirring, remove solvent, and reagents on a vacuum manifold equipped with liquid N_2 cold finger. Once evaporated, leave the product (**6**) to dry for at least a further 2 h on the vacuum line.

(5) Turn off the vacuum and fill the flask back with nitrogen using the dual manifold of the vacuum line or with a nitrogen-filled balloon. Remove the vacuum adaptor, replace septum, and ensure that from this point the product remains under positive nitrogen pressure (or under vacuum) until usage.

3.1.5 Protocol 5: Coupling of N-Fmoc Quinoline Carboxylic Acid Chloride (6) to Resin Bound Amine (7)

Note: This procedure can also be used for the loading of Sieber amide resin. Time required: allow 20–40 min.

(1) On the vacuum station thoroughly rinse the inside surfaces of the reaction vessel and the resin with anhydrous THF (approx. 10 mL) using a dry syringe, ensuring that no DMF remains. Remove all remaining THF, then remove the vessel from the vacuum station, and replace end cap.

(2) To the round bottomed flask containing acid chloride **6** (quantity corresponding to 0.114 mmol, 6 equiv.) add 2 mL anhydrous THF via dry syringe through the septum.

(3) Add 0.3 mL anhydrous THF to the resin using a dry syringe, followed by DIEA (20 µL, 0.114 mmol, 6 equiv.) using a micropipette. Note that it is **essential** that DIEA is added prior to the addition of the acid chloride.

(4) Using a dry syringe, remove 1 mL of the acid chloride solution (ie, corresponding to 0.057 mmol, 3 equiv.) and add to the resin.

(5) Insert fiber optic probe into the reaction vessel (ensuring the tip is immersed in the reaction mixture) and then place the vessel into the microwave cavity with the appropriate insert. Treat with microwaves: 50 W, ramp to 50°C. For acid chlorides with poor THF solubility (eg, **6**) we recommend setting the hold time to 15 min. Otherwise hold time should be set at 5 min.

(6) Remove reaction vessel from microwave cavity, remove end cap, and remove reagent solution on vacuum station. Using a dry syringe, rinse the inside surfaces of the reaction vessel with anhydrous THF (5 mL) to

remove reagents, and ensure all displaced resin is moved down to the resin bed. Remove THF on vacuum station and fit end cap to the vessel.
(7) Repeat steps (3) to (5).
(8) Remove reaction vessel from microwave cavity, remove fiber optic probe and end cap, then remove reagent solution on vacuum station. Using a dry syringe, rinse the inside surfaces of the reaction vessel, and the resin itself with anhydrous THF (approx. 10 mL). Then, using DMF, thoroughly rinse the inside surfaces of the reaction vessel (approx. 10 mL DMF) and the resin itself (approx. 10 mL DMF) to ensure all reagents have been removed.

3.1.6 Protocol 6: Assessment of Coupling Completion Using a Modified DESC Test

Note: DESC (Fig. 5) should be synthesized using the procedure reported in the literature (Claerhout, Ermolat'ev, & Van der Eycken, 2008).

Time required: 7 min.

(1) Remove a very small quantity of resin from the reaction vessel (approx. 20–30 beads) using a Pasteur pipette or spatula, and place in a small sample vial.
(2) Using a Pasteur pipette, add to the sample vial of resin five drops of a 0.1 M solution of DESC in DMF, followed by two drops of a 20% (v/v) solution of DIEA in DMF.
(3) Heat the sample vial at 60°C for 5 min.
(4) Carefully remove supernatant using a Pasteur pipette and add fresh DMF. Repeat in this manner until the supernatant is colorless.
(5) Visualize the resin beads under a microscope. Any red–orange coloration indicates that coupling is incomplete (ie, presence of aromatic amine). The coupling procedure should thus be repeated until no orange coloration is observed with this test.

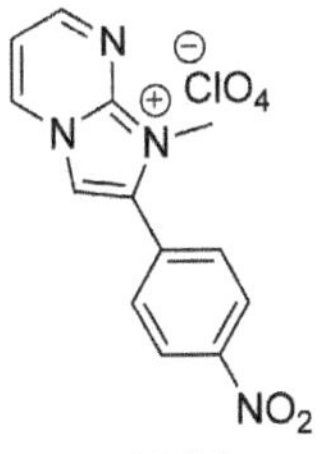

Fig. 5 Structure of DESC color test agent.

3.1.7 Protocol 7: Acetylation of the N-Terminal Aromatic Amine

Time required: 40 min.

(1) On the vacuum station thoroughly rinse the inside surface of the reaction vessel and the resin-bound foldamer **8** with anhydrous THF (approx. 10 mL) using a dry syringe, ensuring that no DMF remains. Remove all remaining THF, then remove the vessel from the vacuum station, and replace end cap.

(2) Using a dry syringe, add 1.3 mL anhydrous THF to the resin, followed by DIEA (20 µL, 0.114 mmol, 6 equiv.) using a micropipette. Then add a diluted acetyl chloride solution (10%, v/v) in anhydrous THF (41 µL, 0.057 mmol, 3 equiv.) using a dry syringe.

(3) Insert fiber optic probe into the reaction vessel (ensuring the tip is immersed in the reaction mixture) and then place the vessel into the microwave cavity with the appropriate insert. Treat with microwaves: 50 W, ramp to 60°C with a 15 min hold time.

(4) Remove reaction vessel from microwave cavity, remove end cap, and remove reagent solution on vacuum station. Using a dry syringe, rinse the inside surfaces of the reaction vessel with anhydrous THF (5 mL) to remove reagents and ensure all displaced resin is moved down to the resin bed. Remove THF on vacuum station and fit end cap to the vessel.

(5) Repeat steps (2) and (3).

(6) Remove reaction vessel from microwave cavity, remove fiber optic probe and end cap, then remove reagent solution on vacuum station. Using a dry syringe, rinse the inside surfaces of the reaction vessel, and the resin itself with anhydrous THF (approx. 10 mL). Then, using DMF, thoroughly rinse the inside surfaces of the reaction vessel (approx. 10 mL DMF), and the resin itself (approx. 10 mL DMF) to ensure all reagents have been removed.

3.1.8 Protocol 8: Cleavage of Foldamer from the Resin

Note: This procedure can be used for cleavage of the foldamer from both Wang and Sieber amide resins.

Time required: Drying resin: 15 h; cleavage: 2 h; workup: 2.5 h.

(1) Using the vacuum station, rinse the resin sequentially with DMF (10 mL), CH_2Cl_2 (10 mL), and (1:1) CH_2Cl_2/MeOH (10 mL). Use the vacuum station to dry the resin briefly (approx. 5 min), then transfer the reaction vessel to a desiccator attached to a vacuum manifold and dry the resin under vacuum for approx. 15 h.

(2) Transfer the resin to a clean, dry 10 mL round bottomed flask equipped with a magnetic stirring bar.

(3) Prepare 1 mL of TFA/TIS/H$_2$O (95:2.5:2.5, v/v/v) solution, and add to the resin. Stopper the flask, and stir the mixture for 2 h at room temperature.
(4) Filter the reaction mixture using a clean, dry sintered glass filter funnel and collect the supernatant. Rinse the resin three times with TFA (3 × 0.5 mL) and combine the supernatants.
(5) Concentrate the TFA solution on a rotary evaporator with a bath temperature of 40°, to obtain a viscous oil. Add diethyl ether to precipitate the product (**9**) and triturate. Filter the product using a clean, dry sintered glass filter funnel, dry and desiccate (approx. 2 h on a vacuum manifold). An alternative to filtration consists in centrifugating, removing the supernatant, redissolving the sticky solid in water or a water–acetonitrile mixture, and freeze-drying.
(6) Purify the product using preparative RP-HPLC.

3.2 7-Amino-8-fluoro-2-quinolinecarboxylic Acid-Based Oligoamides

Advice on SPS of 7-amino-8-fluoro-2-quinolinecarboxylic acid-based oligoamides

Note: SPS of **12** (Fig. 6) can be carried out using the same methods as described in protocols for 8-amino-2-quinolinecarboxylic acid-based oligoamides, with the following exceptions and advice:

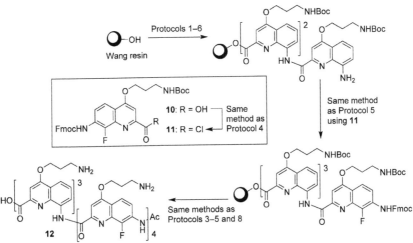

Fig. 6 SPS of 7-amino-8-fluoro-2-quinolinecarboxylic acid-based oligoamides exemplified by synthesis of the antiparallel double helix-forming sequence **12** (Shang et al., 2014).

(1) *Protocol 6*: In contrast with the 8-amino-quinoline, the 7-amino-8-fluoro-quinoline was found to afford a pale yellow coloration using the DESC test, which was difficult to observe. Therefore, we advise that coupling completion should be confirmed by cleaving a small aliquot of resin and analyzing the resulting product(s) via ^{1}H NMR and LC-MS.

(2) *Protocol 8*: The fluoroquinoline amine was found to be particularly susceptible to trifluoroacetylation when undergoing cleavage from the resin (Shang et al., 2014). We therefore advise that if possible, this functionality is blocked (for example, via acetylation, see Protocol 7) during this step.

(3) *Purification/characterization*: It should be noted that products from SPS of these sequences are potentially capable of self- and cross-hybridization (both parallel and antiparallel) after cleavage from the resin. Therefore to avoid difficulties in purification it is recommended that:

a. Each coupling step should be verified as complete in order to avoid as much as possible the presence of deletion sequences (see also point 1).

b. Where possible, the sequence should be designed with a quinoline trimer motif at one terminus which codes for a single helical segment. Since this has no propensity to self-associate, it will force exclusively antiparallel duplex formation.

Final purification should be carried out by RP-HPLC as for the 8-amino-2-quinolinecarboxylic acid-based oligoamides.

3.3 Quinoline/α-Amino Acid Hybrid Oligoamides

Note: Methods described in Protocols 1–8 can be used unchanged for the synthesis of quinoline/α-amino acid hybrid oligoamides (Fig. 7). Coupling of quinoline monomers to the α-amino acid aliphatic amine can be carried out using Protocols 4 and 5 even though other activation methods work as well. Note that for these sequences, all protocols have only been currently validated using Wang resin. They are also not suitable for use with cysteine or 2-aminoisobutyric acid, due to racemization and poor reactivity, respectively.

3.3.1 Protocol 9: Coupling of α-Amino Acid to the Quinoline Amine via In Situ Acid Chloride Formation

Time required: 40 min.

(1) On the vacuum station thoroughly rinse the inside surfaces of the reaction vessel and the resin with anhydrous THF (approx. 10 mL) using a

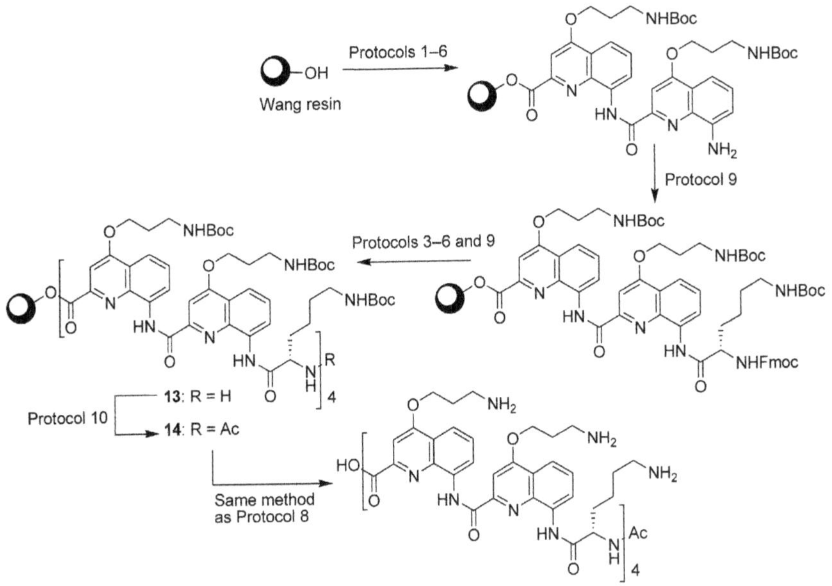

Fig. 7 SPS of quinoline/α-amino acid hybrid oligoamides exemplified by the XQ$_2$-based sequence Ac – (KQ$^{Orn}_2$)$_4$ – OH.

dry syringe, ensuring that no DMF remains. Remove remaining THF, then remove the vessel from the vacuum station and replace end cap.

(2) Add 0.3 mL anhydrous THF to the resin using a dry syringe, followed by 2,4,6-collidine (21.8 μL, 0.165 mmol, 8.7 equiv.) using a micropipette.

(3) Prepare a clean dry 10 mL round bottomed flask and equip with septum and nitrogen balloon.

(4) To the flask, add Fmoc-Lys(Boc)-OH (17.8 mg, 0.038 mmol, 2 equiv. relative to resin loading) and PPh$_3$ (54.8 mg, 0.21 mmol, 11 equiv.) followed by 1 mL anhydrous THF via syringe through the septum. While stirring, add TCAN (16.5 μL, 0.165 mmol, 8.7 equiv.) via dry microsyringe through the septum. Using a dry syringe, immediately transfer the resulting mixture to the resin.

(5) Insert optical fiber probe into the reaction vessel (ensuring the tip is immersed in the reaction mixture) and then place the vessel into the microwave cavity with the appropriate insert. Treat with microwaves: 50 W, ramp to 50°C with a 15 min hold time.

(6) Remove reaction vessel from microwave cavity, remove end cap, and remove reagent solution on vacuum station. Using DMF, thoroughly

rinse the inside surfaces of the reaction vessel (approx. 10 mL DMF) and the resin itself (approx. 10 mL DMF) to ensure all reagents have been removed. Then, using a dry syringe, rinse the inside surfaces of the reaction vessel and the resin itself again with anhydrous THF (approx. 10 mL) to remove the remaining DMF. Remove THF on vacuum station and fit end cap to the vessel.
(7) Repeat steps (2–6). Note: the number of repeat cycles may depend on the Fmoc α-amino acid used. Coupling completion can be determined using Protocol 6. For further details, see Hu et al. (2016).
(8) Remove reaction vessel from microwave cavity, remove fiber optic probe and end cap, then remove reagent solution on vacuum station. Using DMF, thoroughly rinse the inside surfaces of the reaction vessel (approx. 10 mL DMF) and the resin itself (approx. 10 mL DMF) to ensure all reagents have been removed.

3.3.2 Protocol 10: Acetylation of Aliphatic N-Terminal Amine
Time required: 50 min.
(1) On the vacuum station thoroughly rinse the inside surface of the reaction vessel and the resin with anhydrous DMF (approx. 10 mL) using a dry syringe. Remove residual anhydrous DMF from the reaction vessel using the vacuum station. Then remove the vessel from the vacuum station and replace end cap.
(2) Using a dry syringe, add 1.3 mL anhydrous DMF to the resin, followed by DIEA (66.7 μL, 0.38 mmol, 20 equiv.) using a micropipette. Then add acetic anhydride (17.9 μL, 0.19 mmol, 10 equiv.) using a micropipette.
(3) Insert optical fiber probe into the reaction vessel (ensuring the tip is immersed in the reaction mixture) and then place the vessel into the microwave cavity with the appropriate insert. Treat with microwaves: 25 W, ramp to 25°C with a 20 min hold time.
(4) Remove reaction vessel from microwave cavity, remove end cap, and remove reagent solution on vacuum station. Using a dry syringe, rinse the inside surfaces of the reaction vessel with anhydrous DMF (5 mL) to remove reagents and ensure all displaced resin is moved down to the resin bed. Remove residual anhydrous DMF on vacuum station and fit end cap to the vessel.
(5) Repeat steps (2) and (3).
(6) Remove reaction vessel from microwave cavity, remove fiber optic probe and end cap, then remove reagent solution on vacuum station. Using DMF, thoroughly rinse the inside surfaces of the reaction vessel

(approx. 10 mL DMF) and the resin itself (approx. 10 mL DMF) to ensure all reagents have been removed.

4. CHARACTERIZATION OF FOLDAMERS

Section 4.1 shows analytical RP-HPLC and ^{1}H NMR characterization of examples of 8-amino-2-quinolinecarboxylic acid-based oligoamides, 7-amino-8-fluoro-2-quinolinecarboxylic acid-based oligoamides, and quinoline/α-amino acid hybrid oligoamides. Analytical RP-HPLC was carried out on a Macherey–Nagel Nucleodur C_{18} gravity column (4.6 × 100 mm^2, 3 μm) at 1.5 mL/min, running solvents: MilliQ water containing 0.1% (v/v) TFA (solvent 1), CH_3CN containing 0.1% (v/v) TFA (solvent 2). Gradients were as follows: for 8-amino-2-quinolinecarboxylic acid-based oligoamides, 20% to 28% solvent 2 over 10 min (System A); for 7-amino-8-fluoro-2-quinolinecarboxylic acid-based oligoamides, 5% to 100% solvent 2 over 15 min (System B); and for quinoline/α-amino acid hybrid oligoamides, 13% to 18% solvent 2 over 10 min (System C).

4.1 8-Amino-2-quinolinecarboxylic Acid-Based Oligoamides
See Figs. 8 and 9.

4.2 7-Amino-8-fluoro-2-quinolinecarboxylic Acid-Based Oligoamides
See Figs. 10 and 11.

4.3 Quinoline/α-Amino Acid Hybrid Oligoamides
See Figs. 12 and 13.

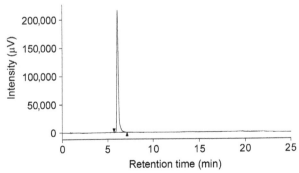

Fig. 8 Analytical RP-HPLC chromatogram (System A) of the 12mer Ac-($Q^{Asp}(Q^{Orn})_2$ $Q^{Ala}Q^{Orn})_2Q^{Asp}Q^{Orn}$-OH after purification by preparative RP-HPLC.

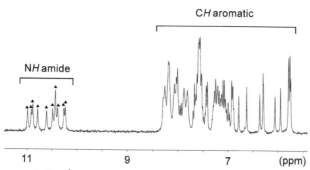

Fig. 9 Excerpt of the ^{1}H NMR spectrum (300 MHz) at 298 K of Ac-$(Q^{Asp}(Q^{Orn})_2 Q^{Ala}Q^{Orn})_2 Q^{Asp}Q^{Orn}$-OH in DMSO-$d_6$. *Solid triangles* indicate the 11 amide NH signals of the 12mer.

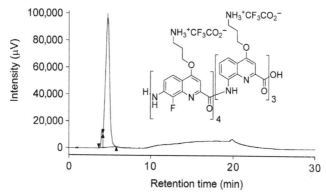

Fig. 10 Analytical RP-HPLC chromatogram (System B) of the 7mer $(Q_F^{Orn})_4(Q^{Orn})_3$ – OH (structure shown in insert) after purification by preparative RP-HPLC.

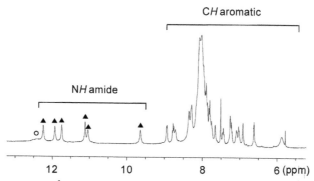

Fig. 11 Excerpt of the ^{1}H NMR spectrum (400 MHz) at 298 K of $H_2N - (Q_F^{Orn})_4(Q^{Orn})_3 -$ OH in DMSO-d_6. *Solid triangles* and the *blank circle* indicate the six amide signals and the carboxylic acid signal of the 7mer, respectively. Note in this solvent, the oligomer is in single helical form (Shang et al., 2014).

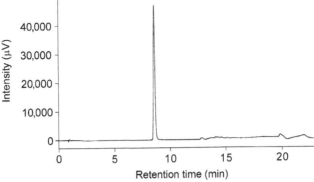

Fig. 12 Analytical RP-HPLC chromatogram (System C) of Ac-K(Q^{Orn})$_2$A(Q^{Orn})$_2$D(Q^{Orn})$_2$S(Q^{Orn})$_2$-OH after purification by preparative RP-HPLC.

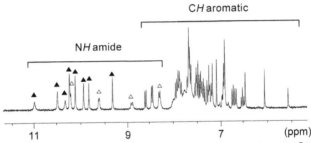

Fig. 13 Excerpt of the ^{1}H NMR spectrum (800 MHz) at 298 K of Ac-K(Q^{Orn})$_2$A(Q^{Orn})$_2$D(Q^{Orn})$_2$S(Q^{Orn})$_2$-OH in D$_2$O/H$_2$O (1:9). *Solid* and *hollow triangles* indicate amide signals from eight aromatic and four aliphatic amines of the 12mer, respectively.

REFERENCES

Appella, D. H., Christianson, L. A., Karle, I. L., Powell, D. R., & Gellman, S. H. (1996). β-peptide foldamers: Robust helix formation in a new family of β-amino acid oligomers. *Journal of the American Chemical Society, 118,* 13071–13072.

Bao, C., Kauffmann, B., Gan, Q., Srinivas, K., Jiang, H., & Huc, I. (2008). Converting sequences of aromatic amino acid monomers into functional three-dimensional structures: Second-generation helical capsules. *Angewandte Chemie International Edition, 47,* 4153–4156.

Baptiste, B., Douat-Casassus, C., Laxmi-Reddy, K., Godde, F., & Huc, I. (2010). Solid phase synthesis of aromatic oligoamides: Application to helical water-soluble foldamers. *The Journal of Organic Chemistry, 75,* 7175–7185.

Barnard, A., Long, K., Martin, H. L., Miles, J. A., Edwards, T. A., Tomlinson, D. C., et al. (2015). Selective and potent proteomimetic inhibitors of intracellular protein–protein interactions. *Angewandte Chemie International Edition, 54,* 2960–2965.

Bassani, D. M., Lehn, J.-M., Baum, G., & Fenske, D. (1997). Designed self-generation of an extended helical structure from an achiral polyheterocylic strand. *Angewandte Chemie International Edition in English, 36,* 1845–1847.

Berl, V., Huc, I., Khoury, R., Krische, M., & Lehn, J.-M. (2000). Interconversion of single and double helices formed from synthetic molecular strands. *Nature, 407,* 720–723.

Buratto, J., Colombo, C., Stupfel, M., Dawson, S. J., Dolain, C., Langlois d'Estaintot, B., et al. (2014). Structure of a complex formed by a protein and a helical aromatic oligoamide foldamer at 2.1 Å resolution. *Angewandte Chemie International Edition, 53*, 883–887.

Chandramouli, N., Ferrand, Y., Lautrette, G., Kauffmann, B., Mackereth, C. D., Laguerre, M., et al. (2015). Iterative design of a helically folded aromatic oligoamide sequence for the selective encapsulation of fructose. *Nature Chemistry, 7*, 334–341.

Claerhout, S., Ermolat'ev, D. S., & Van der Eycken, E. V. (2008). A new colorimetric test for solid-phase amines and thiols. *Journal of Combinatorial Chemistry, 10*, 580–585.

Delaurière, L., Dong, Z., Laxmi-Reddy, K., Godde, F., Toulmé, J.-J., & Huc, I. (2012). Deciphering aromatic oligoamide foldamer–DNA interactions. *Angewandte Chemie International Edition, 51*, 473–477.

Dolain, C., Grélard, A., Laguerre, M., Jiang, H., Maurizot, V., & Huc, I. (2005). Solution structure of quinoline- and pyridine-derived oligoamide foldamers. *Chemistry A European Journal, 11*, 6135–6144.

Ferrand, Y., Chandramouli, N., Kendhale, A. M., Aube, C., Kauffmann, B., Grélard, A., et al. (2012). Long-range effects on the capture and release of a chiral guest by a helical molecular capsule. *Journal of the American Chemical Society, 134*, 11282–11288.

Gan, Q., Bao, C., Kauffmann, B., Grélard, A., Xiang, J., Liu, S., et al. (2008). Quadruple and double helices of 8-fluoroquinoline oligoamides. *Angewandte Chemie International Edition, 47*, 1715–1718.

Garric, J., Léger, J.-M., & Huc, I. (2005). Molecular apple peels. *Angewandte Chemie International Edition, 44*, 1954–1958.

Gillies, E., Deiss, F., Staedel, C., Schmitter, J.-M., & Huc, I. (2007). Development and biological assessment of fully water-soluble helical aromatic amide foldamers. *Angewandte Chemie International Edition, 46*, 4081–4084.

Guichard, G., & Huc, I. (2011). Synthetic foldamers. *Chemical Communications, 47*, 5933–5941.

Hamuro, Y., Geib, S. J., & Hamilton, A. D. (1996). Oligoanthranilamides. Non-peptide subunits that show formation of specific secondary structure. *Journal of the American Chemical Society, 118*, 7529–7541.

Hu, X., Dawson, S. J., Kudo, M., Nagaoka, Y., Tanatani, A., & Huc, I. (2016). Solid-phase synthesis of water-soluble helically folded hybrid α-amino acid/quinoline oligoamides. *The Journal of Organic Chemistry, 81*, 1137–1150.

Huc, I. (2004). Aromatic oligoamide foldamers. *European Journal of Organic Chemistry*, 17–29.

Iriondo-Alberdi, J., Laxmi-Reddy, K., Bouguerne, B., Staedel, C., & Huc, I. (2010). Cellular internalization of water-soluble helical aromatic amide foldamers. *ChemBioChem, 11*, 1679–1685.

Jiang, H., Léger, J.-M., Dolain, C., Guionneau, P., & Huc, I. (2003a). Aromatic deltapeptides: Design, synthesis, and structural studies of helical, quinoline-derived oligoamide foldamers. *Tetrahedron, 59*, 8365.

Jiang, H., Léger, J.-M., & Huc, I. (2003b). Aromatic delta-peptides. *Journal of the American Chemical Society, 125*, 3448–3449.

König, H. M., Abbel, R., Schollmeyer, D., & Kilbinger, A. F. (2006). Solid-phase synthesis of oligo(p-benzamide) foldamers. *Organic Letters, 8*, 1819–1822.

Kudo, M., Maurizot, V., Kauffmann, B., Tanatani, A., & Huc, I. (2013). Folding of a linear array of α-amino acids within a helical aromatic oligoamide frame. *Journal of the American Chemical Society, 135*, 9628–9631.

Kudo, M., Maurizot, V., Masu, H., Tanatani, A., & Huc, I. (2014). Structural elucidation of foldamers with no long range conformational order. *Chemical Communications, 50*, 10090–10093.

Li, X., & Yang, D. (2006). Peptides of aminoxy acids as foldamers. *Chemical Communications, 2006*, 3367–3379.

Lokey, R. S., & Iverson, B. L. (1995). Synthetic molecules that fold into a pleated secondary structure in solution. *Nature, 375*, 303–305.

Morales, G. A., Corbett, J. W., & DeGrado, W. F. (1998). Solid-phase synthesis of benzopiperazinones. *The Journal of Organic Chemistry, 63*, 1172–1177.

Müller, S., Laxmi-Reddy, K., Jena, P. V., Baptiste, B., Dong, Z., Godde, F., et al. (2014). Targeting DNA G-quadruplexes with helical small molecules. *ChemBioChem, 15*, 2563–2570.

Murphy, N. S., Prabhakaran, P., Azzarito, V., Plante, J. P., Hardie, M. J., Kilner, C. A., et al. (2013). Solid-phase methodology for synthesis of O-alkylated aromatic oligoamide inhibitors of α-helix-mediated protein–protein interactions. *Chemistry A European Journal, 19*, 5546–5550.

Nair, R. V., Vijayadas, K. N., Roy, A., & Sanjayan, G. J. (2014). Heterogeneous foldamers from aliphatic–aromatic amino acid building blocks: Current trends and future prospects. *European Journal of Organic Chemistry*, 7763–7780.

Nelson, J. C., Saven, J. G., Moore, J. S., & Wolynes, P. G. (1997). Solvophobically driven folding of nonbiological oligomers. *Science, 277*, 1793–1796.

Nielsen, P. E., Egholm, M., Berg, R. H., & Buchardt, O. (1991). Sequence-selective recognition of DNA by strand displacement with a thymine-substituted polyamide. *Science, 254*, 1497–1500.

Puckett, J. W., Green, J. T., & Dervan, P. B. (2012). Microwave assisted synthesis of Py-Im polyamides. *Organic Letters, 14*, 2774–2777.

Qi, T., Maurizot, V., Noguchi, H., Charoenraks, T., Kauffmann, B., Takafuji, M., et al. (2012). Solvent dependence of helix stability in aromatic oligoamide foldamers. *Chemical Communications, 48*, 6337–6339.

Salaun, A., Potel, M., Roisnel, T., Gall, P., & Le Grel, P. (2005). Crystal structures of aza-beta3-peptides, a new class of foldamers relying on a framework of hydrazinoturns. *The Journal of Organic Chemistry, 70*, 6499–6502.

Seebach, D., Overhand, M., Kühnle, F. N. M., Martinoni, B., Oberer, L., Hommel, U., et al. (1996). β-peptides: Synthesis by *Arndt-Eistert* homologation with concomitant peptide coupling. Structure determination by NMR and CD spectroscopy and by X-ray crystallography. Helical secondary structure of a β-hexapeptide in solution and its stability towards pepsin. *Helvetica Chimica Acta, 79*, 913–941.

Semetey, V., Rognan, D., Hemmerlin, C., Graff, R., Briand, J.-P., Marraud, M., et al. (2002). Stable helical secondary structure in short-chain N,N'-linked oligoureas bearing proteinogenic side chains. *Angewandte Chemie International Edition, 41*, 1893–1895.

Shang, J., Gan, Q., Dawson, S. J., Rosu, F., Jiang, H., Ferrand, Y., et al. (2014). Self-association of aromatic oligoamide foldamers into double helices in water. *Organic Letters, 16*, 4992–4995.

Shi, R., Wang, F., & Yan, B. (2007). Site–site isolation and site–site interaction—Two sides of the same coin. *International Journal of Peptide Research and Therapeutics, 13*, 213–219.

Simon, R. J., Kania, R. S., Zuckermann, R. N., Huebner, V. D., Jewell, D. A., Banville, S., et al. (1992). Peptoids: A modular approach to drug discovery. *Proceedings of the National Academy of Sciences of the United States of America, 89*, 9367–9371.

Singleton, M. L., Pirotte, G., Kauffmann, B., Ferrand, Y., & Huc, I. (2014). Increasing the size of an aromatic helical foldamer cavity by strand intercalation. *Angewandte Chemie International Edition, 53*, 13140–13144.

White, P. D., & Chan, W. C. (2000). Basic procedures. In P. D. White & W. C. Chan (Eds.), *Fmoc solid phase synthesis: A practical approach*.Oxford: Oxford University Press. p. 44.

Wurtz, N. R., Turner, J. M., Baird, E. E., & Dervan, P. B. (2001). Fmoc solid phase synthesis of polyamides containing pyrrole and imidazole amino acids. *Organic Letters, 3*, 1201–1203.

Zhu, J., Parra, R. D., Zeng, H., Skrzypczac-Jankun, E., Zeng, X. C., & Gong, B. (2000). A new class of folding oligomers: Crescent oligoamides. *Journal of the American Chemical Society, 122*, 4219–4220.

CHAPTER FOURTEEN

Conformational Restriction of Peptides Using Dithiol Bis-Alkylation

L. Peraro, T.R. Siegert, J.A. Kritzer[1]
Tufts University, Medford, MA, United States
[1]Corresponding author: e-mail address: joshua.kritzer@tufts.edu

Contents

1. Introduction	304
1.1 Stabilizing Alpha-Helical Structures	304
1.2 Stabilizing Nonhelical Structures	308
2. Using Thiol Alkylation to Constrain Peptides	309
2.1 Libraries of Peptides That Are Constrained Through Alkylated Cysteines	309
2.2 Rational Design Using Cysteine Alkylation	310
2.3 Thiol Bis-Alkylation Is Ideal for Constraining Epitopes into Diverse Conformations	311
3. Protocols for Peptide Synthesis and Cross-Linking	312
4. Applications, Tips, and Troubleshooting	316
4.1 Applying Dithiol Bis-Alkylation for Constraining Loop Epitopes	316
4.2 Monitoring Progression of Thiol Bis-Alkylation	317
4.3 Reaction Scope and Versatility	317
4.4 Investigating the Bis-Alkylation Mechanism	322
4.5 Optimizing and Troubleshooting Bis-Alkylation Reactions	325
5. Future Directions	327
References	327

Abstract

Macrocyclic peptides are highly promising as inhibitors of protein–protein interactions. While many bond-forming reactions can be used to make cyclic peptides, most have limitations that make this chemical space challenging to access. Recently, a variety of cysteine alkylation reactions have been used in rational design and library approaches for cyclic peptide discovery and development. We and others have found that this chemistry is versatile and robust enough to produce a large variety of conformationally constrained cyclic peptides. In this chapter, we describe applications, methods, mechanistic insights, and troubleshooting for dithiol bis-alkylation reactions for the production of cyclic peptides. This method for efficient solution-phase macrocyclization is highly useful for the rapid production and screening of loop-based inhibitors of protein–protein interactions.

1. INTRODUCTION

In the last two decades, early-stage drug discovery has expanded to include targets outside the traditionally druggable classes of enzymes and cell surface receptors (Arkin, Tang, & Wells, 2014). Classically "undruggable" protein–protein interactions can make viable drug targets, but often have large interaction surfaces that are difficult for small molecules to bind with high affinity (Thompson, Dugan, Gestwicki, & Mapp, 2012). Peptides are an attractive option for targeting protein–protein interactions, as they are intermediate in size between small molecules and large biologics and offer many advantages over both (Kawamoto et al., 2012). Peptides are synthetically tractable, they can be optimized to high affinity and selectivity, and they often have good safety and tolerability profiles in animals and humans (Fosgerau & Hoffmann, 2014). These strengths have led to the FDA approval of several peptide-based therapeutics including linaclotide for gastrointestinal disorders, peginesatide for anemia, and carfilzomib for multiple myeloma (Kaspar & Reichert, 2013). While peptides hold many distinct advantages, they also have some inherent weaknesses including their short half-life and sensitivity to proteases (Fosgerau & Hoffmann, 2014). Also, short peptides are often poorly structured in aqueous solution, which can limit their affinity for their targets. One of the largest limitations of peptide drugs is poor membrane permeability, making delivery to intracellular targets difficult (Hill, Shepherd, Diness, & Fairlie, 2014). Chemical modification of peptides, such as isosteric backbone replacement and N-methylation, has substantial promise as a solution to many of these drawbacks (Avan, Hall, & Katritzky, 2014; Bockus et al., 2015; Brust et al., 2013; Chatterjee, Gilon, Hoffman, & Kessler, 2008; Yamagishi et al., 2011). Many modification strategies have been pursued in the effort to transform linear peptides into more natural product-like molecules with high potency, selectivity, cell penetration, and even oral bioavailability (Bock, Gavenonis, & Kritzer, 2013; Clark et al., 2010; Dahan, Tsume, Sun, Miller, & Amidon, 2011; Renukuntla, Vadlapudi, Patel, Boddu, & Mitra, 2013).

1.1 Stabilizing Alpha-Helical Structures

One solution to the inherent limitations of peptides is macrocyclization, which is typically applied as a structure-promoting conformational constraint. The effects of conformational constraints on peptide structure and function have been particularly well-studied in the field of alpha-helix stabilization (Henchey, Jochim, & Arora, 2008). A stable helical structure is difficult to achieve for a short peptide in aqueous solution, and according to the Zimm–Bragg model for helix–coil transition, the primary barrier is nucleation of an initial alpha-helical turn (Zimm & Bragg, 1959). Thus, a primary motivation for developing new peptide cyclization strategies has been stabilization of a nucleating alpha-helical turn, which can then propagate to stabilize an overall helical structure.

There are a host of cyclization strategies that have been developed for alpha-helix stabilization (Table 1). One of the most successful examples is hydrocarbon side-chain "stapling" using olefin metathesis (Miller et al., 1996; Walensky et al., 2004). Specific combinations of alpha-methyl groups and covalent cross-linking of $(i, i+3)$, $(i, i+4)$, or $(i, i+7)$ residues with carbon chains of different lengths have been used to stabilize a variety of helical peptides (Bird et al., 2014; Kim, Kutchukian, & Verdine, 2010; Schafmeister et al., 2000). This method was used to stabilize an alpha-helix that mimics the hDM2-binding epitope of p53 (Chang et al., 2013), and a "stapled helix" of this type is currently in Phase I clinical trials for advanced hematologic and solid malignancies with wild-type p53 (Khoo, Hoe, Verma, & Lane, 2014). Another approach for alpha-helix stabilization using ring-closing olefin metathesis is the "hydrogen-bond surrogate" approach, which replaces the characteristic alpha-helix hydrogen bond between the N-terminal amino acid and the $i+4$ carbonyl with a carbon–carbon bond (Patgiri, Jochim, & Arora, 2008). Lactam bridges between lysine and aspartate residues located at $(i, i+4)$ positions can also dramatically stabilize helical structures and promote potent protein binding (Shepherd et al., 2005). Copper-catalyzed Huisgen cycloaddition, or "click chemistry," has also been used to join azide and alkyne side-chains at $(i, i+4)$ positions (Kawamoto et al., 2012) to stabilize overall helical structure.

Several chemistries are available that capitalize on the reactivity of cysteine residues. Thioether ligation was used to link a cysteine to a bromoacetylated ornithine, yielding an alternative to lactam bridge formation (Brunel & Dawson, 2005). A recent study also presented a thiol-based alternative to ring-closing metathesis, where peptides were stapled using thiol-ene chemistry between two cysteines and a diene linker (Wang & Chou, 2015). Cross-linking two or three unprotected cysteines can also be achieved using arylation reactions with perfluoroaryl groups (Spokoyny et al., 2013), or using alkylation reactions with bis-bromomethyl or tris-bromomethyl linkers (Jo et al., 2012; Timmerman et al., 2005). The diverse chemistries for nucleating alpha-helical structure in short peptides have led

Table 1 Some Common Reactions Used for Intramolecular Peptide Cross-Linking

Cross-Linking Reaction	Reaction Conditions[a]	Synthesis Mode	References
	10 mM Grubbs catalyst in 1,2-dichloroethane, 2 h	On resin	Miller, Blackwell, and Grubbs (1996) and Schafmeister, Po, and Verdine (2000)
	1.5 equiv. BOP, 2 equiv. DIPEA, DMF, 24 h	On resin	Shepherd, Hoang, Abbenante, and Fairlie (2005)
	Aqueous 6 M Gdm-HCl, pH 8.0, 2 h	In solution	Brunel and Dawson (2005)
	50:50 CH$_3$CN/H$_2$O with 5 mM NH$_4$HCO$_3$ pH 8.0, m-dibromoxylene, 15–120 min	In solution and on resin	Jo et al. (2012) and Timmerman, Beld, Puijk, and Meloen (2005)

Reactants	Conditions	Phase	Reference
(azide + alkyne peptide)	2:1 H$_2$O/t-BuOH, 4.4 equiv. CuSO$_4$·5H$_2$O, 30–90 min	In solution and on resin	Kawamoto et al. (2012) and White and Yudin (2011)
(bis-thiol peptide + hexafluorobenzene)	Aqueous 50 mM Tris, DMF, hexafluorobenzene, 4.5 h	In solution	Spokoyny et al. (2013)
(bis-thiol peptide + diene)	DMF, radical initiator, 365 nm UV, 1 h	In solution	Wang and Chou (2015)

[a] All reactions listed were run at room temperature for the times indicated. *BOP*, (benzotriazol-1-yloxy) tris(dimethylamino)phosphonium hexafluorophosphate; *DIPEA*, N,N-diisopropylethylamine; *DMF*, N,N-dimethylformamide; *Gdm–HCl*, guanadinium hydrochloride.

to the successful development of many helical inhibitors of protein–protein interactions and hold promise for the development of many more therapeutic leads (Robertson & Jamieson, 2015).

In all cases of alpha-helix stabilization, the intramolecular cross-link promotes helix-nucleating hydrogen-bonding patterns, thus stabilizing overall alpha-helix structure. Another effective result is that these cyclic peptides are generally more stable to proteolytic degradation (Bird et al., 2014; Bock et al., 2013; Kawamoto et al., 2012; Shepherd et al., 2005). In some cases, cyclic peptides have increased cytosolic penetration compared to linear peptides (Bock et al., 2013; Chang et al., 2013; Muppidi, Wang, Li, Chen, & Lin, 2011; Qian et al., 2013). This sought-after property is essential for targeting intracellular protein–protein interactions.

1.2 Stabilizing Nonhelical Structures

Work on helical peptides has demonstrated the major benefits of macrocyclization for promoting peptide structure and function, but applying cyclic constraints to nonhelical structures has not been as straightforward. While the results from screening large, unbiased libraries of cyclic peptides clearly indicate that this is a valuable chemical space for protein inhibitors (Gao, Amar, Pahwa, Fields, & Kodadek, 2015; Gartner et al., 2004; Passioura, Katoh, Goto, & Suga, 2014; Xiao et al., 2010), rational design of small, nonhelical cyclic peptides is still largely trial and error. The rational design approach is greatly assisted by starting with a peptide epitope derived from a known protein binding partner. Ideally, this epitope accounts for a majority of the binding energy of the interaction by comprising the most important "hot spot" residues (Clackson & Wells, 1995). These hot spots can be identified either through experimental mutagenesis or through computational mutagenesis methods (Kortemme & Baker, 2002; Kortemme, Kim, & Baker, 2004). Several groups have used computational techniques to comprehensively identify peptide epitopes from the Protein Data Bank, focusing on continuous segments or more specifically on helix, sheet, or loop structures (Gavenonis, Sheneman, Siegert, Eshelman, & Kritzer, 2014; Henchey et al., 2008; London, Raveh, Movshovitz-Attias, & Schueler-Furman, 2010; Watkins & Arora, 2014; Wuo, Mahon, & Arora, 2015). Translating these epitopes to effective inhibitory peptides requires replacing the entire protein tertiary structure with a synthetic linker that stabilizes the epitope's highest-affinity 3D structure. For helices, sheets, and beta turns, generalizable approaches for structural stabilization have been developed that can be immediately applied to many epitopes (Yin, 2012). Other, less common 3D structures are loosely grouped as "loop" structures. These require a more general strategy since one cross-link chemistry, spacing and length cannot be used to stabilize the large variety of loop structures that mediate protein–protein interactions (Gavenonis et al., 2014).

To date, computational methods for the prediction of cyclic peptide structure have only begun to be rigorously designed and tested (Damas et al., 2013; Razavi, Wuest, & Voelz, 2014; Wakefield, Wuest, & Voelz, 2015; Yu & Lin, 2015). This makes it very challenging to predictively design specific cross-links to stabilize a desired loop structure. Currently, identifying the proper linker chemistry, length, and positioning can only be done in an iterative process (Berthelot, Gonçalves, Laïn, Estieu-Gionnet, & Déléris, 2006; Cudic, Wade, & Otvos, 2000; Hayouka et al., 2012; Kamens, Eisert, Corlin, Baleja, & Kritzer, 2014; Qvit et al., 2009). One way to accelerate, this process is to introduce diverse conformational constraints at a late stage of synthesis. In this chapter, we provide protocols for a robust, efficient method for late-stage conformational diversification of peptide epitopes using thiol bis-alkylation chemistry (originally reported by Timmerman et al., 2005). This allows for rapid preparation and screening of many conformations of a given loop using a panel of linkers, experimentally searching for the highest-affinity conformation.

2. USING THIOL ALKYLATION TO CONSTRAIN PEPTIDES

2.1 Libraries of Peptides That Are Constrained Through Alkylated Cysteines

Thiol bis-alkylation has been applied in several studies to constrain peptides and introduce structure. This chemistry has been particularly useful for introducing conformational constraints within large, randomized peptide libraries. In a landmark paper, Heinis and coworkers generated a trillion-member phage-display library of bicyclic peptides containing three reactive cysteines which were cyclized using tris-(bromomethyl)benzene (linker **tmb**, Fig. 1) (Heinis, Rutherford, Freund, & Winter, 2009). This library was applied in iterative selections for binding to the plasma protease kallikrein, ultimately producing a low-nanomolar inhibitor which was highly specific for kallikrein over other plasma proteases. Using the same strategy, Heinis and coworkers also discovered a submicromolar bicyclic ligand for the negative regulatory region of the Notch receptor (Urech-Varenne, Radtke, & Heinis, 2015). Thiol bis-alkylation has also been applied to

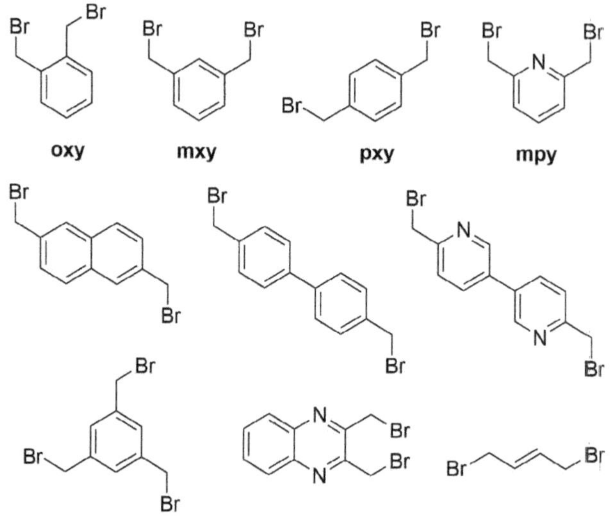

Fig. 1 Common linkers used for thiol cross-linking. *Top row*: Dibromo-*o*-xylene (**oxy**), dibromo-*m*-xylene (**mxy**), dibromo-*p*-xylene (**pxy**), and 2,6-bis(bromomethyl) pyridine (**mpy**). *Second row*: 2,6-bis(bromomethyl)naphthalene, 4,4′-bis(bromomethyl)biphenyl and 6,6′-bis-bromomethyl-[3,3′]bipyridine. *Bottom row*: 1,3,5-tris(bromomethyl)benzene (**tmb**), 2,3-bis(bromomethyl)quinoxaline, and *trans*-1,4-dibromo-2-butene.

mRNA display libraries (Schlippe, Hartman, Josephson, & Szostak, 2012). In one such selection, all the selected peptides were cyclic, even those containing a single cysteine; the **mxy** linker unexpectedly formed a small ring with the N-terminal methionine (White et al., 2015). The rapid cross-linking of cysteine and methionine using dibromoxylene linkers is an observation that we have confirmed in our lab (see later).

More recent work on thiol-alkylated libraries has pushed this chemistry even further. A novel method was reported by the Suga lab where they synthesized tricyclic peptides using flexizyme-assisted translation (Goto, Katoh, & Suga, 2011) and the **tmb** linker (Bashiruddin, Nagano, & Suga, 2015). With this technology, they replaced the N-terminal methionine with a chloroacetyl-containing amino acid, which spontaneously macrocyclizes by forming a thioether bond with a nonadjacent cysteine. The three remaining cysteines were then cyclized using the **tmb** linker to form tricyclic peptides (Bashiruddin et al., 2015).

2.2 Rational Design Using Cysteine Alkylation

As discussed earlier, cysteine alkylation has been used for the generation of libraries via phage or mRNA display, yielding cyclic, bicyclic, or tricyclic peptides with increased conformational rigidity. Alternatively, this chemistry has also been used to induce specific secondary structures in peptides. DeGrado, Greenbaum, and coworkers used thiol bis-alkylation to find the linker that would yield the cleanest cyclization reaction and induce alpha-helical structure in a peptide inhibitor of the protease calpain (Jo et al., 2012). The cysteines were placed at (i, $i+4$) positions within a model peptide that had moderate helicity. A panel of 24 different linkers was screened, and the hits were compared using circular dichroism, from which they determined that cross-linking with dibromo-m-xylene (linker **mxy**, Fig. 1) showed the highest increase in helical structure. Then, they scanned the peptide for the best (i, $i+4$) staple location by moving the cysteines and then cross-linking with linker **mxy** (Jo et al., 2012). Similarly, Muppidi et al. designed stapled helices containing cysteines in (i, $i+7$) positions, using the longer linkers 4,4′-bis-bromomethyl-biphenyl and 6,6′-bis-bromomethyl-[3,3′]bipyridine (Muppidi et al., 2011, 2014) (see Fig. 1). They showed that this cross-linking chemistry led to increased helicity and bioactivity, and even increased cell permeability for a series of BH3-peptide-derived ligands for MCL-1.

2.3 Thiol Bis-Alkylation Is Ideal for Constraining Epitopes into Diverse Conformations

We considered all the above cyclization chemistries when searching for methods for rapid translation of loop epitopes into useful protein–protein interaction inhibitors (Gavenonis et al., 2014). Thiol bis-alkylation has rapid kinetics and broad sequence tolerance, as evidenced by several discrete studies and by its many applications in cyclic peptide libraries (Jo et al., 2012; Smeenk, Dailly, Hiemstra, van Maarseveen, & Timmerman, 2012; Timmerman et al., 2005; Todorova-Balvay, Stoilova, Gargova, & Vijayalakshmi, 2007). Also, studies that used diverse linkers to search for cross-links that specifically stabilize helical conformations indicated that this chemistry is optimal for late-stage conformational diversification (Jo et al., 2012; Muppidi et al., 2011). This chapter describes adaptations and extensions of these prior studies toward general methods for the rapid production of cyclic peptides with diverse conformational constraints.

Late-stage conformational diversification is introduced by including two thiol-containing amino acids at positions known to be nonessential for target binding. Such peptides are readily cross-linked into macrocycles using a wide variety of different linkers. Fig. 1 lists commercially available or readily synthesized linkers that have been used to cross-link cysteine-containing peptides, which range from simple *ortho*, *meta*, and *para* dibromomethylxylenes (**oxy**, **mxy**, and **pxy**, respectively) to 2,3-bis (bromomethyl)quinoxaline (Jo et al., 2012; Muppidi et al., 2011, 2014; Timmerman et al., 2005). While nearly all prior reports have used L-cysteine, one reported application used D-cysteine as the thiol-containing amino acid (Muppidi et al., 2011). We have found that many thiol-containing amino acids are compatible with this approach. To maximize the variety of conformational constraints, we have applied various combinations of L- and D-cysteine, L- and D-homocysteine, and L- and D-penicillamine, all without reduction in cross-linking efficiency. Dithiol-containing amino acids could also potentially be used (Chen et al., 2014), since the rigidity of the linkers should preclude bis-alkylation of both thiols on one side chain. A major advantage of using this chemistry is that it is orthogonal to natural amino acid functional groups including lysine and free N-termini. The exception is methionine, as discussed below. By strategically varying the thiol-containing amino acids, the relative positioning of the thiol-containing amino acids, and the linkers, it is possible to use this chemistry to prepare a large library of peptides with diverse 3D conformations from a single peptide epitope.

3. PROTOCOLS FOR PEPTIDE SYNTHESIS AND CROSS-LINKING

We use standard fluorenylmethoxycarbonyl (Fmoc) solid-phase peptide synthesis (SPPS) to synthesize the linear precursor peptides (Fields & Noble, 1990; Moss, 2005). For an amidated C-terminus, we use Rink Amide Resin (100–200 mesh) with a loading of 0.3–0.6 mmol/g (Han & Kim, 2004). In order to produce a panel of cross-linked peptides, we begin with the synthesis of one parent linear peptide at a scale of 50–100 μmol. Peptides can be synthesized by hand or using an automated synthesizer. After synthesizing the linear sequence, the N-terminus can be capped or left as a free amine, and the peptide is cleaved off the resin. The peptide is precipitated using cold ether to separate it from protecting groups and cleavage reagents, particularly scavengers such as ethanedithiol (EDT). The crude linear peptide can either be purified using reverse-phase high-performance liquid chromatography (RP-HPLC) or directly used in thiol bis-alkylation reactions. The linear peptide is divided into multiple reaction vessels and reacted with different linkers in a 50:50 mixture of acetonitrile (CH$_3$CN) and water buffered at pH 8.0. The reaction is typically complete within

1 h at room temperature. After bis-alkylation, solvents can be concentrated by lyophilizing the reaction and resuspending in a smaller volume of CH_3CN/H_2O. The crude reaction is purified by RP-HPLC to obtain the final cyclic product. An abbreviated procedure is provided later, and a scheme is shown in Fig. 2.

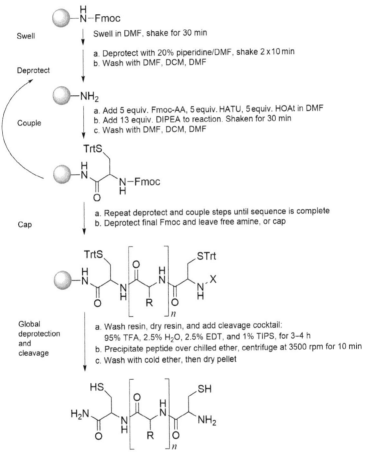

Fig. 2 Scheme for SPPS of linear peptides using Fmoc chemistry. DMF is N,N-dimethylformamide, DCM is dichloromethane, AA refers to the protected monomer form of the desired amino acid, HATU is 1-[bis(dimethylamino)methylene]-1H-1,-2,3-triazolo[4,5-b]pyridinium 3-oxid hexafluorophosphate, HOAt is 1-hydroxy-7-azabenzotriazole, DIPEA is N,N-diisopropylethylamine, TFA is trifluoroacetic acid, EDT is ethanedithiol, TIPS is triisopropylsilane. Though the peptide in this example has cysteines at the N- and C-termini, it is possible to use many other thiol-containing amino acids and to place the thiol-containing amino acids at any positions within the peptide.

(1) Swell resin in 5–10 mL of DMF for at least 30 min with shaking.
(2) Deprotect the resin using 5–10 mL of 20% piperidine in DMF for 2 × 7 min.
(3) Wash the resin with 5–10 mL of DMF, 2 × 30 s, DCM 2 × 30 s, DMF 2 × 30 s.
   The presence of a free amine can be confirmed using a Kaiser test (Shelton & Jensen, 2013).
(4) Dissolve 5 equiv. of the Fmoc-AA-OH, 5 equiv. of coupling reagent, 1-[bis(dimethylamino)methylene]-1H-1,2,3-triazolo[4,5-b] pyridinium 3-oxid hexafluorophosphate (HATU), and 5 equiv. of the coupling additive 1-hydroxy-7-azabenzotriazole (HOAt) in 5–10 mL of DMF or NMP. Add to resin and also add 13 equiv. of DIPEA. Shake at room temp for 30 min. The completion of the reaction can also be checked with a negative Kaiser test.
(5) Wash the resin extensively as in step 3.
(6) Repeat Fmoc deprotection and coupling steps 2–5 until the final amino acid has been coupled to the growing peptide chain.
(7) After coupling the last amino acid, remove the last Fmoc group as described in step 2. Then, if an acetylated N-terminus is desired, cap the resin using 5–10 mL 10% acetic anhydride/10% 2,6-lutidine/80% DMF for 2 × 10 min.
(8) Wash the resin extensively using DMF and DCM, finishing with a methanol wash. Dry out the resin completely using vacuum or dry nitrogen or argon gas.
(9) For global deprotection and cleavage, use 1–4 mL of a standard cleavage cocktail: 95% trifluoroacetic acid (TFA), 2.5% 1,2-ethanedithiol (EDT), 2.5% H_2O, and 1% triisopropylsilane (TIPS). Allow to deprotect for 3–4 h depending on the amino acids in the sequence.
(10) Chill 40 mL of diethyl ether on dry ice for 15 min.
(11) Once the cleavage is complete, filter the cleavage solution to separate it from the resin and add dropwise to chilled ether. You should see your peptide crashing out in the ether and the solution should become opaque.
(12) Centrifuge at 3500 rpm for 10 min. Decant the ether and wash the pellet three times with 40 mL of freshly chilled diethyl ether to ensure removal of cleavage cocktail components. Centrifuge again at 3500 rpm for 10 min.
(13) Decant the ether and dry the pellet under dry argon or nitrogen gas. A precipitated pellet of crude peptide is shown in Fig. 3.

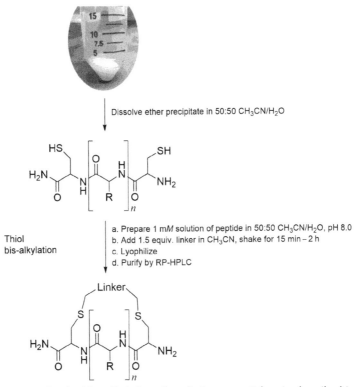

Fig. 3 Scheme for thiol bis-alkylation of crude linear peptides. As described in Fig. 2, the identities of the thiol-containing amino acids and their relative positions within the peptide can be varied with only minor effects on overall yield.

At this point, the linear peptide can be purified by reverse-phase HPLC. However, in most cases, the crude peptide is relatively pure and can be immediately bis-alkylated in solution. We highly recommend skipping HPLC purification of most linear peptides prior to bis-alkylation, unless major byproducts are present. We have had good results by performing the bis-alkylation as follows (Fig. 3):

(14) Dissolve the ether-precipitated pellet (or purified linear peptide) in 50:50 CH_3CN/H_2O.

(15) If the linear peptide has a Tyr or Trp, estimate the concentration by UV–vis spectrophotometry using UV absorbance at 280 nm. If the concentration cannot be estimated by UV, we have more roughly estimated the maximum amount of peptide present by assuming a 100% yield of the overall solid-phase synthesis. This works well because one

would rather have the peptide more dilute than expected for the bis-alkylation reaction than more concentrated.

(16) Prepare a 1 m*M* solution of peptide in 50:50 solution of CH$_3$CN and H$_2$O buffered with 20 m*M* ammonium bicarbonate, pH 8.0.
Note: At this step, check the pH of the solution before you add the linker, especially if you are performing the alkylation on a crude peptide. It is possible for TFA to carry over from ether precipitation or RP-HPLC, which could potentially lower the pH of the reaction.

(17) Dissolve 1.5 equiv. of linker in 1–2 mL of CH$_3$CN and add to the peptide. The reaction is typically completed in under 1 h. You can monitor the formation of the product using mass spectrometry. For instance, adding the **oxy**, **mxy**, or **pxy** linker will result in a cyclic peptide product that is 102 Daltons higher in mass than the linear peptide, as observed in Fig. 5. We have observed that, for most peptides, the appearance of the product peak by MALDI-TOF coincides with the disappearance of the starting material, and that the mass spectrometry peak for the starting material frequently becomes tiny or unobservable after 1 h.

(18) The reaction can be stopped by lowering the pH with HCl or TFA and can be immediately purified. We often freeze the reaction, lyophilize, and redissolve in a smaller volume prior to purification.

(19) Once the cyclic peptide is purified, it can be stored as lyophilized powder at −20°C or directly used in assays.
Note: While dimethyl sulfoxide (DMSO) is a common solvent for concentrated stocks of purified peptides, DMSO can also accelerate the oxidation of sulfides. If you store dithiol bis-alkylated peptides in DMSO, you should monitor their purities and masses over time to ensure sulfur oxidation has not occurred. In a minority of cases, we have observed sulfide oxidation and loss of bioactivity of bis-alkylated peptides that had been stored in DMSO for longer than 4 weeks.

4. APPLICATIONS, TIPS, AND TROUBLESHOOTING

4.1 Applying Dithiol Bis-Alkylation for Constraining Loop Epitopes

We have found thiol bis-alkylation to be an efficient and robust method to add conformational constraints to peptide epitopes. In one application, we designed peptide **1**, which contains a loop epitope derived from the Eps15–stonin2 interaction (Gavenonis et al., 2014; Rumpf et al., 2008). Peptide **1** was synthesized as described earlier, and crude peptide was divided into fractions that were separately cyclized using **oxy**, **mxy**, and **pxy** linkers. Each reaction was then lyophilized and purified by preparative RP-HPLC. Fig. 4 shows the preparative-scale chromatograms along with the analytical reinjections of the major peaks. For each reaction, one pass on the HPLC produced cyclic peptides with purity well over 95%. These peptides can be used directly in biochemical assays and/or cell-based assays to screen for the desired activity.

4.2 Monitoring Progression of Thiol Bis-Alkylation

Model peptide **2**, whose sequence was adapted from the autophagy regulator Beclin 1 (Shoji-Kawata et al., 2013), was bis-alkylated using the **mxy** linker (Fig. 5). The linear peptide was purified and lyophilized, then brought up in a 1:1 mixture of acetonitrile and water (water was buffered to pH 8.0 to a final concentration of 20 mM ammonium bicarbonate). Three equivalents of the linker were dissolved in 500 μL of acetonitrile and added to the reaction, which was then placed on a shaker. At different time points over the course of 1 h, 100 μL were removed from the reaction mixture and quenched with 50 μL of 0.1 M HCl to stop the reaction. Samples at 0, 5, 15, 30, and 60 min were then analyzed by analytical RP-HPLC to monitor the conversion of linear to cyclic peptide. Fig. 6 shows how a peptide bis-alkylation reaction can be monitored for progression and efficiency. The excess linker can be seen eluting at 20 min. The corresponding MALDI traces show that the product (expected mass 1364.61 Da, observed mass m/z 1364.81) has a molecular weight of 102 Da larger than the linear peptide (expected mass 1262.47 Da, observed m/z 1262.42 Da) (Fig. 5).

4.3 Reaction Scope and Versatility

We have obtained excellent yields with a wide variety of dithiol peptides and linkers. Based on this experience, we provide some general observations about the scope and versatility of this macrocyclization reaction.

- *In most peptides, varying the positions of the thiol-containing amino acids does not affect the efficiency of the reaction.* In general, we have observed that relative positions from $(i, i+3)$ to $(i, i+10)$ can be readily cyclized using dithiol bis-alkylation. For example, we prepared model peptides **3**, **4**, and **5** (see Table 2) and observed no difference in cyclization efficiencies among these $(i, i+4)$, $(i, i+5)$, and $(i, i+6)$ macrocyclization reactions. For each of these peptides, the desired bis-alkylated peptide (calculated m/z

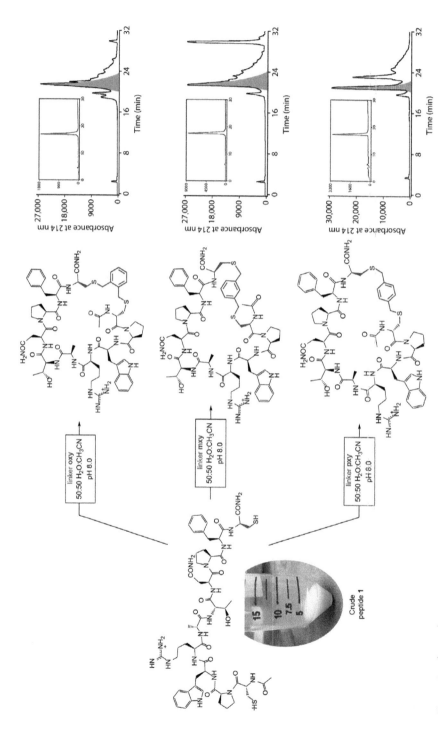

Fig. 4 See legend on opposite page.

1326.56, observed m/z 1327.32 Da) was the major product after only 2 min (Fig. 7A). Minor formation of xylene cross-linked dimer with one linker (calculated m/z 2550.99, observed m/z 2549.60 Da) and two linkers (calculated m/z 2653.12, observed m/z 2654.97 Da) was also observed by mass spectrometry at different time points.

- *Various linkers can be used, though some require additional optimization.* In our hands, bis-alkylation of dithiol-containing peptides has been consistently most efficient with the xylene-based linkers and 2,6-bis(bromomethyl) pyridine (linkers **oxy**, **mxy**, **pxy**, and **mpy**, Fig. 1). Biphenyl and napthyl linkers can also be used, but obtaining high yields of macrocyclized product with these linkers sometimes requires case-by-case optimization to find the most favorable reaction solvents, stoichiometries, times, and conditions. Under similar reaction conditions, we have also used allyl bromide and benzyl bromide to generate bis-alkylated analogs with no macrocyclic constraint. In our hands, linkers lacking benzylic bromides such as 1,4-dibromo-2,3-butanedione, 1,2-dibromoethane, and 1,4-dibromobutane have seldom yielded cleanly cyclized peptides.

- *Other thiol-containing amino acids undergo bis-alkylation with similar efficiency.* Our lab has used L- and D-cysteine, L- and D-homocysteine, and L- and D-penicillamine in various permutations at the N-terminal and C-terminal cyclization positions. We have seen no decrease in the efficiency of cyclization even with the more hindered penicillamines, though we have observed a slight decrease in solubility which can be addressed by reducing peptide concentration in the bis-alkylation reaction. We have also used cysteamine 2-chlorotrityl resin to introduce an achiral, thiol-containing moiety to the C-terminus in place of one cysteine. This cysteine-to-cysteamine substitution at the C-terminus produces peptides that cyclize robustly with a variety of linkers.

Fig. 4 Crude peptide **1** was redissolved from an ether-precipitated pellet, split into three fractions, and each was bis-alkylated as described in Fig. 3. This produced three peptides with identical epitopes but three different conformational constraints. Shown at the right are representative preparative-scale HPLC chromatograms, with peaks corresponding to the product cyclic peptides highlighted in gray. Preparative-scale HPLC was performed using a Agilent ZORBAX SB-C8 Vydac Maxisorb column with a flow rate of 20 mL/min and a gradient of 5% solvent B to 100% solvent B over 30 min (solvent A is H_2O with 0.2% TFA, solvent B is CH_3CN with 0.2% TFA). *Insets* in these chromatograms show analytical HPLC chromatograms of the highlighted fractions, demonstrating greater than 95% purity after one pass. Analytical-scale HPLC was performed using an Agilent ZORBAX Eclipse Plus C18 Vydac Maxisorb column with a flow rate of 1 mL/min and a gradient of 5% solvent B to 100% solvent B over 30 min.

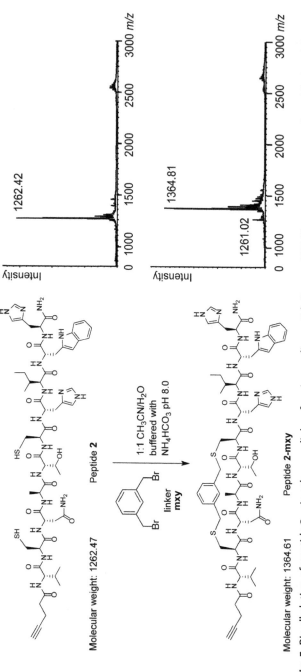

Fig. 5 Bis-alkylation of peptide **2** using the **mxy** linker, forming cyclic peptide **2-mxy**. MALDI mass spectrometry of peptide **2** and the reaction mixture after 60 min are shown at the *right*.

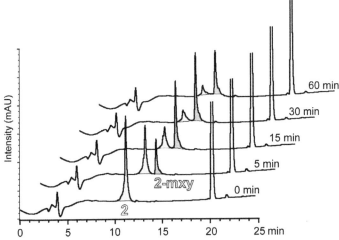

Fig. 6 Monitoring reaction progress by HPLC. Aliquots from the reaction shown in Fig. 5 were removed and quenched at 5, 15, 30, and 60 min and analyzed by analytical HPLC. This stacked plot of HPLC chromatograms shows progression of the reaction from peptide **2** (peaks highlighted in *yellow* (*light gray* in the print version)) to cyclic peptide **2-mxy** (peaks highlighted in *orange* (*gray* in the print version)).

Table 2 Primary Sequences of Model Peptides

Peptide	Sequence	Calculated *m/z*	Observed *m/z*
1	CPWRATNPFC	1235.45	1234.91
2	pa-VCNATCHIWH	1262.47	1262.42
3	VCNATCHIWH	1224.43	1224.57
4	VCNATHCIWH	1224.43	1223.26
5	VCNATIHCWH	1224.43	1223.26
6	VCNATSHIWH	1208.36	1208.54
7	VCNATMHIWH	1252.48	1252.59
8	VMNATMHIWH	1280.53	1280.73

All peptides were prepared with C-terminal amides and N-terminal acetyl groups, except for peptide 2 which was prepared with an N-terminal pentynyl group (denoted pa-) as shown in Fig. 5. Sulfur-containing amino acids capable of participating in macrocyclization reactions are shown in bold. Observed *m/z* are reported as major peaks from MALDI mass spectra.

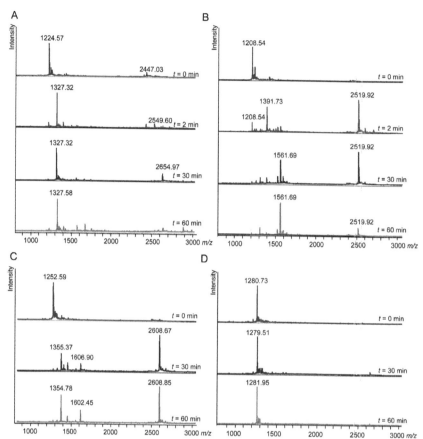

Fig. 7 Peptides listed in Table 2 were incubated at 1 mM with 1.5 mM **mxy** linker and 200 μM TCEP in 50% CH$_3$CN, 50% H$_2$O with 20 mM NH$_4$HCO$_3$, pH 8.0. At time points between 2 and 60 min, aliquots of the reactions were removed, quenched with HCl, and analyzed qualitatively by MALDI mass spectrometry. Data are shown for selected time points for (A) peptide **3**, (B) peptide **6**, (C) peptide **7**, and (D) peptide **8**.

4.4 Investigating the Bis-Alkylation Mechanism

The thiol bis-alkylation reaction is a series of two nucleophilic substitutions. In each, a nucleophilic thiolate attacks the benzylic bromide, and the bromide ion is the leaving group. The robustness of the bis-alkylation reaction with benzylic and allylic bromides compared to alkyl bromides suggests the possibility of some S$_N$1 character in either or both substitutions. In order to investigate the role of the nucleophiles, we carried out test reactions using peptides **3**, **6**, **7**, and **8** (see Table 2 for sequences). These peptides are identical except that peptide **3** contains two cysteines at (i, $i+4$) positions, peptide **6** substitutes one cysteine with serine, peptide **7** substitutes one cysteine with methionine, and peptide **8** substitutes both cysteines with methionine. Each peptide was incubated at 1 mM with 1.5 mM **mxy** linker for 1 h in the presence of 200 µM tris(2-carboxyethyl)phosphine (TCEP), and results at different time points were monitored by MALDI mass spectrometry. Based on prior observations and the results of this limited study (Fig. 7), we conclude the following:

- *The full extent of the reaction usually occurs within the first 30 min*, with no significant change typically observed after 60 min. We note that more time may be necessary for peptides with unusually hindered structures, and for bis-alkylation conducted on-resin (Jo et al., 2012).
- *A peptide containing one cysteine will mostly form cross-linked dimer*. Under these conditions, peptide **6** reacts sluggishly, and primarily forms a xylene-cross-linked dimer. At 2 min, we observe the xylene-cross-linked dimer (calculated m/z 2518.86, observed m/z 2519.92) as well as a peak which corresponds to mono-alkylated product where the linker still has one bromine (calculated m/z 1391.41, observed m/z 1391.73) (Fig. 7B). We note that this mono-alkylated intermediate is detected for peptide **6**, but not for any peptides that bis-alkylate. This indicates that the second substitution step occurs very rapidly for those peptides that can readily bis-alkylate. By $t=30$ min, a new peak at 1561.69 Da is observed, which corresponds to mono-alkylated peptide lacking bromine plus the mass of TCEP (calculated m/z 1562.70). Noncovalent, charge-paired TCEP adducts can be observed by mass spectrometry for some peptides. The loss of bromine could be due to an attack by the serine hydroxyl, histidine imidazole, or alkylated cysteine thioether on the second benzylic bromide. The resulting sulfonium or imidazolium ion would be stabilized by forming a salt with TCEP. Overall, the slow reaction and many difficult-to-characterize products are typically observed only for reactions where rapid intramolecular bis-alkylation is not possible.
- *A peptide-containing one cysteine and one methionine will bis-alkylate*. Fig. 7C shows the results over time for bis-alkylation of peptide **7** under the same conditions. At 30 and 60 min, the major products are bis-alkylated product (calculated m/z 1355.62, observed m/z 1355.37 Da) and cross-linked dimer (calculated m/z 2607.09, observed m/z 2608.67). The peak at 1606.90 Da is a noncovalent TCEP adduct of mono-alkylated peptide (calculated m/z 1606.82).
- *A peptide with no cysteines is not altered under these conditions*. Peptide **8**, which has methionines in place of both cysteines, does not undergo

any reaction under these conditions (Fig. 7D). Notably, starting material is relatively unchanged and little dimeric product is observed. This shows that functional groups such as the histidine imidazole and methionine sulfide are not nucleophilic enough to react under these conditions, at least in the absence of any thiol groups.

The above observations provide some clues as to the overall mechanism. Clearly, a first step must be nucleophilic substitution of a thiolate on the benzyl bromide, but the direct product of this step can only be observed transiently and for peptides that do not rapidly undergo a second alkylation. The second, intramolecular substitution is accelerated due to the high effective concentration of the second bromide in the vicinity of the second nucleophile. Electronic effects at the benzylic bromides may also play an important role. Specifically, the presence of the sulfide directly above or below the benzene ring likely makes the bromomethyl benzyl sulfide even more electrophilic and may even promote a more S_N1-like mechanism for the second bis-alkylation (see Fig. 8). This explains why peptide **7**, with cysteine and methionine in (i, $i+4$) positions, bis-alkylated so rapidly even though peptide **8**, with two methionines, did not. This would also explain why the **oxy, mxy, pxy,** and **mpy** linkers are so reactive toward bis-alkylation, whereas the other linkers require more optimization. These electronic effects would also explain the high reactivity of the monoalkylated peptide to side reactions when a second strong nucleophile is not present in the peptide.

Thiol bis-alkylation appears to form intramolecular cross-links with minimal dependence on the sequence or structure of the intervening

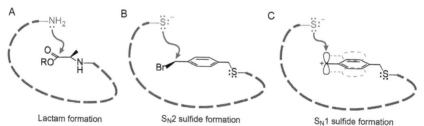

Fig. 8 Nucleophilic attacks for closing peptide macrocycles. (A) Lactam formation is widely used in macrocyclization reactions, but requires attack of a primary amine on an activated ester. (B) If the second substitution is S_N2-like, the ring closure step for thiol bis-alkylation will involve attack by a thiolate on the benzylic bromide. (C) If the second substitution is S_N1-like, the ring closure step for thiol bis-alkylation will involve attack by a thiolate on a benzylic carbocation. Note that the nucleophiles shown in (B) and (C) could also be sulfides (as in the case of methionine).

peptide. This is generally not the case with other macrocyclization approaches (Table 1). We surmise that this major advantage is due to the high reactivity of the bromomethyl benzyl sulfide, as well as an increased tolerance in the angle of the approach of the nucleophile in the macrocyclization step. To illustrate this point, Fig. 8 contrasts lactam formation (where a carboxylic-acid-derived ester is attacked by a primary amine) with a thiolate attacking a benzylic bromide (S_N2-like mechanism) and a thiolate attacking a carbocation (S_N1-like mechanism). While more in-depth investigation is needed, this model illustrates why this chemistry is so robust and versatile for the formation of peptide macrocycles with varied 3D structures.

4.5 Optimizing and Troubleshooting Bis-Alkylation Reactions

Though this cross-linking strategy is very robust and can be easily used for any peptide containing reactive thiols, optimal conditions may vary for different peptides. Here we briefly discuss different conditions with respect to troubleshooting reactions with low yield, or optimizing reactions to maximize total yield.

4.5.1 Concentration

We and others (Dewkar, Carneiro, & Hartman, 2009; Timmerman et al., 2005) have shown that the peptide concentration for this reaction is optimal between 0.01 and 1 mM, but can also be achieved at higher concentrations. One of the byproducts encountered if the peptide is too concentrated is the cross-linked dimer. Trace amounts of this byproduct can often be observed even in highly successful reactions—see, for instance, the tiny peak at $m/z = 2520$ in the mass spectrum of the reaction that produced peptide **2-mxy** (Fig. 5). This is rarely a major product at concentrations 1 mM or below, and it can easily be avoided with a more dilute reaction mixture.

4.5.2 Relative Amount of Linker

While initial reports of cysteine bis-alkylation used a marginal excess (1.05 equiv.) of linker (Timmerman et al., 2005), we have found that using 1.5-fold excess to threefold excess generally results in clean reactions and rapid reaction times. At higher concentrations, using a threefold excess can increase the amount of cross-linked dimer. We have never observed higher yields or faster reactions above threefold excess of linker.

4.5.3 Solvent Selection

The optimal solvent for the reaction is very much dependent on the solubility of the peptide and the linker. A 1:1 mixture of buffered water and acetonitrile has worked well in our hands for most peptides and most linkers, particularly the **oxy**, **mxy**, **pxy**, and **mpy** linkers. Increasing the amount of acetonitrile can be useful for more hydrophobic peptide sequences and larger, more hydrophobic linkers, but it slows the overall reaction. The reaction can also be performed using a 1:1 mixture of buffered water and DMF, which would help with solubility (Hacker, Almohaini, Anbazhagan, Ma, & Hartman, 2015). Since reactions are more easily concentrated if they are run in acetonitrile/water, we have not extensively explored the use of DMF as a cosolvent.

4.5.4 Presence of Reducing Agents

One might assume that a major competing reaction would be disulfide formation, either intermolecular to form a disulfide-linked dimer or intramolecular to form a disulfide-bridged macrocycle. Using degassed, deionized water and paying careful attention to pH (see later), we have rarely observed disulfide formation outcompeting bis-alkylation at the optimized conditions described earlier. In rare cases where bis-alkylation reactions are slow and are incubated overnight, we have observed some disulfide formation. One preventative measure is to use TCEP, a nonnucleophilic, nonsulfur-containing reducing agent which is not reactive with the linker. If disulfide formation has been observed, 100–200 μM TCEP can be added to the peptide prior to cyclization. As observed in Fig. 7, TCEP adducts (+250 Da) can sometimes be observed by mass spectrometry following the reaction, but these are not covalent adducts and typically disappear after acidification and HPLC purification. We do not recommend using TCEP routinely, as the thiol bis-alkylation is typically much faster than oxidation. Rather, we recommend using degassed aqueous buffers and, in general, storing all thiol-containing and sulfide-containing peptides as lyophilized powders rather than DMSO stocks. Overall, we have had excellent success using freshly prepared peptides in bis-alkylation reactions without reducing agents, compared to peptides that have been stored for prolonged periods of time.

4.5.5 pH Dependence

In very few cases, we have observed no product formation with a dithiol-containing peptide. In these cases, the major problem has typically been pH. For instance, TFA can be carried over with the peptides following ether precipitation, altering the resulting pH of the reaction solution. We check reaction pH immediately prior to adding the linker, usually by spotting a small drop onto pH paper, in order to verify the pH is still around 8.0. Overall, we have found that maintaining the pH above 7.6 is critical for clean, fast bis-alkylation reactions. At lower pH, reactions are sluggish or do not progress, typically yielding only starting material. In systematic testing, we have not observed significant improvements in yields or reaction times at pH 8.5, 9, or 9.5.

5. FUTURE DIRECTIONS

This chapter describes a methodology that can be used to produce a large synthetic library of loops as inhibitors for protein–protein interactions. By varying the thiol-containing amino acids, their relative positions, and the linkers, it is possible to lock any individual epitope into a large variety of conformations and rapidly screen them for a desired function. Though these macrocycles make great chemical biology tools for studying specific protein–protein interactions, they are not considered "drug-like" molecules. The cross-link affords structural stability, which generally increases metabolic stability and cell penetration compared to linear peptides (Bock et al., 2013). However, an anticipated problem for thiol bis-alkylated peptides is that benzyl sulfides may be readily oxidized in vivo. Still, these molecules could be regarded as a starting point for drug development, and medicinal chemistry could rapidly identify a suitable isosteric replacement for the thioethers, if necessary. Thus, for chemical biology and drug development, this strategy represents an efficient, rapid, and robust method for the discovery of macrocyclic inhibitors of protein–protein interactions.

REFERENCES

Arkin, M. R., Tang, Y., & Wells, J. A. (2014). Small-molecule inhibitors of protein-protein interactions: Progressing toward the reality. *Chemistry and Biology, 21*(9), 1102–1114. http://doi.org/10.1016/j.chembiol.2014.09.001.

Avan, I., Hall, C. D., & Katritzky, A. R. (2014). Peptidomimetics via modifications of amino acids and peptide bonds. *Chemical Society Reviews, 43*(10), 3575–3594. http://doi.org/10.1039/c3cs60384a.

Bashiruddin, N. K., Nagano, M., & Suga, H. (2015). Bioorganic Chemistry Synthesis of fused tricyclic peptides using a reprogrammed translation system and chemical modification. *Bioorganic Chemistry, 61*, 45–50. http://doi.org/10.1016/j.bioorg.2015.06.002.

Berthelot, T., Gonçalves, M., Laïn, G., Estieu-Gionnet, K., & Déléris, G. (2006). New strategy towards the efficient solid phase synthesis of cyclopeptides. *Tetrahedron, 62*(6), 1124–1130. http://doi.org/10.1016/j.tet.2005.10.080.

Bird, G. H., Irimia, A., Ofek, G., Kwong, P. D., Wilson, I. A., ... Walensky, L. D. (2014). Stapled HIV-1 peptides recapitulate antigenic structures and engage broadly neutralizing antibodies. *Nature Structural & Molecular Biology, 21*(12), 1058–1067. http://doi.org/10.1038/nsmb.2922.

Bock, J. E., Gavenonis, J., & Kritzer, J. A. (2013). Getting in shape: Controlling peptide bio-activity and bioavailability using conformational constraints. *ACS Chemical Biology, 8*(3), 488–499. http://doi.org/10.1021/cb300515u.

Bockus, A. T., Schwochert, J. A., Pye, C. R., Townsend, C. E., Sok, V., Bednarek, M. A., & Lokey, R. S. (2015). Going out on a limb: Delineating the effects of beta-branching, N-methylation, and side chain size on the passive permeability, solubility, and flexibility of sanguinamide a analogues. *Journal of Medicinal Chemistry, 58*(18), 7409–7418. http://doi.org/10.1021/acs.jmedchem.5b00919.

Brunel, F. M., & Dawson, P. E. (2005). Synthesis of constrained helical peptides by thioether ligation: Application to analogs of gp41. *Chemical Communications (Cambridge, England),* 2552–2554. http://doi.org/10.1039/b419015g.

Brust, A., Wang, C.-I. A., Daly, N. L., Kennerly, J., Sadeghi, M., Christie, M. J., ... Alewood, P. F. (2013). Vicinal disulfide constrained cyclic peptidomimetics: A turn mimetic scaffold targeting the norepinephrine transporter. *Angewandte Chemie, 125*(46), 12242–12245. http://doi.org/10.1002/ange.201304660.

Chang, Y. S., Graves, B., Guerlavais, V., Tovar, C., Packman, K., To, K.-H., ... Sawyer, T. K. (2013). Stapled α-helical peptide drug development: A potent dual inhibitor of MDM2 and MDMX for p53-dependent cancer therapy. *Proceedings of the National Academy of Sciences of the United States of America, 110*(36), E3445–E3454. http://doi.org/10.1073/pnas.1303002110.

Chatterjee, J., Gilon, C., Hoffman, A., & Kessler, H. (2008). N-methylation of peptides: A new perspective in medicinal chemistry. *Accounts of Chemical Research, 41*(10), 1331–1342. http://doi.org/10.1021/ar8000603.

Chen, S., Gopalakrishnan, R., Schaer, T., Marger, F., Hovius, R., Bertrand, D., ... Heinis, C. (2014). Dithiol amino acids can structurally shape and enhance the ligand-binding properties of polypeptides. *Nature Chemistry, 6*(11), 1009–1016. http://doi.org/10.1038/nchem.2043.

Clackson, T., & Wells, J. A. (1995). A hot spot of binding energy in a hormone-receptor interface. *Science, 267*(5196), 383–386. http://doi.org/10.1126/science.7529940.

Clark, R. J., Jensen, J., Nevin, S. T., Callaghan, B. P., Adams, D. J., & Craik, D. J. (2010). The engineering of an orally active conotoxin for the treatment of neuropathic pain. *Angewandte Chemie International Edition, 49*(37), 6545–6548. http://doi.org/10.1002/anie.201000620.

Cudic, M., Wade, J. D., & Otvos, L., Jr. (2000). Convenient synthesis of a head-to-tail cyclic peptide containing an expanded ring. *Tetrahedron Letters, 41*(23), 4527–4531. http://doi.org/10.1016/S0040-4039(00)00629-8.

Dahan, A., Tsume, Y., Sun, J., Miller, J. M., & Amidon, G. L. (2011). Oral bioavailability of peptide and peptidomimetic drugs. *Amino acids, peptides and proteins in organic chemistry: Vol. 4* (pp. 277–292). Weinheim, Germany: Wiley-VCH Verlag GmbH & Co. KGaA. http://doi.org/10.1002/9783527631827.ch8.

Damas, J. M., Filipe, L. C. S., Campos, S. R. R., Lousa, D., Victor, B. L., Baptista, A. M., & Soares, C. M. (2013). Predicting the thermodynamics and kinetics of helix formation in a cyclic peptide model. *Journal of Chemical Theory and Computation, 9*(11), 5148–5157. http://doi.org/10.1021/ct400529k.

Dewkar, G. K., Carneiro, P. B., & Hartman, M. C. T. (2009). Synthesis of novel peptide linkers: Simultaneous cyclization and labeling. *Organic Letters, 11*(20), 4708–4711. http://doi.org/10.1021/ol901662c.

Fields, G., & Noble, R. (1990). Solid phase peptide synthesis utilizing 9-fluorenylmethoxycarbonyl amino acids. *International Journal of Peptide and Protein Research*, *35*(3), 161–214. http://doi.org/10.1111/j.1399-3011.1990.tb00939.x.

Fosgerau, K., & Hoffmann, T. (2014). Peptide therapeutics: Current status and future directions. *Drug Discovery Today*, *20*(1), 122–128. http://doi.org/10.1016/j.drudis.2014.10.003.

Gao, Y., Amar, S., Pahwa, S., Fields, G., & Kodadek, T. (2015). Rapid lead discovery through iterative screening of one bead one compound libraries. *ACS Combinatorial Science*, *17*, 49–59.

Gartner, Z. J., Tse, B. N., Grubina, R., Doyon, J. B., Snyder, T. M., & Liu, D. R. (2004). DNA-templated organic synthesis and selection of a library of macrocycles. *Science (New York, N.Y.)*, *305*(September), 1601–1605. http://doi.org/10.1126/science.1102629.

Gavenonis, J., Sheneman, B. A., Siegert, T. R., Eshelman, M. R., & Kritzer, J. A. (2014). Comprehensive analysis of loops at protein-protein interfaces for macrocycle design. *Nature Chemical Biology*, *10* (9), 1–8. http://doi.org/10.1038/nchembio.1580.

Goto, Y., Katoh, T., & Suga, H. (2011). Flexizymes for genetic code reprogramming. *Nature Protocols*, *6*(6), 779–790. http://doi.org/10.1038/nprot.2011.331.

Hacker, D. E., Almohaini, M., Anbazhagan, A., Ma, Z., & Hartman, M. C. T. (2015). Peptide and peptide library cyclization via bromomethylbenzene derivatives. *Methods in molecular biology*: Vol. 1248. New York: Springer. 278 pp. http://doi.org/10.1007/978-1-4939-2020-4.

Han, S.-Y., & Kim, Y.-A. (2004). Recent development of peptide coupling reagents in organic synthesis. *Tetrahedron*, *60*(11), 2447–2467. http://doi.org/10.1016/j.tet.2004.01.020.

Hayouka, Z., Levin, A., Hurevich, M., Shalev, D. E., Loyter, A., Gilon, C., & Friedler, A. (2012). A comparative study of backbone versus side chain peptide cyclization: Application for HIV-1 integrase inhibitors. *Bioorganic and Medicinal Chemistry*, *20*(10), 3317–3322. http://doi.org/10.1016/j.bmc.2012.03.039.

Heinis, C., Rutherford, T., Freund, S., & Winter, G. (2009). Phage-encoded combinatorial chemical libraries based on bicyclic peptides. *Nature Chemical Biology*, *5*(7), 502–507. http://doi.org/10.1038/nchembio.184.

Henchey, L. K., Jochim, A. L., & Arora, P. S. (2008). Contemporary strategies for the stabilization of peptides in the α-helical conformation. *Current Opinion in Chemical Biology*, *12*(6), 692–697. http://doi.org/10.1016/j.cbpa.2008.08.019.

Hill, T. A., Shepherd, N. E., Diness, F., & Fairlie, D. P. (2014). Constraining cyclic peptides to mimic protein structure motifs. *Angewandte Chemie International Edition*, *53*(48), 13020–13041. http://doi.org/10.1002/anie.201401058.

Jo, H., Meinhardt, N., Wu, Y., Kulkarni, S., Hu, X., Low, K. E., … Greenbaum, D. C. (2012). Development of α-helical calpain probes by mimicking a natural protein-protein interaction. *Journal of the American Chemical Society*, *134*(42), 17704–17713. http://doi.org/10.1021/ja307599z.

Kamens, A. J., Eisert, R. J., Corlin, T., Baleja, J. D., & Kritzer, J. A. (2014). Structured cyclic peptides that bind the EH domain of EHD1. *Biochemistry*, *53*, 4758–4760. http://doi.org/10.1021/bi500744q.

Kaspar, A. A., & Reichert, J. M. (2013). Future directions for peptide therapeutics development. *Drug Discovery Today*, *18*(17–18), 807–817. http://doi.org/10.1016/j.drudis.2013.05.011.

Kawamoto, S. A., Coleska, A., Ran, X., Yi, H., Yang, C.-Y., & Wang, S. (2012). Design of triazole-stapled BCL9 α-helical peptides to target the β-catenin/B-cell CLL/lymphoma 9 (BCL9) protein-protein interaction. *Journal of Medicinal Chemistry*, *55*(3), 1137–1146. http://doi.org/10.1021/jm201125d.

Khoo, K. H., Hoe, K. K., Verma, C. S., & Lane, D. P. (2014). Drugging the p53 pathway: Understanding the route to clinical efficacy. *Nature Reviews Drug Discovery*, *13*(3), 217–236. http://doi.org/10.1038/nrd4236.

Kim, Y.-W., Kutchukian, P. S., & Verdine, G. L. (2010). Introduction of all-hydrocarbon i, i+3 staples into alpha-helices via ring-closing olefin metathesis. *Organic Letters*, *12*(13), 3046–3049. http://doi.org/10.1021/ol1010449.

Kortemme, T., & Baker, D. (2002). A simple physical model for binding energy hot spots in protein-protein complexes. *Proceedings of the National Academy of Sciences of the United States of America*, *99*(22), 14116–14121. http://doi.org/10.1073/pnas.202485799.

Kortemme, T., Kim, D. E., & Baker, D. (2004). Computational alanine scanning of protein-protein interfaces. *Science's STKE: Signal Transduction Knowledge Environment*, *2004*(219). pl2, http://doi.org/10.1126/stke.2192004pl2.

London, N., Raveh, B., Movshovitz-Attias, D., & Schueler-Furman, O. (2010). Can self-inhibitory peptides be derived from the interfaces of globular protein-protein interactions? *Proteins: Structure, Function, and Bioinformatics*, *78*(15), 3140–3149. http://doi.org/10.1002/prot.22785.

Miller, S., Blackwell, H., & Grubbs, R. (1996). Application of ring-closing metathesis to the synthesis of rigidified amino acids and peptides. *Journal of the American Chemical Society*, *49*(50), 9606–9614. Retrieved from, http://pubs.acs.org/doi/abs/10.1021/ja9616261.

Moss, J. (2005). Guide for resin and linker selection in solid-phase peptide synthesis. *Current Protocols in Protein Science*, 18.7.1–18.7.19. Chapter 18, http://doi.org/10.1002/0471140864.ps1807s40.

Muppidi, A., Wang, Z., Li, X., Chen, J., & Lin, Q. (2011). Achieving cell penetration with distance-matching cysteine cross-linkers: A facile route to cell-permeable peptide dual inhibitors of Mdm2/Mdmx. *Chemical Communications (Cambridge, England)*, *47*(33), 9396–9398. http://doi.org/10.1039/c1cc13320a.

Muppidi, A., Zhang, H., Curreli, F., Li, N., Debnath, A. K., & Lin, Q. (2014). Design of antiviral stapled peptides containing a biphenyl cross-linker. *Bioorganic & Medicinal Chemistry Letters*, *24*(7), 1748–1751. http://doi.org/10.1016/j.bmcl.2014.02.038.

Passioura, T., Katoh, T., Goto, Y., & Suga, H. (2014). Selection-based discovery of druglike macrocyclic peptides. *Annual Review of Biochemistry*, *83*, 727–752. http://doi.org/10.1146/annurev-biochem-060713-035456.

Patgiri, A., Jochim, A. L., & Arora, P. S. (2008). A hydrogen bond surrogate approach for stabilization of short peptide sequences in α-helical conformation. *Accounts of Chemical Research*, *41*(10), 1289–1300. http://doi.org/10.1021/ar700264k.

Qian, Z., Liu, T., Liu, Y.-Y., Briesewitz, R., Barrios, A. M., Jhiang, S. M., & Pei, D. (2013). Efficient delivery of cyclic peptides into mammalian cells with short sequence motifs. *ACS Chemical Biology*, *8*(2), 423–431. http://doi.org/10.1021/cb3005275.

Qvit, N., Hatzubai, A., Shalev, D. E., Friedler, A., Ben-Neriah, Y., & Gilon, C. (2009). Design and synthesis of backbone cyclic phosphorylated peptides: The IkB model. *Biopolymers*, *91*(2), 157–168. http://doi.org/10.1002/bip.21098.

Razavi, A. M., Wuest, W. M., & Voelz, V. A. (2014). Computational screening and selection of cyclic peptide hairpin mimetics by molecular simulation and kinetic network models. *Journal of Chemical Information and Modeling*, *54*(5), 1425–1432. http://doi.org/10.1021/ci500102y.

Renukuntla, J., Vadlapudi, A. D., Patel, A., Boddu, S. H. S., & Mitra, A. K. (2013). Approaches for enhancing oral bioavailability of peptides and proteins. *International Journal of Pharmaceutics*, *447*(1–2), 75–93. http://doi.org/10.1016/j.ijpharm.2013.02.030.

Robertson, N. S., & Jamieson, A. G. (2015). Regulation of protein-protein interactions using stapled peptides. *Reports in Organic Chemistry*, *5*, 65–74. http://doi.org/10.2147/ROC.S68161.

Rumpf, J., Simon, B., Jung, N., Maritzen, T., Haucke, V., Sattler, M., & Groemping, Y. (2008). Structure of the Eps15-stonin2 complex provides a molecular explanation for EH-domain ligand specificity. *The EMBO Journal, 27*(3), 558–569. http://doi.org/10.1038/sj.emboj.7601980.

Schafmeister, C. E., Po, J., & Verdine, G. L. (2000). An all-hydrocarbon cross-linking system for enhancing the helicity and metabolic stability of peptides. *Journal of the American Chemical Society, 122*(6), 5891–5892.

Schlippe, Y. V., Hartman, M. C., Josephson, K., & Szostak, J. W. (2012). In vitro selection of highly modified cyclic peptides that act as tight binding inhibitors. *Journal of the American Chemical Society, 134*(25), 10469–10477. http://doi.org/10.1021/ja301017y.

Shelton, P. T., & Jensen, K. J. (2013). Linkers, resins, and general procedures for solid-phase peptide synthesis. *Methods in Molecular Biology 1047*, 23–41.

Shepherd, N. E., Hoang, H. N., Abbenante, G., & Fairlie, D. P. (2005). Single turn peptide alpha helices with exceptional stability in water. *Journal of the American Chemical Society, 127*(9), 2974–2983. http://doi.org/10.1021/ja0456003.

Shoji-Kawata, S., Sumpter, R., Leveno, M., Campbell, G. R., Zou, Z., Kinch, L., ... Levine, B. (2013). Identification of a candidate therapeutic autophagy-inducing peptide. *Nature, 494*(7436), 201–206. http://doi.org/10.1038/nature11866.

Smeenk, L. E. J., Dailly, N., Hiemstra, H., van Maarseveen, J. H., & Timmerman, P. (2012). Synthesis of water-soluble scaffolds for peptide cyclization, labeling, and ligation. *Organic Letters, 14*(5), 1194–1197. http://doi.org/10.1021/ol203259a.

Spokoyny, A., Zou, Y., Ling, J., Yu, H., Lin, Y.-S., & Pentelute, B. (2013). A perfluoroaryl-cysteine SNAr chemistry approach to unprotected peptide stapling. *Journal of the American Chemical Society, 135*, 5946–5949. Retrieved from, http://pubs.acs.org/doi/abs/10.1021/ja400119t.

Thompson, A. D., Dugan, A., Gestwicki, J. E., & Mapp, A. K. (2012). Fine-tuning multiprotein complexes using small molecules. *ACS Chemical Biology, 7*(8), 1311–1320. http://doi.org/10.1021/cb300255p.

Timmerman, P., Beld, J., Puijk, W. C., & Meloen, R. H. (2005). Rapid and quantitative cyclization of multiple peptide loops onto synthetic scaffolds for structural mimicry of protein surfaces. *Chembiochem: A European Journal of Chemical Biology, 6*(5), 821–824. http://doi.org/10.1002/cbic.200400374.

Todorova-Balvay, D., Stoilova, I., Gargova, S., & Vijayalakshmi, M. A. (2007). An efficient two step purification and molecular characterization of beta-galactosidases from *Aspergillus oryzae*. *Journal of Molecular Recognition: JMR, 19*(4), 299–304. http://doi.org/10.1002/jmr.

Urech-Varenne, C., Radtke, F., & Heinis, C. (2015). Phage selection of bicyclic peptide ligands of the Notch1 receptor. *ChemMedChem, 10*(10), 1754–1761. http://doi.org/10.1002/cmdc.201500261.

Wakefield, A. E., Wuest, W. M., & Voelz, V. A. (2015). Molecular simulation of conformational pre-organization in cyclic RGD peptides. *Journal of Chemical Information and Modeling, 55*(4), 806–813. http://doi.org/10.1021/ci500768u.

Walensky, L. D., Kung, A. L., Escher, I., Malia, T. J., Barbuto, S., Wright, R. D., ... Korsmeyer, S. J. (2004). Activation of apoptosis in vivo by a hydrocarbon-stapled BH3 helix. *Science (New York, N.Y.), 305*(2002), 1466–1470. http://doi.org/10.1126/science.1099191.

Wang, Y., & Chou, D. H.-C. (2015). A thiol-ene coupling approach to native peptide stapling and macrocyclization. *Angewandte Chemie International Edition, 54* (37), 10931–10934. http://doi.org/10.1002/anie.201503975.

Watkins, A. M., & Arora, P. S. (2014). Anatomy of β-strands at protein–protein interfaces. *ACS Chemical Biology, 9*, 1747–1754.

White, E. R., Sun, L., Ma, Z., Beckta, J. M., Danzig, B. A., Hacker, D. E., … Hartman, M. C. T. (2015). Peptide library approach to uncover phosphomimetic inhibitors of the BRCA1 C-terminal domain. *ACS Chemical Biology, 10*, 1198–1208. http://doi.org/10.1021/cb500757u.

White, C. J., & Yudin, A. K. Contemporary strategies for peptide macrocyclization. *Nature Chemistry 3(7),* 509–524.

Wuo, M. G., Mahon, A. B., & Arora, P. S. (2015). An effective strategy for stabilizing minimal coiled coil mimetics. *Journal of the American Chemical Society, 137*(36), 11618–11621. http://doi.org/10.1021/jacs.5b05525.

Xiao, W., Wang, Y., Lau, E. Y., Luo, J., Yao, N., Shi, C., … Lam, K. S. (2010). The use of one-bead one-compound combinatorial library technology to discover high-affinity alphavbeta3 integrin and cancer targeting arginine-glycine-aspartic acid ligands with a built-in handle. *Molecular Cancer Therapeutics, 9*(10), 2714–2723. http://doi.org/10.1158/1535-7163.MCT-10-0308.

Yamagishi, Y., Shoji, I., Miyagawa, S., Kawakami, T., Katoh, T., Goto, Y., & Suga, H. (2011). Natural product-like macrocyclic N-methyl-peptide inhibitors against a ubiquitin ligase uncovered from a ribosome-expressed de novo library. *Chemistry & Biology, 18*(12), 1562–1570. http://doi.org/10.1016/j.chembiol.2011.09.013.

Yin, H. (2012). Constrained peptides as miniature protein structures. *ISRN Biochemistry, 2012,* 1–15. http://doi.org/10.5402/2012/692190.

Yu, H., & Lin, Y.-S. (2015). Toward structure prediction of cyclic peptides. *Physical Chemistry Chemical Physics, 17*(6), 4210–4219. http://doi.org/10.1039/C4CP04580G.

Zimm, B. H., & Bragg, J. K. (1959). Theory of the phase transition between helix and random coil in polypeptide chains. *The Journal of Chemical Physics, 31*(2), 526–535. http://doi.org/10.1063/1.1730390.

CHAPTER FIFTEEN

Engineering Short Preorganized Peptide Sequences for Metal Ion Coordination: Copper(II) a Case Study

L.M.P. Lima*, O. Iranzo[†,1]
*Instituto de Tecnologia Química e Biológica António Xavier, Universidade Nova de Lisboa, Lisbon, Portugal
[†]Aix Marseille Université, Centrale Marseille, CNRS, iSm2 UMR 7313, Marseille, France
[1]Corresponding author: e-mail address: olga.iranzo@univ-amu.fr

Contents

1. Introduction — 334
   1.1 Designing Short Preorganized Peptide Sequences for Cu(II) Coordination — 337
2. Design of Preorganized Peptidic Scaffolds for Metal Ion Coordination — 340
3. Synthesis and Characterization of the Peptidic Scaffolds — 341
   3.1 Synthesis of Linear Scaffolds — 342
   3.2 Synthesis of Cyclic Scaffolds — 344
4. Preparation of Analytical Stock Solutions — 345
   4.1 Peptidic Scaffolds — 345
   4.2 Metal Cation Stock Solutions — 347
   4.3 Electrolyte Stock Solution — 348
   4.4 Base Titrant Solution — 348
   4.5 Acid Titrant Solution — 349
5. Study of the Metal Ion Coordination Properties — 349
   5.1 Potentiometric Titrations: Determining Stability Constants and Speciation Diagrams — 349
   5.2 Characterization of Major Species by Spectroscopy — 356
6. Concluding Remarks — 360
Acknowledgments — 360
References — 361

Abstract

Peptides are multidentate chiral ligands capable of coordinating different metal ions. Nowadays, they can be obtained with high yield and purity, thanks to the advances on peptide/protein chemistry as well as in equipment (peptide synthesizers). Based on the identity and length of their amino acid sequences, peptides can present different degrees of flexibility and folding. Although short peptide sequences (<20 amino acids) usually lack structure in solution, different levels of structural preorganization can be induced by introducing conformational constraints, such as β-turn/loop template sequences and backbone cyclization. For all these reasons, and the fact that one is not restricted to use proteinogenic amino acids, small peptidic scaffolds constitute a simple and versatile platform for the development of inorganic systems with tailor-made properties and functions. Here we outline a general approach to the design of short preorganized peptide sequences (10–16 amino acids) for metal ion coordination. Based on our experience, we present a general scheme for the design, synthesis, and characterization of these peptidic scaffolds and provide protocols for the study of their metal ion coordination properties.

1. INTRODUCTION

Peptides are an interesting family of ligands for metal ion coordination. From the perspective of coordination chemistry, peptides can provide several donor atoms in a single molecule and in a chiral environment. Therefore, they can be considered as multidentate chiral ligands. From the practical point of view, well-established solid-phase and solution synthetic methodologies allow nowadays their synthesis in high yields and purities. In addition, their modular nature facilitates the modification of both the structure and the spatial distribution of functional groups. All these properties make peptides very attractive for the design of metal ion coordinating chiral ligands. Nonetheless, peptides usually lack structure in solution unless specific sequence motifs, associated with secondary structures, are introduced. This is fundamental to design small peptidic scaffolds with the desired rearrangement of the metal ion-binding units. One of the most common strategies to obtain peptides with potential to adopt a particular folding is the introduction of amino acid sequences known to induce β-turns (Blanco, Ramírez-Alvarado, & Serrano, 1998; Chou & Fasman, 1977; Gellman, 1998; Hughes & Waters, 2006; Hutchinson & Thornton, 1994; Nesloney & Kelly, 1996; Ramírez-Alvarado, Kortemme, Blanco, & Serrano, 1999; Rose, Gierasch, & Smith, 1985). Statistical studies on proteins have revealed that certain amino acids exist preferentially in this type of secondary structures (eg, Pro, Gly, Asn, Asp, Ser) and usually located in specific position within the tetrapeptide turn (eg, X-Pro-Gly-X, X-Ser-Asn-X). This approach has been used by different research groups in the design of peptides with specific metal ion coordination properties (Daugherty, Wasowicz, Gibney, & DeRose, 2002; Imperiali & Kapoor, 1993; Natale et al., 2008; Niedźwiecka, Cisnetti, Lebrun, & Delangle,

2012; Rama et al., 2012; Shults, Pearce, & Imperiali, 2003). However, these sequences were usually not enough to induce specific structures in the apopeptides (in the absence of the metal ions). Another motif reported to induce β-turns is the dipeptide *d*Pro-Pro (*d*Pro stands for D-Pro). This simple unit has strong propensity to induce β-turns (Bean, Kopple, & Peishoff, 1992; Nair, Vijayan, Venkatachalapathi, & Balaram, 1979), and it has been employed to stabilize β-sheet and β-hairpin structures upon introduction in short amino acid sequences (Robinson, 2008; Schneider et al., 2002; Sénèque et al., 2008; Späth, Stuart, Jiang, & Robinson, 1998). Recently, our group showed how a four amino acid peptide containing the *d*Pro-Pro motif in the middle and two Cys in the terminal positions (Ac-Cys-*d*Pro-Pro-Cys-NH$_2$) is a strong chelator for Hg^{2+} (Pires et al., 2012). The *d*Pro-Pro unit placed both Cys residues in a favorable position for metal ion coordination, and indeed, the replacement of *d*Pro by Pro (Pro-Pro unit) generated a peptide with lower affinity for Hg^{2+}. This result highlights the important role of the stereochemical configuration of the first Pro (*d*Pro) when L-amino acids are used (Bean et al., 1992).

Higher constrained peptides can be achieved by cyclization of their backbone. Backbone cyclization restricts their mobility reducing the landscape of possible structures in solution. Cyclic peptides adopt therefore more rigid structures than their linear counterparts, and it is this restricted flexibility that allows for a better control in the spatial distribution of functional groups. This enhances their selectivity in molecular interactions and recognition. This strategy has been used to design protein epitope mimetics, where the structural and conformational properties of the chosen epitopes were recapped on small cyclic scaffolds to obtain biologically active molecules with therapeutic potential (Robinson, 2008). Cyclization also plays a crucial role in fine-tuning and determining the metal ion coordination properties of peptides. Cyclic peptides have been successfully exploited to design mimics of biologically relevant metal centres as well as metal ion chelating and sensing systems (Barba et al., 2012; Butler & Jolliffe, 2011; Farkas, Vass, Hanssens, Majer, & Hollósi, 2005; Fattorusso et al., 1995; Kotynia, Bielinska, Kamysz, & Brasuń, 2012; Neupane, Aldous, & Kritzer, 2013; Ngu-Schwemlein, Gilbert, Askew, & Schwemlein, 2008; Ngyen et al., 2008; Pappalardo, Impellizzeri, & Campagna, 2004). Cyclization represents therefore an attractive way to design small preorganized peptidic ligands with tailor-made metal ion coordination properties. An interesting type of cyclic peptides is the RAFT family (regioselectively addressable functionalized templates) (Dumy et al., 1995; Dumy, Eggleston, Esposito,

Nicula, & Mutter, 1996). These cyclic scaffolds contain two proline-glycine (Pro-Gly) β-turn inducer units that are postulated to constrain the conformation of the peptidic backbone into a β-hairpin or antiparallel β-sheet generating peptides with two potential independent faces. Different functionalities can be introduced by modifying the side chain of Lys (Fig. 1) (Boturyn et al., 2008). Additionally, other amino acids containing amine side chain (diaminopropionic acid, diaminobutyric acid, and ornithine) and also Cys and Ser can be used for this purpose. Initial designs retained still significant backbone flexibility and not well-defined upper and lower faces (as determined by the plane defined by the main peptide backbone), and subsequent modifications were introduced in order to restrict the conformation to the target structure (Dumy et al., 1996; Nikiforovich, Mutter, & Lehmann, 1999; Peluso et al., 2001). The RAFT prototype has been employed to design metal (Cu^+, Zn^{2+}, Cd^{2+}, Hg^{2+}, Pb^{2+}, lanthanides, and actinides) coordinating ligands with interesting properties and functionalities (Bonnet et al., 2009; Lebrun, Starck, Gathu, Chenavier, & Delangle, 2014; Pujol et al., 2011; Rousselot-Pailley et al., 2006; Sénèque et al., 2004; Starck et al., 2015; Tuchscherer, Lehmann, & Mathieu, 1998; Yang et al., 2015). In these cases, the initial Lys residues were replaced either by proteinogenic (Asp, Glu, His, Cys, Asn, Gln, Ser) or nonproteinogenic amino acids with suitable metal ion coordination properties to generate the desired metal complexes.

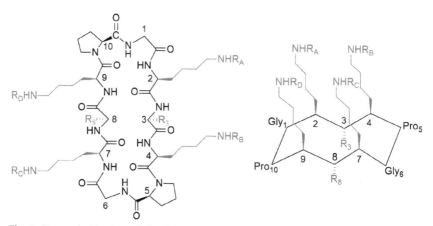

Fig. 1 *Top* and *side* view of the RAFT prototype showing four (R_A, R_B, R_C, R_D) and two (R_3, R_8) potential functionalization sites in the *upper* and *lower* face, respectively.

1.1 Designing Short Preorganized Peptide Sequences for Cu(II) Coordination

The tolerance of this design to amino acid substitutions and its tuneable flexibility prompted us to use the RAFT template as a starting point to generate short preorganized peptide sequences for Cu^{2+} coordination. Cu^{2+} is a d^9 metal ion with borderline Lewis acid properties based on the Pearson's theory of "Hard and Soft Acids and Bases" (Pearson, 1963) and therefore, capable of binding to different donor atoms in the peptides [N (imidazole, amine, amide bond, amide side chain), O (carbonyl, terminal or side chain carboxylic acid, alcohol), S (thiol, thioether)]. Additionally, its coordination number in complexes varies from four to six and includes four coordinate square-planar, five-coordinate trigonal bipyramidal, five-coordinate square-pyramidal, and six-coordinate octahedral geometries. The properties of both the peptidic scaffolds and the Cu^{2+} ion allow for a great assortment in the design since different copper(II) systems can be envisioned with similar scaffolds.

We were interested in using His-containing scaffolds because this amino acid is found as coordinating ligand in the copper centre of numerous metalloproteins and metalloenzymes implicated in a large variety of biological functions (Gaggelli, Kozłowski, Valensin, & Valensin, 2006; Holm, Kennepohl, & Solomon, 1996; Solomon et al., 2014). Hence, such copper systems could mimic the structural and catalytic features of these biological copper sites and be very appealing to engineer miniaturized copper proteins with potential redox and hydrolytic activities. However, designing short His-containing peptides capable of binding the Cu^{2+} ion exclusively through the imidazole side chain and forming a single species stable over a wide pH range is a hard task, and such systems are indeed still scarce (Fragoso, Lamosa, Delgado, & Iranzo, 2013). The His residue plays a crucial role as a nucleation site in the Cu^{2+} coordination properties of peptides, and its number and location within the peptidic sequence strongly determines their coordination ability and the stability of the final complexes (Kozłowski, Kowalik-Jankowska, & Jeżowska-Bojczuk, 2005; Migliorini, Porciatti, Łuczkowski, & Valensin, 2012; Sigel & Martin, 1982). However, as pH values increase and come close to or higher than neutral, Cu^{2+} promotes the deprotonation of the main chain amide nitrogens and binds to them. The formation of fused chelate rings is often the driving force for this type of coordination than can promote amide nitrogen deprotonation and coordination even at acidic pH values. A rational design of the peptidic scaffold is thus needed to stabilize copper(II) species in which the metal ion is bound to the imidazole groups and to prevent the coordination of the main chain amide nitrogens.

Keeping this in mind, we started designing a decapeptide (C-Asp; Fig. 2; Table 1) containing the Pro-Gly unit as β-turn inducing element and having one face dedicated to Cu^{2+} coordination (three His and one Asp) and the other face containing a Trp as a spectroscopic probe (determination of the peptide concentration) and a Lys to avoid peptide aggregation (Fig. 2), and studied its Cu^{2+} coordination properties (Fragoso, Lamosa, et al., 2013). Systematic modifications were introduced afterward to gain insights into the key factors controlling the Cu^{2+} coordination abilities of these scaffolds as well as the properties of the different copper(II) species

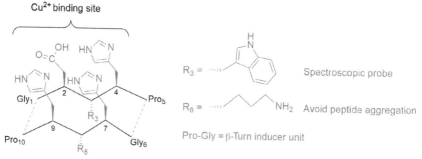

Fig. 2 Schematic representation of the C-Asp peptide in its neutral form.

Table 1 Amino Acid Sequences of Decapeptides[a], Main Copper(II) Complexes, and Their Stability Constants (log $K^{*}_{CuH_hL}$)[b]

Peptide	Sequence ($X_1 \ldots X_{10}$)	Copper(II) Complex	log $K^{*}_{CuH_hL}$
C-Asp	Cyclo(G**D**W**H**PG**H**K**H**P)	[CuH(C-Asp)]$^{2+}$	10.79
O-Asp	Ac-G**D**W**H**PG**H**K**H**G-CONH$_2$	[CuH(O-Asp)]$^{2+}$	9.28
O$_{dPro}$-Asp	Ac-G**D**W**H**^{d}P**P**H**K**H**G-CONH$_2$[c]	[CuH(O$_{dPro}$-Asp)]$^{2+}$	9.28
C-Asn	Cyclo(G**N**W**H**PG**H**K**H**P)	[CuH(C$_{Asn}$-Asp)]$^{3+}$	9.11
O-Asn	Ac-G**N**W**H**PG**H**K**H**G-CONH$_2$	[CuH(O$_{Asn}$-Asp)]$^{3+}$	8.17

[a]Amino acids for Cu^{2+} coordination are highlighted in bold.
[b]log $K^{*}_{CuH_hL}$ = log β_{CuH_hL} − log β_{H_hL} (Fragoso et al., 2015; Fragoso, Delgado, et al., 2013; Fragoso, Lamosa, et al., 2013).
[c] ^{d}P refers to D-Proline (dPro).

generated. Namely, the following elements were tested: replacement of Asp by Asn, use of dPro-Pro unit as β-turn inducer, and opening of the peptidic backbone by deleting one of the β-turn inducing units and introducing capping groups at the N- and C-terminals (Table 1) (Fragoso et al., 2015; Fragoso, Delgado, & Iranzo, 2013; Fragoso, Lamosa, et al., 2013). A thorough characterization of these systems and of their copper(II) complexes using different methodologies (potentiometry, mass spectrometry, spectroscopy, and cyclic voltammetry) showed that, indeed, all these scaffolds were well tuned to obtain copper(II) species in which the metal ion was coordinated exclusively to the side chain groups of His and Asp residues ($[CuH(C-Asp)]^{2+}$, $[CuH(O-Asp)]^{2+}$, $[CuH(O_{dPro}-Asp)]^{2+}$, $[CuH(C_{Asn}-Asp)]^{3+}$, and $[CuH(O_{Asn}-Asp)]^{3+}$; see Fragoso et al., 2015; Fragoso, Delgado, et al., 2013; Fragoso, Lamosa, et al., 2013, for nomenclature of species). The intrinsic nature and different flexibility of the peptidic scaffolds (Table 1) had an important impact on their formation constants, their stability over pH, their Cu^{2+} exchange rates as well as their redox potentials (ascribed to the process Cu^{2+}/Cu^{+}). We have now in hand an interesting family of peptides that have different degrees of flexibility but bind Cu^{2+} through the same amino acids and form very similar main copper(II) complexes. The fact that analogous coordination environments can be achieved using peptidic scaffolds with different structural constrains is very attractive to study the impact of conformational flexibility on the catalytic activity of metal centres, important factors not deeply explored.

This approach can be implemented to design short preorganized peptide sequences for the coordination of different metal ions. One can take advantage of the unique opportunities that become available by simply changing the nature of the amino acids involved in metal ion coordination as well as in β-turn or loop formation. Considering that nonproteinogenic amino acids as well as related building blocks, nonpeptidic β-turn mimetics (Nesloney and Kelly, 1996; Robinson, 2000), organic ligands, and specific probes can be introduced, one has in hand a simple and versatile peptidic platform for the development of inorganic systems with tailor-made properties and functions.

The scope of this chapter is to discuss the rational design of small preorganized peptide scaffolds for the coordination of metal ions. Different synthetic strategies and methodologies for their characterization and study of their metal ion coordination properties will be described with special focus on copper systems.

2. DESIGN OF PREORGANIZED PEPTIDIC SCAFFOLDS FOR METAL ION COORDINATION

The general strategy for the design is shown in Fig. 3. Different elements can be modified to generate a family of linear and cyclic peptidic scaffolds with different sizes, metal ion coordinating abilities, physical properties, and flexibility by simply modifying the number of amino acids

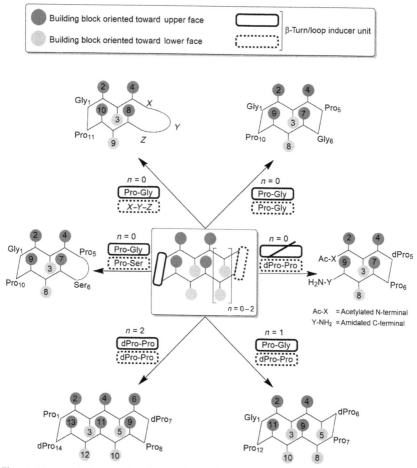

Fig. 3 General scheme for the design of the peptidic scaffolds. Building block denotes proteinogenic and nonproteinogenic amino acids as well as related derivatives. X–Y–Z represent any building block that will generate a very flexible and/or unstructured turn/loop.

or building blocks, their chemical identity as well as the type of β-turn/loop inducing units.

The first decision to be taken is the face in which the metal ion-binding site will be crafted as well as the number of coordination sites (donor atoms). Once this is done, one can place the binding units (proteinogenic and non-proteinogenic amino acids and related counterparts) as needed to promote the desired coordination sphere. In our specific case, three His and one Asp were introduced (Fig. 2) to generate a N_3O first coordination sphere for Cu^{2+}. Depending on the physical characteristics and activities pursued, the opposite face can be employed for fine-tuning different physical properties such as solubility and aggregation (Fig. 2), and for introducing new functionalities such as binding monitoring (eg, sensors development), homing of specific targets (eg, delivery of the peptidic scaffold to specific cells), as well as immobilizing the systems on surfaces or nanoparticles, to mention a few. The functionalization of these peptidic scaffolds can be achieved by using directly amino acids and related building blocks or by functionalizing in a regioselective manner the side chain of amino acids such as Lys (Fig. 1), Cys, or Ser.

At this point it is worth noting that linear peptides will generally lack a defined structure in solution, although features of turns/loops could be observed depending on the constraints arising from the β-turn/loop unit used (Fragoso et al., 2015; Fragoso, Delgado, et al., 2013; Fragoso, Lamosa, et al., 2013). For cyclic peptides, the more flexible the β-turn/loop unit and the larger the peptide scaffold, the higher is the potential of having ill-defined faces since the presence of different conformers in solution will increase. Additionally, a right- or left-handed twist of the peptidic backbone could be observed for larger systems (Athanassiou et al., 2007; Robinson, 2008). This disruption could also be prompted by the chemical identity of the building block side chain, its protonation state, and its location within the scaffold (Campos, Iranzo, & Baptista, 2016). Nonetheless, upon metal ion coordination these systems are prone to adopt the desired folding.

3. SYNTHESIS AND CHARACTERIZATION OF THE PEPTIDIC SCAFFOLDS

These peptidic scaffolds can be prepared by combining solid-phase and solution Fmoc-based synthetic methodologies (Chan & White, 2000). Their synthesis can be carried out manually or using a peptide synthesizer, and in both cases, microwave assisted methods can be used to speed up the successive coupling and deprotection steps. The standard protocols for Fmoc strategy are well described elsewhere (Chan & White, 2000), and if a peptide synthesizer is employed, the Fmoc protocols should be adapted to the specific characteristics of the instrument following the manufacturer recommendations. The synthetic strategy for the regioselective functionalization of Lys (RAFT methodology) is not covered in this section since it has been previously well described (Boturyn et al., 2008). Coupling reactions can be monitored by quantifying either primary amines, using the Kaiser (Kaiser, Colescott, Bossinger, & Cook, 1970) or TNBS (Hancock & Battersby, 1976) tests, or secondary amines, using the acetaldehyde/chloranil (Vojkovsky, 1995) test. In this section, only the general strategy for the synthesis of the peptidic scaffolds will be described. A slightly different approach is followed to obtain the linear (Fig. 4) and cyclic derivatives (Fig. 5).

3.1 Synthesis of Linear Scaffolds

The target peptides have capped N-terminal and C-terminal groups to avoid interference with metal ion binding. The general synthetic strategy is shown in Fig. 4.

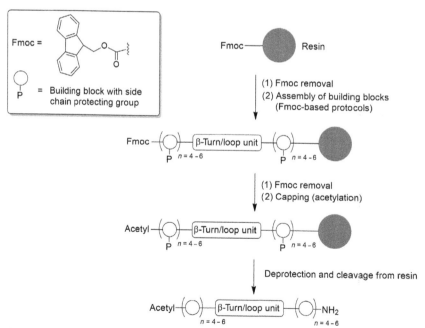

Fig. 4 General strategy for the synthesis of linear peptidic scaffolds.

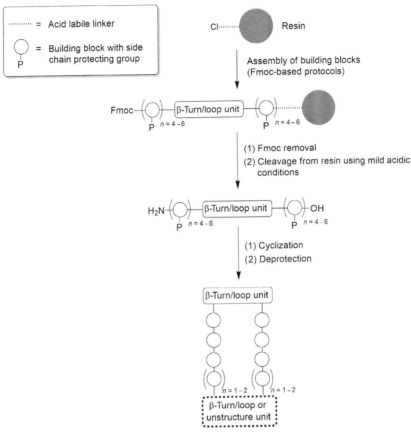

Fig. 5 General strategy for the synthesis of cyclic peptidic scaffolds.

1. Assembly of the linear protected precursor on a Rink amide resin (to obtain peptide amides) using Fmoc-building blocks and standard Fmoc-based protocols.
2. Removal of the end terminal Fmoc group (piperidine/dimethylformamide, v/v, 1:4).
3. Acetylation of the N-terminal (acetic anhydride/dimethylformamide, v/v, 1:9 or acetic anhydride/pyridine/dimethylformamide, v/v/v, 1:2:7).
4. Simultaneous deprotection and cleavage from the resin by treatment with the mixture trifluoroacetic acid (TFA)/triisopropylsilane/H_2O (v/v/v, 95:2.5:2.5, 20 mL/g of resin) for 2 h at room temperature and under nitrogen. If Cys and/or Met are present in the sequence, the mixture should be replaced by TFA/thioanisole/ethanedithiol/anisole (v/v/v/v, 90:5:3:2, 20 mL/g of resin).

5. Filter out and rinse the resin with TFA (~10 mL). Combine filtrate and rinses and concentrate under a nitrogen stream to a crude oil. Add this oil slowly to cold diethyl ether (~20 mL) to precipitate the peptide. Wash with cold diethyl ether and dry. Dissolve the crude peptide in H_2O (alternatives depending on solubility: acetic acid/H_2O (v/v, 1:9), mixtures of H_2O/acetonitrile or H_2O/dimethyl sulfoxide (minimum amount needed)) and lyophilize.
6. Purify the crude peptide by preparative reversed-phase HPLC using the solvent system acetonitrile–H_2O–TFA and characterize by mass spectrometry (ESI or MALDI-TOF). Determine purity using analytical reversed-phase HPLC.

3.2 Synthesis of Cyclic Scaffolds

The location of the β-turn/loop inducing units in the central region of the target amino acid sequence increases the final yields since the end-to-tail cyclization steps can be favored. The general synthetic strategy is shown in Fig. 5.

1. Assembly of the linear protected precursor on a 2-chlorotrityl chloride resin (to obtain fully protected peptide acids, see step 3) using Fmoc-building blocks and standard Fmoc-based protocols. It is recommended to follow the protocol described by the resin manufacturer to load the first amino acid. Alternatively, preloaded resins can be purchased.
2. Removal of the end terminal Fmoc group (piperidine/dimethylformamide, v/v, 1:4).
3. Cleavage of the peptide from the resin using mild acidic conditions (TFA/methylene chloride, v/v, 0.5–1:99.5–99). It is recommended to use the cleavage protocol and work-up procedures described by the resin manufacturer. This generates the fully protected peptide with a N-terminal amino and a C-terminal carboxylic acid groups.
4. Cyclization in solution. The fully protected peptide is dissolved in methylene chloride (≤ 0.5 M, high dilution conditions) and reacted with 1-[bis(dimethylamino)methylene]-1H-1,2,3-triazolo[4,5-b]pyridinium 3-oxid hexafluorophosphate (HATU) or (benzotriazol-1-yloxy)tripyrrolidinophosphonium hexafluorophosphate (PyBOP) (1.0–1.5 equiv.) and N,N-diisopropylethylamine (DIEA) (12 equiv.) at room temperature and under nitrogen atmosphere. Alternatively, concentrated stock solutions of DIEA, peptide (dimethyl sulfoxide may be needed for solubility) and HATU or PyBOP can be prepared in methylene chloride

and added stepwise (15–30 min intervals) to the methylene chloride solution to ensure high dilution conditions. The cyclization reaction is followed by analytical reversed-phase HPLC, and it is usually completed after 2–3 h. Concentrate the reaction mixture in a rotary evaporator and extract with H_2O. Evaporate the methylene chloride of the resulting organic phase and dry it under vacuum to obtain the protected cyclic peptide as a powder. Note: Besides HATU and PyBOP, other coupling reagents can be used.

5. Deprotection of the protected cyclic peptide by treatment with the mixtures TFA/triisopropylsilane/H_2O described for the linear scaffolds.
6. Follow steps 5 and 6 described for the linear scaffolds.

4. PREPARATION OF ANALYTICAL STOCK SOLUTIONS

The water used in all analytical studies, including for the preparation of stock solutions, should be of ultrapure (type I) grade. All commercial compounds used should be of the highest available purity. Similarly, all the synthesized compounds employed should be purified until judged analytically pure.

4.1 Peptidic Scaffolds

The stock solutions of the peptidic scaffolds should be freshly prepared in water before use, and its accurate concentration determined by UV spectroscopy using the protocols described later. Establishing the concentration of the solutions based on the molecular weight of the lyophilized peptidic scaffolds is not accurate because there may be significant quantities of water, salts, or counterions depending on the peptidic sequence, the synthetic procedure and conditions employed for purification. It should be also noted that the use of β-turn/loop inducing units containing aromatic molecules could often lead to additional absorbance in the UV–vis region of interest (see later). In these cases, special attention is required to subtract this absorbance from the final value. Since the determination of the peptidic scaffolds concentrations is supposed to be quantitative, the use of syringes (eg, Hamilton), not pipettes, is recommended, especially for small volumes and dilute solutions.

4.1.1 Trp/Tyr/Disulfide-Containing Scaffolds (Edelhoch, 1967; Gill & von Hippel, 1989; Pace, Vajdos, Fee, Grimsley, & Gray, 1995)

1. Record an UV spectrum of the background solution (eg, 1 mL of water) in a 1-cm pathlength quartz cuvette and measure the absorbance at 280 nm ($Abs_{(O,280)}$).

2. Add 5–10 μL of your peptidic scaffold stock solution, record the UV spectrum, and measure the absorbance at 280 nm ($Abs_{(P,280)}$). The amount of solution added should be adjusted to have a good absorbance intensity (typically 0.1–1.0 absorbance in the linear range).
3. Apply the following equations to calculate the concentration (M) of your stock solution:

$$[\text{Peptidic scaffold}]_{\text{stock sln}} = \frac{(Abs_{(P,280)} - Abs_{(O,280)})}{\varepsilon} \times DF$$

$$\varepsilon = (n_{Trp} \times (5690) + n_{Tyr} \times (1280) + n_{S-S} \times (120))M^{-1}cm^{-1}$$

where DF is the dilution factor, ε is the extinction coefficient of the peptidic scaffold, and n_{Trp}, n_{Tyr}, and n_{S-S} are the number of Trp, Tyr, and disulfide bridges present in the structure, respectively.
4. Repeat the process at least three times and calculate the average concentration.
5. *Note*: the calculation of the ε is based on the extended unfolded conformation of peptides/proteins. Therefore, if Trp and Tyr residues are expected to be buried from solvent in some conformations, their respective ε will be affected and the determination of Abs_{280} will not be accurate. For these cases, it is recommended to measure the absorbance in a denaturing solution such as 6.0 M guanidine hydrochloride.

4.1.2 Cys-Containing Scaffolds (Ellman, 1959)

All solutions must be purged with argon or nitrogen before use to minimize the chances of oxidation of Cys and formation of disulfide bonds.
1. Prepare the following buffer solution: 100 mM phosphate buffer pH 8.0.
2. Prepare a 50 mM stock solution of 5,5′-dithiobis(2-nitrobenzoic acid) (DTNB, Ellman's reagent) using this phosphate buffer. It is recommended to keep this solution in the freezer divided into small aliquots to avoid repeatedly thawing and freezing the original stock solution.
3. Record the UV–vis spectrum of a solution containing 940 μL of phosphate buffer and 50 μL of DTNB solution in a 1-cm pathlength quartz cuvette (background). Measure the absorbance at 412 nm ($Abs_{(O,412)}$).
4. Add 10 μL of the peptidic scaffold solution (total final volume in cuvette = 1000 μL) and mix carefully.
5. Incubate at room temperature for 15 min.
6. Record the UV–vis spectrum and measure the absorbance at 412 nm ($Abs_{(P,412)}$). The amount of solution added should be adjusted to have

a good absorbance intensity and therefore, the quantities of phosphate buffer in step 3 to have a final volume of 1000 µL.

7. Apply the following equations to calculate the concentration (M) of your stock solution:

$$[\text{Peptidic scaffold}]_{\text{stock sln}} = \frac{\left(\dfrac{(\text{Abs}_{(P,412)} - \text{Abs}_{(O,412)})}{14,150 M^{-1} \text{cm}^{-1}} \times \text{DF}\right)}{n_{\text{Cys}}}$$

where DF is the dilution factor, $14,150\ M^{-1}\text{cm}^{-1}$ (Collier, 1973; Riddles, Blakeley, & Zerner, 1983) is the extinction coefficient of the chromophore produced upon reaction of the thiol (Cys) with the Ellman's reagent, and n_{Cys} is the number of Cys present in the peptidic scaffold.

8. Repeat the process at least three times and calculate the average concentration.

9. *Note*: sometimes slow reaction has been observed between the Ellman's reagent and the Cys, especially with nonaccessible Cys, and accurate absorbance values could not be obtained even for longer incubation times. Although most likely this will not be the situation for this type of designed peptide scaffolds, an alternative protocol using 4,4'-dipyridyl disulfide as reagent can be employed (Grassetti & Murray, 1967; Mantle, Stewart, Zayas, & King, 1990).

4.1.3 Scaffolds with No Trp, Tyr, and Cys (Scopes, 1974)

The concentration of peptidic scaffolds that do not contain aromatic amino acids neither Cys residues can be determined by measuring the absorbance at 205 nm. However, this measurement is more prone to errors since many chemicals and solvents absorb at this wavelength. As alternative method, analytical HPLC with monitoring at 205 nm could be employed to simultaneously separate peptide from other constituents absorbing at this wavelength and determine concentration. In this case, a calibration curve at this wavelength using pure peptide solutions of known concentration is needed to obtain accurate results.

4.2 Metal Cation Stock Solutions

The stock solutions of metal cations should be prepared in water at ca. $0.05\ M$ from suitable metal salts and standardized by titration with K_2H_2EDTA using standard methods well described in the literature

(Schwarzenbach & Flaschka, 1969). In our copper(II) studies we use either $Cu(NO_3)_2$ or $CuCl_2$ analytical grade salts, and we standardize the stock solutions with K_2H_2EDTA using murexide or 1-(2-pyridylazo)-2-naphthol (PAN) as complexometric indicators.

4.3 Electrolyte Stock Solution

The stock solution of the chosen electrolyte should be prepared in water from a commercially sourced dry salt. The electrolyte is selected taking into account the purpose of the studies and the solubilities of the compounds employed and is used in a concentration at least 50 times higher than that of the analyte to keep the ionic strength of the media approximately constant throughout the measurements. Usual electrolytes are salts of strong acids with strong bases that are not likely to interfere in the formation equilibria of species. For the latter two reasons, special care must be taken not to use salts that contain interfering cations (such as other transition metals) in amounts above trace levels. Commonly used electrolytes are KNO_3, KCl, or NaCl solutions at 0.10 or 0.15 M concentration, and in our groups we usually employ a 0.10 M KNO_3 electrolyte. The electrolyte stock solution should be at least 10 times more concentrated than the desired concentration in the analytical samples to be studied, to allow preparation of the analytical solutions by dilution of the concentrated electrolyte. Additionally, the selected electrolyte will determine which base and acid titrant solutions will be used, as these should contain the same cation and anion present in the electrolyte.

4.4 Base Titrant Solution

The base titrant solution can be prepared from commercially available analytical ampoules to minimize the phenomenon of carbonation (solubilization of atmospheric carbon dioxide that forms carbonic acid and affects the concentration and use of the base). Carbonation will happen slowly on any basic solution, so it should be freshly prepared or else stored in a flask containing a CO_2 absorption material or kept under constant inert atmosphere. Preparation of this titrant by dissolution of a solid base form or by dilution of stored concentrated solutions is not recommended because both methods inevitably lead to excessive amounts of carbonation and thus cause significant errors in any determination using such a base. In our groups we usually prepare ca. 0.10 M KOH solutions from commercial ampoules and store them in flasks adapted with a CO_2 absorption material. The accurate concentration of any prepared base (even from an ampoule) must be determined by standardization with an acid standard. This can be done initially by neutralization titration of a standard acid solution prepared from a solid compound like potassium hydrogen phtalate dried to constant weight (Vogel, 1989). However, as the base concentration will slightly change over time, a base titrant that is used over a longer period should be regularly standardized. This can be done routinely by potentiometric titration of a standard acid solution (such as described in Section 5.1.1).

4.5 Acid Titrant Solution

The acid titrant solution can be prepared from commercially available analytical ampoules or by dilution of concentrate solutions, and its accurate concentration can be determined by neutralization titration with a freshly prepared and standardized base titrant solution (see earlier). In our groups we usually prepare 0.10 M HNO_3 solutions from ampoules, which after standardization can be stored well closed and used over a long period.

5. STUDY OF THE METAL ION COORDINATION PROPERTIES

The metal ion coordination properties of these peptidic scaffolds can be studied using spectroscopic techniques as well as mass spectrometry and potentiometric titrations. The global analysis of the data obtained in these experiments will give crucial insights into the suitability of these scaffolds to coordinate the target metal ion, the formation and concentration of different metal ion complexes (speciation), and their structure in solution. The selection of the techniques will depend on the metal ion of interest and the type of peptidic scaffold. Herein, we will mainly focus on the study of copper(II) systems, and the different techniques will be illustrated by specific copper(II) examples developed in our group.

5.1 Potentiometric Titrations: Determining Stability Constants and Speciation Diagrams

A metallopeptide system will be completely characterized from the point of view of composition when both the stoichiometry of the formed species and their stability constants are known. Therefore, solution equilibria studies on the peptidic scaffolds in the presence and absence of metal ions are important to obtain detailed information of the different species that exist in solution as a function of pH as well as of their stabilities and concentrations. Once the speciation model has been determined, one can calculate the concentration of all the species present in solution at any given concentration of the system components (metal ion, peptidic scaffold, and pH).

In the framework of the systems described in this chapter, the reactions of complex formation are described by the following general equilibrium and equation:

$$mM + hH + pP \rightleftarrows M_mH_hP_p \quad \beta_{M_mH_hP_p} = \frac{[M_mH_hP_p]}{[M]^m[H]^h[P]^p} \quad (1)$$

where M is the metal ion, P is the peptidic scaffold, H is the proton, β is the overall formation constant, terms in square brackets refer to molar concentration of the species, and charges and solvent are omitted for the sake of simplicity. It is worth to mention that the coefficient h can be negative if the number of proton dissociated from P exceeds the maximum number of protons dissociated in the absence of metal ion. This usually includes the deprotonation of the amide bond nitrogen and the formation of hydroxocomplex species. Because changes in proton concentration are different when the peptidic scaffolds are analyzed in the absence or presence of metal cations, and these changes are a consequence of metal ion coordination (complexation), such experiments allow the monitoring of the reaction progress (Eq. 1).

Initially, a potentiometric titration of the peptidic scaffold is performed to determine its overall protonation constants (Eq. 2). This titration is done under temperature control, constant ionic strength, and usually in the pH range 2.5–11.5.

$$H + P \rightleftarrows HP \quad \beta_{HP} = \frac{[HP]}{[H][P]}$$

$$2H + P \rightleftarrows H_2P \quad \beta_{H_2P} = \frac{[H_2P]}{[H]^2[P]} \quad (2)$$

$$hH + P \rightleftarrows H_hP \quad \beta_{H_hP} = \frac{[H_hP]}{[H]^h[P]}$$

Afterward, potentiometric titrations of the peptidic scaffold in the presence of different proportions of metal ion are carried out under the same experimental conditions to determine the overall complexation constants (exemplified by Eq. 3)

$$M + P \rightleftarrows MP \quad \beta_{MP} = \frac{[MP]}{[M][P]}$$

$$M + H + P \rightleftarrows MHP \quad \beta_{MHP} = \frac{[MHP]}{[M][H][P]}$$

$$M + 2P \rightleftarrows MP_2 \quad \beta_{MP_2} = \frac{[MP_2]}{[M][P]^2}$$

$$M + H + 2P \rightleftarrows MHP_2 \quad \beta_{MHP_2} = \frac{[MHP_2]}{[M][H][P]^2} \quad (3)$$

$$2M + P \rightleftarrows M_2P \quad \beta_{M_2P} = \frac{[M_2P]}{[M]^2[P]}$$

$$2M + H + P \rightleftarrows M_2HP \quad \beta_{M_2HP} = \frac{[M_2HP]}{[M]^2[H][P]}$$

Another equilibrium that should be considered is the self-ionization of water leading to the ionic product (Eq. 4). This also depends on the ionic strength and temperature, and thus, it needs to be adequate for the experimental conditions employed. Although the value of K_w can be experimentally determined, there are accurate values available in the literature for the most common experimental conditions that can be used instead for simplicity (Kron, Marshall, May, Hefter, & Königsberger, 1995).

$$H_2O \rightleftarrows H^+ + OH^- \quad K_W = [H^+][OH^-] \quad (4)$$

From the experimental point of view, pH potentiometric titrations are one of the most relevant and accurate analytical methods to study the formation of metal ion complexes. Changes in proton concentration upon addition of base are monitored by measuring the changes in potential (emf) or pH of a hydrogen-selective electrode or pH-sensitive glass electrode. The simplest experimental setup requires only a combined pH electrode (combining both pH sensing and reference electrodes into one body) connected to a potentiometer to provide the pH or potential readings, with titrations being performed by manual titrant additions. More advanced systems normally use automatic burettes for a more precise volume dispensing, while integrated titrators for combined measurement/titration/control are also available for a fully automatic operation. Temperature should be kept stable throughout all measurements preferably by using a thermostatic control equipment. Potentiometric titrations are very reliable for experiments in the pH range of ca. 2.5–11.5, although they can also be used at pH < 2.5 after corrections.

5.1.1 Calibration of the pH Electrode

The calibration of a pH electrode can be easily made by using two or more standard buffer solutions with pH values in the range of interest. However, for more accurate and frequent measurements of potential a more rigorous method of electrode calibration is recommended. The best way is to perform a calibration titration of a standard acid solution (like the prepared acid titrant) with the base titrant, under the same experimental conditions (analyte concentration, temperature, and ionic strength medium) as those required for the subsequent measurements. This procedure allows the calibration of the electrode response in terms of hydrogen ion concentration.

In the following calibration example, as well as in the titrations described in Section 5.1.2, the analytical solutions used are 0.10 M HNO_3 acid titrant, ca. 0.10 M KOH base titrant, and 1.0 M KNO_3 electrolyte stock solution.

1. Prepare the calibration solution by accurately measuring 0.5 mL of acid titrant, 3 mL of electrolyte solution, and water to make a total volume of 30 mL.
2. Perform the titration by successive additions of 0.05 mL of base titrant, up to a final volume of 1.5 mL. Ideally, the experimental data should contain at least 10 points in each of the pH ranges 2.5–4 and 10.8–11.5.
3. Register the measured potential at each point after a stabilization period (eg, 5 min) upon each titrant addition.

A special software is required for fitting these calibration data, such as the Glee freeware tool (Gans & O'Sullivan, 2000) of the HyperQuad suite of programs (see Section 5.1.3). This software refines the values of both the standard electrode potential (E^0) and the concentration of base titrant. The titration endpoint (thus the concentration of base) may also be calculated by the Gran's method (Rossotti & Rossotti, 1965).

5.1.2 Potentiometric Titrations

For potentiometric titrations, the concentration of analyte (peptide in this case) should be in the range of 0.5–1.0 mM to minimize measurement errors. In special cases it is possible to work at lower concentrations without compromising too much the accuracy of the results, but not lower than 0.2 mM in any case. In the next titration examples, the analytical solutions used are 1.0 mM peptide stock solution and 0.050 M $Cu(NO_3)_2$ metal solution. If such peptide solution concentrations cannot be reached, the protocols must be adapted accordingly by decreasing the volumes of metal solution and decreasing the addition volumes of titrant in order to obtain enough titration points for a good fitting.

For a typical protonation titration:
1. Prepare the analyte solution by accurately measuring 25 mL of peptide stock solution and 3 mL of electrolyte solution. Check the pH and, if necessary, add acid titrant solution until pH≈2.5. Then, add water to make a total volume of 30 mL.
2. Perform the titration by successive additions of 0.05 mL of base titrant, up to pH≈11.5. Ideally, the experimental data should contain at least 10 points for each protonation constant to be determined. If necessary, the titrant addition volume may be decreased to allow for more points.
3. Register the measured potential (or pH) values at each point after a stabilization period (eg, 5 min) upon each titrant addition.

For a typical complexation titration:
1. Prepare the analyte solution by accurately measuring 25 mL of peptide stock solution and 3 mL of electrolyte solution. Check the pH and, if necessary, add acid titrant solution until pH≈2.5. Then, add 0.25 mL of 0.050 M $Cu(NO_3)_2$ solution (corresponding to 0.5 equiv. of peptide concentration). Finally, add water to make a total volume of 30 mL.
2. Perform the titration by successive additions of 0.05 mL of base titrant, up to pH≈11.5. Ideally, the experimental data should contain at least 10 points for each complexation constant to be determined, so the titrant addition volume may be decreased if necessary.
3. Register the measured potential (or pH) values at each point only after the reading is completely stable, which can take longer time than on protonation. It is recommended to start by a stabilization period of 10 min, and increase it as needed.
4. At the end of this titration, perform a back titration of the final solution using the acid titrant solution and a similar method to the one used in the previous forward titration. This back titration should be compared with the forward titration to know if the latter was fully stabilized, or if not, it should be repeated using longer stabilization periods to attain equilibrium.
5. Prepare and run additional complexation titrations using increasing amounts of $Cu(NO_3)_2$ solution (as many equivalents of metal as relevant for your system) by repeating steps 1–4. This allows for monitoring the formation of polynuclear complex species.

5.1.3 Potentiometric Data Treatment

Different computer programs are currently available to fit the potentiometric data from protonation and complexation titrations (with potential or pH measurements) by solving the collection of equations thorough nonlinear least-squares methods. Our groups have been using the HyperQuad suite of programs (http://www.hyperquad.co.uk/) (Gans, Sabatini, & Vacca, 1996). Although all data fitting programs of this suite are commercial, there are a few additional tools that may be used freely. Alternatively, there is a completely free program for academic purposes based on a Microsoft Excel spreadsheet with macros, CurTiPot (http://www.iq.usp.br/gutz/Curtipot).

Initially, the protonation titration data should be fitted to obtain the protonation constants that fully describe the peptidic system under study. Following, complexation titration data should be fitted while using the previously determined protonation constants as fixed values in the refinement. When several complexation titrations were performed in the presence of different ratios of metal ion, each should be fitted independently at first to find the correct model of species existing for each case. Afterward, all complexation titrations should be combined in a single fitting to determine all existing constants more accurately. The fitting process renders the values for all the overall ($\log \beta$) protonation constants of the peptidic scaffold and all the overall metal ion complex formation constants that describe the system under study. Since overall constants can always be expressed as the product of stepwise constants, the stepwise ($\log K$) protonation constants and complex formation constants (or stability constants) of the peptidic scaffold can be calculated using these data. For example, the formation of the species MP_2 could also be described in two steps:

$$M + P \rightleftarrows MP \quad K_{MP} = \frac{[MP]}{[M][P]} = \beta_{MP}$$

$$MP + P \rightleftarrows MP_2 \quad K_{MP_2} = \frac{[MP_2]}{[MP][P]}$$

and thus, the value of K_{MP_2} calculated

$$\beta_{MP_2} = \frac{[MP_2]}{[M][P]^2} = K_{MP} \times K_{MP_2} \quad K_{MP_2} = \frac{\beta_{MP_2}}{K_{MP}}.$$

Moreover, the overall protonation and metal ion complex formation constants can be also computed to draw speciation and competition diagrams using the HySS freeware tool (Alderighi et al., 1999).

The methodology described in Section 5.1 was employed to characterize the binding of Cu^{2+} to our designed scaffolds. Table 2 summarizes the

Table 2 Overall ($\log \beta_i^H$ and $\log \beta_{Cu_mH_hL_l}$) and Stepwise ($\log K_i^H$ and $\log K_{Cu_mH_hL_l}$) Protonation Constants of C-Asp and Stability Constants of Its Copper(II) Complexes in Aqueous Solution at $T = 298.2 \pm 0.1$ K and $I = 0.10 \pm 0.01$ M in KNO_3

Equilibrium Reaction[a]	$\log \beta_i^H$[b]	Equilibrium Reaction	$\log K_i^H$
$L^- + H^+ \rightleftarrows HL$	10.11(1)	$L^- + H^+ \rightleftarrows HL$	10.11
$L^- + 2H^+ \rightleftarrows H_2L^+$	17.33(1)	$HL + H^+ \rightleftarrows H_2L^+$	7.22
$L^- + 3H^+ \rightleftarrows H_3L^{2+}$	23.73(1)	$H_2L^+ + H^+ \rightleftarrows H_3L^{2+}$	6.40
$L^- + 4H^+ \rightleftarrows H_4L^{3+}$	29.34(1)	$H_3L^{2+} + H^+ \rightleftarrows H_4L^{3+}$	5.61
$L^- + 5H^+ \rightleftarrows H_5L^{4+}$	32.203(1)	$H_4L^{3+} + H^+ \rightleftarrows H_5L^{4+}$	2.87

Equilibrium Reaction[a]	$\log \beta_{Cu_mH_hL_l}$[b]	Equilibrium Reaction	$\log K_{Cu_mH_hL_l}$
$Cu^{2+} + 2H^+ + L^- \rightleftarrows [CuH_2L]^{3+}$	24.65(2)	$[CuHL]^{2+} + H^+ \rightleftarrows [CuH_2L]^{3+}$	3.76
$Cu^{2+} + H^+ + L^- \rightleftarrows [CuHL]^{2+}$	20.892(9)	$[CuL]^+ + H^+ \rightleftarrows [CuHL]^{2+}$	8.69
$Cu^{2+} + L^- \rightleftarrows [CuL]^+$	12.20(8)	$Cu^{2+} + L^- \rightleftarrows [CuL]^+$	12.20
$Cu^{2+} + L^- \rightleftarrows [CuH_{-1}L] + H^+$	3.67(8)	$[CuH_{-1}L] + H^+ \rightleftarrows [CuL]^+$	8.53
$Cu^{2+} + L^- \rightleftarrows [CuH_{-2}L]^- + 2H^+$	−6.2(1)	$[CuH_{-2}L]^- + H^+ \rightleftarrows [CuH_{-1}L]$	9.87
$Cu^{2+} + L^- \rightleftarrows [CuH_{-3}L]^{2-} + 3H^+$	−16.3(1)	$[CuH_{-3}L]^{2-} + H^+ \rightleftarrows [CuH_{-2}L]^-$	10.1
$2Cu^{2+} + L^- \rightleftarrows [Cu_2L]^{3+}$	17.2(1)	$[CuL]^+ + Cu^{2+} \rightleftarrows [Cu_2L]^{3+}$	5.0
$2Cu^{2+} + L^- \rightleftarrows [Cu_2H_{-2}L]^+ + 2H^+$	4.28(4)		
$2Cu^{2+} + L^- \rightleftarrows [Cu_2H_{-3}L] + 3H^+$	−2.45(4)	$[Cu_2H_{-3}L] + H^+ \rightleftarrows [Cu_2H_{-2}L]^+$	6.73
$2Cu^{2+} + L^- \rightleftarrows [Cu_2H_{-4}L]^- + 4H^+$	−11.67(5)	$[Cu_2H_{-4}L]^- + H^+ \rightleftarrows [Cu_2H_{-3}L]$	9.22
$2Cu^{2+} + L^- \rightleftarrows [Cu_2H_{-5}L]^{2-} + 5H^+$	[−21.6][c]	$[Cu_2H_{-5}L]^{2-} + H^+ \rightleftarrows [Cu_2H_{-4}L]^-$	10.0
$2Cu^{2+} + L^- \rightleftarrows [Cu_2H_{-6}L]^{3-} + 6H^+$	−31.94(5)	$[Cu_2H_{-6}L]^{3-} + H^+ \rightleftarrows [Cu_2H_{-5}L]^{2-}$	10.3

[a] L^- indicates the completely deprotonated form of the C-Asp peptide.
[b] Values in parenthesis are standard deviations in the last significant figure.
[c] It is an approximate value because it was obtained in the calculated model with a standard deviation larger than the allowed by Hyperquad program; however, the fitting improved (the σ values decrease) with the inclusion of this constant.

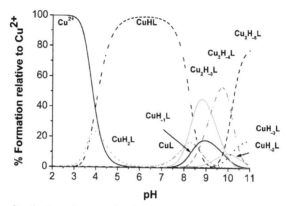

Fig. 6 Species distribution diagrams for the copper(II) complexes of the C-Asp peptide in aqueous solution, at 298.2 K, $I=0.1\ M$ KNO$_3$, and [L=C-Asp]=[Cu(NO$_3$)$_2$]=$1.0\times 10^{-3}\ M$. Charges are omitted for simplicity.

overall protonation constants obtained for the C-Asp peptidic scaffold and the respective overall complex formation constants with Cu^{2+} (Fragoso, Lamosa, et al., 2013). Fig. 6 shows the speciation diagram calculated using these data at 1 mM concentration and at 1:1 C-Asp:Cu^{2+} stoichiometry.

To compare the performance of different metallopeptide systems, it is not correct to directly compare the stability constants for complexes of different peptides as they do not take into account the variable acid–base (protonation) properties of each peptide alone. For a sustained comparison, the effective stability constants (K_{eff}, also known as apparent or conditional constants) at fixed pH may be calculated from the full set of overall equilibrium (both protonation and stability) constants describing each system, for example, using the HySS tool. These effective constants can be obtained for specific pH values or even plotted for a given pH range and allow a direct comparison between different systems (Fig. 7). Such results are also helpful, for example, in understanding at which pH the peptides exhibits a more efficient metal ion complexation or how it changes with pH.

5.2 Characterization of Major Species by Spectroscopy

The major metal ion complexes in solution can be further characterized at pH values for which there are no significant contributions of other species. Based on the metal ion and probes present in the peptidic scaffolds, different

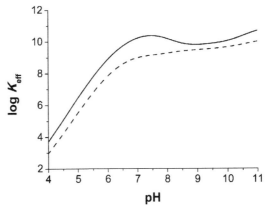

Fig. 7 The log K_{eff} in function of the pH for the copper(II) complexes of C-Asp (*solid line*) and O-Asp (*dashed line*) peptides calculated for 1:1 Cu^{2+}:peptide at [peptide] = 1.0×10^{-3} M. $K_{eff} = \Sigma[Cu_mH_{i+j}L_i]/\Sigma[H_iL]\cdot\Sigma[H_jCu]$, values calculated using the data in Table 2 (Fragoso, Lamosa, et al., 2013) and the HYSS program.

spectroscopic methods can be used, to cite the most common ones: ultraviolet–visible (UV–vis) spectroscopy, fluorescence spectroscopy, circular dichroism (CD) spectroscopy, electron paramagnetic resonance (EPR) spectroscopy, nuclear magnetic resonance (NMR) spectroscopy, and X-ray spectroscopy. These data allow one to propose potential structures in solution for the different metal ion complexes. A description of the experimental procedures for the different spectroscopic methodologies is out of the scope of this chapter. Using the $[CuH(C\text{-}Asp)]^{2+}$ complex as an example (CuHL species in Fig. 6), a general description of the type of data obtained and the solution structure postulated are given.

Cu^{2+} is a spectroscopically rich metal ion and different methods can be used to obtain complementary information (Solomon et al., 2014). Shown in Fig. 8 are the spectra obtained for the $[CuH(C\text{-}Asp)]^{2+}$ complex using UV–vis (Fig. 7A), CD (Fig. 7B), and EPR (Fig. 7C) spectroscopies. The spectroscopic parameters are collected in Table 3. All the data are consistent with the coordination of Cu^{2+} exclusively by the side chain of the three His (imidazole rings) and one Asp (carboxylic group) residues and with the formation of square-planar or square-pyramidal geometry (Fragoso, Lamosa, et al., 2013). Additionally, the solution structure of the C-Asp peptide was evaluated by CD spectroscopy in the absence and presence of Cu^{2+} (Fig. 7D). In this case, the Far-UV region of the spectra was

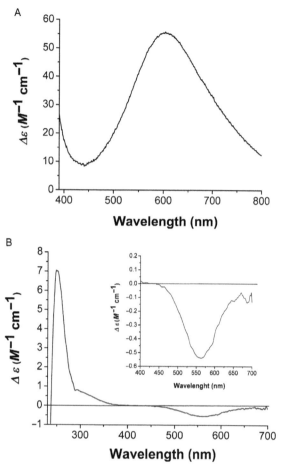

Fig. 8 (A) UV–vis spectrum of the [CuH(C-Asp)]$^{2+}$ complex at 1.0×10^{-3} M, 298.2 K, and pH 6.2. (B) CD spectrum of the [CuH(C-Asp)]$^{2+}$ complex at 1.0×10^{-3} M, 298.2 K, and pH 6.3. The *inset diagram* shows the spectral window 400–700 nm.

(Continued)

analyzed to see if this scaffold presented or not any features of secondary structure, and how this will change upon Cu^{2+} coordination. In the absence of Cu^{2+} the C-Asp scaffold presented a positive band with maximum ellipticity at ~210 nm consistent with the presence of a type II β-turn. Upon metal ion binding, conformational changes were observed. Two bands with positive ellipticity at 210 and 245 nm and one band with negative ellipticity at 225 nm appeared.

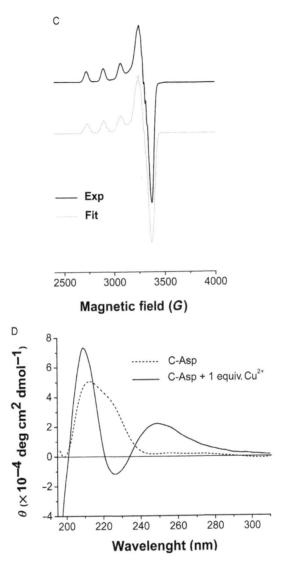

Fig. 8—Cont'd (C) X-band EPR spectrum of the [CuH(C-Asp)]$^{2+}$ complex at 1.0×10^{-3} M in 100 m*M* Mes buffer at pH 6.0 and 92.2 K. (D) Far-UV CD spectra of the C-Asp peptide at 5.0×10^{-6} M, 298.2 K, and pH 6.0 in the absence (*dashed line*) and in the presence (*solid line*) of 1 equiv. of Cu^{2+} (Cu(NO$_3$)$_2$).

Table 3 Spectroscopic Data for the [CuH(C-Asp)]$^{2+}$ Complex[a]

Spectroscopy	CuH(C-Asp)]$^{2+}$
UV–vis	
λ_{max}(nm), $\Delta\varepsilon$ (M^{-1} cm^{-1})	603, 56[b]
CD	
λ_{max}(nm), $\Delta\varepsilon$ (M^{-1} cm^{-1})	248, +7.050[c]
	557, −0.538[b]
EPR	
A_z (×10^{-4} cm^{-1})	177.82
A_x (×10^{-4} cm^{-1})	3.92
A_y (×10^{-4} cm^{-1})	21.82
g_z	2.277
g_y	2.069
g_x	2.045

[a]Values obtained from spectra in Fig. 8A–C.
[b]Cu^{2+} d–d transitions.
[c]N$_{imidazole}$-His → Cu^{2+} charge transfer/intramolecular transition.

6. CONCLUDING REMARKS

Peptidic scaffolds have been shown to be a powerful platform for the design of inorganic systems with tailor-made properties and functions. In this chapter, we provided a detailed methodology for engineering and synthesizing short preorganized peptide sequences for metal ion coordination and showed how potentiometric titrations can enable their full characterization from the point of view of composition (number and stoichiometry of the formed species and their stability constants). An overview of different spectroscopic methods that can give insight into the solution structure of the major metal ion species was given and illustrated by using one of our designed copper(II) systems.

ACKNOWLEDGMENTS

We thank Emmanuel Oheix for helpful comments. This work was supported by Fundação para a Ciência e a Tecnologia (PTDC/QUI-BIQ/098406/2008) and the European Commission (Marie Curie Actions FP7-IRG-230896). L.M.P. Lima acknowledges Fundação para a Ciência e a Tecnologia for a postdoctoral fellowship (SFRH/BPD/73361/2010).

REFERENCES

Alderighi, L., Gans, P., Ienco, A., Peters, D., Sabatini, A., & Vacca, A. (1999). Hyperquad simulation and speciation (HySS): A utility program for the investigation of equilibria involving soluble and partially soluble species. *Coordination Chemistry Reviews, 184*, 311–318.

Athanassiou, Z., Patora, K., Dias, R. L. A., Moehle, K., Robinson, J. A., & Varani, G. (2007). Structure-guided peptidomimetic design leads to nanomolar beta-hairpin inhibitors of the Tat-TAR interaction of bovine immunodeficiency virus. *Biochemistry, 46*, 741–751.

Barba, M., Sobolev, A. P., Zobnina, V., Bonaccorsi di Patti, M. C., Cervoni, L., Spiezia, M. C., et al. (2012). Cupricyclins, novel redox-active metallopeptides based on conotoxins scaffold. *PLoS One, 7*, e30739.

Bean, J. W., Kopple, K. D., & Peishoff, C. E. (1992). Conformational analysis of cyclic hexapeptides containing the D-Pro-L-Pro sequence to fix β-turn positions. *Journal of the American Chemical Society, 114*, 5328–5334.

Blanco, F., Ramírez-Alvarado, M., & Serrano, L. (1998). Formation and stability of β-hairpin structures in polypeptides. *Current Opinion in Structural Biology, 8*, 107–111.

Bonnet, C. S., Fries, P. H., Crouzy, S., Sénèque, O., Cisnetti, F., Boturyn, D., et al. (2009). A gadolinium-binding cyclodecapeptide with a large high-field relaxivity involving second-sphere water. *Chemistry - A European Journal, 15*, 7083–7093.

Boturyn, D., Defrancq, E., Dolphin, G. T., Garcia, J., Labbe, P., Renaudet, O., et al. (2008). RAFT nano-constructs: Surfing to biological applications. *Journal of Peptide Science, 14*, 224–240.

Butler, S. J., & Jolliffe, K. A. (2011). Synthesis of a family of cyclic peptide-based anion receptors. *Organic & Biomolecular Chemistry, 9*, 3471–3483.

Campos, S. R. R., Iranzo, O., & Baptista, A. M. (2016). Constant-pH MD simulations portray the protonation and structural behavior of four decapeptides designed to coordinate Cu^{2+}. *The Journal of Physical Chemistry. B, 120*, 1080–1091.

Chan, W. C., & White, P. D. (2000). *Fmoc solid phase peptide synthesis: A practical approach*. New York: Oxford University Press.

Chou, P. Y., & Fasman, G. D. (1977). β-Turns in proteins. *Journal of Molecular Biology, 115*, 135–175.

Collier, H. B. (1973). Letter: A note on the molar absorptivity of reduced Ellman's reagent, 3-carboxylato-4-nitrothiophenolate. *Analytical Biochemistry, 56*, 310–311.

Daugherty, R. G., Wasowicz, T., Gibney, B. R., & DeRose, V. J. (2002). Design and spectroscopic characterization of peptide models for the plastocyanin copper-binding loop. *Inorganic Chemistry, 41*, 2623–2632.

Dumy, P., Eggleston, I. M., Cervigni, S., Sila, U., Sun, X., & Mutter, M. (1995). A convenient synthesis of cyclic peptides as regioselectively addressable functionalized templates (RAFT). *Tetrahedron Letters, 36*, 1255–1258.

Dumy, P., Eggleston, I. M., Esposito, G., Nicula, S., & Mutter, M. (1996). Solution structure of regioselectively addressable functionalized templates: An NMR and restrained molecular dynamics investigation. *Biopolymers, 39*, 297–308.

Edelhoch, H. (1967). Spectroscopic determination of tryptophan and tyrosine in proteins. *Biochemistry, 6*, 1948–1954.

Ellman, G. L. (1959). Tissue sulfhydryl groups. *Archives of Biochemistry and Biophysics, 82*, 70–77.

Farkas, V., Vass, E., Hanssens, I., Majer, Z., & Hollósi, M. (2005). Cyclic peptide models of the Ca2+-binding loop of α-lactalbumin. *Bioorganic & Medicinal Chemistry, 13*, 5310–5320.

Fattorusso, R., Morelli, G., Lombardi, A., Nastri, F., Maglio, O., D'Auria, G., et al. (1995). Design of metal ion binding peptides. *Biopolymers, 37*, 401–410.

Fragoso, A., Carvalho, T., Rousselot-Pailley, P., Correia dos Santos, M. M., Delgado, R., & Iranzo, O. (2015). Effect of the peptidic scaffold in copper(II) coordination and the redox properties of short histidine-containing peptides. *Chemistry - A European Journal, 21*, 13100–13111.

Fragoso, A., Delgado, R., & Iranzo, O. (2013). Copper(II) coordination properties of decapeptides containing three His residues: The impact of cyclization and Asp residue coordination. *Dalton Transactions, 42*, 6182–6192.

Fragoso, A., Lamosa, P., Delgado, R., & Iranzo, O. (2013). Harnessing the flexibility of peptidic scaffolds to control their copper(II)-coordination properties: A potentiometric and spectroscopic study. *Chemistry - A European Journal, 19*, 2076–2088.

Gaggelli, E., Kozłowski, H., Valensin, D., & Valensin, G. (2006). Copper homeostasis and neurodegenerative disorders (Alzheimer's, prion, and Parkinson's diseases and amyotrophic lateral sclerosis). *Chemical Reviews, 106*, 1995–2044.

Gans, P., & O'Sullivan, B. (2000). GLEE, a new computer program for glass electrode calibration. *Talanta, 51*, 33–37.

Gans, P., Sabatini, A., & Vacca, A. (1996). Investigation of equilibria in solution. Determination of equilibrium constants with the HYPERQUAD suite of programs. *Talanta, 43*, 1739–1753.

Gellman, S. H. (1998). Minimal model systems for β sheet secondary structure in proteins. *Current Opinion in Chemical Biology, 2*, 717–725.

Gill, S. C., & von Hippel, P. H. (1989). Calculation of protein extinction coefficients from amino acid sequence data. *Analytical Biochemistry, 182*, 319–326.

Grassetti, D. R., & Murray, J. F., Jr. (1967). Determination of sulfhydryl groups with 2,2'- or 4,4'-dithiodipyridine. *Archives of Biochemistry and Biophysics, 119*, 41–49.

Hancock, W. S., & Battersby, J. E. (1976). A new micro-test for the detection of incomplete coupling reactions in solid-phase peptide synthesis using 2,4,6-trinitrobenzene-sulphonic acid. *Analytical Biochemistry, 71*, 260–264.

Holm, R. H., Kennepohl, P., & Solomon, E. I. (1996). Structural and functional aspects of metal sites in biology. *Chemical Reviews, 96*, 2239–2314.

Hughes, R. M., & Waters, M. L. (2006). Model systems for β-hairpins and β-sheets. *Current Opinion in Structural Biology, 16*, 514–524.

Hutchinson, E. G., & Thornton, J. M. (1994). A revised set of potentials for beta-turn formation in proteins. *Protein Science, 12*, 2207–2216.

Imperiali, B., & Kapoor, T. M. (1993). The reverse turn as a template for metal coordination. *Tetrahedron, 49*, 3501–3510.

Kaiser, E., Colescott, R. L., Bossinger, C. D., & Cook, P. I. (1970). Color test for detection of free terminal amino groups in the solid-phase synthesis of peptides. *Analytical Biochemistry, 34*, 595–598.

Kotynia, A., Bielinska, S., Kamysz, W., & Brasuń, J. (2012). The coordination abilities of the multiHis-cyclopeptide with two metal-binding centers—Potentiometric and spectroscopic investigation. *Dalton Transactions, 41*, 12114–12120.

Kozłowski, H., Kowalik-Jankowska, T. M., & Jeżowska-Bojczuk, M. (2005). Chemical and biological aspects of Cu^{2+} interactions with peptides and aminoglycosides. *Coordination Chemistry Reviews, 249*, 2323–2334.

Kron, I., Marshall, S. L., May, P. M., Hefter, G., & Königsberger, E. (1995). The ionic product of water in highly concentrated aqueous electrolyte solutions. *Monatshefte für Chemie, 126*, 819–837.

Lebrun, C., Starck, M., Gathu, V., Chenavier, Y., & Delangle, P. (2014). Engineering short peptide sequences for uranyl binding. *Chemistry - A European Journal, 20*, 16566–16573.

Mantle, M., Stewart, G., Zayas, G., & King, M. (1990). The disulphide-bond content and rheological properties of intestinal mucins from normal subjects and patients with cystic fibrosis. *Biochemical Journal, 266*, 597–604.

Migliorini, C., Porciatti, E., Łuczkowski, M., & Valensin, D. (2012). Structural characterization of Cu^{2+}, Ni^{2+} and Zn^{2+} binding sites of model peptides associated with neurodegenerative diseases. *Coordination Chemistry Reviews*, 256, 352–368.

Nair, C. M., Vijayan, M., Venkatachalapathi, Y. V., & Balaram, P. (1979). X-ray crystal structure of pivaloyl-D-Pro-L-Pro-L-Ala-N-methylamide; observation of a consecutive β-turn conformation. *Journal of the Chemical Society, Chemical Communications*, 1183–1184.

Natale, G. D., Damante, C. A., Nagy, Z., Ősz, K., Pappalardo, G., Rizzarelli, E., et al. (2008). Copper(II) binding to two novel histidine-containing model hexapeptides: Evidence for a metal ion driven turn conformation. *Journal of Inorganic Biochemistry*, 102, 2012–2019.

Nesloney, C. L., & Kelly, J. W. (1996). Progress towards understanding β-sheet structure. *Bioorganic & Medicinal Chemistry*, 4, 739–766.

Neupane, K. P., Aldous, A. R., & Kritzer, J. A. (2013). Macrocyclization of the ATCUN motif controls metal binding and catalysis. *Inorganic Chemistry*, 52, 2729–2735.

Ngu-Schwemlein, M., Gilbert, W., Askew, K., & Schwemlein, S. (2008). Thermodynamics and fluorescence studies of the interactions of cyclooctapeptides with Hg^{2+}, Pb^{2+}, and Cd^{2+}. *Bioorganic & Medicinal Chemistry*, 16, 5778–5787.

Ngyen, H., Orlamuender, M., Pretzel, D., Agricola, I., Sternberg, U., & Reissmann, S. (2008). Transition metal complexes of a cyclic pseudo hexapeptide: Synthesis, complex formation and catalytic activities. *Journal of Peptide Science*, 14, 1010–1021.

Niedźwiecka, A., Cisnetti, F., Lebrun, C., & Delangle, P. (2012). Femtomolar Ln(III) affinity in peptide-based ligands containing unnatural chelating amino acids. *Inorganic Chemistry*, 51, 5458–5464.

Nikiforovich, G. V., Mutter, M., & Lehmann, C. (1999). Molecular modeling and design of regioselectively addressable functionalized templates with rigidified three-dimensional structures. *Biopolymers*, 50, 361–372.

Pace, C. N., Vajdos, F., Fee, L., Grimsley, G., & Gray, T. (1995). How to measure and predict the molar absorption coefficient of a protein. *Protein Science*, 4, 2411–2423.

Pappalardo, G., Impellizzeri, G., & Campagna, T. (2004). Copper(II) binding of prion protein's octarepeat model peptides. *Inorganica Chimica Acta*, 357, 185–194.

Pearson, R. G. (1963). Hard and soft acids and bases. *Journal of the American Chemical Society*, 85, 3533–3539.

Peluso, S., Rückle, T., Lehmann, C., Mutter, M., Peggion, C., & Crisma, M. (2001). Crystal structure of a synthetic cyclodecapeptide for template-assembled synthetic protein design. *ChemBioChem*, 2, 432–437.

Pires, S., Habjanič, J., Sezer, M., Soares, C. M., Hemmingsen, L., & Iranzo, O. (2012). Design of a peptidic turn with high affinity for Hg(II). *Inorganic Chemistry*, 51, 11339–11348.

Pujol, A. M., Cuillel, M., Renaudet, O., Lebrun, C., Charbonnier, P., Cassio, D., et al. (2011). Hepatocyte targeting and intracellular copper chelation by a thiol-containing glycocyclopeptide. *Journal of the American Chemical Society*, 133, 286–296.

Rama, G., Ardá, A., Maréchal, J.-D., Gamba, I., Ishida, H., Jiménez-Barbero, J., et al. (2012). Stereoselective formation of chiral metallopeptides. *Chemistry - A European Journal*, 18, 7030–7035.

Ramírez-Alvarado, M., Kortemme, T., Blanco, F. J., & Serrano, L. (1999). β-Hairpin and β-sheet formation in designed linear peptides. *Bioorganic & Medicinal Chemistry*, 7, 93–103.

Riddles, P. W., Blakeley, R. L., & Zerner, B. (1983). Reassessment of Ellman's reagent. *Methods in Enzymology*, 91, 49–60.

Robinson, J. A. (2000). The design, synthesis and conformation of some new β-hairpin mimetics: Novel reagents for drug and vaccine discovery. *Synlett*, 4, 429–441.

Robinson, J. A. (2008). β-Hairpin peptidomimetics: Design, structures and biological activities. *Accounts of Chemical Research, 41*, 1278–1288.

Rose, G. D., Gierasch, L. M., & Smith, J. A. (1985). Turns in peptides and proteins. *Advances in Protein Chemistry, 37*, 1–109.

Rossotti, F. J. C., & Rossotti, H. (1965). Potentiometric titrations using Gran plots. A textbook omission. *Journal of Chemical Education, 42*, 375–378.

Rousselot-Pailley, P., Sénèque, O., Lebrun, C., Crouzy, S., Boturyn, D., Dumy, P., et al. (2006). Model peptides based on the binding loop of the copper metallochaperone Atx1: Selectivity of the consensus sequence MxCxxC for metal ions Hg(II), Cu(I), Cd(II), Pb(II), and Zn(II). *Inorganic Chemistry, 45*, 5510–5520.

Schneider, J. P., Pochan, D. J., Ozbas, B., Rajagopal, K., Pakstis, L., & Kretsinger, J. (2002). Responsive hydrogels from the intramolecular folding and self-assembly of a designed peptide. *Journal of the American Chemical Society, 124*, 15030–15037.

Schwarzenbach, G., & Flaschka, W. (1969). *Complexometric titrations*. London: Methuen & Co.

Scopes, R. K. (1974). Measurement of protein by spectrophotometry at 205 nm. *Analytical Biochemistry, 59*, 277–282.

Sénèque, O., Bourlès, E., Lebrun, V., Bonnet, E., Dumy, P., & Latour, J.-M. (2008). Cyclic peptides bearing a side-chain tail: A tool to model the structure and reactivity of protein zinc sites. *Angewandte Chemie, International Edition, 47*, 6888–6891.

Sénèque, O., Crouzy, S., Boturyn, D., Dumy, P., Ferrand, M., & Delangle, P. (2004). Novel model peptide for Atx1-like metallochaperones. *Chemical Communications*, 770–771.

Shults, M. D., Pearce, D. A., & Imperiali, B. (2003). Modular and tunable chemosensor scaffold for divalent zinc. *Journal of the American Chemical Society, 125*, 10591–10597.

Sigel, H., & Martin, R. B. (1982). Coordinating properties of the amide bond. Stability and structure of metal ion complexes of peptides and related ligands. *Chemical Reviews, 82*, 385–426.

Solomon, E. I., Heppner, D. E., Johnston, E. M., Ginsbach, J. W., Cirera, J., Qayyum, M., et al. (2014). Copper active sites in biology. *Chemical Reviews, 114*, 3659–3853.

Späth, J., Stuart, F., Jiang, L., & Robinson, J. A. (1998). Stabilization of a β-hairpin conformation in a cyclic peptide using the templating effect of a heterochiral diproline unit. *Helvetica Chimica Acta, 81*, 1726–1738.

Starck, M., Sisommay, N., Laporte, F. A., Oros, S., Lebrun, C., & Delangle, P. (2015). Preorganized peptide scaffolds as mimics of phosphorylated proteins binding sites with a high affinity for uranyl. *Inorganic Chemistry, 54*, 11557–11562.

Tuchscherer, G., Lehmann, C., & Mathieu, M. (1998). New protein mimetics: The zinc finger motif as a locked-in tertiary fold. *Angew. Chem. Int. Ed., 37*, 2990–2993. Angewandte Chemie, 110, 3160–3164.

Vogel, A. I. (1989). *Vogel's textbook of quantitative chemical analysis* (5th ed.). Harlow: Longman Scientific & Technical.

Vojkovsky, T. (1995). Detection of secondary amines on solid phase. *Peptide Research, 8*, 236–237.

Yang, C.-T., Han, J., Gu, M., Liu, J., Li, Y., Huang, Z., et al. (2015). Fluorescent recognition of uranyl ions by a phosphorylated cyclic peptide. *Chemical Communications, 51*, 11769–11772.

CHAPTER SIXTEEN

De Novo Construction of Redox Active Proteins

C.C. Moser, M.M. Sheehan, N.M. Ennist, G. Kodali, C. Bialas, M.T. Englander, B.M. Discher, P.L. Dutton[1]

University of Pennsylvania, Philadelphia, PA, United States
[1]Corresponding author: e-mail address: dutton@mail.med.upenn.edu

Contents

1. Introduction	366
2. Sequence Selection: Binary Patterning	367
3. Sequence Selection: Cofactor Self-Assembly	370
4. Binding Heme Redox Cofactors	371
5. Electron-Transfer Reactions of Heme Maquettes	374
6. Binding Light-Activated Zn Tetrapyrrole Cofactors	377
7. Iron–Sulfur Cluster Cofactors	378
8. Redox Metal Binding Sites	379
9. Flavin Cofactor Binding and Electron Transfer	380
10. Translating Aqueous Redox Maquette Designs into Membranes	382
References	384

Abstract

Relatively simple principles can be used to plan and construct de novo proteins that bind redox cofactors and participate in a range of electron-transfer reactions analogous to those seen in natural oxidoreductase proteins. These designed redox proteins are called maquettes. Hydrophobic/hydrophilic binary patterning of heptad repeats of amino acids linked together in a single-chain self-assemble into 4-alpha-helix bundles. These bundles form a robust and adaptable frame for uncovering the default properties of protein embedded cofactors independent of the complexities introduced by generations of natural selection and allow us to better understand what factors can be exploited by man or nature to manipulate the physical chemical properties of these cofactors. Anchoring of redox cofactors such as hemes, light active tetrapyrroles, FeS clusters, and flavins by His and Cys residues allow cofactors to be placed at positions in which electron-tunneling rates between cofactors within or between proteins can be predicted in advance. The modularity of heptad repeat designs facilitates the construction of electron-transfer chains and novel combinations of redox cofactors and new redox cofactor assisted functions. Developing de novo designs that can support cofactor incorporation upon expression in a cell is needed to support a synthetic biology advance that integrates with natural bioenergetic pathways.

Methods in Enzymology, Volume 580
ISSN 0076-6879
http://dx.doi.org/10.1016/bs.mie.2016.05.048

© 2016 Elsevier Inc.
All rights reserved.

1. INTRODUCTION

While the natural amino acids Trp, Tyr, and Cys are important participants in radical electron transfer under appropriate conditions (Stubbe & van der Donk, 1998), the vast majority of natural oxidoreductases exploit redox active cofactors to guide electron tunneling between redox centers and manage the electron-transfer reactions at sites of catalysis (Page, Moser, Chen, & Dutton, 1999). By constructing cofactor binding de novo proteins (maquettes) analogous to natural redox proteins, we can better understand the default properties of these cofactors in a protein matrix independent of the complexities introduced by repeated cycles of natural selection and gain insight into how redox proteins may have operated early in the evolution of life. Such designs also help us to isolate and understand the means by which protein can be engineered to manipulate cofactor properties. De novo design also supports the combination of protein cofactors in novel ways to explore functions unseen in natural systems.

A wide range of redox active centers have been successfully incorporated into de novo designed proteins (for review, see Prabhulkar, Tian, Wang, Zhu, & Li, 2012; Zastrow & Pecoraro, 2013). With a view of potential synthetic biology engineering (Akhtar & Jones, 2014), this work will concentrate on the design and construction of de novo proteins that bind natural redox cofactors and offer the potential for being exploited for intracellular expression and redox cofactor assembly to interface with electron-transfer pathways within the cell.

To be most useful, these de novo protein frames need to be large enough to surround the redox cofactors within a stable protein fold that is tolerant to multiple, diverse changes in amino acid sequence. Furthermore, to make interpretation of the effects of the protein environment on cofactor function as clear as possible, the folding and amino acid sequence should be relatively elementary, selected according to basic principles of design with minimal reference to complex natural protein sequences. In our experience, de novo designed 4-helix bundles based on a binary patterning of hydrophobic and hydrophilic amino acids fulfills these requirements for a large number of redox cofactors, ranging from redox active iron tetrapyrroles such as heme, light active zinc tetrapyrroles, iron sulfur clusters, flavins, and quinones (Farid et al., 2013; Gibney, Mulholland, Rabanal, & Dutton, 1996; Hay, Westerlund, & Tommos, 2007; Lichtenstein et al., 2015; Robertson et al., 1994; Sharp, Moser, Rabanal, & Dutton, 1998). By linking all helices into a single chain via short, flexible loops, bundle assembly under in vivo conditions is simplified.

2. SEQUENCE SELECTION: BINARY PATTERNING

First-principles protein folding design exploits the hydrogen-bonding pattern of the peptide backbone in an alpha helix that completes two full helical turns for every seven residues. Imposing a binary hydrophobic/polar pattern on each heptad of amino acids in a de novo sequence enables helices to fold and stabilize by self-associating with other helices to bury the nonpolar residues in the bundle core (Fig. 1). Traditionally, based on the sequence of the early structurally resolved fibrous 2-helix coiled–coil protein tropomyosin (McLachlan & Stewart, 1975), the heptad positions are lettered sequentially from *a* through *g*, with the *a* and *d* positions associated with core hydrophobic residues.

When the heptads include a third hydrophobic residue, in either the *e* or *g* position, either HPPHHPP or HPPHPPH where H are hydrophobic residues and P polar residues, 4-helix bundles are stabilized with either positions *ade* or *adg* in the core (Liu et al., 2007). The extremely elementary sequence of the three amino acids glutamate (E), leucine (L), and lysine (K) in a 25 amino acid LEELLKK heptad repeat, has a seven turn, 40 Å helix length appropriate for core burial of multiple redox cofactors extending the sequence of a relatively simple earlier de novo bundle forming helix sequence (Ho & DeGrado, 1987). The binary patterned helices are linked together into a single chain with glycine (G) and serine (S) rich loops such as GGSGSGSGG. Single-chain designs have strong advantages over previous bundle designs of symmetric tetramers and dimers, not only because they favor exclusive folding into 4-helix assemblies, but also because they afford greater design freedom in asymmetric sequence changes while engineering the environment of incorporated redox cofactors. Simple sequences such as these can provide extremely robust helical bundle frames that resist unfolding even at near boiling temperatures. For spectroscopic convenience of monitoring maquette concentration, a Trp residue with absorbance at 280 nm is usually included in the sequence, for example replacing the middle S within the second loop.

In the assembly of a single-chain 4-helix bundle, we employ two distinct topologies of helical threading that pack nonpolar residues in the core. Loops connecting helices can proceed circumferentially around the 4-helix cylinder (Fig. 1, *left*) such that helices in contact have an antiparallel arrangement

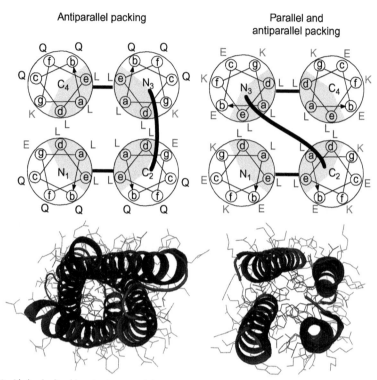

Fig. 1 Alpha helical heptad repeat binary patterning leads to 4-helix bundle association with two distinct topologies: a fully antiparallel arrangement (*left*) or a combination of parallel and antiparallel adjacent helices (*right*) with positive and negatively charged K and E residues shown in *blue* (*light gray* in the print version) and *red* (*gray* in the print version), respectively. 4-Helix bundle topology is viewed from the amino (N_1) end of the first of four helices connected by loops (*black lines*) with the heptad positions *a*, *d*, and *e* comprised predominantly of nonpolar residues (*purple* (*gray* in the print version)) that self-associate to form the bundle core. Bottom right: an X-ray crystal structure of an extended antiparallel bundle for four redox centers based on a LQQLLQX motif; left a mixed parallel/antiparallel packing based on an LEELLKK motif (Huang, Gibney, Stayrook, Dutton, & Lewis, 2003).

and residues at position *d* in the heptad form potential packing interactions with corresponding *d* residues in adjacent helices offset along the bundle long axis nearly a full helical turn. Similarly, *e* position residues interact with adjacent *e* residues. A binary patterned heptad repeat that obeys this pattern is LQQLLQX where Q is Gln and X is E for one sequential pair of helices and K for the other pair. This fully antiparallel arrangement is common in natural helical coiled–coil proteins. Knobs and holes or ridges and grooves formed by large and small protruding core residues in adjacent helices help to stabilize the bundle structure as helices twist around one another in a left-handed super-helical pitch down the central axis of the bundle. This left-handed twist supports a helical pitch of 3.5 residues per alpha helical turn.

In our alternative single-chain topology, there is a mix of parallel and antiparallel helical neighbors (Fig. 1, *right*). The second loop does not proceed along the perimeter of the bundle, but crosses the central axis with a zigzag threading. This topology tends to be favored with the binary patterned heptad repeat LEELLKK. Helical interfaces are more heterogeneous. Instead of the two types of helical interfaces, as in the all antiparallel bundle, there are now three types of helical interfaces. Knob and hole packing in the parallel helices 1 and 3 as well as 2 and 4 have core residues in the *d* position interacting with *e* position residues in the adjacent helix (Huang et al., 2003). Abundant Glu and Lys residues can form stabilizing, charge neutralizing intrahelical salt bridges. However, this topology also supports stabilizing interhelical salt bridges between oppositely charged *b* and *g* position residues. Between antiparallel helices 1 and 2 the *d* residues interact with each other. Between helices 3 and 4 the *e* residues are close enough to interact. Helices in this bundle topology are observed to be straighter with less coiling than a traditional coiled coil (Huang et al., 2003; Skalicky, Gibney, Rabanal, Bieber Urbauer, & Dutton, 1999). With less coiling, the alpha helical pitch increases from 3.5 to about 3.6 residues per turn. At this pitch, if enough turns are added to each helix, the bundle core heptad residues will gradually shift from *adg* to *abe*.

For any maquette sequence that obeys binary patterning, negative design can make the difference between folding into one or the other helical threading (Betz & DeGrado, 1996; Marsh & DeGrado, 2002; Summa, Rosenblatt, Hong, Lear, & DeGrado, 2002), for example, by arranging for favorable E–K salt bridges between adjacent helices in one threading and/or electrostatic repulsion between similarly charged residue in another threading or by including core residues of different sizes that will pack differentially in parallel and antiparallel orientations. Fig. 2 illustrates the conflicting electrostatic interactions between helices with swapped helical threading of the heptad repeats of Fig. 1. Negative design can also suppress multiple interchangeable conformations on the residue level and support folding into singular structures that simplify experimental interpretation of the manipulation of cofactor properties.

Maquette helical bundle structures are generally highly tolerant to changes in amino acid sequence. For example, exterior polar residues can be changed to adjust the net charge from highly negative (-14) to near

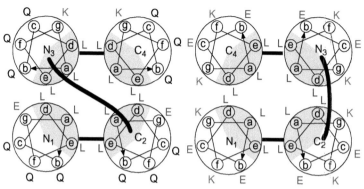

Fig. 2 Negative design, such as electrostatic clash between nearby similarly charged Lys (K) or Glu (E) residues energetically disfavors bundles shown in Fig. 1 from threading into alternate helical bundle topologies shown here.

neutral to positive (+11), which can be exploited to electrostatically manipulate the redox midpoint potential of bound cofactors or to diversify the polar amino acids to simplify NMR structure analysis. Maquette bundles are also generally forgiving to changes in core residues, tolerating departures from strict binary pattering while still displaying robust folding and self-assembly. Most importantly, maquettes are readily adapted by introducing residues to covalently attach redox cofactors and have been shown to be highly malleable to the introduction of cofactors of a wide range of shapes and sizes.

3. SEQUENCE SELECTION: COFACTOR SELF-ASSEMBLY

Cells often employ ligases and accessory maturation proteins to insert cofactors into specific locations with a specific geometry in natural proteins. Although we have observed accessory proteins recognizing and interfacing with unnatural maquette partners (for example, Anderson et al., 2014), in general maquettes are intentionally designed without sequence similarity to any natural protein, and productive interaction with accessory proteins inside the cell cannot be taken for granted. Instead de novo redox proteins are designed with an underlying physical chemistry that supports cofactor self-assembly, both in vivo and in vitro.

For sufficiently nonpolar cofactors, such as heme, simple hydrophobic partitioning is adequate to insert cofactors into a helical bundle core (Solomon, 2013; Solomon, Kodali, Moser, & Dutton, 2014). However, in practice, redox maquettes exploit the chemistry of nitrogen and sulfur of His and Cys to covalently secure the cofactors into specific designed positions.

4. BINDING HEME REDOX COFACTORS

Natural proteins frequently use His to anchor the central metal redox active iron in heme tetrapyrroles, either in a bis-His ligation, as in cytochromes b, or a single His ligation, as in hemoglobin (Reedy & Gibney, 2004). Indeed, during in vivo expression in *Escherichia coli*, maquettes can bind cellular heme, especially if heme synthesis is stimulated by addition of heme precursor aminolevulinic acid to the grown medium. When relatively nonpolar tetrapyrrole cofactors such as heme are added to maquettes in vitro, cofactor aggregation can compete with cofactor binding. To minimize aggregation, tetrapyrrole cofactors are added in successive 0.2 equivalent aliquots from an ~ 1 mM stock solution in DMSO at intervals of 10 min with stirring. Maquette concentrations are typically tens of μM. After an excess of heme is added, unbound heme cofactor and DMSO are then removed by passing through a PD-10 G25 desalting column (GE). Bis-His heme binding is confirmed by conspicuous absorption shifts of the Soret and alpha bands, greater than the subtle bandshifts seen on heme hydrophobic partitioning. For binding affinities measurements, stock heme concentration is confirmed by the hemochrome assay (Berry & Trumpower, 1987). For tetrapyrrole binding titrations in general, bandshift spectra on binding, even if subtle, can be analyzed via singular value decomposition analysis fit to a binding model for accurate K_D values.

Many histidine heme binding de novo helical bundle designs have been explored, from symmetric tetramers (Robertson et al., 1994) to dimers (Ghirlanda et al., 2004) to combinatorial libraries (Moffet et al., 2000). To help minimize sequence complexity, we apply two fundamental principles to single-chain helical bundles based on the elementary LEELLKK and LQQLLQX motifs to secure heme cofactor binding. The first is to simply replace the first interior *a* position leucine with His (H) on each helix to provide the thermodynamically favorable iron ligation by histidine nitrogen. In the elementary maquette sequence based solely on the LEELLKK heptad with four identical helices joined by loops, this in itself is sufficient to secure μM heme binding affinity with no attempt to provide a preformed cavity. The second principle is to provide a primitive accommodation for bound cofactor bulk in the core by shortening Leu to Ala in adjacent helices near the cofactor ligating residues while bulking up core Leu to Phe in remote positions. This second step increases heme affinity by more than a thousand.

In contrast to the fully antiparallel helical topology, bis-His heme ligation between helices 1 and 3 and between 2 and 4 in the mixed antiparallel/parallel helix threading requires helical rotation to appropriately orient *a* position His residues toward heme iron (Fig. 3). With this rotation, some binary patterned residues are partly buried while others are partly exposed, putting strain on the iron ligating histidines. While such strain may be useful for designs that encourage ligand exchange, such as His displacing O_2 binding in oxygen transport (Farid et al., 2013; Koder et al., 2009), we normally avoid excessive hydrophobic residue exposure by replacing *D* position Leu near the end of helices 1 and 2 with a polar residue such as Glu.

In these designs, packing of the maquette core adjusts around bis-His ligated hemes with a diversity of peripheral groups in addition to heme B protoporphyrin IX. These include mesoporphyrin IX, deuteroporphyrin IX, etioporphyrin, isohematoporphyrin, 2,6-diacetyl deuteroporphyrin IX, 2,6 dinitrile porphyrin, and even bulkier groups such as in tetracarboxyphenyl porphyrin or heme A with a farnesyl polyisoprene tail (Gibney et al., 2000; Solomon et al., 2014). By placing a periplasm export tag on the N-terminus of mixed parallel/antiparallel maquettes and using a CXXCH sequence motif for one of the two His groups at a heme binding site, during *E. coli* expression this maquette will interact with the natural cytochrome *c* maturation system and covalently link the Cys groups to heme vinyls creating a heme C (Anderson et al., 2014).

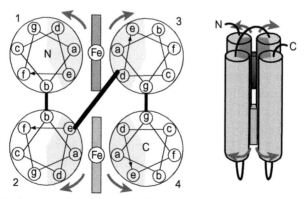

Fig. 3 In mixed parallel/antiparallel bundles, helices are rotated to orient *a* position histidines for bis-His ligation of heme iron. Some binary patterned residues are partly buried while others are partly exposed, putting strain on the iron ligating histidines.

Maquettes also provide a robust frame for creating alternative six-coordinate heme-metal ligation patterns in otherwise similar protein environments by changing just one or two amino acids. Table 1 and Fig. 4 show a variety of coordinate ligations of heme between diagonal helices 1 and 3 in the antiparallel maquette of Fig. 1. These include His-Cys, His-Met, His-Tyr, His-Lys, and His-Ala (Table 1 and Fig. 3, *left*). In general, sites designed to be single His and five coordinate ligation bind heme only weakly. This discrimination can be exploited for site specificity when creating mixed Fe and Zn tetrapyrrole designs as the two tetrapyrroles, respectively, prefer bis-His and single His ligation patterns. On the other hand, in the absence of His, placing a single Cys in maquette cores opposed by a non-coordinating group, akin to the Cys-Ala pattern seen in natural P-450 enzymes, leads to secure heme binding (Fig. 4, *right*).

Table 1 A Single Maquette Frame Allows Change of Heme Ligating Amino Acid While Leaving the Rest of the Sequence Unchanged

Ligation Pattern	Natural Protein	Abs Peaks Natural (nm)	Abs Peaks Maquette (nm)	Em (mV)
His — His	Neuroglobin	426, 560 (red) 413, 529 (ox)	426, 560 (red) 412, 529 (ox)	−260 to −150
His — Cys	Rr CooA, Cysathione b-synthase	422, 540, 570 360, 424, 535	421, 540, 570 360, 424, 535	−290 to −150
His — Met	b562 Cyt c	430, 562 406, 534	429, 562 409, 529	70
His — Tyr	Bovine liver catalase	436, 558 406, 534, 620	427, 558 407, 530, 615	−180
His — Lys	Cyt f (amino)	421, 554 412	427, 532, 560 414	−65
His — Ala	Myoglobin	435, 555 409, 505, 635	428, 530, 562 416, 531	−110
Cys — Ala	P450, NOS	410, 551 391, 635	430, 555 387, 635	−190

Several natural protein heme iron ligation motifs are compared with maquette counterparts. The His-Lys pattern resembles the His-amino terminus pattern of cytochrome *f* (Ponamarev & Cramer, 1998).

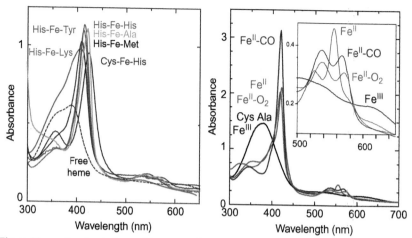

Fig. 4 Maquettes ligate heme B in various ligation schemes, using both amino acids and diatomic ligands. (*Left*) Spectra of oxidized heme B ligated proximally with His and distally ligated His, Cys, Met, Tyr, or Lys in the antiparallel bundle of Fig. 1. For comparison, unbound heme spectrum is dashed. (*Right*) Spectra of reduced heme B proximally ligated with Cys with the distal position occupied by Ala, a nonligating amino acid residue, binds a range of diatomic ligands as in natural protein gas sensors; shown here are CO and O_2.

Unlike the mixed parallel/antiparallel design, the diagonal ligation of bis-His sites in the antiparallel bundle does not lead to helical strain on heme binding and is relatively resistant to displacement of His by added diatomic ligands such as CO and O_2. Both bis-His heme ligation in a mixed parallel/antiparallel bundle with helical strain (Farid et al., 2013), and Cys-Ala in either the mixed parallel/antiparallel bundle or the antiparallel helical bundle without strain are open to binding diatomic ligands to the Fe sixth coordination site (Fig. 4, *right*).

5. ELECTRON-TRANSFER REACTIONS OF HEME MAQUETTES

At rates that vary considerably according to details of maquette design and the redox potential of the heme, reduced heme will bind and eventually reduce O_2 (Farid et al., 2013). Indeed, when a continuous supply of reducing equivalents is supplied to the bis-His heme site of the mixed parallel/antiparallel bundle by anchoring the maquette to a gold electrode surface and performing cyclic voltammetry, a clear catalytic wave of multiple

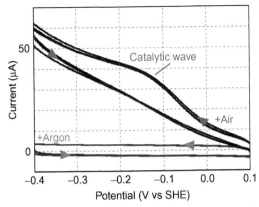

Fig. 5 Cyclic voltammetry of an O_2 binding bis-His heme maquette immobilized on an electrode reveals a catalytic wave of O_2 reduction near the redox midpoint potential of the heme after air is introduced. Here dithiobismaleimidoethane (DTME) is used to anchor a Cys in the heme maquette loop to a 1 cm^2 gold electrode surface. The sweep rate is 0.01 V/s.

turnover O_2 reduction can be seen (Fig. 5). Peak catalytic reduction current occurs as expected around the redox midpoint potential of the heme with at least 50 catalytic turnovers per maquette.

Interprotein electron transfer takes place when heme maquettes are mixed with natural redox proteins such as cytochrome *c* (Fry, Solomon, Dutton, & Moser, 2016) or with other heme maquettes (Solomon, 2013). Electron transfer upon stopped-flow mixing is easily monitored when the donor and acceptor hemes have conspicuously different visible spectra, such as when a reduced lower redox potential heme B maquette is mixed with a higher potential heme C containing cytochrome *c*, or when a reduced heme B maquette is mixed with a red shifted, high potential 2,6-diacetyl deuteroporphyrin IX maquette.

In the absence of obligatory and rate limiting proton-transfer reactions, intraprotein electron transfer between fixed cofactors placed within a single maquette is often controlled by the rate of electron tunneling. Because this rate has an exponential dependence on the edge-to-edge distance between cofactors (Moser, Keske, Warncke, Farid, & Dutton, 1992), the dynamics of an electron-transfer system can be engineered by selecting the position of the cofactor anchoring amino acid along the bundle helices (Fig. 6). For porphyrin tetrapyrrole cofactors anchored with *a* position residues four helical turns apart, this is about 16 Å edge-to-edge and leads to ∼1 ms electron-tunneling times at low driving force, analogous to transmembrane heme-to-heme

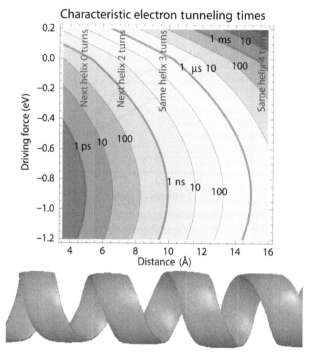

Fig. 6 Expected characteristic times for intraprotein electron tunneling for a range of driving forces, given by the difference in redox midpoint potentials between donor and acceptor, and for a range of edge-to-edge distances between redox cofactors, which is modulated in maquettes by the positions of the cofactor anchoring amino acids along the alpha helix (*bottom*). Here the reorganization energy for electron transfer is a typical 0.9 eV.

electron transfer in cytochrome bc_1. Larger driving force electron transfer at this distance can be as fast as 10 μs. At three helical turns, this distance shrinks to around 9 Å and rates can be expected to be four orders of magnitude faster. To position tetrapyrrole cofactors closer together for more rapid electron tunneling, cofactors need to be anchored to different helices to avoid steric clash.

Both intraprotein and interprotein heme maquette electron-transfer rates are resolved by light activation. For example, photolysis of CO bound to reduced maquette heme frees the heme for electron transfer with redox partners (Fry et al., 2016). With somewhat less control over electron-transfer distances, externally attached, light activatable redox centers such as bipyridyl ruthenium have been attached to external His or Cys. Best control is offered by light activation of redox cofactors within the maquette core, such as a flavin (Sharp, Moser, et al., 1998) or a Zn tetrapyrrole (Farid et al., 2013).

6. BINDING LIGHT-ACTIVATED ZN TETRAPYRROLE COFACTORS

Light activatable Zn tetrapyrroles, including Zn porphyrins, chlorins, and bacteriochlorins, need only one amino acid ligand for ligation to the maquette frame. In natural proteins, most light activatable chlorins and bacteriochlorins are either a noncovalently bound free base, or ligate a Mg metal, which is more abundant than Zn under nonacidic conditions, using His or Lys. Zn bacteriochlorins are found naturally in the acidophilic photosynthetic bacterium *Acidiphilium rubrum* and various Zn tetrapyrroles can be found in other organisms with altered or absent Mg-chelatases (Jaschke, Hardjasa, Digby, Hunter, & Beatty, 2011). Indeed, there is speculation that heme containing cytochromes *b* may have an evolutionary link with early photosynthesis (Xiong & Bauer, 2002) and that in the absence of a Mg-chelatase, Zn replacement of iron may have supported a light-activated role in early life (Jaschke et al., 2011).

Zn tetrapyrroles are bound to maquettes in vitro from stock DMSO solutions much the same as the procedure reported above with Fe tetrapyrroles. Bis-His sites can bind one or two Zn tetrapyrroles as well (Cohen-Ofri et al., 2011; Sharp, Diers, Bocian, & Dutton, 1998). For this reason, in mixed iron/zinc tetrapyrrole maquette designs, a reasonably tight binding heme should be added to fill a bis-His site to a near 1:1 stoichiometry before binding Zn tetrapyrrole. Similarly, maquette interiors accommodate a wide range of peripheral substituents and ring sizes for Zn tetrapyrroles. As a practical matter, the general pattern for rapid and strong binding of Zn tetrapyrroles to maquettes while avoiding counterproductive cofactor aggregation is to have a mix of hydrophobic and hydrophilic peripheral substituents, preferably arranged in an opposing, amphiphilic pattern.

In redox cofactor dyad maquettes with a bis-His site first bound with heme and a second His site four turns away bound with a Zn porphyrin, light activation results in photoreduction of the heme on the 1 ms timescale (Farid et al., 2013). Light-activated electron transfer from Zn chlorin is faster, ~50 μs, presumably because of the change in driving force and shorter electron-tunneling distance with the larger chlorin conjugated ring system (Farid et al., 2013). In these dyads, charge separation has nearly the same energy as charge recombination, which leads to comparable rates and transient populations of the charge-separated state of about 30%. Like natural systems, we introduce a third redox cofactor for increased distance of charge separation to further stabilize light-activated charge-separated states.

7. IRON–SULFUR CLUSTER COFACTORS

Another common and apparently evolutionary ancient type of redox center are iron–sulfur clusters (Blöchl, Keller, Wachtershäuser, & Stetter, 1992). In our experience, Fe_4S_4 cubane clusters readily insert into Cys rich maquette loops containing a ferredoxin consensus motif CIGCGAC. Indeed, a short and flexible 16 amino acid sequence KLCEGGCIACGACGGW will spontaneously incorporate Fe_4S_4 clusters (Gibney et al., 1996) as well as other metals such as Co (Kennedy, Petros, & Gibney, 2004). This sequence was inspired by short hexadecapeptide (Frausto da Silva & Williams, 2001) akin to sequences found in natural proteins.

Fe_4S_4 cluster insertion into apo-maquette proteins is carried out under anaerobic, reducing conditions (added 2-mercaptoethanol) such that all Cys residues are reduced. $FeCl_3$ and Na_2S are slowly added to sixfold stoichiometric excess. Excess reagents are then removed by dialysis or PD-10 G25 buffer exchange column. Redox potentials of these clusters are pH dependent and in the -225 to -375 mV range (Kennedy & Gibney, 2002). Loop insertion leaves the helices free to bind other cofactors such as heme (Gibney et al., 1996).

In nature, an FeS cluster ligating Cys sulfur is sometimes used to ligate additional metals to form extended clusters, manipulating redox potentials, and facilitating catalysis. This strategy was used to modify the maquette frame just described by introducing a metal ligating cluster of three His residues in the core adjacent the Cys rich loop. This enabled the Fe_4S_4 cluster to bridge to a Ni atom upon addition of $NiCl_2$ (Laplaza & Holm, 2001, 2002) to make a cofactor analogous to carbon monoxide dehydrogenase.

Fe_4S_4 clusters insert into de novo designs with single-strand beta-sheet structures analogous to rubredoxin (Nanda et al., 2005). Redox potential of these clusters are relatively high, about 55 mV. Even though most FeS clusters in nature form in loops or beta structures, FeS clusters also assemble in the core, as seen in a computationally designed single-chain 4-helix bundle (Grzyb et al., 2010). A pair of core Fe_4S_4 clusters also form in dimeric 3-helix bundle designs (Roy, Sarrou, Vaughn, Astashkin, & Ghirlanda, 2013; Roy et al., 2014). When the designed edge-to-edge distance between FeS clusters is narrowed to around 9 Å, Fig. 6 suggests intraprotein electron-tunneling time in this dimer should be about 100 μs, fast enough in principle for two-electron transfer to an appropriate acceptor. Intraprotein electron transfer is observed on mixing with the single-electron acceptor oxidized cyt c_{550} (Roy et al., 2014).

8. REDOX METAL BINDING SITES

Besides binding FeS clusters, Co and Ni, Cys and His groups buried in close proximity in de novo helical bundles can bind a host of other redox active metals (for a review, see Tegoni, 2014). Single-chain 4-helix bundle copper centers are promising for future in vivo work. Using a combinatorial approach of template-assisted synthetic protein (TASP) in which four alpha helices were assembled on a cyclic peptide base with various combinations of a Cys and several His, it was found that about a third of the combinations bound Cu^{2+} and most bound Co^{2+} (Schnepf et al., 2001). Spectroscopic evidence suggested a Cu ligation geometry intermediate between the tetragonal type 2 natural copper proteins and the trigonal type 1 copper proteins, called type 1.5. Subsequently, a single-chain 4-helix coiled coil with two buried *a* position His and one *d* position Cys binds the Cu of added $CuCl_2$ under anaerobic conditions, displaying a thiolate to Cu charge transfer absorption responsible for the blue color of type 1 copper centers (Shiga et al., 2010). Nearby core Leu residues replaced Ala to provide space for the metal ions. This redox center has a K_D of ~2 μM and a midpoint potential of 328 mV. In another variant of this frame, the burial of four *a* and *d* position His on two helices four residues apart and two Cys on a third helix also four residues apart, lead to the binding of Cu in a binuclear purple copper center (Shiga et al., 2012) akin to the Cu site of cytochrome *c* oxidase.

Besides His and Cys, Glu is an effective ligand for redox active metals in a de novo helical bundle core. For example, a His-His-Glu variant of the blue copper protein just described binds Cu with similar μM affinity in an apparently distorted equatorial geometry (Shiga et al., 2009). Burial of two His with four Glu in a dimeric 4-helix bundle supports the assembly of a dimetal binding site (Faiella et al., 2009), including redox active Fe. Electron-transfer activity is evident as diferrous centers bind and reduce O_2 through an oxobridged intermediate. However, stability and activity of this liganding assembly requires relatively sophisticated design of second shell residues capable of hydrogen-bonding interactions with the primary liganding residues.

9. FLAVIN COFACTOR BINDING AND ELECTRON TRANSFER

Quinones and flavins are common one- or two-electron redox cofactors in natural biological systems. In natural quinone and flavin redox proteins, these organic redox cofactors are usually noncovalently bound. Binding interactions often employ nonredox active portions of the cofactor, such as the quinone polyisoprene tail or the bases, sugars, or phosphates of natural flavins FAD or FMN. Quinones have been covalently secured in de novo helical bundles by means of native chemical ligation using an unnatural quinone amino acid (Lichtenstein et al., 2015). However, most de novo quinoprotein designs exploit the ready ability of the chemically active unsubstituted ring carbons of p-benzoquinones and naphthoquinone to form a covalent thioester bond with Cys residues (Snell & Weissberger, 1939). These include a TASP bundle (Li, Hellwig, Ritter, & Haehnel, 2006) and a 3-helix bundle (Hay et al., 2007). Yet in natural proteins, covalently attached quinones are not anchored via thioester bonds in natural proteins. Instead covalent quinones are usually formed by modifications of Tyr or Trp, such as the tryptophan tryptophylquinone cofactor formed by cross-linking two Trp residues in methylamine dehydrogenase. In situ generation of quinone or perhaps specific incorporation of quinones via orthogonal tRNA-aminoacyl tRNA (Alfonta, Zhang, Uryu, Loo, & Schultz, 2003; Wang & Schultz, 2005) may be the best path for trying to engineer in vivo biogenesis of de novo quinone proteins.

On the other hand, some natural flavoproteins do covalently attach flavins via His, Tyr, or more commonly, a thioether linkage with Cys at one of several positions on the flavin cofactor (Edmondson & Newton-Vinson, 2001). For example, monoamine oxidase has a thioether linkage at the flavin 8α methyl position while dimethylamine dehydrogenase has a thioether linkage at the C6 position (Mewies, McIntire, & Scrutton, 1998). These patterns suggest that biogenesis of de novo thioester flavoproteins may be tractable. Flavins display a rich variety of natural reactions, including light-activated electron-transfer reactions in catalysis and light and magnetic field sensing (Conrad, Manahan, & Crane, 2014) which make them attractive cofactors for incorporating in de novo designs.

We have experience with the thioester linkage of flavins to Cys residues in de novo helical bundles both directly to the flavin ring at the 8 position (Farid et al., 2013) and via an acetyl at the flavin C7 position (Sharp, Moser,

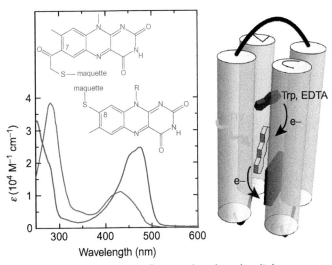

Fig. 7 (*Left*) Oxidized lumiflavin or riboflavin with a thioether linkage to a core Cys at either a C7 (*blue* (*gray* in the print version)) or C8 (*red* (*dark gray* in the print version)) position display distinct UV–vis spectra. (*Right*) Light activation of flavin allows for oxidation of a core Trp or the sacrificial electron donor EDTA. The subsequent flavin semiquinone can reduce neighboring heme.

et al., 1998) (Fig. 7). These two covalent linkages generate flavins with conspicuously different optical properties. Linkage via an acetyl at the 7 position of the flavin is achieved by brominating 7-acetyl-10-methylisoalloxazine (Levine & Kaiser, 1978; Sharp & Dutton, 1999). Maquettes with an *a* position Cys in the core are added to a fivefold excess of brominated flavin while dissolved in 50% DMF in water for 3 h and then dialyzed to remove DMF and unreacted flavin, purified by reverse-phase HPLC (C18 column, H_2O +0.1% TFA aqueous phase and acetonitrile +0.1% TFA organic phase), flash frozen in liquid nitrogen and lyophilized (Sharp & Dutton, 1999). Bis-His sites spaced one heptad away from the Cys are free to bind DMSO solubilized hemes or other tetrapyrroles.

Light-activated flavin is a powerful oxidant. It can extract electrons from Trp, Tyr, or amines such as EDTA to form one-electron reduced flavin semiquinone. Under prolonged illumination, the flavin becomes fully reduced. However, because the redox midpoint potential of the coupled seven acetyl flavin (-95 mV) is more positive than the midpoint potential of ordinary heme (-153 mV), in order to observe intraprotein photoreduction of heme by flavin, the bis-His sites are bound with a higher potential heme, 1-methyl-2-oxomesoheme XIII (-30 mV) (Sharp, Moser, et al.,

1998). Upon light-induced abstraction of an electron from EDTA, the flavin semiquinone intermediate reduces the heme in 100 ns (Sharp, Moser, et al., 1998). This time is consistent with the expected second-order interaction between excited flavin triplet and EDTA while significantly slower than the ~1 ns electron tunneling expected for a heme and a flavin packed in the bundle core.

Direct attachment of a maquette bundle *a* position Cys sulfur to the flavin ring at position 8 is accomplished by first synthesizing 8-bromo-riboflavin (Mansurova, Simon, Salzmann, Marian, & Gärtner, 2013). A typical 5 ml coupling reaction contains ~300 μM maquette that was expressed in *E. coli* and cleaved of its His tag after Ni-NTA resin purification in 20 mM NaH$_2$PO$_4$, 0.5 M NaCl, 10 mM imidazole buffer. A 3–5 M excess of 8-bromoriboflavin (in DMF, final DMF concentration not to exceed 20%) is added with TCEP at a final concentration of 1.25 mM. Cys reductants such as beta mercaptoethanol or dithiothreitol are avoided as they displace bromine and couple directly to flavin. The pH of the solution was adjusted to nine with HCl to assure Cys deprotonation, and stirred overnight at 50°C while protected from light. The flavomaquette is purified by reverse-phase HPLC (C18 column, H$_2$O +0.1% TFA aqueous phase and acetonitrile +0.1% TFA organic phase), flash frozen in liquid nitrogen and lyophilized. After dissolving in 20 mM KH$_2$PO$_4$, 100 mM KCl, pH 7.5, flavin incorporation is confirmed by mass spectrometry and UV-visible spectroscopy (absorption maximum at 475 nm). Typical redox potentials of these flavins are around −100 mV. These flavomaquettes show a multitude of electron-transfer reactions. They are reduced by NADH and catalyze electron transfer from NADH to O$_2$ under aerobic conditions. In the presence of bound heme and EDTA they also photoreduce heme under continuous illumination (Farid et al., 2013). If Trp residues are placed close (~7 Å) to the light-activated flavin, they can be photooxidized in tens of ps (Fig. 7).

10. TRANSLATING AQUEOUS REDOX MAQUETTE DESIGNS INTO MEMBRANES

Redox maquette designs can be adapted for a membrane environment. This offers the possibility to integrate with bioenergetic membranes in synthetic biology or act as novel neuronal membrane potential reporters. However, it is a challenge to translate the method of binary patterning to drive aqueous single-chain 4-helix bundle assembly into membrane spanning electron-transfer proteins (Discher, Moser, Koder, & Dutton, 2003;

Discher et al., 2005). The exterior polar interactions that stabilize the water-soluble four-helix bundles are energetically very unfavorable within the hydrophobic core of lipid membranes and need to be replaced by hydrophobic interactions. Compared to water-soluble proteins, membrane spanning residues between helices tend to be small; without an added hydrophobic effect of core burial within the membrane, the entropic advantage of packing small side chains with few rotatable bonds becomes more conspicuous (Walters & DeGrado, 2006).

The most common helical assembly motif in natural transmembrane proteins is an antiparallel packing with a small left-handed crossing angle between helices (Walters & DeGrado, 2006). Hydrogen bonding plays a role in stabilizing transmembrane helical bundles (Adamian & Liang, 2002) but not a dominant one (Bowie, 2011). Membrane protein design often relies on computational optimization of tight interhelical packing (Bender et al., 2007) or molecular simulations (Korendovych et al., 2010). However, if additional helix associating design elements are added (Goparaju et al., 2016), the transmembrane helices can be simply built from "generic" transmembrane leucine rich sequences with a 4:1 ratio of Leu to Ala and with aromatic residues that have lower insertion energies (Wimley & White, 1996) placed at the membrane interface. Our helix associating elements are a binary patterned extramembrane region to constrain the helical assembly and the enthalpic forces of redox cofactor covalent binding, such as bis-His ligation of heme (Goparaju et al., 2016) (Fig. 8).

For ease of purification, the desired sequence for each maquette is usually extended at the N-terminus with 6×-Histag and TEV protease cleavage site. Sequence encoding a GGDG loop is typically added between the His-tag and TEV protease cleavage sites to enhance TEV cleavage efficiency. The final amino acid sequence then undergoes DNA 2.0 algorithms for codon optimization, back translation, and primer design for insertion into plasmids. We select plasmids that produce fusions with maltose-binding protein (MBP) (pMal vectors) or Δ^5-3-ketosteroidisomerase (KSI) (pET31 vectors). Fusing transmembrane maquettes with KSI results in aggregation of the fusion protein into inclusion bodies and leads to high expression yields. MBP-maquette fusion results in maquette expression within the bacterial membrane, although with low expression yields.

Membrane localization of de novo redox proteins provides a good opportunity for integrating with the oxidants and reductants of natural bioenergetic pathways of respiration and photosynthesis located in cell and organelle membranes. However, some care may have to be taken with

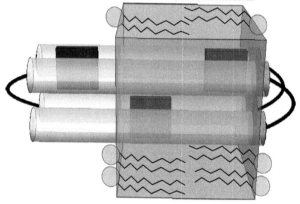

Fig. 8 To aid in helical bundle assembly in a transmembrane design, a binary patterned single-chain maquette is extended with a generic, Leu and Ala rich sequence including helix associating bis-His heme binding sites. Hemes are shown as *dark brown squares* (*gray* in the print version) and the membrane spanning region is *light brown* (*dark gray* in the print version).

maquette designs intended to indiscriminately promote heme-mediated transmembrane electron transfer as there are indications that expression of some of these designs can become deleterious to cell health.

REFERENCES

Adamian, L., & Liang, J. (2002). Interhelical hydrogen bonds and spatial motifs in membrane proteins: Polar clamps and serine zippers. *Proteins, 47,* 209–218.

Akhtar, M. K., & Jones, P. R. (2014). Cofactor engineering for enhancing the flux of metabolic pathways. *Frontiers in Bioengineering and Biotechnology, 2,* 1–6.

Alfonta, L., Zhang, Z., Uryu, S., Loo, J. A., & Schultz, P. G. (2003). Site-specific incorporation of a redox-active amino acid into proteins. *Journal of the American Chemical Society, 125,* 14662–14663.

Anderson, J. L. R., Armstrong, C. T., Kodali, G., Lichtenstein, B. R., Watkins, D. W., Mancini, J. A., et al. (2014). Constructing a man-made c-type cytochrome maquette in vivo: Electron transfer, oxygen transport and conversion to a photoactive light harvesting maquette. *Chemical Science, 5,* 507–514.

Bender, G. M., Lehmann, A., Zou, H., Cheng, H., Fry, H. C., Engel, D., et al. (2007). De novo design of a single-chain diphenylporphyrin metalloprotein. *Journal of the American Chemical Society, 129,* 10732–10740.

Berry, E. A., & Trumpower, B. L. (1987). Simultaneous determination of hemes a, b, and c from pyridine hemochrome spectra. *Analytical Biochemistry, 161,* 1–15.

Betz, S. F., & DeGrado, W. F. (1996). Controlling topology and native-like behavior of de novo-designed peptides: Design and characterization of antiparallel four-stranded coiled coils. *Biochemistry, 35*, 6955–6962.

Blöchl, E., Keller, M., Wachtershäuser, G., & Stetter, K. O. (1992). Reactions depending on iron sulfide and linking geochemistry with biochemistry. *Proceedings of the National Academy of Sciences of the United States of America, 89*, 8117–8120.

Bowie, J. U. (2011). Membrane protein folding: How important are hydrogen bonds? *Current Opinion in Structural Biology, 21*, 42–49.

Cohen-Ofri, I., van Gastel, M., Grzyb, J., Brandis, A., Pinkas, I., Lubitz, W., et al. (2011). Zinc-bacteriochlorophyllide dimers in de novo designed four-helix bundle proteins. A model system for natural light energy harvesting and dissipation. *Journal of the American Chemical Society, 133*, 9526–9535.

Conrad, K. S., Manahan, C. C., & Crane, B. R. (2014). Photochemistry of flavoprotein light sensors. *Nature Chemical Biology, 10*, 801–809.

Discher, B. M., Moser, C. C., Koder, R. L., & Dutton, P. L. (2003). Hydrophilic to amphiphilic design in redox protein maquettes. *Current Opinion in Chemical Biology, 7*, 741–748.

Discher, B. M., Noy, D., Strzalka, J., Ye, S., Moser, C. C., Lear, J. D., et al. (2005). Design of amphiphilic protein maquettes: Controlling assembly, membrane insertion, and cofactor interactions. *Biochemistry, 44*, 12329–12343.

Edmondson, D. E., & Newton-Vinson, P. (2001). The covalent FAD of monoamine oxidase: Structural and functional role and mechanism of the flavinylation reaction. *Antioxidants and Redox Signaling, 3*, 789–806.

Faiella, M., Andreozzi, C., de Rosales, R. T. M., Pavone, V., Maglio, O., Nastri, F., et al. (2009). An artificial di-iron oxo-protein with phenol oxidase activity. *Nature Chemical Biology, 5*, 882–884.

Farid, T. A., Kodali, G., Solomon, L. A., Lichtenstein, B. R., Sheehan, M. M., Fry, B. A., et al. (2013). Elementary tetrahelical protein design for diverse oxidoreductase functions. *Nature Chemical Biology, 9*, 826–833.

Frausto da Silva, J. J. R., & Williams, R. J. P. (2001). *The biological chemistry of the elements* (2nd ed.). Oxford: Oxford University Press.

Fry, B. A., Solomon, L. A., Dutton, P. L., & Moser, C. C. (2016). Design and engineering of a man-made diffusive electron-transport protein. *Biochimica Et Biophysica Acta, 1857*, 513–521.

Ghirlanda, G., Osyczka, A., Liu, W., Antolovich, M., Smith, K. M., Dutton, P. L., et al. (2004). De novo design of a D2-symmetrical protein that reproduces the diheme four-helix bundle in cytochrome bc1. *Journal of the American Chemical Society, 126*, 8141–8147.

Gibney, B. R., Isogai, Y., Rabanal, F., Reddy, K. S., Grosset, A. M., Moser, C. C., et al. (2000). Self-assembly of heme A and heme B in a designed four-helix bundle: Implications for a cytochrome c oxidase maquette. *Biochemistry, 39*, 11041–11049.

Gibney, B. R., Mulholland, S. E., Rabanal, F., & Dutton, P. L. (1996). Ferredoxin and ferredoxin-heme maquettes. *Proceedings of the National Academy of Sciences of the United States of America, 93*, 15041–15046.

Goparaju, G., Fry, B. A., Chobot, S. E., Wiedman, G., Moser, C. C., Dutton, P. L., et al. (2016). First principles design of a core bioenergetic transmembrane electron transfer protein. *Biochimica Et Biophysica Acta, 1857*, 503–512.

Grzyb, J., Xu, F., Weiner, L., Reijerse, E. J., Lubitz, W., Nanda, V., et al. (2010). De novo design of a non-natural fold for an iron-sulfur protein: Alpha-helical coiled-coil with a four-iron four-sulfur cluster binding site in its central core. *Biochimica Et Biophysica Acta, 1797*, 406–413.

Hay, S., Westerlund, K., & Tommos, C. (2007). Redox characteristics of a de novo quinone protein. *The Journal of Physical Chemistry. B, 111*, 3488–3495.

Ho, S. P., & DeGrado, W. F. (1987). Design of a 4-helix bundle protein: Synthesis of peptides which self-associate into a helical protein. *Journal of the American Chemical Society, 109*, 6751–6758.

Huang, S. S., Gibney, B. R., Stayrook, S. E., Dutton, P. L., & Lewis, M. (2003). X-ray structure of a maquette scaffold. *Journal of Molecular Biology, 326*, 1219–1225.

Jaschke, P. R., Hardjasa, A., Digby, E. L., Hunter, C. N., & Beatty, J. T. (2011). A BchD (magnesium chelatase) mutant of rhodobacter sphaeroides synthesizes zinc bacteriochlorophyll through novel zinc-containing intermediates. *The Journal of Biological Chemistry, 286*, 20313–20322.

Kennedy, M. L., & Gibney, B. R. (2002). Proton coupling to [4Fe-4S](2+/+) and [4Fe-4Se](2+/+) oxidation and reduction in a designed protein. *Journal of the American Chemical Society, 124*, 6826–6827.

Kennedy, M. L., Petros, A. K., & Gibney, B. R. (2004). Cobalt(II) and zinc(II) binding to a ferredoxin maquette. *Journal of Inorganic Biochemistry, 98*, 727–732.

Koder, R. L., Anderson, J. L. R., Solomon, L. A., Reddy, K. S., Moser, C. C., & Dutton, P. L. (2009). Design and engineering of an O2 transport protein. *Nature, 458*, 305–309.

Korendovych, I. V., Senes, A., Kim, Y. H., Lear, J. D., Fry, H. C., Therien, M. J., et al. (2010). De novo design and molecular assembly of a transmembrane diporphyrin-binding protein complex. *Journal of the American Chemical Society, 132*, 15516–15518.

Laplaza, C. E., & Holm, R. H. (2001). Helix-loop-helix peptides as scaffolds for the construction of bridged metal assemblies in proteins: The spectroscopic A-cluster structure in carbon monoxide dehydrogenase. *Journal of the American Chemical Society, 123*, 10255–10264.

Laplaza, C. E., & Holm, R. H. (2002). Stability and nickel binding properties of peptides designed as scaffolds for the stabilization of Ni(II)-Fe(4)S(4) bridged assemblies. *JBIC, Journal of Biological Inorganic Chemistry, 7*, 451–460.

Levine, H. L., & Kaiser, E. T. (1978). Oxidation of dihydronicotinamides by flavopapain. *Journal of the American Chemical Society, 100*, 7670–7677.

Li, W.-W., Hellwig, P., Ritter, M., & Haehnel, W. (2006). De novo design, synthesis, and characterization of quinoproteins. *Chemistry: A European Journal, 12*, 7236–7245.

Lichtenstein, B. R., Bialas, C., Cerda, J. F., Fry, B. A., Dutton, P. L., & Moser, C. C. (2015). Designing light-activated charge-separating proteins with a naphthoquinone amino acid. *Angewandte Chemie, International Edition, 127*, 13830–13833.

Liu, J., Zheng, Q., Deng, Y., Li, Q., Kallenbach, N. R., & Lu, M. (2007). Conformational specificity of the lac repressor coiled-coil tetramerization domain. *Biochemistry, 46*, 14951–14959.

Mansurova, M., Simon, J., Salzmann, S., Marian, C. M., & Gärtner, W. (2013). Spectroscopic and theoretical study on electronically modified chromophores in LOV domains: 8-Bromo- and 8-trifluoromethyl-substituted flavins. *Chembiochem: A European Journal of Chemical Biology, 14*, 645–654.

Marsh, E., & DeGrado, W. F. (2002). Noncovalent self-assembly of a heterotetrameric diiron protein. *Proceedings of the National Academy of Sciences of the United States of America, 99*, 5150–5154.

McLachlan, A. D., & Stewart, M. (1975). Tropomyosin coiled–coil interactions: Evidence for an unstaggered structure. *Journal of Molecular Biology, 98*, 293–304.

Mewies, M., McIntire, W. S., & Scrutton, N. S. (1998). Covalent attachment of flavin adenine dinucleotide (FAD) and flavin mononucleotide (FMN) to enzymes: The current state of affairs. *Protein Science, 7*, 7–20.

Moffet, D. A., Certain, L. K., Smith, A. J., Kessel, A. J., Beckwith, K. A., & Hecht, M. H. (2000). Peroxidase activity in heme proteins derived from a designed combinatorial library. *Journal of the American Chemical Society, 122*, 7612–7613.

Moser, C. C., Keske, J. M., Warncke, K., Farid, R. S., & Dutton, P. L. (1992). Nature of biological electron-transfer. *Nature, 355*, 796–802.

Nanda, V., Rosenblatt, M. M., Osyczka, A., Kono, H., Getahun, Z., Dutton, P. L., et al. (2005). De novo design of a redox-active minimal rubredoxin mimic. *Journal of the American Chemical Society, 127*, 5804–5805.

Page, C. C., Moser, C. C., Chen, X., & Dutton, P. L. (1999). Natural engineering principles of electron tunnelling in biological oxidation-reduction. *Nature, 402*, 47–52.

Ponamarev, M. V., & Cramer, W. A. (1998). Perturbation of the internal water chain in cytochrome f of oxygenic photosynthesis: Loss of the concerted reduction of cytochromes f and b6. *Biochemistry, 37*, 17199–17208.

Prabhulkar, S., Tian, H., Wang, X., Zhu, J.-J., & Li, C.-Z. (2012). Engineered proteins: Redox properties and their applications. *Antioxidants & Redox Signaling, 17*, 1796–1822.

Reedy, C. J., & Gibney, B. R. (2004). Heme protein assemblies. *Chemical Reviews, 104*, 617–650.

Robertson, D. E., Farid, R. S., Moser, C. C., Mulholland, S. E., Urbauer, J. L., Pidikiti, R., et al. (1994). Design and synthesis of multi-heme proteins. *Nature, 368*, 425–431.

Roy, A., Sarrou, I., Vaughn, M. D., Astashkin, A. V., & Ghirlanda, G. (2013). De novo design of an artificial bis[4Fe-4S] binding protein. *Biochemistry, 52*, 7586–7594.

Roy, A., Sommer, D. J., Schmitz, R. A., Brown, C. L., Gust, D., Astashkin, A., et al. (2014). A de novo designed 2[4Fe-4S] ferredoxin mimic mediates electron transfer. *Journal of the American Chemical Society, 136*, 17343–17349.

Schnepf, R., Hörth, P., Bill, E., Wieghardt, K., Hildebrandt, P., & Haehnel, W. (2001). De novo design and characterization of copper centers in synthetic four-helix-bundle proteins. *Journal of the American Chemical Society, 123*, 2186–2195.

Sharp, R. E., Diers, J. R., Bocian, D. F., & Dutton, P. L. (1998). Differential binding of iron(III) and zinc(II) protoporphyrin IX to synthetic four-helix bundles. *Journal of the American Chemical Society, 120*, 7103–7104.

Sharp, R. E., & Dutton, P. L. (1999). Flavin synthesis and incorporation into synthetic peptides. *Methods in Molecular Biology (Clifton, N.J.), 131*, 195–206.

Sharp, R. E., Moser, C. C., Rabanal, F., & Dutton, P. L. (1998). Design, synthesis, and characterization of a photoactivatable flavocytochrome molecular maquette. *Proceedings of the National Academy of Sciences of the United States of America, 95*, 10465–10470.

Shiga, D., Funahashi, Y., Masuda, H., Kikuchi, A., Noda, M., Uchiyama, S., et al. (2012). Creation of a binuclear purple copper site within a de novo coiled–coil protein. *Biochemistry, 51*, 7901–7907.

Shiga, D., Nakane, D., Inomata, T., Funahashi, Y., Masuda, H., Kikuchi, A., et al. (2010). Creation of a type 1 blue copper site within a de novo coiled–coil protein scaffold. *Journal of the American Chemical Society, 132*, 18191–18198.

Shiga, D., Nakane, D., Inomata, T., Masuda, H., Oda, M., Noda, M., et al. (2009). The effect of the side chain length of Asp and Glu on coordination structure of $Cu2+$ in a de novo designed protein. *Biopolymers, 91*, 907–916.

Skalicky, J. J., Gibney, B. R., Rabanal, F., Bieber Urbauer, R. J., & Dutton, P. L. (1999). Solution structure of a designed four-α-helix bundle maquette scaffold. *Journal of the American Chemical Society, 121*, 4941–4951.

Snell, J. M., & Weissberger, A. (1939). The reaction of thiol compounds with quinones. *Journal of the American Chemical Society, 61*, 450–453.

Solomon, L. A. (2013). *The physical chemistry underlying the assembly and midpoint potential control in a series of designed protein-maquettes*. Thesis, Philadelphia, PA: University of Pennsylvania.

Solomon, L. A., Kodali, G., Moser, C. C., & Dutton, P. L. (2014). Engineering the assembly of heme cofactors in man-made proteins. *Journal of the American Chemical Society, 136,* 3192–3199.

Stubbe, J. A., & van der Donk, W. A. (1998). Protein radicals in enzyme catalysis. *Chemical Reviews, 98,* 705–762.

Summa, C. M., Rosenblatt, M. M., Hong, J. K., Lear, J. D., & DeGrado, W. F. (2002). Computational de novo design, and characterization of an A(2)B(2) diiron protein. *Journal of Molecular Biology, 321,* 923–938.

Tegoni, M. (2014). De novo designed copper α-helical peptides: From design to function. *European Journal of Inorganic Chemistry, 2014,* 2177–2193.

Walters, R. F. S., & DeGrado, W. F. (2006). Helix-packing motifs in membrane proteins. *Proceedings of the National Academy of Sciences of the United States of America, 103,* 13658–13663.

Wang, L., & Schultz, P. G. (2005). Expanding the genetic code. *Angewandte Chemie (International Ed. in English), 44,* 34–66.

Wimley, W. C., & White, S. H. (1996). Experimentally determined hydrophobicity scale for proteins at membrane interfaces. *Nature Structural & Molecular Biology, 3,* 842–848.

Xiong, J., & Bauer, C. E. (2002). A cytochrome b origin of photosynthetic reaction centers: An evolutionary link between respiration and photosynthesis. *Journal of Molecular Biology, 322,* 1025–1037.

Zastrow, M. L., & Pecoraro, V. L. (2013). Designing functional metalloproteins: From structural to catalytic metal sites. *Coordination Chemistry Reviews, 257,* 2565–2588.

CHAPTER SEVENTEEN

Design Strategies for Redox Active Metalloenzymes: Applications in Hydrogen Production

R. Alcala-Torano, D.J. Sommer, Z. Bahrami Dizicheh, G. Ghirlanda[1]
School of Molecular Sciences, Arizona State University, Tempe, AZ, United States
[1]Corresponding author: e-mail address: giovanna.ghirlanda@asu.edu

Contents

1. Introduction	390
2. Design of FeS Clusters	392
2.1 Use of Natural Sequences	393
2.2 Computational Design of Model Proteins	395
2.3 Incorporation into Prestructured Proteins	396
3. Design of [FeFe]-Hydrogenase Mimics	399
3.1 Use of Peptide Scaffolds	400
3.2 Use of Natural Sequences	402
4. Design of Porphyrin Redox Sites	404
4.1 Redesign of Natural Scaffolds	404
4.2 Use of Peptide Scaffolds	408
5. Conclusion and Future Outlook	409
References	411

Abstract

The last decades have seen an increased interest in finding alternative means to produce renewable fuels in order to satisfy the growing energy demands and to minimize environmental impact. Nature can serve as an inspiration for development of these methodologies, as enzymes are able to carry out a wide variety of redox processes at high efficiency, employing a wide array of earth-abundant transition metals to do so. While it is well recognized that the protein environment plays an important role in tuning the properties of the different metal centers, the structure/function relationships between amino acids and catalytic centers are not well resolved. One specific approach to study the role of proteins in both electron and proton transfer is the biomimetic design of redox active peptides, binding organometallic clusters in well-understood protein environments. Here we discuss different strategies for the design of peptides incorporating redox active FeS clusters, [FeFe]-hydrogenase organometallic mimics, and porphyrin centers into different peptide and protein environments in order to understand natural redox enzymes.

1. INTRODUCTION

One of the most important and urgent challenges facing society is the development of carbon-free or carbon-neutral sustainable fuels. Hydrogen is an attractive carbon-free candidate as it can be utilized in fuel cells, while direct carbon dioxide reduction to carbon monoxide, formate, or methanol is being explored as a mean to achieve carbon-neutral fuels compatible with the current infrastructure. Nature has provided the blueprints for both of these conversions; in photosynthesis, solar energy is utilized as a driving force to push forward the conversion of water and carbon dioxide to oxygen, hydrogen, and carbohydrate molecules (McEvoy & Brudvig, 2006). In this process, electrons extracted from water by the highly oxidizing Photosystem II via light captured by antenna systems are utilized by a series of enzymes for reductive chemistry, finally storing solar energy as carbohydrates. The same highly reducing electrons can also be used to reduce protons to molecular hydrogen via the hydrogenase family of metalloenzymes (Nicolet, Lemon, Fontecilla-Camps, & Peters, 2000).

Inspired by this highly efficient natural process, artificial photosynthesis aims to generate clean and renewable fuels by capturing solar light through artificial antenna system and mimicking the oxygen-evolving complex, hydrogen-evolving reaction, or other reducing processes using organometallic catalysts (Concepcion, House, Papanikolas, & Meyer, 2012; Gust, Moore, & Moore, 2009, 2012). Although progress in this area has been made through the design and study of organometallic catalysts, they tend to function in organic solvents, at high overpotential, and with much less efficiency than the natural enzymes. While organometallic complexes rely on modulation of the redox potential, fine tuning of the donor/acceptor character of the first-sphere coordination ligands, and control of the ligand geometry to tweak the inherent reactivity of the metal center, the optimization of reactivity provided by the second sphere and outer sphere coordination environment of a protein can be difficult to recapitulate in a small molecule. Thus, the chemical principles by which enzymes at the reducing end of photosynthesis, such as hydrogenases, accomplish reversible proton reduction utilizing earth-abundant transition metals remain an important scientific challenge.

Designed metalloenzymes that interface an organometallic catalyst with a protein scaffold have been used as a means to provide the long-range interactions missing in small molecules and thus modulate the activity of the catalyst, with an eye to impart stereospecificity to a reaction (Dürrenberger & Ward, 2014; Hamels & Ward, 2013; Lewis, 2013; Ozaki, Matsui, Roach, & Watanabe, 2000; Ozaki, Roach, Matsui, & Watanabe, 2001; Wilson & Whitesides, 1978). In recent years, this approach has begun to find applications in the design of metalloenzymes dedicated to catalyze energy transformation processes (Caserta, Roy, Atta, Artero, & Fontecave, 2015; Faiella, Roy, Sommer, & Ghirlanda, 2013; Ginovska-Pangovska, Dutta, Reback, Linehan, & Shaw, 2014; Sommer, Alcala-Torano, Bahrami Dizicheh, & Ghirlanda, In Press). Hydrogenases, in particular, have emerged as an attractive target for the development of artificial protein-based mimics (Lubitz, Ogata, Rüdiger, & Reijerse, 2014). These enzymes catalyze the reversible reduction of protons to molecular hydrogen; both sides of the reaction are of interest in a sustainable energy economy. A number of drawbacks have hampered their applications and spurred an intense quest for biomimetic as well as functional organometallic mimics. Although no perfect catalyst has been found, the focus on the bioinorganic chemistry of these mimics provides an extensive body of knowledge on the properties of these complexes (Artero, Chavarot-Kerlidou, & Fontecave, 2011; DuBois, 2014; Simmons, Berggren, Bacchi, Fontecave, & Artero, 2014).

Our group and others have sought to interface an organometallic mimic of the hydrogenase active site with a protein or peptide scaffold with the goal of providing second sphere and long-range interactions that stabilize the catalytic site as well as facilitate proton and electron transfer (Dutta et al., 2013; Dutta, Roberts, & Shaw, 2014; Jones, Lichtenstein, Dutta, Gordon, & Dutton, 2007; Roy, Madden, & Ghirlanda, 2012; Sano, Onoda, & Hayashi, 2011, 2012). These hybrid systems present a number of advantages over their organometallic counterparts. First, functional groups in the protein component can be easily modified by rational redesign of the environment surrounding the catalytic site, either by solid phase peptide synthesis (SPPS) or by mutagenesis. Second, the organometallic site itself can be easily modified by synthetic variation of the ligands. Third, for many of these systems it will be possible to optimize the catalytic activity by directed evolution of the protein scaffold.

The same approach can be extended to the production of carbon-based fuels obtained by reduction of carbon dioxide. Here, the natural enzymes that provide the intellectual inspiration are carbon monoxide dehydrogenase and formate dehydrogenase. Due to the complexity of their active sites, however, existing work has focused on the use of cobalt porphyrins, which have inherent activity in the reduction of carbon dioxide to carbon monoxide and formate.

Finally, the development of robust, redox active metalloenzymes adept at technological applications will ultimately require the engineering of appropriate electron transfer chains to shuttle electrons to/from the active site, and to connect the catalyst to either an electrode or to water soluble electron carriers, such as ferredoxins. This need has led our group to the development of miniaturized proteins containing two iron–sulfur clusters that can easily interface with other proteins (Roy et al., 2014). Recent work on the optimization of the electron transfer rates to a catalytic site as a means to improve activity of an artificial enzyme provide the underpinning for this line of work (Yu et al., 2015).

In this chapter, we review recent work by our group and others in the development of peptide-based electron transfer modules and hydrogenase mimics, with an emphasis on the methods used in the design of these metalloenzymes.

2. DESIGN OF FES CLUSTERS

Iron–sulfur clusters are one of the most utilized metallocenters found in nature, playing roles in catalysis, regulation, and storage of iron, and ligand transport within cells (Beinert, 2000). Despite containing only iron, sulfur, and proteinogenic ligands, the clusters are able to span the entire biological redox range, from −700 to +450 mV (Liu et al., 2014). Proteins utilize multiple structures to modulate function, ranging from a single iron atom to [2Fe2S], [3Fe4S], [4Fe3S], and [4Fe4S] clusters; a number of higher-order clusters also exist that carry out specialized reactions in vivo (Koay, Antonkine, Gärtner, & Lubitz, 2008). Notably, the [4Fe4S] cluster is able to span nearly the whole range itself by using two different redox transitions, both +3/+2 (HiPiP) and +2/+1 (ferredoxin-type). With these clusters having implications in both energy and health, a large literature base has been amassed studying the natural function of these clusters, particularly the oxygen-tolerant [2Fe2S] forms (Fukuyama, 2004; Johnson, Dean, Smith, & Johnson, 2005). Work on natural systems is often times challenging, as these clusters are generally found in large, multidomain proteins that house multiple metallocenters and are sensitive to mutagenesis, complicating the spectroscopic and functional characterization of the bound clusters (Faiella et al., 2013). In order to simplify their study, a number of groups have utilized protein design to obtain artificial FeS cluster constructs (specifically binding [4Fe4S]), aiming to mimic natural clusters of significance and to develop novel materials for artificial redox pathways. There have been three major approaches utilized to design these types of proteins; the design features of each approach are described herein.

2.1 Use of Natural Sequences

The first approach for design of iron–sulfur proteins is to utilize the sequences that are found in natural iron–sulfur cluster proteins and to adopt them into either natural proteins or designed model proteins. This approach has been successful in designing two types of cluster-binding proteins, one that mimics ferredoxin-type [4Fe4S] clusters and the other inspired by the three clusters that play a key role in photosynthesis: F_X, F_A, and F_B. Early bioinformatic work on ferredoxin iron–sulfur proteins identified a conserved motif taking the form of **CXXCXXC…C**, where the fourth cysteine is contributed from a remote portion of the sequence (Koay et al., 2008). Further refinement of the sequence established **CIACGAC** as the minimal conserved sequence, with pseudosymmetric folds enabling the small proteins to bind two iron–sulfur clusters at once. Recent surveys of known iron–sulfur cluster structures identified major structural motifs that accommodate iron–sulfur clusters (Nanda et al., 2016). Unlike other metallocenters or cofactors such as heme, which sit in well-defined protein structural elements, iron–sulfur clusters often reside in transition points of protein structures between β-sheets and α-helices.

Gibney, Mulholland, Rabanal, and Dutton (1996) used this approach to design a protein mimicking Complex II of the electron transport chain, succinate dehydrogenase (Fig. 1A). The native ferredoxin conserved sequence was installed into the unstructured loop of a designed heme-binding helix–loop–helix construct. Following a well-known in vitro incorporation procedure (Antonkine et al., 2009), where inorganic precursors prearranged by β-mercaptoethanol ligands are transferred into the apopeptide via an entropically driven exchange, the designed peptide bundle incorporated a [4Fe4S] cluster while maintaining heme-binding abilities. However, the designed loop was not optimal for the geometric constraints imposed by [4Fe4S] clusters, resulting in higher-order oligomeric states as multiple peptides were recruited to satisfy the preferred pseudotetrahedral state of the cluster. This preference for oligomerization is also seen in nature, where many clusters sit at the interface between two protein subunits and contribute to the stability of the interface (Nanda et al., 2016). Subsequent work aimed to overcome

A *P. aerogenes* ferredoxin   NH$_2$-AYVINDS**CIACGAC**KPECPVNIIEGSITYAIDADSCIACGSCASV**C**PVPVGAPNPED-COOH
Ferredoxin maquette   NH$_2$-KL**C**EGG**C**IA**C**GA**C**GGW-CONH$_2$
Ferredoxin-heme maquette   Ac-LKKLREEHLKLLEEFKKLLEEHLKL**C**EGG**C**IA**C**GA**C**GGGGELWKLHEELLKFEELLKLHEERLKKL-CONH$_2$

    ⟵ **Helical region** ⟶   ⟵ **[4Fe4S] Loop** ⟶   ⟵ **Helical region** ⟶

B Syn$_{7002}$ F$_B$ loop   SHSVKI**YDTC** **IGCTQC**V**RAC** **PL**DVLEMVPW
F$_B$ maquette   ******YDT**C** IG**C**TQ**C**KPE**C** PW*******

Syn$_{7002}$ F$_A$ loop   DGCKAGQIAS SPRT**ED**C**VGC** **KRCETAC**P**T**D
F$_A$ maquette   ******************TED**C**VG**C** KR**C**KPE**C**PW*

Fig. 1 (A) Sequence alignment of the *P. aerogenes* ferredoxin with the minimalist peptide and helical bundle used by Dutton (Gibney et al., 1996), highlighting key structural factors. (B) Alignments of the natural-binding sequences found in PSI for F$_A$ and F$_B$ with the minimalist models used by Lubitz to mimic their properties (Koay et al., 2008).

oligomerization by introducing the conserved sequence for the F_X cluster into two separate loops of a designed four-helix bundle (Scott & Biggins, 1997). The F_X cluster sits at the interface between PsaA and PsaB subunits of PSI, with both subunits containing half of the F_X site sequence, CDGPGRGGTC. Placing this binding site into the loops of the four-helix bundle has the double function of stabilizing the intended fold, resulting in efficient incorporation of an intact [4Fe4S] cluster that accurately mimics the EPR and spectroscopy of the F_X cluster; however, the redox potentials did not match well (-705 mV vs SHE in PSI, -422 mV in the modeled cluster).

In both early models, the designed scaffold structures proved to be unnecessary, either not affecting or harming the overall design of the cluster site. To establish if the structural elements imposed by the peptides were required for cluster incorporation, further work aimed to explore how unstructured peptides would handle cluster incorporation. By taking the native sequences of both the F_A and F_B cluster-binding sites and appending a fourth cysteine via a flexible linker (mimicking the distant-binding Cys found in vivo), two 16-residue peptides were generated as models (Fig. 1B). The small, flexible peptides readily incorporated [4Fe4S] clusters and showed spectroscopic signatures that correspond well to the F_A and F_B clusters. Further work attempted to utilize this approach to mimic the F_X clusters, although cluster properties were not replicated accurately (Koay et al., 2008). In all the constructs discussed thus far, the clusters exist as solvent-exposed centers, which play a dominant role in the determination of cluster properties.

2.2 Computational Design of Model Proteins

One relatively unexplored, albeit potentially promising method to design iron–sulfur clusters is the use of computational methodology. This so-called metal first approach builds on computational parameterization of the cluster and aims to extend the inherent symmetry found in [4Fe4S] cluster to the ligating protein. Recent work from the groups of Nanda and Noy exemplified this approach (Grzyb et al., 2010, 2012). A search of the PDB identified a natural [4Fe4S] cluster, found at the interface of three-helical bundles, as containing a minimal helical element containing the CXXC motif and displaying two additional Cys on neighboring helices. Starting from this helical CXXC element a library of C_2-symmetric protein backbones was computationally generated. The library was searched for optimal backbone rotation, bundle radius, offset displacement between helices, and angle between helices, until a single-optimized backbone was chosen for further study. The backbone was converted into a four-helix bundle by linking together the helices with designed loops and by optimizing side chain identity with a combination of the protCAD and ROSETTA computational packages.

The designed protein was able to incorporate a [4Fe4S] cluster, as evidenced by UV–vis and EPR traces. As the cluster is contained within the core of the bundle, the protein was unstructured in the absence of cluster, but transitions to an α-helical structure upon incorporation. As with previous constructs, the designed protein tended to form higher-order oligomeric states.

2.3 Incorporation into Prestructured Proteins

Peptides and proteins that have a prearranged structure optimal inclusion of binding sites can also be used as starting point for the design of iron–sulfur cluster proteins. While the strategies discussed earlier utilized either highly unstructured peptides, allowing for flexibility to accommodate the desired cluster, or peptides whose structure was dependent on the cluster incorporation, resulting in poor control of the oligomerization state of the peptides, proteins that have well-defined protein contacts and structure may avoid these problems. Furthermore, use of well-structured proteins opens the possibility of repurposing natural proteins by generating modules containing iron–sulfur clusters that function in tailor-made redox pathways.

We have utilized this approach to design a series of peptides that bind two iron–sulfur clusters within the same construct at once (Roy, Sarrou, Vaughn, Astashkin, & Ghirlanda, 2013; Roy et al., 2014; Sommer, Roy, Astashkin, & Ghirlanda, 2015). While in nature iron–sulfur clusters are generally found in chains of two or more clusters spaced within 15 Å of one another, previous designs incorporated only a single, electronically isolated cluster (Beinert, 2000). Our design started with a dimeric three-helix bundle (domain-swapped dimer, DSD, PDB 1G6U), in which two identical helix–loop–helix peptides pair with twofold symmetry (Ogihara et al., 2001). The helices are held together by a leucine-rich hydrophobic core, as well as designed salt bridges at helix termini to impart selectivity onto the peptide. DSD possesses 2-D symmetry reminiscent of the ferredoxin fold, in which two iron–sulfur cluster-binding sites are comprised of three contiguous cysteines and a fourth distal cysteine related by internal symmetry. We modeled a [4Fe4S] cluster taken from tryptophanyl-tRNA synthetase into the leucine-rich core by overlapping the helical CXXC motif with a LXXL motif and identified two additional Leu residues that were compatible with the positioning of Cys in the first-coordination sphere of the native cluster (Fig. 2A) (Han et al., 2010). This set of four Cys residues recapitulates the pseudotetrahedral geometry preference of the iron–sulfur cluster; further, three cysteines are contributed from one monomer and the fourth from the second monomer similar to the natural ferredoxin-binding site arrangement. Due to the inherent symmetry in this scaffold, the binding site designed at one end of the bundle was mirrored on the other side.

Mutation of core Leu to Cys residues resulted in a significant destabilization compared to the parent peptide due to the presence of hydrophilic pockets in the peptide. Cluster incorporation was shown to rescue this destabilization, increasing the overall stability of the designed peptide as the protein–metal contacts stabilized the fold. Unlike previous constructs, the DSD-bis[4Fe4S] remained the same oligomeric state as the parent peptide, a single-dimeric species. In order to confirm the presence of two electronically coupled clusters, we used the pulse electron–electron double-resonance technique to measure the distance between the clusters via magnetic dipole interactions. With this technique we measured a cluster–cluster distance of 29–34 Å, which agreed well with the modeled distance of ∼30 Å. This construct is the first example of a de novo iron–sulfur cluster protein that binds two electronically coupled clusters (Roy et al., 2013).

In a second generation design, deemed DSD-Fdm, we utilized the heptad repeat pattern of coiled coils to move the cluster-binding sites from 30 Å

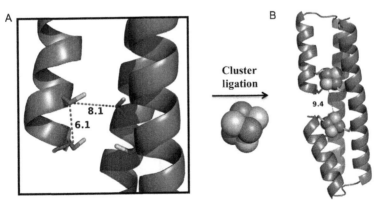

Fig. 2 (A) Model of the DSD family iron–sulfur cluster-binding site, showing key α-carbon distances (in Å) compatible with incorporation of clusters. (B) Intercluster distance in the DSD-Fdm construct, mimicking natural ferredoxin cluster distances.

to within 12 Å of one another; iron–sulfur cluster chains in nature have intercluster distances of 15 Å or less (Fig. 2B) (Roy et al., 2014). As with the previous design, cluster incorporation proceeded readily with the desired stoichiometry and resulted in stabilization of the bound protein. DSD-Fdm was also shown to be functional, as clusters bound by DSD-Fdm are able to accept energy from photosensitized dyes, a process reminiscent of the electron accepting ability of ferredoxins from Photosystem I. Furthermore, the reduced clusters were able to interact with a natural redox protein, cytochrome c_{550}, reducing the bound heme with a stoichiometry of 2 cyt c_{550}: 1 DSD-Fdm, confirming the ability of the designed peptide to carry two reducing equivalents, a major step forward in the quest toward design of artificial redox pathways.

Iron–sulfur-binding pockets can be carved into well-folded natural proteins, endowing them with novel functionality. Recently, protein-based bacterial microcompartments (BMCs) were reengineered to make them competent for electron transfer by introducing an iron–sulfur-binding site at the subunits interface (Fig. 3A) (Aussignargues et al., 2016). BMCs are protein macrostructures comprised of multiple subunits and are often implicated for applications as bionanoreactors. The protein shell contains a hydrophilic trimeric face with three Ser residues facing toward a pore in a trigonal planar orientation. Mutation of these residues to Cys provided an iron–sulfur-binding site, with the fourth ligand for the iron provided by the solvent (Fig. 3B). The designed protein was expressed in vivo and found to

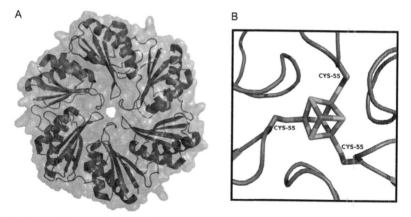

Fig. 3 (A) Crystal structure (PDB 5DII) of the BMC protein showing the trimeric structure and pore in the center of the structure. (B) Engineered iron–sulfur cluster-binding site at the trimer interface, showing the incorporated [4Fe4S] cluster.

spontaneously incorporate an iron–sulfur cluster by spectroscopic characterization. Further, structure determination confirmed the presence of an iron–sulfur cluster at the designed interface, giving the first structural confirmation of a designed iron–sulfur cluster-binding site. This work demonstrates that four binding residues in a pseudotetrahedral orientation may not be necessary for successful cluster incorporation, if the protein scaffold provides a preorganized-binding site.

3. DESIGN OF [FeFe]-HYDROGENASE MIMICS

One of the most important applications of redox active peptides and proteins is as catalysts for the sustainable production of fuels, as part of a global search for renewable energy sources. In recent years, hydrogen has emerged as a good candidate in this area due to its high heat of combustion and its suitability as green fuel. However, current processes for the production of this gas are far from being cost-effective for its widespread utilization, as they often require precious metals or high energy processes. In contrast, hydrogen metabolism in nature is regulated by hydrogenases, enzymes that are capable of reversibly reducing protons to hydrogen with high efficiency under mild conditions while utilizing earth-abundant metals. Of particular interest are the [FeFe]-hydrogenases, which are biased toward the production of hydrogen gas. The active site in these hydrogenases, the H-cluster, is comprised of a [4Fe4S] cluster tethered by a cysteine residue to the proximal iron atom of a peculiar diiron complex. The metal atoms in the latter are coordinated by carbonyl and biologically unique cyanide ligands; the metal atoms are linked together by an azadithiolate bridge (Fig. 4).

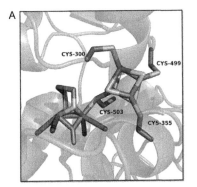
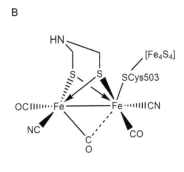

Fig. 4 Modeled H-cluster of *C. pasteurianum* [FeFe]-hydrogenase, PDB code: 1FEH (A) and its corresponding line structure representation (B).

Many organometallic mimics of the H-cluster have been reported to date, but none have shown the high turnover numbers seen for natural hydrogenases, much less in aqueous systems. The reason behind these shortcomings is generally attributed to the lack of second sphere and long-range interactions, which in the natural enzymes is provided by the protein environment around the active site. These interactions help stabilize important intermediates during the course of the reaction and alter their redox potential. The protein environment also provides a proton channel that controls the movement of substrates and product to and from the redox center. With this in mind, few research groups have ventured into exploring the utilization of protein and peptide scaffolds to mimic the role of protein during catalysis and study proton reduction by this diiron complex.

The diiron complex has been incorporated into protein or peptide systems either by forming protein contacts with both metals, thus mimicking the dithiolate bridge, or by coordinating only one of the iron atoms. Evidently, linking the organometallic compound through a cysteine residue coordinating one of the iron atoms would be the most accurate mimic of the native H-cluster. However, in the native enzyme the diiron center is attached to this cysteine *after* incorporation of the [4Fe4S] cluster, possibly because a distinct electron density at the coordinating thiol is required for this type of bond. For these reasons, the exact type of bond between the cluster and the diiron site may be difficult to replicate in bioinorganic mimics, leaving the dithiolate bridge option as the most explored one. Previous studies on organometallic complexes have shown that two thiol moieties held at a distance of in a CXXC motif can template the assembly of a biomimetic diiron hexacarbonyl cluster. In the context of a peptide, this arrangement of the dithiol moiety can be achieved either via two cysteine residues or by using unnatural amino acids.

3.1 Use of Peptide Scaffolds

The first example of incorporation of a diiron cluster mimic into a peptide was achieved using an alanine-rich helical 36-mer peptide, in which two residues were mutated to cysteines to coordinate a diiron hexacarbonyl mimic of hydrogenase. The cysteines were placed at i and $i+3$ positions of an α-helix to achieve the same thiolate–thiolate distance observed in the azadithiol bridging ligand of [FeFe] hydrogenases (Jones et al., 2007).

A related approach took advantage of the disulfide bond of octreotide, a macrocyclic octapeptide linked through residues Cys2 and Cys7, providing a new alternative for choosing and/or designing peptide scaffolds other than α-helices (Apfel et al., 2010).

Using cysteine side chains, however, introduces design limitations related to correct placement of these amino acids either within a helix or within a cyclic scaffold. These limitations are alleviated by using artificial amino acids bearing a propanedithiol moiety, which can serve as bridge for templating the formation of diiron hexacarbonyl mimic. This moiety can also be introduced within simple peptides by modifying an existing side chain such as lysine. For example, the peptide WASKLPSG was synthesized through SPPS with the lysine bearing an orthogonal protecting group (iVDde, 1-(4,4-dimethyl-2,6-dioxocyclohex-1-ylidene)-3-methylbutyl). This enabled orthogonal on-resin deprotection of the amino group and coupling to 3-(acetylthio)-2-(acetylthiomethyl)propanoic acid. Cleavage and deprotection of the peptide from the resin provides the modified peptide bearing two free thiols optimally spaced to incorporate the desired diiron hexacarbonyl mimic with a propanedithiolate bridge (Fig. 5A) (Roy et al., 2011). Modification of the lysine residue, however, results in a long chain that brings the functional group far from the peptide backbone and interferes with the design of second sphere interactions.

Our group utilized an unnatural aminoacid to directly anchor a diiron hexacarbonyl center into any synthetic protein. The approach utilizes an

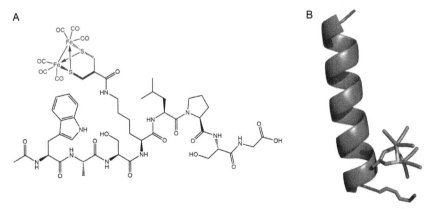

Fig. 5 (A) Octapeptide with a modified lysine residue to incorporate a diiron hexacarbonyl mimic (Roy, Shinde, Hamilton, Hartnett, & Jones, 2011). (B) Model of the single-helix peptide with an artificial amino acid bearing the diiron hexacarbonyl moiety (Roy et al., 2012).

Fmoc-protected, SPPS-compatible unnatural amino acid bearing a propanedithiol moiety. This amino acid was incorporated in a 19-mer helical peptide and used to template the diiron hexacarbonyl site after peptide purification. A lysine residue included at position 19 ($i+3$) could potentially aid catalysis by acting as a pendant base (Fig. 5B). The construct was active for the photocatalytic hydrogen production in water at pH 4.5 using [Ru(bpy)$_3$]$^{2+}$ as photosensitizer and ascorbic acid as sacrificial electron donor, showing a TON of 84 after 2.3 h (Roy et al., 2012).

Mimics of the diiron cluster can also be incorporated into a peptide scaffold through coordination of only one of the iron atoms. This was achieved in two similar fashions: one by modification of a lysine residue (akin to the work previously described) and the second one by utilization of unnatural amino acids compatible with SPPS. By introducing an amino acid with a phosphine side chain through the latter approach, the diiron mimic was incorporated into the peptide by exchanging a carbonyl for the better electron donating phosphine ligand. This binding mode not only resembles the native enzyme, but also it provides the opportunity to modify the properties of the dithiolate bridge, such as the azadithiolate that is seen in hydrogenases which is likely to result in an increase in catalytic activity (Roy, Nguyen, Gan, & Jones, 2015).

3.2 Use of Natural Sequences

Early work on methodology to incorporate diiron mimics into proteins was done on relatively simple, small peptides that provided no solvent shielding or complex secondary structure elements to protect and stabilize the cluster. Inspired by the success using the bis-Cys moiety in peptides, Sano et al. (2011) exploited a similar CXXC motif in horse heart cytochrome *c* (cyt *c*) which is normally responsible for forming thioether linkages with heme (Fig. 6). Incorporation of the diiron hexacarbonyl cluster resulted in photochemical hydrogen production using [Ru(bpy)$_3$]$^{2+}$ as photosensitizer and ascorbate as sacrificial electron donor at pH 4.7 with a TON of ~80 after 2 h and TOF of 2.1 min^{-1}, comparable to helical peptides. Interestingly, a heptapeptide fragment sequence, YKCAQCH, gave rise to a TON of ~10 and TOF of 0.47 min^{-1} under the same conditions, clearly suggesting that either secondary interactions or structural elements provided by the cytochrome scaffold strongly affect catalysis.

Simultaneous incorporation of the diiron hexacarbonyl moiety through the CXXCH motif and of a Ru complex through ligation by the histidine

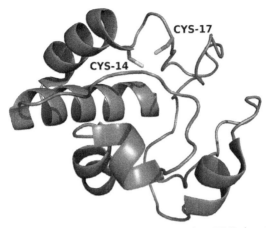

Fig. 6 Structure of horse heart cytochrome *c* (PDB code: 1HRC) showing the cysteine residues used for incorporation of a [FeFe]-hydrogenase mimic.

imidazole ring in a fragment of cyt c_{556} comprising 18 residues was meant to increase the electron rate transfer upon illumination. However, the experiments in the presence of ascorbate as sacrificial electron donor at pH 8.5 showed a TON of 9 after 2 h and TOF of 0.19 min^{-1} (Sano et al., 2012).

Positioning cysteine residues correctly in an existing protein scaffold is inherently more difficult than on a peptide model, due to the constraints imposed by the allowed cysteine side chains. Alternatively, proteins may be covalently modified using biorthogonal chemistry to incorporate organometallic mimics. For example, the diiron hexacarbonyl was interfaced with a maleimide-modified propanedithiolate bridge, allowing for site-specific functionalization of proteins using the well-known maleimide-thiol chemistry (Fig. 7A). The selected protein for incorporation, nitrobindin, is a β barrel protein normally binds heme for NO transport in vivo; removal of the heme leaves a hydrophobic pocket optimal for incorporation (Fig. 7B). A mutation to Cys in the hydrophobic pocket, Q96C, positions the metallic center at the entrance of the barrel. Light-driven hydrogen production yielded TON of ~130 with a TOF of ~2.3 min^{-1} in a Tris–HCl buffer at pH 4.0. These values are similar to the free complex in 4% v/v THF. However, this work does provide a different way to incorporate hydrogenase mimics into protein scaffolds (Onoda et al., 2014).

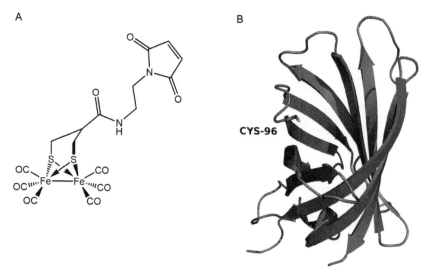

Fig. 7 (A) Maleimide-modified diiron hexacarbonyl mimic. (B) Crystal structure of mutant nitrobindin showing the engineered cysteine residue for inclusion of the mimic through maleimide-thiol chemistry (Onoda, Kihara, Fukumoto, Sano, & Hayashi, 2014).

4. DESIGN OF PORPHYRIN REDOX SITES

Metalloporphyrins are found as cofactors in natural proteins, bound via both covalent and noncovalent interactions. Porphyrin-binding proteins cover a vast range of function, including reversible binding of gases, small molecule detection, oxygen activation, intracellular electron transport, and catalysis. Due to the crucial roles that metalloporphyrins play in nature, a large body of research on designing porphyrin-binding proteins has been developed. More recently investigators have started to exploit the intrinsic activity of cobalt porphyrins in hydrogen reduction and carbon dioxide reduction, using either redesign of natural scaffolds, which creates specific functionality into an existing protein, and de novo design, which designs both the scaffold and binding site from scratch. This section will discuss general strategies for both approaches by highlighting recent advances in the field.

4.1 Redesign of Natural Scaffolds

Natural heme-binding proteins can be repurposed by exchanging the native iron for cobalt and by reengineering the secondary coordination shell. The first example of this approach examined the hydrogen production activity of the cobalt derivative of microperoxidase-11 (MP11), a proteolytic fragment

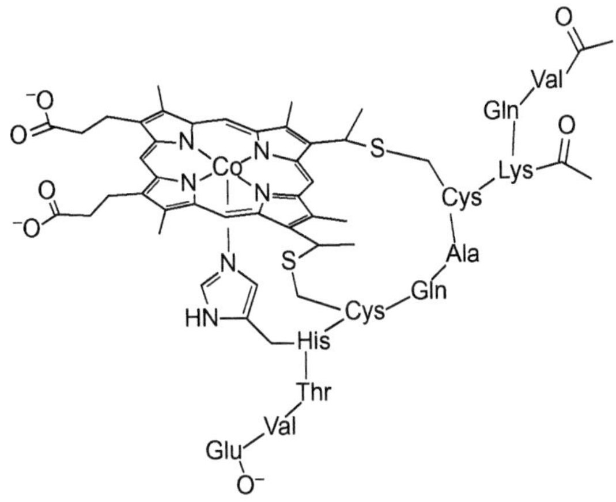

Fig. 8 Bisacetylated cobalt microperoxidase-11.

of horse cytochrome c (Kleingardner, Kandemir, & Bren, 2014). MP11 comprises 11 residues and contains a covalently linked heme, coordinated at one of the axial positions by a histidine residue, leaving an open coordination site for the substrate (Fig. 8). CoMP11 catalyzes hydrogen production electrochemically with TON over 20,000 in aqueous solution at pH 7 and in the presence of oxygen. However, the activity was sustained for only 15 min; this was attributed to porphyrin degradation that is exposed to the solution.

Incorporation into a folded protein should shelter the porphyrin from side reactions. We employed myoglobin, which holds the porphyrin ring within a well-defined hydrophobic pocket represented in Fig. 9 (Sommer, Vaughn, & Ghirlanda, 2014). The cobalt-substituted protein, CoMyo, exhibited catalytic proton reduction potential at −1 V vs SHE. In order to overcome this large overpotential, we looked at photoinduced hydrogen production in the presence of $[Ru(bpy)_3]^{2+}$ as photosensitizer and sodium ascorbate as a sacrificial electron donor, which provides a driving force of −1.26 V vs SHE. Under these conditions, CoMyo has a TON of 518 at pH 7. We then investigated whether mutations in the active site could modulate the activity by replacing the ionizable histidine residues with nonionizable alanines, producing three mutants, H64A, H97A, and H64/97A. We found that the double mutant increased the porphyrin activity 2.5-fold compared to the wild-type construct. This design approach exemplifies the importance of secondary coordination shell of the artificial H_2-evolving active site, showing that optimizing protein contacts can result in a significant increase in activity.

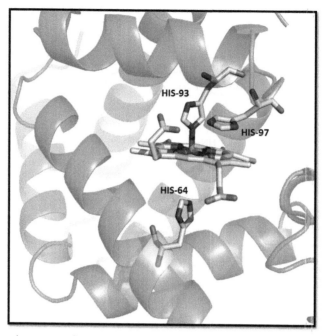

Fig. 9 Crystal structure of CoMyo active site (PDB 1YOI), showing H64, H93, H97 residues.

We applied this approach to cytochrome b_{562} (cyt b_{562}) in order to investigate the effect of dual axial coordination to cobalt protoporphyrin IX (CoPP(IX)) on catalytic activity (Sommer et al., 2016). The natural protein coordinates a heme within its hydrophobic cavity through axial ligation by His102 and Met7. We investigated the effect of loss of one coordination site (M7A), or of exchange of the coordinated ligand (M7D, M7E), and found that the activity of the bound cobalt porphyrin was modulated. The M7D variants had increased affinity comparing to wild type (K_d of 3.5 vs 8.9 µM), while the other two were essentially unchanged. Under the same conditions as CoMyo, M7A showed the highest activity with TON of 310, which supports the role of having an open axial coordination site accessible for protons. Introduction of a carboxylic acid into the axial site in M7E and M7D variants also increased the catalytic activity compared to the wild type, with TON of 195 and 270, respectively. The increased activity of M7D and M7E toward wild type attributed to increased structural flexibility induced by mutations. This work exemplifies the impact of primary coordination sphere of a cofactor on its catalytic activity, while highlighting the nature of the axial ligands as one of the critical factors that affect heme-binding proteins reactivity.

An elegant approach to change the reactivity of natural porphyrin-binding proteins consists in adding a second metal-binding site in the vicinity, thus generating heteronuclear metalloenzymes. These sites can be stabilized by genetically incorporating nonnatural amino acids in the protein sequence. This has been recently demonstrated by mimicking the structural features of the copper-heme-binding sites of cytochrome c oxidase (CcO) within a myoglobin scaffold (Yang et al., 2015). One of these features replicates the role of tyrosine, which can donate an electron and a proton during O_2 reduction by CcO. By genetically replacing Phe33 in Cu_BMyoglobin (Cu_BMb) with 3-methoxy tyrosine (OMeY), a Phe33OMeYCu_BMb mutant can mimic the Tyr–His pair seen in Fig. 10. Cu-Myo-OMeY33 shows the structural model of Cu-Myo-OMeY33, based on the crystal structure of Cu-Myo The rate of oxygen reduction catalyzed by Phe33OMeYCu_BMb, in the presence of ascorbate as reductant and tetramethyl-p-phenylenediamine dihydrochloride a redox mediator, displayed a TON of 1100, which is twofold higher than

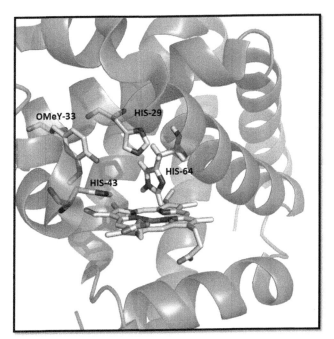

Fig. 10 Model of Cu-myoglobin active site, highlighting replacement of Phe33 with OMeY. The model is constructed based on the crystal structure of Phe33Tyr-Cu$_B$Mb (PDB 4FWX).

the wild type's TON of 500, demonstrating how genetic code expansion offers a novel approach for protein design.

The activity of designed redox enzymes can be increased considerably by optimizing the rate of electron transfer by their redox partner (Yu et al., 2015). Redesigning the interface between an artificial oxidase within myoglobin and its natural redox partner, cyt b_5, resulted in an increase of 400-fold in the electron transfer rate. In turn, this improvement is reflected in oxygen reduction rate of 52 s^{-1}, comparable to natural oxidases.

4.2 Use of Peptide Scaffolds

The observation that heme prosthetic groups are often found at the interface between α-helical bundles has led to the de novo design of α-helical heme-binding peptides. Heme centers were first incorporated into four-helix bundles in order to dissect factors contributing to the electron transfer properties of the membrane protein cytochrome bc_1, for which at the time no structure was available (Lichtenstein et al., 2012; Robertson et al., 1994). The design process begins with a generic protein sequence encoding a pattern of hydrophobic and hydrophilic residues that results in a helical bundle topology; additional features such as loops, cofactor-binding sites, etc. are added iteratively onto this minimalistic model. This approach has been used to impart function ranging from binding different redox cofactors, including covalent c-type heme binding (Anderson et al., 2014), coordination of unnatural cofactors (Fry, Lehmann, Saven, DeGrado, & Therien, 2010; Fry et al., 2013; Korendovych et al., 2010; Polizzi et al., 2016), membrane solubility (Cordova, Noack, Hilcove, Lear, & Ghirlanda, 2007; Korendovych et al., 2010), and reversible oxygen binding (Osyczka, Moser, Daldal, & Dutton, 2004) to a series of designed maquettes (Cochran et al., 2005; Ogihara et al., 2001). The sequence patterning concept has been applied to generate libraries of heme-binding proteins with various functions (Hecht, Das, Go, Bradley, & Wei, 2004). The use of computational programs to predict optimal sequences to accommodate complex artificial cofactors has proven particularly powerful to investigate the electron transfer properties of proteins (Bender et al., 2007). Below we highlight a few examples.

Recently, Farid et al. (2013) designed and characterized a monomeric, four-helix bundle maquette protein malleable enough to accommodate various light- and redox active cofactors and with an exterior tolerating extensive charge patterning for modulation of redox potentials and environmental. The net charge of the bundle was raised from −16 to +11 by changing amino acid residues of the construct, and the reduction potential of both heme sites changed from −290 to −150 mV. Therefore, by simply increasing the net charge of the protein the redox potential of the heme could be modulated. The results of this work demonstrate the importance of protein's charge and electrostatic interactions in fine tuning the redox activity of the designed variants.

An elegant application of de novo designed proteins accommodating artificial cofactors was recently used to demonstrate changes in dielectric constant during electron transfer reactions (Polizzi et al., 2016). The authors mapped the electron transfer dynamics after photoexcitation of a donor-bridge-acceptor molecule free in solution to those of the molecule embedded into a de novo designed four-helix bundle and concluded that the dielectric constant of the protein decreases from $\varepsilon_s = 8$ to $\varepsilon_s = 3$ following the initial charge separation event.

Currently, catalysis of hydrogen production and CO_2 reduction by protein-embedded porphyrins has been achieved by repurposing natural heme-binding proteins, which have the advantage of robustness and of built-in electron transfer chains to other biomolecules. However, a limitation of this approach is that the active site is limited to CoPP(IX), which has limited intrinsic activity. In contrast, artificial porphyrins with much higher activity have been obtained by modulating the substituents on the porphyrin ring. De novo design of proteins able to bind artificial cobalt porphyrins holds promise to generate new artificial enzymes with increased catalytic prowess.

5. CONCLUSION AND FUTURE OUTLOOK

The design of novel metalloenzymes by incorporation of an unnatural organometallic active site within a protein scaffold is emerging as a powerful approach to expand the range of catalytic activities intrinsic to natural enzymes. Here, we discussed strategies used to generate novel redox metalloenzymes, with a focus on artificial hydrogenases and electron transfer modules. The underlying hypothesis is that the protein scaffold would serve to protect the organometallic catalyst from rapid deactivation, while simultaneously providing second sphere and long-range interactions to modulate the catalytic activity, for example, by providing a proton relay during the catalytic cycle, or by stabilizing the relevant redox state of the metal center.

In all the systems surveyed here, embedding the catalytic site in a polypeptide scaffold resulted in increased lifetime for the active species under photocatalytic conditions, thus resulting in increased turnover numbers. Similar effects are observed with encasement of diiron hexacarbonyl centers within biopolymers or nonbiological scaffolds, such as cyclodextrins, surfactants, dendrimers, or metal-organic frameworks, suggesting that the increased lifetime may be due in part to supramolecular confinement (Caserta et al., 2015; Jian et al., 2013; Onoda & Hayashi, 2015; Pullen, Fei, Orthaber, Cohen, & Ott, 2013; Quentel, Passard, & Gloaguen, 2012a, 2012b; Singleton, Reibenspies, & Darensbourg, 2010; Wang et al., 2010; Yu et al., 2013). On the other hand, mutations in the cavity surrounding the porphyrin active site resulted in further increase in activity in two systems, CoMyo and Co-cyt b_{562}. We found that increase in activity was linked mainly to modulation of the primary coordination sphere and of the redox potential of the metal center. These results suggest that designed hybrid systems containing explicit mutations to tune certain interactions and properties hold promise to increase the catalytic prowess. In particular, mutations that stabilize the catalytically active redox state may result in reduction in the overpotential for the reaction.

Building on the extensive knowledge of the mechanism of action of the organometallic active sites, two yet unexplored paths hold promise in the future for the design of artificial hydrogenases. First, computational design has proven successful in the design of artificial enzymes for a handful of well-understood reactions, such as the Kemp elimination and the Diels–Alder cycloaddition. This approach builds on the detailed understanding of the mechanism of the reactions and utilizes protein design methods to stabilize the "theozyme," eg, the ensemble of substrate and key protein contact in the transition state calculated by DFT. Second, directed evolution has been remarkably successful in tweaking the activity of existing enzymes, either natural or de novo designed ones (Blomberg et al., 2013; Giger et al., 2013; Khersonsky et al., 2012, 2010, 2011; Moroz et al., 2015; Röthlisberger et al., 2008). This approach allows exploring a large number of variants that can be screened for activity. When applied to de novo designed enzymes, directed evolution resulted in improvements of over 10^3 in k_{cat}/K_M (Blomberg et al., 2013) through mutations that remodel the active site and introduce additional interactions to the catalytic amino acids, not predicted in the initial design. These results highlight the power of directed evolution in optimizing our rudimental understanding of all the steps involved in the catalytic mechanism, including product release.

We foresee that application of computationally driven methods complemented by directed evolution to the design of novel hybrid hydrogenase mimics will result in improved catalytic properties while deepening our understanding of the catalytic mechanism. The different strategies surveyed here for the incorporation of a variety of organometallic compounds on protein scaffolds will serve as blueprints in this endeavor.

Finally, the lessons learned from the two-electron process of proton reduction will certainly prove useful in the design of more complex multielectron redox processes such as carbon dioxide reduction and nitrogen fixation.

REFERENCES

Anderson, J. L., Armstrong, C. T., Kodali, G., Lichtenstein, B. R., Watkins, D. W., Mancini, J. A., ... Dutton, P. L. (2014). Constructing a man-made c-type cytochrome maquette *in vivo*: Electron transfer, oxygen transport and conversion to a photoactive light harvesting maquette. *Chemical Science*, 5(2), 507–514. http://dx.doi.org/10.1039/C3SC52019F.

Antonkine, M. L., Koay, M. S., Epel, B., Breitenstein, C., Gopta, O., Gärtner, W., ... Lubitz, W. (2009). Synthesis and characterization of de novo designed peptides modelling the binding sites of [4Fe–4S] clusters in photosystem I. *Biochimica et Biophysica Acta (BBA) - Bioenergetics*, 1787(8), 995–1008. http://dx.doi.org/10.1016/j.bbabio.2009.03.007.

Apfel, U., Rudolph, M., Apfel, C., Robl, C., Langenegger, D., Hoyer, D., ... Weigand, W. (2010). Reaction of $Fe_3(CO)_{12}$ with octreotide-chemical, electrochemical and biological investigations. *Dalton Transactions*, 39(12), 3065–3071. http://dx.doi.org/10.1039/B921299J.

Artero, V., Chavarot-Kerlidou, M., & Fontecave, M. (2011). Splitting water with cobalt. *Angewandte Chemie International Edition*, 50(32), 7238–7266. http://dx.doi.org/10.1002/anie.201007987.

Aussignargues, C., Pandelia, M. E., Sutter, M., Plegaria, J. S., Zarzycki, J., Turmo, A., ... Kerfeld, C. A. (2016). Structure and function of a bacterial microcompartment shell protein engineered to bind a [4Fe-4S] cluster. *Journal of the American Chemical Society*, 138, 5262–5270. http://dx.doi.org/10.1021/jacs.5b11734.

Beinert, H. (2000). Iron-sulfur proteins: Ancient structures, still full of surprises. *Journal of Biological Inorganic Chemistry*, 5(1), 2–15. http://dx.doi.org/10.1007/s007750050002.

Bender, G. M., Lehmann, A., Zou, H., Cheng, H., Fry, H. C., Engel, D., ... DeGrado, W. F. (2007). De novo design of a single-chain diphenylporphyrin metalloprotein. *Journal of the American Chemical Society*, 129(35), 10732–10740. http://dx.doi.org/10.1021/ja071199j.

Blomberg, R., Kries, H., Pinkas, D. M., Mittl, P. R. E., Grütter, M. G., Privett, H. K., ... Hilvert, D. (2013). Precision is essential for efficient catalysis in an evolved Kemp eliminase. *Nature*, 503(7476), 418–421. http://dx.doi.org/10.1038/nature12623.

Caserta, G., Roy, S., Atta, M., Artero, V., & Fontecave, M. (2015). Artificial hydrogenases: Biohybrid and supramolecular systems for catalytic hydrogen production or uptake. *Current Opinion in Chemical Biology*, 25, 36–47. http://dx.doi.org/10.1016/j.cbpa.2014.12.018.

Cochran, F. V., Wu, S. P., Wang, W., Nanda, V., Saven, J. G., Therien, M. J., & DeGrado, W. F. (2005). Computational de novo design and characterization of a four-helix bundle protein that selectively binds a nonbiological cofactor. *Journal of the American Chemical Society, 127*(5), 1346–1347. http://dx.doi.org/10.1021/ja044129a.

Concepcion, J. J., House, R. L., Papanikolas, J. M., & Meyer, T. J. (2012). Chemical approaches to artificial photosynthesis. *Proceedings of the National Academy of Sciences, 109*(39), 15560–15564. http://dx.doi.org/10.1073/pnas.1212254109.

Cordova, J. M., Noack, P. L., Hilcove, S. A., Lear, J. D., & Ghirlanda, G. (2007). Design of a functional membrane protein by engineering a heme-binding site in glycophorin A. *Journal of the American Chemical Society, 129*(3), 512–518. http://dx.doi.org/10.1021/ja057495i.

DuBois, D. L. (2014). Development of molecular electrocatalysts for energy storage. *Inorganic Chemistry, 53*(8), 3935–3960. http://dx.doi.org/10.1021/ic4026969.

Dürrenberger, M., & Ward, T. R. (2014). Recent achievments in the design and engineering of artificial metalloenzymes. *Current Opinion in Chemical Biology, 19*, 99–106. http://dx.doi.org/10.1016/j.cbpa.2014.01.018.

Dutta, A., Lense, S., Hou, J., Engelhard, M. H., Roberts, J. A. S., & Shaw, W. J. (2013). Minimal proton channel enables H_2 oxidation and production with a water-soluble nickel-based catalyst. *Journal of the American Chemical Society, 135*(49), 18490–18496. http://dx.doi.org/10.1021/ja407826d.

Dutta, A., Roberts, J. A. S., & Shaw, W. J. (2014). Arginine-containing ligands enhance H_2 oxidation catalyst performance. *Angewandte Chemie International Edition, 53*(25), 6487–6491. http://dx.doi.org/10.1002/anie.201402304.

Faiella, M., Roy, A., Sommer, D. J., & Ghirlanda, G. (2013). De novo design of functional proteins: Toward artificial hydrogenases. *Biopolymers, 100*(6), 558–571. http://dx.doi.org/10.1002/bip.22420.

Farid, T. A., Kodali, G., Solomon, L. A., Lichtenstein, B. R., Sheehan, M. M., Fry, B. A., ... Dutton, P. L. (2013). Elementary tetrahelical protein design for diverse oxidoreductase functions. *Nature Chemical Biology, 9*(12), 826–833. http://dx.doi.org/10.1038/nchembio.1362.

Fry, H. C., Lehmann, A., Saven, J. G., DeGrado, W. F., & Therien, M. J. (2010). Computational design and elaboration of a de novo heterotetrameric α-helical protein that selectively binds an emissive abiological (porphinato)zinc chromophore. *Journal of the American Chemical Society, 132*(11), 3997–4005. http://dx.doi.org/10.1021/ja907407m.

Fry, H. C., Lehmann, A., Sinks, L. E., Asselberghs, I., Tronin, A., Krishnan, V., ... Therien, M. J. (2013). Computational de novo design and characterization of a protein that selectively binds a highly hyperpolarizable abiological chromophore. *Journal of the American Chemical Society, 135*(37), 13914–13926. http://dx.doi.org/10.1021/ja4067404.

Fukuyama, K. (2004). Structure and function of plant-type ferredoxins. *Photosynthesis Research, 81*(3), 289–301. http://dx.doi.org/10.1023/B:PRES.0000036882.19322.0a.

Gibney, B. R., Mulholland, S. E., Rabanal, F., & Dutton, P. L. (1996). Ferredoxin and ferredoxin-heme maquettes. *Proceedings of the National Academy of Sciences of the United States of America, 93*(26), 15041–15046. http://dx.doi.org/10.1073/pnas.93.26.15041.

Giger, L., Caner, S., Obexer, R., Kast, P., Baker, D., Ban, N., & Hilvert, D. (2013). Evolution of a designed retro-aldolase leads to complete active site remodeling. *Nature Chemical Biology, 9*(8), 494–498. http://dx.doi.org/10.1038/nchembio.1276.

Ginovska-Pangovska, B., Dutta, A., Reback, M. L., Linehan, J. C., & Shaw, W. J. (2014). Beyond the active site: The impact of the outer coordination sphere on electrocatalysts for hydrogen production and oxidation. *Accounts of Chemical Research, 47*(8), 2621–2630. http://dx.doi.org/10.1021/ar5001742.

Grzyb, J., Xu, F., Nanda, V., Luczkowska, R., Reijerse, E., Lubitz, W., & Noy, D. (2012). Empirical and computational design of iron-sulfur cluster proteins. *Biochimica et Biophysica Acta, 1817*(8), 1256–1262. http://dx.doi.org/10.1016/j.bbabio.2012.02.001.

Grzyb, J., Xu, F., Weiner, L., Reijerse, E. J., Lubitz, W., Nanda, V., & Noy, D. (2010). De novo design of a non-natural fold for an iron-sulfur protein: Alpha-helical coiled-coil with a four-iron four-sulfur cluster binding site in its central core. *Biochimica et Biophysica Acta, 1797*(3), 406–413. http://dx.doi.org/10.1016/j.bbabio.2009.12.012.

Gust, D., Moore, T. A., & Moore, A. L. (2009). Solar fuels via artificial photosynthesis. *Accounts of Chemical Research, 42*(12), 1890–1898. http://dx.doi.org/10.1021/ar900209b.

Gust, D., Moore, T. A., & Moore, A. L. (2012). Realizing artificial photosynthesis. *Faraday Discussions, 155*, 9–26. http://dx.doi.org/10.1039/C1FD00110H.

Hamels, D. R., & Ward, T. R. (2013). Biomacromolecules as ligands for artificial metalloenzymes. In J. Poeppelmeier & K. Reedijk (Eds.), *Comprehensive inorganic chemistry II: Vol. 6.* (2nd ed., pp. 737–761). Amsterdam: Elsevier.

Han, G. W., Yang, X. L., McMullan, D., Chong, Y. E., Krishna, S. S., Rife, C. L., ... Wilson, I. A. (2010). Structure of a tryptophanyl-tRNA synthetase containing an iron-sulfur cluster. *Acta Crystallographica. Section F, Structural Biology and Crystallization Communications, 66*, 1326–1334. http://dx.doi.org/10.1107/S1744309110037619.

Hecht, M. H., Das, A., Go, A., Bradley, L. H., & Wei, Y. (2004). De novo proteins from designed combinatorial libraries. *Protein Science, 13*(7), 1711–1723. http://dx.doi.org/10.1110/ps.04690804.

Jian, J.-X., Liu, Q., Li, Z.-J., Wang, F., Li, X.-B., Li, C.-B., ... Wu, L.-Z. (2013). Chitosan confinement enhances hydrogen photogeneration from a mimic of the diiron subsite of [FeFe]-hydrogenase. *Nature Communications, 4*. Art no. 2695, http://dx.doi.org/10.1038/ncomms3695.

Johnson, D. C., Dean, D. R., Smith, A. D., & Johnson, M. K. (2005). Structure, function, and formation of biological iron-sulfur clusters. *Annual Review of Biochemistry, 74*, 247–281. http://dx.doi.org/10.1146/annurev.biochem.74.082803.133518.

Jones, A. K., Lichtenstein, B. R., Dutta, A., Gordon, G., & Dutton, P. L. (2007). Synthetic hydrogenases: Incorporation of an iron carbonyl thiolate into a designed peptide. *Journal of the American Chemical Society, 129*(48), 14844–14845. http://dx.doi.org/10.1021/ja075116a.

Khersonsky, O., Kiss, G., Röthlisberger, D., Dym, O., Albeck, S., Houk, K. N., ... Tawfik, D. S. (2012). Bridging the gaps in design methodologies by evolutionary optimization of the stability and proficiency of designed Kemp eliminase KE59. *Proceedings of the National Academy of Sciences, 109*(26), 10358–10363. http://dx.doi.org/10.1073/pnas.1121063109.

Khersonsky, O., Röthlisberger, D., Dym, O., Albeck, S., Jackson, C. J., Baker, D., & Tawfik, D. S. (2010). Evolutionary optimization of computationally designed enzymes: Kemp eliminases of the KE07 series. *Journal of Molecular Biology, 396*(4), 1025–1042. http://dx.doi.org/10.1016/j.jmb.2009.12.031.

Khersonsky, O., Röthlisberger, D., Wollacott, A. M., Murphy, P., Dym, O., Albeck, S., ... Tawfik, D. S. (2011). Optimization of the in-silico-designed Kemp eliminase KE70 by computational design and directed evolution. *Journal of Molecular Biology, 407*(3), 391–412. http://dx.doi.org/10.1016/j.jmb.2011.01.041.

Kleingardner, J. G., Kandemir, B., & Bren, K. L. (2014). Hydrogen evolution from neutral water under aerobic conditions catalyzed by cobalt microperoxidase-11. *Journal of the American Chemical Society, 136*(1), 4–7. http://dx.doi.org/10.1021/ja406818h.

Koay, M. S., Antonkine, M. L., Gärtner, W., & Lubitz, W. (2008). Modelling low-potential [Fe4S4] clusters in proteins. *Chemistry & Biodiversity, 5*, 1571–1587. http://dx.doi.org/10.1002/cbdv.200890145.

Korendovych, I. V., Senes, A., Kim, Y. H., Lear, J. D., Fry, H. C., Therien, M. J., ... Degrado, W. F. (2010). De novo design and molecular assembly of a transmembrane diporphyrin-binding protein complex. *Journal of the American Chemical Society, 132*(44), 15516–15518. http://dx.doi.org/10.1021/ja107487b.

Lewis, J. C. (2013). Artificial metalloenzymes and metallopeptide catalysts for organic synthesis. *ACS Catalysis*, *3*(12), 2954–2975. http://dx.doi.org/10.1021/cs400806a.

Lichtenstein, B. R., Farid, T. A., Kodali, G., Solomon, L. A., Anderson, J. L., Sheehan, M. M., ... Dutton, P. L. (2012). Engineering oxidoreductases: Maquette proteins designed from scratch. *Biochemical Society Transactions*, *40*(3), 561–566. http://dx.doi.org/10.1042/BST20120067.

Liu, J., Chakraborty, S., Hosseinzadeh, P., Yu, Y., Tian, S., Petrik, I., ... Lu, Y. (2014). Metalloproteins containing cytochrome, iron-sulfur, or copper redox centers. *Chemical Reviews*, *114*(8), 4366–4469. http://dx.doi.org/10.1021/cr400479b.

Lubitz, W., Ogata, H., Rüdiger, O., & Reijerse, E. (2014). Hydrogenases. *Chemical Reviews*, *114*(8), 4081–4148. http://dx.doi.org/10.1021/cr4005814.

McEvoy, J. P., & Brudvig, G. W. (2006). Water-splitting chemistry of photosystem II. *Chemical Reviews*, *106*(11), 4455–4483. http://dx.doi.org/10.1021/cr0204294.

Moroz, Y. S., Dunston, T. T., Makhlynets, O. V., Moroz, O. V., Wu, Y., Yoon, J. H., ... Korendovych, I. V. (2015). New tricks for old proteins: Single mutations in a non-enzymatic protein give rise to various enzymatic activities. *Journal of the American Chemical Society*, *137*(47), 14905–14911. http://dx.doi.org/10.1021/jacs.5b07812.

Nanda, V., Senn, S., Pike, D. H., Rodriguez-Granillo, A., Hansen, W. A., Khare, S. D., & Noy, D. (2016). Structural principles for computational and de novo design of 4Fe-4S metalloproteins. *Biochimica et Biophysica Acta*, *1857*, 531–538. http://dx.doi.org/10.1016/j.bbabio.2015.10.001.

Nicolet, Y., Lemon, B. J., Fontecilla-Camps, J. C., & Peters, J. W. (2000). A novel FeS cluster in Fe-only hydrogenases. *Trends in Biochemical Sciences*, *25*(3), 138–143. http://dx.doi.org/10.1016/S0968-0004(99)01536-4.

Ogihara, N. L., Ghirlanda, G., Bryson, J. W., Gingery, M., DeGrado, W. F., & Eisenberg, D. (2001). Design of three-dimensional domain-swapped dimers and fibrous oligomers. *Proceedings of the National Academy of Sciences of the United States of America*, *98*(4), 1404–1409. http://dx.doi.org/10.1073/pnas.98.4.1404.

Onoda, A., & Hayashi, T. (2015). Artificial hydrogenase: Biomimetic approaches controlling active molecular catalysts. *Current Opinion in Chemical Biology*, *25*, 133–140. http://dx.doi.org/10.1016/j.cbpa.2014.12.041.

Onoda, A., Kihara, Y., Fukumoto, K., Sano, Y., & Hayashi, T. (2014). Photoinduced hydrogen evolution catalyzed by a synthetic diiron dithiolate complex embedded within a protein matrix. *ACS Catalysis*, *4*(8), 2645–2648. http://dx.doi.org/10.1021/cs500392e.

Osyczka, A., Moser, C. C., Daldal, F., & Dutton, P. L. (2004). Reversible redox energy coupling in electron transfer chains. *Nature*, *427*(6975), 607–612. http://dx.doi.org/10.1038/nature02242.

Ozaki, S.-i, Matsui, T., Roach, M. P., & Watanabe, Y. (2000). Rational molecular design of a catalytic site: Engineering of catalytic functions to the myoglobin active site framework. *Coordination Chemistry Reviews*, *198*(1), 39–59. http://dx.doi.org/10.1016/S0010-8545(00)00234-4.

Ozaki, S.-i, Roach, M. P., Matsui, T., & Watanabe, Y. (2001). Investigations of the roles of the distal heme environment and the proximal heme iron ligand in peroxide activation by heme enzymes via molecular engineering of myoglobin. *Accounts of Chemical Research*, *34*(10), 818–825. http://dx.doi.org/10.1021/ar9502590.

Polizzi, N. F., Eibling, M. J., Perez-Aguilar, J. M., Rawson, J., Lanci, C. J., Fry, H. C., ... Therien, M. J. (2016). Photoinduced electron transfer elicits a change in the static dielectric constant of a de novo designed protein. *Journal of the American Chemical Society*, *138*(7), 2130–2133. http://dx.doi.org/10.1021/jacs.5b13180.

Pullen, S., Fei, H., Orthaber, A., Cohen, S. M., & Ott, S. (2013). Enhanced photochemical hydrogen production by a molecular diiron catalyst incorporated into a metal–organic framework. *Journal of the American Chemical Society*, *135*(45), 16997–17003. http://dx.doi.org/10.1021/ja407176p.

Quentel, F., Passard, G., & Gloaguen, F. (2012a). A binuclear iron–thiolate catalyst for electrochemical hydrogen production in aqueous micellar solution. *Chemistry – A European Journal*, *18*(42), 13473–13479. http://dx.doi.org/10.1002/chem.201201884.

Quentel, F., Passard, G., & Gloaguen, F. (2012b). Electrochemical hydrogen production in aqueous micellar solution by a diiron benzenedithiolate complex relevant to [FeFe] hydrogenases. *Energy & Environmental Science*, *5*(7), 7757–7761. http://dx.doi.org/10.1039/C2EE21531D.

Robertson, D. E., Farid, R. S., Moser, C. C., Urbauer, J. L., Mulholland, S. E., Pidikiti, R., ... Dutton, P. L. (1994). Design and synthesis of multi-haem proteins. *Nature*, *368*(6470), 425–432. http://dx.doi.org/10.1038/368425a0.

Röthlisberger, D., Khersonsky, O., Wollacott, A. M., Jiang, L., DeChancie, J., Betker, J., ... Baker, D. (2008). Kemp elimination catalysts by computational enzyme design. *Nature*, *453*(7192), 190–195. http://dx.doi.org/10.1038/nature06879.

Roy, A., Madden, C., & Ghirlanda, G. (2012). Photo-induced hydrogen production in a helical peptide incorporating a [FeFe] hydrogenase active site mimic. *Chemical Communications*, *48*(79), 9816–9818. http://dx.doi.org/10.1039/C2CC34470J.

Roy, S., Nguyen, T.-A. D., Gan, L., & Jones, A. K. (2015). Biomimetic peptide-based models of [FeFe]-hydrogenases: Utilization of phosphine-containing peptides. *Dalton Transactions*, *44*(33), 14865–14876. http://dx.doi.org/10.1039/C5DT01796C.

Roy, A., Sarrou, I., Vaughn, M. D., Astashkin, A. V., & Ghirlanda, G. (2013). De novo design of an artificial bis[4Fe-4S] binding protein. *Biochemistry*, *52*(43), 7586–7594. http://dx.doi.org/10.1021/bi401199s.

Roy, S., Shinde, S., Hamilton, G. A., Hartnett, H. E., & Jones, A. K. (2011). Artificial [FeFe]-hydrogenase: On resin modification of an amino acid to anchor a hexacarbonyldiiron cluster in a peptide framework. *European Journal of Inorganic Chemistry*, *2011*(7), 1050–1055. http://dx.doi.org/10.1002/ejic.201000979.

Roy, A., Sommer, D. J., Schmitz, R. A., Brown, C. L., Gust, D., Astashkin, A., & Ghirlanda, G. (2014). A de novo designed 2[4Fe-4S] ferredoxin mimic mediates electron transfer. *Journal of the American Chemical Society*, *136*(49), 17343–17349. http://dx.doi.org/10.1021/ja510621e.

Sano, Y., Onoda, A., & Hayashi, T. (2011). A hydrogenase model system based on the sequence of cytochrome c: Photochemical hydrogen evolution in aqueous media. *Chemical Communications*, *47*(29), 8229–8231. http://dx.doi.org/10.1039/C1CC11157D.

Sano, Y., Onoda, A., & Hayashi, T. (2012). Photocatalytic hydrogen evolution by a diiron hydrogenase model based on a peptide fragment of cytochrome c_{556} with an attached diiron carbonyl cluster and an attached ruthenium photosensitizer. *Journal of Inorganic Biochemistry*, *108*, 159–162. http://dx.doi.org/10.1016/j.jinorgbio.2011.07.010.

Scott, M. P., & Biggins, J. (1997). Introduction of a [4Fe-4S (S-cys)4]$^{+1,+2}$ iron-sulfur center into a four-α helix protein using design parameters from the domain of the F_X cluster in the photosystem I reaction center. *Protein Science*, *6*(2), 340–346. http://dx.doi.org/10.1002/pro.5560060209.

Simmons, T. R., Berggren, G., Bacchi, M., Fontecave, M., & Artero, V. (2014). Mimicking hydrogenases: From biomimetics to artificial enzymes. *Coordination Chemistry Reviews*, *270-271*, 127–150. http://dx.doi.org/10.1016/j.ccr.2013.12.018.

Singleton, M. L., Reibenspies, J. H., & Darensbourg, M. Y. (2010). A cyclodextrin host/guest approach to a hydrogenase active site biomimetic cavity. *Journal of the American Chemical Society*, *132*(26), 8870–8871. http://dx.doi.org/10.1021/ja103774j.

Sommer, D. J., Alcala-Torano, R., Bahrami Dizicheh, Z., & Ghirlanda, G. (In Press). Design of electron transfer peptides: Towards functional materials. In T. Z. Grove & A. L. Cortajarena (Eds.), *Protein-based engineered nanostructures*. Springer.

Sommer, D. J., Roy, A., Astashkin, A., & Ghirlanda, G. (2015). Modulation of cluster incorporation specificity in a *de novo* iron-sulfur cluster binding peptide. *Biopolymers, 104*(4), 412–418. http://dx.doi.org/10.1002/bip.22635.

Sommer, D. J., Vaughn, M. D., Clark, B. C., Tomlin, J., Roy, A., & Ghirlanda, G. (2016). Reengineering cyt b_{562} for hydrogen production: A facile route to artificial hydrogenases. *Biochimica et Biophysica Acta, 1857*, 598–603. http://dx.doi.org/10.1016/j.bbabio.2015.09.001.

Sommer, D. J., Vaughn, M. D., & Ghirlanda, G. (2014). Protein secondary-shell interactions enhance the photoinduced hydrogen production of cobalt protoporphyrin IX. *Chemical Communications, 50*(100), 15852–15855. http://dx.doi.org/10.1039/c4cc06700b.

Wang, W.-G., Wang, F., Wang, H.-Y., Si, G., Tung, C.-H., & Wu, L.-Z. (2010). Photocatalytic hydrogen evolution by [FeFe] hydrogenase mimics in homogeneous solution. *Chemistry – An Asian Journal, 5*(8), 1796–1803. http://dx.doi.org/10.1002/asia.201000087.

Wilson, M. E., & Whitesides, G. M. (1978). Conversion of a protein to a homogeneous asymmetric hydrogenation catalyst by site-specific modification with a diphosphinerhodium(I) moiety. *Journal of the American Chemical Society, 100*(1), 306–307. http://dx.doi.org/10.1021/ja00469a064.

Yang, Y., Zhou, Q., Wang, L., Liu, X., Zhang, W., Hu, M., ... Wang, J. (2015). Significant improvement of oxidase activity through the genetic incorporation of a redox-active unnatural amino acid. *Chemical Science, 6*(7), 3881–3885. http://dx.doi.org/10.1039/C5SC01126D.

Yu, Y., Cui, C., Liu, X., Petrik, I. D., Wang, J., & Lu, Y. (2015). A designed metalloenzyme achieving the catalytic rate of a native enzyme. *Journal of the American Chemical Society, 137*(36), 11570–11573. http://dx.doi.org/10.1021/jacs.5b07119.

Yu, T., Zeng, Y., Chen, J., Li, Y.-Y., Yang, G., & Li, Y. (2013). Exceptional dendrimer-based mimics of diiron hydrogenase for the photochemical production of hydrogen. *Angewandte Chemie International Edition, 52*(21), 5631–5635. http://dx.doi.org/10.1002/anie.201301289.

CHAPTER EIGHTEEN

Equilibrium Studies of Designed Metalloproteins

B.R. Gibney*,†,1
*Brooklyn College, Brooklyn, NY, United States
†Ph.D. Programs in Chemistry and Biochemistry, The Graduate Center of the City University of New York, New York, NY, United States
[1]Corresponding author: e-mail address: BGibney@brooklyn.cuny.edu

Contents

1. Introduction	418
2. Metalloprotein Thermodynamics	419
2.1 The Influence of Protons on Metal-Ion Affinity	420
2.2 The Thermodynamic Relationship Between Heme Electrochemistry and Affinity	422
2.3 The Influence of Protons on Heme Electrochemistry	424
2.4 The Thermodynamic Relationship Between Heme Electrochemistry and Protein Folding	426
3. Essential Equilibrium Measurements	426
3.1 Cofactor Binding Equilibria	426
3.2 Protonation Equilibrium	427
3.3 Electrochemical Equilibrium	428
4. Thermodynamic Analysis of a Heme Binding Four Helix Bundle, [Δ7-His]$_2$	430
4.1 Heme Affinity	430
4.2 Electrochemistry	431
4.3 Proton Competition	432
4.4 The PCET Event	434
5. Experimental Procedures	435
5.1 Protein Synthesis and Purification	435
5.2 Equilibrium Measurements	435
5.3 Kinetic Evaluation	436
6. Summary	436
Acknowledgment	437
References	437

Abstract

Complete thermodynamic descriptions of the interactions of cofactors with proteins via equilibrium studies are challenging, but are essential to the evaluation of designed metalloproteins. While decades of studies on protein–protein interaction thermodynamics provide a strong underpinning to the successful computational design of novel protein folds and de novo proteins with enzymatic activity, the corresponding paucity of data on metal–protein interaction thermodynamics limits the success of computational metalloprotein design efforts. By evaluating the thermodynamics of metal–protein interactions via equilibrium binding studies, protein unfolding free energy determinations, proton competition equilibria, and electrochemistry, a more robust basis for the computational design of metalloproteins may be provided. Our laboratory has shown that such studies provide detailed insight into the assembly and stability of designed metalloproteins, allow for parsing apart the free energy contributions of metal–ligand interactions from those of porphyrin–protein interactions in hemeproteins, and even reveal their mechanisms of proton-coupled electron transfer. Here, we highlight studies that reveal the complex interplay between the various equilibria that underlie metalloprotein assembly and stability and the utility of making these detailed measurements.

1. INTRODUCTION

Life is a nonequilibrium situation, however, chemical equilibria permeate, and are of fundamental importance to, biological systems. The vast majority of chemical reactions exist in equilibrium between reactants and products where the forward and reverse reactions occur simultaneously. Reversible chemical reactions such as these can be found throughout biochemistry including in metal-ion transport and trafficking (Fu, Chang, & Giedroc, 2014; Reddi, Jensen, & Culotta, 2009) as well as energy conversion processes (Dempsey, Winkler, & Gray, 2010). However, nonequilibrium events are also critical to biochemical function. In the case of metal-ion trafficking, equilibrium binding constants of various proteins contribute to metal-ion distribution throughout the cell, and it may be that kinetic traps are critical for specific metal-ion delivery. For instance, the Zn(II) in carbonic anhydrase is kinetically trapped which allows formation of the holoenzyme despite strong competitors (Kiefer & Fierke, 1994). In the case of energy conversion, the mitochondrial electron transport chain uses the energy provided by NADH to drive proton pumps that create the chemiosmotic pH gradient, a nonequilibrium situation. The enzyme ATP synthase attempts to restore the equilibrium state and in the process produces chemical energy in the form of ATP (Suzuki, Tanaka, Wakabayashi, Saita, & Yoshida, 2014).

Reactant and products in a reversible chemical reaction exist in a dynamic equilibrium that maintains their activities over time. Once equilibrium is established, the concentrations of reactants and products can be determined to directly measure the equilibrium constant, K_{eq}, which provides insight into the thermodynamics of the chemical reaction, ie, $\Delta G_{rxn} = -RT \ln K_{eq}$. Equilibrium constants can also be determined by measuring the kinetics of the approach to equilibrium, $k_{forward}/k_{reverse} = K_{eq}$. The thermodynamic approach is often limited by the concentration of the reactants and products required to achieve an accurately measurable spectroscopic signal. The rapid establishment of equilibrium prior to kinetic measurements can limit the kinetic approach to K_{eq} determination.

Once determined, K_{eq} values can be used to provide powerful insight into the thermodynamics of the given reaction. Metal-ion binding constants have long been used as sensitive probes for reporting the thermodynamics of protein structural changes (Magyar & Arnold-Godwin, 2003). In fact, one of the earliest β-sheet propensity scales was developed based on the Co(II) binding constants of consensus peptide 1, a designed zinc finger protein with a ββα fold (Kim & Berg, 1993). Herein, we will discuss the application of metal-ion binding equilibrium studies to de novo designed metalloproteins. We will show how metal-ion binding equilibria are interdependent with ligand deprotonation equilibria, metal oxidation/reduction equilibria, and global protein folding/unfolding equilibria. The data will show that the deceptively simple concept of a metal-ion binding equilibrium constant belies the complexity that is biological systems. Our systematic approach to measuring each of these interrelated equilibrium constants provides fundamental insight into the design of metalloproteins from first principles as well as their basis for proton-coupled electron transfer (PCET) function.

2. METALLOPROTEIN THERMODYNAMICS

Metalloproteins are in essence highly elaborated inorganic coordination compounds (Karlin, 1993). Our approach to de novo metalloprotein design is based on delineating the equilibrium thermodynamics of our designed proteins to reveal their fundamental engineering (Reddi, Guzman, Breece, Tierney, & Gibney, 2007). We view this as a necessary step toward both the rational and the computational design of metalloproteins and metalloenzymes from first principles (Cangelosi, Deb, Penner-Hahn, & Pecoraro, 2014). This assertion is based on the success of computation protein design that is firmly rooted in decades of protein–protein interaction thermodynamics studies (Bryson et al., 1995; Dahiyat & Mayo, 1997; Jiang et al., 2008). At the current stage, these measurements allow for an evaluation of the success of our individual designs and for the delineation of the effects of structural perturbation within them. This has allowed us to provide insight into the role of the porphyrin substituents of heme *a*, the heme found in human cytochrome *c* oxidase (Zhuang, Amoroso, et al., 2006; Zhuang, Reddi, et al., 2006).

Our studies begin with detailed measurements of the metal-ion binding constants, also called stability constants or association constants, ie, K_a values which are not to be confused with the acid dissociation constants in pK_a values. Each metal-ion binding constant is determined under a set of experimental conditions that affect its value (temperature, solution pH, ionic strength, potential, etc.). Thus, the individual measurement is a conditional metal-ion binding constant. These are typically expressed in the biochemical literature as conditional dissociation constants, or K_d values, in units of concentration (M). It is useful to keep in mind that for a 1:1 binding reaction, a sample that contains a equimolar amount of protein and metal-ion, at the K_d value concentration generates 38% product, at 10 times the K_d value concentration the same reaction stoichiometry generates 73% product, at 100 times the K_d value it generates 90% product, and 1000 times the K_d value concentration is necessary to generate 99% product. Each K_d value provides a measurement of the binding reaction free energy using the relationship $\Delta G_{rxn} = -RT \ln K_d$.

2.1 The Influence of Protons on Metal-Ion Affinity

Since the binding of metal-ions to protein-based ligands is often in competition with proton binding, the solution pH is a major effector of conditional binding constants. For example, let us consider the most prevalent de novo hemeprotein design, the coordination of a ferric heme, iron(III)(protoporphyrin IX) to a bis-His binding site (Reedy & Gibney, 2004). Fig. 1 shows a plot of the $-\log K_d$ values vs pH for one such a system, monoheme-[Δ7-His]$_2$ (vide infra), which demonstrate the pH dependence of the heme affinity. Depending on solution conditions, the binding reaction can be described by a combination of the following reactions:

$$\text{Heme} + \text{Protein} - (\text{HisH}^+)_2 \rightleftharpoons \text{Hemeprotein} - (\text{HisH}^+)_2 \quad \text{(i)}$$
$$\text{Heme} + \text{Protein} - (\text{HisH}^+)_2 \rightleftharpoons \text{Hemeprotein} - (\text{His})_2 + 2\text{H}^+ \quad \text{(ii)}$$
$$\text{Heme} + \text{Protein} - (\text{His})_2 \rightleftharpoons \text{Hemeprotein} - (\text{His})_2 \quad \text{(iii)}$$

Reaction (i) is pH-independent and a conditional binding constant measured under these conditions reflects the formation constant of the

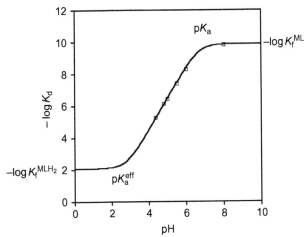

Fig. 1 Plot of $-\log K_d$ values vs pH for ferric monoheme-[Δ7-His]$_2$ is used to determine the formation constant, K_f^{ML}. Individual K_d measurements are shown as *open squares*. The location of the critical variables (pK_a, pK_a^{eff}, $-\log K_f^{MLH_2}$, and $-\log K_f^{ML}$) are noted in the graph.

protonated metal–ligand complex, $K_f^{MLH_2}$. This formation constant may be measured at pH values below both the intrinsic pK_a of histidine in the protein and the effective pK_a^{eff} of the hemeprotein histidine, eg, pH 1.5. Since the HisH$^+$, the imidazolium form, does not coordinate the heme iron, the free energy of this reaction, $\Delta G_{rxn1} = -RT \ln K_f^{MLH_2}$, may be ascribed to porphyrin–protein interactions, or a hydrophobic association. Reaction (iii) is also pH-independent and the conditional binding constant measured under these conditions represents the formation constant of the deprotonated metal–ligand complex, K_f^{ML}. In this case, the imidazole form of histidine coordinates the heme iron and the free energy of this reaction, $\Delta G_{rxn3} = -RT \ln K_f^{ML}$ includes the free energy of both the iron coordination and porphyrin–protein interactions. The value of $K_f^{MLH_2}$ may be measured at pH values above the intrinsic pK_a of histidine in the protein, eg, pH 9.0. Lastly, reaction (ii) represents the pH-dependent conditional dissociation constant, K_d, observed between pH 2 and 8.

$$K_f^{MLH_2} = [\text{Hemeprotein} - (\text{HisH}^+)_2]/[\text{Heme}][\text{Protein} - (\text{HisH}^+)_2] \quad (1)$$

$$K_a = [\text{Hemeprotein}][\text{H}^+]^2/[\text{Heme}][\text{Protein}] \quad (2)$$

$$K_d = [\text{Heme}][\text{Protein}]/[\text{Hemeprotein}][\text{H}^+]^2$$
$$= [\text{Heme}][\text{Protein}]/[\text{Hemeprotein}]\,(10^{-2\text{pH}}) \tag{3}$$
$$K_f^{ML} = [\text{Hemeprotein} - (\text{His})_2]/[\text{Heme}][\text{Protein} - (\text{His})_2]. \tag{4}$$

The conditional binding constant, K_a or K_d, shows a $[\text{H}^+]^2$ dependence; in other words, a unit change in the solution pH value changes the K_a or K_d value by a factor of 100. Thus, it is critical to measure and effectively buffer the pH at which the conditional value is measured.

The two formation constants and the conditional dissociation constants are related by the intrinsic and effective pK_a values of the histidine ligands. The intrinsic pK_a value of histidine in the protein is the dissociation of Protein-$(\text{HisH}^+)_2$ to form Protein-$(\text{His})_2 + 2\text{H}^+$, which is equivalent to the difference between Reaction (i) and (ii). Likewise, the difference between Reaction (ii) and Reaction (iii), Hemeprotein-$(\text{HisH}^+)_2$ to form Hemeprotein-$(\text{His})_2 + 2\text{H}^+$, is the effective pK_a^{eff} of the histidine coordinated to the heme iron.

Ultimately, the conditional dissociation constants, K_d values; the formation constants for the protonated, $K_f^{MLH_2}$, and deprotonated, K_f^{ML}, forms of the hemeprotein; the intrinsic, pK_a, and effective, pK_a^{eff}, acid dissociation constant values of the histidine ligands can be combined into a single equilibrium expression. In the case of the designed hemeprotein $[\Delta 7\text{-His}]_2$, the pH dependence of $-\log K_d$ is shown in Fig. 1. The fit to the data is given by the following expression:

$$-\log K_d = -\log K_f * \frac{10^{(-2\text{pH}+2pK_a)} + 10^{(-\text{pH}+pK_a)}}{10^{(-2\text{pH}+2pK_a^{\text{eff}})} + 10^{(-\text{pH}+pK_a^{\text{eff}})}}.$$

At basic pH values, above the intrinsic histidine pK_a, the data plateaus at the value of $-\log K_f^{ML}$ due to the dominance of reaction (iii). As the solution pH is lowered below the intrinsic pK_a value of histidine, the observed $-\log K_d$ value is attenuated with a slope of 2. This attenuation ceases below the effective pK_a^{eff} value of histidine, and plateaus at the value of $-\log K_f^{MLH_2}$. This clearly shows that the simple metal-ion binding constant is more complex than it seemed upon first inspection.

2.2 The Thermodynamic Relationship Between Heme Electrochemistry and Affinity

The situation is even more complex, as the description earlier is for heme in a single oxidation state, and solution potential is also a major factor in heme binding constants since it affects the heme iron oxidation state. Reaction (iv) expresses the half-cell reaction for hemeprotein reduction, the simplest electrochemical reaction. The Nernst equation relates the midpoint reduction potential of a redox-active species, E_m, to its standard reduction potential, E^0; the number of electrons in the reduction reaction, n; the Faraday constant, F; and the activities of the oxidized and reduced states of the redox-active species, α^{ox} and α^{red}, respectively.

$$\text{Ferric} - \text{Hemeprotein} - (\text{His})_2 + e^-$$
$$\rightleftharpoons \text{Ferrous} - \text{Hemeprotein} - (\text{His})_2 \quad \text{(iv)}$$

$$E_m = E^0 + \frac{RT}{nF} \ln \frac{K_d^{red}}{K_d^{ox}}. \quad (5)$$

For our purposes, it is convenient to convert the activities into the hemeprotein dissociation constants, K_d^{ox} and K_d^{red} as given in Eq. (5). Thus, the midpoint reduction potential, E_m, reflects the *ratio* of the hemeprotein dissociation constants in the two oxidation state and can be converted to a free energy using the relationship $\Delta G = -nFE_m$. Factors that differentially affect the oxidized or reduced heme binding constants, K_d^{ox} and K_d^{red} values, have an impact on the measured midpoint reduction potential value since the free energy difference between oxidized and reduced heme binding, $\Delta G^{ox} - \Delta G^{red} = \left(-RT \ln K_d^{ox} + RT \ln K_d^{red}\right) = -nFE_m$. Thus, the midpoint reduction potential, E_m, is a measure of the differential stability of oxidized and reduced hemeproteins, factors that stabilize ferric heme binding more than ferrous make the midpoint reduction potential more negative, whereas those that preferentially stabilize ferrous heme binding over ferric make the reduction potential more positive. If one knows the standard reduction potential, E^0, of the redox cofactor, then knowledge of the binding constant in one oxidation state and the midpoint reduction potential can be used to deduce the binding constant in the other oxidation state.

In the case of hemeproteins, the standard reduction potential of heme, E^0, has not been determined, likely due to the tendency of heme to oligomerize under aqueous conditions (De Villiers, Kashula, Egan, & Marques, 2007). However, it may be possible to determine the formal reduction potential of heme, $E^{0\prime}$, under specific conditions from knowledge of the ferric and ferrous heme dissociation constant ratio, K_d^{red}/K_d^{ox}, and the midpoint reduction potential of the hemeprotein, E_m. The E_m values extant in the biochemical literature demonstrate an 1100 mV range, from the −450 mV value of the heme acquisition A (HasA) hemeprotein to the +550 mV value of diheme cytochrome *c* peroxidase (Reedy, Elvekrog, &

Gibney, 2008). Since $\Delta\Delta G = -nF\Delta E_m$, the observed 1100 mV range (ΔE_m) represents a 25.5 kcal/mol difference in the ratio of the ferric and ferrous heme dissociation constants since 43 mV is equivalent to 1 kcal/mol at 25°C. While heme dissociation constants are relatively rare in the literature, the electrochemical data indicate a 6×10^{18} change in the ratio of the ferric and ferrous heme dissociation constants between HasA and diheme cytochrome c peroxidase based on the relationship $\Delta\Delta G = -RT \ln \Delta K_{eq}$.

2.3 The Influence of Protons on Heme Electrochemistry

As discussed earlier, the binding of heme to the typical bis-His site in a designed hemeprotein is a pH-dependent reaction. The conditional dissociation constants for heme binding *in each oxidation state* are impacted by the intrinsic pK_a and effective pK_a^{eff} values of the histidine ligands and the formation constant, K_f^{ML}. While the value of the intrinsic pK_a for the histidine ligands is not oxidation state dependent, the values of the effective pK_a^{eff} and the formation constant, K_f^{ML} are. Thus, the K_d^{red}/K_d^{ox} ratio and the resulting E_m value are pH dependent as shown for monoheme-[Δ7-His]$_2$ in Fig. 2.

Reaction (iv) shows the simplest electrochemical reaction for a hemeprotein with bis-His ligation. This reduction potential, E_m value, requires that neither the oxidized state nor the reduced state of hemeprotein has

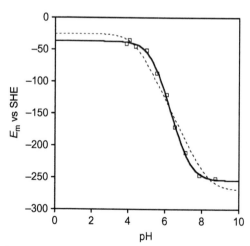

Fig. 2 Pourbaix diagram of monoheme-[Δ7-His]$_2$. Individual measurements of the reduction potential, E_m values, are given as *open squares*. The *solid line* shows that the data are well fit to a 2H$^+$/1e$^-$ process. The *dashed fit* is a 1H$^+$/1e$^-$ process for comparison.

protons bound to the His residues. As such, this E_m value is observed at pH values above the intrinsic pK_a values of the histidine ligands in both oxidation states. We shall refer to this E_m value as E_m^{base} as it is observed under basic conditions as shown in Fig. 2. As the solution pH becomes more acidic, the reduction potential becomes pH dependent. The reaction represented at this point is not purely electrochemical, but rather involves proton binding in a PCET mechanism (Cukier & Nocera, 1998; Hammes-Schiffer & Stuchebrukhov, 2010):

$$\text{Ferric} - \text{Hemeprotein} - (\text{His})_2 + e^- + 2\text{H}^+ \rightleftharpoons \text{Ferrous} - \text{Hemeprotein} - (\text{HisH}^+)_2. \tag{v}$$

Fig. 2 shows the monoheme-[Δ7-His]$_2$ Pourbaix diagram, a plot of reduction potential vs solution pH that demonstrates the pH dependence. The data are best fit with a slope of 120 mV, ie, $(-2RT/nF)$, as a function of pH (solid line fit) indicative of a $2\text{H}^+/1e^-$ PCET event (Reddi, Reedy, Mui, & Gibney, 2007). A fit with a slope of 60 mV per pH unit indicative of a $1\text{H}^+/1e^-$ PCET event (dashed line fit) is also shown to demonstrate that it does not describe the data accurately.

It should be noted that the pH dependence in the Pourbaix diagram should continue until the solution pH is below the $pK_a^{\text{eff-ox}}$ value where the electrochemical reaction returns to a simple electron transfer mechanism as represented by reaction (vi).

$$\text{Ferric} - \text{Hemeprotein} - (\text{HisH}^+)_2 + e^- \rightleftharpoons \text{Ferrous} - \text{Hemeprotein} - (\text{HisH}^+)_2. \tag{vi}$$

However, data were not collected in this pH regime for practical reasons and the fits shown do not reflect that this additional equilibrium.

This example of a $2\text{H}^+/1e^-$ PCET event is very rare outside of de novo hemeproteins, it is far more common to observe $1\text{H}^+/1e^-$ PCET events in biochemical hemeprotein systems. These more common $1\text{H}^+/1e^-$ PCET usually involve the protonation of a local amino acid side chain which changes its pK_a value upon reduction of the heme. In a Pourbaix diagram, the data will evince a 60 mV per pH unit slope indicative of a $1\text{H}^+/1e^-$ PCET event. This redox-Bohr behavior has been observed in one de novo designed hemeprotein, heme-[H10A24]$_2$, where it involves a glutamic acid residue and not the histidine ligands to heme iron (Shifman, Moser, Kalsbeck, Bocian, & Dutton, 1998).

2.4 The Thermodynamic Relationship Between Heme Electrochemistry and Protein Folding

The thermodynamic hypothesis for protein folding forwarded by Anfinsen states that under physiological conditions the native state of a protein in stable since it lies at a global free energy minima (Anfinsen, 1973). The reaction of any cofactor with a protein that results in a cofactor–protein complex stabilizes the global fold of said complex relative to the *apo*-protein. Otherwise the reaction is thermodynamically unfavorable and the cofactor would not be observed to bind. The free energy of heme binding to a protein is the net free energy contribution to the stability of the global protein fold. In other words, the global free energy of a hemeprotein is the free energy of the *apo*-protein plus the free energy of heme binding, ie, $\Delta G_{folding}^{holo} = \Delta G_{folding}^{apo} + \Delta G_{rxn}$. It should be noted that the value of ΔG_{rxn} includes any free energy cost of protein folding required to accommodate the cofactor. Indeed, we have been able to use Zn(II) metalloprotein K_d values to determine the free energy cost of protein folding in zinc finger proteins with metal-induced protein folding events (Reddi, Guzman, et al., 2007). As discussed previously, the ΔG_{rxn} value for heme binding is dependent on heme oxidation state, solution pH, temperature, ionic strength, etc. The free energy of protein folding can be measured using chaotropic agents (guanidinium chloride, urea), pressure, and temperature. To date we have not found conditions suitable for $\Delta G_{folding}^{holo}$ measurements for our designed hemeproteins because the unfolding reactions proved less than reversible. This is likely due to partially irreversible heme aggregation from the unfolded state. However, our determinations of K_d values, and the resulting ΔG_{rxn} values, allow for the estimation of $\Delta G_{folding}^{holo}$ values from measurements of $\Delta G_{folding}^{apo}$ values.

3. ESSENTIAL EQUILIBRIUM MEASUREMENTS

3.1 Cofactor Binding Equilibria

Analyzing the heme binding thermodynamics for a designed hemeprotein begins with an evaluation of the hemeprotein binding equilibria in the two relevant iron oxidation states. As with all measurements of this type, the system should be evaluated to ensure that equilibrium has been established prior to measurement. In addition, it is critical that the fraction of hemeprotein complex formed at equilibrium is between 20% and 80% so that accurate values of the dissociation constant can be derived from a fit to the data. Above 80% fractional saturation, the equilibrium constant approaches a tight binding regime where the K_d is underdetermined and only a weak-limit may be placed on the K_d value. Below 20% fractional saturation, the K_d is difficult to accurately determine due to the small hemeprotein signal relative to the free heme signal.

The binding of ferric heme to a designed protein is typically evaluated first since it can be performed aerobically. Typically, the system is first evaluated to determine the time required to reach equilibrium before a full titration is performed. In a standard titration, aliquots of 0.1 equiv. of hemin in DMSO or sodium hydroxide are added to buffered solutions of the designed protein with gentle stirring and allowed the time necessary to reach equilibrium before spectroscopic measurement. Initial determinations of the hemeprotein stoichiometry are made at relatively high concentrations of protein, 1000-fold higher than the K_d value concentration, or at basic pH values where competition between protons and the heme iron for the ligands is minimized, cf. reaction (iii). Once the reaction stoichiometry is determined, titrations into protein at lower concentrations, and/or buffered at more acidic pH values are performed until the equilibrium allows for the accurate determination of the K_d value. As stated earlier, it is critical to measure the pH at the start and end of the titration to know the conditions of the measurement and ensure that the buffer held the pH constant.

The binding of ferrous heme to a protein requires anaerobic conditions to prevent autoxidation of the heme and hemeprotein to the ferric state. All solutions required for the titrations are subjected to repetitive freeze–pump–thaw cycles under a nitrogen atmosphere to remove dissolved oxygen. These solutions are then stored in an anaerobic chamber with low ppm O_2 levels. The processes of determining the hemeprotein stoichiometry and K_d values are analogous to that for ferric heme; however, they are performed under strictly anaerobic conditions and with a slight excess (1.1 equiv.) of the reductant sodium dithionite added to protein solution in order to maintain the heme iron in the ferrous state throughout the titration. It is generally observed that equilibration times for ferrous heme are shorter than for ferric heme. Additionally, the K_d values for ferrous heme are typically weaker than the corresponding values for ferric heme due to the prevalent use of bis-His heme binding sites in designed proteins to date.

3.2 Protonation Equilibrium

Heme binding to protein-based ligands often involves proton competition below the intrinsic pK_a values of the ligands as discussed earlier. It is possible to measure K_d values as a function of pH to ascertain the pH dependence. However, it is simpler to determine the protonation equilibria by potentiometric titration. In the case of the *apo*-protein, potentiometric titrations using a pH meter could in theory be used to determine the intrinsic pK_a values of the ligands. However, this is not practical for proteins of even moderately small size, as they may contain 10–20 titratable residues. Thus, assignment of a particular pK_a value to the metal-ion binding residue is very challenging. Potentiometric titrations using multidimensional NMR, with residue specific assignments, allow for direct determination of the ligand intrinsic pK_a values.

The determination of the effective pK_a values of the ligands with the metal-ion bound is straightforward in the case of hemes due to their strong optical signals. The hemeprotein may be placed in a cuvette and the solution pH value adjusted with strong acid as its spectrum is recorded. This process can be done for the ferric and ferrous hemeprotein to reveal the $pK_a^{\text{eff-ox}}$ and $pK_a^{\text{eff-red}}$ values, respectively. The resulting plot the spectroscopic signal as a function of solution pH is fit to a standard pK_a value equation:

$$S_{\text{meas}} = S_0 + \sum \left\{ \Delta S^* \frac{1}{10^{(-x\text{pH} + x\text{p}K_a^{\text{eff}})} + 1} \right\}, \quad (5)$$

where S_{meas} is the measured signal at a particular pH value, S_0 is the signal at the original pH value, ΔS is the signal change due to the protonation event, pH is the solution pH, pK_a^{eff} is the effective pK_a value, and x is the proton stoichiometry. For hemeprotein systems, the best fit should be determined from a single one-proton pK_a event (one pK_a^{eff} value, $x=1$), two single one-proton pK_a events (two pK_a^{eff} values, both $x=1$, with an observable intermediate), and a single two-proton pK_a event (one pK_a^{eff} value, $x=2$). The slope of the plot of signal vs pH is the value of x, and therefore discriminates between the $x=1$ and $x=2$ events. This is shown in Fig. 3, where the slope of the transition for the single pK_a^{eff} $x=2$ event is easily discriminated from the single pK_a^{eff} (or double pK_a^{eff}) $x=1$ fits. If $x=1$, the first two can be discriminated between based on the observation (or not) of the monoprotonated species intermediate.

3.3 Electrochemical Equilibrium

Once the stoichiometry of heme binding has been established in both oxidation states, conditions have be identified where both oxidized and reduced heme are bound to the protein scaffold before an analysis of the heme

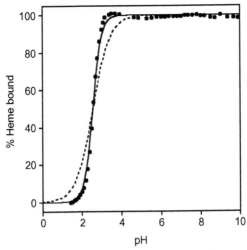

Fig. 3 Potentiometric titration of monoheme-[Δ7-His]$_2$. The data are well fit to a single 2H$^+$ coupled event (*solid line*). A fit to a 1H$^+$ event is shown for comparison (*dotted line*).

electrochemistry can be undertaken, cf. reaction (iv). Otherwise, the electrochemical measurement does not reflect simple electron transfer, but rather a more complex EC mechanism where electron transfer is followed by a chemical reaction. Such EC mechanisms may be studied, and may be quite informative as in the case of PCET reactions, but the experimentalist should be cognizant of their existence before the measurements are interpreted. There are a multitude of electrochemical techniques suitable for evaluating hemeprotein reduction potentials including an array of voltammetric techniques (Bard & Faulkner, 2001). One time honored method is redox potentiometry, as described in this series by Dutton (1978). In redox potentiometry, the potential of a buffered hemeprotein solution containing redox mediators is adjusted and the spectrum recorded. Analysis of the spectral changes as a function of solution potential is then used to derive the reduction potential, E_m, and the number of electrons in the reaction, n.

$$E_m = E^0 - \frac{RT}{nF} \ln \frac{[\text{red}]}{[\text{ox}]}. \qquad (6)$$

Redox potentiometry offers the advantages that it is based on simple laboratory equipment (spectrophotometer, pH meter), allows accurate determination of the reduction potential and identifies the moiety being oxidized/reduced in the experiment through the collected spectra. It is slower than most voltammetry methods and requires more material, but accurate E_m values (±5 mV) can be determined in an afternoon with ~7 mL of a hemeprotein solution whose optical density is ~0.4. It is important to perform the experiment in both the reductive and oxidative directions on the same sample to ensure there is no hysteresis in the data. Reduction potential measurements may be taken at a single pH value to determine the redox activity of the heme, or may be performed over a pH range to demonstrate the presence of PCET, ie, a redox-Bohr effect.

4. THERMODYNAMIC ANALYSIS OF A HEME BINDING FOUR HELIX BUNDLE, [Δ7-HIS]$_2$

4.1 Heme Affinity

Using a stably folded four-helix bundle [Δ7-His]$_2$ of our design, we were able to delineate the thermodynamic cycle relating heme affinity and electrochemistry in a protein for the first time, as shown in Fig. 4 (Reedy, Kennedy, & Gibney, 2003). Using equilibrium titrations of heme into *apo*-protein, the affinity of [Δ7-His]$_2$ for ferric heme was determined at pH 8.0. The data showed that binding of the first ferric heme was too tight to measure by optical methods, K_{d1}^{ox} value tighter than 1 nM. The weaker

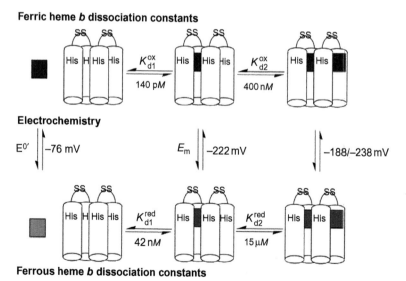

Fig. 4 Thermodynamic scheme relating ferric and ferrous heme affinity with the related electrochemistry.

binding of the second ferric heme, K_{d2}^{ox} value of 400 nM, allowed for its direct determination by titration. The binding of the second ferric heme is thermodynamically favorable by 8.7 kcal/mol, $\Delta G_{rxn} = -RT \ln K_{d2}^{ox}$. In the ferrous state, the K_{d1}^{red} and K_{d2}^{red} values were weaker than the ferric values which allowed both to be determined directly. The K_{d1}^{red} value was measured to be 42 nM, $\Delta G = 10.0$ kcal/mol, and the K_{d2}^{red} value was measured to be 15 μM, $\Delta G = 6.6$ kcal/mol. The tighter affinity for ferric heme over ferrous heme, K_{d1} values of <1 and 42 nM, respectively, was as expected for a bis-His heme binding site based on small-molecule studies. Since ferrous heme is formally charge neutral, ie, $[Fe^{2+}por^{2-}]^0$, the observed 300-fold ($\Delta\Delta G = 3.4$ kcal/mol) difference in heme affinity between the first and second ferrous hemes likely represents steric interactions between the two porphyrin macrocycles. This steric interaction is expected to contribute 3.4 kcal/mol to the difference between K_{d1}^{ox} and K_{d2}^{ox} in the ferric hemeprotein as well.

4.2 Electrochemistry

The addition monoheme- and diheme-[Δ7-His]$_2$ electrochemical values identified the other contributor to the difference in ferric K_{d1}^{ox} and K_{d2}^{ox} values and allowed for deduction of the K_{d1}^{ox} value, the affinity of the first ferric heme. In the single heme-bound state, monoheme-[Δ7-His]$_2$, the electrochemistry showed a single heme reduction potential of −222 mV vs SHE, a value typical of bis-His designed hemeproteins. The electrochemistry of the two heme-bound state of [Δ7-His]$_2$ showed two distinct midpoint reduction potentials at −188 and −238 mV vs SHE. Thus, in diheme-[Δ7-His]$_2$, one heme is easier to reduce, and the other harder to reduce than in monoheme-[Δ7-His]$_2$. This is ascribed to an electrostatic interaction between the two formally cationic ferric hemes, ie, $[Fe^{3+}por^{2-}]^+$. The observed 50 mV split in the reduction potentials represents 1.2 kcal/mol. Combining the 3.4 kcal/mol steric interaction observed in the ferrous hemeprotein with the 1.2 kcal/mol electrostatic interaction from the ferric hemeprotein, shows that the difference in heme affinity between the first and second oxidized heme must be 4.6 kcal/mol or about 2800-fold. Since the ferric K_{d2}^{ox} value is 400 nM, the ferric K_{d1}^{ox} value is deduced to be 140 pM, $\Delta G = 13.4$ kcal/mol. In addition, these data allow the formal reduction potential of heme, to be evaluated at −76 mV vs SHE which completes this thermodynamic scheme. This exercise clearly demonstrates the power of thermodynamics to provide values that are not practical to measure directly.

The thermodynamic scheme presented in Fig. 4 contains three interconnected thermodynamic cycles. The first is the relationship between the K_{d1}^{ox} value, the K_{d1}^{red} value, the E_m of the monoheme-$[\Delta 7\text{-His}]_2$ and the $E^{0\prime}$ value, the left most box. The second thermodynamic cycle is comprised of the K_{d2}^{ox} value, the K_{d2}^{red} value, the E_m of the monoheme-$[\Delta 7\text{-His}]_2$ state and the two E_m values of the diheme-$[\Delta 7\text{-His}]_2$, the right most box. Lastly, there is the thermodynamic cycle that includes both the K_{d1}^{ox} and K_{d2}^{ox} values, the K_{d1}^{red} and K_{d2}^{red} values, the E_m values of diheme-$[\Delta 7\text{-His}]_2$, and the $E^{0\prime}$ value, the outer box. The accuracy of these values are each confirmed by the remaining values in these thermodynamic cycles which ascribes great confidence in the resulting thermodynamic description of this system at this single pH value.

4.3 Proton Competition

The competition between metal-ion binding and proton binding to the His ligands modulates the affinity of the hemes for the protein scaffold and was determined both directly and indirectly. Proton competition for the metal-ion binding His residues was measured directly using potentiometric titrations followed by UV–vis spectroscopy. Potentiometric titrations involved titrating minimally buffered hemeprotein solutions, monoheme-$[\Delta 7\text{-His}]_2$, with submicroliter aliquots of strong acid. Proton competition was also evident in determinations of the pH dependence of the K_d values and redox activity. In these cases of indirect measurement, individual K_d values or E_m values were measured over a pH range.

Potentiometric titrations in the ferric state were followed by spectroscopic changes in the absorbance of the Soret, α-bands and β-bands of the heme porphyrin. The data revealed a coupled two-proton event, $x=2$, with a single pK_a^{eff-ox} value of 2.6 as shown in Fig. 3. The data are consistent with a process where two HisH$^+$, or two histidine imidazolium ions, coordinate ferric heme iron with the concomitant loss of two protons. The shift in the pK_a value of HisH$^+$ from the intrinsic pK_a value of 6.5 to the effective pK_a^{eff-ox} value of 2.6 represents 10.6 kcal/mol in free energy contributed by ferric iron coordination by the two His residues in $[\Delta 7\text{-His}]_2$, $\Delta\Delta G = -RT \ln \left(10^{2*pK_a} - 10^{2*pK_a^{eff-ox}}\right)$. This is a major contributor to the 13.4 kcal/mol free energy of ferric heme binding observed in the value of K_{d1}^{ox}. This suggests that the hydrophobic interactions between the porphyrin and the protein contribute the remaining 2.8 kcal/mol of heme affinity.

The corresponding ferrous state potentiometric titration, performed under anaerobic conditions, revealed a single $pK_a^{\text{eff-red}}$ value of 5.7 that also reflects a cooperative two-proton event, $x=2$. In the case of the ferrous heme, the difference between the intrinsic and effective pK_a values of the ligands, 6.5 and 5.7, is smaller than observed in the ferric heme case due to the weaker binding of ferrous heme to the protein. The difference between the effective and intrinsic pK_a values of the ligands indicate that ferrous heme coordination is only 4.4 kcal/mol favorable. This free energy contribution of ferrous iron coordination, 4.4 kcal/mol is less than half the total free energy of ferrous heme binding, 10.0 kcal/mol. Thus, the hydrophobic interaction between ferrous heme and the [Δ7-His]$_2$ protein contributes more to heme binding than iron coordination to the two His residues. This shows how thermodynamic analyses can be used to parse apart the individual contributors to metalloprotein stability.

Fig. 5 shows the K_{d1} values of ferric and ferrous heme bound to [Δ7-His]$_2$ as a function of pH over the range of 4.0–8.0. These data reveal the effects of proton competition on heme affinity indirectly. The K_{d1} values of ferric and ferrous heme measured at pH 8.0, where histidine is predominantly in the imidazole form, reflect reaction (iii). This is the maximal affinity and the pH-independent formation constant for oxidized and reduced heme binding to [Δ7-His]$_2$, $K_f^{\text{ML-ox}}$ and $K_f^{\text{ML-red}}$, respectively.

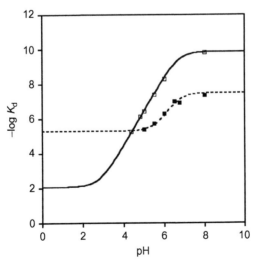

Fig. 5 Plot of $-\log K_d$ values vs pH for monoheme-[Δ7-His]$_2$ in the ferric (*open squares*) and ferrous (*filled squares*) oxidation states.

At more acidic pH values, proton competition for the His ligands begins to attenuate the observed K_{d1} values. The onset of this occurs in both oxidation states at the intrinsic pK_a values of the His ligands, pH 6.5. Importantly, the slope of a plot of K_{d1} value vs pH shows a slope of 2.0, indicating that heme binding is in competition with two protons in both oxidation states, cf. reaction (ii). This is as expected based on the potentiometric titration, both oxidation states show single $x=2$ events and is consistent with the bis-His coordination motif designed into [Δ7-His]$_2$.

In the oxidized state, this attenuation continues to until the pH is more acidic than the $pK_a^{\text{eff-ox}}$ value of 2.6. Below pH 2.6, oxidized heme binding is pH independent, cf. reaction (i), and the K_{d1}^{ox} value is the $K_f^{MLH_2-ox}$ value, the formation constant of ferric heme associating with the protein hydrophobic core but not binding to the histidine imidazolium ions.

In the reduced state, the attenuation of the K_{d1}^{red} value stops after the $pK_a^{\text{eff-red}}$ value is reached at a pH of 5.7, below which ferrous heme binding to the protein becomes pH independent. This K_{d1}^{red} value represents the formation constant of ferrous heme binding to the protein with histidine imidazolium ions, $K_f^{MLH_2-red}$.

In both oxidation states, the observed $\Delta\Delta G$ value between K_f^{ML} and $K_f^{MLH_2}$, ($-RT \ln \Delta K_f$), is thermodynamically equivalent to the $\Delta\Delta G$ value derived from the intrinsic and effective pK_a values of the two ligands, ie, $\Delta\Delta G = -RT \ln \left(10^{2^*pK_a} - 10^{2^*pK_a^{\text{eff}}} \right)$. This is a critical check when evaluating proton competition in metalloproteins as it combines the results of several experimental determinations into a single equilibrium model. Furthermore, Fig. 5 shows that the values of K_{d1}^{ox} and K_{d1}^{red} are equivalent at pH 4.5 which indicates that the formal reduction potential of heme, $E^{0\prime}$, at pH 4.5 is -38 mV vs SHE. This is within 1 kcal/mol of the $E^{0\prime}$ value of -76 mV value measured at pH 8.0.

4.4 The PCET Event

Analysis of the pH dependence of the midpoint reduction potential between pH 4.0 and 8.0 was also used as an indirect evaluation of the proton competition equilibrium and it revealed a rare $2H^+/1e^-$ PCET event. The Pourbaix diagram in Fig. 2 demonstrates the pH dependence of the reduction potential of a single heme bound to [Δ7-His]$_2$. At basic pH values above the pK_a of the His ligands, the reduction potential is pH independent, as expected, at a value of $E_m^{\text{base}} = -222$ mV vs SHE, cf. reaction (iv). Between pH 7.0 and pH 5.0, the reduction potential rises with decreasing pH value with a slope of 120 mV per pH unit, indicative of a $2H^+/1e^-$ coupled reaction. Ultimately, the reduction potential becomes pH independent at acidic pH values, $E_m^{acid} = -38\,\text{mV}$ vs SHE, cf. reaction (v). Since 43 mV represents 1.0 kcal/mol at 25 °C for an $n=1$ reduction, the 184 mV difference between the E_m values at pH 4.5 and 8.0 represents 4.3 kcal/mol difference in the free energy between reaction (iv) and reaction (v). The difference between reaction (iv) and (v) is also at the difference between pK_a^{eff} and $pK_a^{eff-red}$ values. Thus, the free energy difference between $E_{m4.5}$ and E_m^{base} is equivalent to the free energy difference between pK_a^{eff} and $pK_a^{eff-red}$, $\Delta\Delta G = -nF\Delta\left(E_m^{acid} - E_m^{base}\right) = -RT\ln\left(10^{2*pK_a} - 10^{2*pK_a^{eff-red}}\right)$. Again, this thermodynamic equivalence provides a further validation of all the equilibrium measurements made and the model developed.

5. EXPERIMENTAL PROCEDURES

5.1 Protein Synthesis and Purification

The $[\Delta 7\text{-His}]_2$ protein was prepared by solid-phase peptide synthesis methods using Fmoc/tBu chemistry at 0.2 mmol scale (Amblard, Fehrentz, Martinez, & Subra, 2006). Single 30 min coupling cycles with HATU activation and the following side chain protecting groups were used: ^{t}Boc (Lys); O^{t}Bu (Glu); Trt (Cys); Pmc (Arg). The N-terminus was manually acetylated with acetic anhydride followed by thorough washing with DMF, MeOH, and CH$_2$Cl$_2$. The peptide was cleaved from the resin and the side chains deprotected using 90:8:2 (v/v/v) trifluoroacetic acid: ethanedithiol:water. Crude peptide was precipitated and triturated with cold ether, followed by dissolution in water (0.1% (v/v) TFA), and lyophilization. Reversed-phase C$_{18}$ HPLC employing aqueous acetonitrile gradients containing 0.1% (v/v) TFA was used to purify the peptide. The N-terminal cysteine residues of the purified peptide were air oxidized to the symmetric disulfides in 100 mM ammonium carbonate buffer, pH 9.5 (4–7 h), which was followed by analytical HPLC. Electrospray ionization mass spectrometry was used to confirm the identity of the peptide.

5.2 Equilibrium Measurements

As described in Reddi, Reedy, et al. (2007), the binding of heme to the *apo*-protein was followed using UV–vis spectroscopy on a Varian Cary 100 or Cary 300 spectrophotometer. The concentrations of *apo*-peptide were determined using the intrinsic absorbance of the tryptophan residues (ε_{280} of 5600 M^{-1} cm^{-1} Trp^{-1}). 1.0 cm path length semi-microfluorescence cuvettes (sample volume 1.0 mL, protein concentrations 1–7 μM) were used for K_d determinations in the 0.5–20 μM range. Turned 90 degrees, the same cuvette has a 0.4-cm path length and was used for K_d determinations from 5 to 100 μM. The determination of K_d values in the range of 10–5 μM required the use of a 10-cm path length cell (sample volume 30 mL, protein concentration of 100–700 nM). Anaerobic samples were prepared in an inert atmosphere glove box. These samples include 1.1 equiv. of sodium dithionite to maintain the heme in the reduced state. The cuvettes used to ferrous heme K_d determinations had screw caps with septa to allow introduction of μL volumes of heme in DMSO or 0.1 N NaOH.

5.3 Kinetic Evaluation

It is critical to know that the system has reached equilibrium prior to measurement. This requires an examination of the kinetic approach to equilibrium. This is typically done by following the growth of the bound heme at λ^{max} as a function of time. Equilibration times may be in the minutes timeframe, as generally observed for heme b, hours timeframe, as observed for heme o and heme a, and can be unmeasurably slow, as observed for the ferric state of outer mitochondrial membrane cytochrome b_5 (Silchenko et al., 2000). The latter is indicative of kinetic trapping of the heme that may play an important role in directing heme transport within the cell.

6. SUMMARY

We have shown that detailed thermodynamic analysis of designed metalloproteins offers significant insight into their assembly, stability, and nascent function. The analysis of simple chemical equilibria allows the complexity that is biochemistry to be studied in great detail. The complex interrelationships between protein stability, cofactor affinity, proton binding, and electron transfer can all be delineated in a rigorous treatment of the overlapping equilibrium expressions. In addition, these thermodynamic studies can be used to identify kinetic traps that may be important to metal-ion trafficking. The biophysical techniques used and the analyses employed require equipment and software that is readily available in most biochemical laboratories. Wider application of thermodynamic analyses to natural and designed metalloproteins is necessary to provide the thermodynamic data to improve computational design of metalloproteins from first principles.

This will provide a most rigorous constructive test of our understanding of metallobiochemistry.

ACKNOWLEDGMENT
This work was supported by the American Heart Association (0455900T).

REFERENCES
Amblard, M., Fehrentz, J. A., Martinez, J., & Subra, G. (2006). Methods and protocols of modern solid phase peptide synthesis. *Molecular Biotechnology, 33*, 239–254.

Anfinsen, C. B. (1973). Principles that govern the folding of protein chains. *Science, 181*, 223–230.

Bard, A. J., & Faulkner, L. R. (2001). *Electrochemical methods: Fundamentals and applications* (2nd ed.). . New York: Wiley.

Bryson, J. W., Betz, S. F., Lu, H. S., Suich, D. J., Zhou, H. X. X., O'Neil, K. T., et al. (1995). Protein design: A hierarchical approach. *Science, 270*, 935–941.

Cangelosi, V. M., Deb, A., Penner-Hahn, J. E., & Pecoraro, V. L. (2014). A de novo designed metalloenzyme for the hydration of CO_2. *Angewandte Chemie, International Edition, 53*, 7900–7903.

Cukier, R., & Nocera, D. G. (1998). Proton coupled electron transfer. *Annual Review of Physical Chemistry, 49*, 337–369.

Dahiyat, B. I., & Mayo, S. L. (1997). De novo protein design: Fully automated sequence selection. *Science, 278*, 82–87.

De Villiers, K. A., Kashula, C. H., Egan, T. J., & Marques, H. M. (2007). Speciation and structure of ferriprotoporphyrin IX in aqueous solution: Spectroscopic and diffusion measurements demonstrate dimerization, but no μ-oxo dimer formation. *Journal of Biological Inorganic Chemistry, 12*, 101–117.

Dempsey, J. L., Winkler, J. R., & Gray, H. B. (2010). Proton-coupled electron flow in protein redox machines. *Chemical Reviews, 110*, 7024–7039.

Dutton, P. L. (1978). Redox potentiometry: Determination of midpoint reduction potentials of oxidation-reduction components of biological electron-transfer systems. *Methods in Enzymology, 54*, 411–435.

Fu, Y., Chang, F.-M. J., & Giedroc, D. P. (2014). Copper transport and trafficking at the host-bacterial pathogen interface. *Accounts of Chemical Research, 47*, 3605–3613.

Hammes-Schiffer, S., & Stuchebrukhov, A. A. (2010). Theory of coupled electron and proton transfer reactions. *Chemical Reviews, 110*, 6939–6960.

Jiang, L., Altoff, E. A., Clemente, F. R., Doyle, L., Röthlisberger, D., Zanghellini, A., et al. (2008). De novo computational design of retro-aldol enzymes. *Science, 319*, 1387–1391.

Karlin, K. D. (1993). Metalloenzymes, structural motifs, and inorganic models. *Science, 261*, 701–708.

Kiefer, L. L., & Fierke, C. A. (1994). Functional characterization of human carbonic anhydrase II variants with altered zinc binding sites. *Biochemistry, 33*, 15233–15240.

Kim, W. A., & Berg, J. M. (1993). Thermodynamic β-sheet propensities measured using a zinc-finger host peptide. *Nature, 362*, 267–270.

Magyar, J. S., & Arnold-Godwin, H. (2003). Spectrophotometric analysis of metal binding to structural zinc-binding sites: Accounting quantitatively for pH and metal ion buffering effects. *Analytical Biochemistry, 320*, 39–54.

Reddi, A. R., Guzman, T., Breece, R. M., Tierney, D. L., & Gibney, B. R. (2007). Deducing the energetic cost of protein folding in zinc finger proteins using designed metallopeptides. *Journal of the American Chemical Society, 129*, 12815–12827.

Reddi, A. R., Jensen, L. T., & Culotta, V. C. (2009). Manganese homeostasis in saccharomyces cerevisiae. *Chemical Reviews, 109*, 4722–4732.

Reddi, A. R., Reedy, C. J., Mui, S., & Gibney, B. R. (2007). Thermodynamic investigation into the mechanisms of proton-coupled electron transfer in heme protein maquettes. *Biochemistry, 46*, 291–305.

Reedy, C. J., Elvekrog, M. M., & Gibney, B. R. (2008). Development and analysis of a heme protein structure-electrochemical function database. *Nucleic Acids Research, 36*, D307–D313.

Reedy, C. J., & Gibney, B. R. (2004). Heme-protein assemblies. *Chemical Reviews, 104*, 617–649.

Reedy, C. J., Kennedy, M. L., & Gibney, B. R. (2003). Thermodynamic characterization of ferric and ferrous haem binding to a designed four-α-helix protein. *Chemical Communications*, 570–571.

Shifman, J. M., Moser, C. C., Kalsbeck, W. A., Bocian, D. F., & Dutton, P. L. (1998). Functionalized de novo designed proteins: Mechanisms of proton coupling to oxidation/reduction in heme protein maquettes. *Biochemistry, 37*, 16815–16827.

Silchenko, S., Sippel, M. L., Kuchment, O., Benson, D. R., Mauk, A. G., Altuve, A., et al. (2000). Hemin is kinetically trapped in cytochrome b_5 from rat outer mitochondrial membrane. *Biochemical and Biophysical Research Communications, 5*, 467–472.

Suzuki, T., Tanaka, K., Wakabayashi, C., Saita, E.-I., & Yoshida, M. (2014). Chemomechanical coupling of human mitochondrial F1-ATPase motor. *Nature Chemical Biology, 10*, 930–936.

Zhuang, J., Amoroso, J. H., Kinloch, R., Dawson, J. H., Baldwin, M. J., & Gibney, B. R. (2006). Evaluation of electron-withdrawing group effects on heme binding in designed proteins: Implications for heme *a* in cytochrome *c* oxidase. *Inorganic Chemistry, 45*, 4685–4694.

Zhuang, J., Reddi, A. R., Wang, Z., Khodaverdian, B., Hegg, E. L., & Gibney, B. R. (2006). Evaluating the roles of the heme *a* side chains in cytochrome *c* oxidase using designed heme proteins. *Biochemistry, 45*, 12530–12538.

CHAPTER NINETEEN

Reconstitution of Heme Enzymes with Artificial Metalloporphyrinoids

K. Oohora*,†,‡, T. Hayashi*,[1]

*Department of Applied Chemistry, Graduate School of Engineering, Osaka University, Suita, Japan
†Frontier Research Base for Global Young Researchers, Graduate School of Engineering, Osaka University, Suita, Japan
‡PRESTO, Japan Science and Technology Agency, Kawaguchi, Japan
[1]Correponding author: e-mail address: thayashi@chem.eng.osaka-u.ac.jp

Contents

1. Introduction — 440
2. Design and Synthesis of Artificial Metalloporphyrinoids — 442
    2.1 Dianionic Porphyrinoid Ligand — 442
    2.2 Trianionic Porphyrinoid Ligand — 444
    2.3 Monoanionic Porphyrinoid Ligand — 444
3. Reconstitution of Hemoproteins — 446
    3.1 Preparation of Apoproteins — 447
    3.2 Common Protocols for Heme Protein Reconstitution with Artificial Metalloporphyrinoids — 448
    3.3 Characterization of Reconstituted Proteins — 449
4. Representative Characteristics of Reconstituted Hemoproteins — 451
    4.1 Peroxidase Activity of Reconstituted Myoglobin and HRP — 451
    4.2 Hydroxylase Activity of Reconstituted Myoglobin — 452
    4.3 Methionine Synthase Model — 452
Acknowledgments — 453
References — 453

Abstract

An important strategy used in engineering of hemoproteins to generate artificial enzymes involves replacement of heme with an artificial cofactor after removal of the native heme cofactor under acidic conditions. Replacement of heme in an enzyme with a nonnatural metalloporphyrinoid can significantly alter the reactivity of the enzyme. This chapter describes the design and synthesis of three types of artificial metalloporphyrinoid cofactors consisting of mono-, di-, and tri-anionic ligands (tetradehydrocorrin, porphycene, and corrole, respectively). In addition, practical procedures for the preparation of apo-hemoproteins, incorporation of artificial cofactors, and characterization techniques are presented. Furthermore, the representative catalytic activities of artificial enzymes generated by reconstitution of hemoproteins are summarized.

1. INTRODUCTION

Heme proteins are among the most versatile of metalloproteins. They have one or several characteristic iron porphyrins operating as cofactors in the protein matrices (Philips, 2001). In nature, there are a number of different iron porphyrin variants, including, for example, heme a, heme b, heme c, and heme d. In these heme cofactors, heme b, (also known as protoheme IX), is among the most common heme cofactors (Reedy & Gibney, 2004). As shown in Fig. 1, heme b, protoporphyrin IX iron complex, has three different peripheral substituents represented by four methyl groups, two vinyl groups, and two propionate side chains. The cofactor is usually bound within the protein matrix in an area known as the "heme pocket," via Fe–ligand coordination and noncovalent interactions with high affinity of 10^8–10^{15} M^{-1}. The linkage of the two methyl groups at 1- and 3-positions and two vinyl groups at 2- and 4-positions (Fischer nomenclature) of the β-pyrrole rings provides asymmetric structure about the α,γ-*meso* axis of the heme framework. Thus, the cofactor possesses two possible orientations in the heme pocket (La Mar, Pande, Hauksson, Pandey, & Smith, 1989).

Several well-known hemoproteins such as myoglobin, hemoglobin, horseradish peroxidase (HRP), cytochrome b_5, and cytochrome P450 have heme b in the heme pocket. Under the acidic conditions, the native heme

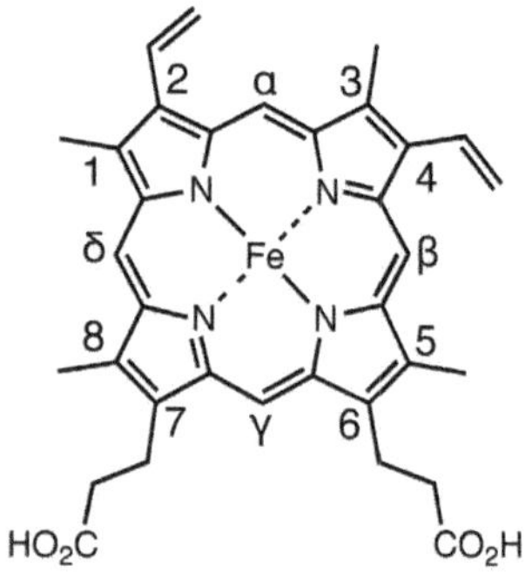

Fig. 1 Molecular structure of heme b with Fischer nomenclature.

cofactor can be removed from the heme pocket to yield the corresponding generally colorless apoprotein. The addition of heme *b* into an apoprotein solution provides the reconstituted hemoprotein (Hayashi, 2013). A series of experiments has indicated that the structures, physicochemical properties, and reactivities of the reconstituted proteins are identical to those of the corresponding native proteins. This indicates that the inserted heme moiety adopts the same orientation in the heme pocket as the native heme moiety (Wagner, Perez, Toscano, & Gunsalus, 1981). This has led to preparation of hemoproteins reconstituted with artificial cofactors (Fig. 2). This strategy is used to generate new cofactor-dependent proteins with catalytic activity. Thus, the reconstitution of hemoproteins gives important insights into structure–function relationships of native proteins as well as generation of an artificial biocatalysts (Hayashi, Sano, & Onoda, 2015) and/or biomaterials (Oohora, Onoda, & Hayashi, 2012).

To construct a reconstituted hemoprotein, selection of an appropriately designed artificial heme molecule is quite important because a given heme pocket provides a relatively limited hydrophobic space. There are at least three strategies employed in the design of an artificial cofactor: (i) metal substitution within the tetrapyrrole macrocyclic ligand (Cai, Li, Jing, & Zhang, 2013), (ii) modification of peripheral substituents of the porphyrin framework (Matsuo, Fukumoto, Watanabe, & Hayashi, 2011), and (iii) utilization of a porphyrinoid framework instead of a porphyrin framework (Hayashi et al., 2002). Reconstitution of hemoproteins with artificial cofactors has been carried out for over 50 years. However, almost all research have included reconstitution of heme proteins using only the first two of these strategies. It is expected that use of a nonnatural porphyrinoid as a cofactor ligand will yield useful results because the nonnatural porphyrinoids provides unique chemical properties distinct from those of the native heme cofactor. For example, the redox potential of a cofactor with a nonnatural porphyrinoid framework will be remarkably shifted, while the nonnatural cofactor resides in the same position of the heme pocket as the native heme

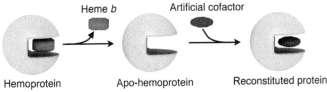

Fig. 2 Schematic representation of reconstitution of a hemoprotein with an artificial cofactor.

cofactor. This chapter focuses on reconstitution of hemoproteins with nonnatural porphyrinoid cofactors and describes the synthesis and chemical properties of reconstituted hemoproteins with artificial metalloporphyrinoids to generate new heme enzyme models.

2. DESIGN AND SYNTHESIS OF ARTIFICIAL METALLOPORPHYRINOIDS

Various artificial porphyrinoids have been designed and synthesized to change the redox and reactivity of a bound metal ion in the porphyrinoid core with the aim of functionally modifying and improving the hemoproteins. Efficient insertion and robust binding of artificial metalloporphyrinoids into the apo-form of hemoprotein require the planar structure of porphyrinoid framework, which is suitable to the heme-binding site. In addition, propionate side chains are useful for anchoring as well as for increasing the solubility of heme in aqueous solutions. In this chapter, the framework structures and the valencies of porphyrinoid ligands are discussed. Porphyrin (a native macrocycle) and porphycene are dianionic ligands with D_{4h} and D_{2h} symmetries, respectively, whereas corrole and tetradehydrocorrin each lack a *meso*-carbon in the tetrapyrrole framework and act as trianionic and monoanionic ligands, respectively (Fig. 3).

2.1 Dianionic Porphyrinoid Ligand

Porphycene, a constitutional isomer of porphyrin, serves as a dianionic porphyrinoid. This porphyrinoid framework was firstly synthesized by Vogel, Köcher, Schmickler, and Lex (1986). Porphycene is prepared by intermolecular McMurry coupling of two bipyrrole units. To date, various porphycene and metalloporphycene derivatives have been prepared and it has been found that several physicochemical properties of porphycenes

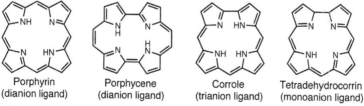

Fig. 3 Comparison of porphyrinoid ligands as shown in this chapter.

are significantly different from those of porphyrins (Fowler et al., 2002). Moreover, it was recognized that a hemoprotein reconstituted with the metalloporphycene would have unique properties. An iron porphycene complex (FePc), 2,7-diethyl-3,6,12,17-tetramethyl-13,16-*bis*-carboxyethylporphycenato iron, was designed (Hayashi et al., 2002). This compound contains two propionate side chains at the appropriate positions and was prepared according to Scheme 1 which includes a total of

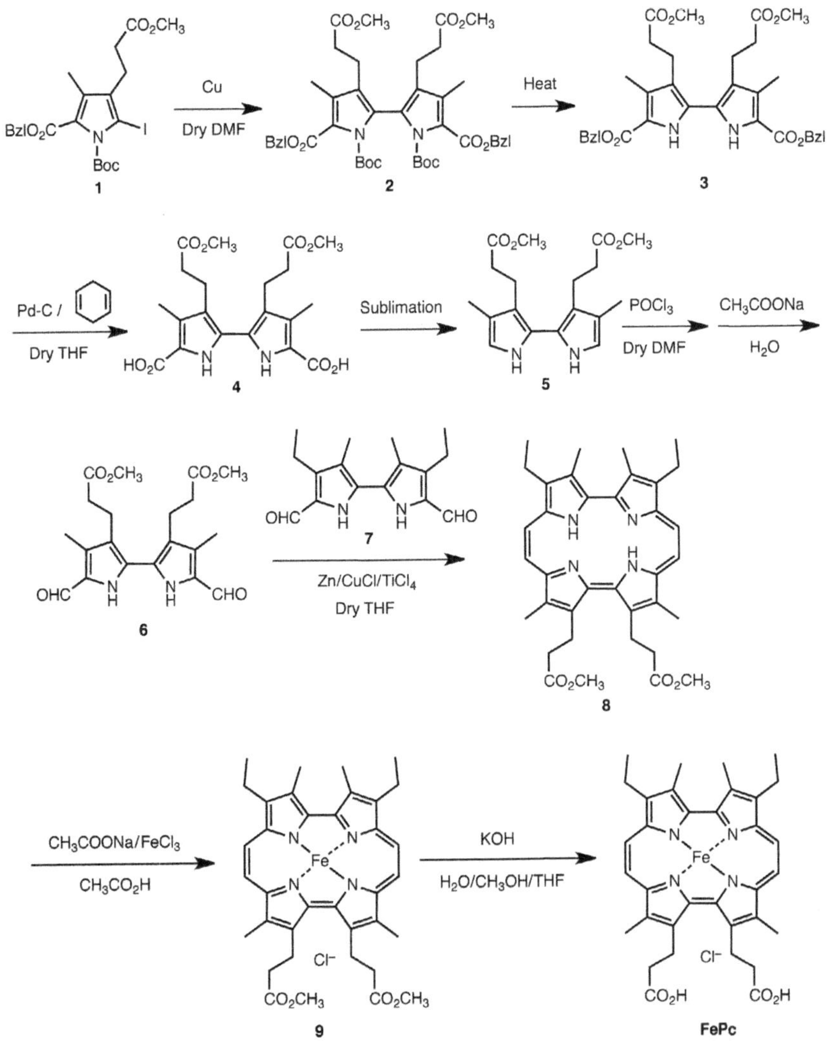

Scheme 1 Synthetic scheme for preparation of FePc.

25 steps. Two types of bipyrrole units, **6** and **7**, are used as precursor of porphycene **8**. Both of the pyrrole units are synthesized by Ullman coupling of α-iodo-substituted pyrrole molecules, in which another α-position and a nitrogen atom are protected by suitable functional groups. After deprotection of each precursor to generate bipyrrole, the formylation of both α-positions is carried out via the Vilsmeier–Haack reaction. McMurry coupling of the two bipyrrole units yields the porphycene ligand with protected propionate side chains. Insertion of iron and followed by hydrolysis of methyl esters yields the desired porphycene cofactor, FePc. Manganese porphycene (MnPc) can also be prepared essentially the same synthetic pathway (Oohora, Kihira, Mizohata, Inoue, & Hayashi, 2013).

2.2 Trianionic Porphyrinoid Ligand

Corrole is a porphyrinoid compound which lacks one *meso* carbon compared to the porphyrin framework (Erben, Will, & Kadish, 2000). Since corrole retains the 18π-conjugated aromatic system despite the contracted structure, it functions as a trianionic ligand. Therefore, a high-valent state of a metal ion carried by the porphyrinoid framework should be stabilized when the corrole framework is used as a ligand (Wasbotten & Ghosh, 2006). Thermostable high-valent metallocorroles equipped with electron-withdrawing *meso*-substituents have been reported and are relatively easy to synthesize. However, a *meso*-unsubstituted metallocorrole was required to be synthesized prior to being investigated as an artificial hemoprotein cofactor. In this context, an iron corrole complex (FeCor), 2,3,17,18-tetraethyl-7,13-dimethyl-8,12-bis(2-carboxyl)corrolate iron, was designed and synthesized according to Scheme 2 (Matsuo et al., 2009). The dipyrromethane unit **11** with protected at the α-positions by benzyl ester moieties was synthesized by the coupling of pyrrole **10** under acidic conditions. After deprotection of the ester, dipyrromethane was modified with α-formylated pyrrole via decarbonation under strongly acidic conditions to form the *a,c*-biladiene **14**. Cyclization of the biladiene **14**, tetrapyrrole compound, in the presence of $FeCl_3$ yields a precursor of the iron corrole complex **15**. After hydrolysis of the ester groups in this precursor, FeCor is obtained.

2.3 Monoanionic Porphyrinoid Ligand

Corrin, a tetrapyrrole compound, has a similar ligand framework which is seen in cobalamin, a native cofactor found in enzymes such as methionine synthase (Kräutler, 2009). The low valent state of a metal ion in the corrinoid

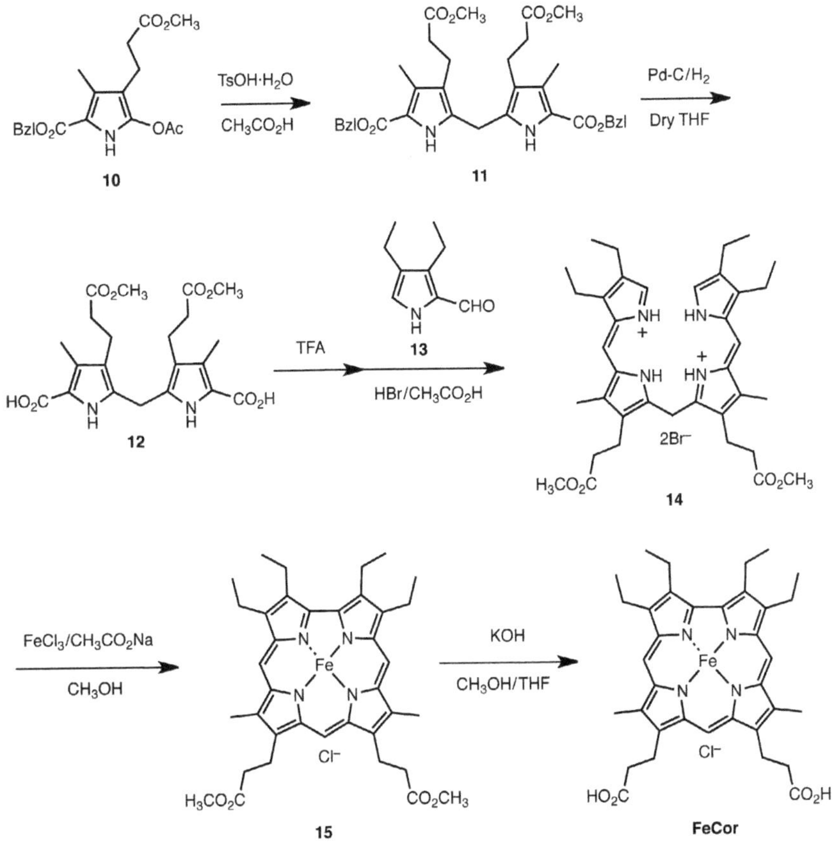

Scheme 2 Synthetic scheme for preparation of FeCor.

core should be stabilized because the corrinoid framework serves as a monoanionic porphyrinoid ligand. Unfortunately, the cobalamin itself is not suitable for use as a cofactor of hemoproteins because it has a flexible and nonplanar structure with long peripheral side chains. To mimic the cobalamin framework, tetradehydrocorrin and didehydrocorrin have been prepared and these cobalt complexes have been investigated as cobalamin-type structures. It remains challenging to prepare a model for cobalamin-dependent enzymes (Murakami, Aoyama, & Tokunaga, 1980). It was recognized that a cobalamin-dependent enzyme model could be prepared using a hemoprotein reconstituted with a tetradehydrocorrin or didehydrocorrin cobalt complex. To insert the cobalt complex into an apoprotein, 8,12-dicarboxyethyl-1,2,3,7,13,17,18,19-octamethyltetradehydrocorrinate cobalt complex (Co(TDHC)) was designed and prepared according

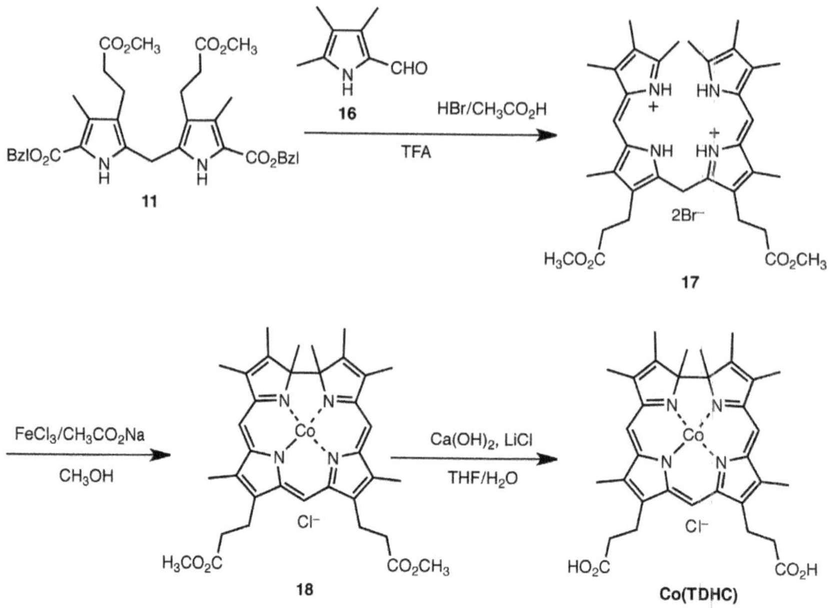

Scheme 3 Synthetic scheme for preparation of Co(TDHC).

to Scheme 3 (Hayashi et al., 2014). The synthetic scheme has some steps which are similar to steps used in synthesis of the corrole metal complex. The protected dipyrromethane unit **11** is the same unit used in corrole synthesis. In this scheme, deprotection and coupling of the dipyrromethane with α-formylated pyrrole were carried out simultaneously under strongly acidic conditions to yield a,c-biladiene **17**. The cobalt tetradehydrocorrin complex can be obtained via cyclization of a,c-biladiene **17** in the presence of $Co(OAc)_2$. Hydrolysis of the ester groups provides the product, Co(TDHC).

3. RECONSTITUTION OF HEMOPROTEINS

Reconstitution of hemoproteins with artificial metalloporphyrinoids is performed in three steps: (i) removal of heme from a hemoprotein to generate the corresponding apoprotein, (ii) insertion of an artificial metalloporphyrinoid into the apoprotein to generate the reconstituted protein, and (iii) characterization of the reconstituted protein. For each step, the general procedures are described below.

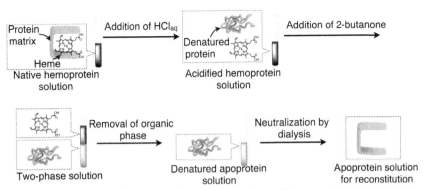

Fig. 4 Schematic route for preparation of the apoform of hemoprotein.

3.1 Preparation of Apoproteins

A typical procedure for the preparation of an apo-hemoprotein is illustrated in Fig. 4. A general procedure for removal of heme from hemoprotein which is known as Teale's 2-butanone method, is outlined below (Teale, 1959). Hemoprotein dissolved in ultrapure water or buffering solution in a glass cuvette is acidified to pH 2.0–2.2 upon addition of 0.1 HCl aqueous solution in an ice bath while monitoring the pH of the solution. To the solution, the same volume of cooled 2-butanone is added. The mixture is gently shaken several times and centrifuged at 4°C to separate the two phases. The aqueous solution is treated with cooled 2-butanone more than three times until the aqueous solution turns colorless. The aqueous phase is then transferred into a dialysis membrane (molecular weight cutoff from 6 to 8 kDa is recommended) and dialyzed against 1 L of potassium phosphate buffer (100 mM, pH 7.0) for 2 h at 4°C. The dialysis process is repeated at least three times to remove 2-butanone. The resulting solution is then stored at 4°C.

In the case of heme-buried hemoproteins such as HRP and cytochrome P450, the pH value of the solution should be lowered when heme is removed in the presence of 2-butanone. Additives such as histidine or guanidine hydrochloride are also helpful in removing heme from hemoproteins which have strong binding interactions between the heme and the hemoprotein.

In most cases, the apoprotein generated by removal of the heme is thermodynamically less stable than the corresponding holoprotein. Thus, the apoprotein should be handled with care and stored at an appropriate temperature (Fandrich et al., 2003). Furthermore, contamination with organic solvent should be avoided because it could lead to irreversible aggregation.

3.2 Common Protocols for Heme Protein Reconstitution with Artificial Metalloporphyrinoids

An artificial cofactor with efficient affinity and specificity for a heme pocket will be spontaneously incorporated into the apoprotein. A general procedure for heme protein reconstitution with an artificial metalloporphyrinoid is now described. To the apo-form of the hemoprotein (10–50 μM) in 100 mM potassium phosphate buffer solution, pH 7.0, at 4°C, is added dropwise the artificial metalloporphyrinoid (>1 mM) in 0.01 M NaOH solution. In general, incorporation of the metalloporphyrinoid is completed within 5 min after each dropwise addition. The incorporation should be monitored by UV–vis spectroscopy. If the artificial metalloporphyrinoid is successfully incorporated, a titration curve will demonstrate a significant transition when one equivalent of artificial metalloporphyrinoid is added (Fig. 5). After mild shaking for a 2-h period to finish the reconstitution, the solution is concentrated using an ultrafiltration membrane to a volume less than 2% of the column volume of a gel-filtration column (such as a

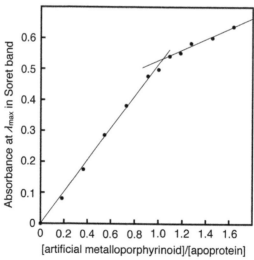

Fig. 5 An example of a plot of absorbance against the equivalent of added artificial metalloporphyrinoid in a reconstitution reaction.

Sephadex G-25 column). The solution is loaded onto the gel-filtration column equilibrated in the elution buffer (100 mM potassium phosphate buffer, pH 7.0) and colored fractions are collected. The UV–vis spectrum of each fraction is measured and the fractions with the highest ratio of absorbances of Soret band and at 280 nm (A_{Soret}/A_{280}) are combined. The purified protein solution is concentrated to greater than 1 mM but less than 3 mM by ultrafiltration and stored in the freezer.

The incorporation of a cofactor into a series of apoproteins should be performed in aqueous solution without any organic solvents, if the cofactor is sufficiently soluble in aqueous media. If an organic solvent such as DMSO, pyridine, or DMF is needed to dissolve the artificial metalloporphyrinoid, its volume should be kept as low as possible.

3.3 Characterization of Reconstituted Proteins

Characterization of reconstituted myoglobin is performed by UV–vis absorption, circular dichroism (CD), and electrospray ionization mass spectroscopic (ESI MS) methods. Additionally, X-ray crystal structure analysis is extremely powerful identification method. Each measurement is carried out as described below.

1. UV–vis absorption spectral measurements and determination of extinction coefficient

    Diluted protein samples in various concentrations (1–50 μM) are prepared. It is confirmed that the samples have the same absorption maxima. Next, the metal concentration for each sample is measured by inductively coupled plasma atomic emission spectrometry to determine the concentration of artificial metalloporphyrinoid in the sample.

2. CD spectral measurements

    To reduce noise, a quartz cell with a short light path (1 mm path length is recommended) is employed. A diluted solution (less than 10 μM) of the reconstituted protein is prepared to verify proper folding in the wavelength region from 190 to 300 nm, where the signals caused by secondary structure of protein provide very high intensity. To verify the signal assigned to absorption by the artificial metalloporphyrinoid, a highly concentrated protein solution (over 100 μM) should be used.

3. ESI MS measurements

    The buffer salts should be exchanged for volatile salts prior to ESI MS measurements. Common buffer components such as potassium phosphate are not vaporized and lead to clogging of the needle in the MS during the ionization operation. A useful buffer for the measurement is a 5 mM NH$_4$OAc or (NH$_4$)$_2$CO$_3$ solution. The buffer is exchanged by ultrafiltration and dilution with NH$_4$OAc buffer at least 10 times. To measure a cofactor-bound protein, a slightly concentrated protein solution (~10 μM) is appropriate. To detect the mass of the cofactor-bound protein, the acceleration voltage in the detector should be as low as possible. Multiple m/z values caused by species with varying degrees of protonation will be observed. If a molecular mass measurement is required, deconvolution with the appropriate software should be performed. An example is shown in Fig. 6.

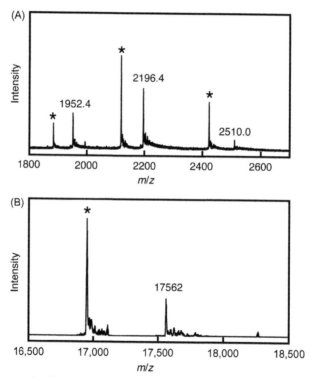

Fig. 6 An example of an ESI mass spectrum of a reconstituted hemoprotein. (A) Raw mass spectrum of reconstituted myoglobin with Co(TDHC). Three characteristic peaks are consistent with the calculated mass numbers as follows. m/z (z): 1952.1 (9+), 2196.1 (8+), and 2509.6 (7+). (B) Deconvoluted mass spectrum based on (A), indicating molecular mass of the reconstituted myoglobin: calculated $m/z = 17561$ (1+). Peaks with asterisks were assigned as ionized species of apomyoglobin.

4. REPRESENTATIVE CHARACTERISTICS OF RECONSTITUTED HEMOPROTEINS

4.1 Peroxidase Activity of Reconstituted Myoglobin and HRP

Myoglobin is an oxygen storage protein containing heme b as its native cofactor. Myoglobin has low peroxidase activity, although both myoglobin and HRP have two histidine residues at the proximal and distal sites. Compared to HRP, myoglobin does not have a suitable H_2O_2 activation system in its heme pocket. One of the structural differences between myoglobin and HRP is the Fe–His(axial ligand) coordination bond; the bond strength of this bond in HRP is larger than that of myoglobin. To effectively increase peroxidase activity of myoglobin, one strategy is to enhance the coordination bond strength in myoglobin by reconstituted with a nonnatural porphyrinoid ligand such as porphycene (Hayashi et al., 2006). Reconstituted myoglobin with iron porphycene, rMb(FePc), is found to accelerate the rate of guaiacol oxidation, which is an indicator of peroxidase activity (Scheme 4A). The initial rate of oxidation promoted by rMb(FePc) is 11-fold faster than that observed by native myoglobin at pH 7.0 at 20 °C.

HRP reconstituted with FePc is also available (Matsuo, Murata, Hisaeda, Hori, & Hayashi, 2007). HRP is indeed a powerful peroxidase. The activity toward the thioanisole oxidation by rHPR(FePc) in the presence of H_2O_2 is

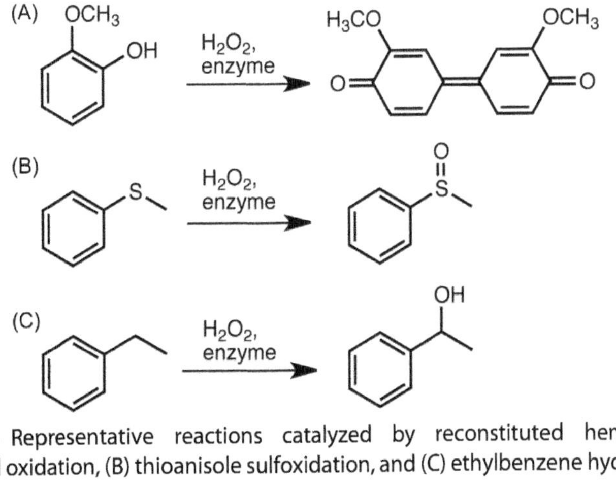

Scheme 4 Representative reactions catalyzed by reconstituted hemoproteins: (A) guaiacol oxidation, (B) thioanisole sulfoxidation, and (C) ethylbenzene hydroxylation.

clearly higher than that of the native enzyme (Scheme 4B). The turnover number frequency (TOF) of rHRP(FePc) is 17 min^{-1} at pH 7.0, 25°C (TOF of native HRP is 1.4 min^{-1} under the same conditions). In addition, the compound I-like species, an oxoferryl porphycene π-cation intermediate, is detectable by transient near-IR spectroscopic measurements and EPR spectroscopy upon addition of H_2O_2 to a solution of rHRP(FePc).

Iron corrole (FeCor) is also an attractive cofactor for myoglobin (Matsuo et al., 2009). The TOF value of rMb(FeCor) is 1.6 s^{-1} for guaiacol oxidation upon addition of H_2O_2, although no significant reaction is catalyzed by native myoglobin under the same conditions. In contrast, the TOF value of rHRP(FeCor) is 0.43 s^{-1}, which is clearly lower than that observed for rMb(FeCor). The trianionic corrole stabilizes the reactive intermediate, allowing the substrate to be efficiently oxidized.

4.2 Hydroxylase Activity of Reconstituted Myoglobin

One of the representative catalytic reactions supported by cytochrome P450 is hydroxylation of an inert C(sp^3)–H bond in alkane substrates. For example, cytochrome P450cam (which has heme *b* bound by a cysteine thiolate) is known to promote the 5-*exo*-hydroxylation of *d*-camphor. Myoglobin also has heme *b* but is incapable of promoting hydroxylation reaction. Hydroxylated products from external substrates have never been obtained by native and mutant myoglobins with heme *b*. In contrast, rMb(MnPc) is found to promote catalytic hydroxylation of toluene, ethylbenzene, and cyclohexane to the corresponding alcohol products catalytically (Scheme 4C) (Oohora et al., 2013). The turnover number of rMb(MnPc) for ethylbenzene to α-hydroxyethylbenzene is 13 at 25°C, pH 8.5. The KIE value determined in isotope effect experiments using toluene is 6.1, indicating that the rate determining step is hydrogen abstraction from the C–H bond by the Mn-oxo species which is produced by H_2O_2. The hydroxylation reaction has never been observed for rMb(Mn-protoporphyrin IX) and rMb(FePc).

4.3 Methionine Synthase Model

Methionine synthase is known to promote methyl group transfer from methylated folate to homocysteine to yield methionine. This reaction is supported by cobalamin as a cofactor in methionine synthase. During this catalytic reaction, there are two key intermediates, nucleophilic cobalt(I) and methylated cobalamin. However, the Co(I) complex has never been detected in the protein and the reaction mechanism is not completely understood because the structures of the enzyme and cofactor are complicated. In contrast, myoglobin reconstituted with Co(TDHC) is a good model for the cobalamin binding domain of methionine synthase (Hayashi et al., 2014). X-ray crystal structure analysis of reduced rMb(Co(TDHC)) reveals the Co(I)(TDHC) is a four-coordinate species without any axial ligands in the heme pocket. Furthermore, the methylated Co(TDHC) complex can be obtained upon the addition of methyl iodide, as confirmed by ESI-TOF MS (Morita et al., 2016). The methyl group bound to the Co(TDHC) is transferred to the imidazole ring of the distal His64. Although the catalytic methylation of external substrates has never been demonstrated by the reconstituted protein, rMb(Co (TDHC)) is the first example of a suitable model which generates two key-intermediates and replicates the methyl transfer reaction of methionine synthase.

ACKNOWLEDGMENTS

The authors thank the financial supports from the Grants-in-Aid (JP16H00837, JP16H06045, JP16K14036, and JP15H05804) provided by JSPS and MEXT. The authors also appreciate the supports from PRESTO and SICORP by JST.

REFERENCES

Cai, Y.-B., Li, X.-H., Jing, J., & Zhang, J.-L. (2013). Effect of distal histidines on hydrogen peroxide activation by manganese reconstituted myoglobin. *Metallomics*, *5*, 828–835.

Erben, C., Will, S., & Kadish, K. M. (2000). Metallocorroles: Molecular structure, spectroscopy and electronic states. In K. M. Kadish, K. M. Smith, & R. Guilard (Eds.), *Heteroporphyrins, expanded porphyrins and related macrocycles*: Vol. 2. *Porphyrin handbook* (pp. 233–300). San Diego: Academic Press.

Fandrich, M., Forge, V., Buder, K., Kittler, M., Dobson, C. M., & Diekmann, S. (2003). Myoglobin forms amyloid fibrils by association of unfolded polypeptide segments. *Proceedings of the National Academy of Sciences of the United States of America*, *100*, 15463–15468.

Fowler, C. J., Sessler, J., Lynch, V. M., Waluk, J., Gebauer, A., Lex, J., et al. (2002). Metal complexes of porphycene, corrphycene, and hemiporphycene: Stability and coordination chemistry. *Chemistry A European Journal*, *8*, 3485–3496.

Hayashi, T. (2013). Generation of functionalized biomolecules using hemoprotein matrices with small protein cavities for incorporation of cofactors. In T. Ueno & Y. Watanabe (Eds.), *Coordination chemistry in protein cages: Principles, design, and applications* (pp. 87–110). Hoboken: Wiley.

Hayashi, T., Dejima, H., Matsuo, T., Sato, H., Murata, D., & Hisaeda, Y. (2002). Blue myoglobin reconstituted with an iron porphycene shows extremely high oxygen affinity. *Journal of the American Chemical Society*, *124*, 11226–11227.

Hayashi, T., Morita, Y., Mizohata, E., Oohora, K., Ohbayashi, J., Inoue, T., et al. (2014). Co(II)/Co(I) reduction-induced axial histidine-flipping in myoglobin reconstituted with a cobalt tetradehydrocorrin as a methionine synthase model. *Chemical Communications*, *50*, 12560–12563.

Hayashi, T., Murata, D., Makino, M., Sugimoto, H., Matsuo, T., Sato, H., et al. (2006). Crystal structure and peroxidase activity of myoglobin reconstituted with iron porphycene. *Inorganic Chemistry, 45*, 10530–10536.

Hayashi, T., Sano, Y., & Onoda, A. (2015). Generation of new artificial metallopoteins by cofactor modification of native hemoproteins. *Israel Journal of Chemistry, 55*, 76–84.

Kräutler, B. (2009). Organometallic chemistry of B_{12} coenzyme. In A. Sigel, H. Sigel, & R. K. O. Sigel (Eds.), *Metal–carbon bonds in enzymes and cofactors: Vol. 6. Metal ions in life science* (pp. 1–51). Cambridge: RSC.

La Mar, G. N., Pande, U., Hauksson, J. B., Pandey, R. K., & Smith, K. M. (1989). Proton nuclear magnetic resonance investigation of the mechanism of the reconstitution of myoglobin that leads to metastable heme orientational disorder. *Journal of the American Chemical Society, 111*, 485–491.

Matsuo, T., Fukumoto, K., Watanabe, T., & Hayashi, T. (2011). Precise design of artificial cofactors for enhancing peroxidase activity of myoglobin: Myoglobin mutant H64D reconstituted with a "single-winged cofactor" is equivalent to native horseradish peroxidase in oxidation activity. *Chemistry, an Asian Journal, 6*, 2491–2499.

Matsuo, T., Hayashi, A., Abe, M., Matsuda, T., Hisaeda, Y., & Hayashi, T. (2009). Meso-unsubstituted iron corrole in hemoproteins: Remarkable differences in effects on peroxidase activities between myoglobin and horseradish peroxidase. *Journal of the American Chemical Society, 131*, 15124–15125.

Matsuo, T., Murata, D., Hisaeda, Y., Hori, H., & Hayashi, T. (2007). Porphyrinoid chemistry in hemoprotein matrix: Detection and reactivities of iron(IV)-oxo species of porphycene incorporated into horseradish peroxidase. *Journal of the American Chemical Society, 129*, 12906–12907.

Morita, Y., Oohora, K., Sawada, A., Doitomi, K., Ohbayashi, J., Kamachi, T., et al. (2016). Intraprotein transmethylation *via* a CH_3–Co(III) species in myoglobin reconstituted with a cobalt corrinoid complex. *Dalton Transactions, 45*, 3277–3284.

Murakami, Y., Aoyama, Y., & Tokunaga, K. (1980). Transition-metal complexes of pyrrole pigments. 16. Cobalt complexes of 1,19-dimethyldehydrocorrins as vitamin B_{12} models. *Journal of the American Chemical Society, 102*, 6736–6744.

Oohora, K., Kihira, Y., Mizohata, E., Inoue, T., & Hayashi, T. (2013). C(sp^3)–H bond hydroxylation catalyzed by myoglobin reconstituted with manganese porphycene. *Journal of the American Chemical Society, 135*, 17282–17285.

Oohora, K., Onoda, A., & Hayashi, T. (2012). Supramolecular assembling systems formed by heme–heme pocket interactions in hemoproteins. *Chemical Communications, 48*, 11714–11726.

Philips, G. N., Jr. (2001). Myoglobin. In A. Messerschmidt, R. Huber, T. Poulos, & K. Wieghardt (Eds.), *Handbook of metalloproteins: Vol. 1.* (pp. 5–15). Chichester: Wiley.

Reedy, C. J., & Gibney, B. R. (2004). Heme protein assemblies. *Chemical Reviews, 104*(2), 617–649.

Teale, F. (1959). Cleavage of the haem–protein link by acid methylethylketone. *Biochimica et Biophysica Acta, 35*, 543.

Vogel, E., Köcher, M., Schmickler, H., & Lex, J. (1986). Porphycene—A novel porphin isomer. *Angewandte Chemie, International Edition, 25*, 257–259.

Wagner, G. C., Perez, M., Toscano, W. A., Jr., & Gunsalus, I. C. (1981). Apoprotein formation and heme reconstitution of cytochrome P-450$_{cam}$. *The Journal of Biological Chemistry, 256*, 6262–6265.

Wasbotten, I., & Ghosh, A. (2006). Theoretical evidence favoring true iron(V)-oxo corrole and corrolazine intermediates. *Inorganic Chemistry, 45*, 4910–4913.

CHAPTER TWENTY

Creation of a Thermally Tolerant Peroxidase

Y. Watanabe*, H. Nakajima[†,1]
*Research Center of Materials Science, Nagoya University, Nagoya, Japan
[†]Graduate School of Science, Osaka City University, Osaka, Japan
[1]Corresponding author: e-mail address: nakajima@sci.osaka-cu.ac.jp

Contents

1. Introduction	456
2. Molecular Design for Cyt c_{552} to Acquire Peroxidase Activity	458
3. Gene Constructs of Recombinant Wild Type and Mutant Cyt c_{552}	459
3.1 Materials	459
3.2 Procedures	459
4. Expression and Purification of the Recombinant Proteins	460
4.1 Materials	460
4.2 Procedures	460
5. Thermal Stability of the Cyt c_{552} Mutants	461
5.1 Materials	462
5.2 Procedures	462
6. Analysis of a Key Intermediate in the Peroxidase Reaction of Cyt c_{552} V49D/M69A Mutant	462
6.1 Materials	464
6.2 Procedures	464
7. Effect of Trp45 on the Active Intermediate of Mutants	465
7.1 Procedures	467
References	468

Abstract

An artificial peroxidase with thermal tolerance and high catalytic activity has been successfully prepared by mutagenesis of an electron transfer protein, cytochrome c_{552} from *Thermus thermophilus*. The mutant enzymes were rationally designed based on the general peroxidase mechanism and spectroscopic analyses of an active intermediate formed in the catalytic reaction. Stopped flow UV–vis spectroscopy and EPR spectroscopy with a rapid freezing sample technique revealed that the initial double mutant, V49D/M69A, which was designed to reproduce the peroxidase mechanism, formed an active oxo-ferryl heme intermediate with a protein radical predominantly localized on Tyr45 during the catalytic reaction. The magnetic power saturation measurement obtained from EPR studies showed little interaction between the oxo-ferryl heme and the tyrosyl radical. Kinetics studies indicated that the isolated oxo-ferryl heme component in the active intermediate was a possible cause of heme degradation during the reaction with H_2O_2. Strong interaction between the oxo-ferryl heme and the radical was achieved by replacing Tyr45 with tryptophan (resulting in the Y45W/V49D/M69A mutant), which was similar to a tryptophanyl radical found in active intermediates of some catalase-peroxidases. Compared to the protein radical intermediates of V49D/M69A mutant, those of the Y45W/V49D/M69A mutant showed higher reactivity to an organic substrate than to H_2O_2. The Y45W/V49D/M69A mutant exhibited improved peroxidase activity and thermal tolerance.

1. INTRODUCTION

Peroxidases are some of the most common and ubiquitous enzymes. Many peroxidases contain an iron–porphyrin derivative (heme) in their active site, and they can catalyze oxidation of a wide variety of organic compounds using hydrogen peroxide (H_2O_2). H_2O_2 being an environmentally low-load oxidant, it has been used in industry and bioremediation for purposes such as food processing and biorefining of oil, and peroxidases have been utilized in some of these processes to enhance the oxidation activity of H_2O_2 (Ayala, Verdin, & Vazquez-Duhalt, 2007; Krieg & Halbhuber, 2003; Sigoillot et al., 2012; Veitch, 2004; Weng, Hendrickx, Maesmans, & Tobback, 1991). Despite the benefits of peroxidases, their full-scale use is still restricted due to their facile inactivation in the presence of H_2O_2 under supraphysiological conditions. In addition, their poor thermal and environmental tolerance reduces the range of practical uses of peroxidases. Many efforts have been devoted to overcoming these problems. Screening peroxidases from thermophiles has been employed to explore robust natural peroxidases (Apitz & van Pee, 2001; Gudelj et al., 2001; Kengen, Bikker, Hagen, de Vos, & van der Oost, 2001; McEldoon, Pokora, & Dordick, 1995). Random mutagenesis based on evolutionary engineering has also improved the stability and enzymatic activity of known peroxidases under catalytic conditions (Cherry et al., 1999; Morawski, Quan, & Arnold, 2001). To date, these biological approaches have been reasonably successful for obtaining peroxidases bearing desirable stability and activity. However, rational understanding of mutants' acquired chemical properties is often lacking, which may present obstacles to further rational modulation of catalytic activity and stability, or introduction of additional functions.

Engineering appropriate scaffold proteins based on the mechanistic aspects of peroxidases could be another approach for obtaining desirable peroxidases (de Lauzon, Desfosses, Mansuy, & Mahy, 1999; Ozaki, Matsui, Roach, & Watanabe, 2000; Watanabe, Nakajima, & Ueno, 2007).

This methodology allows successive molecular modifications leading to further development of desired properties. Despite this advantage, studies on such engineered proteins have been scarce, possibly due to the intrinsic fragility of the scaffold proteins examined. In this study, we decided to exploit a heme-protein from a thermophile as a scaffold for protein engineering.

Cytochrome c_{552} (Cyt c_{552}) from *Thermus thermophilus* HB8 is a small 14-kDa heme-protein expressed in the periplasmic space (Than et al., 1997; Yoshida et al., 1984). As seen in other proteins from thermophiles (Robb, Antranikian, Grogan, & Driessen, 2007), this protein exhibits high stability against heat. Although cytochrome c possesses weak peroxidase activity (Diederix, Busson, Ubbink, & Canters, 2004; Suzumura et al., 2005; Vazquez-Duhalt, 1999), simple mutagenesis to mimic residues essential for the peroxidase reaction conferred enhanced catalytic activity without sacrificing protein stability. The mutagenesis we performed on Cyt c_{552} consisted of two steps. First, an active site on the heme iron was introduced by replacing Met69 with alanine (M69A) and a general acid–base catalyst was deployed by replacing Val49 with aspartic acid (V49D) (Fig. 1). Catalytic activity of the mutant (V49D/M69A), as estimated from initial reaction rates, was highly improved in comparison to the original Cyt c_{552}, and the catalytic activity increased as reaction temperature rose to a maximum of 70°C. This result demonstrated the advantage of using an intrinsically thermostable protein as a scaffold to produce artificial peroxidases capable

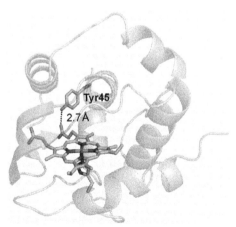

Fig. 1 Locus of Tyr45 in Cyt c_{552}. A hydrogen bond connects Tyr45 with a heme propionate group (*dotted line*) at a distance of 2.7 Å.

of acting at high temperatures. However, the mutant showed insufficient persistence in catalytic activity not due to denaturation, but due to heme degradation during the reaction cycle. Detailed analysis of the reaction of V49D/M69A with H_2O_2 revealed two points concerning the heme degradation. The major active intermediate was an oxo-ferryl heme with a protein radical localized predominantly on Tyr45, which is located close to the heme group and is connected to a heme propionate group through a hydrogen bond. The active intermediate could be transformed into a precursor for heme degradation by reaction with H_2O_2. Extensive mutagenesis of V49D/M69A at Tyr45 revealed that replacing this residue with a tryptophan (Y45W) changed the reactivity of the basal mutant with H_2O_2 and an organic substrate, resulting in improved persistence in the peroxidase activity at high temperatures. In the following sections, we describe the processes of transforming Cyt c_{552} to a thermally tolerant peroxidase based on spectroscopic characterization and kinetic analyses of reaction intermediates. We propose that proteins from thermophiles are useful resources not only for exploring a protein of desired function but also for constructing thermally stable bio-based materials amenable to rational molecular design based on the chemistry of a target function.

2. MOLECULAR DESIGN FOR CYT C_{552} TO ACQUIRE PEROXIDASE ACTIVITY

Two components are essential for heme-based peroxidase activity: one is a 5-coordinated heme species with an axial ligand (usually His), and the other is an acid–base catalyst generally served by an imidazole (His) or carboxyl (Asp or Glu) side chain (Watanabe et al., 2007). The acid–base catalyst should be placed 5–6 Å above the heme iron for the smooth exchange of a proton with H_2O_2 during reaction with the heme to form an oxo-ferryl heme active intermediate species. In order to introduce the above components, the mutagenesis described later was performed on Cyt c_{552} based on the crystal structure of the protein (PDB: 1C52; Fig. 1) (Than et al., 1997).

1. Replacement of Met69 with Ala (M69A): transformation of the original 6-coordinated heme with Met69–His15 axial ligands to a 5-coordinated heme with the His15 axial ligand.
2. Replacement of Val49 with Glu (V49D): introduction of the acid–base catalyst above the heme. The distance from C_γ of Val49 to the heme iron was estimated to be 5.5 Å based on the crystal structure. A simple model structure anticipated that replacement of Val49 with

Asp would place the carboxyl group of Asp at a position appropriate for acting as the acid–base catalyst.

In addition to these mutations essential for peroxidase activity, Tyr45 was replaced with Trp (Y45W) according to detailed kinetics analysis of the V49D/M69A mutant peroxidase reaction. This mutagenesis ameliorated a major inactivation process in the catalytic reaction, consequently improving persistence of the V49D/M69A mutant's catalytic activity (vide infra) (Nakajima, Ramanathan, Kawaba, & Watanabe, 2010).

3. GENE CONSTRUCTS OF RECOMBINANT WILD TYPE AND MUTANT CYT c_{552}

3.1 Materials

Primers for Cyt c_{552} gene cloning
Sense primer: 5′-<u>GGATCC</u>GCAGGCGGACGGGGCCAAGATC-3′
Antisense primer: 5′-<u>GGATCC</u>TTACTTCAGGCCGAGCTTCTTCTTCCG-3′

Primers for mutagenesis
5′-GGCGGTAGGGAG<u>TGG</u>CTCATCCTGGACC-3′ (for replacement of Tyr45 with Trp)
5′-GTACCTCATCCTG<u>GAC</u>CTTCTCTACGGCC-3′ (for replacement of Val49 with Asp)
5′-GAAGTACAACGGCGTC<u>GCG</u>TCCTCCTTCGC-3′ (for replacement of Met69 with Ala)

T. thermophilus strain HB8 (ATCC 27634)
pET-22b(+) vector (GE Health Care Co., PA, USA)
Escherichia coli strain BL21(DE3)
*Bam*HI restriction enzyme
QuikChange Site-Directed Mutagenesis Kit (Agilent Tech. Co., CA, USA)

3.2 Procedures

The thermophilic bacteria, *T. thermophilus*, were grown in ATCC 697 thermus medium at 75°C. Genomic DNA of *T. thermophilus* was recovered from cell lysates by a conventional method using cetyltrimethylammonium bromide (Ausubel et al., 2003). The Cyt c_{552} gene was obtained by PCR amplification using the isolated chromosome and the primers shown above, where underlined bases represent the *Bam*HI restriction site, and double-underlined bases encode the first residue of Cyt c_{552} immediately after a signal peptide for periplasmic localization in *T. thermophilus*. The amplified gene was inserted into the BamHI site of a pET-22b(+) expression vector plasmid to produce pET-cyt c_{552}, where the pelB leader sequence included in the plasmid was attached on the 5′-end of the inserted gene so that the pelB leader served as the signal peptide of the overexpressed protein in *E. coli* (Pritchard et al., 1997). The bacterial strain, *E. coli* BL21(DE3), was transformed with pET-cyt c_{552} and was used for expressing recombinant Cyt c_{552}. The N-terminus of the recombinant protein was preceded by the residual peptide "MDIGINSDP" derived from the pelB sequence. The peptide was not removable from the protein due to the lack of a protease site, but it did not alter the thermal stability of recombinant Cyt c_{552}.

All mutants were constructed using a site-directed mutagenesis kit (QuikChange II, Agilent Tech. Co., CA, USA), using pET-cyt c_{552} as a template. Sequences listed earlier and their reverse compliments were used for sense and antisense primers for the mutagenesis, where the underlined sequences correspond to the codons being altered.

4. EXPRESSION AND PURIFICATION OF THE RECOMBINANT PROTEINS

4.1 Materials

LB medium with 50 μg mL^{-1} ampicillin
Isopropyl-β-D-thiogalactopyranoside (IPTG), 0.5 mM in LB medium
Working buffer solution: 20 mM succinic acid-NaOH buffer (pH 5.0)
Cation exchange column (CM sepharose, GE Healthcare Co., PA, USA)
Size exclusion chromatography column (Sephadex 30 pg, GE Healthcare Co., PA, USA)
Desalting column (HiPrep 26/10 Desalting, GE Healthcare Co., PA, USA)
Regenerated cellulose membrane for ultrafiltration, MWCO = 3000 (Merck Millipore Co., Darmstadt, Germany)
Chemicals for pyridine hemochrome method (1 M aqueous NaOH, pyridine, sodium dithionite)

4.2 Procedures

BL21(DE3) cells expressing pET-cyt c_{552} were cultured in the LB medium with ampicillin at 37°C. When cell density in the culture reached an OD600 of ~1.0, IPTG was added to induce overexpression of the Cyt c_{552} protein. Cells were harvested 8 h after IPTG induction by centrifugation at $8000 \times g$ for 5 min, and pellets were stored at −80°C until use.

The harvested cells were resuspended and sonicated in the working buffer solution on ice. The cell debris was separated from supernatant by centrifugation at 45,000 × g for 45 min at 4°C. Although the recombinant proteins were stable at room temperature, cell-free extracts were handled at 4°C until the heat treatment described later, in case of unexpected proteolysis of the overexpressed protein by an active protease at room temperature. The Cyt c_{552} protein was purified by heating the supernatant at 75°C for 30 min, immediately followed by centrifugation at 15,000 × g for 30 min at 20°C. The resulting solution, which had the characteristic reddish-brown color, was loaded onto a cation exchange column preequilibrated with the working buffer. After washing out unbound proteins from the column, the Cyt c_{552} protein was eluted with a linear gradient of 0–1 M NaCl in the equilibration buffer. Fractions containing the reddish-orange protein were combined and concentrated by ultrafiltration. A size exclusion chromatography column preequilibrated with the working buffer containing 0.1 M NaCl was utilized when further purification was needed. Purified protein samples showing a 410 nm/280 nm absorbance ratio >5.0 were used for subsequent experiments. After the purification process, protein solution was loaded onto a desalting column preequilibrated with 20 mM MES-NaOH buffer (pH 5.0) for buffer exchange. The concentration of the purified protein was adjusted using the extinction coefficient of a soret band in the ferric form of the protein, which was determined by the pyridine hemochrome method (Fuhrhop & Smith, 1975). Identical procedures were adopted for the expression and purification of all proteins used in this study.

The pyridine hemochrome method was carried out as follows: a solution containing the mutants in their ferric form (400 µL) was mixed with pyridine (50 µL) and 1 M NaOH (50 µL). After 30-min incubation at room temperature, an appropriate amount of sodium dithionite powder was added to the solution to completely reduce the ferric c-type heme exposed from the denatured protein. The concentration of the heme was determined using the extinction coefficient of the α-band of a ferrous c-type heme (29.1 mM^{-1} cm^{-1} at 551 nm). The determined concentration was used to calculate the extinction coefficient of the soret band of the mutants in the ferric form, which were 167 and 169 mM^{-1} cm^{-1} for the V49D/M69A and Y45F/V49D/M69A mutants, respectively.

5. THERMAL STABILITY OF THE CYT C_{552} MUTANTS

In order to estimate the thermal stability of the Cyt c_{552} mutants (V49D/M69A and Y45W/V49D/M69A), temperature-dependent circular dichroism (CD) and UV–vis spectroscopy were performed (Nakajima et al., 2008, 2010). Similar to the wild-type protein, the mutants showed no apparent changes in the CD spectra from 20°C to 85°C. Incubation of the mutants at 70°C for 1200 s, which mimicked reaction conditions of present peroxidase assays (vide infra), also showed no effects on the spectra. Likewise, UV–vis spectroscopy indicated no significant change in heme coordination structure of the mutants, even at 80°C. Thus, the mutants retained high thermal stability comparable to the wild type; therefore, thermal denaturation of the mutants was unlikely to occur under the catalytic reaction conditions employed in the present study (vide infra).

5.1 Materials

Cyt c_{552} wild type and mutants, 10 μM
Working buffer solution, 20 mM MES-NaOH buffer (pH 5.0)

5.2 Procedures

A thin optical cell (1-mm path length) with a screw top was used to measure CD spectra. The protein concentration of sample solutions was adjusted to 10 μM using the extinction coefficient of the mutant mentioned earlier. The temperature was raised linearly from 20°C to 85°C at a rate of 1°C/min. All data were corrected at 222 nm, which corresponds to the maximum of a negative cotton effect from α-helices.

6. ANALYSIS OF A KEY INTERMEDIATE IN THE PEROXIDASE REACTION OF CYT C$_{552}$ V49D/M69A MUTANT

Reaction intermediates formed by the Cyt c_{552} V49D/M69A mutant were analyzed by combination of the stopped flow UV–vis technique and EPR spectroscopy with rapidly frozen samples (Nakajima et al., 2008, 2010). Using these approaches, we determined that an intermediate in the catalytic reaction was associated with heme degradation and deactivation, and therefore, we formulated a strategy to avoid deactivation and improve the persistence of the catalytic activity of the mutant.

The stopped flow UV–vis technique revealed that the reaction of the V49D/M69A mutant in the ferric state with excess H_2O_2 quickly formed an oxo-ferryl heme species as an initial product, which resulted in immediate heme degradation. When a substrate of the peroxidase reaction was added to the initial product, heme degradation was suppressed, and the initial product was restored to its original ferric form. These results implied that the oxo-ferryl heme species was associated with the heme degradation in the absence of peroxidase substrate. EPR spectroscopy and site-directed mutagenesis revealed that the oxo-ferryl heme species described earlier was accompanied by a phenoxy radical localized on Tyr45, which was located close to the heme (Fig. 1) and was connected with a heme propionate group through a hydrogen bond; however, microwave power saturation measurements from EPR spectroscopy (Pogni et al., 2005; Schunemann et al., 2004) indicated little interaction between the phenoxy radical and the oxo-ferryl heme species.

A possible catalytic cycle to explain the experimental results and the general peroxidase mechanism is shown in Scheme 1. Several heme degradation mechanisms have been reported for the peroxidase reaction (DePillis, Ozaki, Kuo, Maltby, & deMontellano, 1997; Valderrama, Ayala, & Vazquez-Duhalt, 2002; Villegas, Mauk, & Vazquez-Duhalt, 2000). In this study, detailed kinetic analyses of the V49D/M69A mutant proposed that a peroxy-ferric heme radical species formed by reaction of the oxo-ferryl species with H_2O_2 was a major cause of the heme degradation (Valderrama et al., 2002; Villegas et al., 2000). This mechanism is characteristic of the oxo-ferryl heme species. Therefore, we further altered the V49D/M69A mutant to avoid the formation of the isolated oxo-ferryl species **4** in the catalytic cycle. After extensive mutagenesis of Tyr45, replacement with Trp was found to be most efficient in preventing the formation of the isolated oxo-ferryl heme species by inducing the magnetic interaction with a tryptophanyl radical. An analogous interaction is observed for Class I peroxidases such as cytochrome c peroxidase (Pogni et al., 2005, 2006; Ruiz-Duenas et al., 2009).

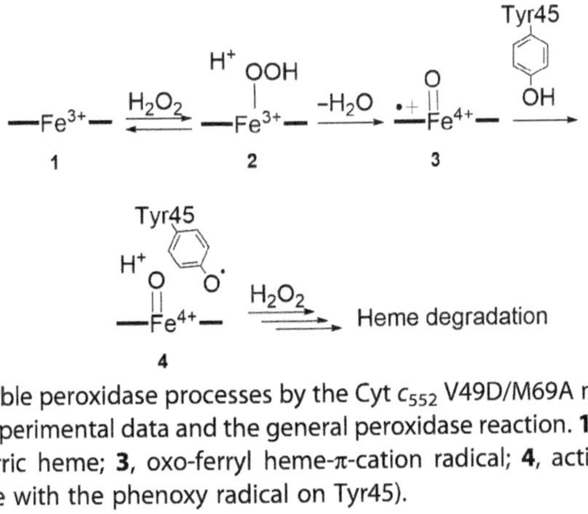

Scheme 1 Possible peroxidase processes by the Cyt c_{552} V49D/M69A mutant, as determined by the experimental data and the general peroxidase reaction. **1**, resting state; **2**, hydroperoxo-ferric heme; **3**, oxo-ferryl heme-π-cation radical; **4**, active intermediate (oxo-ferryl heme with the phenoxy radical on Tyr45).

6.1 Materials

Cyt c_{552} mutants, 10 µM (for stopped flow measurement), 200 µM (for EPR measurement)

H_2O_2, 100–800 µM (for stopped flow measurement), 2 mM (for EPR measurement)

$Cu(NO_3)_2$, 1 mM

Ethylenediamine-N, N, N', N'-tetra acetic acid disodium salt (EDTA·2Na), 1 mM

Working buffer solution, 20 mM MES-NaOH (pH 5.0)

6.2 Procedures

A Cyt c_{552} mutant and H_2O_2 (100–800 µM) were mixed in the working buffer solution on a stopped flow apparatus at room temperature or 70°C. The kinetic traces at 398 nm that correspond to the absorption maxima of the variants were used for determining pseudo first-order rates of initial product formation. The rate constants were given by the slopes of plots of the observed rates vs H_2O_2 concentration.

Typical procedures of sample preparation for X-band EPR measurements were as follows: a solution containing the Cyt c_{552} mutant (200 µM, 100 µL) and H_2O_2 (2 mM, 100 µL) was quickly mixed at a 1:1 ratio at room temperature by using a modified RSP-1000 stopped flow apparatus (Unisoku Co.), and the solution was placed in an EPR tube that was precooled in an acetone/dry ice bath. The reaction mixture was rapidly frozen at the bottom of the EPR tube (inner diameter of 4 mm), and the frozen samples were stored in liquid nitrogen until measurement. All EPR spectra were recorded at 10 K. The modulation amplitude and frequency were set to 0.2 mT and 100 kHz, respectively. Spin quantification was performed alongside Cu(II)-EDTA concentrations of 10, 20, 50, and 100 µM as a standard under nonsaturating conditions. The standard solutions were prepared by mixing appropriate amounts of 1 mM aqueous solutions of $Cu(NO_3)_2$ and EDTA disodium salt at room temperature, and then adjusting the final volume to 200 µL with water. Values obtained by double integration of the signals were divided by a correcting factor that is a function of the principal g values.

EPR saturation data were collected by measuring the EPR absorption derivatives signal as a function of microwave power. The saturation data were fitted to the following equation:

$$SP^{-1/2} = \left(1 + P/P_{1/2}\right)^{-b/2}$$

where S is the EPR derivative signal intensity, P is the microwave power, $P_{1/2}$ is the half-saturation power, and b is the inhomogeneity parameter. Nonlinear least-squares fit to the equation above yielded b and $P_{1/2}$ values. Details about the b value are described elsewhere (Pogni et al., 2005). In short, this parameter is an index of magnetic interaction between the observed spin and other spin systems. A value of 1.0 indicates that the observed spin is under diamagnetic conditions, while magnetic interaction with another spin system reduces the value to less than 1.0. The b values of the V49D/M69A and Y45W/V49D/M69A mutants were determined to be 0.99 and 0.67, respectively.

7. EFFECT OF Trp45 ON THE ACTIVE INTERMEDIATE OF MUTANTS

The active intermediate could be isolated by adding a small amount of catalase to scavenge residual H_2O_2 from the reaction mixture. The isolated intermediate was stable except that prolonged incubation at 25 °C resulted in gradual regeneration of the resting state without significant loss of the heme (Lardinois & de Montellano, 2003). Therefore, subsequent reactions like oxidation of organic substrate or heme degradation could be initiated by manually adding an organic substrate (ferulic acid, a natural substrate of horseradish peroxidase; Peyron, Abecassis, Autran, & Rouau, 2001) or H_2O_2 to the solution, respectively, and this process could be monitored by ordinary UV–vis spectroscopy. Under the experimental conditions employed here, the stoichiometric reactions with ferulic acid and H_2O_2 were traced by simple pseudo first-order kinetics depending on their concentrations. The calculated rate constants for the oxidation of organic substrate (k_r) and heme degradation (k_{hd}) are listed in Table 1. Based on the general peroxidase reaction mechanism, oxidation of the organic substrate by the active intermediate is depicted in Scheme 2. As the protein radical does not affect the UV–vis spectrum of the oxo-ferryl heme, intermediates

Table 1 Rate Constants of k_{hd} and k_r (s^{-1} M^{-1}) at 25 °C for the Prepared Mutants

	V49D/M69A	Y45W/V49D/M69A
k_r	5.8×10^4	9.2×10^4
k_{hd}	2.4×10^4	4.3×10^3
k_{hd}/k_r	0.41	0.05

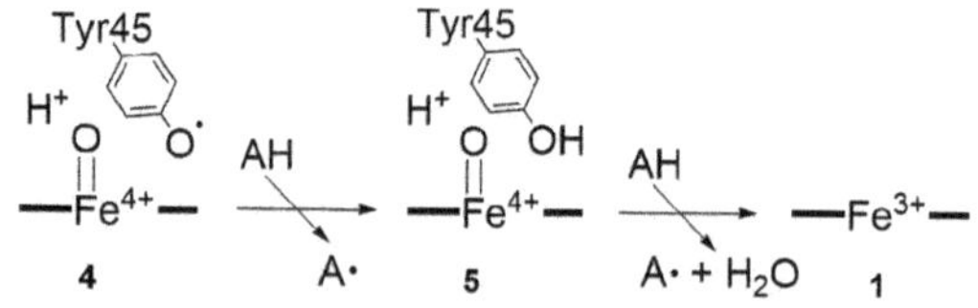

Scheme 2 Model oxidation mechanism of an organic substrate by the active intermediate (**4**). **5**, the oxo-ferryl heme; AH, the organic substrate (ferulic acid in this study).

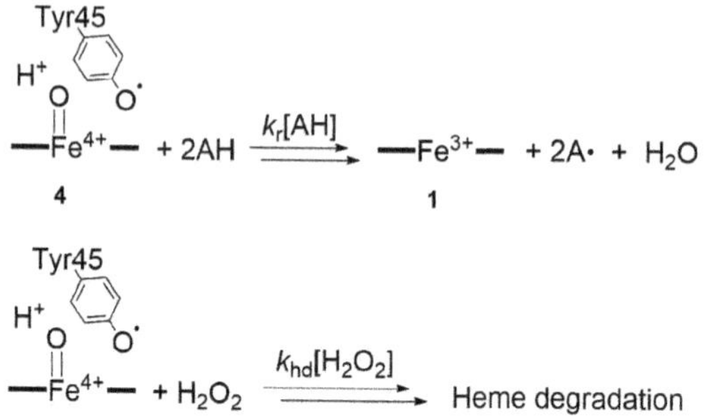

Scheme 3 Assignment of k_r and k_{hd} to the oxidation reaction of the organic substrate (*upper reaction*) and heme degradation (*lower reaction*), respectively, by the active intermediate.

4 and **5** show virtually the same spectra; therefore, the first and second steps in Scheme 2 are spectroscopically indistinguishable.

Accordingly, k_r was assigned to the overall oxidation process by the protein radical intermediate as shown in Scheme 3. The actual heme degradation must be a complex reaction (DePillis et al., 1997; Valderrama et al., 2002); however, the spectroscopic analysis indicated that the reactions simply depended on [H_2O_2] and immediately led to heme bleaching with no observable intermediate. Thus, the heme degradation process kinetics could be represented as $k_{hd}[H_2O_2]$ (Scheme 3).

Since heme degradation (k_{hd}) and substrate oxidation (k_r) are competitive processes in the described peroxidase reaction, the k_{hd}/k_r ratio could be used as an index to deduce the likelihood of the active intermediates leaving the catalytic cycle for the heme degradation pathway. The k_{hd}/k_r values at 25°C were calculated to be 0.41 and 0.05 for V49D/M69A and Y45W/V49D/M69A, respectively. This indicates that the replacement of Tyr45 with Trp is preferable for kinetic control of the heme degradation in the present peroxidase cycle. As a result of the decreased heme degradation, the Y45W/V49D/M69A mutant exhibited improved activity and persistency in the peroxidase reaction compared to the V49D/M69A mutant (Fig. 2).

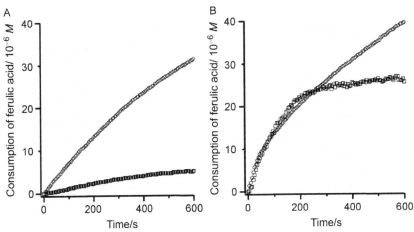

Fig. 2 Time courses of oxidation substrate (ferulic acid) consumption in the catalytic reaction by the mutants at 25°C (A) and 70°C (B) up to 600 s. *Symbols* represent V49D/M69A (□) and Y45W/V49D/M69A (o). Reaction conditions: 2×10^{-7} M enzyme, 2.0×10^{-4} M H_2O_2, and 2.0×10^{-4} M ferulic acid in 20 mM MES-NaOH buffer (pH 5.0).

7.1 Procedures

7.1.1 Kinetic Measurement

Cyt c_{552} mutant, 10 µM

H_2O_2, 20 µM and 100–500 µM

Ferulic acid (a substrate for peroxidase reaction), 100–500 µM

Catalase from bovine liver, 50 µg L^{-1}

Working buffer solution, 20 mM MES-NaOH buffer (pH 5.0)

Active intermediates of the V49D/M69A and Y45W/V49D/M69A mutants were prepared as follows: mutant proteins in working buffer solution were mixed with 1 mM aqueous H_2O_2 to reach a final H_2O_2 concentration of 20 µM at 25°C. Immediately after observing maximal formation of the active intermediate, which was confirmed by complete diminishment of the original visible band of the mutants at 620 nm (characteristic of the high spin ferric heme species), catalase from bovine liver was added to the reaction mixture to scavenge the residual H_2O_2 in the solution. In the absence of H_2O_2, the isolated active intermediate was stable for at least 5 min at 25°C. Then, ferulic acid or H_2O_2 was added to the isolated active intermediate at a final concentration of 100–500 µM. The kinetic traces at 620 nm were used for determining pseudo first-order rates. The rate constants were obtained by the slopes of plots of the observed rates vs the concentration of ferulic acid or H_2O_2. Due to instability of the employed catalase, the reproducible rate constants could not be obtained at 70°C.

7.1.2 Measurement of Catalytic Activity

Cyt c_{552} mutant, 200 μM
Ferulic acid, 200 mM
H_2O_2, 200 mM
Working buffer solution, 20 mM MES-NaOH, pH 5.0

Prior to use in the catalytic activity assay, ferulic acid was heated at 80°C in the working buffer solution for 8 h. This heat treatment is crucial for obtaining a stable ferulic acid absorption spectrum under the assay conditions, otherwise ferulic acid showed a spontaneous decrease in absorbance during the assay at 70°C, leading to overestimation of the peroxidase activity of the mutants. This decrease in absorbance cannot yet be explained, as no chemical alteration of ferulic acid was observed as confirmed by ^{1}H-NMR spectroscopy. After heat treatment, ferulic acid shows stable absorption ($\varepsilon_{287nm} = 8.0 \times 10^3$ cm^{-1} M^{-1}) at both 20°C and 70°C. A 3-mL reaction mixture containing ferulic acid and mutant enzyme in working buffer solution were preincubated at each reaction temperature (20°C or 70°C) for 10 min. To initiate the reaction, 60 mL of 10 M aqueous H_2O_2 kept at 25°C was added to the solution. Since ferulic acid provides various oxidation products in the peroxidase reaction, observation of the reaction product is not appropriate for the quantitative analysis by UV–vis spectroscopy. Therefore, consumption of ferulic acid was monitored to trace the peroxidase activity of the mutants.

REFERENCES

Apitz, A., & van Pee, K. H. (2001). Isolation and characterization of a thermostable intracellular enzyme with peroxidase activity from *Bacillus sphaericus*. *Archives of Microbiology*, *175*, 405–412.

Ausubel, F. M., Brent, R., Kingston, R. E., Moore, D. D., Seidman, J. G., & Smith, J. A., et al. (2003). *Current protocols in molecular biology*. (Ringbou ed.). New York: John Wiley & Sons, Inc.

Ayala, M., Verdin, J., & Vazquez-Duhalt, R. (2007). The prospects for peroxidase-based biorefining of petroleum fuels. *Biocatalysis and Biotransformation*, *25*, 114–129.

Cherry, J. R., Lamsa, M. H., Schneider, P., Vind, J., Svendsen, A., Jones, A., et al. (1999). Directed evolution of a fungal peroxidase. *Nature Biotechnology*, *17*, 379–384.

de Lauzon, S., Desfosses, B., Mansuy, D., & Mahy, J. P. (1999). Studies of the reactivity of artificial peroxidase-like hemoproteins based on antibodies elicited against a specifically designed ortho-carboxy substituted tetraarylporphyrin. *FEBS Letters*, *443*, 229–234.

DePillis, G. D., Ozaki, S., Kuo, J. M., Maltby, D. A., & deMontellano, P. R. O. (1997). Autocatalytic processing of heme by lactoperoxidase produces the native protein-bound prosthetic group. *The Journal of Biological Chemistry, 272,* 8857–8860.

Diederix, R. E. M., Busson, S., Ubbink, M., & Canters, G. W. (2004). Increase of the peroxidase activity of cytochrome c_{550} by the interaction with detergents. *Journal of Molecular Catalysis B: Enzymatic, 27,* 75–82.

Fuhrhop, J. H., & Smith, K. M. (1975). Porphyrins and metalloporphyrins. In K. Smith (Ed.), Amsterdam: Elsevier.

Gudelj, M., Fruhwirth, G. O., Paar, A., Lottspeich, F., Robra, K. H., Cavaco-Paulo, A., et al. (2001). A catalase-peroxidase from a newly isolated thermoalkaliphilic *Bacillus sp* with potential for the treatment of textile bleaching effluents. *Extremophiles, 5,* 423–429.

Kengen, S. W. M., Bikker, F. J., Hagen, W. R., de Vos, W. M., & van der Oost, J. (2001). Characterization of a catalase-peroxidase from the hyperthermophilic archaeon *Archaeoglobus fulgidus. Extremophiles, 5,* 323–332.

Krieg, R., & Halbhuber, K. J. (2003). Recent advances in catalytic peroxidase histochemistry. *Cellular and Molecular Biology, 49,* 547–563.

Lardinois, O. M., & de Montellano, P. R. O. (2003). Intra- and intermolecular transfers of protein radicals in the reactions of sperm whale myoglobin with hydrogen peroxide. *The Journal of Biological Chemistry, 278,* 36214–36226.

McEldoon, J. P., Pokora, A. R., & Dordick, J. S. (1995). Lignin peroxidase-type activity of soybean peroxidase. *Enzyme and Microbial Technology, 17,* 359–365.

Morawski, B., Quan, S., & Arnold, F. H. (2001). Functional expression and stabilization of horseradish peroxidase by directed evolution in *Saccharomyces cerevisiae. Biotechnology and Bioengineering, 76,* 99–107.

Nakajima, H., Ichikawa, Y., Satake, Y., Takatani, N., Manna, S. K., Rajbongshi, J., et al. (2008). Engineering of *Thermus thermophilus* cytochrome c552: Thermally tolerant artificial peroxidase. *ChemBioChem, 9,* 2954–2957.

Nakajima, H., Ramanathan, K., Kawaba, N., & Watanabe, Y. (2010). Rational engineering of *Thermus thermophilus* cytochrome c_{552} to a thermally tolerant artificial peroxidase. *Dalton Transactions, 39,* 3105–3114.

Ozaki, S., Matsui, T., Roach, M. P., & Watanabe, Y. (2000). Rational molecular design of a catalytic site: Engineering of catalytic functions to the myoglobin active site framework. *Coordination Chemistry Reviews, 198,* 39–59.

Peyron, S., Abecassis, J., Autran, J. C., & Rouau, X. (2001). Enzymatic oxidative treatments of wheat bran layers: Effects on ferulic acid composition and mechanical properties. *Journal of Agricultural and Food Chemistry, 49,* 4694–4699.

Pogni, R., Baratto, M. C., Giansanti, S., Teutloff, C., Verdin, J., Valderrama, B., et al. (2005). Tryptophan-based radical in the catalytic mechanism of versatile peroxidase from *Bjerkandera adusta. Biochemistry, 44,* 4267–4274.

Pogni, R., Baratto, M. C., Teutloff, C., Giansanti, S., Ruiz-Duenas, F. J., Choinowski, T., et al. (2006). A tryptophan neutral radical in the oxidized state of versatile peroxidase from *Pleurotus eryngii*—A combined multifrequency EPR and density functional theory study. *The Journal of Biological Chemistry, 281,* 9517–9526.

Pritchard, M. P., Ossetian, R., Li, D. N., Henderson, C. J., Burchell, B., Wolf, C. R., et al. (1997). A general strategy for the expression of recombinant human cytochrome P450s in *Escherichia coli* using bacterial signal peptides: Expression of CYP3A4, CYP2A6, and CYP2E1. *Archives of Biochemistry and Biophysics, 345,* 342–354.

Robb, Frank, Antranikian, Garabed, Grogan, Dennis, & Driessen, Arnold (Eds.), (2007). *Thermophiles: Biology and technology at high temperatures.* New York: CRC Press.

Ruiz-Duenas, F. J., Pogni, R., Morales, M., Giansanti, S., Mate, M. J., Romero, A., et al. (2009). Protein radicals in fungal versatile peroxidase catalytic tryptophan radical in both compound I and compound II and studies on W164Y, W164H, and W164S variants. *The Journal of Biological Chemistry, 284*, 7986–7994.

Schunemann, V., Lendzian, F., Jung, C., Contzen, J., Barra, A. L., Sligar, S. G., et al. (2004). Tyrosine radical formation in the reaction of wild type and mutant cytochrome P450cam with peroxy acids—A multifrequency EPR study of intermediates on the millisecond time scale. *The Journal of Biological Chemistry, 279*, 10919–10930.

Sigoillot, J. C., Berrin, J. G., Bey, M., Lesage-Meessen, L., Levasseur, A., Lomascolo, A., et al. (2012). Fungal strategies for lignin degradation. In L. Jouann & C. Lapierre (Eds.), *Lignins: Biosynthesis, biodegradation and bioengineering* (pp. 263–308). San Diego: Elsevier Academic Press Inc.

Suzumura, A., Paul, D., Sugimoto, H., Shinoda, S., Julian, R. R., Beauchamp, J. L., et al. (2005). Cytochrome c-crown ether complexes as supramolecular catalysts: Cold-active synzymes for asymmetric sulfoxide oxidation in methanol. *Inorganic Chemistry, 44*, 904–910.

Than, M. E., Hof, P., Huber, R., Bourenkov, G. P., Bartunik, H. D., Buse, G., et al. (1997). *Thermus thermophilus* cytochrome c_{552}: A new highly thermostable cytochrome c structure obtained by MAD phasing. *Journal of Molecular Biology, 271*, 629–644.

Valderrama, B., Ayala, M., & Vazquez-Duhalt, R. (2002). Suicide inactivation of peroxidases and the challenge of engineering more robust enzymes. *Chemistry & Biology, 9*, 555–565.

Vazquez-Duhalt, R. (1999). Cytochrome c as a biocatalyst. *Journal of Molecular Catalysis B: Enzymatic, 7*, 241–249.

Veitch, N. C. (2004). Horseradish peroxidase: A modern view of a classic enzyme. *Phytochemistry, 65*, 249–259.

Villegas, J. A., Mauk, A. G., & Vazquez-Duhalt, R. (2000). A cytochrome c variant resistant to heme degradation by hydrogen peroxide. *Chemistry & Biology, 7*, 237–244.

Watanabe, Y., Nakajima, H., & Ueno, T. (2007). Reactivities of oxo and peroxo intermediates studied by hemoprotein mutants. *Accounts of Chemical Research, 40*, 554–562.

Weng, Z. J., Hendrickx, M., Maesmans, G., & Tobback, P. (1991). Immobilized peroxidase—A potential bioindicator for evaluation of thermal-processes. *Journal of Food Science, 56*, 567–570.

Yoshida, T., Lorence, R. M., Choc, M. G., Tarr, G. E., Findling, K. L., & Fee, J. A. (1984). Respiratory proteins from the extremely thermophilic aerobic bacterium, thermus-thermophilus—Purification procedures for cytochrome c_{552}, cytochrome $c_{555,549}$, and cytochrome c_1 aa_3 and chemical evidence for a single subunit cytochrome aa_3. *The Journal of Biological Chemistry, 259*, 112–123.

CHAPTER TWENTY-ONE

Designing Covalently Linked Heterodimeric Four-Helix Bundles

M. Chino*, L. Leone*, O. Maglio*,[†], A. Lombardi*,[1]
*University of Napoli Federico II, Napoli, Italy
[†]Institute of Biostructures and Bioimages-IBB, CNR, Napoli, Italy
[1]Corresponding author: e-mail address: alombard@unina.it

Contents

1. Introduction	472
1.1 The Four-Helix Bundle: A Widespread Structural Motif	472
1.2 Designing Functional Four-Helix Bundle Proteins	473
2. Selection of the Best Docking Hotspot Given a Predefined Anchor Bolt	477
2.1 Structure Preparation of the Target Protein	479
2.2 Structure Preparation of the Linker	479
2.3 Performing the Geometrical Parameter Calculations	479
2.4 Data Analysis	482
2.5 Generation of the Best Candidates for Click Reaction	483
2.6 Evaluation of the Best Candidates for Synthesis	484
3. Broadening the Hotspot and Linker Selections	486
3.1 Structure Preparation of Any Target Protein	488
3.2 Structure Preparation of the Linker Library	488
3.3 Performing the Geometrical Parameter Calculations for Each Residue Pair	488
3.4 Data Analysis	490
3.5 Generation of the Best Candidates for the Identified Residue Pairs	492
3.6 Evaluation of the Best Models Amenable for the Selected Linkers	492
4. Concluding Remarks	493
Acknowledgments	494
References	495

Abstract

De novo design has proven a powerful methodology for understanding protein folding and function, and for mimicking or even bettering the properties of natural proteins. Extensive progress has been made in the design of helical bundles, simple structural motifs that can be nowadays designed with a high degree of precision. Among helical bundles, the four-helix bundle is widespread in nature, and is involved in numerous and fundamental processes. Representative examples are the carboxylate bridged diiron proteins, which perform a variety of different functions, ranging from reversible dioxygen binding to catalysis of dioxygen-dependent reactions, including epoxidation, desaturation, monohydroxylation, and radical formation. The "Due Ferri" (two-irons; DF) family of proteins is the result of a de novo design approach, aimed to reproduce in minimal four-helix bundle models the properties of the more complex natural diiron proteins, and to address how the amino acid sequence modulates their functions. The results so far obtained point out that asymmetric metal environments are essential to reprogram functions, and to achieve the specificity and selectivity of the natural enzymes. Here, we describe a design method that allows constructing asymmetric four-helix bundles through the covalent heterodimerization of two different α-helical harpins. In particular, starting from the homodimeric DF3 structure, we developed a protocol for covalently linking the two $α_2$ monomers by using the Cu(I) catalyzed azide–alkyne cycloaddition. The protocol was then generalized, in order to include the construction of several linkers, in different protein positions. Our method is fast, low cost, and in principle can be applied to any couple of peptides/proteins we desire to link.

1. INTRODUCTION
1.1 The Four-Helix Bundle: A Widespread Structural Motif

The four-helix bundle is a ubiquitous structural motif in nature, as it is found among a wide range of functionally diverse proteins and metalloproteins. For example, four-helix bundles are involved in the RNA-binding process (Banner, Kokkinidis, & Tsernoglou, 1987) and they are found in several proteins, such as growth hormones (De Vos, Ultsch, & Kossiakoff, 1992) and cytokines (Rozwarski et al., 1994). Numerous complex metalloproteins yet contain a simple four-helix bundle at the heart of the protein, where the metal cofactor (such as a heme, a dinuclear iron or copper site) necessary to accomplish functions is housed. A heme site is found in the electron transfer cytochrome c' (Weber et al., 1980) and cytochrome $b562$ (Mathews, Bethge, & Czerwinski, 1979). Diiron sites in the class of carboxylate-bridged diiron proteins are involved in dioxygen binding and activation (Lee & Lippard, 2003; Maglio, Nastri, & Lombardi, 2012). Hemerythrin and myohemerythrin (Stenkamp, 1994) reversibly bind and transport oxygen, whereas ferritins and bacterioferritins are devoted to ferroxidase activity and iron storage within the core of a polymeric four-helix bundle structure (Frolow, Kalb, & Yariv, 1994; Harrison & Arosio, 1996; Wahlgren et al., 2012). Diiron proteins also catalyze a diverse set of dioxygen-dependent reactions, including desaturation (acyl carrier $Δ^9$ desaturase), hydroxylation (catalytic component of bacterial monooxygenases), and radical formation (R2 subunit of ribonucleotide reductase) (Jordan & Reichard, 1998; Lindqvist, Huang, Schneider, & Shanklin, 1996; Sazinsky & Lippard,

2015; Sirajuddin & Rosenzweig, 2015). Dinuclear copper sites, housed into the interior of four-helix bundles, also play important roles in dioxygen binding and activation. Among them, hemocyanins reversibly bind dioxygen, catechol oxidase and tyrosinase further activate dioxygen for substrate hydroxylation or oxidation (Yoon, Fujii, & Solomon, 2009). Recently, a four-helix bundle protein, able to accumulate copper for particulate methane monooxygenase, was isolated from the methanotroph *Methylosinus trichosporium* OB3b (Vita et al., 2015), further expanding the repertoire of fundamental processes played by this protein scaffold. Due to its central role in Nature, numerous attempts have been made to construct artificial four-helix bundles by de novo design, not only to allow a better interpretation of the chemistry supported by the natural systems but also to develop novel proteins with programmed functions (Chino et al., 2015; Samish, MacDermaid, Perez-Aguilar, & Saven, 2011; Slope & Peacock, 2016; Yu et al., 2014).

1.2 Designing Functional Four-Helix Bundle Proteins

The four-helix bundle can be viewed as an α-helical coiled coil, which is, more generally, a super-secondary structure made up of α-helices packed together in a parallel or antiparallel orientation. Coiled coils amino acid sequences are usually described in terms of seven residues (heptad) repeats, since seven residues are present per two turns of the α-helix (Kohn & Hodges, 1998). This scaffold is very robust and thermodynamically stable, since it is able to tolerate multiple residue substitutions without disrupting the global three-dimensional fold. As a consequence, the four-helix bundle scaffold is of great interest in the field of protein design, as it represents a useful template for structure-to-function relationship analysis and for developing novel artificial metalloenzymes (Chino et al., 2015; Peacock, 2016). In principle, active site environment (first and second coordination sphere) can be modified to induce metal-binding selectivity and to finely tune the chemistry of the cofactor to achieve specific functions. This task often involves introducing asymmetry around the metal environment, thus representing a difficult challenge in the de novo design of α-helical coiled coils.

One possible strategy for developing an asymmetric four-helix bundle involves the noncovalent heterodimerization of four single α-helices or two helix–loop–helix (α_2) domains (Fig. 1A and B). This approach requires establishing a large energy gap to stabilize the desired heteromeric form respect to both homooligomeric folds, and any undesired heteromeric

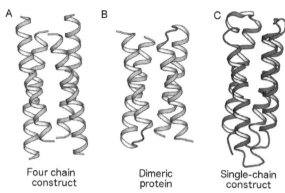

Fig. 1 Possible strategies for developing antiparallel four-helix bundles. (A) Tetramerization of four single α-helices. (B) Dimerization of two helix-loop-helix motifs. (C) Single-chain construct.

topology. Thus, the design methodology should include specific elements of both positive and negative design, to prevent alternate topologies from occurring (Grigoryan, Reinke, & Keating, 2009; Havranek & Harbury, 2003; Hill, Raleigh, Lombardi, & DeGrado, 2000). Even though the "rules" that guide oligomerization are now well established, all the interactions responsible for the pairing specificity are strictly dependent on slight variations of pH, ionic strength, and other physicochemical conditions of the environment (Fairman et al., 1996; Fry, Lehmann, Saven, DeGrado, & Therien, 2010; Marsh & DeGrado, 2002; Zhang et al., 2015). The noncovalently assembled complexes are generally not suitable for structural characterization, since it is difficult to completely avoid the presence of alternatively assembled species. On the other hand, heteromeric systems consisting of disconnected helices, which can be separately synthesized, purified, and combinatorially assembled, are well suited for the production of an array of any desired helical bundles from a significantly smaller number of peptides (Calhoun et al., 2005). A variety of de novo designed heteromeric two-stranded coiled coils (Litowski & Hodges, 2002; Thomas, Boyle, Burton, & Woolfson, 2013), three-helix (Chakraborty, Iranzo, Zuiderweg, & Pecoraro, 2012; Dieckmann et al., 1997), and four-helix bundles (Kaplan & DeGrado, 2004; Summa, Rosenblatt, Hong, Lear, & DeGrado, 2002) have been successfully developed and reported to date.

An alternative strategy to mimic the asymmetry of natural proteins in the context of designed coiled coils uses a single polypeptide chain (Fig. 1C), in which helices are connected by loops (Calhoun et al., 2003; Chakraborty et al., 2011; Smith & Hecht, 2011). Such proteins have generally unambiguous three-dimensional structures, thus greatly facilitating structural analysis. Nevertheless, the design of large proteins requires methods that are computationally intensive. In particular, the choice of interhelical loops is crucial, since it greatly affects both the stability and flexibility of the bundle. Further, the complexity of a single-chain construct limits its applicability for catalytic screening purposes, aimed at evaluating how systematic changes in the sequence affect structure, substrate-binding, and catalytic properties.

A third exploited strategy to obtain heteromeric four-helix bundles involves the covalent binding onto a predefined molecular scaffold. Mutter and colleagues introduced the concept of template assembled synthetic proteins (TASP) (Mutter, 2013; Mutter & Tuchscherer, 1997), which have been successfully adopted as scaffold for recognition and coupling of exogenous ligands (Monien, Drepper, Sommerhalter, Lubitz, & Haehnel, 2007; Rau, DeJonge, & Haehnel, 2000; Rau & Haehnel, 1998). Following the pioneering works of Mutter and coworkers, which adopted a properly designed cyclic decapeptide as template for assembling a variety of tertiary structures (Mutter et al., 1988), Haehnel and coworkers developed modular organized proteins (MOPS), for selectively binding metal cofactors, such as heme and copper ion (Monien et al., 2007; Rau, DeJonge, & Haehnel, 1998; Rau et al., 2000; Rau & Haehnel, 1998; Schnepf, Haehnel, Wieghardt, & Hildebrandt, 2004). A suitable chemoselective synthetic strategy was developed in order to control the identity and directionality of the helical segments, and obtain the desired heteromers.

A further approach for the design of heteromers retraces the way chosen by Nature, by building side chain/side chain covalent ligation through disulfide bond formation. Several computational methods have been developed so far, for predicting which pairs of residues, once mutated to cysteines, are suitable to form a disulfide bond. The used algorithms are derived from the analysis of side chains packing preferences of cysteine pairs involved in disulfide bonds, as found in crystal structures (Burton, Oas, Fterke, & Hunt, 2000; Craig & Dombkowski, 2013; Hazes & Dijkstra, 1988; Pabo & Suchanek, 1986; Sowdhamini et al., 1989). Despite inspired by Nature, the applicability of this strategy is limited, due to the stringent geometrical requirements for disulfide bond formation.

Recently, we implemented a novel design method to obtain an asymmetric four-helix bundle through the covalent heterodimerization of two different α-helical hairpins (Chino et al., 2016). This strategy aims at realizing an easy-to-screen system in a robust covalent framework, thus merging the advantages of using self-assembled monomers and single-chain constructs. We selected an efficient and orthogonal chemistry, to properly bind two different monomers in native conditions. In 2001 Kolb, Finn, and Sharpless published their seminal paper on the application of powerful and selective reactions to join small units through heteroatom links, and coined the term "Click Chemistry" (Kolb, Finn, & Sharpless, 2001). Several research groups explored the different applications of the Click Chemistry in numerous fields, such as drug discovery and synthesis (Galibert et al., 2010; Góngora-Benítez, Cristau, Giraud, Tulla-Puche, & Albericio, 2012; Valverde, Vomstein, Fischer, Mascarin, & Mindt, 2015) and polymer bioconjugation (Marine, Song, Liang, Watson, & Rudick, 2015; Rachel & Pelletier, 2016; Shu, Tan, DeGrado, & Xu, 2008). In particular, the use of Cu(I)-catalyzed azide–alkyne cycloaddition (CuAAC) has been largely employed as amide bond surrogate in the generation of α-helical and β-turn pseudopeptides (Beierle et al., 2009; Horne, Yadav, Stout, & Ghadiri, 2004), in TASP based molecular assemblies (Avrutina et al., 2009), as well as in strategies of peptide stapling, a macrocyclization process, where an intramolecular linkage is introduced to constrain the peptide in the desired α-helical conformation (Jacobsen et al., 2011; Lau, Wu, de Andrade, Galloway, & Spring, 2015; Scrima et al., 2010). Finally, it is also notable the work of Kolmar and coworkers (Empting et al., 2011), who constructed a triazole bridge, in replacement of a disulfide bond, employing the Ru(II)-catalyzed azide–alkyne cycloaddition (RuAAC).

The wide range of Click Chemistry applications prompted us to test this reaction in the selective intermolecular chemical ligation of two functionalized α_2 peptides, to generate heterodimeric proteins. One of the major advantages of the Click Chemistry-based methodology relies on its orthogonality to the chemistry involved in solid-phase peptide synthesis. Once chosen the binding positions, the peptides to be ligated can be easily functionalized during the synthetic step, by introducing noncanonical amino acids bearing the azide and alkyne moieties in their sequences (Fig. 2). CuAAC provides a simple to perform coupling process, leading to a thermally and hydrolytically stable triazole connection between the

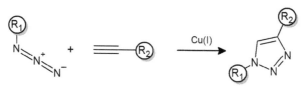

Fig. 2 Copper catalyzed 1,3-dipolar cycloaddition.

peptides. Moreover, the triazole ring of the linker could be introduced as a ligand in the metal-binding site.

Finally, given the large number of commercially available azide and alkyne building blocks, the designer can finely control the length and flexibility of the linker by choosing different pairs of functionalized amino acids.

In this chapter, we describe the developed computational protocol, first applied to the DF3 structure, as a specific case study. As logical extension of the method, a more general protocol is also reported, aimed at including the construction of several linkers, in different protein positions, and at fulfilling as many as possible designer needs.

2. SELECTION OF THE BEST DOCKING HOTSPOT GIVEN A PREDEFINED ANCHOR BOLT

In this section, we define a method for the rational design of a covalent attachment between the two subunits of the de novo designed DF3 protein. DF3 is made up of two identical 48-residue helix–loop–helix (α_2) motifs, able to specifically self-assemble into an antiparallell four-helix bundle, in the presence of metal ions (Faiella et al., 2009). The diiron form of the DF3 protein is able to perform phenol oxidase activity, rivaling natural counterparts in terms of catalytic efficiency. In order to move from oxidase to monooxygenase activity, a careful observation of natural monooxygenases structures, such as methane and toluene monooxygenases, points out that symmetry is broken in proximity of the active site (Friedle, Reisner, & Lippard, 2010). Starting from this observation, we developed a new asymmetric family of DF compounds, named DF-Click (Chino et al., 2016). In order to accomplish this task, we adopted a protocol for the design and synthesis of a covalent linkage between the two α_2 subunits based on Click Chemistry. This allowed us to generate a hybrid variant between a self-assembled heteromer and a single-chain protein, preserving the pros of dimeric derivatives, ie, simplified synthesis and structural data interpretation, ability to quickly generate several compounds for catalytic screening, and to finely tune the active site properties.

Here, we will first trace the steps leading to DF-Click, rationalizing, and elucidating the design process. In the next section, we will generalize our approach both in terms of different linkers and binding positions.

In the first member of DF-Click family, we used propargyl glycine (Pra) and 6-azidohexanoic acid (6aha) as the alkyne and the azide moiety, respectively, for the Click reaction. 6aha is often used to functionalize amino groups (Witte et al., 2013); in DF-Click family, it has been used as N-capping reagent of one subunit, replacing the N-terminal acetyl group. Once located the azide moiety, the next step required searching for the best position to mutate to Pra residue, in order to obtain efficient triazole formation, as well as the lowest perturbation of the global protein fold. This can be obtained when: (a) the interresidue distance is comparable with the distance spanned by the linker; (b) the conformation adopted by the linker is energetically favorable, otherwise the bundle could be strained to allow the linker to reach a more stable conformation. One possible way to address these requirements is to compare the geometrical parameters of the linker with those calculated for each possible binding position on the protein. A key step is to define a suitable description of the binding geometry, as well as to consider all the most favorable conformations of the linker. To accomplish these tasks, we define three binding parameters, which are calculated both for the protein and for the linker. Then, we compare these parameters to find the best candidates for synthesis. Two pivot bonds are chosen to describe the binding geometry, the C1–C2 bond of 6aha and the Cα–Cβ bond of Pra, which are compared to the C–C$_{methyl}$ bond of the terminal acetyl group and the Cα–Cβ bond of each residue from the other subunit, respectively. The geometrical parameters, illustrated in Fig. 3, are: (1) d, the C–C pivotal distance; (2) θ, the angle described by the first pivot bond and the first atom of the second pivot; (3) θ', the angle described by the second pivot bond and the first atom of the first pivot.

Upon generation of the "clicked" model, we evaluate the designed linker conformations in terms of RMSD from the energetically favorable starting conformations. A detailed step-by-step procedure will clarify this general approach.

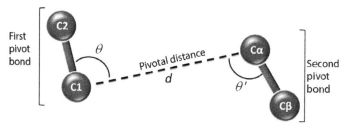

Fig. 3 Geometrical parameters calculated for the linker and for any pair of binding positions on the protein. For the linker, the first and the second pivot are the C1–C2 bond of 6aha and Cα–Cβ bond of Pra, respectively. For the protein, the first pivot is the C–C$_{methyl}$ bond of N-terminal acetyl group, and the second pivot is the Cα–Cβ bond of each selected residue from the other subunit.

2.1 Structure Preparation of the Target Protein

1. Retrieve the coordinates of the protein structure that you wish to adopt as a template. You can choose either an X-ray or a NMR structure, since no limitation is imposed by the method. In the protocol described in the following, we used the NMR model of the di-Zn-DF3, which can be downloaded from PDB (PDB ID: 2kik). We limited the search to the first model of the NMR bundle for the sake of brevity; in principle, the protocol can be adopted for each model of the bundle. It is worth to say that performing the protocol for multiple NMR models may result in a more exhaustive search, since the analysis on alternative conformations could result in a higher number of hotspots.
2. Remove water molecules, if present.
3. Add hydrogen atoms to the structure using the preferred software. In our case, we used Accelrys Discovery Studio 3.0 (DS3) (Accelrys Software Inc., 2012).
4. Save the structure as pdb format, and open it in PyMOL (DeLano, 2002).

2.2 Structure Preparation of the Linker

1. Manually generate the linker coordinates. Specifically, combine 6aha and Pra in the triazole form of the linker. We used DS3 to perform this task.
2. Perform a fast minimization cycle to clean the linker geometry. We adopted a DREIDING force field (Mayo, Olafson, & Goddard, 1990) to perform the minimization. DS3 uses a predefined function, activated by clicking *Structure|Clean Geometry* in the menu bar.
3. Save the linker coordinates in sdf format.
4. Submit this file to the Frog2 web server (http://bioserv.rpbs.univ-paris-diderot.fr/services/Frog2/; Miteva, Guyon, & Tufféry, 2010) to generate the 3D conformation ensemble, which will be analyzed in Sections 2.3 and 2.4. In our example, 50 conformations have been generated in pdb format, by imposing a minimization cycle for each of them.
5. Open the structural ensemble in PyMOL. Fig. 4 shows the Frog2 output we used for this protocol.

2.3 Performing the Geometrical Parameter Calculations

This protocol relies on the evaluation of three geometrical parameters, which have been calculated for: (i) each acetyl/residue pair of the protein and (ii) each linker conformation. Two PyMOL scripts have been produced

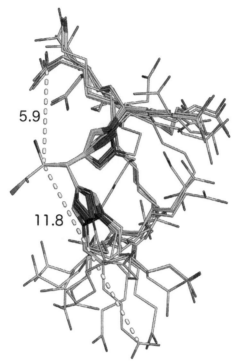

Fig. 4 Bundle of the 50 conformers generated by Frog2 (Miteva et al., 2010) for the Pra/6aha linker, showing the minimum and the maximum distance spanned by this linker. Conformers are fitted toward Pra backbone, for clarity.

to perform the calculation. They can be copied and saved as .py script files and run under the PyMOL environment with the command "*run/path/to/script.py.*"

A first script (script 1 reported below) has been used for the protein; it generates two files: "distca.txt" containing the pivotal distances d for each residue, and "angcbcaca.txt" containing θ, θ' angles.

```
#script 1 starts here
from pymol import cmd
#get the model coordinates for the 4 atoms. caA and cbA are the acyl
carbon and the methyl carbon of acetyl residue of the A subunit. caB
and cbB are the alpha and the beta carbon coming from B subunit.
caA = cmd.get_model("(n. C and c. A and resn ace)")
caB = cmd.get_model("(n. CA and c. B)")
cbA = cmd.get_model("(n. CH3 and c. A and resn ace)")
cbB = cmd.get_model("(n. CB+HA2 and c. B)")
```

```
#this is to generate a file with caA-caB distances
outFile = open("./distca.txt", 'w')
for atA in caA.atom:
        for atB in caB.atom:
                outFile.write( "%s %s %s %s %s\n" %(atA.resn, str(atA.resi), atB.resn, str(atB.resi), str(cmd.get_distance("(id %s)" % atA.id, "(id %s)" % atB.id))))
outFile.close()
#this is to generate a file for theta and theta prime.
outFile = open("./angcbcaca.txt", 'w')

outFile.write("CHAIN A (THETA)\n")

for acaA in caA.atom:
        for acbA in cbA.atom:
                if acaA.resi == acbA.resi:
                        for acaB in caB.atom:
                                outFile.write( "%s %s %s %s %s\n" %(acaA.resn, str(acaA.resi), acaB.resn, str(acaB.resi), str(cmd.get_angle("(id %s)" % acbA.id, "(id %s)" % acaA.id, "(id %s)" % acaB.id))))

outFile.write("CHAIN B (THETA')\n")

for acaA in caA.atom:
        for acaB in caB.atom:
                for acbB in cbB.atom:
                        if acaB.resi == acbB.resi:
                                outFile.write( "%s %s %s %s %s\n" %(acaA.resn, str(acaA.resi), acaB.resn, str(acaB.resi), str(cmd.get_angle("(id %s)" % acbB.id, "(id %s)" % acaB.id, "(id %s)" % acaA.id))))
outFile.close()
```

A second script (script 2 reported below) has been used for the linker.

```
#script 2 starts here
from pymol import cmd

#Before starting the script, generate four selections in pymol GUI, as follows: sele1 (C1 of 6aha), sele2 (C2 of 6aha), sele3 (CA of Pra), sele4 (CB of Pra).
```

```
outFile = open("./distca_linker.txt", 'w')
for i in range(1,51):
    outFile.write("%s %s\n" %(str(i),
cmd.get_distance("sele1","sele3",int(i))))
outFile.close()

outFile = open("./theta_linker.txt", 'w')
for i in range(1,51):
    outFile.write("%s %s %s\n" %(str(i),
cmd.get_angle("sele2","sele1","sele3",int(i)),
cmd.get_angle("sele4","sele3","sele1",int(i))))
outFile.close()
```

Also this script generates two files "distca_linker.txt" and "theta_linker.txt," containing the three selected geometrical parameters (d, θ, and θ', illustrated in Fig. 3). Table 1 reports the statistics of the d parameter calculated for the linker ensemble. It is evident that the linker is able to span a wide range of distances from 5.9 to 11.8 Å, with a mean value of 9.0 Å. The agreement between residue pairs and linker conformations will be evaluated in Section 2.4.

2.4 Data Analysis

Each subunit of DF3 is composed of 48 residues, and for each of them the three geometrical parameters (d, θ, and θ') have been calculated. Each triad of parameters has to be compared with the 50 triads calculated for the linker, for a total of 2400 deviations to be analyzed. Before generating this huge amount of data, it is appropriate to filter only for those residues that fall in the range of distances spanned by the linker (5.9–11.8 Å). Sorting the DF3 residues of one subunit according to the d-value (ie, the distance between each residue of one subunit and the N-terminal acetyl of the other subunit) results in identifying only three residues, within the maximum distance spanned by the Pra/6aha linker: Ala20 (11.5 Å), Tyr23 (10.9 Å), and Thr24 (12.0 Å). For these three residues, deviations for d, and θ, θ' parameters have been calculated as follows:

Table 1 Descriptive Statistics for the Conformers of the Pra/6aha Linker

Total number of conformers	Minimum Distance (Å)	Maximum Distance (Å)	(Mean Distance ± Standard Deviation) (Å)
50	5.9	11.8	9.0 ± 1.5

$$\text{dev}(d) = d_{\text{linker}} - d_{\text{protein}}$$

$$\text{rmsd}(\theta) = \sqrt{\left(\theta_{\text{linker}} - \theta_{\text{protein}}\right)^2 + \left(\theta'_{\text{linker}} - \theta'_{\text{protein}}\right)^2}$$

Table 2 shows the five best conformers in terms of dev(d) for the three evaluated binding positions. Tyr23 gives the best fitness with the linker, in terms of both distance and angles, resulting, for the conformer 28 (numbered as in the Frog2 output), in a dev(d) equal to zero and in a rmsd(θ) of only ~6°. In a refined protocol, it would be desirable to define a threshold for the calculated deviations to discard all the unproductive pairs. However, before taking a general rule, best models for each of the three residue positions have to be evaluated.

2.5 Generation of the Best Candidates for Click Reaction

DS3 has been used to generate the three desired structural models. The procedure described in the following has been performed with the implemented superimposition routine, even though other docking software could be adopted.

Table 2 Deviations of d and θ Parameters for the Five Best Conformers of the Pra/6aha Linker Respect to the Three Selected Binding Positions

Binding Position	Conformer n°	dev(d) (Å)	rmsd(θ) (°)
A20	10	0.2	35.8
A20	20	0.3	48.9
A20	28	−0.6	67.5
A20	27	−0.6	40.9
A20	46	−0.6	42.6
Y23	28	0	5.6
Y23	27	0	25.8
Y23	46	0	23.8
Y23	25	−0.1	6.3
Y23	37	−0.1	8.7
T24	20	−0.2	56.3
T24	10	−0.3	43.8
T24	28	−1.1	71.3
T24	27	−1.1	48.9
T24	46	−1.1	50.2

1. Load the protein structure and the best linker conformer in the same Molecule Window. We loaded conformer 10 for Ala20 hotspot, conformer 28 for Tyr23, and conformer 10 for Thr24 (see Table 2).
2. Add tethers between corresponding atoms, giving the command *Structure|Superimpose|Add Tether* (Fig. 5A). We tethered the main chain atoms of Pra and the carboxyl atoms of 6aha.
3. Set rotatable bonds on the linker, giving the command *Structure|Superimpose|Add Rotatable Bonds* (Fig. 5B). We set all bonds between the two pivotal bonds as free.
4. Perform superimposition with flexible torsions, giving the command *Structure|Superimpose|Molecular Overlay*. Fig. 5C displays the settings adopted to perform the superimposition.
5. Repeat the step 4 until the superimposed linker converges to a final invariable conformation.

2.6 Evaluation of the Best Candidates for Synthesis

In DS3, the molecular overlay with flexible torsions relies on a non-deterministic algorithm; thus, the final results may depend on the starting conformation of the linker adopted. For this reason, careful analysis of the resulting models should be made. In particular, the models could present some forbidden dihedrals, or the final superimposition could be not satisfactory as the tethered atoms are too far from the desired positions.

We found out that the best and more reproducible results are obtained when the final linker conformation is close to the starting one. Thus, the final linker models can be classified according to the RMSD respect to the starting conformer coordinates. This ranking has the double advantage of checking the superimposition task, as well as the goodness of the resulting conformation, since the starting conformers can be considered as minima in the energy landscape of the linker. Table 3 reports the deviations with respect to the starting conformations for the three models as obtained in Section 2.5. Comparing the RMSD values among the three models, it results that the linker bound in position 23 gives the lowest RMSD. A careful inspection of the three models (Fig. 6), further confirms the goodness of the resulting linker structures: the best model in terms of RMSD does not present any violation in the dihedrals.

We have already successfully synthesized a covalent heterodimeric DF analog, named DF-Click1, using the best linker described earlier (Chino et al., 2016), corroborating the design process. Given the success of this

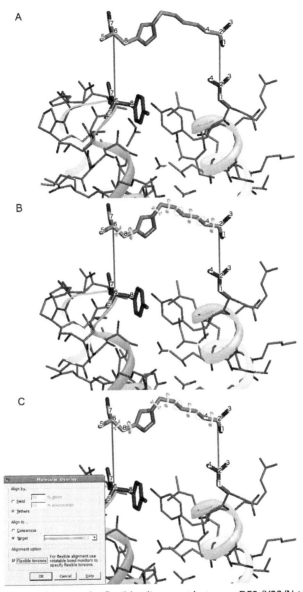

Fig. 5 Discovery Studio steps for flexible alignment between DF3 (Y23/N-terminal acetyl pair) and Pra/6aha linker with rotatable bonds. (A) Tether assignments for the defined fixed atom set. (B) Definition of the rotatable bonds of the linker molecule. (C) Molecular overlay settings.

Table 3 RMSD Values Calculated for the Conformations of the Pra/6aha Linker Before and After Flexible Alignment

Binding Position	Starting Conformer	RMSD (Å)
A20	10	1.08
Y23	28	0.91
T24	20	1.42

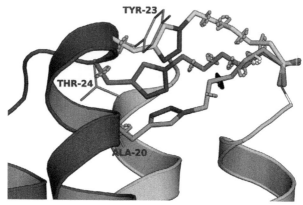

Fig. 6 Stick representation of the three possible docking hotspot positions of Pra/6aha on DF3: A20, Y23, and T24. Eclipsed torsions are highlighted in *black*, gauche torsions in *white*, trans torsions in *gray*.

approach, we have generalized the method, giving the designer the possibility to select both docking hotspots, and a wider set of linker options.

3. BROADENING THE HOTSPOT AND LINKER SELECTIONS

Keeping fixed one docking position (in our case the N-terminus of one subunit) limits the search output to only those positions that are in proximity of the chosen hotspot. This option greatly simplifies the design process; however, it may result reductive, since it may not be generally applicable to any protein scaffold of interest. Furthermore, one may be interested in stapling specific positions at a predefined distance, narrowing down the choice of suitable linkers. To meet these requirements, a generalized method has been defined (Fig. 7), which allows designing the link between any given residue pair and to adopt any linker combination. Three linker combinations have been considered to demonstrate the goodness of this approach. Covalently linked models will be then generated and evaluated, and all the steps for their design will be discussed.

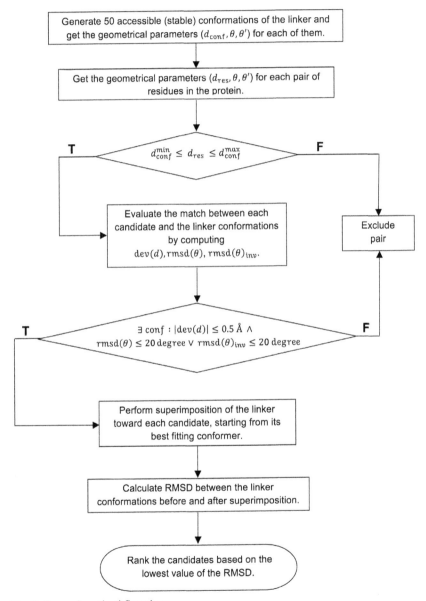

Fig. 7 General method flowchart.

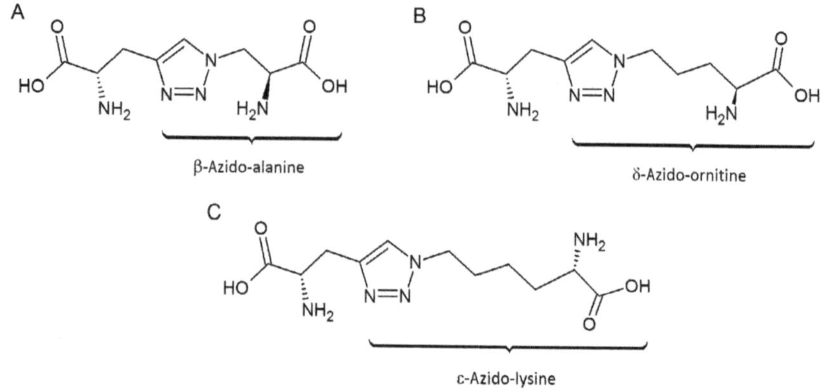

Fig. 8 Chemical structures of the selected linkers. Each of them is composed by a Pra residue and an azido-amino acid of different side chain length. (A) β-Azido-alanine, (B) δ-Azido-ornitine, (C) ε-Azido-lysine.

3.1 Structure Preparation of Any Target Protein

For any protein we wish to use as template, the structure preparation can be performed following the steps described in Section 2.1

3.2 Structure Preparation of the Linker Library

Three linker combinations have been chosen: Pra/β-azido-alanine (azido-Ala), Pra/δ-azido-ornitine (azido-Orn), and Pra/ε-azido-lysine (azido-Lys); their structures are shown in Fig. 8. These linkers have been chosen as they offer a wide range of distances between the binding residues. Steps to generate the set of conformations for each of them are reported in Section 2.2.

3.3 Performing the Geometrical Parameter Calculations for Each Residue Pair

A PyMOL script (script 3) has been developed to calculate the geometrical parameters, when the two pivotal bonds are both Cα–Cβ bonds, each from a different subunit.

```
#script 3 starts here
from pymol import cmd
#get the model coordinates for the 4 atoms. caA and cbA are the alpha
carbon and the beta carbon in the A subunit. caB and cbB are the alpha
and the beta carbon in the B subunit.
caA = cmd.get_model("(n. CA and c. A)")
caB = cmd.get_model("(n. CA and c. B)")
cbA = cmd.get_model("(n. CB+HA2 and c. A)")
```

```
cbB = cmd.get_model("(n. CB+HA2 and c. B)")
#this is to generate a file with caA-caB distances
outFile = open("./distca.txt", 'w')
for atA in caA.atom:
      foratB in caB.atom:
            outFile.write( "%s %s %s %s %s\n" %(atA.resn, str(atA.
resi), atB.resn, str(atB.resi), str(cmd.get_distance("(id %s)"
% atA.id, "(id %s)" % atB.id))))
outFile.close()
#this is to generate a file for theta and theta prime.
outFile = open("./angcbcaca.txt", 'w')

outFile.write("CHAIN A (THETA)\n")

for acaA in caA.atom:
      for acbA in cbA.atom:
            if acaA.resi == acbA.resi:
                  for acaB in caB.atom:
                        outFile.write( "%s %s %s %s %s\n" %(acaA.
resn, str(acaA.resi), acaB.resn, str(acaB.resi), str(cmd.get_angle
("(id %s)" % acbA.id, "(id %s)" % acaA.id, "(id %s)" % acaB.id))))
outFile.write("CHAIN B (THETA')\n")

for acaA in caA.atom:
      for acaB in caB.atom:
            for acbB in cbB.atom:
                  if acaB.resi == acbB.resi:
                        outFile.write( "%s %s %s %s %s\n" %(acaA.resn,
str(acaA.resi), acaB.resn, str(acaB.resi), str(cmd.get_angle("(id
%s)" % acbB.id, "(id %s)" % acaB.id, "(id %s)" % acaA.id))))
outFile.close()
```

The above reported script generates two files with the three geometrical parameters (d, θ, and θ') for each residueA–residueB pair. Given the great amount of data, we suggest ordering the script result in any spreadsheet software.

For each linker ensemble, the same script reported in Section 2.3 can be adopted to calculate the geometrical parameters for the linker. Before running the script, it is needed to create in the PyMOL GUI four selections with the four pivotal atoms.

3.4 Data Analysis

Table 4 summarizes the linker distances of the three ensembles, adopted in this procedure. Only pairs of residues whose Cα–Cα distance falls in the range spanned by the linkers have been taken into account, and for each of them the scores defined in Section 2.4 have been calculated. In this case, also a third score has been calculated, which considers the switch between the alkyne and the azide moieties:

$$\mathrm{rmsd}(\theta)_{\mathrm{inv}} = \sqrt{\left(\theta_{\mathrm{linker}} - \theta'_{\mathrm{protein}}\right)^2 + \left(\theta'_{\mathrm{linker}} - \theta_{\mathrm{protein}}\right)^2}$$

Only residue pairs showing at least one linker conformer meeting the required deviations have been considered as candidates for the design step. In particular, only those conformers whose dev(d) value was equal or lower than 0.5 Å and at least one of rmsd(θ) or rmsd(θ)$_{\mathrm{inv}}$ was lower than 20 degree have been selected. These thresholds have been adopted taking into account the results obtained from the previous designs. Further, since the design protocol does not consider backbone flexibility (Butterfoss & Kuhlman, 2006), it is not suggested to narrow down these thresholds. The numbers of candidate pairs for each linker, resulting from this analysis, are reported in Table 5. Thresholds filtered out 67% of candidates on average, resulting in less than 20 candidates for the Pra/azido-Lys linker. Tables 6–8 display the scores of the best matching conformation for each candidate pair of

Table 4 Descriptive Statistics for the 50 Conformers of Each of the Selected Linkers

Linker	Minimum Distance (Å)	Maximum Distance (Å)	Mean Distance ± Standard deviavtion (Å)
Pra/azido-Ala	5.9	7.0	6.6 ± 0.4
Pra/azido-Orn	6.8	9.4	8.5 ± 0.8
Pra/azido-Lys	5.4	10.7	8.2 ± 1.2

Table 5 Number of Residue Pairs Whose Geometrical Parameters Match with the Low-Energy Conformations of Each Linker

Linker	Matching Pairs (Evaluated Pairs)
Pra/azido-Ala	2 (4)
Pra/azido-Orn	10 (42)
Pra/azido-Lys	19 (81)

Table 6 Scores of the Best Fitting Conformation of the Pra/azido-Ala Linker for Each of the Selected Candidate Pairs of Binding Residues

Binding Positions	Best Conformer	dev(d) (Å)	rmsd(θ) (°)	rmsd(θ)$_{inv}$ (°)
V28–W42	50	0.1	12.9	23.9
I32–H39	6	0.3	1.9	15.7

Table 7 Scores of the Best Fitting Conformation of the Pra/azido-Orn Linker for Each of the Selected Candidate Pairs of Binding Residues

Binding Positions	Best Conformer	dev(d) (Å)	rmsd(θ) (°)	rmsd(θ)$_{inv}$ (°)
D35–H39	1	0.2	16.6	18.0
K31–W42	44	0.4	15.4	17.3
Y2–Q16	41	−0.2	18.6	14.8
I32–K38	46	−0.2	11.5	12.8
L6–A20	8	−0.2	16.4	14.6
N26–I46	10	0.2	10.8	18.2
K31–H39	46	−0.3	17.1	16.0
T24–I46	50	0.4	18.5	39.7
E36–E36	46	−0.5	16.4	16.8
Y2–A20	50	0.3	44.2	19.6

Table 8 Scores of the Best Fitting Conformation of the Pra/azido-Lys Linker for Each of the Selected Candidate Pairs of Binding Residues

Binding Positions	Best Conformer	dev(d) (Å)	rmsd(θ) (°)	rmsd(θ)$_{inv}$ (°)
D35–H39	40	−0.1	11.5	19.6
K31–W42	40	0.1	21.4	13.9
V28–T45	3	0.0	13.7	55.9
T24–L43	23	−0.1	19.5	67.4
E5–Q16	12	−0.3	38.1	17.1
K31–D35	33	0.0	28.6	13.7
Y2–Q16	38	0.1	7.2	11.1
I32–L43	34	−0.3	16.4	43.4

Continued

Table 8 Scores of the Best Fitting Conformation of the Pra/azido-Lys Linker for Each of the Selected Candidate Pairs of Binding Residues—cont'd

Binding Positions	Best Conformer	dev(d) (Å)	rmsd(θ) (°)	rmsd(θ)$_{inv}$ (°)
K31–K38	12	−0.4	17.8	7.2
L3–T24	33	0.4	15.0	5.4
Y2–A20	34	−0.3	8.7	35.4
A20–L43	12	−0.3	13.6	26.2
E10–G13	22	0.1	4.3	10.1
E10–I32	30	0.3	42.3	7.5
L6–Q16	34	−0.2	40.0	13.1
T24–I46	31	−0.2	30.9	5.2
Y2–I19	30	0.3	49.1	6.4
I32–W42	34	0.2	35.0	13.8
K31–H39	22	−0.4	21.3	7.7

residues. The reported conformers have been chosen to perform superimposition in the design step.

3.5 Generation of the Best Candidates for the Identified Residue Pairs

The desired models can be generated following the steps described in Section 2.5, with the only exception that in this case tethers for the superimposition can be imposed in the main chain for both ends of the linker.

3.6 Evaluation of the Best Models Amenable for the Selected Linkers

As discussed in Section 2.6, RMSD between the starting and the modeled linker coordinates can be adopted to rank the designed models. For the sake of brevity, Table 9 summarizes the results for the two best models for each designed linker. The average RMSD value of 1.1 Å is in line with the values obtained from the previous designs (see Table 3). The three best designs, one for each linker, are shown in Fig. 9. As expected by the RMSDs, the Pra/azido-Lys linker gives the best designed structure in terms of dihedrals, even though all of them may be considered as good candidates for synthesis.

Table 9 RMSD Values Calculated for the Conformations Before and After Flexible Alignment, for Each of the Three Linkers

Linker	Binding Positions	Starting Conformer	RMSD (Å)
Pra/azido-Ala	H39-I32'	6	1.22
Pra/azido-Ala	V28-W42'	3	1.35
Pra/azido-Orn	Y2-A20'	50	1.07
Pra/azido-Orn	L6-A20'	1	1.25
Pra/azido-Lys	V28-T45'	3	0.65
Pra/azido-Lys	T24-L43'	23	1.26

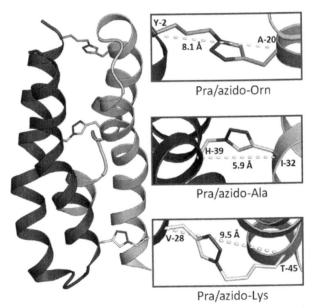

Fig. 9 Representation of the three best linker models (in terms of RMSD), along the bundle structure. *Inlets* show the details for each designed linker.

4. CONCLUDING REMARKS

In this chapter, we have described the steps leading to the generation of heterodimeric DF proteins through the rational design of covalent linkage between the two subunits. Asymmetrization of the active site in DF proteins has been already achieved in the tetrameric construct by self-assembly

(Kaplan & DeGrado, 2004) and in the single-chain construct (Reig et al., 2012), resulting in the modulation of the catalytic properties. The presented design methodology fills the gap between these two extremes, as it allows designing asymmetric models in the framework of the dimeric constructs, by means of a very simple and reliable approach. Oligomerization of two or more smaller subunits is frequently preferred in Nature to achieve complex protein structures, as supermolecular assembly is relatively simple and economical (Boersma & Roelfes, 2015). The protocol here described keeps the advantages given by the oligomerization of small subunits by fusing them in rationally designed positions that do not alter the global folding of the four-helix bundle. These advantages have been proven particularly remarkable in the study of the first DF-Click analog, as it showed complete reduction of dioxygen, coupled to the oxidation of a phenolic substrate leading to only one specific product (Chino et al., 2016).

The described linking moieties are based on the widely adopted CuAAC, which holds the advantage to be orthogonal to peptide chemistry; nonetheless this method can be efficiently applied to any class of linker, widening the applicability of the methodology. We kept this method as simple as possible, largely using simple scripts, with easily accessible softwares. This is particularly favorable as this protocol does not rely specifically on the four-helix bundle scaffold, and it can be freely applied to covalently link any protein/protein interface through a structure-based approach.

It is worth to say that there is still room for further improvements to make the methodology more accessible and reliable. In the actual implementation, this methodology does not consider explicitly backbone flexibility. One possible way to circumvent this limitation is to apply the method to a previously generated ensemble of protein structures, or to evaluate the protein flexibility of the finally designed dimers through molecular dynamics simulations. In the next future, we aim to integrate all the design tasks in an open-source environment, with the creation of a freely accessible web server.

ACKNOWLEDGMENTS

We wish to thank Flavia Nastri for critically reading the manuscript, and Fabrizia Sibillo for helping with editing.

The STRAIN PROJECT (POR Campania FSE 2007/2013 CUP B25B0900000000), which has provided a postdoctoral fellowship to M.C., and the Cost Action CM1003 (Biological Oxidation Reactions—Mechanisms and Design of New Catalysts) are kindly acknowledged.

REFERENCES

Accelrys Software Inc. (2012). *Discovery Studio, Release 3.0*. San Diego: Accelrys Software Inc.

Avrutina, O., Empting, M., Fabritz, S., Daneschdar, M., Frauendorf, H., Diederichsen, U., et al. (2009). Application of copper(i) catalyzed azide–alkyne [3+2] cycloaddition to the synthesis of template-assembled multivalent peptide conjugates. *Organic & Biomolecular Chemistry, 7*(20), 4177–4185.

Banner, D. W., Kokkinidis, M., & Tsernoglou, D. (1987). Structure of the ColE1 rop protein at 1.7 A resolution. *Journal of Molecular Biology, 196*(3), 657–675.

Beierle, J. M., Horne, W. S., van Maarseveen, J. H., Waser, B., Reubi, J. C., & Ghadiri, M. R. (2009). Conformationally homogeneous heterocyclic pseudotetrapeptides as three-dimensional scaffolds for rational drug design: Receptor-selective somatostatin analogues. *Angewandte Chemie International Edition, 48*(26), 4725–4729.

Boersma, A. J., & Roelfes, G. (2015). Protein engineering: The power of four. *Nature Chemistry, 7*(4), 277–279.

Burton, R. E., Oas, T. G., Fterke, C. A., & Hunt, J. A. (2000). Novel disulfide engineering in human carbonic anhydrase II using the PAIRWISE side-chain geometry database. *Protein Science, 9*(4), 776–785.

Butterfoss, G. L., & Kuhlman, B. (2006). Computer-based design of novel protein structures. *Annual Review of Biophysics and Biomolecular Structure, 35*(1), 49–65.

Calhoun, J. R., Kono, H., Lahr, S., Wang, W., DeGrado, W. F., & Saven, J. G. (2003). Computational design and characterization of a monomeric helical dinuclear metalloprotein. *Journal of Molecular Biology, 334*(5), 1101–1115.

Calhoun, J. R., Nastri, F., Maglio, O., Pavone, V., Lombardi, A., & DeGrado, W. F. (2005). Artificial diiron proteins: From structure to function. *Peptide Science, 80*(2–3), 264–278.

Chakraborty, S., Iranzo, O., Zuiderweg, E. R. P., & Pecoraro, V. L. (2012). Experimental and theoretical evaluation of multisite cadmium(ii) exchange in designed three-stranded coiled-coil peptides. *Journal of the American Chemical Society, 134*(14), 6191–6203.

Chakraborty, S., Yudenfreund Kravitz, J., Thulstrup, P. W., Hemmingsen, L., DeGrado, W. F., & Pecoraro, V. L. (2011). Design of a three-helix bundle capable of binding heavy metals in a triscysteine environment. *Angewandte Chemie International Edition, 50*(9), 2049–2053.

Chino, M., Leone, L., Maglio, O., Pavone, V., Nastri, F., & Lombardi, A. (2016). *Enhancing the selectivity of de novo due ferri proteins through heterodimer formation*. Manuscript in preparation.

Chino, M., Maglio, O., Nastri, F., Pavone, V., DeGrado, W. F., & Lombardi, A. (2015). Artificial diiron enzymes with a de novo designed four-helix bundle structure. *European Journal of Inorganic Chemistry, 2015*(21), 3371–3390.

Craig, D. B., & Dombkowski, A. A. (2013). Disulfide by Design 2.0: A web-based tool for disulfide engineering in proteins. *BMC Bioinformatics, 14*, 346.

De Vos, A. M., Ultsch, M., & Kossiakoff, A. A. (1992). Human growth hormone and extracellular domain of its receptor: Crystal structure of the complex. *Science, 255*(5042), 306–312.

DeLano, W. L. (2002). *The PyMOL molecular graphics system*. San Carlos, CA, USA: De Lano Scientific.

Dieckmann, G. R., McRorie, D. K., Tierney, D. L., Utschig, L. M., Singer, C. P., O'Halloran, T. V., et al. (1997). De novo design of mercury-binding two- and three-helical bundles. *Journal of the American Chemical Society, 119*(26), 6195–6196.

Empting, M., Avrutina, O., Meusinger, R., Fabritz, S., Reinwarth, M., Biesalski, M., et al. (2011). "Triazole bridge": Disulfide-bond replacement by ruthenium-catalyzed formation of 1,5-disubstituted 1,2,3-triazoles. *Angewandte Chemie International Edition, 50*(22), 5207–5211.

Faiella, M., Andreozzi, C., de Rosales, R. T. M., Pavone, V., Maglio, O., Nastri, F., et al. (2009). An artificial di-iron oxo-protein with phenol oxidase activity. *Nature Chemical Biology, 5*(12), 882–884.

Fairman, R., Chao, H.-G., Lavoie, T. B., Villafranca, J. J., Matsueda, G. R., & Novotny, J. (1996). Design of heterotetrameric coiled coils: Evidence for increased stabilization by Glu−Lys+ ion pair interactions. *Biochemistry, 35*(9), 2824–2829.

Friedle, S., Reisner, E., & Lippard, S. J. (2010). Current challenges of modeling diiron enzyme active sites for dioxygen activation by biomimetic synthetic complexes. *Chemical Society Reviews, 39*(8), 2768–2779.

Frolow, F., Kalb, A. J., & Yariv, J. (1994). Structure of a unique twofold symmetric haem-binding site. *Nature Structural Biology, 1*(7), 453–460.

Fry, H. C., Lehmann, A., Saven, J. G., DeGrado, W. F., & Therien, M. J. (2010). Computational design and elaboration of a de novo heterotetrameric α-helical protein that selectively binds an emissive abiological (porphinato)zinc chromophore. *Journal of the American Chemical Society, 132*(11), 3997–4005.

Galibert, M., Sancey, L., Renaudet, O., Coll, J.-L., Dumy, P., & Boturyn, D. (2010). Application of click-click chemistry to the synthesis of new multivalent RGD conjugates. *Organic & Biomolecular Chemistry, 8*(22), 5133.

Góngora-Benítez, M., Cristau, M., Giraud, M., Tulla-Puche, J., & Albericio, F. (2012). A universal strategy for preparing protected C-terminal peptides on the solid phase through an intramolecular click chemistry-based handle. *Chemical Communications, 48*(17), 2313.

Grigoryan, G., Reinke, A. W., & Keating, A. E. (2009). Design of protein-interaction specificity gives selective bZIP-binding peptides. *Nature, 458*(7240), 859–864.

Harrison, P. M., & Arosio, P. (1996). The ferritins: Molecular properties, iron storage function and cellular regulation. *Biochimica et Biophysica Acta (BBA)-Bioenergetics, 1275*(3), 161–203.

Havranek, J. J., & Harbury, P. B. (2003). Automated design of specificity in molecular recognition. *Nature Structural & Molecular Biology, 10*(1), 45–52.

Hazes, B., & Dijkstra, B. W. (1988). Model building of disulfide bonds in proteins with known three-dimensional structure. *Protein Engineering, 2*(2), 119–125.

Hill, R. B., Raleigh, D. P., Lombardi, A., & DeGrado, W. F. (2000). De novo design of helical bundles as models for understanding protein folding and function. *Account of Chemical Research, 33*(11), 745–754.

Horne, W. S., Yadav, M. K., Stout, C. D., & Ghadiri, M. R. (2004). Heterocyclic peptide backbone modifications in an α-helical coiled coil. *Journal of the American Chemical Society, 126*(47), 15366–15367.

Jacobsen, Ø., Maekawa, H., Ge, N.-H., Görbitz, C. H., Rongved, P., Ottersen, O. P., et al. (2011). Stapling of a 310-helix with click chemistry. *The Journal of Organic Chemistry, 76*(5), 1228–1238.

Jordan, A., & Reichard, P. (1998). Ribonucleotide reductases. *Annual Review of Biochemistry, 67*(1), 71–98.

Kaplan, J., & DeGrado, W. F. (2004). De novo design of catalytic proteins. *Proceedings of the National Academy of Sciences of the United States of America, 101*(32), 11566–11570.

Kohn, W. D., & Hodges, R. S. (1998). De novo design of α-helical coiled coils and bundles: Models for the development of protein-design principles. *Trends in Biotechnology, 16*(9), 379–389.

Kolb, H. C., Finn, M. G., & Sharpless, K. B. (2001). Click chemistry: Diverse chemical function from a few good reactions. *Angewandte Chemie International Edition, 40*(11), 2004–2021.

Lau, Y. H., Wu, Y., de Andrade, P., Galloway, W. R. J. D., & Spring, D. R. (2015). A two-component "double-click" approach to peptide stapling. *Nature Protocols, 10*(4), 585–594.

Lee, D., & Lippard, S. J. (2003). 8.13—Nonheme di-iron enzymes. In J. A. McCleverty & T. J. Meyer (Eds.), *Comprehensive coordination chemistry II* (pp. 309–342). Oxford: Pergamon.

Lindqvist, Y., Huang, W., Schneider, G., & Shanklin, J. (1996). Crystal structure of delta9 stearoyl-acyl carrier protein desaturase from castor seed and its relationship to other di-iron proteins. *The EMBO Journal, 15*(16), 4081–4092.

Litowski, J. R., & Hodges, R. S. (2002). Designing heterodimeric two-stranded α-helical coiled-coils. Effects of hydrophobicity and α-helical propensity on protein folding, stability, and specificity. *Journal of Biological Chemistry, 277*(40), 37272–37279.

Maglio, O., Nastri, F., & Lombardi, A. (2012). Structural and functional aspects of metal binding sites in natural and designed metalloproteins. In A. Ciferri & A. Perico (Eds.), *Ionic interactions in natural and synthetic macromolecules* (pp. 361–450). Hoboken, NJ: John Wiley & Sons, Inc.

Marine, J. E., Song, S., Liang, X., Watson, M. D., & Rudick, J. G. (2015). Bundle-forming α-helical peptide–dendron hybrid. *Chemical Communications, 51*(76), 14314–14317.

Marsh, E. N. G., & DeGrado, W. F. (2002). Noncovalent self-assembly of a heterotetrameric diiron protein. *Proceedings of the National Academy of Sciences, 99*(8), 5150–5154.

Mathews, F. S., Bethge, P. H., & Czerwinski, E. W. (1979). The structure of cytochrome b562 from Escherichia coli at 2.5 A resolution. *Journal of Biological Chemistry, 254*(5), 1699–1706.

Mayo, S. L., Olafson, B. D., & Goddard, W. A. (1990). DREIDING: A generic force field for molecular simulations. *The Journal of Physical Chemistry, 94*(26), 8897–8909.

Miteva, M. A., Guyon, F., & Tufféry, P. (2010). Frog2: Efficient 3D conformation ensemble generator for small compounds. *Nucleic Acids Research, 38*(Suppl. 2), W622–W627.

Monien, B. H., Drepper, F., Sommerhalter, M., Lubitz, W., & Haehnel, W. (2007). Detection of heme oxygenase activity in a library of four-helix bundle proteins: Towards the de novo synthesis of functional heme proteins. *Journal of Molecular Biology, 371*(3), 739–753.

Mutter, M. (2013). Four decades, four places and four concepts. *CHIMIA International Journal for Chemistry, 67*(12), 868–873.

Mutter, M., Altmann, E., Altmann, K.-H., Hersperger, R., Koziej, P., Nebel, K., et al. (1988). The construction of new proteins. Part III. Artificial folding units by assembly of amphiphilic secondary structures on a template. *Helvetica Chimica Acta, 71*(4), 835–847.

Mutter, M., & Tuchscherer, G. (1997). Non-native architectures in protein design and mimicry. *Cellular and Molecular Life Sciences CMLS, 53*(11–12), 851–863.

Pabo, C. O., & Suchanek, E. G. (1986). Computer-aided model-building strategies for protein design. *Biochemistry, 25*(20), 5987–5991.

Peacock, A. F. (2016). Recent advances in designed coiled coils and helical bundles with inorganic prosthetic groups—From structural to functional applications. *Current Opinion in Chemical Biology, 31*, 160–165.

Rachel, N. M., & Pelletier, J. N. (2016). One-pot peptide and protein conjugation: A combination of enzymatic transamidation and click chemistry. *Chemical Communications, 52*(12), 2541–2544.

Rau, H. K., DeJonge, N., & Haehnel, W. (1998). Modular synthesis of de novo-designed metalloproteins for light-induced electron transfer. *Proceedings of the National Academy of Sciences, 95*(20), 11526–11531.

Rau, H. K., DeJonge, N., & Haehnel, W. (2000). Combinatorial synthesis of four-helix bundle hemoproteins for tuning of cofactor properties. *Angewandte Chemie International Edition, 39*(1), 250–253.

Rau, H. K., & Haehnel, W. (1998). Design, synthesis, and properties of a novel cytochrome b model. *Journal of the American Chemical Society, 120*(3), 468–476.

Reig, A. J., Pires, M. M., Snyder, R. A., Wu, Y., Jo, H., Kulp, D. W., et al. (2012). Altering the O_2-dependent reactivity of de novo due ferri proteins. *Nature Chemistry*, 4(11), 900–906.

Rozwarski, D. A., Gronenborn, A. M., Clore, G. M., Bazan, J. F., Bohm, A., Wlodawer, A., et al. (1994). Structural comparisons among the short-chain helical cytokines. *Structure*, 2(3), 159–173.

Samish, I., MacDermaid, C. M., Perez-Aguilar, J. M., & Saven, J. G. (2011). Theoretical and computational protein design. *Annual Review of Physical Chemistry*, 62(1), 129–149.

Sazinsky, M. H., & Lippard, S. J. (2015). Methane monooxygenase: Functionalizing methane at iron and copper. In P. M. H. Kroneck & M. E. S. Torres (Eds.), *Sustaining life on planet earth: Metalloenzymes mastering dioxygen and other chewy gases* (pp. 205–256): Switzerland: Springer International Publishing.

Schnepf, R., Haehnel, W., Wieghardt, K., & Hildebrandt, P. (2004). Spectroscopic identification of different types of copper centers generated in synthetic four-helix bundle proteins. *Journal of the American Chemical Society*, 126(44), 14389–14399.

Scrima, M., Le Chevalier-Isaad, A., Rovero, P., Papini, A. M., Chorev, M., & D'Ursi, A. M. (2010). CuI-catalyzed azide–alkyne intramolecular i-to-(i+4) side-chain-to-side-chain cyclization promotes the formation of helix-like secondary structures. *European Journal of Organic Chemistry*, 2010(3), 446–457.

Shu, J. Y., Tan, C., DeGrado, W. F., & Xu, T. (2008). New design of helix bundle peptide–polymer conjugates. *Biomacromolecules*, 9(8), 2111–2117.

Sirajuddin, S., & Rosenzweig, A. C. (2015). Enzymatic oxidation of methane. *Biochemistry*, 54(14), 2283–2294.

Slope, L. N., & Peacock, A. F. A. (2016). De novo design of xeno-metallo coiled coils. *Chemistry—An Asian Journal*, 11(5), 660–666.

Smith, B. A., & Hecht, M. H. (2011). Novel proteins: From fold to function. *Current Opinion in Chemical Biology*, 15(3), 421–426.

Sowdhamini, R., Srinivasan, N., Shoichet, B., Santi, D. V., Ramakrishnan, C., & Balaram, P. (1989). Stereochemical modeling of disulfide bridges. Criteria for introduction into proteins by site-directed mutagenesis. *Protein Engineering*, 3(2), 95–103.

Stenkamp, R. E. (1994). Dioxygen and hemerythrin. *Chemical Reviews*, 94(3), 715–726.

Summa, C. M., Rosenblatt, M. M., Hong, J.-K., Lear, J. D., & DeGrado, W. F. (2002). Computational de novo design, and characterization of an A2B2 diiron protein. *Journal of Molecular Biology*, 321(5), 923–938.

Thomas, F., Boyle, A. L., Burton, A. J., & Woolfson, D. N. (2013). A set of de novo designed parallel heterodimeric coiled coils with quantified dissociation constants in the micromolar to sub-nanomolar regime. *Journal of the American Chemical Society*, 135(13), 5161–5166.

Valverde, I. E., Vomstein, S., Fischer, C. A., Mascarin, A., & Mindt, T. L. (2015). Probing the backbone function of tumor targeting peptides by an amide-to-triazole substitution strategy. *Journal of Medicinal Chemistry*, 58(18), 7475–7484.

Vita, N., Platsaki, S., Baslé, A., Allen, S. J., Paterson, N. G., Crombie, A. T., et al. (2015). A four-helix bundle stores copper for methane oxidation. *Nature*, 525(7567), 140–143.

Wahlgren, W. Y., Omran, H., von Stetten, D., Royant, A., van der Post, S., & Katona, G. (2012). Structural characterization of bacterioferritin from blastochloris viridis. *PloS One*, 7(10), e46992.

Weber, P. C., Bartsch, R. G., Cusanovich, M. A., Hamlin, R. C., Howard, A., Jordan, S. R., et al. (1980). Structure of cytochrome c': A dimeric, high-spin haem protein. *Nature*, 286(5770), 302–304.

Witte, M. D., Theile, C. S., Wu, T., Guimaraes, C. P., Blom, A. E. M., & Ploegh, H. L. (2013). Production of unnaturally linked chimeric proteins using a combination of sortase-catalyzed transpeptidation and click chemistry. *Nature Protocols*, 8(9), 1808–1819.

Yoon, J., Fujii, S., & Solomon, E. I. (2009). Geometric and electronic structure differences between the type 3 copper sites of the multicopper oxidases and hemocyanin/tyrosinase. *Proceedings of the National Academy of Sciences, 106*(16), 6585–6590.

Yu, F., Cangelosi, V. M., Zastrow, M. L., Tegoni, M., Plegaria, J. S., Tebo, A. G., et al. (2014). Protein design: Toward functional metalloenzymes. *Chemical Reviews, 114*(7), 3495–3578.

Zhang, Y., Bartz, R., Grigoryan, G., Bryant, M., Aaronson, J., Beck, S., et al. (2015). Computational design and experimental characterization of peptides intended for pH-dependent membrane insertion and pore formation. *ACS Chemical Biology, 10*(4), 1082–1093.

CHAPTER TWENTY-TWO

Design of Heteronuclear Metalloenzymes

A. Bhagi-Damodaran[1], P. Hosseinzadeh[1], E. Mirts, J. Reed, I.D. Petrik, Y. Lu[2]

University of Illinois at Urbana-Champaign, Urbana, IL, United States
[2]Corresponding author: e-mail address: yi-lu@illinois.edu

Contents

1. Step 1: Computational Design of Heteronuclear Metal-Binding Sites in Mb or CcP — 505
   1.1 Structure-Guided Design of Primary Ligands — 506
   1.2 Design of Secondary Coordination Sphere Interactions — 508
   1.3 Computational Analysis of the Design — 510
2. Step 2: Purification and Structural Characterization of Rationally Designed Proteins — 513
   2.1 UV–Vis Spectroscopy — 514
   2.2 Mass Spectrometry (MS) — 516
   2.3 EPR — 516
   2.4 X-Ray Crystallography — 517
3. Step 3: Functional Characterization of Designed Heterobinuclear Metalloenzymes — 518
   3.1 Activity Assays — 519
   3.2 Redox Potential Measurement — 521
   3.3 Kinetic and Mechanistic Studies — 522
   3.4 Capture and Characterization of Reaction Intermediates — 524
4. Step 4: Further Improvement of Designed Heteronuclear Metalloenzymes: Case Studies — 527
   4.1 Improving ET to the Heme Center — 527
   4.2 Tuning the Heme Reduction Potential — 527
   4.3 Increasing the Binding Affinity of Mn^{2+} in MnCcP — 527
5. Conclusions — 529
Acknowledgments — 529
References — 529

---

[1] These authors contributed equally.

Methods in Enzymology, Volume 580
ISSN 0076-6879
http://dx.doi.org/10.1016/bs.mie.2016.05.050

Abstract

Heteronuclear metalloenzymes catalyze some of the most fundamentally interesting and practically useful reactions in nature. However, the presence of two or more metal ions in close proximity in these enzymes makes them more difficult to prepare and study than homonuclear metalloenzymes. To meet these challenges, heteronuclear metal centers have been designed into small and stable proteins with rigid scaffolds to understand how these heteronuclear centers are constructed and the mechanism of their function. This chapter describes methods for designing heterobinuclear metal centers in a protein scaffold by giving specific examples of a few heme–nonheme bimetallic centers engineered in myoglobin and cytochrome c peroxidase. We provide step-by-step procedures on how to choose the protein scaffold, design a heterobinuclear metal center in the protein scaffold computationally, incorporate metal ions into the protein, and characterize the resulting metalloproteins, both structurally and functionally. Finally, we discuss how an initial design can be further improved by rationally tuning its secondary coordination sphere, electron/proton transfer rates, and the substrate affinity.

Metal ions have long been known to play important roles in biology as essential structural elements to stabilize protein tertiary structure, as multivalent redox-active elements to promote long-range electron transfer (ET), or as active sites to catalyze a variety of otherwise difficult reactions (Gray & Winkler, 1996, 2010; Liu et al., 2014; Lu, Garner, & Zhang, 2009; Lu, Yeung, Sieracki, & Marshall, 2009; Malmström, 1990). An important class of metalloenzymes contains active sites consisting of two or more distinct metal centers, called heteronuclear metalloenzymes (Table 1) (Collman,

Table 1 Examples of Heteronuclear Metalloenzymes

Enzyme	Metal-Binding Site
Heme–copper oxidase	$(N_{His})Heme-Cu(N_{His})_3$
Nitric oxide reductase	$(N_{His})Heme-Fe(N_{His})_3(O_{Glu})$
Sulfite/nitrite reductase	$Heme-(\mu_2-S_{Cys})-(Fe_4S_4)$
Nitrogenase P-clusters	$[Fe_4S_3]-(\mu_2-S_{Cys})_2(\mu_{4or6}-S)-[Fe_4S_3]$
Nitrogenase FeMoCo	$[Fe_4S_3]-(\mu_2-S)_3(\mu_6-C)-[MoFe_3S_3]$
[NiFe]-hydrogenase	$(S_{Cys})_2Ni(\mu_2-S_{Cys})_2Fe(CN)_2CO$
CO dehydrogenase	$[NiFe_3S_3]-(\mu_3-S)-[Fe(N_{His})(S_{Cys})]$

Ligands are shown in parenthesis with the first letter being the coordinating atom and the subscript being the amino acid harboring that atom. The number after indicates the number of ligands of a specified type that coordinate to the metal ion. Symbol μ represents a bridging ligand and the subscript on μ indicates the number of metal centers bridged.

Boulatov, Sunderland, & Fu, 2004; Collman & Wang, 1999; Holm, 1995; Kim, Chufan, Kamaraj, & Karlin, 2004).

Heteronuclear metalloenzymes are responsible for catalyzing some of the most fundamentally interesting and practically useful reactions found in *Nature*, including the oxygen reduction reaction catalyzed by heme–copper oxidase (HCO) that is known to be important for bioenergetics and fuel cells research (Gao et al., 2012; Garcia-Horsman, Barquera, Rumbley, Ma, & Gennis, 1994), NO reduction to N_2O catalyzed by nitric oxide reductase (NOR) that contributes to the global nitrogen cycle (Flock, Watmough, & Ädelroth, 2005; Richardson & Watmough, 1999), and Mn^{2+} oxidation to Mn^{3+} catalyzed by manganese peroxidases (MnP) that is critical to degrading lignin, the second most abundant biopolymer on earth (Hofirchter, 2002) (Fig. 1). The importance of the reactions catalyzed by these and similar enzymes (Table 1) have rendered them attractive targets for biochemical and biophysical investigations and for biomimetic studies to design artificial catalysts or enzymes with similar active site structures and activities (Fontecave & Pierre, 1998; Mahadevan, Gebbink, & Stack, 2000).

Despite the importance of these heteronuclear metalloenzymes in catalysis, they are generally more difficult to study because many of them (eg, HCO and NOR) are membrane enzymes that are relatively difficult to purify to homogeneity in high yields and contain other metal-binding sites, making it difficult to focus on spectroscopic study of the heteronuclear metal-binding sites. In the case of MnP, even though it is a water-soluble protein containing only a heme-Mn center, it has been difficult to construct and express mutations. To overcome these limitations, much effort has been devoted to making synthetic models using small organic molecules as ligands. The study of these models has provided deeper insight into the physical

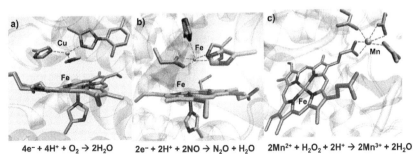

Fig. 1 Crystal structures of the active sites of cbb_3 oxidase (PDB ID: 3MK7), cNOR (PDB ID: 3O0R), and MnP (PDB ID: 1MNP) and the reaction catalyzed by each enzyme.

underpinnings of the chemical reactions facilitated by these enzymes by allowing researchers to isolate oxidation states, reactive intermediates, and find structural details in the primary coordination geometry of catalytic metal sites (Fontecave & Pierre, 1998; Koziol et al., 2012; Mahadevan et al., 2000; Punniyamurthy, Velusamy, & Iqbal, 2005; Que & Tolman, 2008; Sanders, 1999; Tard & Pickett, 2009). Despite the progress made, the majority of these synthetic models tend to exhibit low activity (Lilie, 2003; Petrik, Liu, & Lu, 2014).

To complement the study of native enzymes and synthetic models, we and others have employed a biosynthetic approach of designing heteronuclear metalloenzymes using small, stable scaffold proteins that can already be expressed in high yields with exceptional purity through well-established purification schemes (Chakraborty, Hosseinzadeh, & Lu, 2014; Chakraborty, Reed, et al., 2014; Degrado, Summa, Pavone, Nastri, & Lombardi, 1999; Degrado, Wasserman, & Lear, 1989; Gibney, Rabanal, Skalicky, Wand, & Dutton, 1999; Lu et al., 2013; Lu, Garner, & Zhang, 2009; Lu et al., 1999; Lu, Yeung, et al., 2009; Marshall, Miner, Wilson, & Lu, 2013; Reedy & Gibney, 2004; Richter, Leaver-Fay, Khare, Bjelic, & Baker, 2011; Thomas & Ward, 2005; Zanghellini et al., 2006; Zastrow & Pecoraro, 2013). Protein scaffolds are innately optimized to function under physiological conditions, offer a broad ligand set that can be expanded through the use of noncanonical amino acids, and comprise a genetically programmable platform for bottom-up engineering of metal-binding sites. A key advantage of protein-based biosynthetic models is the ability to construct and tune the noncovalent secondary sphere interactions around the primary coordination sphere using the much more rigid protein scaffold.

Study of native enzymes and their models has shown the importance of the secondary sphere interactions in tuning the enzymatic activities (Butland, Spiro, Watmough, & Richardson, 2001; Marshall et al., 2009; Miller, 2008; Varadarajan, Zewert, Gray, & Boxer, 1989). The secondary interactions can confer their roles in many ways: engaging in a hydrogen bond or salt bridge to the primary ligands, changing the overall electrostatic environment of the metal-binding site by changing charge or hydrophobicity, engaging in proton/electron transfer, or providing space for the metal cofactors (Hosseinzadeh & Lu, 2015). The stability and reactivity of heteronuclear metal-binding sites depend heavily on secondary sphere interactions that may contribute to electronic coupling between metal cofactors, the redox potential ($E^{o\prime}$) of one or both metals, and the stabilization of substrates and reactive intermediates (Berry, Baker, & Reardon, 2010;

Jackson & Brunold, 2004; Marshall et al., 2009; New, Marshall, Hor, Xue, & Lu, 2012; Petrik et al., 2016; Shook & Borovik, 2010; Yikilmaz, Xie, Brunold, & Miller, 2002). Engineering such sites into easily expressible protein scaffolds makes the process of studying an isolated heteronuclear site as well as its secondary coordination sphere less complicated by excluding the complex global effects present in native enzymes while working on a stable, three-dimensional platform that can be altered rationally.

In this chapter, we will describe methods of engineering heteronuclear, bimetallic active sites into small, stable protein scaffolds sperm whale myoglobin (swMb), and yeast cytochrome c peroxidase (CcP) from the incipient stages that make use of computational methods to identify and design metal-binding sites, through expression and quality-assessment of the biosynthetic models, and finally to specialized characterization techniques that take advantage of the physiologically relevant nature of biosynthetic models to probe their structural, electronic, and catalytic properties.

1. STEP 1: COMPUTATIONAL DESIGN OF HETERONUCLEAR METAL-BINDING SITES IN MB OR CcP

In this step, we describe the computational design process of a heteronuclear center beginning with choosing a scaffold, continuing through designing secondary sphere interactions for tuning activity, and finally to in silico generation and assessment of the models. An overall flow diagram of the described process is shown in Fig. 2.

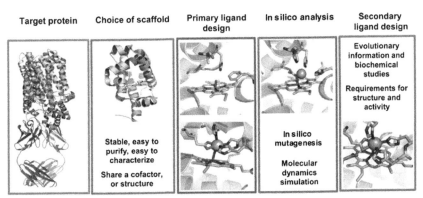

Fig. 2 General process of computational design of heteronuclear metalloproteins.

1.1 Structure-Guided Design of Primary Ligands

The design process begins with the choice of an appropriate protein scaffold and finding residues to mutate as primary ligands. As a general rule, the chosen scaffold should be relatively small, stable, easy to express recombinantly in large quantities, easy to crystallize, and devoid of any additional ligand or cofactor unless required as part of the design. It is also advisable to choose a scaffold that can achieve some of the chemistry of the reaction of interest. For example, for the design of a Cu_B–heme center for investigating oxygen reduction activity, the scaffold protein should contain a heme cofactor and bind oxygen but not natively catalyze its reduction. SwMb is a small, very well-studied protein that can be easily crystallized and expressed in large quantities in *Escherichia coli* (Springer & Sligar, 1987). It possesses a histidine coordinated heme group (similar to HCOs) that is known to bind to oxygen, making it an excellent scaffold for studying the reactivity of the Cu_B site in HCOs (Fig. 3A and B) (Varadarajan et al., 1989). Similarly, MnP (Annele, Taina, Martin, & Pekka, 2003; Hofirchter, 2002) shares strong structural similarity and shares enzymatic intermediates with yeast C*c*P (Finzel, Poulos, & Kraut, 1984; Sundaramoorthy, Kishi, Gold, & Poulos, 1994), making C*c*P an excellent starting point for mimicking MnP activity (Wilcox et al., 1998; Yeung, Wang, Sigman, Petillo, & Lu, 1997).

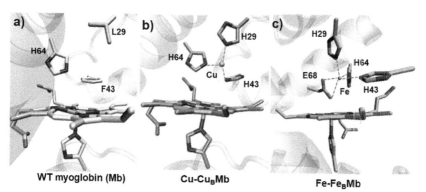

Fig. 3 X-ray crystal structures of Mb-based models. (A) The histidine coordinated heme center of WT-swMb (PDB ID: 1MBO) that already contains one of the histidines (H64) required to coordinate copper. (B) The Cu bound structure of a Cu_BMb mutant (PDB ID: 4FWY) that was obtained by mutating L29 and F43 to histidines. The three histidines coordinate copper in trigonal geometry. (C) The nonheme Fe added Fe_BMb (PDB ID: 3M38) that was obtained by incorporating a glutamate (68E) in the nonheme metal-binding center of Cu_BMb. The three histidines and a glutamate bind nonheme Fe in distorted octahedral geometry.

1. Choosing scaffolds: There are several methods and tools to perform an advanced search to find appropriate scaffold(s). The protein data bank (PDB, www.rscb.org) (Berman et al., 2000) contains over 100,000 biological macromolecular structures and has a very straightforward and versatile advanced search module that can narrow down proteins based on size, desired ligands, and even expression system. In addition to PDB, several other servers such as FATCAT (http://fatcat.burnham.org) (Ye & Godzik, 2004), MarkUs (http://wiki.c2b2.columbia.edu/honiglab_public/index.php/Software:Mark-Us) (Fischer et al., 2011), Dali (http://ekhidna.biocenter.helsinki.fi/dali_server) (Holm & Rosenstrom, 2010), and SwissProt (http://www.expasy.ch/sprot/ and http://www.ebi.ac.uk/swissprot/) (Bairoch & Apweiler, 2000) also provide a variety of tools that can assist in searching the scaffolds.

   The next step is to design the primary ligands of the second metal-binding site. This process can be carried out in one of several ways depending on the similarity between the target protein and the scaffold and the availability of initial structural data.

2. Finding primary ligand positions: (a) homologous target and scaffold: If the two proteins share high structural similarity, the first method is to perform structural alignment and identify the residues in the scaffold protein that occupy the same location as the metal ligands in the target protein. These residues can then be directly mutated to the corresponding residues of the target protein. This strategy was successfully applied to design a manganese-binding site near the heme in CcP based on structural similarity with MnP (Fig. 4). Several programs and online services can perform structure-based alignment including VMD (Humphrey, Dalke, & Schulten, 1996), Dali (http://ekhidna.biocenter.helsinki.fi/dali_server) (Holm & Rosenstrom, 2010), FATCAT (http://fatcat.burnham.org) (Ye & Godzik, 2004), SSAP (http://v3-4.cathdb.info/cgi-bin/SsapServer.pl) (Sillitoe et al., 2015), CLUSTAL omega (http://www.clustal.org/omega/) (Sievers et al., 2011), and many more.

3. Finding primary ligand positions: (b) nonhomologous target and scaffold: If the structures of the target and scaffold are known and contain a common cofactor but are not closely related, the two proteins can be aligned relative to this cofactor followed by visual inspection of the scaffold protein for positions that can potentially accommodate ligands of the target metal site in the appropriate geometry. This strategy was successfully used to design a Cu_B-heme center in swMb by aligning the heme of swMb with the catalytic heme of HCO

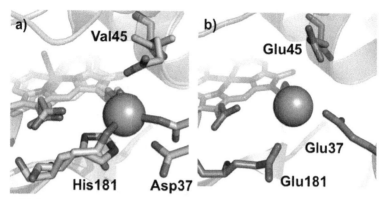

Fig. 4 (A) Structural alignment of C*c*P (PDB ID: 2CYP, *green*) and MnP (PDB ID: 1MNP, *gray*) indicates the possible position of primary ligands to be designed in C*c*P for Mn-binding site. (B) Final MnCcP.1 model binds Mn(II) (PDB ID: 5D6M, *orange*). (See the color plate.)

(Sigman, Kwok, Gengenbach, & Lu, 1999; Sigman, Kwok, & Lu, 2000; Sigman, Wang, & Lu, 2000). This strategy can also be extended to model the active sites of proteins with unsolved structures. For example, the sequence similarity within HCOs and NORs was used to model the Fe$_B$-heme center in Mb at a time when the X-ray crystal structure of NOR had not been solved (Fig. 3C) (Yeung et al., 2009). If the ligands of the desired metal-binding site are mostly within a loop, loop-directed mutagenesis is a good strategy for incorporation of the ligands (Hay et al., 1998). In this method, a loop in the scaffold protein is replaced with the metal-binding loop of the target protein.

Tip. Often in the design of heteronuclear bimetallic sites, the chosen scaffold should already contain one of the desired metal centers, eg, heme in the examples discussed in the review, to eliminate the additional difficult step of designing a heme-binding site.

Tip. For more complicated cases, the Rosetta software suite—specifically the Matcher program (Richter et al., 2011)—can be used to find sites within protein scaffolds that can satisfy constraints of the target site. While this method is powerful in searching for large cofactors, it still has limitations for small cofactors, such as metal ions, especially if there is no a priori information about the expected position of the metal center.

1.2 Design of Secondary Coordination Sphere Interactions

While the design of the primary metal-binding ligands is the first necessary step in achieving functional biosynthetic models, it has been observed in many cases that these features are not sufficient to recapitulate the complete activity of the target system (Sigman et al., 1999; Sigman, Kwok, et al., 2000; Sigman, Wang, et al., 2000; Wilcox et al., 1998; Yeung et al., 2009, 1997) in part due to the essential role of secondary coordination sphere interactions. There are several approaches to design these secondary interactions successfully. We briefly describe each approach here and refer the readers to more comprehensive reviews and articles.

1. Using evolutionary information and biochemical studies as a guide: Use of evolutionary relationships between members of a class of native metalloproteins combined with biochemical studies of the conserved residues is one of the most effective approaches to identify the secondary sphere interactions essential for the activity of interest. It is known that conserved residues found upon sequence or structural alignment of homologous proteins with similar functions play important roles in the observed activity (Dib & Carbone, 2012; Worth, Gong, & Blundell, 2009) and that mutation of these residues usually abolishes the activity of the enzyme under study (Sollewijn Gelpke, Moenne-Loccoz, & Gold, 1999). Thus, it is expected that adding these residues to the design will impart or improve the desired activity. A substantial amount of software is available for alignment of proteins with similar structure and function (homologous proteins) at the DNA sequence, amino acid sequence, or structural levels. The proteins in the alignment can be further separated and clustered based on stricter homology criteria to identify subfamilies. Each subfamily usually has a common feature that separates it from others. The residues that are conserved within members of each subfamily and that differ from others are potential targets for design. For instance, a Glu in the secondary coordination sphere of NORs was suggested to be important by biochemical studies (Butland et al., 2001), and incorporation of an additional Glu into the NOR model Fe$_B$Mb increased the NOR activity by two-fold (Zhao, Yeung, Russell, Garner, & Lu, 2006). Similarly, presence of a Tyr in HCOs has long been known to be essential for their activity based on mutagenesis studies (Das, Pecoraro, Tomson, Gennis, & Rousseau, 1998). Addition of Tyr in the right orientation in Cu$_B$Mb imparted oxygen reduction activity within a factor of 300 of a native HCO and with 1000 turnovers (Miner et al., 2012). Similarly, addition of an imidazole-Tyr residue resulted in substantial improvements in activity (Liu et al., 2012).

Tip. Sometimes general features rather than residues are conserved—eg, positive charge in a certain position—and in those cases the residue that satisfies this feature and fits the design scaffold best should be chosen.

This method has been successfully used to enhance the MnP activity in a designed mimic called MnC*c*P (Hosseinzadeh et al., 2016). It has also been used to increase the potential of Fe/Mn superoxide dismutase (Miller, 2008).

2. Structural requirements: After the primary ligands are found and put in place, the structure should be carefully inspected to find out whether the ligands fit well into the protein structure and whether there are any clashes or unsatisfied buried charges to be removed. For example, by incorporating a Gly in the loop that contains primary Mn-binding ligands into MnC*c*P and giving the ligands more flexibility, this designed mimic of MnP could achieve binding affinities for Mn comparable to the native system (Hosseinzadeh et al., 2016). Removing a disfavored hydrogen bond to another Mn ligand in the same mimic increased the activity and binding affinity (Hosseinzadeh et al., 2016).

3. Activity requirements: Information about the activity can be of prime importance in figuring out what secondary interactions should be designed. Biological reactions in heteronuclear metalloproteins are usually highly coupled to proton and/or electron transfer (Blomberg & Siegbahn, 2006; Chang, Chang, Damrauer, & Nocera, 2004; Fukuda et al., 2016; Gray & Winkler, 2010; Liu et al., 2014; Malmström, 1990; Reece & Nocera, 2009; Weinberg et al., 2012). The metal centers need to shuttle electrons and protons from the environment through redox-active amino acids or hydrogen bonding networks involving water and/or charged/polar amino acids. In order to improve the activity, similar residues should be incorporated as well (Weinberg et al., 2012).

1.3 Computational Analysis of the Design

Design hypotheses and methods often suggest several if not many possible solutions. Before experimental analyses of all of these designs, they should first be analyzed computationally. In this section, we briefly explain the steps for generating in silico models and analyzing them.

1. Generation of in silico designs: After a decision has been made of which residues to mutate, a molecular model containing those mutations should be generated. There are several ways to achieve this starting from an experimentally determined structure—the reader may use tools they are most familiar with—but the easiest methods are offered by programs like chimera (Pettersen et al., 2004), pymol, and coot (Emsley, Lohkamp, Scott, & Cowtan, 2010) which enable easy replacement of a residue with another residue and even a choice of rotamers in coot or pymol.

Tip. For loop-directed mutagenesis, manual alignment and splicing of the loop structure into the scaffold structure can be done, but an easy way to carry out this design is to feed the amino acid sequence into a protein structure prediction servers like I-TASSER (http://zhanglab.ccmb.med.umich.edu/I-TASSER) (Roy, Kucukural, & Zhang, 2010; Yang et al., 2015; Zhang, 2008) or use the remodel app in the Rosetta suite (explained in detail in Rosetta documentation) (Hu, Wang, Ke, & Kuhlman, 2007).

Assuming that the initial mutagenesis reveals no major problems, such as geometric clashes after incorporation of the desired feature, the designs should be further evaluated for feasibility. One easy way to evaluate the designs and eliminate those that are unlikely to work is to subject them to 1–5 ns of molecular dynamics simulation. These simulations can be carried out using one of several software packages, such as CHARMM (Brooks et al., 2009), NAMD (Phillips et al., 2005), Amber (Case et al., 2005), and GROMACS (Pronk et al., 2013), which implement various macromolecular force fields, such as CHARMM, AMBER, GROMOS, and OPLS.

The authors are most familiar with NAMD employing the CHARMM force field and VMD for simulation preparation and will give general procedural advice for this package.

2. Preparing models for molecular dynamics simulation: Initialization of molecular dynamics simulation requires the definition of two related properties: (1) the exact connectivity of the atoms in the simulation, known as the topology and (2) the physical constants of the atoms in the molecule, such as mass, charge, and interatomic force constants for bonds, three-atom angles, and four-atom torsional values. In CHARMM, the topologies of individual residues of a polymer and other monomers, such as water, lipids, and many exogenous cofactors, have been determined and are available with the default force field parameter package. In VMD/NAMD, the overall topology of macromolecules is generated by a script called "psfgen." Instructions for general use of this script to produce topologies for macromolecules are readily available on the NAMD/VMD website. However, the authors would like to point out that special actions must usually be taken when additional cofactors are present. For instance, while parameters for heme are included in the standard CHARMM force field, bonds between His or Cys residues and the heme iron must be explicitly defined using patches. Even when the heme is unligated, a patch must be applied to ensure proper planarity of the Fe in the ring.

Tip. For some cofactors, such as FeS clusters, topology, and force field parameters are not included. In such cases, parameters can sometimes be found in the literature or from the websites of various biophysical research labs (Chang & Kim, 2009). More often, these properties are not readily available for the desired force field. There are several avenues to pursue in these cases: (1) if parameters are available from another force field, they can sometimes be translated into parameters for the desired force field; (2) if parameters for similar bonds exist, and quantitative precision is not critical, the parameters can be approximated from existing parameters; and (3) otherwise, the ligand and its interactions with the macromolecule must be parameterized by quantum mechanical calculations. Guidance for such parameterization is provided by the developers of the relevant force fields. Depending on the model, it may be necessary to constrain the metal atom to prevent it from moving out of the protein pocket.

3. Simulation parameters: After defining the macromolecular topologies and corresponding parameters, the structure should be solvated in explicitly modeled water, such as TIP3 parameterized water, and the charge balanced with ions of choice. Further general considerations about simulation parameters, including size of the simulation box and ensemble are discussed in literature and software documentation. As a start, the authors suggest 1000–2000 steps of minimization and a 2-ns simulation (2 fs/step) in an NPT ensemble with a box that is at least twice as long in each direction as the longest protein dimension. More complicated methods for simulation are also available that are not within the scope of this chapter.

4. Analysis of the simulation: After the simulation is over, the ".dcd" file (or any file that contains the simulation trajectory) should be loaded and the overall movement of the protein should be evaluated, especially in the designed area to make sure no undesired movements of the ligand, changes in the structure, or solvation of the metal site is observed. General structural RMSDs should be calculated with respect to the starting scaffold to ensure that there are no major structural changes. The VMD package has several useful extensions for these analyses such as a RMSD calculator, hydrogen bond, and electrostatic interaction predictor, and a pK_a calculator that the user can apply when required.

Tip. Make sure to align all the scaffolds in the whole trajectory before running RMSD measurements.

Tip. It is important to note that simulations cannot guarantee that the designed variants will work in an experimental setting, but they are extremely useful to filter those variants that are likely to fail. Overall, if a variant causes drastic changes in the protein structure after 2 ns of simulation, it probably is very destabilizing and should not be further pursued.

Tip. It is also important to note that all force fields, including CHARMM, are currently limited in handling electrostatic interactions to evaluate metal binding and thus we do not suggest using the simulation as a way to assess metal affinity. However, the results from simulation can be a guide to get some ideas about factors that can enhance metal-binding affinity.

2. STEP 2: PURIFICATION AND STRUCTURAL CHARACTERIZATION OF RATIONALLY DESIGNED PROTEINS

In this step, we provide a detailed description of rationally designed heteronuclear metalloprotein characterization aimed at identifying binding events and interactions of designed metal-binding sites. Single mutation site-directed mutagenesis is typically performed using the QuikChange Site Directed Mutagenesis Kit (Sigman et al., 1999). Where applicable, Gibson assembly (Gibson, 2011) and Mega-Primer PCR (Tyagi, Lai, & Duggleby, 2004) can be also used. Detailed descriptions of these methods are beyond the scope of this chapter and have been covered previously (Casini, Storch, Baldwin, & Ellis, 2015; Nakayama & Shimamoto, 2014; Séraphin & Kandels-Lewis, 1996; Weissensteiner, Nolan, Bustin, Griffin, & Griffin, 2003). The presence of desired mutations in the DNA sequence is confirmed by DNA sequencing. General purification of our designed heteronuclear metalloproteins will not be discussed in great detail here, but generally falls into two categories: inclusion body and cytosolic purification. The inclusion body protocol is used extensively in models such as Cu_BMb and Fe_BMb and has been covered previously in those works (Bhagi-Damodaran, Petrik, Marshall, Robinson, & Lu, 2014; Chakraborty, Hosseinzadeh, et al., 2014; Chakraborty, Reed, et al., 2014; Lin, Yeung, Gao, Miner, Lei, et al., 2010; Lin, Yeung, Gao, Miner, Tian, et al., 2010; Miner et al., 2012; Petrik et al., 2016; Pfister et al., 2005; Sigman, Kim, Zhao, Carey, & Lu, 2003; Sigman, Kwok, et al., 2000; Sigman, Wang, et al., 2000; Springer & Sligar, 1987; Yeung et al., 2009; Zhao et al., 2006; Zhao, Yeung, Wang, Guo, & Lu, 2005). Additionally, cytosolic purification of C*c*P and its rationally designed mutants have also been described in great detail by our group (Feng et al., 2003; Hosseinzadeh et al., 2016; Miner et al., 2014; Pfister et al., 2007; Sigman et al., 1999; Yeung et al., 1997). After purification, the protein mass is determined by SDS-PAGE (Brunelle & Green, 2014) and ESI-MS (Strupat, 2005). A pyridine hemochromogen assay is used to determine the molar extinction coefficient of the Soret band in the designed mutants (Berry & Trumpower, 1987). Additionally, circular dichroism (CD) is a very useful technique to assess the overall folding state of the designed proteins (Greenfield, 2006). Using these techniques, it is possible to immediately assess whether the protein is correctly folded.

A crucial component of designing heteronuclear metalloenzymes is being able to perform a quick check of metal ion binding at the designed metal center, a procedure for which a number of techniques are helpful.

2.1 UV–Vis Spectroscopy

Metalloenzymes containing transition metal ions typically exhibit characteristic UV–vis signals that are of great advantage in detecting the binding of metal ions. For example, the titration of Cu(II) in apo-WT azurin (a copper containing ET protein) results in an obvious increase in broad absorption bands centered at ~450 and ~850 nm due to ligand to metal charge transfer and d–d transitions, respectively (Hadt et al., 2012). However, detecting metal binding using UV–vis is not as straightforward for heme-based metalloenzymes; heme exhibits strong UV–vis signals from 400 to 650 nm. As a result, we monitor the changes in the heme UV–vis signals as metal ions are titrated into the protein (Fig. 5A). These are

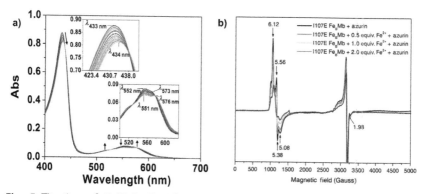

Fig. 5 Titration of Fe(II) in Fe$_B$Mb variants as observed by UV–vis (A) and EPR (B) spectroscopy. *Adapted with permission from Chakraborty, S., Reed, J., Ross, M., Nilges, M. J., Petrik, I. D., Ghosh, S., et al. (2014). Spectroscopic and computational study of a nonheme iron nitrosyl center in a biosynthetic model of nitric oxide reductase. Angewandte Chemie International Edition, 126, 2449–2453; Lin, Y.-W., Yeung, N., Gao, Y.-G., Miner, K. D., Tian, S., Robinson, H., et al. (2010). Roles of glutamates and metal ions in a rationally designed nitric oxide reductase based on myoglobin.* Proceedings of the National Academy of Sciences of the United States of America, 107, 8581–8586, *respectively.*

small but significant changes resulting from alteration in the heme electronic environment upon addition of a new metal ion in its vicinity (Yeung & Lu, 2008). It should be noted that if no change is observed, it does not preclude the possibility of metal still being bound to the site. In addition, it should be verified that the changes in the UV–vis signals are not due to (1) unfolding of the protein, which will be apparent by an increase in the absorbance at 280 nm or (2) removal of heme from the protein, which will be apparent by an increase in 360/280 nm signals due to free heme. We discuss the protocol for titration of nonheme Fe in Fe_BMb below (Yeung et al., 2009).

2.1.1 UV–Vis Spectroscopic Titration of Nonheme Fe(II) in E-Fe_BMb

1. Degas 1000 μL of 7 μM protein solution in 50 mM Bis-Tris (bis (2-hydroxyethyl)iminotris(hydroxymethyl)methane) pH 7 buffer on a Schlenk line using standard freeze–pump–thaw techniques.
2. Bring the degassed empty Fe_BMb (that is Fe_BMb with no nonheme metal added called E-Fe_BMb) solution into an anaerobic chamber.
3. Reduce the protein using a small amount of solid dithionite, leaving an excess to keep the protein reduced during titration.
4. Record an initial spectrum of the protein solution.
5. Prepare a solution of 50 mM $FeCl_2$ in the anaerobic chamber by dissolving 6.33 mg solid $FeCl_2$ in 1 mL degassed water.
6. Dilute 28 μL of the 50 mM $FeCl_2$ solution to 1.4 mM by adding 972 μL degassed water, total volume 1 mL.
7. Add 5 μL of $FeCl_2$ solution to the protein solution with rapid stirring and incubate for 5 min.
8. Collect a UV–vis spectrum of the metal-added protein.
9. Repeat steps 7 and 8 for a total of 2 molar equivalents nonheme metal added.
10. Dilute 56 μL of the 50 mM $FeCl_2$ solution to 2.8 mM by adding 944 μL degassed water, total volume 1 mL.
11. Repeat step 9 for a total of 5 molar equivalents nonheme metal added.
12. Changes in the absorbance profile of the Soret (433 nm → 434 nm redshift) and visible (formation of split peak at 550 and 572 nm) regions are indicative of nonheme metal binding (Yeung et al., 2009).
13. Dissociation constants based on difference spectra (ΔAbs vs λ) can be calculated using previously described protocols (Bidwai et al., 2003).

Tip. Vary the concentration of your protein so that the Soret absorption is ~1 or within an acceptable range for the particular spectrometer.

Tip. If the titration is incomplete after addition of 5 M equivalents of nonheme metal, adjust the concentration of $FeCl_2$ such that a 5 µL addition is a larger molar equivalent (eg, 1, 2, 4, etc.).

Tip. Run a control experiment with WT protein (WT-swMb in this case) to make sure that the change in the UV–vis spectrum is due to binding of metal in the designed binding site and not an undesired endogenous site close to the heme iron.

Tip. If no binding was observed, the protein should be treated with chelating agents (eg, EDTA) to make sure there is no exogenous metal binding that prevents the binding of the desired metal ion.

Tip. While it is difficult in the glovebag, the titration of metal ions should ideally be performed with the protein at ~4°C. This practice helps stabilize some mutant proteins that are not very stable at room temperature.

2.2 Mass Spectrometry (MS)

The addition of metal ions to the protein, given that the binding is strong, will lead to an increase in mass which can be detected using MS. Care should be taken that the protocol used for MS does not lead to protein disruption or loss of metal ions. In order to do so, a milder electrospray ionization (ESI) method is used which is called "syringe pump ESI." In this technique, the protein is exchanged into water or buffer with very low ionic strength. The lines of the ESI instrument are washed with water thoroughly and then the protein is loaded into the instrument using a syringe at very low flow rates (~5 µL/min) without addition of formic acid. There are also reports of using matrix-assisted laser desorption/ionization time-of-flight MS (MALDI-TOF MS) with specific resins to find the mass of metal-bound species (Zeng et al., 2007). In general, MS has been useful in detecting metal binding for metalloproteins with high binding affinities to the metal ion such as azurin, but for proteins that exhibit low binding affinities for nonheme metal ions (K_D ~10 µM) such as Mb and CcP, it is more difficult to obtain accurate MS results by this method.

While UV–vis and MS are excellent methods to perform initial characterization of the designed metalloenzymes, more specific methods are required for their complete characterization of properties including coordination, geometry, oxidation state, and spin state. Next, we describe methods that can be used to determine them.

2.3 EPR

Monitoring the changes in the spin state of metal ions is another approach through which metal binding can be readily detected. For example, addition of nonheme Fe(II) close to high spin ($S=5/2$) heme iron(III) leads to spin coupling between the two metals that causes attenuation of the $g=6$ signal (Fig. 5B). The spin coupling between nonheme Fe and heme Fe in I107E Fe$_B$Mb will be described as an example. The nonheme Fe(II) addition was performed in the reduced form of Fe$_B$Mb, which is electron paramagnetic resonance (EPR) silent. After metal addition, the resulting Fe-Fe$_B$Mb was oxidized using azurin.

1. Prepare a solution of degassed protein as described in "Metal Titration, steps 1–3."
2. Remove the excess dithionite by passing the protein solution down a size exclusion PD-10 column (GE Healthcare, Buckinghamshire, UK) preequilibrated with 50 mM Bis-Tris pH 7.
3. Add varying molar equivalents of FeCl$_2$ (eg, 0.5, 1.0, 1.5, 2.0) in one addition to the protein solution(s) with rapid stirring for 20 min.
4. Add 3 M equivalents of blue copper Cu(II)-azurin to the nonheme metal-added protein solution(s) with rapid stirring for 5 min.
5. Add glycerol to 10% final concentration as a glassing agent.
6. Transfer the protein solution(s) into separate EPR tubes and flash freeze using liquid N$_2$.

Tip. Final protein concentrations before flash freezing should be 0.5 mM Fe$_B$Mb and 1.5 mM azurin.

Tip. Other metal ions, like Cu(II) and Zn(II), can bind Fe$_B$Mb in the oxidized state and neither need to be reduced nor kept in an anaerobic environment (Lin, Yeung, Gao, Miner, Lei, et al., 2010; Lin, Yeung, Gao, Miner, Tian, et al., 2010).

2.4 X-Ray Crystallography

Crystallization is a standard technique for metalloprotein characterization to obtain overall structure, metal coordination, and geometry. As mentioned, one criterion for choosing the scaffold proteins for designing heteronuclear metalloenzymes is ease of crystallization; thus, extensive screening is often not required to obtain crystals of their mutants. For example, to model HCOs and NORs, we have used swMb, which was the first protein whose X-ray crystal structure was obtained. Thus, crystallization conditions for myoglobin are well known and standardized. Here, we discuss the protocol to obtain the structure of nonheme metal-added Fe$_B$Mb variants.

1. Start with the crystallization of E-Fe$_B$Mb and screen for crystallization conditions. Details and protocols used for screening are described in

the following reviews (Luft, Newman, & Snell, 2014; Skarina, Xu, Evdokimova, & Savchenko, 2014).
2. For Mb variants, the following crystallization conditions obtain diffraction-quality crystals with the variation in the pH of Tris/MES buffer and polyethylene glycol (PEG) molecular weights: well buffer: 0.1 M Tris/MES, 0.2 M sodium acetate, 30% PEG 10,000 kDa MW; protein: ~1 mM in 20 mM potassium phosphate buffer.
3. Once the crystallization conditions for E-Fe$_B$Mb are finalized, we prepare the Fe-Fe$_B$Mb protein as described previously and setup the crystal trays in an anaerobic chamber at 4°C using conditions determined for E-Fe$_B$Mb. The resulting crystals have obtained the structure of Fe-Fe$_B$Mb at 1.7 Å resolution (Yeung et al., 2009).
4. Alternatively, for certain variants of Fe$_B$Mb, crystals obtained for empty protein are transferred to the anaerobic chamber, reduced using 20 mM dithionite in crystallization buffer, and soaked with crystallization buffer containing ~10 mM FeCl$_2$ for ~1 h to obtain the nonheme Fe containing crystal structure.
5. For metal ions like Cu(II) and Zn(II) that are stable in the presence of oxygen (unlike Fe(II) which tends to oxidize to Fe(III) and precipitate out), crystal trays can be set in aerobic conditions. The protein is mixed with 1, 2, and 3 molar equivalents of ZnCl$_2$ or CuSO$_4$, and conditions similar to E-Fe$_B$Mb are used for crystallization (Lin, Yeung, Gao, Miner, Lei, et al., 2010; Lin, Yeung, Gao, Miner, Tian, et al., 2010; Miner et al., 2012).
6. At the time of X-ray data collection, diffraction should be obtained at the metal edge (for example at Cu-edge if the nonheme metal is copper) to be sure that the electron density at the designed metal center is indeed due to the presence of a metal ion and not, for example, a water molecule.

Tip. Avoid any crystallization buffers that may chelate the metal ion out of the protein, eg, citrate or cacodylate buffer.

3. STEP 3: FUNCTIONAL CHARACTERIZATION OF DESIGNED HETEROBINUCLEAR METALLOENZYMES

In this section, we describe methods for functional characterization of designed metalloenzymes that include activity assays, mechanistic studies, redox potential ($E^{o\prime}$), and characterization of reaction intermediates.

3.1 Activity Assays

There are three measures for assessing the functional efficiency of a designed metalloenzyme, namely the rate of reaction, the product selectivity, and the turnover number. As an example, we will detail how the designed Mb-based HCO mimics (Cu_BMb) were tested for all of these measures in the presence or absence of nonheme metal ions.

3.1.1 Catalytic Efficiency of the Reaction

The activity assays used for testing designed metalloenzymes should ideally be the same as that of native enzymes. For example, an oxygen electrode is the technique commonly utilized for measuring oxygen reduction rates in native HCOs and was also employed to characterize Cu_BMb variants. The reduction of oxygen requires a constant supply of electrons, thus a mixture of TMPD/ascorbate was used for electron donation, again the same system used for native HCOs. The rate of oxygen reduction is measured as a slope of concentration of oxygen with respect to time. To further optimize the reaction conditions, similar experiments should be performed at different pH values and different concentrations of reductant. Finally, the reaction should be repeated in presence of the nonheme metal ion of interest (copper in this case). Additionally, redox-inactive metal ions such as Ag(I) or Zn(II) should also be used as controls (Fig. 6A).

3.1.2 Product Selectivity

One important feature of an enzyme-catalyzed reaction is its high selectivity and specificity with minimal byproduct formation. Hence, the products should be analyzed to evaluate the quality of the designed protein as a model compared with its native counterpart and whether it can recapitulate the native activity. The assessment of products can be performed with methods similar to those applied to the native enzyme. For example, a hallmark of an oxidase is that it performs complete four-electron reduction of oxygen to water. An incomplete reduction of oxygen can produce a multitude of reactive oxygen species (ROS), namely O_2^-, O_2^{2-}, and $OH^{\bullet}$. Thus, it is important to obtain the ratio of water with respect to ROS produced during oxygen reduction by Mb-based HCO mimics. To measure this ratio, the reaction is repeated in the presence of superoxide dismutase (that reacts with superoxide converting it to oxygen and peroxide) and catalase (that reacts with peroxide converting it to oxygen). Thus, if superoxide or peroxide is produced during the reaction, the rate of oxygen reduction is decreased.

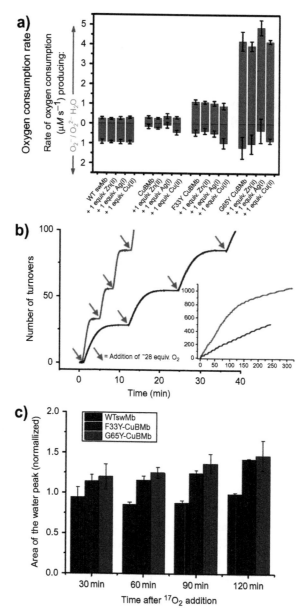

Fig. 6 Characterization of the O_2 reduction: (A) rates of O_2 reduction to form either water or superoxide/peroxide with 18 μM WTswMb, Cu_BMb, F33Y-Cu_BMb, or G65Y-Cu_BMb and (B) number of turnovers of O_2 reduction by E-F33Y-Cu_BMb and E-G65Y-Cu_BMb; *arrows* indicate addition of approximately 28 equiv. O_2. *Inset*, region of a high number of turnovers. (C) Quantitation of $H_2^{17}O$ product by ^{17}O NMR spectroscopy: area of the $H_2^{17}O$ signal normalized to an external standard for WTswMb, E-F33Y-Cu_BMb, and E-G65Y-Cu_BMb at 30, 60, 90, and 120 min. *Figures are adapted with permission from Miner, K. D., Mukherjee, A., Gao, Y.-G., Null, E. L., Petrik, I. D., Zhao, X., et al. (2012). A designed functional metalloenzyme that reduces O_2 to H_2O with over one thousand turnovers. Angewandte Chemie International Edition, 51, 5589–5592, PMC3461324.*

These assays when performed for swMb converted oxygen almost exclusively to ROS, while Cu$_B$Mb variants performed selective reduction of oxygen to water (Fig. 6B) (Bhagi-Damodaran et al., 2014; Miner et al., 2012, Yu, Lv, et al., 2015; Yu et al., 2014). Other methods such as nuclear magnetic resonance (NMR), high performance liquid chromatography, and gas chromatography–MS (GC–MS) can also be used to identify and quantify the products of a reaction (Hosseinzadeh et al., 2016; Yeung et al., 2009).

3.1.3 Turnover Numbers

Native enzymes often exhibit turnover numbers in the order of thousands while most synthetic small-molecule models exhibit turnover numbers less than 10. Thus, turnover number is an important parameter to evaluate the efficiency of designed proteins. There are several ways to assess the turnover number of a reaction, but all follow a similar procedure of continuously providing substrate under reaction conditions until the reaction ceases. For example, to measure turnover number of oxygen reduction by G65Y-Cu$_B$Mb, after the oxygen in the reaction chamber was exhausted, the chamber was vented with pure oxygen to boost the oxygen concentration to ~0.6 mM. Over the course of the reaction, the sacrificial reductant (ascorbate) is also consumed, so its concentration must also be maintained. These experiments showed that G65Y-Cu$_B$Mb could perform up to 1000 turnovers without any degradation (Miner et al., 2012).

Tip. It is a good idea to perform a secondary assay to confirm findings, especially when measuring the product selectivity where a number of variables are at play. For example, the assay for oxidase activity involved two enzymes—SOD and catalase—that could be inhibited upon addition of metal ions. Thus, we used $^{17}O_2$ NMR spectroscopy to analyze the product of $^{17}O_2$ reduction in parallel to the quantitative oxygen electrode experiments described earlier (Fig. 6C).

3.2 Redox Potential Measurement

The redox potentials ($E^{\circ\prime}$) of metalloenzymes provide significant insights into the electronic nature of metal ions, their ligands, and environment. For example, E-Fe$_B$Mb possesses a heme $E^{\circ\prime}$ of -159 mV (vs SHE), which is ~200 mV lower than that of Cu$_B$Mb, suggesting that the presence of negatively charged Glu68 in the distal pocket of Fe$_B$Mb increases the electron density on the heme iron to favor its Fe(III) state over Fe(II) state (Yeung et al., 2009). Upon addition of nonheme Fe(II), the $E^{\circ\prime}$ of Fe$_B$Mb increases to −60 mV due to increased distance between the negatively charged Glu68, now coordinated to the nonheme Fe, and the heme iron and due also to the presence of the positively charged nonheme Fe close to the heme center. Thus, changes (generally increase) in $E^{o\prime}$ upon addition of a second metal ion can be an indication of the proximity of two metal centers in the protein. Heme proteins in particular exhibit strong UV–vis spectroscopic signals characteristic to their oxidized and reduced states that are useful to deduce their heme $E^{o\prime}$. Protocols for measurement of heme $E^{o\prime}$ for $Fe^{3+/2+}$ couple through spectroelectrochemistry are fairly standardized and discussed extensively in other reviews (Hosseinzadeh & Lu, 2015; Taboy, Bonaventura, & Crumbliss, 1999). Cyclic voltammetry (CV) can also be used to measure $E^{o\prime}$ of a heterobinuclear center. Along with determining $E^{o\prime}$, a CV setup can also be utilized to perform electrocatalysis that yields mechanistic information, specifically the number of electrons involved in the reaction as well as the rate of reaction (Armstrong, 2002; Evans & Armstrong, 2014). Recently, in situ electrocatalysis has been coupled with vibrational spectroscopy to detect reaction intermediates (Mukherjee et al., 2015). As the $Fe^{3+/2+}$ transition of heme in swMb is accompanied by the loss of a water ligand, CV usually does not produce reversible signals in Mb variants and hence most studies on these models used spectroelectrochemistry (Van Dyke, Saltman, & Armstrong, 1996). C*c*P models, however, produce reversible CV signals (Becker, Watmough, & Elliott, 2009).

Tip. Typically an electrolyte is added to make the protein solution conducting for spectroelectrochemical titrations. It should be made sure that the electrolyte does not react with the nonheme metal of interest. For example, when measuring the heme $E^{o\prime}$ of Ag(I)-Cu_BMb, $NaNO_3$ should be used as an electrolyte and not NaCl which will precipitate any Ag(I) bound to the protein.

Tip. In order to get a high quality CV signal, vary the scan rates and the initial potential. Pyrolytic graphite electrodes are usually well suited for protein samples. However, other electrodes such as glassy carbon or gold may yield preferable behavior at the electrode surface, which is variable depending on the protein.

3.3 Kinetic and Mechanistic Studies

Respiratory enzymes like HCOs and NORs contain a number of metal cofactors, thus it is difficult to selectively study their heterobinuclear catalytic center. Designed heteronuclear metalloenzymes offer an easy alternative route to study reaction mechanism under physiological conditions.

Initial mechanistic studies are performed using a stopped-flow apparatus wherein the protein is mixed with the reactant and the reaction is monitored in a time-resolved fashion using either UV–vis spectroscopy or vibrational spectroscopy. The strong and well-characterized UV–vis signals of heme proteins, in particular, make UV–vis stopped-flow spectroscopy a direct and easy method to probe their mechanism. Here, we describe how the mechanism of NO reduction by I107E-Fe$_B$Mb (an NOR mimic) was investigated using UV–vis stopped-flow spectroscopy (Matsumura, Hayashi, Chakraborty, Lu, & Moënne-Loccoz, 2014). The reduction of NO to N$_2$O involves the cleavage of an N–O bond and formation of an N–N bond. This reaction can happen through three possible pathways: the *cis*-heme b_3, *cis*-Fe$_B$, or the *trans* mechanisms (Fig. 7A).

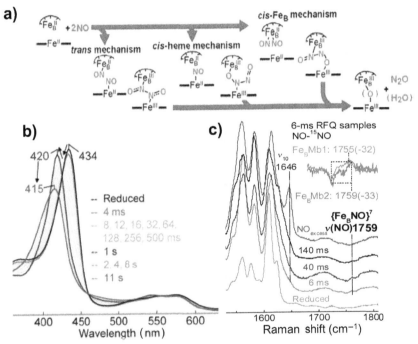

Fig. 7 (A) Possible mechanisms for NO reduction in Fe$_B$Mb model. (B) Stopped-flow UV–vis absorption spectra of the reaction of Fe$_B$Mb with 60 μM NO at 4.0°C. (C) The rR spectra of RFQ samples of the reaction of reduced I107E-Fe$_B$Mb with excess NO. *Inset* is an overlay of the ^{14}NO–^{15}NO differential signal for the nonheme ν(NO) modes in the 6-ms RFQ samples of Fe$_B$Mb and I107E-Fe$_B$Mb. *Figures are adapted from Matsumura, H., Hayashi, T., Chakraborty, S., Lu, Y., & Moënne-Loccoz, P. (2014). The production of nitrous oxide (N$_2$O) by the heme/non-heme diiron center of engineered myoglobins (FeBMbs) proceeds through a trans iron-nitrosyl dimer. Journal of the American Chemical Society, 136, 2420–2431.* © 2014 American Chemical Society.

1. To investigate the reaction mechanism, stopped-flow experiments were performed with an SX20 apparatus (Applied Photophysics) with a 1-cm path length cell equilibrated at 4°C inside an anaerobic chamber.
2. Briefly, 10 μM of Fe(II)-I107E-Fe$_B$Mb in 50 mM Bis-Tris pH 7.3 buffer was mixed with 0.8 mM NO in the same buffer in 1:1 ratio and the reaction was investigated with exponentially increasing time range for up to 50 s.
3. The results obtained (shown in Fig. 7B) showed two distinct intermediates depicted in *red (gray* in the print version) (4 ms) and *blue (dark gray* in the print version) (128 ms) followed by a final species in *green (gray* in the print version) (11 s).
4. A comparison of the UV–vis spectra obtained at 128 ms and 11 s with literature values revealed them to be heme Fe(II)–NO and heme Fe(III)–NO species, respectively. These results strongly suggest the *trans*-mechanism of NO reduction for Fe$_B$Mb variants (Fig. 7A).

Tip. To get a better understanding of the role of nonheme metal in the reaction mechanism, experiments should be repeated with empty protein or redox-inactive metal controls like Zn(II). For example, when E-I107E-Fe$_B$Mb was mixed with NO under conditions described earlier, the reaction stopped at the formation of heme Fe(II)–NO, suggesting the importance of nonheme Fe for N–N bond formation and subsequent reactions.

3.4 Capture and Characterization of Reaction Intermediates

For a deeper insight into the reaction mechanism of the designed metalloenzymes, it is necessary to capture and identify the intermediates involved in the reaction. The use of transition metal ions with rich spectroscopic features provides researchers with a wealth of techniques with which they can characterize these intermediates. Techniques such as X-ray absorption spectroscopy (XAS), EPR, Mössbauer, NMR, magnetic circular dichroism (MCD), resonance Raman (rR), nuclear resonance vibrational spectroscopy (NRVS), and other techniques have been extensively used to characterize heteronuclear metal centers. Table 2 shows a list of these techniques, the information gained from each, and some studies in which these techniques have been used on designed heteronuclear metalloenzymes.

Usually, the presence of an intermediate can be observed through stopped-flow UV–vis experiments. These intermediates can be captured at millisecond time points using rapid freeze-quench (RFQ)/chemical-

Table 2 Techniques for Characterization of Metalloproteins and Their Intermediates

Spectroscopy	Obtained Information	References
Electronic absorption (UV–vis)	M–ligand, geometry, stoichiometry, affinity, electronic structure	Lin, Yeung, Gao, Miner, Lei, Robinson, et al. (2010), Lin, Yeung, Gao, Miner, Tian, Robinson, et al. (2010), Chakraborty, Hosseinzadeh, et al. (2014), and Chakraborty, Reed, et al. (2014)
Electron paramagnetic resonance (EPR)	Ligand, geometry, stoichiometry, spin state	Lin, Yeung, Gao, Miner, Lei, Robinson, et al. (2010), Lin, Yeung, Gao, Miner, Tian, Robinson, et al. (2010), Chakraborty, Hosseinzadeh, et al. (2014), Chakraborty, Reed, et al. (2014), Yu, Lv, et al. (2015), and Yu et al. (2014)
Electron-nuclear double resonance (ENDOR)	Ligand and metal interactions Reaction intermediates (w/cryoreduction)	Shanmugam, Xue, Que, and Hoffman (2012) and Petrik et al. (2016)
Magnetic circular dichroism (MCD)	M–ligand, geometry, stoichiometry, affinity, and electronic structure	Sarangi et al. (2008) and Hadt et al. (2012)
X-ray absorption spectroscopy (XAS)	Ligands, metal–ligand distances (EXAFS) Oxidation states (XANES)	Sarangi et al. (2008) and Hadt et al. (2012)
FTIR/resonance Raman (rR)	Reaction intermediates of O_2/NO reduction	Hayashi, Lin, Chen, Fee, and Moënne-Loccoz (2007), Lu, Zhao, Lu, Rousseau, and Yeh (2010), and Matsumura et al. (2014)
Nuclear resonance vibrational spec (NRVS)	M–ligand vibrations, bond strengths, spin state, coordination	Chakraborty et al. (2015)
Mössbauer	Oxidation and spin states of iron	Chakraborty, Hosseinzadeh, et al. (2014) and Chakraborty, Reed, et al. (2014)
Paramagnetic NMR (PNMR)	M–ligand, binding stoichiometry	Wang and Lu (1999)

quench (Matsumura & Moenne-Loccoz, 2014) techniques and further characterized using various spectroscopic techniques described in Table 1. To give an overview of this process, we describe the capture and vibrational characterization of intermediates obtained during reaction of Fe-Fe$_B$Mb with NO below.

1. RFQ was used to capture the reaction mixture at early time points. Protocols for preparation of RFQ samples and their rR analysis have been discussed in detail in this chapter (Matsumura & Moenne-Loccoz, 2014).
2. Briefly, glass syringes (1 or 2 mL) were loaded with protein solutions (0.6 mM reduced Fe$_B$Mbs in 50 mM Bis-Tris, pH 7.0) and NO solutions (2 mM ^{14}NO or ^{15}NO in 50 mM Bis-Tris, pH 7.0) inside an anaerobic chamber before mounting them to the System 1000 Chemical/Freeze Quench Apparatus (Update Instruments).
3. Reaction times were controlled by varying the syringe displacement rate or by varying the length of the reactor hose after the mixer.
4. Mixed volumes of 250 μL were ejected into a glass funnel attached to NMR tubes filled with liquid ethane at or below −120°C. The frozen samples were packed into the tube as the assembly set within a Teflon block cooled to −120°C. Liquid ethane was subsequently removed by incubation of samples at −80°C for 2 h.

The rR spectra recorded on all of the RFQ samples shown in Fig. 6C display a prominent high frequency vibration at 1755 cm^{-1} corresponding to ν(NO) of nonheme Fe(II)–NO with a continuous increase in vibration at 1660 cm^{-1} with time corresponding to ν(NO) of heme Fe(II)–NO (Fig. 7C). These results suggest that the 6 ms species obtained in UV–vis stopped-flow spectroscopy corresponds to a nonheme Fe(II)–NO species. Thus, NO binds first to nonheme iron following which it binds to the heme iron. Overall, stopped-flow experiments in tandem with RFQ-rR experiments strongly support the *trans*-mechanism for NO reduction for Mb-based NOR mimic (Matsumura et al., 2014).

Tip. Generally, the strong spectroscopic signature of heme iron along with its high affinity makes characterization of the nonheme metal ion challenging. To overcome this challenge, the heme can be replaced with Zn protoporphyrin IX. This replacement has been successfully done in Fe$_B$Mb protein for the complete characterization of nonheme Fe(II)–NO species via UV–vis, EPR, and Mössbauer spectroscopy (Chakraborty, Hosseinzadeh, et al., 2014; Chakraborty, Reed, et al., 2014).

4. STEP 4: FURTHER IMPROVEMENT OF DESIGNED HETERONUCLEAR METALLOENZYMES: CASE STUDIES

Here, we discuss a few examples of how the information obtained from structural, functional, and mechanistic studies of heteronuclear metalloenzymes was used to further improve their activity and robustness.

4.1 Improving ET to the Heme Center

Increasing the concentration of an electron donor (TMPD/ascorbate) led to a direct increase in the oxygen reduction rates of G65Y-Cu$_B$Mb, suggesting that ET is the rate-limiting step in oxygen reduction of Mb-based HCO mimics (Miner et al., 2012). Thus, an efficient method to increase the overall oxidase activity of HCO mimics is to enhance ET rates to the catalytic heme center. In order to enhance ET rates, a triple lysine mutant of G65Y-Cu$_B$Mb called G65Y-Cu$_B$Mb(+6) was created that displayed very strong interactions with the native redox partner of Mb, cytochrome b_5 (Livingston, Mclachlan, La Mar, & Brown, 1985). The G65Y-Cu$_B$Mb(+6) mutant in the presence of cytochrome b_5 (Fig. 8A) reduced oxygen at a rate (52 s^{-1}) comparable to that of a native HCO (cytochrome cbb_3 oxidase, 50 s^{-1}) (Yu, Cui, et al., 2015).

4.2 Tuning the Heme Reduction Potential

The heme $E^{o\prime}$ of F33Y-Cu$_B$Mb is rather low (95 mV) relative to native HCO, and hence has a higher overpotential for biofuel cell applications, which require oxygen reduction catalysts with higher E^o (close to 0.8 V at pH 7). To increase the heme $E^{o\prime}$ of the Mb-based HCO mimic, we used several strategies including tuning of the H-bonding interactions of the proximal histidine and introducing nonnative heme cofactors which contain electron-withdrawing carbonyl group conjugated to the porphyrin macrocycle. Four F33Y-Cu$_B$Mb variants were obtained with systematically increasing heme $E^{o\prime}$ (Fig. 8B). Interestingly, the HCO mimics with higher heme $E^{o\prime}$ showed a corresponding increase in oxygen reduction rates (Bhagi-Damodaran et al., 2014).

4.3 Increasing the Binding Affinity of Mn^{2+} in MnCcP

The initial activity assays on MnCcP mimics showed much lower rates of manganese oxidation than native MnPs. One contributor to this effect is

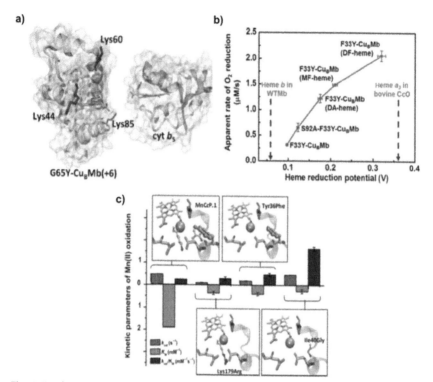

Fig. 8 Further improvement of designed heteronuclear metalloenzymes. (A) Structures of G65Y-Cu$_B$Mb(+6), showing the engineered lysines in *blue* (*black* in the print version), and cyt b5 (PDB IDs: 1CYO for cyt b5) and 4FWY for F33Y-Cu$_B$Mb. (B) Correlation between the heme $E^{o\prime}$ and catalytic oxygen reduction reactivity of HCO mimics. (C) Enhancement of binding affinity and catalytic efficiency of MnCcP mimics by mutation secondary coordination sphere residues. *Figures are adapted with permission from Yu, Y., Cui, C., Liu, X., Petrik, I. D., Wang, J., & Lu, Y. (2015). A designed metalloenzyme achieving the catalytic rate of a native enzyme.* Journal of the American Chemical Society, 137, 11570–11573; Bhagi-Damodaran, A., Petrik, I. D., Marshall, N. M., Robinson, H., & Lu, Y. (2014). Systematic tuning of heme redox potentials and its effects on O$_2$ reduction rates in a designed oxidase in myoglobin. Journal of the American Chemical Society, 136, 11882–11885; Hosseinzadeh, P., Mirts, E. N., Pfister, T. D., Gao, Y.-G., Mayne, C., Robinson, H., et al. (2016). Enhancing Mn(II)-binding and manganese peroxidase activity in a designed cytochrome c peroxidase through fine-tuning secondary-sphere interactions. Biochemistry, 55, 1494–502, respectively. © 2014 American Chemical Society.

the lower affinity for Mn^{2+} in the designed variants compared with their native counterparts. A series of mutations in the secondary coordination sphere of the Mn-binding site resulted in variants with higher activities. In particular, the Ile40Gly variant showed K_M values comparable with native MnPs (Fig. 8C) (Hosseinzadeh et al., 2016).

5. CONCLUSIONS

The process of designing biosynthetic models of heteronuclear metalloenzymes is intricate and requires simultaneous consideration of many variables. Scaffold selection should favor the desired properties of the target metal-binding site while maintaining a useful level of plasticity for investigative purposes; this initial choice carries forward for subsequent design considerations, and sometimes a promising first choice may not ultimately be the best. The presence of and ability to rationally tune secondary sphere interactions (and beyond) is a powerful tool for accessing new regimes of catalysis or for generating states found in nature that cannot be stabilized by primary interactions alone. The design space for these outer sphere interactions increases correspondingly, and predicting optimal changes to achieve the desired outcome is not trivial. Fortunately, a large assortment of tools is available to assist in the design process, and the versatility and accuracy of computational modeling suites are already quite high and constantly improving: the task need not be daunting. A rich knowledge base exists for physical and spectroscopic characterization of metalloprotein active sites, much of which has been honed so that meaningful results can be obtained while working with dilute samples at varying levels of purity. A major advantage of the biosynthetic approach is the ability to make use of these techniques with a large amount of highly pure samples. Experiments can target individual parameters such as redox potential, pK_a, or catalytic kinetic parameters such as K_M to both probe and improve upon nature, and all of these changes benefit from direct comparison through the same scaffold. Biosynthetic models have already contributed greatly to our understanding of heteronuclear metalloprotein catalysis, and as the techniques in design and genetic manipulation continue to improve, they are likely to continue to grow in importance for increasingly complex designed biological and biomimetic systems.

ACKNOWLEDGMENTS

We wish to thank all the former and current Lu group members who have contributed to the development of the protocols and obtaining results published in papers cited in this chapter. The Lu group research described in papers cited in this chapter has been supported by the US National Institute of Health (R01GM06211) and National Science Foundation (CHE 14-13328).

REFERENCES

Annele, H., Taina, L., Martin, H., & Pekka, M. (2003). Manganese peroxidase and its role in the degradation of wood lignin. In J. N. Saddler (Ed.), *Applications of enzymes to lignocellulosics*. Washington, DC: American Chemical Society.

Armstrong, F. A. (2002). Protein film voltammetry: Revealing the mechanisms of biological oxidation and reduction. *Russian Journal of Electrochemistry, 38*, 49–62.

Bairoch, A., & Apweiler, R. (2000). The SWISS-PROT protein sequence database and its supplement TrEMBL in 2000. *Nucleic Acids Research, 28*, 45–48.

Becker, C. F., Watmough, N. J., & Elliott, S. J. (2009). Electrochemical evidence for multiple peroxidatic heme states of the diheme cytochrome c peroxidase of Pseudomonas aeruginosa. *Biochemistry, 48*, 87–95.

Berman, H. M., Westbrook, J., Feng, Z., Gilliland, G., Bhat, T. N., Weissig, H., et al. (2000). The protein data bank. *Nucleic Acids Research, 28*, 235–242.

Berry, S. M., Baker, M. H., & Reardon, N. J. (2010). Reduction potential variations in azurin through secondary coordination sphere phenylalanine incorporations. *Journal of Inorganic Biochemistry, 104*, 1071–1078.

Berry, E. A., & Trumpower, B. L. (1987). Simultaneous determination of hemes a, b, and c from pyridine hemochrome spectra. *Analytical Biochemistry, 161*, 1–15.

Bhagi-Damodaran, A., Petrik, I. D., Marshall, N. M., Robinson, H., & Lu, Y. (2014). Systematic tuning of heme redox potentials and its effects on O_2 reduction rates in a designed oxidase in myoglobin. *Journal of the American Chemical Society, 136*, 11882–11885.

Bidwai, A., Witt, M., Foshay, M., Vitello, L. B., Satterlee, J. D., & Erman, J. E. (2003). Cyanide binding to cytochrome c peroxidase (H52L). *Biochemistry, 42*, 10764–10771.

Blomberg, M. R., & Siegbahn, P. E. (2006). Different types of biological proton transfer reactions studied by quantum chemical methods. *Biochimica et Biophysica Acta, 1757*, 969–980.

Brooks, B. R., Brooks, C. L., 3rd., Mackerell, A. D., Jr., Nilsson, L., Petrella, R. J., Roux, B., et al. (2009). CHARMM: The biomolecular simulation program. *Journal of Computational Chemistry, 30*, 1545–1614.

Brunelle, J. L., & Green, R. (2014). One-dimensional SDS-polyacrylamide gel electrophoresis (1D SDS-PAGE). *Methods in Enzymology, 541*, 151–159.

Butland, G., Spiro, S., Watmough, N. J., & Richardson, D. J. (2001). Two conserved glutamates in the bacterial nitric oxide reductase are essential for activity but not assembly of the enzyme. *Journal of Bacteriology, 183*, 189–199.

Case, D. A., Cheatham, T. E., 3rd., Darden, T., Gohlke, H., Luo, R., Merz, K. M., Jr., et al. (2005). The Amber biomolecular simulation programs. *Journal of Computational Chemistry, 26*, 1668–1688.

Casini, A., Storch, M., Baldwin, G. S., & Ellis, T. (2015). Bricks and blueprints: Methods and standards for DNA assembly. *Nature Reviews. Molecular Cell Biology, 16*, 568–576.

Chakraborty, S., Hosseinzadeh, P., & Lu, Y. (2014). Metalloprotein design and engineering. In R. A. Scott (Ed.), *Encyclopedia of inorganic and bioinorganic chemistry* (pp. 1–51). Chichester: John Wiley and Sons, Ltd.

Chakraborty, S., Reed, J., Ross, M., Nilges, M. J., Petrik, I. D., Ghosh, S., et al. (2014). Spectroscopic and computational study of a nonheme iron nitrosyl center in a biosynthetic model of nitric oxide reductase. *Angewandte Chemie, International Edition, 126*, 2449–2453.

Chakraborty, S., Reed, J., Sage, J. T., Branagan, N. C., Petrik, I. D., Miner, K. D., et al. (2015). Recent advances in biosynthetic modeling of nitric oxide reductases and insights gained from nuclear resonance vibrational and other spectroscopic studies. *Inorganic Chemistry, 54*, 9317–9329.

Chang, C. J., Chang, M. C., Damrauer, N. H., & Nocera, D. G. (2004). Proton-coupled electron transfer: A unifying mechanism for biological charge transport, amino acid radical initiation and propagation, and bond making/breaking reactions of water and oxygen. *Biochimica et Biophysica Acta, 1655*, 13–28.

Chang, C. H., & Kim, K. (2009). Density functional theory calculation of bonding and charge parameters for molecular dynamics studies on [FeFe] hydrogenases. *Journal of Chemical Theory and Computation, 5*, 1137–1145.

Collman, J. P., Boulatov, R., Sunderland, C. J., & Fu, L. (2004). Functional analogues of cytochrome *c* oxidase, myoglobin, and hemoglobin. *Chemical Reviews, 104,* 561–588.

Collman, J. P., & Wang, Z. (1999). Synthetic heme chemistry: The cradle of functional molecules and catalysts. *Chemtracts, 12,* 229–263.

Das, T. K., Pecoraro, C., Tomson, F. L., Gennis, R. B., & Rousseau, D. L. (1998). The post-translational modification in cytochrome *c* oxidase is required to establish a functional environment of the catalytic site. *Biochemistry, 37,* 14471–14476.

Degrado, W. F., Summa, C. M., Pavone, V., Nastri, F., & Lombardi, A. (1999). De novo design and structural characterization of proteins and metalloproteins. *Annual Review of Biochemistry, 68,* 779–819.

Degrado, W. F., Wasserman, Z. R., & Lear, J. D. (1989). Protein design, a minimalist approach. *Science, 243,* 622–628.

Dib, L., & Carbone, A. (2012). Protein fragments: Functional and structural roles of their coevolution networks. *PloS One, 7,* e48124.

Emsley, P., Lohkamp, B., Scott, W. G., & Cowtan, K. (2010). Features and development of Coot. *Acta Crystallographica. Section D, Biological Crystallography, 66,* 486–501.

Evans, R. M., & Armstrong, F. A. (2014). Electrochemistry of metalloproteins: Protein film electrochemistry for the study of E. coli [NiFe]-hydrogenase-1. *Methods in Molecular Biology, 1122,* 73–94.

Feng, M., Tachikawa, H., Wang, X., Pfister, T. D., Gengenbach, A. J., & Lu, Y. (2003). Resonance Raman spectroscopy of cytochrome c peroxidase variants that mimic manganese peroxidase. *Journal of Biological Inorganic Chemistry, 8,* 699–706.

Finzel, B. C., Poulos, T. L., & Kraut, J. (1984). Crystal structure of yeast cytochrome c peroxidase refined at 1.7-A resolution. *The Journal of Biological Chemistry, 259,* 13027–13036.

Fischer, M., Zhang, Q. C., Dey, F., Chen, B. Y., Honig, B., & Petrey, D. (2011). MarkUs: A server to navigate sequence–structure–function space. *Nucleic Acids Research, 39,* W357–W361.

Flock, U., Watmough, N. J., & Ädelroth, P. (2005). Electron/proton coupling in bacterial nitric oxide reductase during reduction of oxygen. *Biochemistry, 44,* 10711–10719.

Fontecave, M., & Pierre, J.-L. (1998). Oxidations by copper metalloenzymes and some biomimetic approaches. *Coordination Chemistry Reviews, 170,* 125–140.

Fukuda, Y., Tse, K. M., Nakane, T., Nakatsu, T., Suzuki, M., Sugahara, M., et al. (2016). Redox-coupled proton transfer mechanism in nitrite reductase revealed by femtosecond crystallography. *Proceedings of the National Academy of Sciences of the United States of America, 113,* 2928–2933.

Gao, Y., Meyer, B., Sokolova, L., Zwicker, K., Karas, M., Brutschy, B., et al. (2012). Heme-copper terminal oxidase using both cytochrome c and ubiquinol as electron donors. *Proceedings of the National Academy of Sciences of the United States of America, 109,* 3275–3280.

Garcia-Horsman, J. A., Barquera, B., Rumbley, J., Ma, J., & Gennis, R. B. (1994). The superfamily of heme-copper respiratory oxidases. *Journal of Bacteriology, 176,* 5587–5600.

Gibney, B. R., Rabanal, F., Skalicky, J. J., Wand, A. J., & Dutton, P. L. (1999). Iterative protein redesign. *Journal of the American Chemical Society, 121,* 4952–4960.

Gibson, D. G. (2011). Enzymatic assembly of overlapping DNA fragments. *Methods in Enzymology, 498,* 349–361.

Gray, H. B., & Winkler, J. R. (1996). Electron transfer in proteins. *Annual Review of Biochemistry, 65,* 537–561.

Gray, H. B., & Winkler, J. R. (2010). Electron flow through metalloproteins. *Biochimica et Biophysica Acta (BBA)—Bioenergetics, 1797,* 1563–1572.

Greenfield, N. J. (2006). Using circular dichroism spectra to estimate protein secondary structure. *Nature Protocols, 1,* 2876–2890.

Hadt, R. G., Sun, N., Marshall, N. M., Hodgson, K. O., Hedman, B., Lu, Y., et al. (2012). Spectroscopic and DFT studies of second-sphere variants of the type 1 copper site in azurin: Covalent and nonlocal electrostatic contributions to reduction potentials. *Journal of the American Chemical Society, 134*, 16701–16716.

Hay, M. T., Ang, M. C., Gamelin, D. R., Solomon, E. I., Antholine, W. E., Ralle, M., et al. (1998). Spectroscopic characterization of an engineered purple Cu_A center in azurin. *Inorganic Chemistry, 37*, 191.

Hayashi, T., Lin, I. J., Chen, Y., Fee, J. A., & Moënne-Loccoz, P. (2007). Fourier transform infrared characterization of a CuB–nitrosyl complex in cytochrome ba3 from Thermus thermophilus: Relevance to NO reductase activity in heme–copper terminal oxidases. *Journal of the American Chemical Society, 129*, 14952–14958.

Hofirchter, M. (2002). Review: Lignin conversion by manganese 682 peroxidase (MnP). *Enzyme and Microbial Technology, 30*, 454–466.

Holm, R. H. (1995). Chemical approaches to bridged biological metal assemblies. *Pure and Applied Chemistry, 67*, 217–224.

Holm, L., & Rosenstrom, P. (2010). Dali server: Conservation mapping in 3D. *Nucleic Acids Research, 38*, W545–W549.

Hosseinzadeh, P., & Lu, Y. (2015). Design and fine-tuning redox potentials of metalloproteins involved in electron transfer in bioenergetics. *Biochimica et Biophysica Acta, Bioenergetics, 187*(5), 557–581.

Hosseinzadeh, P., Mirts, E. N., Pfister, T. D., Gao, Y.-G., Mayne, C., Robinson, H., et al. (2016). Enhancing Mn(II)-binding and manganese peroxidase activity in a designed cytochrome c peroxidase through fine-tuning secondary-sphere interactions. *Biochemistry, 55*, 1494–1502.

Hu, X., Wang, H., Ke, H., & Kuhlman, B. (2007). High-resolution design of a protein loop. *Proceedings of the National Academy of Sciences of the United States of America, 104*, 17668–17673.

Humphrey, W., Dalke, A., & Schulten, K. (1996). VMD: Visual molecular dynamics. *Journal of Molecular Graphics, 14*(33–38), 27–28.

Jackson, T. A., & Brunold, T. C. (2004). Combined spectroscopic/computational studies on Fe- and Mn-dependent superoxide dismutases: Insights into second-sphere tuning of active site properties. *Accounts of Chemical Research, 37*, 461–470.

Kim, E., Chufan, E. E., Kamaraj, K., & Karlin, K. D. (2004). Synthetic models for heme-copper oxidases. *Chemical Reviews, 104*, 1077–1133.

Koziol, L., Valdez, C. A., Baker, S. E., Lau, E. Y., Floyd, W. C., 3rd., Wong, S. E., et al. (2012). Toward a small molecule, biomimetic carbonic anhydrase model: Theoretical and experimental investigations of a panel of zinc(II) aza-macrocyclic catalysts. *Inorganic Chemistry, 51*, 6803–6812.

Lilie, H. (2003). Designer proteins in biotechnology. International Titisee Conference on protein design at the crossroads of biotechnology, chemistry and evolution. *EMBO Reports, 4*, 346–351.

Lin, Y.-W., Yeung, N., Gao, Y.-G., Miner, K. D., Lei, L., Robinson, H., et al. (2010a). Introducing a 2-His-1-Glu nonheme iron center into myoglobin confers nitric oxide reductase activity. *Journal of the American Chemical Society, 132*, 9970–9972. PMC2917603.

Lin, Y.-W., Yeung, N., Gao, Y.-G., Miner, K. D., Tian, S., Robinson, H., et al. (2010b). Roles of glutamates and metal ions in a rationally designed nitric oxide reductase based on myoglobin. *Proceedings of the National Academy of Sciences of the United States of America, 107*, 8581–8586.

Liu, J., Chakraborty, S., Hosseinzadeh, P., Yu, Y., Tian, S., Petrik, I. D., et al. (2014). Metalloproteins containing cytochrome, iron-sulfur or copper redox centers. *Chemical Reviews, 114*, 4366–4469.

Liu, X., Yu, Y., Hu, C., Zhang, W., Lu, Y., & Wang, J. (2012). Significant increase of oxidase activity through the genetic incorporation of a tyrosine-histidine cross-link in a myoglobin model of heme-copper oxidase. *Angewandte Chemie, International Edition, 51*, 4312–4316. PMC3511862.

Livingston, D. J., Mclachlan, S. J., La Mar, G. N., & Brown, W. D. (1985). Myoglobin:cytochrome b5 interactions and the kinetic mechanism of metmyoglobin reductase. *The Journal of Biological Chemistry, 260*, 15699–15707.

Lu, Y., Chakraborty, S., Miner, K. D., Wilson, T. D., Mukherjee, A., Yu, Y., et al. (2013). Metalloprotein design. In J. Reedijk & K. Poeppelmeier (Eds.), *Comprehensive inorganic chemistry II* (pp. 565–593). Amsterdam: Elsevier Science.

Lu, Y., Garner, D. K., & Zhang, J.-L. (2009). Artificial metalloproteins: Design and engineering. In T. P. Begley (Ed.), *Wiley encyclopedia of chemical biology*. New York: John Wiley & Sons, Inc.

Lu, Y., Wang, X., Sigman, J. A., Gengenbach, A., Berry, S. M., & Brueshoff, P. J. (1999). Redesigning metalloproteins. In *Book of abstracts, 218th ACS national meeting, New Orleans, Aug 22–26*. INOR-403.

Lu, Y., Yeung, N., Sieracki, N., & Marshall, N. M. (2009). Design of functional metalloproteins. *Nature, 460*, 855–862. PMC153914.

Lu, C., Zhao, X., Lu, Y., Rousseau, D. L., & Yeh, S.-R. (2010). Role of copper ion in regulating ligand binding in a myoglobin-based cytochrome c oxidase model. *Journal of the American Chemical Society, 132*, 1598–1605. PMC3761084.

Luft, J. R., Newman, J., & Snell, E. H. (2014). Crystallization screening: The influence of history on current practice. *Acta Crystallographica. Section F, Structural Biology Communications, 70*, 835–853.

Mahadevan, V., Gebbink, R. K., & Stack, T. D. (2000). Biomimetic modeling of copper oxidase reactivity. *Current Opinion in Chemical Biology, 4*, 228–234.

Malmström, B. G. (1990). Structural control of electron-transfer properties in metalloproteins. *Biology of Metals, 3*, 64–66.

Marshall, N. M., Garner, D. K., Wilson, T. D., Gao, Y.-G., Robinson, H., Nilges, M. J., et al. (2009). Rationally tuning the reduction potential of a single cupredoxin beyond the natural range. *Nature, 462*, 113–116.

Marshall, N. M., Miner, K. D., Wilson, T. D., & Lu, Y. (2013). Rational design of protein cages for alternative enzymatic functions. In T. Ueno & Y. Watanabe (Eds.), *Coordination chemistry in protein cages* (pp. 111–150). New York: John Wiley & Sons, Inc.

Matsumura, H., Hayashi, T., Chakraborty, S., Lu, Y., & Moënne-Loccoz, P. (2014). The production of nitrous oxide (N_2O) by the heme/non-heme diiron center of engineered myoglobins (Fe_BMbs) proceeds through a trans iron-nitrosyl dimer. *Journal of the American Chemical Society, 136*, 2420–2431.

Matsumura, H., & Moenne-Loccoz, P. (2014). Characterizing millisecond intermediates in hemoproteins using rapid-freeze-quench resonance Raman spectroscopy. *Methods in Molecular Biology, 1122*, 107–123.

Miller, A. F. (2008). Redox tuning over almost 1 V in a structurally conserved active site: Lessons from Fe-containing superoxide dismutase. *Accounts of Chemical Research, 41*, 501–510.

Miner, K. D., Mukherjee, A., Gao, Y.-G., Null, E. L., Petrik, I. D., Zhao, X., et al. (2012). A designed functional metalloenzyme that reduces O_2 to H_2O with over one thousand turnovers. *Angewandte Chemie, International Edition, 51*, 5589–5592. PMC3461324.

Miner, K. D., Pfister, T. D., Hosseinzadeh, P., Karaduman, N., Donald, L. J., Loewen, P. C., et al. (2014). Identifying the elusive sites of tyrosyl radicals in cytochrome *c* peroxidase: Implications for oxidation of substrates bound at a site remote from the heme. *Biochemistry, 53*, 3781–3789.

Mukherjee, S., Mukherjee, M., Dey, A., Mukherjee, A., Bhagi-Damodaran, A., & Lu, Y. (2015). A biosynthetic model of cytochrome c oxidase as an electrocatalyst for oxygen reduction. *Nature Communications, 6*, 8467.

Nakayama, H., & Shimamoto, N. (2014). Modern and simple construction of plasmid: Saving time and cost. *Journal of Microbiology, 52*, 891–897.

New, S. Y., Marshall, N. M., Hor, T. S. A., Xue, F., & Lu, Y. (2012). Redox tuning of two biological copper centers through non-covalent interactions: Same trend but different magnitude. *Chemical Communications, 48*, 4217–4219.

Petrik, I. D., Davydov, R., Ross, M., Zhao, X., Hoffman, B., & Lu, Y. (2016). Spectroscopic and crystallographic evidence for the role of a water-containing H-bond network in oxidase activity of an engineered myoglobin. *Journal of the American Chemical Society, 138*, 1134–1137.

Petrik, I. D., Liu, J., & Lu, Y. (2014). Recent advances in functional metalloprotein design. *Current Opinion in Structural Biology, 19*, 67–75.

Pettersen, E. F., Goddard, T. D., Huang, C. C., Couch, G. S., Greenblatt, D. M., Meng, E. C., et al. (2004). UCSF chimera—A visualization system for exploratory research and analysis. *Journal of Computational Chemistry, 25*, 1605–1612.

Pfister, T. D., Mirarefi, A. Y., Gengenbach, A. J., Zhao, X., Danstrom, C., Conatser, N., et al. (2007). Kinetic and crystallographic studies of a redesigned manganese-binding site in cytochrome *c* peroxidase. *Journal of Biological Inorganic Chemistry, 12*, 126–137.

Pfister, T. D., Ohki, T., Ueno, T., Hara, I., Adachi, S., Makino, Y., et al. (2005). Monooxygenation of an aromatic ring by F43W/H64D/V68I myoglobin mutant and hydrogen peroxide. Myoglobin mutants as a model for P450 hydroxylation chemistry. *The Journal of Biological Chemistry, 280*, 12858–12866.

Phillips, J. C., Braun, R., Wang, W., Gumbart, J., Tajkhorshid, E., Villa, E., et al. (2005). Scalable molecular dynamics with NAMD. *Journal of Computational Chemistry, 26*, 1781–1802.

Pronk, S., Pall, S., Schulz, R., Larsson, P., Bjelkmar, P., Apostolov, R., et al. (2013). GROMACS 4.5: A high-throughput and highly parallel open source molecular simulation toolkit. *Bioinformatics, 29*, 845–854.

Punniyamurthy, T., Velusamy, S., & Iqbal, J. (2005). Recent advances in transition metal catalyzed oxidation of organic substrates with molecular oxygen. *Chemical Reviews, 105*, 2329–2363.

Que, L., Jr., & Tolman, W. B. (2008). Biologically inspired oxidation catalysis. *Nature, 455*, 333–340.

Reece, S. Y., & Nocera, D. G. (2009). Proton-coupled electron transfer in biology: Results from synergistic studies in natural and model systems. *Annual Review of Biochemistry, 78*, 673–699.

Reedy, C. J., & Gibney, B. R. (2004). Heme protein assemblies. *Chemical Reviews, 104*, 617–649.

Richardson, D. J., & Watmough, N. J. (1999). Inorganic nitrogen metabolism in bacteria. *Current Opinion in Chemical Biology, 3*, 207–219.

Richter, F., Leaver-Fay, A., Khare, S. D., Bjelic, S., & Baker, D. (2011). De novo enzyme design using Rosetta3. *PloS One, 6*, e19230.

Roy, A., Kucukural, A., & Zhang, Y. (2010). I-TASSER: A unified platform for automated protein structure and function prediction. *Nature Protocols, 5*, 725–738.

Sanders, J. K. M. (1999). Supramolecular catalysis in transition. In R. Ungaro & E. Dalcanale (Eds.), *Supramolecular science: Where it is and where it is going* (pp. 273–286). Dordrecht, Netherlands: Springer.

Sarangi, R., Gorelsky, S. I., Basumallick, L., Hwang, H. J., Pratt, R. C., Stack, T. D. P., et al. (2008). Spectroscopic and density functional theory studies of the blue-copper site in

M121SeM and C112SeC azurin: Cu-Se versus Cu-S bonding. *Journal of the American Chemical Society, 130,* 3866.

Séraphin, B., & Kandels-Lewis, S. (1996). An efficient PCR mutagenesis strategy without gel purification [correction of "purificiation"] step that is amenable to automation. *Nucleic Acids Research, 24,* 3276–3277.

Shanmugam, M., Xue, G., Que, L., Jr., & Hoffman, B. M. (2012). 1H-ENDOR evidence for a hydrogen-bonding interaction that modulates the reactivity of a nonheme Fe(IV)=O unit. *Inorganic Chemistry, 51,* 10080–10082.

Shook, R. L., & Borovik, A. S. (2010). Role of the secondary coordination sphere in metal-mediated dioxygen activation. *Inorganic Chemistry, 49,* 3646–3660.

Sievers, F., Wilm, A., Dineen, D., Gibson, T. J., Karplus, K., Li, W., et al. (2011). Fast, scalable generation of high-quality protein multiple sequence alignments using Clustal Omega. *Molecular Systems Biology, 7,* 539.

Sigman, J. A., Kim, H. K., Zhao, X., Carey, J. R., & Lu, Y. (2003). The role of copper and protons in heme-copper oxidases: Kinetic study of an engineered heme-copper center in myoglobin. *Proceedings of the National Academy of Sciences of the United States of America, 100,* 3629–3634.

Sigman, J. A., Kwok, B. C., Gengenbach, A., & Lu, Y. (1999). Design and creation of a Cu(II)-binding site in cytochrome *c* peroxidase that mimics the Cu_B-heme center in terminal oxidases. *Journal of the American Chemical Society, 121,* 8949–8950.

Sigman, J. A., Kwok, B. C., & Lu, Y. (2000). From myoglobin to heme-copper oxidase: Design and engineering of a Cu_B center into sperm whale myoglobin. *Journal of the American Chemical Society, 122,* 8192.

Sigman, J. A., Wang, X., & Lu, Y. (2000). Coupled oxidation of heme by myoglobin is mediated by exogenous peroxide. *Journal of the American Chemical Society, 123,* 6945–6946.

Sillitoe, I., Lewis, T. E., Cuff, A., Das, S., Ashford, P., Dawson, N. L., et al. (2015). CATH: Comprehensive structural and functional annotations for genome sequences. *Nucleic Acids Research, 43,* D376–D381.

Skarina, T., Xu, X., Evdokimova, E., & Savchenko, A. (2014). High-throughput crystallization screening. *Methods in Molecular Biology, 1140,* 159–168.

Sollewijn Gelpke, M. D., Moenne-Loccoz, P., & Gold, M. H. (1999). Arginine 177 is involved in Mn(II) binding by manganese peroxidase. *Biochemistry, 38,* 11482–11489.

Springer, B. A., & Sligar, S. G. (1987). High-level expression of sperm whale myoglobin in Escherichia coli. *Proceedings of the National Academy of Sciences of the United States of America, 84,* 8961–8965.

Strupat, K. (2005). Molecular weight determination of peptides and proteins by ESI and MALDI. *Methods in Enzymology, 405,* 1–36.

Sundaramoorthy, M., Kishi, K., Gold, M. H., & Poulos, T. L. (1994). The crystal structure of manganese peroxidase from Phanerochaete chrysosporium at 2.06-A resolution. *The Journal of Biological Chemistry, 269,* 32759–32767.

Taboy, C. H., Bonaventura, C., & Crumbliss, A. L. (1999). Spectroelectrochemistry of heme proteins: Effects of active-site heterogeneity on Nernst plots. *Bioelectrochemistry and Bioenergetics, 48,* 79–86.

Tard, C., & Pickett, C. J. (2009). Structural and functional analogues of the active sites of the [Fe]-, [NiFe]-, and [FeFe]-hydrogenases. *Chemical Reviews, 109,* 2245–2274.

Thomas, C. M., & Ward, T. R. (2005). Design of artificial metalloenzymes. *Applied Organometallic Chemistry, 19,* 35–39.

Tyagi, R., Lai, R., & Duggleby, R. G. (2004). A new approach to "megaprimer" polymerase chain reaction mutagenesis without an intermediate gel purification step. *BMC Biotechnology, 4,* 2.

Van Dyke, B. R., Saltman, P., & Armstrong, F. A. (1996). Control of myoglobin electron-transfer rates by the distal (nonbound) histidine residue. *Journal of the American Chemical Society, 118*, 3490–3492.

Varadarajan, R., Zewert, T. E., Gray, H. B., & Boxer, S. G. (1989). Effects of buried ionizable amino acids on the reduction potential of recombinant myoglobin. *Science, 243*, 69–72.

Wang, X., & Lu, Y. (1999). Proton NMR investigation of the heme active site structure of an engineered cytochrome *c* peroxidase that mimics manganese peroxidase. *Biochemistry, 38*, 9146–9157.

Weinberg, D. R., Gagliardi, C. J., Hull, J. F., Murphy, C. F., Kent, C. A., Westlake, B. C., et al. (2012). Proton-coupled electron transfer. *Chemical Reviews, 112*, 4016–4093.

Weissensteiner, T., Nolan, T., Bustin, S., Griffin, H., & Griffin, A. (2003). *PCR technology: Current innovations* (3rd ed.). Boca Raton, FL: CRC Press.

Wilcox, S. K., Putnam, C. D., Sastry, M., Blankenship, J., Chazin, W. J., Mcree, D. E., et al. (1998). Rational design of a functional metalloenzyme: Introduction of a site for manganese binding and oxidation into a heme peroxidase. *Biochemistry, 37*, 16853–16862.

Worth, C. L., Gong, S., & Blundell, T. L. (2009). Structural and functional constraints in the evolution of protein families. *Nature Reviews. Molecular Cell Biology, 10*, 709–720.

Yang, J., Yan, R., Roy, A., Xu, D., Poisson, J., & Zhang, Y. (2015). The I-TASSER suite: Protein structure and function prediction. *Nature Methods, 12*, 7–8.

Ye, Y., & Godzik, A. (2004). FATCAT: A web server for flexible structure comparison and structure similarity searching. *Nucleic Acids Research, 32*, W582–W585.

Yeung, N., Lin, Y.-W., Gao, Y.-G., Zhao, X., Russell, B. S., Lei, L., et al. (2009). Rational design of a structural and functional nitric oxide reductase. *Nature (London, United Kingdom), 462*, 1079–1082.

Yeung, N., & Lu, Y. (2008). One heme, diverse functions: Using biosynthetic myoglobin models to gain insights into heme-copper oxidases and nitric oxide reductases. *Chemistry & Biodiversity, 5*, 1437–1454. PMC153901.

Yeung, B. K., Wang, X., Sigman, J. A., Petillo, P. A., & Lu, Y. (1997). Construction and characterization of a manganese-binding site in cytochrome *c* peroxidase: Towards a novel manganese peroxidase. *Chemistry and Biology, 4*, 215–221.

Yikilmaz, E., Xie, J., Brunold, T. C., & Miller, A. F. (2002). Hydrogen-bond-mediated tuning of the redox potential of the non-heme Fe site of superoxide dismutase. *Journal of the American Chemical Society, 124*, 3482–3483.

Yu, Y., Cui, C., Liu, X., Petrik, I. D., Wang, J., & Lu, Y. (2015). A designed metalloenzyme achieving the catalytic rate of a native enzyme. *Journal of the American Chemical Society, 137*, 11570–11573.

Yu, Y., Lv, X., Li, J., Zhou, Q., Cui, C., Mukherjee, A., et al. (2015). Defining the role of tyrosine and rational tuning of oxidase activity by genetic incorporation of unnatural tyrosine analogs. *Journal of the American Chemical Society, 137*(14), 4594–4597.

Yu, Y., Mukherjee, A., Nilges, M. J., Hosseinzadeh, P., Miner, K. D., & Lu, Y. (2014). Direct observation of a tyrosyl radical in a functional oxidase model in myoglobin during both H_2O_2 and O_2 reactions by EPR. *Journal of the American Chemical Society, 138*, 3869–3875.

Zanghellini, A., Jiang, L., Wollacott, A. M., Cheng, G., Meiler, J., Althoff, E. A., et al. (2006). New algorithms and an in silico benchmark for computational enzyme design. *Protein Science, 15*, 2785–2794.

Zastrow, M. L., & Pecoraro, V. L. (2013). Designing functional metalloproteins: From structural to catalytic metal sites. *Coordination Chemistry Reviews, 257*, 2565–2588.

Zeng, J., Geng, M., Jiang, H., Liu, Y., Liu, J., & Qiu, G. (2007). The IscA from Acidithiobacillus ferrooxidans is an iron-sulfur protein which assemble the [Fe4S4] cluster with intracellular iron and sulfur. *Archives of Biochemistry and Biophysics, 463*, 237–244.

Zhang, Y. (2008). I-TASSER server for protein 3D structure prediction. *BMC Bioinformatics*, *9*, 40.

Zhao, X., Yeung, N., Russell, B. S., Garner, D. K., & Lu, Y. (2006). Catalytic reduction of NO to N2O by a designed heme copper center in myoglobin: Implications for the role of metal ions. *Journal of the American Chemical Society*, *128*, 6766–6767. PMC63152.

Zhao, X., Yeung, N., Wang, Z., Guo, Z., & Lu, Y. (2005). Effects of metal ions in the Cu_B center on the redox properties of heme in heme-copper oxidases: Spectroelectrochemical studies of an engineered heme-copper center in myoglobin. *Biochemistry*, *44*, 1210–1214.

CHAPTER TWENTY-THREE

Periplasmic Screening for Artificial Metalloenzymes

M. Jeschek*, S. Panke*, T.R. Ward[†,1]
*ETH Zurich, Basel, Switzerland
[†]University of Basel, Basel, Switzerland
[1]Corresponding author e-mail address: thomas.ward@unibas.ch

Contents

1. Introduction	540
2. Anchoring Strategies	541
3. Increasing the Throughput	543
4. Advantages of Periplasmic Screening	544
5. Periplasmic Screening for ArMs	546
5.1 Choice of Expression Strain and Translocation Strategy	546
5.2 Establishment of a Screening Workflow	549
6. Summary	553
References	554

Abstract

Artificial metalloenzymes represent an attractive means of combining state-of-the-art transition metal catalysis with the benefits of natural enzymes. Despite the tremendous recent progress in this field, current efforts toward the directed evolution of these hybrid biocatalysts mainly rely on the laborious, individual purification of protein variants rendering the throughput, and hence the outcome of these campaigns feeble. We have recently developed a screening platform for the directed evolution of artificial metalloenzymes based on the streptavidin–biotin technology in the periplasm of the Gram-negative bacterium *Escherichia coli*. This periplasmic compartmentalization strategy comprises a number of compelling advantages, in particular with respect to artificial metalloenzymes, which lead to a drastic increase in the throughput of screening campaigns and additionally are of unique value for future in vivo applications. Therefore, we highlight here the benefits of this strategy and intend to propose a generalized guideline for the development of novel transition metal-based biocatalysts by directed evolution in order to extend the natural enzymatic repertoire.

1. INTRODUCTION

Over the past decades, the field of biocatalysis has evolved from a discovery-based science to an application-driven engineering discipline. Accordingly, the biocatalytic scope has exceeded the mere harnessing of the natural enzymatic repertoire toward the directed evolution of biological entities ranging from single enzymes to entire organisms (Bornscheuer et al., 2012). However, these efforts have so far mainly concentrated on repurposing natural enzymes to achieve desired biotransformations (Renata, Wang, & Arnold, 2015).

In stark contrast to this "canonic" chemistry found in Nature, transition metal catalysis offers a wide range of reaction mechanisms that are absent from the enzymatic repertoire (Hyster & Ward, 2016), presumably due to the scarcity of such precious metals in the earth crust which has led to their exclusion from natural evolution. Only recently methods for the creation of artificial metalloenzymes (hereafter referred to as "ArMs") are being developed which allow to introduce this versatile reaction repertoire of precious metals into biocatalysis (Heinisch & Ward, 2010; Hyster & Ward, 2016). These recent endeavors are accompanied by mutual benefits for the two fields: homogenous catalysis will benefit from the mild reaction conditions as well as the evolvability of the second coordination sphere provided by the proteinaceous scaffold (Hyster & Ward, 2016). By contrast, biological systems could be complemented by a novel and bioorthogonal reaction toolbox facilitating new-to-nature reaction cascades and xenobiotic metabolism.

However, most efforts in tailoring and improving ArMs have so far relied on laborious purification procedures for the selected protein mutants (Reetz, Peyralans, Maichele, Fu, & Maywald, 2006; Sauer et al., 2015; Srivastava, Yang, Ellis-Guardiola, & Lewis, 2015). This severely limits the screening throughput and consequently leads to a rather unsatisfactory efficiency of these methods. Here, we summarize current basic strategies in the design and evolution of ArMs with a particular focus on the compatibility of the different methods with directed evolution strategies, which largely rely on the capability to screen large collections of mutants. More precisely, we focus on different strategies for anchoring the catalytically active metal within the protein scaffold which lead to profound differences in the compatibility of the different systems with high-throughput methodologies and in vivo screening. We introduce the periplasm of the bacterium *Escherichia coli* as a promising compartmentalization strategy that is of particular interest for ArMs for numerous reasons and zoom in on the underlying experimental design and procedures of this method in order to identify generalized guidelines for future projects.

2. ANCHORING STRATEGIES

In order to introduce and localize the respective metal cofactor into a desired protein scaffold and therefore endow it with novel catalytic properties, three different anchoring strategies have been pursued based on (i) dative binding of the metal by Lewis-basic amino acid residues, (ii) covalent cofactor attachment to reactive amino acids side chains, or (iii) high-affinity noncovalent cofactor binding (Hyster & Ward, 2016; Ilie & Reetz, 2015; Lewis, 2013; Yu et al., 2014). These different methods can be viewed as complementary to a certain degree and have all recently been assisted by computational design algorithms (Heinisch et al., 2015; Khare et al., 2012; Song & Tezcan, 2014; Zastrow, Peacock, Stuckey, & Pecoraro, 2012).

While all of these strategies have been shown to lead to the emergence of interesting novel enzymatic reactivities using purified protein samples, the methods differ significantly in the underlying coupling characteristics and the properties of the resulting complexes or conjugates. This has a profound impact on their suitability for directed evolution and their applicability in vivo (Table 1). Covalent anchoring, for instance, on the one hand provides a desirable strong and stable interaction between the protein and the metal-

Table 1 Key Characteristics of Different Metal Anchoring Strategies

Metal Anchoring	Dative	Covalent	Noncovalent
Coupling mechanism	Spontaneous	Reactive	Spontaneous
Coupling efficiency	High/quant.	Low intermediate	High/quant.
Off-target binding	Moderate	Likely	Unlikely
Bond stability	Intermediate	High	Variable
In vivo coupling	Possible	Challenging	Straightforward

containing ligand (cofactor) but on the other hand requires a coupling step with a specific amino acid residues (eg, cysteine) in order to anchor the catalyst (Hilvert & Kaiser, 1987; Philippart et al., 2013; Qi, Tann, Haring, & Distefano, 2001). This step is often prone to off-target reactivity (with other residues in the scaffold protein or even with other proteins) rendering application in multiprotein systems and in vivo challenging. Moreover, the coupling in many cases is not quantitative and thus requires multiple equivalents of cofactor vs scaffold protein to ensure (near) quantitative loading (Philippart et al., 2013).

On the contrary, dative noncovalent binding offers an attractive means to overcome these limitations since incorporation is spontaneous and reversible. However, creating metal-binding sites de novo is a challenging endeavor especially when rare and inert metals (eg, ruthenium, rhodium, iridium, etc.) are desired which lack exemplar binding sites in the natural enzymes' repertoire.

A promising noncovalent approach is the exploitation of the (strept-) avidin–biotin technology due to the very specific and highly stable binding of biotin to the protein ($K_D < 10^{-13}$ M) (Miyamoto & Kollman, 1993; Weber, Ohlendorf, Wendoloski, & Salemme, 1989). Due to its high affinity, this "molecular velcro" technology allows for the specific and quantitative placement of the biotinylated cofactors in a "quasi-covalent" manner even at subnanomolar concentrations as can be deduced from law of mass action (see Fig. 1). Additionally it benefits from the remarkable chemical- and physical stability of either streptavidin or avidin (Gonzalez, Argarana, & Fidelio, 1999; Hofmann, Wood, Brinton, Montibeller, & Finn, 1980). In the context of ArMs, this concept was first introduced by Wilson and Whitesides (1978) and has since then been applied in a number of compelling cases (Dürrenberger et al., 2011; Heinisch et al., 2015; Hyster, Knorr, Ward, & Rovis, 2012; Reetz et al., 2006). The use of biotinylated metal cofactors in this technology allows for the spontaneous assembly of the ArM simply by mixing the two components (ie, (strept-) avidin and metal cofactor) in solution. It is thus ideally suited for in vivo application where covalent coupling strategies are difficult to realize due to unwanted side reactivity. This is, in principle, also the case for other anchors (ie, inhibitors) which noncovalently bind to a desired protein scaffold of interest but the exceptionally high affinity between (strept-)avidin and biotin, which is the strongest noncovalent interaction known in Nature (Weber et al., 1989), sets this particular technology apart from alternative methods.

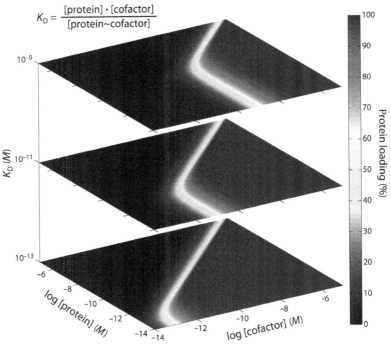

Fig. 1 Model for cofactor binding. Protein loading (ie, percentage of binding sites occupied by the cofactor) is derived from the law of mass action and displayed as a function of the concentration of the protein and the cofactor as well as the dissociation constant K_D. For high affinities (low K_D) between the cofactor and the protein as for instance for the streptavidin–biotin interaction ($K_D \sim 10^{-13}$ M) quantitative occupation (>99%) of the binding sites occurs even at subnanomolar concentrations for both binding partners. (See the color plate.)

3. INCREASING THE THROUGHPUT

Despite the rapid progress in the field of ArMs, efforts tackling the systematic directed evolution of the hybrid biocatalysts remain scarce (Ilie & Reetz, 2015; Song & Tezcan, 2014). The main reason for this can be traced back to the fact that the available methods mainly rely on purification of the host protein since many transition metal reactions are frequently inhibited by cellular components and are therefore not compatible with in vivo conditions (Wilson, Dürrenberger, Nogueira, & Ward, 2014). Due to this limitation, the directed evolution of ArMs was so far limited to few targeted amino acid exchanges since the throughput of the corresponding in vitro assays with purified mutants is low. On the contrary, screening of large mutant libraries may lead to a substantial improvement of the catalytic properties and is therefore highly desirable.

Song and Tezcan recently evolved an artificial metallo-β-lactamase on the basis of a structurally and functionally unrelated zinc enzyme that hydrolyses ampicillin with high efficiency (Song & Tezcan, 2014). This study is particularly noteworthy since directed evolution has been achieved via a periplasmic screening strategy in *E. coli* without the need to undergo laborious purification strategies and improved mutants could be retrieved simply by selection on a medium containing ampicillin. The periplasm has proven versatile for the recombinant expression of various proteins of interest (Choi & Lee, 2004; Skerra & Pluckthun, 1988; Thie, Schirrmann, Paschke, Dubel, & Hust, 2008; Voss & Skerra, 1997) and ensures a tight linkage between genotype and phenotype which is indispensable for the efficient screening of mutant libraries (Chen et al., 2001; Song & Tezcan, 2014; Sroga & Dordick, 2002). Building up on these studies, we recently developed an in vivo screening platform for ArMs based on the streptavidin–biotin technology exploiting the *E. coli* periplasm. Relying on a fluorescent model reaction, this system allowed for the screening of several thousand streptavidin mutants directed toward the evolution of an ArM capable of catalyzing an organometallic transformation absent from the natural enzymatic repertoire. Since this study emphasizes the high potential of the periplasmic environment as a favorable reaction compartment for transition metal-catalyzed reactions, we highlight the particular advantages of this strategy in the context of ArMs later.

4. ADVANTAGES OF PERIPLASMIC SCREENING

A major bottleneck for the evolution of ArMs has so far been the incompatibility of the catalysts with reaction conditions in the cytoplasm in vivo or even in crudely purified cell extracts. As a result, it is required to purify protein mutants and subsequently examine those for their catalytic properties. Even though in many cases the exact mechanisms of inhibition in more complex, cell-like environments remain elusive, several potential reasons for catalyst inactivation come to mind including: (i) undesired interaction of the metal-complexes with reactive amino acid residues or metabolites as well as (ii) metal complexation for instance by nucleic acids to name but a few. In this context, a related study had revealed that glutathione (GSH), the most abundant low-molecular thiol in living cells (Meister & Anderson, 1983), is a major source of inhibition when performing transition metal catalysis in cell-free extracts. This limitation can be overcome by the addition of suitable neutralizing agents (Wilson et al., 2014). Based on these observations, we hypothesized that the periplasm might offer a propitious reaction environment for ArMs since it contains GSH in much lower quantities than the cytoplasm and mainly in the oxidized, noninhibiting GSSG form (Meister & Anderson, 1983; Smirnova & Oktyabrsky, 2005). Additionally, the lower background of potentially disturbing proteins, both in terms of absolute protein content and number of different proteins (Han et al., 2011; Pooley, Merchante, & Karamata, 1996), as well as the absence of nucleic acids also speaks in favor of the periplasmic space as opposed to the cytoplasm where the bulk of the host cell's natural metabolic household is executed (Fig. 2).

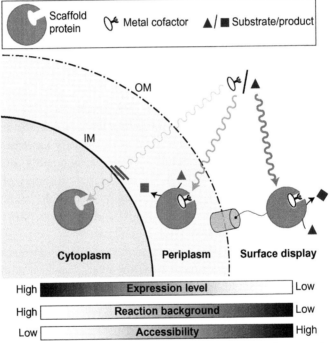

Fig. 2 Comparison of different compartmentalization strategies for screening of ArM variants in vivo. Three basic strategies for the production of the scaffold protein in E. coli are displayed: the conventional cytoplasmic expression, secretion across the inner membrane (IM) into the periplasmic space and surface display on the outer membrane (OM). In the context of ArMs, these strategies diverge significantly with respect to their suitability for high-throughput screening and in vivo applicability. The most important characteristics are differences in achievable expression levels, the reaction background stemming from host enzymes and metabolites as well as the accessibility of the scaffold protein for externally added agents as for instance the metal cofactor as well as the designated target substrate for the reaction of the ArM.

Notably, in the context of ArMs, additional considerations regarding the accessibility of the scaffold protein come into play. In order to assemble the holoenzyme, the desired abiotic cofactor has to gain access to its proteinaceous counterpart and cellular cofactor uptake can become a limiting step in the screening even if the considerations about inhibitory factors discussed earlier did not apply. In the case of cytoplasmic expression of the scaffold protein, for instance, this translates to the mandatory passage of two distinct cellular membranes including the more selective inner membrane, which usually requires specific uptake mechanisms that are not available for artificial cofactors. On the contrary due to the presence of large unselective pores (eg, porins) (Delcour, 2009), the outer membrane is more promiscuous in its uptake characteristics and consequently the periplasm might be a "more accessible" choice in many cases. A similar argument may be made for the uptake of designated substrates for the ArM reactions which, in many cases, represent nonnatural substrates and consequently lack specific uptake mechanisms that allow transit through the inner membrane.

The argument of low reaction background and good accessibility likewise applies for different surface display methods like autodisplay by autotransporters (Maurer, Jose, & Meyer, 1997; Park et al., 2011) or equivalent techniques (Fig. 2). However, the achievable cell-specific amount of recombinant protein is comparably low due to capacity limitations of the outer membrane (Bannwarth & Schulz, 2003). Consequently, only highly active catalysts that produce enough (ie, detectable) amounts of product can be analyzed. Additionally, the requirement to fuse the target protein to a vehicle protein that dictates the export and insertion into the outer membrane may not be a suitable choice in all cases.

Hence, exporting the target protein to the periplasm represents a valuable trade-off between respectable expression levels, low background reactivity, and facile accessibility for cofactors and/or substrates in comparison to other localization strategies. We therefore portray this method in a detailed and application-oriented manner later. At the same time, we would like to point out that we only intend to give a general guideline without claiming universality of the proposed methods.

5. PERIPLASMIC SCREENING FOR ArMs

5.1 Choice of Expression Strain and Translocation Strategy

In principle, most common laboratory strains of *E. coli* can be used for periplasmic expression proteins and the strains JM83, KS272, HM125, W3110,

BL21, and TOP10 have been reported to deliver good results (Choi & Lee, 2004). However, we have observed significant differences in the apparent export capacity of different strains and overall strain fitness and growth behavior was severely compromised in some cases. Moreover, the used strains diverge strongly in their efficiency of transformation and the ease of genome modification which should be considered if the generation of large mutant libraries for screening or the editing of the genome is intended. In our case, this led to the selection of TOP10 over JM83 at the cost of a lower expression level in the periplasm for the benefit of ease of genetic modification and efficiency of transformation. Nevertheless, the strain selection remains a project-specific choice and an evaluation of a range of candidate strains is advisable.

In order to export the target protein to the periplasmic space, different pathways are available in *E. coli* that either secrete the protein in a fully or partially folded state (twin-arginine translocation pathway; TAT) or in an unfolded manner with subsequent folding in the periplasm (Sec translocon) (Natale, Bruser, & Driessen, 2008). Depending on the protein of interest, the selection of the export mechanism is an important consideration. Unfolded export (Sec) is the method of choice in cases where the protein of interest contains structural disulfide bonds (Georgiou & Segatori, 2005) or is toxic in the cytosol, as for instance reported for streptavidin, which depletes the essential intracellular biotin pool (Wetzel, Muller, Flaschel, Friehs, & Risse, 2016). On the contrary, some proteins like beta-galactosidase or the green fluorescent protein do not readily fold in the periplasm and are therefore rendered inactive upon Sec translocation. In this case, activity can be restored using TAT export pathways or specifically engineered strains with increased chaperone activity in the periplasm (Santini et al., 2001; Schlapschy, Grimm, & Skerra, 2006).

The designated export pathway is charted by the selection of the N-terminal fusion peptide that directs the cargo into the periplasmic space and in many cases is cleaved off after transport. This cleavage is a very important benefit of periplasmic compartmentalization since it leaves the protein of interest in the desired compartment without any remnant of the export signal (authentic N-terminus). Several homo- and heterologous leader peptides have been successfully employed in *E. coli* both for Sec (eg, PelB, OmpA, StII, PhoA, LamB, etc.) and TAT mediated (eg, TorA, Tap, etc.) (Choi & Lee, 2004) translocation and signal peptides can be identified as such using prediction algorithms for Sec (Petersen, Brunak, von Heijne, & Nielsen, 2011) or TAT (Bendtsen, Nielsen, Widdick, Palmer, & Brunak, 2005). Detailed descriptions and application examples of frequently used signal peptides can be found elsewhere (Choi & Lee, 2004; Schlapschy & Skerra, 2011) and are therefore not detailed herein. We have generally had good experiences with OmpA and PelB for the export of streptavidin and other recombinant proteins. Ideally however, the suitable leader peptide should be identified for each protein of interest individually.

Once a suitable set of candidate signal peptides has been identified and fused to the cargo protein in the designated plasmid vector, it is important to validate whether the fusion is expressed properly and more importantly if the cargo is efficiently exported into the periplasmic space. Whereas in most cases, the sole expression of the protein can be readily determined by appropriate methods (eg, SDS-PAGE, Western blot, activity assay, etc.), verification of the correct compartmentalization needs more carful assessment. We generally use commercially available reagents (eg, Peripreps Periplasting Kit, Madison, USA) that selectively lyse the outer membrane and can therefore be used for fractionation of cellular protein into a periplasmic and a cytoplasmic fraction. Other protocols likewise deliver satisfying results (Schlapschy & Skerra, 2011). These fractions can then be analyzed separately for the presence of the protein of interest. However, the purity of the fractions should be analyzed in order to ensure proper separation. This can for instance be done by detection of cyto- and/or periplasmic marker proteins in both fractions, either based on mere presence of the marker or its activity. We have successfully used beta-galactosidase activity to determine impurities of fractions, but other proteins for the cytoplasm (eg, ribosomal proteins and GroEL) or periplasm (eg, alkaline phosphatase PhoA and beta-lactamase) may serve equally well as markers for the respective compartment. Given a correct localization of the cargo protein in the periplasm, it is advisable to examine whether the cleavage of the leader peptide upon export is successful and quantitative in cases where an authentic N-terminus is required for activity of the protein. An adequate method to confirm the latter is mass spectrometry but SDS-PAGE of the periplasmic protein fraction in comparison to a purified sample of the untagged target protein will suffice in most cases since the added molecular weight due to the leader peptide (usually more than 2 kDa) allows for in-gel discrimination between cleaved and uncleaved forms.

These aforementioned quality controls are to be understood as general recommendations for good practice and are to be adapted for the specific cases. Once the correct synthesis and localization of the protein is confirmed the screening workflow can be established which we try to elucidate in a generalized manner later.

5.2 Establishment of a Screening Workflow

This section is aimed at providing general guidelines on how to set up a valid screening workflow for the evolution of ArMs. We intend to give recommendations for library generation and screening format, the applied cultivation conditions, improvement of cofactor uptake, and the final activity assay as well as the retrieval of beneficial mutants (Fig. 3). While we generally had very good experience with the following practices, it should be emphasized that each of these steps needs careful consideration for individual screening campaigns and the practices are hence to be understood only as inspirational suggestions without imperative character.

5.2.1 Library Generation

In order to generate genetic diversity in the scaffold protein different in vitro mutagenesis strategies can be pursued. These include but are not limited to (i) random mutation of the entire gene of interest or specific parts thereof (eg, error-prone PCR), (ii) randomization of individual or multiple residues (eg, site-specific saturation mutagenesis), and (iii) recombination between different mutants of interest (eg, DNA shuffling), all potentially combined with computational design algorithms to improve library quality. A detailed description of the available methods was recently reviewed (Packer & Liu, 2015) and shall not be discussed in detail here. Generally speaking, the method should be selected in accordance to the screening capacity of the subsequent assay and ideally be based on structural information and/or prior mutagenesis studies if available. For instance, full randomization of the target protein is a valuable method if large numbers of mutants ($>100,000$) can be screened (eg, by flow cytometry) and may even lead to the identification of "counter-intuitive" residues spread across the entire protein. By contrast, readouts with comparably low throughput (eg, UPLC) require more focused libraries in order to achieve a sufficient coverage of the mutational space to be able to retrieve improved variants. In the latter case, saturation mutagenesis of residues close to the active site provides a useful method for diversification and beneficial mutations can be combined in order to gradually improve the desired properties by appropriate heuristics like iterative saturation mutagenesis (ISM) (Reetz & Carballeira, 2007).

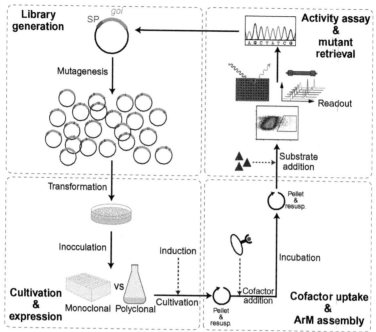

Fig. 3 Generalized workflow of the periplasmic screening for artificial metalloenzymes. Starting from an expression plasmid containing the *gene of interest* (*goi*) fused to the sequence of a signal peptide for periplasmic export (SP), a mutant library is created and transformed into the screening strain. Clones are cultivated, either in a monoclonal or in a polyclonal way, and expression is induced and continued for a sufficient time before harvesting by centrifugation. Cells are resuspended in an appropriate buffer containing the cofactor and incubated in order to allow uptake of the cofactor and concomitant assembly of the ArM in the periplasm. Afterwards, the buffer containing excess cofactor is removed and cells are resuspended in the desired reaction buffer containing the target substrate. Reactions can be tracked by an appropriate readout method, either directly in the culture (eg, formation of a fluorescent product) or after workup (eg, solvent extraction and subsequent (U)HPLC analysis). Finally, improved variants are isolated and subjected to sequencing to identify relevant mutations and/or further screening rounds.

Another important consideration is the question whether to include the signal sequence for periplasmic export into the diversification. In most cases, it is advisable to exclude this part from mutagenesis and thereby ensure that mutants with impaired or improved export behavior do not degrade library quality or emerge as false positives. However, if high export efficiency as such is a desired trait and should concomitantly be evolved, the signal peptide can be included for the respective campaign.

5.2.2 Cultivation and Expression

After in vitro library generation, the diversified plasmid pool is used to transform the respective screening strain for instance by electroporation and clones are recovered on selective agar plates from where cultivation can be subsequently started. A suitable cultivation format has to be selected in accordance to the final activity assay and cultures can be maintained in either a monoclonal (eg, multiwell plates, droplets, etc.) or a polyclonal (ie, growth of all library members in bulk) manner. While the latter is a convenient way to keep the experimental effort at a minimum, it requires a tight genotype–phenotype linkage without cross-talk between single mutants as well as a functional single-cell readout to be able to retrieve improved mutants after all. For ArMs, however, the analyzed trait will in most cases be the production of a low-molecular compound that readily diffuses across cell membranes and between cells. Therefore, library members have to be cultivated individually (ie, monoclonal) and analyzed separately in order to reliably identify variants with improved traits.

The growth medium, either complex or chemically defined, can be selected mainly according to the requirements of the respective screening strain. However, if possible, media should be kept free of components that might interfere with the reaction of the ArM. Therefore, chemically defined media that allow for the specific exchange of inhibitory components may be the superior choice in some cases. Likewise, cultivation conditions (temperature, pH, etc.) should be in agreement with the strains requirements.

To initiate expression and export of the scaffold protein the respective inducer is added to the culture. We generally use IPTG/lactose-inducible expression systems which offer the possibility to choose between concomitant induction of all variants by addition of the inducer at a desired time point or depending on cell density using autoinduction media (Studier, 2005). Other inducible systems as well as constitutive expression should work equally well and may be chosen depending on the specific requirements and/or availability. However, it is critical to optimize the respective expression conditions in order to avoid overloading of the cellular export machinery which can have detrimental effects on the amount of active protein produced in the periplasmic space and overall strain fitness.

Lowering of inducer concentrations as well as expression at lower temperatures can represent helpful measures to improve strain behavior.

5.2.3 Cofactor Uptake and ArM Assembly

After the induction period, cells containing the scaffold protein variants in the periplasm can be harvested by centrifugation and resuspended in a buffer solution containing the artificial cofactor. Ideally, the buffer should be selected according to the cells' requirements (ie, pH ~7.0, physiological salt concentration, etc.) and should not contain compounds interfering with the metal complex. If specific medium components exhibit inhibitory effects on the reaction, washing steps in buffer lacking the cofactor can be applied to remove all medium components prior to incubation in presence of the cofactor. During the incubation period, the cofactor should be able to pass the outer membrane and bind to the scaffold protein. This assembly of the ArM largely depends on the uptake kinetics of the cofactor which can be analyzed by methods that quantify the rare metal content of the cells (eg, ICP-OES, inductively coupled plasma optical emission spectrometry). We have identified a range of critical parameters for cofactor uptake that need to be optimized prior to screening including cofactor concentration, incubation temperature and duration, as well as buffer composition. For a specific case we found that incubation on ice for 30 min is optimal for the uptake of a biotinylated ruthenium cofactor and that addition of high salt concentrations (eg, 9% NaCl) can significantly improve uptake. The latter has previously been reported to improve the uptake of bulky compounds into the periplasm without compromising cellular integrity significantly (Chen et al., 2001).

After incubation, the cells are pelleted and the buffer containing excess cofactor is removed. In order to ensure full removal of unbound cofactor, washing steps in plain buffer can be applied prior to the subsequent activity assessment of mutants.

5.2.4 Activity Assay and Mutant Retrieval

To evaluate variants for their reactivity, cells are resuspended in an appropriate reaction buffer containing the substrate to be transformed by the ArM reaction. The reaction is allowed to proceed for an appropriate amount of time and under the desired conditions and can be analyzed either in a kinetic manner (eg, for reactions with an optical readout) or using end-point analysis. Wherever possible, references should be included in each screening round that enable reliable comparison of variants' performance throughout longer campaigns of directed evolution compensating for parameters that inevitably vary over time. For instance, we by default add multiple replicates of a "mock control" (empty vector) as well as a positive reference (parent variant prior to mutagenesis) in every plate throughout 96-well screening campaigns which allows to compensate for variability (plate-to-plate, batch-to-batch, day-to-day, etc.). Moreover, including the parent variant as a control is particularly advisable since the applied workflow will be frequently adjusted and improved based on new findings throughout the course of a screening campaign and therefore a proper normalization of results is essential.

Improved variants can be isolated and further analyzed prior to sequencing. Since the proposed workflow is based on an in vivo assay, which is subject to biological variability, it is recommended to include a verification round in order to confirm improved performance of the respective mutants. This can either be done by repeating the described workflow for promising candidates in multiple replicates or, if feasible, by purifying the protein variants and performing a detailed in vitro analysis.

Finally mutations can be analyzed, potentially in combination with structural information of the scaffold protein, and subsequently combined by appropriate methods (eg, ISM) in order to achieve further improvement. Moreover, several rounds of directed evolution can be conducted in an iterative manner to improve overall performance.

6. SUMMARY

In this work, we summarized important considerations for the directed evolution of ArMs and introduced a periplasmic screening platform designed to increase the throughput of the respective campaigns. From our point of view, periplasmic compartmentalization offers a number of compelling advantages including adequate reaction conditions and facile accessibility for cofactor and substrates. The proposed blueprint workflow elucidates crucial aspects to be considered before and during screening, including the choice of export mechanism and strain, library generation, growth conditions, cofactor uptake, and final assessment of variants for improved traits. With this we hope to provide a useful guideline for future projects in the field as well as for the scientific community.

REFERENCES

Bannwarth, M., & Schulz, G. E. (2003). The expression of outer membrane proteins for crystallization. *Biochimica et Biophysica Acta—Biomembranes, 1610*(1), 37–45. http://dx.doi.org/10.1016/S0005-2736(02)00711-3.

Bendtsen, J. D., Nielsen, H., Widdick, D., Palmer, T., & Brunak, S. (2005). Prediction of twin-arginine signal peptides. *BMC Bioinformatics, 6*, 167. http://dx.doi.org/10.1186/1471-2105-6-167.

Bornscheuer, U. T., Huisman, G. W., Kazlauskas, R. J., Lutz, S., Moore, J. C., & Robins, K. (2012). Engineering the third wave of biocatalysis. *Nature, 485*(7397), 185–194. http://dx.doi.org/10.1038/nature11117.

Chen, G., Hayhurst, A., Thomas, J. G., Harvey, B. R., Iverson, B. L., & Georgiou, G. (2001). Isolation of high-affinity ligand-binding proteins by periplasmic expression with cytometric screening (PECS). *Nature Biotechnology, 19*(6), 537–542. http://dx.doi.org/10.1038/89281.

Choi, J. H., & Lee, S. Y. (2004). Secretory and extracellular production of recombinant proteins using Escherichia coli. *Applied Microbiology and Biotechnology, 64*(5), 625–635. http://dx.doi.org/10.1007/s00253-004-1559-9.

Delcour, A. H. (2009). Outer membrane permeability and antibiotic resistance. *Biochimica et Biophysica Acta—Proteins and Proteomics, 1794*(5), 808–816. http://dx.doi.org/10.1016/j.bbapap.2008.11.005.

Dürrenberger, M., Heinisch, T., Wilson, Y. M., Rossel, T., Nogueira, E., Knorr, L., ... Ward, T. R. (2011). Artificial transfer hydrogenases for the enantioselective reduction of cyclic imines. *Angewandte Chemie (International ed. in English), 50*(13), 3026–3029. http://dx.doi.org/10.1002/anie.201007820.

Georgiou, G., & Segatori, L. (2005). Preparative expression of secreted proteins in bacteria: Status report and future prospects. *Current Opinion in Biotechnology, 16*(5), 538–545. http://dx.doi.org/10.1016/j.copbio.2005.07.008.

Gonzalez, M., Argarana, C. E., & Fidelio, G. D. (1999). Extremely high thermal stability of streptavidin and avidin upon biotin binding. *Biomolecular Engineering, 16*(1-4), 67–72. http://dx.doi.org/10.1016/S1050-3862(99)00041-8.

Han, M. J., Yun, H., Lee, J. W., Lee, Y. H., Lee, S. Y., Yoo, J. S., ... Hur, C. G. (2011). Genome-wide identification of the subcellular localization of the *Escherichia coli* B proteome using experimental and computational methods. *Proteomics, 11*(7), 1213–1227. http://dx.doi.org/10.1002/pmic.201000191.

Heinisch, T., Pellizzoni, M., Dürrenberger, M., Tinberg, C. E., Kohler, V., Klehr, J., ... Wardt, T. R. (2015). Improving the catalytic performance of an artificial metalloenzyme by computational design. *Journal of the American Chemical Society, 137*(32), 10414–10419. http://dx.doi.org/10.1021/jacs.5b06622.

Heinisch, T., & Ward, T. R. (2010). Design strategies for the creation of artificial metalloenzymes. *Current Opinion in Chemical Biology, 14*(2), 184–199. http://dx.doi.org/10.1016/j.cbpa.2009.11.026.

Hilvert, D., & Kaiser, E. T. (1987). Semisynthetic enzymes—Design of flavin-dependent oxidoreductases. *Biotechnology & Genetic Engineering Reviews, 5*, 297–318.

Hofmann, K., Wood, S. W., Brinton, C. C., Montibeller, J. A., & Finn, F. M. (1980). Iminobiotin affinity columns and their application to retrieval of streptavidin. *Proceedings of the National Academy of Sciences of the United States of America—Biological Sciences, 77*(8), 4666–4668. http://dx.doi.org/10.1073/pnas.77.8.4666.

Hyster, T. K., Knorr, L., Ward, T. R., & Rovis, T. (2012). Biotinylated Rh(III) complexes in engineered streptavidin for accelerated asymmetric C-H activation. *Science, 338*(6106), 500–503. http://dx.doi.org/10.1126/science.1226132.

Hyster, T. K., & Ward, T. R. (2016). Genetic optimization of metalloenzymes: Enhancing enzymes for non-natural reactions. *Angewandte Chemie (International ed in English), 55*, 2–16. http://dx.doi.org/10.1002/anie.201508816.

Ilie, A., & Reetz, M. T. (2015). Directed evolution of artificial metalloenzymes. *Israel Journal of Chemistry*, *55*(1), 51–60. http://dx.doi.org/10.1002/ijch.201400087.

Khare, S. D., Kipnis, Y., Greisen, P. J., Takeuchi, R., Ashani, Y., Goldsmith, M., ... Baker, D. (2012). Computational redesign of a mononuclear zinc metalloenzyme for organophosphate hydrolysis. *Nature Chemical Biology*, *8*(3), 294–300.

Lewis, J. C. (2013). Artificial metalloenzymes and metallopeptide catalysts for organic synthesis. *ACS Catalysis*, *3*(12), 2954–2975. http://dx.doi.org/10.1021/cs400806a.

Maurer, J., Jose, J., & Meyer, T. F. (1997). Autodisplay: One-component system for efficient surface display and release of soluble recombinant proteins from Escherichia coli. *Journal of Bacteriology*, *179*(3), 794–804.

Meister, A., & Anderson, M. E. (1983). Glutathione. *Annual Review of Biochemistry*, *52*, 711–760. http://dx.doi.org/10.1146/annurev.bi.52.070183.003431.

Miyamoto, S., & Kollman, P. A. (1993). Absolute and relative binding free-energy calculations of the interaction of biotin and its analogs with streptavidin using molecular-dynamics free-energy perturbation approaches. *Proteins: Structure, Function, and Genetics*, *16*(3), 226–245. http://dx.doi.org/10.1002/prot.340160303.

Natale, P., Bruser, T., & Driessen, A. J. M. (2008). Sec- and Tat-mediated protein secretion across the bacterial cytoplasmic membrane—Distinct translocases and mechanisms. *Biochimica et Biophysica Acta—Biomembranes*, *1778*(9), 1735–1756. http://dx.doi.org/10.1016/j.bbamem.2007.07.015.

Packer, M. S., & Liu, D. R. (2015). Methods for the directed evolution of proteins. *Nature Reviews. Genetics*, *16*(7), 379–394. http://dx.doi.org/10.1038/nrg3927.

Park, M., Jose, J., Thommes, S., Kim, J. I., Kang, M. J., & Pyun, J. C. (2011). Autodisplay of streptavidin. *Enzyme and Microbial Technology*, *48*(4–5), 307–311. http://dx.doi.org/10.1016/j.enzmictec.2010.12.006.

Petersen, T. N., Brunak, S., von Heijne, G., & Nielsen, H. (2011). SignalP 4.0: Discriminating signal peptides from transmembrane regions. *Nature Methods*, *8*(10), 785–786. http://dx.doi.org/10.1038/nmeth.1701.

Philippart, F., Arlt, M., Gotzen, S., Tenne, S. J., Bocola, M., Chen, H. H., ... Okuda, J. (2013). A hybrid ring-opening metathesis polymerization catalyst based on an engineered variant of the beta-barrel protein FhuA. *Chemistry—A European Journal*, *19*(41), 13865–13871.

Pooley, H. M., Merchante, R., & Karamata, D. (1996). Overall protein content and induced enzyme components of the periplasm of Bacillus subtilis. *Microbial Drug Resistance—Mechanisms, Epidemiology and Disease*, *2*(1), 9–15. http://dx.doi.org/10.1089/mdr.1996.2.9.

Qi, D. F., Tann, C. M., Haring, D., & Distefano, M. D. (2001). Generation of new enzymes via covalent modification of existing proteins. *Chemical Reviews*, *101*(10), 3081–3111. http://dx.doi.org/10.1021/cr000059o.

Reetz, M. T., & Carballeira, J. D. (2007). Iterative saturation mutagenesis (ISM) for rapid directed evolution of functional enzymes. *Nature Protocols*, *2*(4), 891–903. http://dx.doi.org/10.1038/nprot.2007.72.

Reetz, M. T., Peyralans, J. J. P., Maichele, A., Fu, Y., & Maywald, M. (2006). Directed evolution of hybrid enzymes: Evolving enantioselectivity of an achiral Rh-complex anchored to a protein. *Chemical Communications*, *41*, 4318–4320.

Renata, H., Wang, Z. J., & Arnold, F. H. (2015). Expanding the enzyme universe: Accessing non-natural reactions by mechanism-guided directed evolution. *Angewandte Chemie (International ed. in English)*, *54*(11), 3351–3367. http://dx.doi.org/10.1002/anie.201409470.

Santini, C. L., Bernadac, A., Zhang, M., Chanal, A., Ize, B., Blanco, C., & Wu, L. F. (2001). Translocation of jellyfish green fluorescent protein via the Tat system of Escherichia coli and change of its periplasmic localization in response to osmotic up-shock. *Journal of Biological Chemistry*, *276*(11), 8159–8164. http://dx.doi.org/10.1074/jbc.C000833200.

Sauer, D. F., Himiyama, T., Tachikawa, K., Fukumoto, K., Onoda, A., Mizohata, E., ... Okuda, J. (2015). A highly active biohybrid catalyst for olefin metathesis in water: Impact of a hydrophobic cavity in a β-barrel protein. *ACS Catalysis*, 7519–7522. http://dx.doi.org/10.1021/acscatal.5b01792.

Schlapschy, M., Grimm, S., & Skerra, A. (2006). A system for concomitant overexpression of four periplasmic folding catalysts to improve secretory protein production in Escherichia coli. *Protein Engineering, Design and Selection*, 19(8), 385–390. http://dx.doi.org/10.1093/protein/gzl018.

Schlapschy, M., & Skerra, A. (2011). Periplasmic chaperones used to enhance functional secretion of proteins in E. coli. *Methods in Molecular Biology*, 705, 211–224. http://dx.doi.org/10.1007/978-1-61737-967-3_12.

Skerra, A., & Pluckthun, A. (1988). Assembly of a functional immunoglobulin-Fv fragment in Escherichia coli. *Science*, 240(4855), 1038–1041. http://dx.doi.org/10.1126/science.3285470.

Smirnova, G. V., & Oktyabrsky, O. N. (2005). Glutathione in bacteria. *Biochemistry (Moscow)*, 70(11), 1199–1211. http://dx.doi.org/10.1007/s10541-005-0248-3.

Song, W. J., & Tezcan, F. A. (2014). A designed supramolecular protein assembly with in vivo enzymatic activity. *Science*, 346(6216), 1525–1528.

Srivastava, P., Yang, H., Ellis-Guardiola, K., & Lewis, J. C. (2015). Engineering a dirhodium artificial metalloenzyme for selective olefin cyclopropanation. *Nature Communications*, 6, 7789.

Sroga, G. E., & Dordick, J. S. (2002). A strategy for in vivo screening of subtilisin E reaction specificity in E. coli periplasm. *Biotechnology and Bioengineering*, 78(7), 761–769. http://dx.doi.org/10.1002/bit.10269.

Studier, F. W. (2005). Protein production by auto-induction in high-density shaking cultures. *Protein Expression and Purification*, 41(1), 207–234. http://dx.doi.org/10.1016/j.pep.2005.01.016.

Thie, H., Schirrmann, T., Paschke, M., Dubel, S., & Hust, M. (2008). SRP and Sec pathway leader peptides for antibody phage display and antibody fragment production in E. coli. *New Biotechnology*, 25(1), 49–54. http://dx.doi.org/10.1016/j.nbt.2008.01.001.

Voss, S., & Skerra, A. (1997). Mutagenesis of a flexible loop in streptavidin leads to higher affinity for the Strep-tag II peptide and improved performance in recombinant protein purification. *Protein Engineering*, 10(8), 975–982.

Weber, P. C., Ohlendorf, D. H., Wendoloski, J. J., & Salemme, F. R. (1989). Structural origins of high-affinity biotin binding to streptavidin. *Science*, 243(4887), 85–88. http://dx.doi.org/10.1126/science.2911722.

Wetzel, D., Muller, J. M., Flaschel, E., Friehs, K., & Risse, J. M. (2016). Fed-batch production and secretion of streptavidin by Hansenula polymorpha: Evaluation of genetic factors and bioprocess development. *Journal of Biotechnology*, 225, 3–9. http://dx.doi.org/10.1016/j.jbiotec.2016.03.017.

Wilson, Y. M., Dürrenberger, M., Nogueira, E. S., & Ward, T. R. (2014). Neutralizing the detrimental effect of glutathione on precious metal catalysts. *Journal of the American Chemical Society*, 136(25), 8928–8932. http://dx.doi.org/10.1021/ja500613n.

Wilson, M. E., & Whitesides, G. M. (1978). Conversion of a protein to a homogeneous asymmetric hydrogenation catalyst by site-specific modification with a diphosphinerhodium(I) moiety. *Journal of the American Chemical Society*, 100(1), 306–307.

Yu, F. T., Cangelosi, V. M., Zastrow, M. L., Tegoni, M., Plegaria, J. S., Tebo, A. G., ... Pecoraro, V. L. (2014). Protein design: Toward functional metalloenzymes. *Chemical Reviews*, 114(7), 3495–3578. http://dx.doi.org/10.1021/cr400458x.

Zastrow, M. L., Peacock, A. F. A., Stuckey, J. A., & Pecoraro, V. L. (2012). Hydrolytic catalysis and structural stabilization in a designed metalloprotein. *Nature Chemistry*, 4(2), 118–123. http://dx.doi.org/10.1038/Nchem.1201.

CHAPTER TWENTY-FOUR

De Novo Designed Imaging Agents Based on Lanthanide Peptides Complexes

A.F.A. Peacock[1]

School of Chemistry, University of Birmingham, Edgbaston, United Kingdom
[1]Corresponding author: e-mail address: a.f.a.peacock@bham.ac.uk

Contents

1. Introduction	558
2. Lanthanide Coordination Chemistry	558
3. Lanthanide Imaging Agents	558
4. Lanthanide Peptides and Proteins	560
4.1 Inspired by Native Ca(II)-Binding Loops	560
4.2 Other Lanthanide-Binding Sites	562
5. Strategies for Designing CCs for Lanthanide Coordination	562
5.1 CC Design	562
5.2 Lanthanide-Binding Site Design	565
5.3 Controlling Water Coordination and Access	568
5.4 Lanthanide Selectivity	571
5.5 Controlling Stability	571
5.6 MRI Efficiency of Gadolinium CCs	574
6. Conclusions and Perspectives	576
Acknowledgments	576
References	577

Abstract

Herein are discussed a selection of lanthanide peptide/protein complexes in view of their potential applications as imaging agents, both in terms of luminescence detection and magnetic resonance imaging. Though this chapter covers a range of different peptides and protein, if focuses specifically on the opportunities afforded by the de novo design of coiled coils, miniature protein scaffolds, and the development on lanthanide-binding sites into these architectures. The requirements for lanthanide coordination and the challenges that need to be addressed when preparing lanthanide peptides with a view to their potential adoption as clinical imaging applications, will be highlighted.

1. INTRODUCTION

The opportunities afforded by progress in the characterization of biomolecules over the last decade, has had a huge impact on the design of new, artificial, and miniaturized proteins. With the advent of *Synthetic Biology*, there is much interest in the development of functional systems as a result of interfacing biology with disciplines such as engineering, chemistry, and physics. Herein we will discuss the opportunities afforded by combining protein design and inorganic chemistry, in an effort to develop functional novel metalloproteins. Specifically, this chapter will address the use of lanthanide metal ions, with no known biological role (apart from one example of a lanthanide dependent enzyme; Nakagawa et al., 2012), but with attractive photophysical and magnetic properties, that make their use highly attractive for applications in luminescence and magnetic resonance imaging (MRI). This chapter will not be an exhaustive review of lanthanide peptides and proteins, but will instead focus specifically on the de novo design of miniature artificial proteins scaffolds based on the coiled coil (CC) motif, and lanthanide complexes thereof, with a view to their potential use as novel imaging agents.

2. LANTHANIDE COORDINATION CHEMISTRY

The lanthanide ions exist overwhelmingly in the +3 oxidations state and are hard lewis acids, with their coordination chemistry dominated by multidentate ligands which feature hard donor atoms such as oxygen. Resulting complexes are held together primarily by electrostatic interactions, with ligand sterics determining coordination geometries. Lanthanide complexes tend to be 9- or 8-coordinate, depending on the size of the lanthanide ion, which gradually decreases as you go across the period due to the lanthanide contraction. Much of the chemistry of the lanthanides is due to the 4f-orbitals being effectively shielded by electrons in the 5s and 5p orbitals, one consequence of this is that electrons located in the 4f-orbital are rather insensitive to the ligands.

3. LANTHANIDE IMAGING AGENTS

The majority of the lanthanides display attractive photophysical properties including large Stokes' shifts, narrow emission lines, and long emission life-times, the latter due to these transitions being Laporte forbidden. Therefore, lanthanide complexes are often employed as luminescence imaging agents, which can span the visible, near-infrared, and infrared regions, and which can take advantage of time resolved spectroscopy. However, as the f–f transitions responsible for luminescence are Laporte forbidden, the lanthanides are weak absorbers and instead an organic chromophore is required to sensitize lanthanide luminescence via the antenna effect. However, sensitization efficiency drops off rapidly with increasing distance between the chromophore and the lanthanide ion.

Of relevance to this chapter are the antennas that have been adopted in lanthanide peptides and proteins, with the native amino acid Trp, and to a lesser extent Tyr, commonly exploited as natural sensitizers. However, much effort has been invested in the use of nonnatural amino acids, in which the side chain instead features an organic chromophore capable of lanthanide sensitization. This not only allows the excitation wavelength to be shifted to lower energy, but also allows one to explore the luminescence of a wider range of lanthanide ions, as Trp is largely limited to the efficient sensitization of Tb(III) and Eu(III). Examples of chromophores that have been introduced as nonnatural amino acids for lanthanide sensitization include carbostyril 124, acridone (Reynolds, Sculimbrene, & Imperiali, 2008), and naphthalimide (Bonnet, Devocelle, & Gunnlaugsson, 2008) (Fig. 1).

Paramagnetic complexes are routinely used to enhance the contrast between diseased and normal tissue in MRI, by (normally) shortening the longitudinal and transverse relaxation time of bulk water. Gd(III) complexes are often employed as MRI contrast agents, due to the seven unpaired f-electrons, coupled with its long electronic relaxation time. Relaxivity or contrast agent efficiency, is dependent on the number of water molecules coordinated to the Gd(III); the Gd(III)–water distance (ie, inner vs outer sphere); the rate of exchange of these waters with bulk water; and the rotational correlation time of the complex, ie, its tumbling rate in solution (Caravan, 2006). The latter is dependent on complex size, with gadolinium proteins having a possible advantage over small molecule complexes in this regard.

A key distinction between the coordination chemistry requirements of lanthanides for applications in luminescence and MRI, is related to the extent of inner sphere water directly coordinated to the lanthanide. Though important for efficient MRI contrast agents, the O–H vibration quenches lanthanide luminescence, and so lanthanide sites are often designed for applications in either luminescence or MRI, but generally not both.

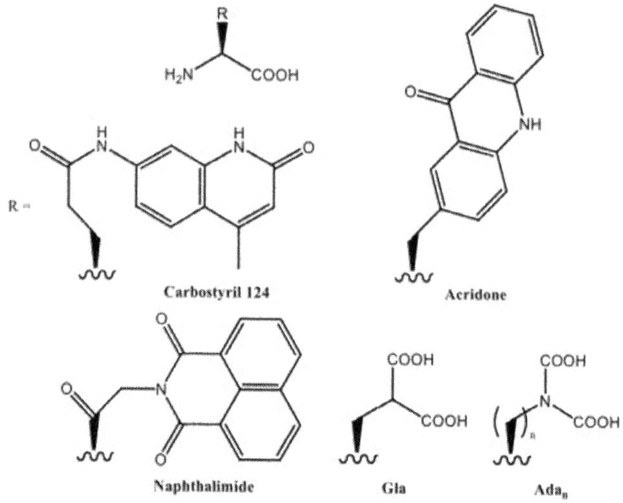

Fig. 1 Structures of selected nonnatural amino acids for lanthanide sensitization (carbostyril 124, acrdione, and naphthalimide) or coordination (Gla=γ-carboxyglutamic acid, Ada$_n$=aminodiacetate with side chains of varying length n).

4. LANTHANIDE PEPTIDES AND PROTEINS

There is much interest in coupling the exciting chemistry/imaging described earlier, with the opportunities afforded by the use of proteins as ligands. Resulting designs could be biologically compatible, retain biological recognition and targeting properties, features that are challenging to introduce into small molecule lanthanide complexes. The opportunities to make "smart" and biologically responsive imaging agents are therefore endless. Though there are many examples of lanthanide peptide/protein hybrid complexes in which a small molecule lanthanide complex has been conjugated to a peptide/protein for use in imaging (Lauffer, 1987; Franano et al., 1995), this chapter will instead focus on complexes in which the peptide/protein is the ligand which coordinates directly to the lanthanide ion.

4.1 Inspired by Native Ca(II)-Binding Loops

Due to the similarities between the coordination chemistry of lanthanide ions and Ca(II), much work has involved the substitution of lanthanide ions into native calcium-binding loops, such as the work by Franklin and coworkers, who generated a chimeric peptide which featured a DNA-binding domain from one protein, and the Ca(II)-binding loop from calmodulin (Welch, Sirish, Lindstrom, & Franklin, 2001). The resulting lanthanide complexes were investigated for catalysis (Welch et al., 2001) and the Gd(III) complex found to be a sequence-specific DNA responsive MRI contrast agent, with DNA binding reducing complex tumbling and enhancing MRI relaxivity (Caravan, Greenwood, Welch, & Franklin, 2003) (P3W, Table 2).

Similarly, Imperiali and coworkers have developed high affinity lanthanide-binding tags (LBTs), short peptide sequences (≥20 residues) inspired originally by native Ca(II)-binding sites, capable of binding one (Franz, Nitz, & Imperiali, 2003) or two lanthanide ions with high affinity (Martin et al., 2007). These can be introduced toward either the N- (Franz et al., 2003) or C-terminus of a protein (Martin et al., 2007), and in some cases inserted into existing loops and turns (Barthelmes et al., 2011). These LBTs have found applications as luminescence probes (Franz et al., 2003), can provide distance information in NMR (Martin et al., 2007; Wöhnert, Franz, Nitz, Imperiali, & Schwalbe, 2003) and can aid in phasing protein X-ray crystal structures (Silvage, Martin, Schwalbe, Imperiali, & Allen, 2007). More recently, a LBT was redesigned to bind Gd(III) with access for one inner sphere water molecule and investigated as a MRI contrast agent tag which could be readily introduced into a range of proteins (Daughtry et al., 2012) (Fig. 2A) (sSE3 and m-sSE3, Table 2).

A Gd(III)-binding site was designed, aided by computational methods, inspired by native Ca(II)-binding sites, on to the surface of a rat cell-adhesion

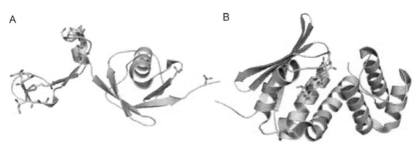

Fig. 2 Structure of a (A) double LBT Gd-ubiquitin adducts (PDB code 3VTZ, Daughtry et al., 2012) and (B) a manganese substituted porphyrin in the H-NOX protein (PDB code 4IT2, Winter et al., 2013), for which the related gadolinium complex was also investigated.

protein (Yang et al., 2008). The resulting Gd(III) complex was postulated to contain two coordinated water molecules, and displayed promising MRI relaxivity (Yang et al., 2008) (CA1.CD2, Table 2). Furthermore, it offers potential as a targeted imaging agent, supported by targeted HER2 imaging by a fusion construct with a HER2 antibody (Qiao et al., 2011).

4.2 Other Lanthanide-Binding Sites

Inspired by zinc finger proteins, lanthanide fingers have been designed which feature the same ββα structure motif, but in which the Cys$_2$His$_2$ Zn(II)-binding site has been replaced with a Asp$_2$Glu$_2$ Ln(III)-binding site, which binds with micromolar affinity (am Ende, Meng, Ye, Pandey, & Zondlo, 2010).

Small cyclic peptides have been developed by Delangle and coworkers as rigid scaffolds onto which lanthanide-binding sites, with, eg, Asp$_2$Glu$_2$ side chains directed on the same face of the scaffold, can be engineered. The Gd(III) complexes have been shown to be promising MRI contrast agents (Bonnet et al., 2009), with both an inner and outer sphere mechanism contributing to relaxivity (Bonnet, Fries, Crouzy, & Delangle, 2010) (PA, Table 2). This work has been extended to the use of nonnatural amino acids for the generation of higher affinity sites, and will be discussed further in Section 5.5.3.

Though the large number of examples focus on coordinating lanthanides via carboxylate side chains, there are also examples in which Gd(III) has substituted iron in a porphyrin, as has been done with the H-NOX protein (Fig. 2B). The resulting Gd(III) derivative (alongside a Mn(II/III) derivative) was found to display high MRI relaxivity (Table 2), however, the amount of inner sphere water was not determined (Winter et al., 2013).

5. STRATEGIES FOR DESIGNING CCS FOR LANTHANIDE COORDINATION

5.1 CC Design

The term de novo is defined as "anew" or "from the beginning," and though this can be applied to the examples described previously in which the lanthanide-binding sites are engineered de novo, a truer version of de novo metallopeptide/protein design, involves both the design of the lanthanide-binding site, and the peptide/protein scaffold from first principles. Despite being a daunting task, much effort has been invested into the latter, and a number of key peptide design rules established. Though a variety of peptide folds have been reproduced de novo, including β-sheets (Venkatraman, Naganagowda, Sudha, & Balaram, 2001), β-hairpins (Platt, Chung, & Searle, 2001), γ-turns (Bonomo et al., 1997), and mixed αβ motifs (am Ende et al., 2010; Krizek, Merkle, & Berg, 1993), the large majority of work has focused on the α-helix, and CCs thereof. In which the CC is a supercoil of two or more α-helices, twisted around one another (akin to a coil of rope) in a left-handed fashion.

CC designs are based on the heptad repeat approach, a seven amino acid repeat (approximately two turns of an α-helix) with amino acid positions identified as *abcdefg*, which is repeated a minimum of three times in order to generate a folded CC (Woolfson, 2005). Amino acids are selected which have a high propensity to form an α-helix, and tend to include Leu, Glu, and Lys, with Ala demonstrating the highest α-helical propensity (Pace & Scholtz, 1998). The major driving force for assembly is the generation of a hydrophobic core, achieved by burying hydrophobic residues which are most commonly located in the *a* and *d* positions. The CC assembly can also be directed by the formation of stabilizing interhelical salt bridges, formed between positively and negatively charged residues located commonly in the *e* and *g* positions. External sites such as the *f* position, tend to contain hydrophilic residues to enhance the water solubility of the CC (Fig. 3).

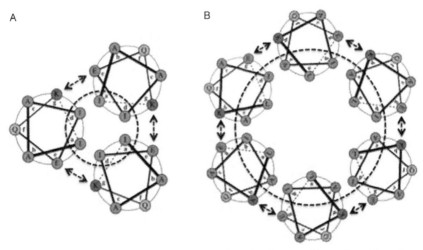

Fig. 3 Helical wheel diagrams illustrating the absolute position of residues in a (A) parallel three-stranded homo-CC or (B) parallel six-stranded homo-CC. The hydrophobic core and favorable electrostatic interactions are identified.

The number of α-helices within a CC can be controlled by either altering the packing of the hydrophobic core, or by extending the hydrophobic surface of the α-helix. In the case of the former, in the *a* and *d* positions contain hydrophobic residues, and are located on the same face of the α-helix (consistent with 3.6 residues per turn), thereby generating an amphipathic α-helix. These are normally Ile and Leu residues, and to a lesser extent Ala and Val, rather than aromatic containing side chains. These pack in the hydrophobic core and it is the sterics associated with this which determines the oligomeric state of the CC. Generally the combination of Leu in a *a* site and Ile in a *d* site forms a dimer, when both sites contain the same residue one obtains a trimer, and when the order is reversed (eg, Ile in a *a* site and Leu in a *d* site) this commonly yields a tetramer (Apostolovic, Danial, & Klok, 2010; Woolfson, 2005). However, care must be taken, as these are merely guidelines, and the presence of other amino acids can alter the oligomerization state. For example, inclusion of polar amino acids such as Lys in *a/d* sites, can destabilize higher-order oligomers instead favoring dimers in which the polar side chain can more readily protrude from the core and interact with the solvent. In order to generate higher-order oligomers, such as a six-stranded CC (Fig. 3B), one has to extend the hydrophobic face of the α-helix across more residues, eg, the *a*, *d*, and *e* sites, while at the same time moving the favorable salt bridges further out (eg, *b* and *g* positions) (Zaccai et al., 2011). Similar strategies can be adopted to generate larger oligomers (Thomson et al., 2014).

Another important consideration is whether the α-helices are aligned in a parallel or antiparallel fashion, and if they are all the same (homo-CC) or different (hetero-CC). Both features are largely dependent on ionic interactions. For example, the use of positively charged residues in both the *e* and *g* positions would destabilize homo-CC formation due to unfavorable electrostatics. However, the addition of a complementary peptide with negatively charged residues in the *e* and *g* positions would trigger the formation of a hetero-CC (Keating, Malashkevich, Tidor, & Kim, 2001). Similarly peptides which feature positively charged residues on one half of the peptide, and negatively charged residues on the other half, can self-associate in an antiparallel fashion (Apostolovic et al., 2010; Woolfson, 2005). The symmetry of the CC is an important consideration when engineering metal ion-binding sites into the structure, and in addition to the structures described earlier, asymmetry can be achieved in helical bundles. In these a single peptide chain folds into helix-loop-helix motifs, in which adjacent helices are aligned in a mixture of parallel and antiparallel arrangements.

Metal-binding sites can be engineered into CC scaffolds at various locations, including on its exterior (generally via the *f* position), at the α-helical interface or more commonly within the hydrophobic core. The latter is normally achieved by replacing a core hydrophobic residue, with a natural, or nonnatural amino acid capable of coordinating metal ions (eg, Cys or His) (Gamble & Peacock, 2014).

5.2 Lanthanide-Binding Site Design
5.2.1 At the Helical Interface

Lanthanide-binding sites have been engineered within de novo designed CCs, with the first report involving their introduction at the α-helical interface. Work by Hodges and coworkers, involved the design of a two-stranded CC, held together at one end by a disulfide bridge, in an effort to compensate for unfavorable and destabilizing ionic interhelical repulsion between Glu residues located in adjacent *e* and *g* sites on the two α-helices (Kohn, Kay, & Hodges, 1998) ($E_2(20,22)$, Table 1). Lanthanum binding, in which the two Glu residues are "bridged" and the charge effectively neutralized, is accompanied by peptide folding and enhanced stability.

Table 1 Sequences of the Peptide CCs Described Herein

Peptide	Sequence
$E_2(20,22)$[a]	Ac-Q C*GALQKQ VGALQKQ VGAL<u>E</u>K<u>E</u> VGALQKQ VGALQK-NH$_2$
Gla$_2$Nx[b]	Ac-Q C*GALQKQ VGALQK<u>X</u> VGAL<u>X</u>KQ <u>N</u>GALQKQ VGALQK-NH$_2$
Pep1[c]	Ac-YGGEEK IAAIEKK IAA<u>A</u>EEK <u>X</u>AAIEKK IAAIEEK GGY-NH$_2$
Pep2[c]	Ac-YGGEEK IAAIEKK <u>W</u>AA<u>A</u>EEK <u>X</u>AAIEKK IAAIEEK GGY-NH$_2$
Pep3[c]	Ac-YGGEEK IAAIEKK <u>A</u>AA<u>A</u>EEK <u>X</u>AAIEKK IAAIEEK GGY-NH$_2$
MB1-1[d]	Ac-G IAA<u>N</u>E<u>W</u>K <u>D</u>AAIEQK IAAIEQK IAAIEQK IAAIEQK G-NH$_2$
MB1-2[d,e]	Ac-G IAAIEQK IAA<u>N</u>E<u>W</u>K <u>D</u>AAIEQK IAAIEQK IAAIEQK G-NH$_2$
MB1-3[d]	Ac-G IAAIEQK IAAIEQK IAA<u>N</u>E<u>W</u>K <u>D</u>AAIEQK IAAIEQK G-NH$_2$
MB1-4[d]	Ac-G IAAIEQK IAAIEQK IAAIEQK IAA<u>N</u>E<u>W</u>K <u>D</u>AAIEQK G-NH$_2$
CS1-1[d]	Ac-G IAAIE<u>W</u>K <u>D</u>AAIEQK IAAIEQK IAAIEQK IAAIEQK G-NH$_2$

[a]Kohn, Kay, and Hodges (1998).
[b]Kohn, Kay, Sykes, and Hodges (1998).
[c]Kashiwada, Ishida, and Matsuda (2007).
[d]Berwick et al. (2016).
[e]Berwick et al. (2014).
C*: dimerized via a disulfide bond to generate a two-stranded CC; X = γ-carboxyglutamic acid (Gla). Residues bold and underlined are essential for generating the lanthanide binding site.

The design of a lanthanum site with three Glu residues in place of two, displayed even greater stability (Kohn, Kay, & Hodges, 1998).

In an effort to further enhance lanthanide-binding affinity, the nonnatural amino acid γ-carboxyglutamic acid (Gla), similar to Glu but with two carboxylate groups attached to the β-carbon (Fig. 1), was introduced at the interhelical interface (Gla$_2$Nx, Table 1). However, the enhancement was somewhat modest (Kohn, Kay, & Hodges, 1998).

5.2.2 Within the Hydrophobic Core

Instead of generating the binding site at the helical interface, sites buried within the hydrophobic core, where the majority of metal-binding sites are located in proteins, allows one to more readily control the coordination chemistry of the metal ion site, as well as its hydration state and solvent access. The first example of a lanthanide site engineered within a hydrophobic core was reported by Kashiwada and coworkers (Kashiwada et al., 2007), who introduced a Gla layer in a *a* site of a parallel three-stranded homo-CC. In order to accommodate the bulk of the Gla residue, the adjacent Ile layer was replaced with a sterically less demanding Ala layer (Pep1, Table 1). The substitution of two adjacent hydrophobic layers and introduction of a bulky and negatively charged Gla layer, resulted in peptide unfolding. However, CC folding was again found to be induced on binding Eu(III), Tb(III), or Ce(III). In order to probe Ln(III) binding directly by luminescence, a single Trp sensitizer was introduced two layers above the Gla. This required the formation of an AAB heterotrimer with an AlaAlaTrp layer (Pep2/3, Table 1), so as to accommodate the bulk of the Trp within the hydrophobic core (Kashiwada et al., 2007).

Despite the design of lanthanide CCs, this work had focused on lanthanide binding inducing a folding event, but had not exploited any of the imaging opportunities afforded by the use of lanthanides, with no gadolinium derivatives reported. We therefore chose to design a lanthanide-binding site within the hydrophobic core of a de novo designed CC, specifically for the coordination of gadolinium, with the view to the resulting complexes being investigated for their potential use as MRI contrast agents.

In view of the trivalent charge of the lanthanides, as well as the optimal Ln–O bond distances, we selected a three-stranded CC based on the IAAIEQK heptad repeat, where the Ile residues in the *a* and *d* sites select for a trimer. In the interior of this we introduced a layer of Asp residues in *a* sites. These would provide the overall −3 charge that would be effectively neutralized on Ln(III) coordination, as well as providing three-bidentate carboxylate ligands. However, as this only presents a maximum of six oxygen donor atoms for Ln(III) coordination, a neutral layer, so as to not alter the overall charge of the site, of Asn residues was introduced directly above in d sites, capable of providing an additional three oxygen donors to achieve an overall nine-coordinate site. It was essential that the Asn was introduced into a d site, as its positioning in an a site has been found to promote the formation of two-stranded CCs (Fletcher et al., 2012). The disruption of two adjacent hydrophobic layers with polar and negatively charged residues, was believed to be sufficiently destabilizing, that a total of five heptads were employed to compensate for this. The final peptide sequence, MB1–2, is presented in Table 1 (Berwick et al., 2014).

The resulting peptide was poorly folded, however, as observed previously (Kohn, Kay, & Hodges, 1998; Kohn, Kay, Sykes, et al., 1998; Kashiwada et al., 2007), the addition of a Ln(III) promoted folding and CC formation. In addition to the templating role, Ln(III) coordination was accompanied by stabilization as determined from both chemical and thermal denaturation experiments, with the lanthanide CC remaining folded at biologically relevant temperatures. Titrations and MS evidence were consistent with the formation of the desired Ln(MB1–2)$_3$ species, and a dissociation constant in the micromolar range, which for a first design using natural amino acids was as expected. A short MD simulation of the proposed lanthanide CC, was consistent with the intended design of a Ln(Asn$_3$Asp$_3$) site (Berwick et al., 2014) (Fig. 4).

Intriguingly the MD simulation contains a water molecule bound to the Ln(III) ion. This is despite the luminescence decay experiments for the Tb(III) complex being consistent with 0.1 water molecules (Berwick et al., 2014). Therefore, the experimental evidence is consistent with a complex which is coordinatively saturated with all the ligands being provided by the peptide scaffold, consistent with the original design, and a Ln(III)-binding site engineered within a hydrophobic core. However, the MD simulation certainly suggests that there is adequate space to accommodate a water molecule within the first coordination sphere of a coordinated Ln(III) ion. Clearly this feature is crucial given the potential imaging applications of these complexes, and especially given the general assumption that a coordinated water is required for gadolinium-based MRI contrast agents.

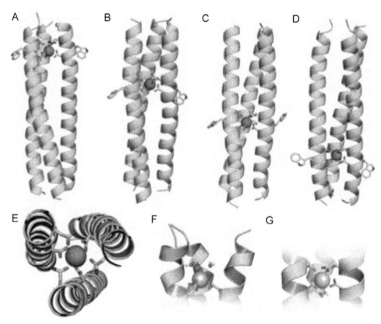

Fig. 4 Cartoon representation of a series of lanthanide coiled coils in which the AsnAsp-binding site is translated linearly along the CC: (A) Ln(MB1–1)$_3$, (B) Ln(MB1–2)$_3$, (C) Ln(MB1–3)$_3$, and (D) Ln(MB1–4)$_3$. (E) A top-down view of a core-binding site, and side-on views of (F) top and (G) bottom lanthanide-binding sites. *Adapted from Berwick, M. R., Slope, L., N., Smith, C. F., King, S., Newton, S. L., Gillis, R. B., et al. (2016). Location dependent coordination chemistry and MRI relaxivity, in* de novo *designed lanthanide coiled coils. Chemical Science, 7, 2207–2216.*

5.3 Controlling Water Coordination and Access

The coordination of small molecules, such as water, to designed metallopeptides, is of much interest as sites such as these can be catalytic active or involved in small molecule transport (such as oxygen), natural biological processes in which there is much interest in fully understanding and ultimately mimicking the activity. However, in the case of lanthanide CCs, the hydration state of the lanthanide has a large impact on the potential imaging applications, with luminescent and MRI favored by dehydrated and hydrated lanthanide sites, respectively. Within this section, methods by which the hydration state of a metallo-CC can be altered, will be discussed.

5.3.1 Cavity

Hodges and coworkers proposed that the introduction of a cavity within the hydrophobic core of a CC, could be used to accommodate water molecules

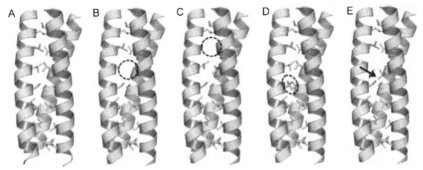

Fig. 5 Cartoon representation of a series of coiled coils with a (A) Cys$_3$-binding site with differing degrees of sterics and as a result water coordination to bound metal ions. Relieving steric bulk (B) adjacent or (C) further removed from the active site can enhance the metal hydration state. Introducing steric bulk in (D) the coordinating plane or (E) in the layer above, can reduce the metal hydration state.

(Monera, Sonnichsen, Hicks, Kay, & Hodges, 1996). This was subsequently confirmed by Pecoraro and coworkers, who replaced a hydrophobic Leu residue directly above a cadmium thiolate site with a sterically less demanding amino acids, such as the helix destabilising Gly, or the better tolerated helix stabilizing Ala (Pace & Scholtz, 1998) (Fig. 5B), enhanced the hydration state of the bound Cd(II) (Lee, Cabello, Hemmingsen, Marsh, & Pecoraro, 2006; Lee, Matzapetakis, Mitra, Marsh, & Pecoraro, 2004). Using a similar strategy, but instead introducing the cavity below the trigonal thiolate site, had a similar effect and was also able to slightly enhance the hydration state of the coordinated Cd(II) (Lee et al., 2004). The introduction of a cavity further away in the CC (Fig. 5C), separated from the metal-binding site by intervening hydrophobic layers, can also impact on the hydration state of a coordinated metal. However, the location of this cavity is imperative with one example of a cavity toward the N-terminus leading to metal hydration, whereas the same cavity introduced toward the C-terminus having no impact on the metal hydration state (Iranzo, Jakusch, Lee, Hemmingsen, & Pecoraro, 2009).

5.3.2 Enhanced Steric Bulk

In contrast is was similarly reasoned that if removing steric bulk could lead to an increase in hydration state of a coordinated metal ion, that enhancing the steric bulk in close proximity to the binding site, could instead exclude the presence and coordination of exogenous water. This was shown to be the case when a dehydrated cadmium trigonal thiolate-binding site was generated, by introducing the nonnatural amino acid L-penicillamine (Pen) in place of Cys (Iranzo, Cabello, & Pecoraro, 2007; Lee et al., 2006) (Fig. 5D), where Pen is a bulkier analogue of Cys with sterically demanding methyl groups in place of the β-methylene protons (Peacock, Stuckey, & Pecoraro, 2009) and was first reported for use in de novo metallopeptide design in 2004 by Gibney and coworkers (Petros, Shaner, Costello, Tierney, & Gibney, 2004).

Though this example has focused on altering sterics within the metal-binding plane, one can also take advantage of the sterics afforded by noncoordinating second-sphere residues to alter the hydration of the coordinated metal ion. For example, by altering the chirality of a Leu directly above the metal-binding site, in which the side chain in the L-form is directed away from the metal toward the N-terminus, to the D-enantiomer, instead directs the bulky isopropyl side chain toward the C-terminus and metal ion-binding plane, preventing the coordination of an exogenous water molecule (Peacock, Hemmingsen, & Pecoraro, 2008) (Fig. 5E). A number of reports now exist which successfully use D-amino acids in de novo peptide design (Aravinda, Shamala, Desirajub, & Balaram, 2002; Fairman, Anthony-Cahill, & DeGrado, 1992; Karle, Gopi, & Balaram, 2003; Sia & Kim, 2001).

5.3.3 Location Within the CC

Pecoraro and coworkers reported that solvent access to a catalytically active ZnHis$_3$O site is dependent on the linear location of the metal-binding site along a CC, with decreased water access when located more buried in the center of the CC (Zastrow & Pecoraro, 2013). Likewise the percentage of water coordinated to cadmium thiolate sites, was found to be slightly sensitive to the binding site location along the CC (Iranzo et al., 2009).

Given the above it is not surprising that the lanthanide hydration state, when bound in a CC hydrophobic core, was found to be location dependent. Binding sites generated within the middle of the CC are coordinatively saturated, with all nine-donor atoms being provided by the peptide ligand (Berwick et al., 2014). Designs such as these would therefore be ideal for use as luminescence imaging agents. In contrast, translating the binding site toward either termini, led to increased water coordination (Fig. 4) (Berwick et al., 2016), essential for MRI contrast agents. For example, Tb(III) binding to a AsnAsp site located toward the C-terminus (bottom, MB1–4) of the CC, coordinates to an additional two exogenous water molecules, whereas the identical ligand set located toward the N-terminus (top, MB1–1) of the

CC, binds Tb(III) with three exogenous water molecules (Fig. 4F) (Berwick et al., 2016). These observations could be consistent with altered local peptide structure associated with fraying and unwinding of the CC toward the termini, or ease of water penetration, with fewer intervening hydrophobic layers between the bulk solvent and the lanthanide-binding site.

The presence of three-coordinated water molecules when bound to MB1–1 suggests that only six donor atoms are presented by the peptide scaffold. This hypothesis was tested by preparing the related peptide, CS1–1, which differs only in that the Asn layer is missing, retaining the essential Asp layer (overall −3 charge and six oxygen donor atoms). Despite the missing Asn layer, the resulting peptide was still capable of lanthanide binding and with a similar hydration state. However, the Asn residue appears essential for effective lanthanide coordination to the three remaining binding sites (Berwick et al., 2016).

5.4 Lanthanide Selectivity

Given that the coordination chemistry of calcium is not dissimilar to that of lanthanide ions, and that calcium is ubiquitous in biology, being the most abundant metal in the human body, it is imperative that binding sites with strong preferences for lanthanides over calcium, are designed, if lanthanide peptides are to find applications as biological imaging agents.

For lanthanide-binding sites generated at the helical interface of dimers, the metal dissociation constant for trivalent lanthanides was micromolar. In contrast, the affinity for divalent calcium was substantially higher and in the millimolar range (Kohn, Kay, & Hodges, 1998; Kohn, Kay, Sykes, et al., 1998). This was similarly observed for the binding sites generated within the hydrophobic core of a three-stranded CC (Berwick et al., 2014; Kashiwada et al., 2007), and in the case of MB1–2, 10 μM terbium was not displaced on addition of as much as 10 mM calcium (Berwick et al., 2014). This selectivity for a trivalent lanthanide is proposed to be related to both a complementary charge of the designed site, as well as preferences for bidentate coordination within confined coordination environments (Dudev & Lim, 2007).

5.5 Controlling Stability

Stability is an important safety consideration if de novo lanthanide peptides are to be translated into the clinic for real-life use. There are three different types of stability that need to be considered: stability of the peptide fold with respect to denaturation and unfolding; its proteolytic stability; and the gadolinium dissociation constant.

5.5.1 Peptide Folding

The degree of CC folding, and stability with respect to unfolding (triggered by either chemical or thermal denaturation) is dependent on both the nature of the metal-binding site and its relative location. Introduction of a metal-binding site within the hydrophobic core involves the replacement of a stabilizing hydrophobic layer (in either a *a* or *d* site), with a destabilizing residue that may differ in sterics, polarity, and even charge. We previously described how the presence of some amino acids can in fact program for a different oligomeric state (Section 5.1) and some residues may be more or less helix inducing (Pace & Scholtz, 1998).

For the introduction of a lanthanide-binding site within the hydrophobic core, it is generally necessary to disturb two adjacent hydrophobic layers and to introduce repulsive negatively charged residues. This either involved the use of the nonnatural amino acid Gla, which requires two layers to accommodate its bulky side chain (Kashiwada et al., 2007), or the introduction of two different residues, Asn and Asp, to fully satisfy lanthanide coordination (Berwick et al., 2014, 2016). In both cases, these changes were highly destabilizing leading to CC unfolding. However, lanthanide coordination nucleated CC assembly (Berwick et al., 2014, 2016; Kashiwada et al., 2007).

Notably, translating the AsnAsp-binding site linearly along the CC has a large impact on folding and stability of these isomeric peptides, Fig. 4. When the binding site is located toward the N-terminus (top) of the CC, the apo peptide is well folded, but when located more centrally within core heptads, is poorly folded (Berwick et al., 2016). Similarly translating the binding site toward the top of the CC stabilizes the lanthanide bound CC by c.5 kcal mol^{-1} compared to the more central isomer (Berwick et al., 2016). In fact the location of metal-binding site introduction is crucial, with modifications to core heptads being three times more destabilizing than terminal heptads (de Crescenzo, Litowski, Hodges, & O'Connor-McCourt, 2003).

CC stability is related to the number of heptads in the design, with a minimum of three required, and more routinely four. Perhaps not surprising in view of the destabilizing nature of lanthanide-binding sites, all of the lanthanide CCs discussed herein contain five heptads, with the fifth heptad presumably necessary to compensate for introduction of the destabilizing lanthanide-binding site (Berwick et al., 2014, 2016; Kashiwada et al., 2007; Kohn, Kay, & Hodges, 1998; Kohn, Kay, Sykes, et al., 1998).

5.5.2 Proteolytic Stability

One of the challenges associated with the use of peptides as therapeutics, is related to their short-lived plasma half-lives. Strategies to circumvent this challenge have involved cyclizing the peptide scaffold or stabilizing of secondary structure (eg, through stapling or clipping) (Fosgerau & Hoffmann, 2015) as well as the generation of peptidomimetics, including the use of D-amino acids in the sequence (Van Regenmortel & Muller, 1998). These strategies can enhance the proteolytic stability of peptides, which is important if these are to be employed in therapeutic applications.

5.5.3 Gadolinium Dissociation Constant

For lanthanide complexes to be adopted for clinical applications it is essential that the lanthanide ion must be bound sufficiently tightly to avoid its release, which in turn is associated with toxicity when lanthanides interfere with calcium processes and end up irreversibly associated in bone (Darrah et al., 2009). In fact retention of gadolinium, most commonly in patients with impaired kidney function, following the repeated use of gadolinium-based MRI contrast agents, can lead to Nephrogenic Systemic Fibrosis which can be a life-threatening condition (Grobner, 2006).

Therefore it is of concern that lanthanide-binding sites generated from natural amino acids, generally have micromolar affinity, though some optimized examples have achieved nanomolar affinity (Nitz, Franz, Maglathlin, & Imperiali, 2003; Nitz et al., 2004). However, these remain substantially poorer than the affinity associated with the small molecule gadolinium-based MRI contrast agents currently used in the clinic, which instead have log K values of around 23. We have previously mentioned the use of the nonnatural Gla, a higher affinity nonnatural amino acid derivative of Glu, and more recently the group of Delangle and coworkers have begun investigated the use of nonnatural amino acids which feature polyaminocarboxylate side chains (Fig. 1), ranging from tridentate to tetradentate donor ligands, so as to enhance lanthanide affinity further (Ancel, Niedźwiecka, Lebrun, Gateau, & Delangle, 2013). Coupled with the enhanced affinity of gadolinium complexes for the more basic donor atoms (Bravard, Rosset, & Delangle, 2004), they were able, by careful peptide design, to successfully design lanthanide peptide-binding sites with picomolar (Cisnetti, Gateau, Lebrun, & Delangle, 2009), and when they employed pentadentate derivatives, record breaking femtomolar affinity (Niedźwiecka, Cisnetti, Lebrun, & Delangle, 2012).

However, an important consequence of these high affinity sites is that the lanthanide ions tended to be dehydrated (not contain any first coordination sphere water), and therefore despite these sites being of interest for luminescence imaging applications, they are less likely to be attractive for use in MRI contrast agents.

5.6 MRI Efficiency of Gadolinium CCs

The MRI relaxivity (r), efficiency at altering the longitudinal and transverse relaxation time of bulk water, of the gadolinium CCs described in Section 5.2.2, was investigated at 7 T. The latter being chosen as the trend in clinical MRI scanners is toward higher field instruments, despite higher field strengths normally being associated with lower relaxivity (Table 2). These findings were put into context by comparing the results with those obtained for Dotarem, Gd(DOTA), a clinically employed small molecule gadolinium MRI contrast agent with one molecule of inner sphere water.

For gadolinium CCs with no evidence of inner sphere water molecules (MB1–2 and MB1–3), the longitudinal relaxivities (r_1) were comparable to that of Dotarem (Table 2). However, surprisingly the transverse relaxivities (r_2, Fig. 6) were superior to Dotarem, despite the apparent lack of inner sphere water (Berwick et al., 2014, 2016). This was postulated to be due to a combination of an outer sphere mechanism, which most likely involves a hydrogen bonding network between outer sphere water molecules and the CC external surface, exchange with peptide protons, and reduced tumbling in solution, both of which are intrinsically due to the CC ligand (Berwick et al., 2014) and are the subject of further investigation (Newton & Britton, in preparation).

Changes to the lanthanide coordination chemistry, and specifically its hydration state, on translating the binding site linearly along the CC, were not surprisingly accompanied by an enhancement in relaxivity, by bringing into play, in addition to an outer sphere mechanism, an inner sphere mechanism. On coordination of two water molecules, as is the case when bound to MB1–4, the gadolinium complex displays both enhanced longitudinal and transverse relaxivity (Table 2) (Berwick et al., 2016). This is

Table 2 Relaxivity of Select Gadolinium Peptides and Proteins

Gd Peptide	r_1 (mM^{-1} s^{-1})	r_2 (mM^{-1} s^{-1})	Field Strength (MHz)	#H$_2$O	References
GdP3W	16.2	—	20	2	Caravan et al. (2003)
GdP3W +DNA	29.6	—	20	2	Caravan et al. (2003)
GdP3W	21.1	—	60	2	Caravan et al. (2003)
GdP3W +DNA	42.4	—	60	2	Caravan et al. (2003)
Gd-sSE3	1.2	—	500	0	Daughtry, Martin, Sarraju, Imperiali, and Allen (2012)
Gd-m-sSE3	5.5	—	500	1	Daughtry et al. (2012)
Gd-CA1.CD2	117	129	64	~2	Yang et al. (2008)
Gd-CA1.CD2	34	57	128	~2	Yang et al. (2008)
Gd-PA	29.8	~40	200	2	Bonnet et al. (2009)
Gd-PA	~21	~36	400	2	Bonnet et al. (2009)
Gd-PA	~18	~36	500	2	Bonnet et al. (2009)
Gd-H-NOX	28.7±5.0	39.9±4.9	60	—	Winter, Klemm, Phillips-Piro, Raymond, and Marletta (2013)
Gd(MB1–1)$_3$	10.0±1.5	89.3±16.8	300	3	Berwick et al. (2016)
Gd(MB1–2)$_3$	4.2±1.2	21.3±2.6	300	0	Berwick et al. (2016)
Gd(MB1–3)$_3$	4.0±1.0	20.9±1.0	300	0	Berwick et al. (2016)
Gd(MB1–4)$_3$	7.5±4.1	37.9±4.0	300	2	Berwick et al. (2016)

further enhanced when three water molecules are coordinated, to yield the highest relaxivity gadolinium CC, Gd(MB1–1)$_3$ (Table 2). Translating the gadolinium-binding site linearly (1 nm) along the CC from a central to a N-terminus site, transforms a coordinatively saturated lanthanide site, suitable for use as a luminescence imaging agent, into a highly hydrated site more suitable for use in MRI, with more than a fourfold enhancement in transverse relaxivity (Berwick et al., 2016).

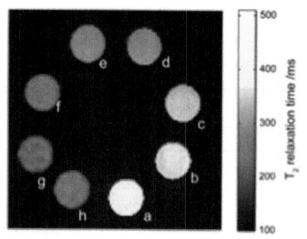

Fig. 6 T_2 weighted MR image of Gd(III) samples on addition of increasing concentrations of MB1–2, and formation of Gd(MB1–2)$_3$. *Reprinted from Berwick, M. R., Lewis, D. J., Jones, A. W., Parslow, R. A., Dafforn, T. R., Cooper, H. J., et al. (2014). De novo design of Ln(III) coiled coils for imaging applications. Journal of the American Chemical Society, 136, 1166–1169.* (See the color plate.)

6. CONCLUSIONS AND PERSPECTIVES

Though relatively early stage work, the use of nonbiological metal ions, such as the lanthanides, in protein design, certainly opens up extensive opportunities to develop functional systems for applications beyond the current repertoire of biology, such as the development of potential imaging agents as described herein. However, it should be noted that the examples discussed are preliminary proof-of-concept studies and have not yet been translated into clinical settings. Furthermore, much of the work has focused on establishing key design rules and principles related to the introduction of appropriate sensitizers for luminescence, control of hydration state, and achieving high dissociation constant lanthanide-binding sites. These important findings, are vital for building on the preliminary exciting reports described herein, and offer much hope for lanthanide peptides and their applications in clinical imaging in the future.

ACKNOWLEDGMENTS

Support from the University of Birmingham, the EPSRC Directed Assembly Grand Challenge Network, and EU COST Action CM1105 are gratefully acknowledged.

REFERENCES

am Ende, C. W., Meng, H. Y., Ye, M., Pandey, A. K., & Zondlo, N. J. (2010). Design of lanthanide fingers: Compact lanthanide-binding metalloproteins. *ChemBioChem, 11*, 1738–1747.

Ancel, L., Niedźwiecka, A., Lebrun, C., Gateau, C., & Delangle, P. (2013). Rational design of lanthanide binding peptides. *Comptes Rendus Chimie, 16*, 515–523.

Apostolovic, B., Danial, M., & Klok, H.-A. (2010). Coiled coils: Attractive protein folding motifs for the fabrication of self-assembled, responsive and bioactive materials. *Chemical Society Reviews, 39*, 3541–3575.

Aravinda, S., Shamala, N., Desirajub, S., & Balaram, P. (2002). A right handed peptide helix containing a central double D-amino acid segment. *Chemical Communications (Cambridge, England)*, 2454–2455.

Barthelmes, K., Reynolds, A. M., Peisach, E., Jonker, H. R. A., DeNunzio, N. J., Allen, K. N., et al. (2011). Engineering encodable lanthanide-binding tags into loop regions of proteins. *Journal of the American Chemical Society, 133*, 808–819.

Berwick, M. R., Lewis, D. J., Jones, A. W., Parslow, R. A., Dafforn, T. R., Cooper, H. J., et al. (2014). De novo design of Ln(III) coiled coils for imaging applications. *Journal of the American Chemical Society, 136*, 1166–1169.

Berwick, M. R., Slope, L. N., Smith, C. F., King, S., Newton, S. L., Gillis, R. B., et al. (2016). Location dependent coordination chemistry and MRI relaxivity, in de novo designed lanthanide coiled coils. *Chemical Science, 7*, 2207–2216.

Bonnet, C. S., Devocelle, M., & Gunnlaugsson, T. (2008). Structural studies in aqueous solution of new binuclear lanthanide luminescent peptide conjugates. *Chemical Communications*, 4552–4554.

Bonnet, C. S., Fries, P. H., Crouzy, S., & Delangle, P. (2010). Outer-sphere investigation of MRI relaxation contrast agents. Example of a cyclodecapeptide gadolinium complex with second-sphere water. *The Journal of Physical Chemistry. B, 114*, 8770–8781.

Bonnet, C. S., Fries, P. H., Crouzy, S., Sénèque, O., Cisnetti, F., Boturyn, D., et al. (2009). A gadolinium-binding cyclodecapeptide with a large high-field relaxivity involving second-sphere water. *Chemistry: A European Journal, 15*, 7083–7093.

Bonomo, R. P., Casella, L., De Gioia, L., Molinari, H., Impellizzeri, G., Jordan, T., et al. (1997). Metal ion and proton stabilisation of turn motif in the synthetic octapeptide histidyltris(glycylhistidyl) glycine. *Journal of the Chemical Society Dalton Transactions, 14*, 2387–2389.

Bravard, F., Rosset, C., & Delangle, P. (2004). Cationic lanthanide complexes of neutral tripodal N, O ligands: Enthalpy versus entropy-driven podate formation in water. *Dalton Transactions*, 2012–2018.

Caravan, P. (2006). Strategies for increasing the sensitivity of gadolinium based MRI contrast agents. *Chemical Society Reviews, 35*, 512–523.

Caravan, P., Greenwood, J. M., Welch, J. T., & Franklin, S. J. (2003). Gadolinium-binding helix–turn–helix peptides: DNA-dependent MRI contrast agents. *Chemical Communications*, 2574–2575.

Cisnetti, F., Gateau, C., Lebrun, C., & Delangle, P. (2009). Lanthanide(III) complexes with two hexapeptides incorporating unnatural chelating amino acids: Secondary structure and stability. *Chemistry: A European Journal, 15*, 7456–7469.

Darrah, T. H., Prutsman-Pfeiffer, J. J., Poreda, R. J., Campbell, M. E., Hauschka, P. V., & Hannigan, R. E. (2009). Incorporation of excess gadolinium into human bone from medical contrast agents. *Metallomics, 1*, 479–488.

Daughtry, K. D., Martin, L. J., Sarraju, A., Imperiali, B., & Allen, K. N. (2012). Tailoring encodable lanthanide-binding tags as MRI contrast agents. *ChemBioChem, 13*, 2567–2574.

de Crescenzo, G., Litowski, J. R., Hodges, R. S., & O'Connor-McCourt, M. D. (2003). Real-time monitoring of the interactions of two-stranded de novo designed coiled-coils: Effect of chain length on the kinetic and thermodynamic constants of binding. *Biochemistry, 42*, 1754–1763.

Dudev, T., & Lim, C. (2007). Effect of carboxylate-binding mode on metal binding/selectivity and function in proteins. *Accounts of Chemical Research, 40*, 85–93.

Fairman, R., Anthony-Cahill, S. J., & DeGrado, W. F. (1992). The helix-forming propensity of D-alanine in a right-handed α-helix. *Journal of the American Chemical Society, 114*, 5458–5459.

Fletcher, J. M., Boyle, A. L., Bruning, M., Bartlett, G. J., Vincent, T. L., Zaccai, N. R., et al. (2012). A basis set of de novo coiled-coil peptide oligomers for rational protein design and synthetic biology. *ACS Synthetic Biology, 1*, 240–250.

Fosgerau, K., & Hoffmann, T. (2015). Peptide therapeutics: Current status and future directions. *Drug Discovery Today, 20*, 122–128.

Franano, F. N., Edwards, W. B., Welch, M. J., Brechbiel, M. W., Gansow, O. A., & Duncan, J. R. (1995). Biodistribution and metabolism of targeted and nontargeted protein-chelate-gadolinium complexes: Evidence for gadolinium dissociation in vitro and in vivo. *Magnetic Resonance Imaging, 13*, 201–214.

Franz, K. J., Nitz, M., & Imperiali, B. (2003). Lanthanide-binding tags as versatile protein coexpression probes. *ChemBioChem, 4*, 265–271.

Gamble, A. J., & Peacock, A. F. A. (2014). De novo design of peptide scaffolds as novel preorganized ligands for metal-ion coordination. In V. Köhler (Ed.), *Methods in molecular biology: Vol. 1216. Protein design* (pp. 211–231). NY: Humana Press.

Grobner, T. (2006). Gadolinium—A specific trigger for the development of nephrogenic fibrosing dermopathy and nephrogenic systemic fibrosis. *Nephrology, Dialysis, Transplantation, 21*, 1104–1108.

Iranzo, O., Cabello, C., & Pecoraro, V. L. (2007). Heterochromia in designed metallopeptides: Geometry-selective binding of Cd^{II} in a de novo peptide. *Angewandte Chemie, International Edition, 46*, 6688–6691.

Iranzo, O., Jakusch, T., Lee, K.-H., Hemmingsen, L., & Pecoraro, V. L. (2009). The correlation of ^{113}Cd NMR and ^{111m}Cd PAC spectroscopies provides a powerful approach for the characterization of the structure of Cd^{II}-substituted Zn^{II} proteins. *Chemistry: A European Journal, 15*, 3761–3772.

Karle, I. L., Gopi, H. N., & Balaram, P. (2003). Crystal structure of a hydrophobic 19-residue peptide helix containing three centrally located D amino acids. *Proceedings of the National Academy of Sciences of the United States of America, 100*, 13946–13951.

Kashiwada, A., Ishida, K., & Matsuda, K. (2007). Lanthanide ion-induced folding of de novo designed coiled-coil polypeptides. *Bulletin of the Chemical Society of Japan, 80*, 2203–2207.

Keating, A. E., Malashkevich, V. N., Tidor, B., & Kim, P. S. (2001). Side-chain repacking calculations for predicting structures and stabilities of heterodimeric coiled coils. *Proceedings of the National Academy of Sciences of the United States of America, 98*, 14825–14830.

Kohn, W. D., Kay, C. M., & Hodges, R. S. (1998). Effects of lanthanide binding on the stability of de novo designed alpha-helical coiled-coils. *The Journal of Peptide Research, 51*, 9–18.

Kohn, W. D., Kay, C. M., Sykes, B. D., & Hodges, R. S. (1998). Metal ion induced folding of a de novo designed coiled-coil peptide. *Journal of the American Chemical Society, 120*, 1124–1132.

Krizek, B. A., Merkle, D. L., & Berg, J. M. (1993). Ligand variation and metal-ion binding specificity in zinc finger peptides. *Inorganic Chemistry, 32*, 937–940.

Lauffer, R. B. (1987). Paramagnetic metal complexes as water proton relaxation agents for NMR imaging: Theory and design. *Chemical Reviews, 87*, 901–927.

Lee, K.-H., Cabello, C., Hemmingsen, L., Marsh, E. N. G., & Pecoraro, V. L. (2006). Using non-natural amino acids to control metal coordination number in three stranded coiled-coils. *Angewandte Chemie, International Edition, 45*, 2864–2868.

Lee, K.-H., Matzapetakis, M., Mitra, S., Marsh, E. N. G., & Pecoraro, V. L. (2004). Control of metal coordination number in de novo designed peptides through subtle sequence modifications. *Journal of the American Chemical Society, 126*, 9178–9179.

Martin, L. J., Hähnke, M. J., Nitz, M., Wöhnert, J., Silvage, N. R., Allen, K. N., et al. (2007). Double-lanthanide-binding tags: Design, photophysical properties, and NMR applications. *Journal of the American Chemical Society, 129*, 7106–7113.

Monera, O. D., Sonnichsen, F. D., Hicks, L., Kay, C. M., & Hodges, R. S. (1996). The relative positions of alanine residues in the hydrophobic core control the formation of two-stranded or four-stranded alpha-helical coiled-coils. *Protein Engineering, 9*, 353–362.

Nakagawa, T., Mitsui, R., Tani, A., Sasa, K., Tashiro, S., Iwama, T., et al. (2012). A catalytic role of XoxF1 as La^{3+}-dependent methanol dehydrogenase in Methylobacterium extorquens strain AM1. *PLoS One, 7*, e50480.

Newton, S. L., Britton, M. M., & Peacock, A. F. A. (manuscript in preparation).

Niedźwiecka, A., Cisnetti, F., Lebrun, C., & Delangle, P. (2012). Femtomolar Ln(III) affinity in peptide-based ligands containing unnatural chelating amino acids. *Inorganic Chemistry, 51*, 5458–5464.

Nitz, M., Franz, K. J., Maglathlin, R. L., & Imperiali, B. (2003). A powerful combinatorial screen to identify high-affinity terbium(III)-binding peptides. *ChemBioChem, 4*, 272–276.

Nitz, M., Sherawat, M., Franz, K. J., Peisach, E., Allen, K. N., & Imperiali, B. (2004). Structural origin of the high affinity of a chemically evolved lanthanide-binding peptide. *Angewandte Chemie, International Edition, 43*, 3682–3685.

Pace, C. N., & Scholtz, J. M. (1998). A helix propensity scale based on experimental studies of peptides and proteins. *Biophysical Journal, 75*, 422–427.

Peacock, A. F. A., Hemmingsen, L., & Pecoraro, V. L. (2008). Using diastereopeptides to control metal ion coordination in proteins. *Proceedings of the National Academy of Sciences of the United States of America, 105*, 16566–16571.

Peacock, A. F. A., Stuckey, J. A., & Pecoraro, V. L. (2009). Switching the chirality of the metal environment alters the coordination mode in designed peptides. *Angewandte Chemie, International Edition, 48*, 7371–7374.

Petros, A. K., Shaner, S. E., Costello, A. L., Tierney, D. L., & Gibney, B. R. (2004). Comparison of cysteine and penicillamine ligands in a Co(II) maquette. *Inorganic Chemistry, 43*, 4793–4795.

Platt, G., Chung, C. W., & Searle, M. S. (2001). Design of histidine-Zn^{2+} binding sites within a beta-hairpin peptide: Enhancement of beta-sheet stability through metal complexation. *Chemical Communications, 13*, 1162–1163.

Qiao, J., Li, S., Wei, L., Jiang, J., Long, R., Mao, H., et al. (2011). HER2 targeted molecular MR imaging using a de novo designed protein contrast agent. *PLoS One, 6*, e18103.

Reynolds, M., Sculimbrene, B. R., & Imperiali, B. (2008). Lanthanide-binding tags with unnatural amino acids: Sensitizing Tb^{3+} and Eu^{3+} luminescence at longer wavelengths. *Bioconjugate Chemistry, 19*, 588–591.

Sia, S. K., & Kim, P. S. (2001). A designed protein with packing between left-handed and right-handed helices. *Biochemistry, 40*, 8981–8989.

Silvage, N. R., Martin, L. J., Schwalbe, H., Imperiali, B., & Allen, K. N. (2007). Double-lanthanide-binding tags for macromolecular crystallographic structure determination. *Journal of the American Chemical Society, 129*, 7114–7120.

Thomson, A. R., Wood, C. W., Burton, A. J., Bartlett, G. J., Sessions, R. B., Brady, R. L., et al. (2014). Computational design of water-soluble alpha-helical barrels. *Science, 346*, 485–488.

Van Regenmortel, M. H. V., & Muller, S. (1998). D-peptides as immunogens and diagnostic reagents. *Current Opinion in Biotechnology, 9*, 377–382.

Venkatraman, J., Naganagowda, G. A., Sudha, R., & Balaram, P. (2001). De novo design of a five-stranded beta-sheet anchoring a metal-ion binding site. *Chemical Communications, 24*, 2660–2661.

Welch, J. T., Sirish, M., Lindstrom, K. M., & Franklin, S. J. (2001). De novo nucleases based on HTH and EF-hand chimeras. *Inorganic Chemistry, 40*, 1982–1984.

Winter, M. B., Klemm, P. J., Phillips-Piro, C. M., Raymond, K. N., & Marletta, M. A. (2013). Porphyrin-substituted H-NOX proteins as high-relaxivity MRI contrast agents. *Inorganic Chemistry, 52*, 2277–2279.

Wöhnert, J., Franz, K. J., Nitz, M., Imperiali, B., & Schwalbe, H. (2003). Protein alignment by a coexpressed lanthanide-binding tag for the measurement of residual dipolar couplings. *Journal of the American Chemical Society, 125*, 13338–13339.

Woolfson, D. N. (2005). The design of coiled-coil structures and assemblies. In D. Parry & J. Squire (Eds.), *Advances in protein chemistry and structural: Vol. 70. Fibrous proteins: Coiled-coils, collagen and elastomers* (1st ed., pp. 79–112). MA: Elsevier and Academic Press.

Yang, J. J., Yang, J., Wei, L., Zurkiya, O., Yang, W., Li, S., et al. (2008). Rational design of protein-based MRI contrast agents. *Journal of the American Chemical Society, 130*, 9260–9267.

Zaccai, N. R., Chi, B., Thomson, A. R., Boyle, A. L., Bartlett, G. J., Bruning, M., et al. (2011). A *de novo* peptide hexamer with a mutable channel. *Nature Chemical Biology, 7*, 935–941.

Zastrow, M., & Pecoraro, V. L. (2013). Influence of active site location on catalytic activity in *de novo*-designed zinc metalloenzymes. *Journal of the American Chemical Society, 135*, 5895–5903.

CHAPTER TWENTY-FIVE

Peptide Binding for Bio-Based Nanomaterials

N.M. Bedford*, C.J. Munro[†], M.R. Knecht[†,1]

*National Institute of Standards and Technology, Boulder, CO, United States
[†]University of Miami, Coral Gables, FL, United States
[1]Corresponding author: e-mail address: knecht@miami.edu

Contents

1. Introduction	582
1.1 General Opening Remarks	582
1.2 Biocombinatorial Approaches to Isolate Peptides with Material Affinity	583
1.3 Peptide-Based Approaches for NP Production	584
1.4 Advantages and Limitations of Peptides for NP Production	584
2. Materials	585
2.1 Materials for Peptide Synthesis	585
2.2 Materials for Peptide Purification	586
2.3 Materials for Peptide-Capped NP Synthesis	586
2.4 Materials for PDF Analysis	586
3. Methods	587
3.1 Peptide Synthesis	587
3.2 Peptide-Capped NP Synthesis	590
3.3 Methods for PDF Analysis	592
4. Conclusions	596
Acknowledgments	596
References	596

Abstract

Peptide-based strategies represent transformative approaches to fabricate functional inorganic materials under sustainable conditions by modeling the methods exploited in biology. In general, peptides with inorganic affinity and specificity have been isolated from organisms and through biocombinatorial selection techniques (ie, phage and cell surface display). These peptides recognize and bind the inorganic surface through a series of noncovalent interactions, driven by both enthalpic and entropic contributions, wherein the biomolecules wrap the metallic nanoparticle structure. Through these interactions, modification of the inorganic surface can be accessed to drive the incorporation of significantly disordered surface metal atoms, which have been found to be highly catalytically active for a variety of chemical transformations. We have employed synthetic, site-directed mutagenesis studies to reveal localized binding effects of the peptide at the metallic nanoparticle structure to begin to identify the biological basis of control over biomimetic nanoparticle catalytic activity. The protocols described herein were used to fabricate and characterize peptide-capped nanoparticles in atomic resolution to identify peptide sequence effects on the surface structure of the materials, which can then be directly correlated to the catalytic activity to identify structure/function relationships.

1. INTRODUCTION

1.1 General Opening Remarks

Peptide-enabled routes for nanomaterial synthesis and assembly comprise an area of increasing interest as a versatile strategy to create materials with enhanced and emergent properties (Bedford, Ramezani-Dakhel, et al., 2015; Coppage et al., 2013; Slocik, Stone, & Naik, 2005). The complexity, specificity, and materials-recognition properties of peptides allow for potential rational design routes that are not readily achieved using conventional nanoparticle (NP) ligand combinations (Coppage et al., 2013, 2010; Li, Tang, Prasad, Knecht, & Swihart, 2014; Ruan et al., 2014). Peptides have demonstrated the capability to direct the nucleation, growth, and organization of inorganic nanomaterials in aqueous solution at room temperature (Brown, Sarikaya, & Johnson, 2000; Chiu et al., 2011; Pacardo, Sethi, Jones, Naik, & Knecht, 2009; Slocik & Naik, 2006; Slocik et al., 2005), while imparting sequence-dependent shape, size, and/or materials property control through specific motif recognition (Chiu et al., 2011; Coppage et al., 2013, 2010; Li et al., 2014).

Although initial prospects are exciting, an in-depth understanding of sequence-dependent structure/function relationships is paramount to establish rational bioinspired design rules. Initial efforts have focused on studying the structure of the biotic/abiotic interface of various inorganic surfaces to determine how the overall ensemble of peptide morphologies at this interface can influence NP properties (Heinz, Vaia, Farmer, & Naik, 2008; Ramezani-Dakhel, Ruan, Huang, & Heinz, 2015; Tang et al., 2013). Though valuable fundamental insights into peptide–nanomaterial interactions were obtained, the underlying assumption in most cases has been that the inorganic component is a perfectly ordered, "locally crystalline" material. For peptide-directed NP synthesis, the presence of different binding motifs of the peptide could potentially influence the atomic scale structure at the particle surface, causing structural differences that can impart sequence-dependent material properties. Indeed, we have recently demonstrated such phenomena with peptide-capped Pd, Au, and PdAu bimetallic materials (Bedford, Hughes, et al., 2016; Bedford, Ramezani-Dakhel, et al., 2015; Merrill et al., 2015). These results indicate the possibility to tune NP properties through peptide sequence manipulation, but more knowledge about this interfacial structure is required to achieve such capabilities.

1.2 Biocombinatorial Approaches to Isolate Peptides with Material Affinity

To obtain peptide sequences with known materials binding affinity, biocombinatorial methods have typically been employed. Commonly, M13 phage display and cell surface display are used (Fig. 1), wherein each library consists of a large number ($\sim 10^9$) of members displaying unique and randomized peptide binding domains. Using phage display as an example, the phage library is diluted into buffer and incubated with a target material. Library members with high affinity to the target will bind, while the

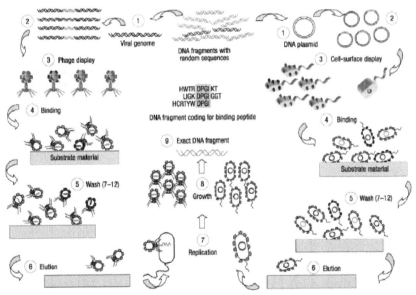

Fig. 1 Scheme of biocombinatorial selection approaches for the isolation of peptides with affinity to inorganic materials. The method for phage display is shown on the *left*, while the process for cell surface display is presented on the *right*. Adapted with permission from Sarikaya, M., Tamerler, C., Jen, A. K. Y., Schulten, K., & Baneyx, F. (2003). Molecular biomimetics: Nanotechnology through biology. Nature Materials, 2, 577–585; copyright 2003, Nature Publishing Group. (See the color plate.)

remaining library members will not. Unbound phage are removed thorough washing procedures, followed by an elution step to remove the bound phage. The elution step usually involves placing the bound phage–target in a buffered solution of lower pH that alters binding through changes in ionic strength and pH. The phage are then amplified in *E. coli* for additional rounds of screening. Typically 2–4 more rounds are performed with increasing stringency, usually by increasing detergent concentration, to remove any nonspecific and marginal target binders. After the final round of biopanning, the phage are isolated and sequenced to obtain the primary structure of the target binding peptide. Several reviews are available that provide a comprehensive description of biocombinatorial approaches for materials binding peptide isolation (Dickerson, Sandhage, & Naik, 2008; Sarikaya, Tamerler, Jen, Schulten, & Baneyx, 2003; Sarikaya, Tamerler, Schwartz, & Baneyx, 2004).

1.3 Peptide-Based Approaches for NP Production

Aqueous-based methods are common for peptide-enabled NP synthesis given the ease at which many peptides can readily dissolve in H_2O (Coppage et al., 2011, 2013, 2010; Li et al., 2014; Slocik & Naik, 2006; Slocik et al., 2005). This requires metal ion precursors and reducing agents that are also soluble in H_2O, which are common as well. In a typical procedure, metal ions are added to a known concentration of peptide in solution to obtain a predetermined peptide:metal ion ratio. After a suitable incubation period, an excess of reducing agent is used (commonly $NaBH_4$) to create peptide-capped NPs. While free peptide in solution can readily cap NP nucleation and growth, peptides bound to supports can also be used to create NP under aqueous conditions. The most well-known example would be the virus-templated nanomaterials from Angela Belcher's laboratory (Lee et al., 2012; Nam et al., 2006, 2010), with other examples including graphene-bound peptides (Kim et al., 2011) and peptide functionalized dendrimers (Bedford et al., 2014).

1.4 Advantages and Limitations of Peptides for NP Production

Traditional approaches for NP synthesis employ indiscriminate, covalent, and monovalent binding of ligands to the inorganic surface (Daniel & Astruc, 2004; Quintanilla et al., 2010; Thomas, Zajicek, & Kamat, 2002). As a result, conventional ligands only interact with the NP at a single site. Furthermore, these methods commonly employ harsh organic solvents and energy intensive high temperatures, whereas the generation of NPs in water at room temperature is ideal to abate energy and environmental concerns. Peptides (in the absence of thiols) bind noncovalently through multiple anchor residues to wrap metal NP surfaces under aqueous conditions. These multiple noncovalent interactions disrupt the metallic structure at the binding sites, generating highly disordered metal atoms at the anchor locations, which can be tuned directly with the amino acid sequence of the peptide. Additionally, peptide capping agents result in small NPs due to the high affinity of the biomolecule to the inorganic surface arresting nucleation and growth (Coppage et al., 2011). The biggest current limitation is the associated cost of peptides, which require time-consuming solid-phase synthesis methods. That said the price of peptides commercially has gone down quite a bit in the past 20 years, making the barrier for technological relevance of peptide-capped NPs achievable.

2. MATERIALS

Unless otherwise noted, all chemicals are used without additional purification.

2.1 Materials for Peptide Synthesis

Protocols for 9-fluorenylmethoxycarbonyl (FMOC) solid-phase peptide synthesis are described; however, other approaches can be employed for peptide synthesis (ie, BOC-based approaches).

1. Solid powders of FMOC- and side chain-protected amino acids. Side chains are protected with various acid-cleavable groups to prevent undesired coupling and byproduct formation.
2. Wang resins substituted with the first amino acid. Wang resins are used to ensure the synthesis of peptides with unmodified C-termini after acid cleavage. Other resins are available that generate peptides with modifications of the C-terminus.
3. Piperidine
4. N,N-diisopropylethylamine (DIPEA)
5. Hydroxybenzotriazole (HoBT hydrate)
6. N,N,N',N'-tetramethyl-O-(1H-benzotriazol-1-yl)uronium hexafluorophosphate (HBTU)
7. Organic solvents including N,N-dimethylformamide (DMF), methanol (MeOH), and diethyl ether
8. Trifluoroacetic acid (TFA)

9. Triisopropylsilane (TIS)
10. 1,2-Ethanedithiol (EDT)
11. Anisole
12. Thioanisole
13. Double-deionized water

2.2 Materials for Peptide Purification
1. HPLC grade acetonitrile
2. HPLC grade 1% TFA water by volume
3. 0.2 μm PTFE syringe filter

2.3 Materials for Peptide-Capped NP Synthesis
1. Metal salts ($HAuCl_4$ and K_2PdCl_4 for Au and Pd NPs, respectively)
2. $NaBH_4$ solid powder
3. Double-deionized water
4. Lyophilized peptide

2.4 Materials for PDF Analysis
To perform PDF analysis, first a high-energy X-ray diffraction (HE-XRD) pattern needs to be taken at sufficiently high reciprocal space values (>25 Å^{-1}, but usually much higher) with very good statistics. Therefore, synchrotron radiation sources are most commonly used to obtain HE-XRD patterns for PDF analysis. For example, the Advanced Photon Source (APS) at Argonne National Laboratory has a series of beam lines capable to perform such measurements, including 1-ID, 6-ID-D, 11-ID-B, and 11-ID-C. To capture large reciprocal space values, area detectors are employed and placed within close proximity the sample. HE-XRD patterns are captured in transmission geometry with the area detector. As with any transmission-based diffraction measurement, a wide range of sample preparation can be implemented. Thin-walled capillaries are the most popular method for preparing samples for PDF analysis, given the associated low background scattering.

1. Beamtime at a HE-XRD beamline with an area detector
2. Lyophilized NP samples
3. Reference material for sample to detector distance calibration. The most commonly used material is the NIST Standard Reference material CeO_2, which is available at most HE-XRD beamlines. In principle, any crystalline material with well-known lattice spacing can be used.

4. Thin-walled capillaries. Kapton capillaries provide low background and are mechanically robust, and thus are the most popular choice for HE-XRD measurements. Glass capillaries may be used as an alternative, particularly for environmentally sensitive samples given the lower gas permeability for glass compared to Kapton. Kapton tubes usually do not come precut or presealed, and therefore require some additional preparation work. Tubes can be cut to the length needed for the experiment, and are commonly sealed with epoxy.
5. Weigh paper for loading capillaries

3. METHODS
3.1 Peptide Synthesis

For the peptide synthesis protocols, all methods are described for a TETRAS model synthesizer manufactured by CreoSalus, Inc. This system is equipped with 108 independent reaction wells to simultaneously synthesize 108 different sequences. Changes in protocols should be employed due to instrument-specific requirements.

3.1.1 Solutions Required for Peptide Synthesis

1. Program the sequence of the peptide being synthesized in the computer controlling the system for each reaction well. The last amino acid in the sequence is attached to the Wang resin and should not be included in calculations of solution volumes required for synthesis.
2. For every time an individual amino acid is added across all peptides being synthesized, 1 mL of a 0.5 M amino acid solution must be made using DMF as the solvent. The additional mass of the FMOC and side chain-protecting groups must be taken into account. For example, if 10 alanine residues are being added over all of the sequences being synthesized, 10 mL of 0.5 M alanine should be prepared. An additional 10 mL of this solution should be made for the instrument priming process.
3. For every amino acid being added in the peptide sequence, prepare 4 mL of 20% by volume piperidine, 1 mL of 1 M DIPEA, and 1 mL of 0.5:0.5 M HoBT:HBTU. All of these solutions are diluted or dissolved to their appropriate concentration using DMF. At least an additional 15 mL of each solution should be made for the instrument priming step. Note that the HoBT:HBTU mixture is light sensitive and should be put in an amber bottle.

3.1.2 Protocols for Automated Synthesis

1. For every reaction well of peptide being made, 150 mg of the appropriate Wang resin should be added into the well. Place the wells in the appropriate carousel positions.
2. Prime the instrument by flushing all stations with the appropriate amount of solution being used at that station (ie, the amino acid being dispensed from that station). The amount of priming solution is dictated by how long the transport tubing is from the solution to the well. This step is very important to minimize byproduct formation due to extraneous materials being in the lines from previous syntheses. Once the stations are primed and the wells are activated, the synthesis is ready to begin.
3. Begin the synthesis by adding 3 mL of DMF to swell the resin. Mix the reaction for 15 min before purging the DMF from the reaction well. Repeat this process four times.
4. Once the resin is swollen, perform a washing step. For this, add 2 mL of DMF and mix for 3 min before purging. Repeat this three times. Next add 2 mL of MeOH and mix for 3 min before purging and repeat this three times. Finally add 2 mL of DMF and mix for 3 min before purging. Repeat this last step three times. This constitutes one full washing cycle, which is used after each coupling and deprotection step.
5. Deprotect the amine of the residue on the resin. For this, add 2 mL of the piperidine solution and mix for 10 min before purging. Next add another 2 mL of the piperidine solution and mix for 15 min before purging. The second piperidine addition is used to ensure maximal FMOC deprotection. Once deprotected, wash the resin using the process of step 4.
6. With the newly exposed amine functionality, coupling of the next amino acid in the sequence can occur. Add 1 mL of the desired amino acid solution, 1 mL of the DIPEA solution, and 1 mL of the HoBT:HBTU solution. Mix the reaction for 120 min before purging. Once this process is complete, wash the resin using the process of step 4.
7. Repeat steps 5 and 6 to couple subsequent amino acids to the growing peptide chain on the resin surface.
8. When the final amino acid is added to the resin, repeat step 5 to deprotect the final amine group of the amino acid. If this does not occur, removal of the FMOC group after cleavage of the peptide from the resin is exceedingly difficult. After the final FMOC deprotection step, wash the resin using the process of step 4.

3.1.3 Cleavage from Resin

1. Cleave the peptide from the resin using an appropriate cleavage cocktail. For peptides containing a cysteine residue, EDT, thioanisole, and anisole are required to ensure thiol protecting group scavenging; however, these reagents are not used if thiols are not present in the sequence.
2. Prepare the appropriate cleavage cocktail at a 4-mL volume per reaction well:

Cleavage Cocktail

	Volume of Solution (mL)	TFA (mL)	EDT (mL)	Thioanisole (mL)	Anisole (mL)	Water (mL)	TIS (mL)
Peptide with cysteine	4	3.60	0.12	0.20	0.08	0.00	0.00
Peptide without cysteine	4	3.80	0.00	0.00	0.00	0.10	0.10

3. First inject 3 mL of the cleavage cocktail into the reaction well, covering the resin. Allow the cocktail to mix with the resin for 2 h before purging into a collection port to collect the liquid that contains the deprotected peptides released from the resin surface.
4. Inject another 1 mL of the cleavage cocktail over the resin and mix for 15 min before it is purged and the solution is collected with the original 3 mL cocktail solution. This second step is used to ensure complete cleavage and deprotection of any remaining peptides on the resin.
5. Empty each collection port (~4 mL of cocktail/peptide solution) into 20 mL of cold diethyl ether to precipitate the peptides.
6. Centrifuge the cloudy peptide dispersion in diethyl ether for 12 min at 5°C and 9000 rpm to pellet the synthesized biomolecule. Decant the ether and disperse the peptide pellet in additional ether. Repeat the ether cleaning cycle two more times for a total volume of 60 mL of ether for each peptide well.
7. Store the solid peptide at −80°C until it can be purified by reverse-phase HPLC.

3.1.4 Purification via Reverse-Phase HPLC

1. If the peptide was at −80°C, allow it to come to room temperature.
2. Once at room temperature add 5 mL of 1% by volume TFA water. This should dissolve most hydrophilic peptides; however, if the peptide is

sufficiently hydrophobic, titrate in a sufficient volume of HPLC grade acetonitrile until all of the solid material is dissolved.
3. Filter the peptide solution using a 0.2-μm PTFE syringe filter.
4. Purify the peptide using a gradient of 100% water to 100% acetonitrile, both containing 1% by volume TFA, over the course of 60 min. at a flow rate of 10 mL per minute using a C18 column. Different instrument systems will require different gradients and flow rates depending upon the column used.
5. Collect the peptide as it elutes from the column based upon biomolecule absorbance at 215 nm.
6. From the eluent, obtain a mass spectrum of the sample to confirm the appropriate molecular weight and sequence.
7. Once confirmed, lyophilize the sample of peptide and store the solid powder at −80°C.

3.2 Peptide-Capped NP Synthesis

Identical approaches are employed for the formation of peptide-capped NPs composed of either Au or Pd. As such, a generic metal ion M^+ is used throughout the text, which represents either Au or Pd (Fig. 2) (Pacardo et al., 2009).

3.2.1 Solutions Required for Peptide-Capped NP Synthesis

1. Prepare a 100 mM metal salt solution in water. This requires measurement of appropriate masses of the metal salt and dissolution of the material in the needed volume of water. These salts are readily soluble, thus formation of the solution is straightforward. For many metal salts, the use of iron-based spatulas should be avoided due to galvanic reactions that can occur upon contact with iron. We use glass pipettes, which provide sufficient space to measure out small amounts of precious metal salts.
2. For the peptide solution, dissolve sufficient mass of the lyophilized peptide prepared earlier in water to reach a final concentration of 1.0 mM. Obtaining an accurate mass of the peptide can be challenging due to the low mass of biomolecule typically used to prepare a solution of 1.0 mL, thus a highly accurate analytical balance is necessary. The peptides isolated from phage display are typically very soluble in water, thus solution formation is relatively easy.
3. The NaBH$_4$ solution should be prepared immediately before use. The reductant readily reacts with water, thus if it is prepared too early,

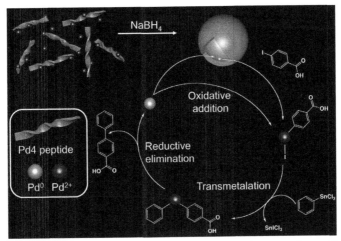

Fig. 2 Scheme for the general synthesis and catalytic application of peptide-capped metal nanoparticles. In general, metal ions are comixed with the peptide in water. After incubation, the metal ions are reduced with an excess of $NaBH_4$ to generate the inorganic core capped on the surface with the biomolecules. The scheme shown is for Pd; however, the approach is readily translatable to other compositions. Once fabricated, the materials can be studied for numerous properties, including catalysis as shown here for Stille coupling. *Adapted with permission from Pacardo, D. B., Sethi, M., Jones, S. E., Naik, R. R., & Knecht, M. R. (2009). Biomimetic synthesis of Pd nanocatalysts for the Stille coupling reaction. ACS Nano, 3, 1288–1296; copyright 2009, American Chemical Society.*

its reducing power is rapidly diminished. For this solution, dissolve the solid $NaBH_4$ powder in ice cold water to reach a concentration of 100 mM. Keep the solution on ice throughout the duration of the experiment.

3.2.2 Protocols for NP Synthesis

In the protocols listed later, a M^+:peptide ratio of 2 is employed; however, this ratio can be varied by increasing the amount of M^+ added to the solution concomitant with a decrease in the amount of water added to reach a final reaction volume of 5.0 mL. In all cases, sufficient $NaBH_4$ must be added to ensure complete reduction of M^+ to M^0.

1. Dilute 500 µL of the 1.0 mM peptide solution with 4.460 mL of water.
2. To the diluted peptide solution, add 10 µL of the prepared aqueous M^+ solution. This solution should become pale yellow upon metal addition.
3. Vigorously stir the solution for 10 min to allow for complex formation between the biomolecule and metal ions.

4. Inject 30 μL of the freshly prepared, ice cold NaBH$_4$ solution. The reductant should be added rapidly while the solution is stirred vigorously. Upon NaBH$_4$ addition, a swift color change should be evident, turning dark red/brown for the Au NPs and brown for the Pd materials.
5. Allow the reaction to continue to reduce for 1 h without stirring at ambient temperature.

3.2.3 Characterization of Peptide-Capped NPs

Two standard characterization methods of the peptide-capped NPs are employed: UV–vis spectroscopy and transmission electron microscopy (TEM), as described later.

For UV–vis spectroscopy, a quartz cuvette with a 1.0-cm pathlength should be used. Quartz is important due to its low spectral absorbance, allowing for accurate measurement of the absorbance properties of the materials in the UV region. Standard UV–vis analyses methods should be used on either a single beam or a photodiode array spectrometer. Should the sample be too concentrated, smaller pathlength cuvettes can be used to diminish the absorbance intensity. Dilution of the sample is usually not employed to avoid material loss due to analysis.

TEM analysis is typically performed using a microscope with an electron beam operating at 200 kV. Such a beam intensity is required to view smaller NPs and achieve high-resolution, lattice-resolved images. To prepare the sample, commercially available metal grids are used, which are coated in a thin layer of carbon. Place the grid such that the carbon surface is displayed at the top. Drop 5.0 μL of the NP solution on the grid surface, allowing the solvent to evaporate. This is usually done inside of a desiccator to ensure complete evaporation of the water to avoid release the solvent inside the TEM.

3.3 Methods for PDF Analysis

3.3.1 Sample Preparation

Filling capillaries with lyophilized NP samples can be accomplished using a variety of methods. The most common is to place the sample on creased weigh paper, then simply run the capillary across the crease until the material is inside the capillary. While most HE-XRD beamlines have X-ray beam sizes on the order of ~100 μm^2, filling the capillaries up to 2–3 cm is preferred to allow for straightforward sample alignment.

3.3.2 HE-XRD Data Acquisition
3.3.2.1 Sample Preparation
Samples can be loaded in the beam in a variety of ways, but the most common sample holder is a multicapillary holder that can hold 30–40 capillaries. This is loaded onto a translational stage that can move vertically and perpendicular to the X-ray beam to allow for sample alignment.

3.3.2.2 Calibrating the Sample to Detector Distance
Standard materials, such as CeO_2, need to be used to calibrate the sample to detector distance. This is commonly accomplished using the program Fit2D, which is also used for subsequent data processing (Hammersley, Svensson, Hanfland, Fitch, & Hausermann, 1996). Standard material HE-XRD patterns are collected at very short exposure times (usually 0.1 s) as they are highly scattering and can leave residual impressions on area detectors. For calibrations using CeO_2 as a standard in Fit2D, import the HE-XRD pattern. In the calibration menu, select CeO_2. Set all possible parameters to "refine" *except* the X-ray wavelength (which is already known). Use a reasonable approximation for the sample to detector distance, otherwise the fitting protocol fails in Fit2D if the distance is too far off. Click on a minimum of four points in the first diffraction feature of CeO_2 evenly spaced around the ring. Once these points are set, click calibrate, which runs a fitting protocol that provides the sample to detector distance. If you move your sample stage parallel to the beam at any time, the calibration needs to be redone.

3.3.2.3 Setting Acquisition Times
When performing HE-XRD measurements for PDF analysis, along with achieving high reciprocal space values (which are determined by the X-ray energy and sample to detector distance), very good statistics, and signal are necessary. In terms of statistics, this involves taking multiple exposures that are summed together to obtain the HE-XRD pattern. More exposures are better, but usually at least 500 exposures are considered to be the minimum number of exposures needed (although opinions vary). There is usually a software limit to how many exposures can be summed in one HE-XRD pattern. In this instance, multiple patterns are averaged together to arrive at the desired amount of exposures. Detector saturation determines how long each exposure will be, where researchers typically aim for an exposure time that yields a number of counts equivalent to ~50–60% saturation of the detector. For example, the Perkin Elmer area

detectors used at APS saturate at ~60,000 counts, so one should aim for ~30,000 counts in a HE-XRD experiment. The exposure time and number of overall exposures ultimately dictates your acquisition time.

3.3.2.4 Data Acquisition
Once the number of exposures and exposure time has been determined, use these parameters to collect HE-XRD patterns on your samples. A background sample also needs to be performed using an identical number of exposures and the same exposure time for background correction.

3.3.3 HE-XRD Pattern Conversion to PDFs
3.3.3.1 HE-XRD Integration
Using Fit2D, import all HE-XRD data from the area detector (usually as .tiff files) and perform an integration. Integrated HE-XRD data are meaningless if not appropriately calibrated, so ensure that a calibration was performed. One caveat to integration is the simple fact that most area detectors are rectangular in shape, while diffraction patterns are circular. This leads to issues in integrating data out to very high reciprocal space values, as intensity will unevenly drop off from the pixels in the corner of the detector. Be sure to pick an appropriate q-range for integration, keeping this effect in mind. Consult with beamline scientists for more specific details.

3.3.3.2 Calculating PDFs
The calculation of a PDF from HE-XRD data can be performed using readily available software. The most commonly used programs are PDF-GetX2 and the newer PDF-GetX3 from Simon Billinge and coworkers (Juhas, Davis, Farrow, & Billinge, 2013; Qiu, Thompson, & Billinge, 2004). Both are freely available online. Another program, RAD, from Valeri Petkov can also convert HE-XRD into PDFs and is intuitive to use (Petkov, 1989). The newest version of this software is not available for free online, although unsupported older versions can be found online free of charge. PDF-GetX2, PDF-GetX3, and RAD all operate using the following general pathway. First, the HE-XRD dataset of interest is background corrected using a background sample (usually an empty capillary). The background corrected data are converted to a total structure function, $S(Q)$, and reduced structure function, $F(Q)$, which are defined as:

$$S(Q) = 1 + \frac{I^{coh}(Q) - \Sigma c_i |f_i(Q)|^2}{|\Sigma c_i f_i(Q)|^2}$$
$$F(Q) = Q[S(Q) - 1]$$

where I^{coh} is the coherent part of the HE-XRD pattern (ie, the background corrected data), and c_i and $f_i(Q)$ are the atomic concentration and X-ray scattering factor of each elemental species of type i, respectively. From here, $F(Q)$ is Fourier transformed in the atomic PDF, $G(r)$ via:

$$G(r) = \left(\frac{2}{\pi}\right) \int_{Q=0}^{Q_{max}} Q[S(Q)-1] \sin(Qr) dQ$$

where $G(r) = 4\pi r(\rho(r) - \rho(0))$, r is the distance between two atom pairs, $\rho(r)$ is local atomic density, and $\rho(0)$ is the average atomic density. Peaks in the PDF represent distances within the material of a high number of atomic pairs and valleys represent distances of little atomic density. The PDF for nanomaterials may not extend to distances longer than the size of the nanomaterial. For more details, see excellent review articles by Simon Billinge and Valeri Petkov (Billinge & Kanatzidis, 2004; Petkov, 2008).

3.3.4 Structural Modeling

Atomic PDFs can be modeled using crystallographic and noncrystallographic approaches. Choosing an approach ultimately depends on how much is known about this system prior to modeling. If the material is anticipated to be moderately disordered, then crystallographic modeling methods may not be serviceable. A good rule of thumb is to first attempt using crystallographic methods, and if unsuccessful due to a lack of structural order, then use noncrystallographic approaches. PDFgui is the most popular software for crystallographic PDF modeling, wherein lattice constants, atomic positions, thermal coefficient, and other parameters are used to describe the atomic PDF data (Farrow et al., 2007). For noncrystallographic approaches, reverse Monte Carlo (RMC) simulations are commonly used, with RMCProfile (Krayzman et al., 2009) and RMC++ (Orsolya & Valeri, 2013) being the most popular options. A new RMC platform, fullRMC, has been recently published (Aoun, 2016), which has a multitude of features found in both RMCProfile and RMC++ while providing options for "molecular moves," such as translations and rotations. In RMC, an atom is moved at random from an initial configuration of atoms and a new structure function (here, the atomic PDF) is calculated. If the new PDF provides a better fit to the experimental data, the move is accepted and the algorithm proceeds. Otherwise, the move is disregarded at a certain percentage of moves to prevent the simulation from getting stuck in a local minimum in configuration space. This process is repeated until the computed PDF converges to the experimental PDF data in very good detail. One advantage of RMC over crystallographic modeling is that it can be performed on systems of any size, directly providing structural models that can be used to assess structure/function relationships.

4. CONCLUSIONS

In conclusion, peptide-based methods are intriguing avenues for the fabrication of functional materials. They have been employed as catalysts and sensors, and for optical applications (Bedford et al., 2014; Bedford, Hughes, et al., 2016; Bedford, Ramezani-Dakhel, et al., 2015; Chen, Zhang, & Rosi, 2008; Chiu et al., 2011; Coppage et al., 2011, 2013, 2010; Lévy et al., 2004; Li et al., 2014; Merrill et al., 2015; Nam et al., 2010; Slocik & Naik, 2006; Slocik et al., 2005; Slocik, Zabinski, Phillips, & Naik, 2008; Tang et al., 2015). The next great challenge is the ability to de novo design peptide sequences that can control the material properties via surface binding. It is anticipated that the peptide could be designed to control the size, shape, surface morphology, and overall properties of the NP to generate functional materials with targeted activities. The ability to derive design rules for this capability require an in-depth understanding of the structure/function relationships of peptide-capped NPs where PDF methods are making major advances in this arena. More work is definitely required, but initial critical steps have been achieved using the methods described herein. Advancing this approach to materials of new compositions and properties is required.

ACKNOWLEDGMENTS
The authors greatly acknowledge support from the Air Force Office of Scientific Research, Grant number FA9550-12-1-0226. Support from the University of Miami is also acknowledged.

REFERENCES
Aoun, B. (2016). Fullrmc, a rigid body reverse Monte Carlo modeling package enabled with machine learning and artificial intelligence. *Journal of Computational Chemistry, 37*, 1102–1111.

Bedford, N. M., Bhandari, R., Slocik, J. M., Seifert, S., Naik, R. R., & Knecht, M. R. (2014). Peptide-modified dendrimers as templates for the production of highly reactive catalytic nanomaterials. *Chemistry of Materials, 26*, 4082–4091.

Bedford, N. M., Hughes, Z. E., Tang, Z., Li, Y., Briggs, B. D., Ren, Y., et al. (2016). Sequence-dependent structure/function relationships of catalytic peptide-enabled gold nanoparticles generated under ambient synthetic conditions. *Journal of the American Chemical Society, 138*, 540–548.

Bedford, N. M., Ramezani-Dakhel, H., Slocik, J. M., Briggs, B. D., Ren, Y., Frenkel, A. I., et al. (2015). Elucidation of peptide-directed palladium surface structure for biologically tunable nanocatalysts. *ACS Nano, 9*, 5082–5092.

Billinge, S. J. L., & Kanatzidis, M. G. (2004). Beyond crystallography: The study of disorder, nanocrystallinity and crystallographically challenged materials with pair distribution functions. *Chemical Communications, 749–760.*

Brown, S., Sarikaya, M., & Johnson, E. (2000). A genetic analysis of crystal growth. *Journal of Molecular Biology, 299*, 725–735.

Chen, C.-L., Zhang, P., & Rosi, N. L. (2008). A new peptide-based method for the design and synthesis of nanoparticle superstructures: Construction of highly ordered gold nanoparticle double helices. *Journal of the American Chemical Society, 130*, 13555–13557.

Chiu, C.-Y., Li, Y., Ruan, L., Ye, X., Murray, C. B., & Huang, Y. (2011). Platinum nanocrystals selectively shaped using facet-specific peptide sequences. *Nature Chemistry, 3*, 393–399.

Coppage, R., Slocik, J. M., Briggs, B. D., Frenkel, A. I., Heinz, H., Naik, R. R., et al. (2011). Crystallographic recognition controls peptide binding for bio-based nanomaterials. *Journal of the American Chemical Society, 133*, 12346–12349.

Coppage, R., Slocik, J. M., Ramezani-Dakhel, H., Bedford, N. M., Heinz, H., Naik, R. R., et al. (2013). Exploiting localized surface binding effects to enhance the catalytic reactivity of peptide-capped nanoparticles. *Journal of the American Chemical Society, 135*, 11048–11054.

Coppage, R., Slocik, J. M., Sethi, M., Pacardo, D. B., Naik, R. R., & Knecht, M. R. (2010). Elucidation of peptide effects that control the activity of nanoparticles. *Angewandte Chemie International Edition, 49*, 3767–3770.

Daniel, M.-C., & Astruc, D. (2004). Gold nanoparticles: Assembly, supramolecular chemistry, quantum-size-related properties, and applications toward biology, catalysis, and nanotechnology. *Chemical Reviews, 104*, 293–346.

Dickerson, M. B., Sandhage, K. H., & Naik, R. R. (2008). Protein- and peptide-directed syntheses of inorganic materials. *Chemical Reviews, 108*, 4935–4978.

Farrow, C. L., Juhas, P., Liu, J. W., Bryndin, D., Božin, E. S., Bloch, J., et al. (2007). PDFfit2 and PDFgui: Computer programs for studying nanostructure in crystals. *Journal of Physics. Condensed Matter, 19*, 335219.

Hammersley, A. P., Svensson, S. O., Hanfland, M., Fitch, A. N., & Hausermann, D. (1996). Two-dimensional detector software: From real detector to idealised image or two-theta scan. *High Pressure Research, 14*, 235–248.

Heinz, H., Vaia, R. A., Farmer, B. L., & Naik, R. R. (2008). Accurate simulation of surfaces and interfaces of face-centered cubic metals using 12−6 and 9−6 Lennard-Jones potentials. *Journal of Physical Chemistry C, 112*, 17281–17290.

Juhas, P., Davis, T., Farrow, C. L., & Billinge, S. J. L. (2013). PDFgetX3: A rapid and highly automatable program for processing powder diffraction data into total scattering pair distribution functions. *Journal of Applied Crystallography, 46*, 560–566.

Kim, S. N., Kuang, Z., Slocik, J. M., Jones, S. E., Cui, Y., Farmer, B. L., et al. (2011). Preferential binding of peptides to graphene edges and planes. *Journal of the American Chemical Society, 133*, 14480–14483.

Krayzman, V., Levin, I., Woicik, J. C., Proffen, T., Vanderah, T. A., & Tucker, M. G. (2009). A combined fit of total scattering and extended X-ray absorption fine structure data for local-structure determination in crystalline materials. *Journal of Applied Crystallography, 42*, 867–877.

Lee, Y., Kim, J., Yun, D. S., Nam, Y. S., Shao-Horn, Y., & Belcher, A. M. (2012). Virus-templated Au and Au-Pt core-shell nanowires and their electrocatalytic activities for fuel cell applications. *Energy & Environmental Science, 5*, 8328–8334.

Lévy, R., Thanh, N. T. K., Doty, R. C., Hussain, I., Nichols, R. J., Schiffrin, D. J., et al. (2004). Rational and combinatorial design of peptide capping ligands for gold nanoparticles. *Journal of the American Chemical Society, 126*, 10076–10084.

Li, Y., Tang, Z., Prasad, P. N., Knecht, M. R., & Swihart, M. T. (2014). Peptide-mediated synthesis of gold nanoparticles: Effects of peptide sequence and nature of binding on physicochemical properties. *Nanoscale, 6*, 3165–3172.

Merrill, N. A., McKee, E. M., Merino, K. C., Drummy, L. F., Lee, S., Reinhart, B., et al. (2015). Identifying the atomic-level effects of metal composition on the structure and catalytic activity of peptide-templated materials. *ACS Nano, 9*, 11968–11979.

Nam, K. T., Kim, D. W., Yoo, P. J., Chiang, C. Y., Meethong, N., Hammond, P. T., et al. (2006). Virus-enabled synthesis and assembly of nanowires for lithium ion battery electrodes. *Science, 312*, 885–888.

Nam, Y. S., Magyar, A. P., Lee, D., Kim, J.-W., Yun, D. S., Park, H., et al. (2010). Biologically templated photocatalytic nanostructures for sustained light-driven water oxidation. *Nature Nanotechnology, 5*, 340–344.

Orsolya, G., & Valeri, P. (2013). Reverse Monte Carlo study of spherical sample under non-periodic boundary conditions: The structure of Ru nanoparticles based on x-ray diffraction data. *Journal of Physics. Condensed Matter, 25*, 454211.

Pacardo, D. B., Sethi, M., Jones, S. E., Naik, R. R., & Knecht, M. R. (2009). Biomimetic synthesis of Pd nanocatalysts for the Stille coupling reaction. *ACS Nano, 3*, 1288–1296.

Petkov, V. (1989). Rad, a program for analysis of X-ray-diffraction data from amorphous materials for personal computers. *Journal of Applied Crystallography, 22*, 387–389.

Petkov, V. (2008). Nanostructure by high-energy X-ray diffraction. *Materials Today, 11*, 28–38.

Qiu, X., Thompson, J. W., & Billinge, S. J. L. (2004). PDFgetX2: A GUI-driven program to obtain the pair distribution function from X-ray powder diffraction data. *Journal of Applied Crystallography, 37*, 678.

Quintanilla, A., Butselaar-Orthlieb, V. C. L., Kwakernaak, C., Sloof, W. G., Kreutzer, M. T., & Kapteijn, F. (2010). Weakly bound capping agents on gold nanoparticles in catalysis: Surface poison? *Journal of Catalysis, 271*, 104–114.

Ramezani-Dakhel, H., Ruan, L., Huang, Y., & Heinz, H. (2015). Molecular mechanism of specific recognition of cubic Pt nanocrystals by peptides and of the concentration-dependent formation from seed crystals. *Advanced Functional Materials, 25*, 1374–1384.

Ruan, L., Ramezani-Dakhel, H., Lee, C., Li, Y., Duan, X., Heinz, H., et al. (2014). A rational biomimetic approach to structure defect generation in colloidal nanocrystals. *ACS Nano, 8*, 6934–6944.

Sarikaya, M., Tamerler, C., Jen, A. K. Y., Schulten, K., & Baneyx, F. (2003). Molecular biomimetics: Nanotechnology through biology. *Nature Materials, 2*, 577–585.

Sarikaya, M., Tamerler, C., Schwartz, D. T., & Baneyx, F. (2004). Materials assembly and formation using engineered polypeptides. *Annual Review of Materials Research, 34*, 373–408.

Slocik, J. M., & Naik, R. R. (2006). Biologically programmed synthesis of bimetallic nanostructures. *Advanced Materials, 18*, 1988–1992.

Slocik, J. M., Stone, M. O., & Naik, R. R. (2005). Synthesis of gold nanoparticles using multifunctional peptides. *Small, 1*, 1048–1052.

Slocik, J. M., Zabinski, J. S., Phillips, D. M., & Naik, R. R. (2008). Colorimetric response of peptide-functionalized gold nanoparticles to metal ions. *Small, 4*, 548–551.

Tang, Z., Lim, C.-K., Palafox-Hernandez, J. P., Drew, K. L. M., Li, Y., Swihart, M. T., et al. (2015). Triggering nanoparticle surface ligand rearrangement via external stimuli: Light-based actuation of biointerfaces. *Nanoscale, 7*, 13638–13645.

Tang, Z., Palafox-Hernandez, J. P., Law, W.-C., Hughes, Z. E., Swihart, M. T., Prasad, P. N., et al. (2013). Biomolecular recognition principles for bionanocombinatorics: An integrated approach to elucidate enthalpic and entropic factors. *ACS Nano, 7*, 9632–9646.

Thomas, K. G., Zajicek, J., & Kamat, P. V. (2002). Surface binding properties of tetraoctylammonium bromide-capped gold nanoparticles. *Langmuir, 18*, 3722–3727.

AUTHOR INDEX

Note: Page numbers followed by "*f*" indicate figures, and "*t*" indicate tables.

A

Aaronson, J., 473–474
Abbel, R., 284
Abbenante, G., 305, 306–307*t*, 308
Abe, M., 444, 452
Abecassis, J., 465–466
Abel-Santos, E., 150–151
Abotsi, E., 150
Adachi, S., 513–514
Adamian, L., 383
Adams, D.J., 304
Adams, J.J., 54
Adams, K.L., 100
Adams, P.D., 142–144, 243–244
Adams, S.-L., 204
Ädelroth, P., 503
Aerni, H.R., 130, 214
Agricola, I., 335–336
Ahmad, I., 128–129
Ahmad, R., 252–253, 260
Ahsan, M.Q., 3–4
Ai, H.W., 92–93, 113
Aiken, M.J., 170–173, 173*f*, 175*t*
Akbari, J., 162
Akhtar, M.K., 366
Al Hashimi, H.M., 252–253, 259
Albeck, S., 204, 410
Alber, T., 257–259
Albericio, F., 475–476
Alberts, I.L., 235–236
Alcala-Torano, R., 390–411
Alderighi, L., 354
Aldous, A.R., 335–336
Alewood, D., 257
Alewood, P.F., 257, 304
Alexander, C.G., 163
Alfonta, L., 113, 124, 380
Al-Hashimi, H.M., 252–254, 265
Aliyan, A., 12
Allemann, R.K., 172–173, 173*f*
Allen, K.N., 561, 561*f*, 573–574, 575*t*
Allen, L., 159

Allen, S.J., 472–473
Almohaini, M., 326
Althoff, E.A., 504
Altmann, E., 475
Altmann, K.-H., 475
Altoff, E.A., 419–420
Altuve, A., 436
am Ende, C.W., 562–563
Amar, S., 308–309
Amblard, M., 435
Ambrogelly, A., 113
Ambroggio, X.I., 225–227, 232–236, 245–247
Amidon, G.L., 304
Amoros, D., 243
Amoroso, J.H., 419–420
Anbazhagan, A., 326
Ancel, L., 573–574
Anderson, E.A., 252–253
Anderson, J.C., 90–91, 111–112
Anderson, J.L.R., 370, 372, 408
Anderson, J.T., 252–255, 256*t*, 257–259
Anderson, M.E., 544–545
Andreozzi, C., 379, 477
Anfinsen, C.B., 426
Ang, M.C., 507
Annele, H., 506–508
Antholine, W.E., 507
Anthony-Cahill, S.J., 570
Antipov, A.N., 116
Antolovich, M., 371–372
Antoniou, C., 170–173, 173*f*, 175*t*, 177, 180
Antonkine, M.L., 392–395, 394*f*
Antos, J.M., 12, 150–151
Antranikian, G., 457–458
Aono, Y., 180–181*t*
Aoun, B., 595–596
Aoyama, K., 96
Aoyama, Y., 444–446
Apfel, C., 400–401
Apfel, U., 400–401
Apitz, A., 456

Apostolov, R., 511
Apostolovic, B., 564
Appel, R.D., 158–159
Appella, D.H., 280
Apweiler, R., 507
Arai, M., 154
Aramburu, I.V., 91, 97
Aravinda, S., 570
Ardá, A., 334–335
Arendall, W.B., 142–144, 143t
Arganda-Carreras, I., 185
Argarana, C.E., 542
Arkin, M.R., 304
Arlt, M., 541–542
Armstrong, C.T., 370, 372, 408
Armstrong, F.A., 521–522
Armstrong, R.N., 225, 231–232
Arnold, F.H., 155–156, 192, 228, 456, 540
Arnold-Godwin, H., 419
Arora, P.S., 304–305, 308–309
Arosio, P., 472–473
Arslan, E., 237–238
Artero, V., 390–391, 410
Ashani, Y., 541
Ashford, P., 507
Askew, K., 335–336
Asselberghs, I., 408
Astashkin, A.V., 378–379, 392, 396–398
Astruc, D., 584–585
Athanassiou, Z., 341
Atta, M., 390–391, 410
Aube, C., 281–282
Ault-Riché, D., 57–58
Aussignargues, C., 398–399
Ausubel, F.M., 459–460
Autran, J.C., 465–466
Avan, I., 304
Avilés, F.X., 212–213
Avrutina, O., 475–476
Ayala, M., 456, 463, 466
Azain, M.J., 150, 159
Aziz, R., 252–253
Azzarito, V., 284

B

Baarlink, C., 170–171, 173–174
Baase, W.A., 254
Baca, M., 48, 61

Bacchi, M., 390–391
Bader, G.D., 204
Bahrami Dizicheh, Z., 390–411
Baiga, T.J., 91–93
Bailey, C.W., 48
Bailey, J.B., 224–247
Bailey, R.L., 159
Baird, E.E., 284
Bairoch, A., 507
Bajjuri, K.M., 90
Baker, D., 128, 224–225, 260, 265, 268, 308–309, 410, 504, 508, 541
Baker, E.N., 151
Baker, M.H., 504–505
Baker, S.E., 503–504
Balaram, P., 334–335, 475, 562–563, 570
Baldwin, E.P., 254
Baldwin, G.S., 513–514
Baldwin, M.J., 419–420
Baldwin, R.L., 254
Bale, J.B., 224–225
Bale, S., 48, 57–58
Baleja, J.D., 252–253, 260, 309
Ball, Z.T., 2–18
Ballard, C.C., 142–144, 243–244
Ban, N., 410
Baneyx, F., 583–584
Banner, D.W., 472–473
Bannwarth, M., 546
Banville, S., 280
Bao, C., 281–282
Baptista, A.M., 309, 341
Baptiste, B., 280–281, 284, 286
Baratto, M.C., 462–465
Barba, M., 335–336
Barbas, C.F., 61
Barbuto, S., 305
Bard, A.J., 428–429
Barnard, A., 283–284
Barnea, I., 82
Barnhill, H.N., 13
Barquera, B., 503
Barra, A.L., 462–463
Barrios, A.M., 308
Barthelmes, K., 561
Bartlett, G.J., 564, 566–567
Bartsch, R.G., 472–473
Bartunik, H.D., 457–459

Bartz, R., 473–474
Bashiruddin, N.K., 310
Baslé, A., 472–473
Bass, S.H., 52–54
Bassani, D.M., 280
Basumallick, L., 525t
Bates, P.A., 29–30
Battersby, J.E., 341–342
Battye, T.G., 243–244
Bauer, A., 204
Bauer, C.E., 377
Baum, G., 280
Bazan, J.F., 472–473
Bean, J.W., 334–335
Bear, J.E., 170–176, 173f, 178–180
Bear, J.L., 3–5
Beasley, J.A.R., 136–138
Beatty, J.T., 377
Beauchamp, J.L., 457–458
Beck, S., 473–474
Becker, C.F., 521–522
Beckta, J.M., 309–310
Beckwith, K.A., 371–372
Bedford, N.M., 582–596
Bednarek, M.A., 304
Begley, T.P., 502–504
Beierle, J.M., 475–476
Beinborn, M., 252–253
Beinert, H., 392–393, 396–397
Belcher, A.M., 584
Beld, J., 305–309, 306–307t, 311–312, 325
Belousov, V.V., 122–123
Bender, G.M., 383, 408
Bendtsen, J.D., 547–548
Benhar, I., 82
Benkovic, S.J., 150–151, 236
Benner, J., 150–151
Ben-Neriah, Y., 309
Bennett, B., 234–235
Bennett, S.P., 150–164
Benson, D.R., 436
Benzinger, D., 170–173, 173f
Berg, J.M., 228–229, 419, 562–563
Berg, R.H., 280
Bergdoll, M., 225, 231–232
Berggren, G., 390–391
Berglund, H., 163
Berl, V., 280

Berman, H.M., 507
Bernadac, A., 547
Berrin, J.G., 456
Berry, E.A., 371, 513–514
Berry, S.M., 504–505
Berson, S.A., 22–24, 26
Berthelot, T., 309
Bertini, I., 228–229
Bertozzi, C.R., 124–125
Bertrand, D., 311–312
Berube, A., 113
Berwick, M.R., 565t, 566–567, 570–575, 575t
Beste, G., 48
Bethge, P.H., 472–473
Betker, J., 410
Betz, S.F., 369, 419–420
Bey, M., 456
Beyermann, M., 90, 99–101, 101f, 103–104
Bezsonova, I., 252
Bhagi-Damodaran, A., 502–529
Bhakta, M.S., 218
Bhandari, R., 584, 596
Bhat, T.N., 507
Bialas, C., 366–384
Bianco, A., 97–99
Biava, H., 252–253
Bidwai, A., 515
Bieber Urbauer, R.J., 369
Bielinska, S., 335–336
Bienert, M., 101
Biesalski, M., 475–476
Biffinger, J.C., 254
Biggins, J., 393–395
Bikker, F.J., 456
Bilgiçer, B., 252–254, 259–260
Bill, E., 379
Billinge, S.J.L., 594–595
Bird, G.H., 305, 308
Biris, N., 57–58
Bischoff, R., 22–23
Bjelic, S., 504, 508
Bjelkmar, P., 511
Bjornsson, R., 151
Blaber, M., 254
Black, R.E., 159
Blackman, M.L., 97–99
Blackwell, H., 305, 306–307t

Blake, D.A., 37–38
Blake, R.C., 37–38
Blakeley, B.D., 207–209
Blakeley, R.L., 347
Blanc, E., 142–144
Blanco, C., 547
Blanco, F.J., 334–335
Blankenship, J., 506–509
Bloch, J., 595–596
Blöchl, E., 378
Blom, A.E.M., 477–478
Blomberg, M.R., 510
Blomberg, R., 192, 194, 410
Blundell, T.L., 509
Bocian, D.F., 377, 425
Bock, J.E., 304, 308, 327
Bockus, A.T., 304
Bocola, M., 541–542
Boddu, S.H.S., 304
Boder, E.T., 26, 28–32, 150–151, 204
Boersma, A.J., 493–494
Bogan, A.A., 61
Bogdanov, A.M., 122–123
Bogomolni, R.A., 180–181t
Bohlke, R.A., 159
Bohm, A., 472–473
Bolduc, J.M., 128
Bolin, J.T., 225, 231–232
Bollong, M., 92–93
Bonaccorsi di Patti, M.C., 335–336
Bonaventura, C., 521–522
Bonge, H.T., 3–4
Bonnet, C.S., 335–336, 559, 562, 575t
Bonnet, E., 334–335
Bonomo, R.P., 562–563
Bonsor, D.A., 204
Bonvin, A.M.J.J., 29–30
Boone, C., 50, 53, 68, 70, 72
Borbas, J., 91
Borinskaya, S., 170–171
Bornscheuer, U.T., 192, 540
Borovik, A.S., 504–505
Borrmann, A., 97–99
Borzilleri, K.A., 158–159
Bösche, M., 204
Bossinger, C.D., 341–342
Boturyn, D., 335–336, 341–342, 475–476, 562, 575t
Bouguerne, B., 280–281

Boulatov, R., 502–503
Bourenkov, G.P., 457–459
Bourlès, E., 334–335
Bovet, C.L., 3, 12
Bowie, J.U., 383
Boxer, S.G., 504–508
Boyana, K., 22–24, 26–27, 29
Boyko, K.M., 116
Boyle, A.L., 473–474, 564, 566–567
Božin, E.S., 595–596
Bracher, A., 154
Bradley, L.H., 408
Bradley, P., 265
Brady, R.L., 564
Bragg, J.K., 304–305
Braithwaite, V., 151
Branagan, N.C., 525t
Brand, L., 265
Brandis, A., 377
Brandl, M., 142–144
Brandts, J.F., 163
Brandts, M., 163
Brasuń, J., 335–336
Brault, P.A., 37–38
Braun, R., 511
Bravard, F., 573–574
Braymer, J.J., 234–235
Brechbiel, M.W., 560
Breece, R.M., 419–420, 426
Breitenstein, C., 393–395
Breitsprecher, D., 163
Bren, K.L., 404–405
Brender, J.R., 252–253, 256t, 265
Brenner, M.D., 151–153, 164
Brenner, S.E., 173f
Brent, R., 459–460
Briand, J.-P., 280
Bricogne, G., 142–144
Briesewitz, R., 308
Briggs, B.D., 582–585, 596
Briggs, J.A., 97–99
Briggs, P., 142–144
Briggs, W.R., 180–181t
Brinton, C.C., 542
Brittain, S.M., 90–91
Brock, A., 90–91, 111–112, 125–127
Brodin, J.D., 225–227, 232–236
Brooks, B.R., 511
Brooks, C.L., 511

Brown, C.L., 378–379, 392, 396–398
Brown, K.D., 113
Brown, R.S.H., 203
Brown, S., 582
Brown, W.D., 527
Bruché, L., 255
Brudvig, G.W., 390
Bruesehoff, P.J., 504
Brunak, S., 547–548
Brunel, F.M., 305–308, 306–307t
Brunelle, J.L., 513–514
Bruning, M., 564, 566–567
Brunold, T.C., 504–505
Bruser, T., 547
Brust, A., 304
Brustad, E., 90–91
Brustad, E.M., 192
Brutschy, B., 503
Bryant, M., 473–474
Bryden, W.L., 159
Bryndin, D., 595–596
Bryson, J.W., 396–397, 408, 419–420
Buchardt, O., 280
Buder, K., 447–448
Budisa, N., 252–253, 255–257
Buer, B.C., 252–254, 256t, 259, 261–263, 265
Bugaj, L.J., 170–171
Buhl, M., 151
Bukau, B., 162–163
Buratto, J., 280–281, 286
Burchell, B., 459–460
Burton, A.J., 473–474, 564
Burton, R.E., 475
Buse, G., 457–459
Bushey, M.L., 90–91
Busson, S., 457–458
Bustin, S., 513–514
Butland, G., 504–505, 509
Butler, S.J., 335–336
Butselaar-Orthlieb, V.C.L., 584–585
Butterfoss, G.L., 490–492
Bylund, D.B., 26, 35, 37

C

Cabello, C., 568–570
Cagney, G., 204
Cai, Y.-B., 441–442
Calhoun, J.R., 473–475

Callaghan, B.P., 304
Callot, H.J., 3–4
Caltabiano, G., 90
Camarero, J.A., 150–151, 153
Cameron, A.D., 225, 231–232
Campagna, T., 335–336
Campbell, G.R., 317
Campbell, M.E., 573
Campos, S.R.R., 309, 341
Campos-Olivas, R., 252–253
Caner, S., 410
Cangelosi, V.M., 419–420, 472–473, 541
Cano, J.P., 22–23
Canters, G.W., 457–458
Caravan, P., 559–561, 575t
Carballeira, J.D., 549
Carbone, A., 509
Cardona, A., 185
Carell, T., 124–125
Carey, J.R., 513–514
Carloni, P., 236
Carneiro, P.B., 325
Carpenter, S., 218
Carter, L.G., 151, 152f
Carvalho, T., 338–339, 338t, 341
Casatta, E., 158
Cascio, D., 138–139
Case, D.A., 511
Casella, L., 562–563
Caserta, G., 390–391, 410
Cash, B.M., 13
Casini, A., 513–514
Cassio, D., 335–336
Cavaco-Paulo, A., 456
Celik, E., 151, 158, 164
Cellitti, S., 90–91
Cellitti, S.E., 114
Cerda, J.F., 366–367, 380
Certain, L.K., 371–372
Cervigni, S., 335–336
Cervoni, L., 335–336
Chakraborty, S., 136–139, 142–144, 392–393, 473–475, 502–504, 510, 513–514, 523–524, 525t, 526
Chan, M.K., 111–112
Chan, S.I., 110–111, 114–115
Chan, W.C., 141, 289, 341–342
Chanal, A., 547
Chandonia, J.M., 173f

Chandramouli, N., 281–282
Chandran, K., 48, 57–58, 60
Chang, C.H., 511–512
Chang, C.J., 510
Chang, F.-M.J., 418
Chang, M.C., 510
Chang, Y.S., 305, 308
Chao, H.-G., 473–474
Chapman, A.M., 207–209
Charbonnier, P., 335–336
Charoenraks, T., 280–281
Chatterjee, A., 90–93, 99
Chatterjee, J., 304
Chavan, M.Y., 3–4
Chavarot-Kerlidou, M., 390–391
Chazin, W.J., 506–509
Cheatham, T.E., 252–253, 511
Chen, B.Y., 507
Chen, C.-L., 596
Chen, D., 170–171
Chen, G., 48, 57–58, 60, 544, 552
Chen, H.H., 541–542
Chen, H.Y., 91
Chen, J., 150, 206–207, 218, 308, 310–312, 410
Chen, P.R., 90–91
Chen, S., 90–91, 311–312
Chen, T.F., 25–26, 30–32
Chen, V.B., 142–144, 143t
Chen, X., 366
Chen, Y., 525t
Chen, Z., 3, 12
Chenavier, Y., 335–336
Cheng, G., 504
Cheng, H., 200–201, 383, 408
Cheng, K.P., 180
Cheng, L., 214
Cheng, R.P., 252–253, 255, 256t, 260
Cheng, Y., 36
Cheresh, D.A., 61
Cherf, G.M., 30–31
Cherry, J.R., 456
Cheung, J., 163
Chew, G.-L., 150–151
Chi, B., 564
Chiang, C.Y., 584
Chin, J.W., 90–93, 97–99
Chino, M., 472–494

Chittock, E.C., 151, 158, 164
Chittuluru, J.R., 90–91
Chiu, C.-Y., 582, 596
Chiu, H.-P., 252–253, 255, 256t, 260
Chiva, G., 255
Cho, J., 252–253
Chobot, S.E., 383
Choc, M.G., 457–458
Choe, S., 138–139
Choi, C., 228
Choi, E.J., 170–173, 173f, 175t
Choi, J.H., 544, 546–548
Choi, S.H., 90
Choinowski, T., 463
Chong, Y.E., 396–397
Choo, Y., 48
Chorev, M., 475–476
Chothia, C., 224–225, 254
Chou, C., 97–99
Chou, D.H.-C., 305–308, 306–307t
Chou, F.-C., 260
Chou, P.Y., 334–335
Christianson, L.A., 280
Christie, J.M., 180–181t
Christie, M.J., 304
Chuang, R.-Y., 155
Chuang, T.-Y., 150–151
Chudakov, D.M., 122–123
Chufan, E.E., 502–503
Chugh, J., 252–253, 265
Chung, C.W., 562–563
Chung, J., 236
Church, G.M., 130, 214
Churchfield, L.A., 224–247
Cierpicki, T., 265
Ciferri, A., 472–473
Cirera, J., 337–338, 357–358
Cisnetti, F., 334–336, 562, 573–574, 575t
Clackson, T., 46, 58–59, 61, 66, 68, 308–309
Claerhout, S., 280–298
Clark, B.C., 406
Clark, G.A., 252–253, 255–257, 260
Clark, J.D., 15
Clark, R.J., 304
Clemente, F.R., 419–420
Clore, G.M., 472–473
Clow, F., 151

Cochran, F.V., 408
Cochran, J.R., 22–42
Cockburn, J.J.B., 61–62
Cohen, S.M., 410
Cohen, S.X., 142–144
Cohen-Ofri, I., 377
Coin, I., 90–104
Colby, D.W., 28, 30–32
Cole, J.L., 242–243
Colescott, R.L., 341–342
Coleska, A., 304–305, 306–307t, 308
Coll, J.-L., 475–476
Collier, H.B., 347
Collman, J.P., 114–115, 502–503
Colombo, C., 280–281, 286
Comeforo, K., 252–253
Conatser, N., 513–514
Concepcion, J.J., 390
Conrad, K.S., 380
Conte, L.L., 224–225
Contreras-Martel, C., 151–152
Contzen, J., 462–463
Cook, P.I., 341–342
Cooper, H.J., 565t, 566–567, 570–574
Cooper, J.C., 3, 9
Coppage, R., 582, 584–585, 596
Corbett, J.W., 286–287
Corchnoy, S.B., 180–181t
Cordova, J.M., 408
Corey, E.J., 3–5
Corlin, T., 309
Cornilescu, C.C., 252–253
Cornilescu, G., 252–253
Correia dos Santos, M.M., 338–339, 338t, 341
Correia, J.J., 242–243
Cortajarena, A.L., 206–207, 209, 213
Costello, A.L., 569–570
Couch, G.S., 510
Coughlin, J.M., 2–3, 9, 11–12, 14–15
Coulibaly, F., 151
Cowtan, K.D., 142–144, 243–244, 510
Cox, D.J., 97–99
Craggs, T.D., 252
Craig, D.B., 475
Craik, D.J., 304
Crain, P.F., 113
Cramer, W.A., 373t

Crane, B.R., 180–181t, 380
Creighton, T.E., 150–151
Crevat-Pisano, P., 22–23
Crisma, M., 335–336
Cristau, M., 475–476
Crombie, A.T., 472–473
Crooks, G.E., 173f
Cropp, T.A., 90–91
Crosson, S., 172
Crouzy, S., 335–336, 562, 575t
Crowder, M.W., 234–236
Crumbliss, A.L., 521–522
Cudic, M., 309
Cuff, A., 507
Cui, C., 124, 392, 408, 519–521, 525t, 527
Cui, W., 252
Cui, Y., 584
Cui, Z.-Q., 207–209
Cuillel, M., 335–336
Cukier, R., 424–425
Culotta, V.C., 418
Cunningham, B.C., 50
Cunningham, F., 158
Curran, D.P., 254
Curreli, F., 310–312
Cusack, S., 113
Cusanovich, M.A., 472–473
Cushing, P.R., 5
Czekelius, C., 252–253, 256t
Czerwinski, E.W., 472–473

D

D'Ursi, A.M., 475–476
Da Silva, G.F., 57–58
Dafforn, T.R., 565t, 566–567, 570–574
Dafik, L., 252–253
Dahan, A., 304
Dahiyat, B.I., 419–420
Dai, Z., 57–58
Dailly, N., 311
Dalcanale, E., 503–504
Daldal, F., 408
Dalke, A., 507
Dalvit, C., 252
Daly, N.L., 304
Damante, C.A., 334–335
Damas, J.M., 309
Damrauer, N.H., 510

D'Andrea, L., 206
Daneschdar, M., 475–476
Danial, M., 564
Daniel, M.-C., 584–585
Danielson, M.A., 252
Danstrom, C., 513–514
Danuser, G., 170–173, 173f
Danzig, B.A., 309–310
Darden, T., 511
Darensbourg, M.Y., 410
Darling, R.J., 37–38
Darrah, T.H., 573
Dary, O., 159
Das, A., 408
Das, R., 260, 265
Das, S., 507
Das, T.K., 509
Daugherty, R.G., 334–335
Daughtry, K.D., 561, 561f, 575t
D'Auria, G., 335–336
Davenport, A.P., 22–23
Davis, L., 97–99
Davis, T., 594–595
Davydov, R., 504–505, 513–514, 525t
Dawson, J.H., 419–420
Dawson, N.L., 507
Dawson, P.E., 305–308, 306–307t
Dawson, S.J., 280–298
de Andrade, P., 475–476
de Benoist, B., 159
de Crescenzo, G., 572
De Gioia, L., 136–138, 562–563
de Jong, L.A.A., 22–23
de la Salud-Bea, R., 252–253, 259
de Lauzon, S., 456–457
de Marco, A., 158
de Montellano, P.R.O., 465–466
de Picciotto, S., 25–26, 30–32
de Rosales, R.T.M., 379, 477
De Villiers, K.A., 423–424
De Vos, A.M., 472–473
de Vos, W.M., 150–151, 456
Dean, D.R., 392–393
Deb, A., 419–420
Deber, C.M., 158
Debnath, A.K., 310–312
DeChancie, J., 410
Decreau, R.A., 114–115

Defrancq, E., 335–336, 341–342
DeGrado, W.F., 136–138, 192–193, 228, 252–253, 255–257, 286–287, 367, 369, 382–383, 396–397, 408, 472–476, 493–494, 504, 570
Deiss, F., 280–281
Deiters, A., 90–91, 97–99
Dejima, H., 441–444
DeJonge, N., 475
Delangle, P., 334–336, 562, 573–574
DeLano, W.L., 479
Delaurière, L., 280–281
Delcour, A.H., 546
Déléris, G., 309
Delgado, R., 337–339, 338t, 341, 354–358, 357f
DeLisa, M.P., 178
Dellas, N., 91
Demeler, B., 136–139
deMontellano, P.R.O., 463, 466
Dempsey, J.L., 418
Deng, Y., 367
DeNunzio, N.J., 561
DePillis, G.D., 463, 466
Der, B.S., 224–225
DeRose, V.J., 334–335
Dervan, P.B., 284
Desfosses, B., 456–457
Desirajub, S., 570
Desmet, J., 48
Deupi, X., 90
Devaraj, N.K., 114–115
Devocelle, M., 559
Dewkar, G.K., 325
Dey, A., 521–522
Dey, F., 507
Di Ventura, B., 170–173, 173f
Dias, R.L.A., 341
Dib, L., 509
Dickerson, M.B., 583–584
Dickinson, D.J., 170–173, 173f
Dieckmann, G.R., 136–138, 473–474
Diederich, F., 252
Diederichsen, U., 475–476
Diederix, R.E.M., 457–458
Diekmann, S., 447–448
Diers, J.R., 377
Digby, E.L., 377

Dijkstra, B.W., 475
DiMagno, S.G., 254
DiMaio, F., 224–225
Dineen, D., 507
Diness, F., 304
Discher, B.M., 366–384
Distefano, M.D., 541–542
Ditri, T.B., 230–231, 238–239
Dixon, H.B., 158–159
Dobson, C.M., 447–448
Dodd, L.E., 91
Dodson, E.J., 142–144, 243–244
Doitomi, K., 452–453
Dolain, C., 280–281, 286
Dolphin, G.T., 335–336, 341–342
Dombkowski, A.A., 475
Dommerholt, J., 97–99
Donald, L.J., 513–514
Dong, J.S., 120–122
Dong, Z., 280–281
Doran, T.M., 252–253
Dordick, J.S., 456, 544
Doty, R.C., 596
Douat-Casassus, C., 284, 286
Dougan, S.K., 150–151
Doyle, L., 419–420
Doyon, J.B., 308–309
Draebing, T., 170–173, 173f
Drake, A.W., 38
Drepper, F., 475
Drescher, M., 91
Drew, K.L.M., 596
Driessen, A.J.M., 547
Driessen, A., 457–458
Drummy, L.F., 582–583, 596
Du, G., 150
Du, K., 238–239
Duan, X., 582
Dubel, S., 544
DuBois, D.L., 390–391
Dudev, T., 571
Duffy, S.P., 110
Dugan, A., 304
Duggleby, R.G., 513–514
Duhr, S., 163
Dumas, P., 225, 231–232
Dumy, P., 334–336, 475–476
Duncan, J.R., 560

Dunston, T.T., 192–193, 200–201, 410
Dürrenberger, M., 116–117, 390–391, 541–545
Dutta, A., 390–391, 400
Dutton, P.L., 366–384
Duvaud, S., 158–159
Dym, O., 204, 410
Dyson, H.J., 236

E

Ebina, W., 114–115
Ebrahimi, M., 162
Edelhoch, H., 345
Edmondson, D.E., 380
Edwards, T.A., 283–284
Edwards, W.B., 560
Egan, T.J., 423–424
Eggleston, I.M., 335–336
Egholm, M., 280
Ehlers, M.D., 170–171
Ehlert, F.J., 36–37
Eibling, M.J., 408–409
Eils, R., 170–173, 173f
Eisenberg, D., 138–139, 396–397, 408
Eisert, R.J., 309
El Mortaji, L., 151–152
Eliceiri, K.W., 180
Elliott, S.J., 521–522
Elliott, T.S., 97–99
Ellis, J., 113
Ellis, T., 513–514
Ellis-Guardiola, K., 540–541
Ellman, G.L., 346
Elsasser, S.J., 91–93, 97–99
Eltis, L.D., 225, 231–232
Elvekrog, M.M., 423–424
Empting, M., 475–476
Emsley, P., 142–144, 243–244, 510
Engel, D., 383, 408
Engelhard, M.H., 391
Englander, M.T., 366–384
Englander, S.W., 154
Ennist, N.M., 366–384
Epel, B., 393–395
Erben, C., 444
Erdbrink, H., 252–253, 256t
Eriksson, A., 254
Erman, J.E., 515

Ermolat'ev, D.S., 291
Ernst, R.J., 97–99, 150–151
Esaki, K., 204–205
Escher, I., 305
Eshelman, M.R., 308–309, 311, 316–317
Espargaró, A., 212–213
Esposito, G., 335–336
Essen, L.O., 170–171
Estieu-Gionnet, K., 309
Evanics, F., 252
Evans, J.N.S., 252–253
Evans, P.R., 142–144, 243–244
Evans, R.M., 521–522
Evans, T.C., 150–151
Evdokimova, E., 517

F

Fabiane, S.M., 236
Fabritz, S., 475–476
Faeh, C., 252
Fägerstam, L., 38
Faiella, M., 379, 390–393, 477
Fairhead, M., 153
Fairlie, D.P., 304–305, 306–307t, 308
Fairman, R., 252–253, 260, 473–474, 570
Falke, J.J., 252
Fan, J.-Y., 207–209
Fandrich, M., 447–448
Fang, X., 91, 176
Faraone-Mennella, J., 225–230, 232, 234–235, 240–242
Farid, R.S., 366–367, 371–372, 375–376, 408
Farid, T.A., 366–367, 372, 374–377, 380–382, 408–409
Farkas, V., 335–336
Farmer, B.L., 582–584
Farrer, B.T., 136–138
Farrow, C.L., 594–596
Fasman, G.D., 334–335
Fass, D., 233–234
Fattorusso, R., 335–336
Faulkner, L.R., 428–429
Fee, J.A., 457–458, 525t
Fee, L., 345
Fehrentz, J.A., 435
Fei, H., 410
Fekner, T., 111–112

Feldman, H.A., 23
Feng, M., 513–514
Feng, Z., 507
Fenske, D., 280
Ferrand, M., 335–336
Ferrand, Y., 281–282, 284, 286, 293f, 294, 298f
Fersht, A.R., 265–266
Fichera, A., 252–257, 256t
Fidelio, G.D., 542
Fields, G., 308–309, 312–316
Fierer, J.O., 150–151, 153–156, 158–160, 163–164
Fierke, C.A., 418
Figueroa, J.S., 230–231, 238–239
Filipe, L.C.S., 309
Filonov, G.S., 207–209
Findling, K.L., 457–458
Finn, F.M., 542
Finn, M.G., 124–125, 475–476
Finzel, B.C., 506–508
Fischer, C.A., 475–476
Fischer, M., 507
Fitch, A.N., 593
Flaschel, E., 547
Flaschka, W., 347–348
Fleishman, S.J., 224–225
Flensburg, C., 142–144
Fletcher, J.M., 566–567
Fletcher, S., 224–225
Flock, U., 503
Floyd, W.C., 503–504
Fontecave, M., 390–391, 410, 503–504
Fontecilla-Camps, J.C., 390
Forest, K.T., 252–253
Forge, V., 447–448
Forman-Kay, J., 252
Forouhar, F., 128
Forrer, P., 46, 178
Fosgerau, K., 304, 573
Foshay, M., 515
Fowler, C.J., 442–444
Fragoso, A., 337–339, 338t, 341, 354–358, 357f
Franano, F.N., 560
Francis, M.B., 12
Franke, J.P., 22–23
Franklin, S.J., 560–561, 575t

Franz, K.J., 561, 573–574
Frasca, V., 163
Frauendorf, H., 475–476
Frausto da Silva, J.J.R., 228–229, 378
Frei, J.C., 46–85
Frenkel, A.I., 582–585, 596
Fretes, N., 61–62
Freund, S., 309–310
Frey, D., 170–171, 173–174
Friedle, S., 477
Friedler, A., 309
Friehs, K., 547
Fries, P.H., 335–336, 562, 575t
Frise, E., 185
Frolow, F., 472–473
Frostell, A., 38
Fruhwirth, G.O., 456
Fry, B.A., 366–367, 372, 374–377, 380–383, 408–409
Fry, H.C., 383, 408–409, 473–474
Fterke, C.A., 475
Fu, L., 502–503
Fu, Y., 418, 540–542
Fuh, G., 46
Fuhrhop, J.H., 460–461
Fujii, S., 472–473
Fukuda, Y., 510
Fukumoto, K., 403, 404f, 441–442, 540–541
Fukumoto, Y., 96
Fukunaga, R., 97–99
Fukuyama, K., 392–393
Funahashi, Y., 379
Furman, J.L., 99
Fushman, D., 150–151, 153
Fustero, S., 255

G

Gaggelli, E., 337–338
Gagliardi, C.J., 510
Gai, S.A., 30–31
Galibert, M., 475–476
Gall, P., 280
Galloway, W.R.J.D., 475–476
Gamba, I., 334–335
Gambhir, S.S., 207–209
Gamble, A.J., 565
Gamelin, D.R., 507
Gan, L., 402
Gan, Q., 281–282, 284, 286, 293f, 294, 298f
Gans, P., 352–354
Gansow, O.A., 560
Gao, F., 123–124, 123f
Gao, J., 252–253
Gao, Y.-G., 308–309, 503–505, 507–510, 513–515, 517–522, 525t, 527–528
Garcia de la Torre, J., 243
Garcia, J., 335–336, 341–342
Garcia-Horsman, J.A., 503
Gardner, K.H., 170–173, 175t, 177, 180, 180–181t
Gargova, S., 311
Garner, D.K., 502–505, 509, 513–514
Garric, J., 281–282
Gärtner, W., 382, 392–395, 394f
Gartner, Z.J., 308–309
Gassaway, B.M., 130, 203–205, 207, 213–214, 216
Gasteiger, E., 158–159
Gateau, C., 573–574
Gathu, V., 335–336
Gatticker, A., 158–159
Gavathiotis, E., 57–58
Gavenonis, J., 304, 308–309, 311, 316–317, 327
Gavin, A.-C., 204
Gawlak, G., 172–173, 175t, 177
Gayet, R.V., 151–153, 164
Ge, N.-H., 475–476
Gebauer, A., 442–444
Gebbink, R.K., 503–504
Geerlof, A., 158
Geib, S.J., 280
Geierstanger, B.H., 90–91, 114
Gellman, S.H., 192–193, 252–253, 260, 280, 334–335
Geng, M., 516
Gengenbach, A.J., 504, 507–509, 513–514
Gennis, R.B., 503, 509
Geoghegan, K.F., 158–159
Georgiou, G., 178, 544, 547, 552
Gerig, J.T., 252
Gerling, U.I.M., 252–253, 256t
Gestwicki, J.E., 304
Getahun, Z., 378–379
Geyer, C.R., 68

Ghadiri, M.R., 228, 475–476
Ghirlanda, G., 252–253, 255–257, 371–372, 378–379, 390–411
Ghosh, A., 444
Ghosh, D., 136–139
Ghosh, I., 207–209, 212–213
Ghosh, S., 504, 513–514, 525t, 526
Giansanti, S., 462–465
Gibney, B.R., 334–335, 366–367, 368f, 369, 371–372, 378, 393–395, 394f, 418–437, 440, 504, 569–570
Gibson, D.G., 155, 513–514
Gibson, T.J., 507
Giedroc, D.P., 234–235, 418
Gierasch, L.M., 334–335
Giger, L., 410
Gilbert, W., 335–336
Gilbreth, R.N., 57–58, 204–205
Gill, S.C., 345
Gillies, E., 280–281
Gilliland, G., 507
Gillis, R.B., 565t, 570–575, 575t
Gilon, C., 304, 309
Gingery, M., 396–397, 408
Ginovska-Pangovska, B., 390–391
Ginsbach, J.W., 337–338, 357–358
Giraud, M., 475–476
Girona, G.E., 91, 97
Gladwin, S.T., 268
Gloaguen, F., 410
Glotzer, M., 170–173, 173f, 175t, 177
Go, A., 408
Goddard, A., 48, 61–62
Goddard, T.D., 510
Goddard, W.A., 252–253, 255–257, 479
Godde, F., 280–281, 284, 286
Godzik, A., 507
Gohlke, H., 511
Gold, M.H., 506–509
Goldenberg, D.P., 150–151
Goldsmith, M., 541
Goldstein, B., 170–173, 173f
Gonçalves, M., 309
Gonen, S., 224–225
Gonen, T., 224–225
Gong, B., 280
Gong, S., 509
Gong, W.M., 120–122

Góngora-Benítez, M., 475–476
Gonzalez, M., 542
Goodman, D.B., 130
Goodman, J.L., 260
Gopalakrishnan, R., 311–312
Goparaju, G., 383
Gopi, H.N., 570
Gopta, O., 393–395
Görbitz, C.H., 475–476
Gordon, G., 391, 400
Gorelsky, S.I., 525t
Gorensek, A.H., 265
Goto, Y., 304, 308–310
Gottler, L.M., 252–253
Gotzen, S., 541–542
Graff, C.P., 28, 30–32
Graff, R., 280
Grandi, P., 204
Grassetti, D.R., 347
Graves, B., 305, 308
Gray, H.B., 122, 227–229, 234–235, 240–242, 418, 502–508, 510
Gray, J.J., 232–233
Gray, T., 345
Green, J.T., 284
Green, R., 513–514
Greenbaum, D.C., 305–308, 306–307t, 310–312, 323
Greenblatt, D.M., 510
Greene, R., 52–53
Greenfield, N.J., 513–514
Greenwood, J.M., 560–561, 575t
Gregg, C.J., 214
Greisen, P.J., 541
Grélard, A., 280–282
Griffin, A., 513–514
Griffin, H., 513–514
Griffiths, A.D., 48
Grigoryan, G., 473–474
Grimm, S., 547
Grimsley, G., 345
Grobner, T., 573
Groemping, Y., 316–317
Groff, D., 90–91
Grogan, D., 457–458
Grome, M.W., 130
Gronenborn, A.M., 252–253, 472–473
Grosse, R., 170–171, 173–174

Grosse-Kunstleve, R.W., 142–144, 243–244
Grosset, A.M., 372
Grossman, J.H., 150–151, 153
Grove, T.Z., 209
Grubbs, R., 305, 306–307t
Grubina, R., 308–309
Gruhler, A., 204
Grunbeck, A., 103–104
Grütter, M.G., 410
Grzyb, J., 377–379, 395–396
Gu, M., 335–336
Gudelj, M., 456
Guerlavais, V., 305, 308
Guichard, G., 280
Guilard, R., 444
Guimaraes, C.P., 477–478
Guionneau, P., 280–281
Guldberg, C.M., 23
Gullickson, D., 252–253, 260
Gumbart, J., 511
Gunasekera, T.S., 234–235
Gunnlaugsson, T., 559
Gunsalus, I.C., 440–441
Guntaka, V.R., 154
Guntas, G., 170–173, 173f, 178–180
Guo, J.T., 90–91, 111–112, 114
Guo, X., 204
Guo, Z., 513–514
Gust, D., 378–379, 390, 392, 396–398
Gutsmiedl, K., 124–125
Guyon, F., 479, 480f
Guzei, I.A., 252–253
Guzman, T., 419–420, 426

H

Habjanič, J., 334–335
Hackel, B.J., 25–26, 30–32
Hacker, D.E., 309–310, 326
Hadley, E.B., 252–253, 260
Hadt, R.G., 514–515, 525t
Haehnel, W., 379–380, 475
Hagan, R.M., 151
Hagen, W.R., 456
Hahn, K.M., 170–174, 173f, 175t
Hähnke, M.J., 561
Haimovich, A.D., 130, 203–205, 207, 213–214, 216

Hajare, H.S., 203
Halavaty, A.S., 170–172
Halbhuber, K.J., 456
Hall, C.D., 304
Hall, T.M.T., 218
Hallett, R.A., 170–176, 173f, 175t, 178–180
Hamels, D.R., 390–391
Hamilton, A.D., 207–209, 212–213, 224–225, 280
Hamilton, G.A., 401, 401f
Hamlin, R.C., 472–473
Hammersley, A.P., 593
Hammes-Schiffer, S., 424–425
Hammond, P.T., 584
Hamuro, Y., 280
Han, G.W., 128, 396–397
Han, I.S., 180–181t
Han, J., 204, 335–336
Han, M.J., 544–545
Han, S.-Y., 312–316
Hanashima, S., 252–253
Hancock, W.S., 341–342
Handa, M., 4
Handel, T.M., 228
Hands, M., 209
Hanfland, M., 593
Hannigan, R.E., 573
Hansen, S., 218
Hansen, T., 3–4
Hansen, W.A., 393–395
Hanssens, I., 335–336
Haque, M.E., 265
Hara, I., 513–514
Harbury, P.B., 257–259, 473–474
Hardie, M.J., 284
Harding, M.M., 225, 228–230
Hardjasa, A., 377
Haring, D., 541–542
Hariton, C., 22–23
Harper, S.M., 170–171, 175t, 180, 180–181t
Harrison, J.S., 57–58, 61
Harrison, P.M., 472–473
Hart, G.T., 204
Hartl, F.U., 162–163
Hartman, K., 252–253, 256t, 265
Hartman, M.C.T., 309–310, 325–326
Hartnett, H.E., 401, 401f
Harvey, B.R., 544, 552

Hatzubai, A., 309
Haucke, V., 316–317
Hauksson, J.B., 440
Hauschka, P.V., 573
Hausermann, D., 593
Havranek, J.J., 473–474
Hawryluk, N.A., 3–4
Hay, M.T., 507
Hay, S., 366–367, 380
Hayashi, A., 444, 452
Hayashi, T., 391, 402–403, 404f, 410, 440–453, 523–524, 525t, 526
Hayhurst, A., 544, 552
Hayles, L., 154
Haymore, B.L., 228
Hayouka, Z., 309
Hazes, B., 475
Headd, J.J., 142–144, 143t
Hecht, M.H., 136–138, 371–372, 408, 474–475
Hedman, B., 514–515, 525t
Heemskerk, A.H., 15
Hefter, G., 351
Hegg, E.L., 419–420
Heilbut, A., 204
Heinemann, I.U., 90, 214
Heinis, C., 309–312
Heinisch, T., 540–542
Heinz, D., 254
Heinz, H., 582–585, 596
Hellwig, P., 380
Helms, G.L., 252–253
Hemmerlin, C., 280
Hemmingsen, L., 136–139, 334–335, 474–475, 568–570
Henchey, L.K., 304–305, 308–309
Henderson, C.J., 459–460
Hendrickx, M., 456
Heppel, L.A., 244–245
Heppner, D.E., 337–338, 357–358
Herberich, B., 111–112, 125–127
Hersperger, R., 475
Hess, K.R., 110
Heydari, A., 162
Hicks, L., 568–569
Hiemstra, H., 311
Higaki, J., 255, 256t
Hilcove, S.A., 408
Hildebrandt, P., 379, 475

Hill, T.A., 304
Hilty, C., 91
Hilvert, D., 192–193, 236, 410, 541–542
Himiyama, T., 540–541
Hirano, H., 162
Hirata, M., 255, 256t
Hisaeda, Y., 441–444, 451–452
Ho, S.P., 367
Ho, Y., 204
Hoang, H.N., 305, 306–307t, 308
Hochstrasser, M., 213
Hodges, R.S., 136–138, 473–474, 565–569, 565t, 571–573
Hodgson, K.O., 514–515, 525t
Hoe, K.K., 305
Hoess, R.H., 48
Hoey, R.J., 57–58, 204–205
Hof, P., 457–459
Hoffman, A., 304
Hoffman, B.M., 504–505, 513–514, 525t
Hoffmann, T., 304, 573
Hofirchter, M., 503, 506–508
Hofmann, K., 542
Hokenson, M., 180–181t
Hol, W.G.J., 225, 231–232
Hollósi, M., 335–336
Holm, L., 507
Holm, R.H., 235–236, 337–338, 378, 502–503
Hommel, U., 280
Hon, G., 173f
Hong, J.-K., 369, 473–474
Hong, S., 204
Honig, B., 507
Hoogenboom, H.R., 48
Hoogland, C., 158–159
Hope, C.M., 170–173, 173f, 175t, 177
Hoppmann, C., 91
Hor, T.S.A., 504–505
Hore, P.J., 252
Hori, H., 451–452
Horikawa, M., 3–4
Horne, W.S., 475–476
Horng, J.-C., 252–253
Horst, R., 265
Hörth, P., 379
Hosseinkhani, S., 162
Hosseinzadeh, P., 124, 392–393, 502–529
Hoth, L.R., 158–159

Hou, J., 391
Houk, K.N., 410
House, R.L., 390
Hovius, R., 311–312
Howard, A., 472–473
Howard, R.I., 150
Howard, S., 150
Howarth, M., 150–164
Hoyer, D., 400–401
Hsieh, K.-H., 255
Hsu, F.C., 224–225
Hu, C., 110–130, 123f, 509
Hu, M.R., 116, 119, 407–408
Hu, X., 280–298, 305–308, 306–307t, 310–312, 323, 511
Hu, Z., 234–235
Huang, C.C., 510
Huang, J., 204–205
Huang, K., 12
Huang, P.-S., 224–225
Huang, S., 218
Huang, S.J., 224–225
Huang, S.S., 368f, 369
Huang, W., 472–473
Huang, X., 48
Huang, Y., 582–583, 596
Huang, Z., 335–336
Huber, R., 440, 457–459
Huber, T., 93, 103–104
Huc, I., 280–298
Huebner, V.D., 280
Hufton, S.E., 48
Hughes, R.M., 170–171, 334–335
Hughes, Z.E., 582–583, 596
Huh, W.-K., 209
Huhmann, S., 252–253
Huisman, G.W., 192, 540
Hull, J.F., 510
Hulme, E.C., 22–24, 26–29, 34
Humphrey, W., 507
Hunt, J.A., 475
Hunter, C.N., 377
Hunter, S., 22–42
Huntley, J.J., 236
Hur, C.G., 544–545
Hurevich, M., 309
Hurrell, R., 159
Hussain, I., 596
Hust, M., 544

Hutchinson, E.G., 334–335
Hutchison, C.A., 155
Huynh, M.H.V., 124
Hwang, H.J., 29–30, 525t
Hyster, T.K., 540–542

I

Ichikawa, Y., 461–462
Ienco, A., 354
Ignatchenko, A., 204
Ilie, A., 541, 543–544
Immormino, R.M., 142–144, 143t
Impellizzeri, G., 335–336, 562–563
Imperiali, B., 334–335, 559, 561, 561f, 573–574, 575t
Inaba, K., 234–235
Inomata, T., 379
Inoue, N., 114–115
Inoue, T., 442–446, 452–453
Iqbal, J., 503–504
Iranzo, O., 136–139, 334–360, 473–474, 568–570
Irimia, A., 305, 308
Iriondo-Alberdi, J., 280–281
Isaacs, F.J., 203–205, 207, 213–214, 216
Ishibashi, K., 96
Ishida, H., 334–335
Ishida, K., 565t, 566–567, 571–573
Ishida, Y., 252–253
Ishihara, S., 255, 256t
Ishii, R., 97–99
Ishitani, R., 113
Isogai, Y., 372
Ito, K., 234–235
Ivarsson, B., 38
Iverson, B.L., 280, 544, 552
Iwai, H., 150–151, 154
Iwakura, M., 154
Iwama, T., 558
Ize, B., 547
Izoré, T., 151–152

J

Jacak, R., 198–199, 265
Jäckel, C., 236, 252–254, 260
Jackrel, M.E., 203, 207, 209, 211, 213
Jackson, C.J., 410
Jackson, J.C., 110
Jackson, S.E., 252, 265–266

Jackson, T.A., 504–505
Jacobsen, Ø., 475–476
Jaehrig, A., 170–171, 173–174
Jakusch, T., 136–138, 568–570
Jamieson, A.C., 48
Jamieson, A.G., 305–308
Janin, J., 224–225
Jansen van Rensburg, E.I., 150
Janssen, D.B., 192
Jaschke, P.R., 377
Jen, A.K.Y., 583–584
Jena, P.V., 280–281
Jensen, J., 304
Jensen, K.J., 314
Jensen, L.T., 418
Jeschek, M., 540–553
Jewell, D.A., 280
Jewett, J.C., 124–125
Jeżowska-Bojczuk, M., 337–338
Jhiang, S.M., 308
Jian, J.-X., 410
Jiang, B., 252, 265
Jiang, H., 280–282, 284, 286, 293f, 294, 298f, 516
Jiang, J., 561–562
Jiang, L., 116, 124, 252, 265, 334–335, 410, 419–420, 504
Jiménez-Barbero, J., 334–335
Jing, J., 441–442
Jo, H., 305–308, 306–307t, 310–312, 323, 493–494
Jochim, A.L., 304–305, 308–309
Johnson, C.M., 163
Johnson, D.C., 392–393
Johnson, E., 582
Johnson, K.A., 151, 152f
Johnson, M.K., 392–393
Johnson, M.L., 265
Johnson, O., 243–244
Johnsson, B., 38
Johnsson, N., 207–209
Johnston, E.M., 337–338, 357–358
Jolliffe, K.A., 335–336
Jones, A., 257, 456
Jones, A.K., 391, 400–402, 401f
Jones, A.W., 565t, 566–567, 570–574
Jones, D.S., 32
Jones, P.R., 366

Jones, S.E., 582, 584, 590
Jonker, H.R.A., 561
Jonson, P.H., 234
Jönsson, U., 38
Jordan, A., 472–473
Jordan, S.R., 472–473
Jordan, T., 562–563
Jose, J., 546
Josephson, K., 309–310
Jouann, L., 456
Juhas, P., 594–596
Julian, R.R., 457–458
Jung, C., 462–463
Jung, N., 316–317
Jung, S., 46

K

Kaboli, M., 3–4
Kadish, K.M., 3–4, 444
Kaiser, E.T., 341–342, 380–381, 541–542
Kajander, T., 206–207
Kalb, A.J., 472–473
Kallenbach, N.R., 367
Kalsbeck, W.A., 425
Kamachi, T., 452–453
Kamaraj, K., 502–503
Kamat, P.V., 584–585
Kamens, A.J., 309
Kamysz, W., 335–336
Kanatzidis, M.G., 594–595
Kandels-Lewis, S., 513–514
Kandemir, B., 404–405
Kane, R.S., 170–171
Kang, H.J., 151
Kang, J.H., 91, 97
Kang, M.J., 546
Kania, R.S., 280
Kanno, A., 162
Kapadnis, P.B., 97–99
Kaplan, J., 136–138, 473–474, 493–494
Kapoor, T.M., 334–335
Kapral, G.J., 142–144, 143t
Kapteijn, F., 584–585
Karaduman, N., 513–514
Karamata, D., 544–545
Karas, M., 503
Karle, I.L., 280, 570
Karlin, K.D., 419–420, 502–503

Karlsson, R., 38
Karplus, K., 507
Kashiwada, A., 565t, 566–567, 571–573
Kashula, C.H., 423–424
Kaspar, A.A., 304
Kast, P., 236, 410
Kastritis, P.L., 29–30
Kataoka, Y., 4
Katoh, T., 304, 308–310
Katona, G., 472–473
Katritch, V., 90, 99–100, 101f, 103–104, 265
Katritzky, A.R., 304
Katz, S.R., 130
Kauffmann, B., 280–284
Kawaba, N., 459, 461–462
Kawada, K., 255, 256t
Kawakami, T., 304
Kawamoto, S.A., 304–305, 306–307t, 308
Kawamoto, T., 4
Kawano, F., 180–181t
Kay, C.M., 565–569, 565t, 571–573
Kaya, E., 124–125
Kaynig, V., 185
Kazlauskas, R.J., 192, 540
Kazmi, M., 93
Ke, H., 511
Keasling, J.D., 110
Keating, A.E., 473–474, 564
Keeble, A.H., 204
Keedy, D.A., 142–144, 143t
Keller, M., 378
Keller, P., 142–144
Kellogg, B.A., 28, 30–32
Kelly, J.W., 334–335, 339
Kelly, S.M., 265–266
Kenakin, T., 24
Kendhale, A.M., 281–282
Kengen, S.W.M., 456
Kennedy, M.J., 170–171
Kennedy, M.L., 378, 430–431
Kennepohl, P., 235–236, 337–338
Kennerly, J., 304
Kenney, J., 57–58
Kent, C.A., 510
Kent, S.B.H., 257
Kerfeld, C.A., 398–399
Kersten, R.J.A., 15

Keske, J.M., 375–376
Kessel, A.J., 371–372
Kessler, H., 304
Keydar, I., 82
Khan, F., 252
Khare, S.D., 128, 194, 393–395, 504, 508, 541
Khavari-Nejad, R.A., 162
Khersonsky, O., 410
Khodaverdian, B., 419–420
Khoo, K.H., 305
Khoury, R., 280
Kiefer, L.L., 418
Kielian, M., 48, 57–58, 61–62
Kiernan, E.A., 180
Kihara, Y., 403, 404f
Kihira, Y., 442–444, 452
Kikuchi, A., 379
Kikuchi, I., 96
Kikuti, C.M., 61–62
Kilbinger, A.F., 284
Kilner, C.A., 284
Kim, D., 252–253
Kim, D.E., 308–309
Kim, D.W., 584
Kim, E., 502–503
Kim, H.J., 204
Kim, H.K., 513–514
Kim, H.W., 254
Kim, J.I., 546
Kim, J.R., 154
Kim, J.-W., 584, 596
Kim, K., 511–512
Kim, P.S., 257–259, 564, 570
Kim, S.-H., 48
Kim, S.N., 584
Kim, W.A., 419
Kim, Y.-A., 312–316
Kim, Y.H., 383, 408
Kim, Y.-W., 305
Kinch, L., 317
King, D.S., 90–91, 111–112
King, M., 347
King, N.P., 224–225
King, S., 565t, 570–575, 575t
Kingston, R.E., 459–460
Kinloch, R., 419–420
Kipnis, Y., 541

Kishi, K., 506–508
Kiss, G., 194, 410
Kitchens, J., 4–5
Kitevski, J.L., 252
Kittler, M., 447–448
Klakamp, S.L., 38
Kleanthous, C., 204
Klehr, J., 541–542
Kleingardner, J.G., 404–405
Klejnot, J., 170–171
Klemm, P.J., 561f, 562, 575t
Klok, H.-A., 564
Kloster, R.A., 3–4
Klotz, I.M., 24
Klug, A., 48
Knecht, M.R., 582–596
Knorr, L., 542
Knudsen, S.E., 3, 10–12
Ko, W.H., 175t, 180, 180–181t
Koay, M.S., 392–395, 394f
Kobayashi, T., 97–99, 113
Köcher, M., 442–444
Kodadek, T., 308–309
Kodali, G., 366–384, 408–409
Koder, R.L., 192, 372, 382–383
Kodibagkar, V.D., 252
Koehler, C., 97–99, 124–125
Koellhoffer, J.F., 48, 57–58, 60
Köhler, V., 253–254, 541–542, 565
Kohn, W.D., 136–138, 473, 565–567, 565t, 571–573
Kohrer, C., 93
Koide, A., 48, 54, 57–58, 204–205
Koide, S., 48, 54, 57–58, 204–205
Kokkinidis, M., 472–473
Kokona, B., 252–253, 260
Koksch, B., 252–254, 256t, 260
Kolb, H.C., 124–125, 475–476
Kolb, V.A., 154
Kollman, P.A., 542
Kommer, A., 154
König, H.M., 284
Königsberger, E., 351
Kono, H., 378–379, 474–475
Kontogiannis, L., 243–244
Kopple, K.D., 334–335
Korendovych, I.V., 192–201, 383, 408, 410
Korsmeyer, S.J., 305

Kortemme, T., 224–225, 308–309, 334–335
Kossiakoff, A.A., 48, 61, 204–205, 472–473
Kosuri, S., 214
Kotynia, A., 335–336
Kowalik-Jankowska, T.M., 337–338
Koziej, P., 475
Koziol, L., 503–504
Kozłowski, H., 337–338
Krantz, B.A., 228
Krause, R., 204
Kraut, J., 506–508
Kräutler, B., 444–446
Krayzman, V., 595–596
Kretsinger, J., 334–335
Kreutzer, M.T., 584–585
Krieg, R., 456
Kries, H., 192, 410
Krische, M., 280
Krishna, M.M.G., 154
Krishna, S.S., 396–397
Krishnaji, S.T., 252–253
Krishnamoorthy, J., 252–253, 256t, 265
Krishnamurthy, V.M., 252–253
Krishnan, V., 408
Kritzer, J.A., 304–327, 335–336
Krizek, B.A., 562–563
Krogan, N.J., 204
Krohn, K.A., 36–37
Krois, A.S., 265
Kron, I., 351
Kroneck, P.M.H., 472–473
Ku, J., 48
Kua, J., 252–253, 255–257
Kuang, Z., 584
Kuc, R.E., 22–23
Kuchment, O., 436
Kucukural, A., 511
Kudo, M., 283–284, 283f, 296
Kuhlman, B., 170–188, 224–227, 231–233, 268, 490–492, 511
Kühnle, F.N.M., 280
Kulkarni, S., 305–308, 306–307t, 310–312, 323
Kulp, D.W., 493–494
Kumar, K., 252–257, 256t, 259–260
Kundu, R., 3, 5, 9
Kung, A.L., 305
Kunkel, T.A., 66, 70

Kunzler, P., 237–238
Kuo, J.M., 463, 466
Kuprov, I., 252
Kuriyan, J., 22–24, 26–27, 29
Kutchukian, P.S., 305
Kuzin, A.P., 194
Kuznetsov, G., 130
Kwakernaak, C., 584–585
Kwok, B.C., 114–115, 507–509, 513–514
Kwon, O.-H., 252–253
Kwong, P.D., 305, 308

L

La Mar, G.N., 440, 527
Labbe, P., 335–336, 341–342
Lacey, V.K., 91
Laguerre, M., 280–282
Lahr, S., 474–475
Lai, J.R., 46–85
Lai, R., 513–514
Laïn, G., 309
Lajoie, M.J., 130, 214
Lakshman, V., 111–112
Lam, K.S., 308–309
Lamosa, P., 337–339, 338t, 341, 354–358, 357f
Lamsa, M.H., 456
Lamzin, V.S., 142–144
Lanci, C.J., 408–409
Lane, D.P., 305
Lang, K., 90–91, 97–99
Lange, O.F., 198–199, 265
Langenegger, D., 400–401
Langer, G.G., 142–144
Langer, T., 162–163
Langlois d'Estaintot, B., 280–281, 286
Lanzetti, A.J., 158–159
Lapierre, C., 456
Laplaza, C.E., 378
Laporte, F.A., 335–336
Lardinois, O.M., 465–466
Larsson, P., 511
Lary, J.W., 242–243
Latour, J.-M., 334–335
Lau, E.Y., 308–309, 503–504
Lau, Y.H., 475–476
Laue, T.M., 242–243
Lauffer, R.B., 560

Lautrette, G., 281–282
Lavoie, T.B., 473–474
Law, W.-C., 582–583
Laxmi-Reddy, K., 280–281, 284, 286
Le Chevalier-Isaad, A., 475–476
Le Grel, P., 280
Lear, J.D., 136–138, 192–193, 369, 382–383, 408, 473–474, 504
Leaver-Fay, A., 198–199, 265, 504, 508
Lebowitz, J., 242–243
Lebrun, C., 334–336, 573–574
Lebrun, V., 334–335
Lee, C., 582
Lee, C.-S., 252–254
Lee, C.V., 46
Lee, D., 57–58, 472–473, 584, 596
Lee, H.S., 90–91, 120
Lee, H.-Y., 252–254, 257–259
Lee, J.C., 234–235
Lee, J.-J., 204
Lee, J.S., 111–112, 114
Lee, J.W., 113, 544–545
Lee, K.F., 91–93
Lee, K.-H., 136–139, 252–254, 257–259, 568–570
Lee, M.M., 111–112
Lee, S., 582–583, 596
Lee, S.-C., 204
Lee, S.Y., 544–548
Lee, Y., 584
Lee, Y.H., 544–545
Léger, J.-M., 280–282
Legzdins, P., 4
Lehmann, A., 383, 408, 473–474
Lehmann, C., 335–336
Lehn, J.-M., 280
Lei, L., 507–509, 513–515, 517–522, 525t
Lei, X.G., 150, 159
Lemke, E.A., 90–91, 97–99, 124–125
Lemon, B.J., 390
Lendzian, F., 462–463
Lense, S., 391
Lentz, D., 256t
Leonard, P.G., 3, 10–12
Leone, L., 472–494
Lerner, A.M., 170–173, 173f
Lesage-Meessen, L., 456
Leslie, A.G., 243–244

Levasseur, A., 456
Leveno, M., 317
Lever, M., 153
Levin, A., 309
Levin, B.J., 252–253, 256t, 261–263, 265
Levin, I., 595–596
Levin, K.B., 204
Levine, B., 317
Levine, H.L., 380–381
Levskaya, A., 170
Lévy, R., 596
Lew, S., 128
Lewis, D.J., 565t, 566–567, 570–574
Lewis, J.C., 390–391, 540–541
Lewis, M.S., 242–243, 368f, 369
Lewis, R.A., 225–227, 229–230, 232–233
Lewis, S.M., 198–199, 265
Lewis, T.E., 507
Lex, J., 442–444
Li, C., 252, 265
Li, C.-B., 410
Li, C.-Z., 366
Li, D.N., 459–460
Li, F., 91, 255–257, 265
Li, J., 150, 255–257, 265, 519–521, 525t
Li, J.S., 119–124, 123f
Li, K., 170–171
Li, N., 310–312
Li, P., 176
Li, Q., 218, 367
Li, S., 57–58, 150, 561–562, 575t
Li, T., 218
Li, W., 507
Li, W.-W., 380
Li, X., 111–112, 280, 308, 310–312
Li, X.-B., 410
Li, X.-H., 441–442
Li, Y., 335–336, 410, 582–584, 596
Li, Y.-Y., 410
Li, Z., 130
Li, Z.-J., 410
Liang, J., 383
Liang, X., 475–476
Lichtenstein, B.R., 366–367, 370, 372, 374–377, 380–382, 391, 400, 408–409
Lifsey, R.S., 3–4
Lilie, H., 503–504
Lim, C., 571

Lim, C.-K., 596
Lim, R.K.V., 124–125
Lim, W.A., 170
Lima, L.M.P., 334–360
Limbird, L.E., 27, 29
Lin, C., 170–171
Lin, I.J., 525t
Lin, Q., 124–125, 308, 310–312
Lin, S.L., 224–225
Lin, X.Q., 3–4
Lin, Y., 170–173, 173f, 175t, 177
Lin, Y.-S., 305–309, 306–307t
Lin, Y.-W., 507–509, 513–515, 517–522, 525t
Lindqvist, Y., 472–473
Lindstrom, K.M., 560–561
Linehan, J.C., 390–391
Ling, J., 90, 305–308, 306–307t
Link, J.M., 36–37
Lipinski, A.K., 265
Lippard, S.J., 228–229, 472–473, 477
Lisiero, D., 170–171
Litowski, J.R., 473–474, 572
Liu, A., 53–54
Liu, C., 154
Liu, C.C., 90, 110, 114, 119–120
Liu, C.L., 225
Liu, D.R., 150, 178, 308–309, 549
Liu, H., 151, 152f, 170–171
Liu, J., 218, 335–336, 367, 392–393, 502–504, 510, 516
Liu, J.J., 265
Liu, J.W., 595–596
Liu, L., 150
Liu, N., 48, 57–58, 60
Liu, Q., 410
Liu, S., 281–282
Liu, T., 308
Liu, T.Y., 209, 213
Liu, W., 371–372
Liu, W.R., 90
Liu, W.S., 120
Liu, X., 252, 265, 392, 407–408, 509, 527
Liu, X.H., 115–116, 119–122, 124, 127f
Liu, Y., 57–58, 218, 231–232, 516
Liu, Y.-Y., 308
Livingston, D.J., 527
Loewen, P.C., 513–514

Löfas, S., 38
Lohkamp, B., 510
Lokey, R.S., 280, 304
Lomascolo, A., 456
Lombardi, A., 136–138, 335–336, 472–494, 504
London, N., 308–309
Long, H., 48, 57–58, 60
Long, K., 283–284
Long, R., 561–562
Long, S., 13
Longair, M., 185
Loo, J.A., 113, 124, 380
Lorence, R.M., 457–458
Lottspeich, F., 456
Lou, Y., 3–5
Louie, G.V., 91–93
Lousa, D., 309
Love, J.J., 224–225
Lovejoy, B., 138–139
Low, K.E., 305–308, 306–307t, 310–312, 323
Lowitz, J., 57–58
Lowman, H.B., 46, 48, 50, 54, 58–59, 66, 68
Loyter, A., 309
Lu, C., 525t
Lu, H., 53–54
Lu, H.S., 419–420
Lu, M., 367
Lu, Y., 114–116, 120, 127f, 136–138, 392–393, 408, 502–529
Lubitz, W., 377–379, 390–396, 394f, 475
Luchette, P.A., 252
Luczkowska, R., 395–396
Łuczkowski, M., 136–138, 337–338
Luft, J.R., 517
Lundh, K., 38
Lungu, O.I., 170–174, 173f, 175t
Luo, J., 308–309
Luo, R., 511
Luo, Z.Y., 254
Lusic, H., 97–99
Lutz, S., 192, 540
Lv, X., 519–521, 525t
Lv, X.X., 123–124, 123f
Lynch, V.M., 442–444
Lyubovitsky, J.G., 234–235

M

Ma, J., 503
Ma, Z., 309–310, 326
MacDermaid, C.M., 472–473
Machius, M., 224–225
Mack, K.L., 195, 200–201
Mackay, J.P., 218
Mackerell, A.D., 511
Mackereth, C.D., 281–282
Madden, C., 391, 401–402, 401f
Madden, D.R., 5
Maeda, Y., 192
Maekawa, H., 475–476
Maesmans, G., 456
Magdassi, S., 204
Maglathlin, R.L., 573–574
Magliery, T.J., 206–213
Maglio, O., 335–336, 379, 472–494
Maguire, J.J., 22–23
Magyar, A.P., 584, 596
Magyar, J.S., 419
Mahadevan, V., 503–504
Mahesh, M., 97–99
Mahon, A.B., 308–309
Mahy, J.P., 456–457
Maichele, A., 540–542
Main, E.R.G., 206–207
Mainz, D.T., 252–253, 255–257
Majer, Z., 335–336
Makhatadze, G.I., 163
Makhlynets, O.V., 192–201, 410
Makino, M., 451
Makino, Y., 513–514
Malashkevich, V.N., 564
Malia, T.J., 305
Malmström, B.G., 502–503, 510
Maltby, D.A., 463, 466
Manahan, C.C., 380
Mancini, J.A., 370, 372, 408
Mandal, K., 57–58
Mandell, D.J., 224–225
Manna, S.K., 461–462
Mansurova, M., 382
Mansuy, D., 456–457
Mantle, M., 347
Manzano, C., 151–152
Mao, H., 561–562

Mapp, A.K., 265, 304
Marcotte, E.M., 204
Marcus, R.A., 122
Maréchal, J.-D., 334–335
Marger, F., 311–312
Marian, C.M., 382
Marine, J.E., 475–476
Maritzen, T., 316–317
Markley, J.L., 252–253
Marletta, M.A., 561f, 562, 575t
Marques, H.M., 423–424
Marraud, M., 280
Marsh, E.N.G., 136–138, 252–272, 369, 473–474, 568–570
Marshall, G.R., 255
Marshall, N.M., 120, 136–138, 502–505, 513–515, 519–521, 525t, 527
Marshall, S.L., 351
Marszalek, P.E., 162–163
Mart, R.J., 172–173, 173f
Martell, A.E., 230
Martí, A.A., 12
Martin, A.B., 110
Martin, H., 506–508
Martin, H.L., 283–284
Martin, L.J., 561, 561f, 575t
Martin, R.B., 337–338
Martin, S.C., 2–18
Martinez, J., 435
Martinoni, B., 280
Marzioch, M., 204
Mascarin, A., 475–476
Mason, R.P., 252
Masu, H., 283–284
Masuda, H., 379
Mate, M.J., 463
Mathews, F.S., 472–473
Mathieu, M., 335–336
Matsuda, K., 565t, 566–567, 571–573
Matsuda, T., 444, 452
Matsueda, G.R., 473–474
Matsui, H., 192
Matsui, T., 390–391, 456–457
Matsumura, H., 523–526, 525t
Matsuo, T., 441–444, 451–452
Matthews, B.W., 142–144, 254
Matzapetakis, M., 136–138, 568–569
Mauk, A.G., 436, 463

Maurer, J., 546
Maurizot, V., 280–281, 283–284
May, P.M., 351
Mayer, B.J., 170–171
Mayne, C., 509–510, 513–514, 519–521, 527–528
Mayo, S.L., 224–225, 419–420, 479
Maywald, M., 540–542
Mazor, Y., 82
McCleverty, J.A., 472–473
McCloskey, J.A., 113
McConnell, S.J., 48
McCoy, A.J., 142–144, 243–244
McEldoon, J.P., 456
McEvoy, J.P., 390
McGuirl, M.A., 234–235
McIntire, W.S., 380
McKee, E.M., 582–583, 596
McKinney, M., 22–23, 36–37
McLachlan, A.D., 367
McLachlan, J., 218
Mclachlan, S.J., 527
McLaughlin, J.M., 195
McMahon, S.A., 151, 152f
McMullan, D., 396–397
McNamara, D.E., 224–225
McNaughton, B.R., 207–209
Mcree, D.E., 506–509
McRorie, D.K., 136–139, 473–474
Meagher, J.L., 252–253, 263
Meah, D., 172–173, 173f
Mecozzi, S., 252–253, 260
Medina-Morales, A., 234–236
Meethong, N., 584
Mehl, R.A., 110
Meiler, J., 504
Meinhardt, N., 305–308, 306–307t, 310–312, 323
Meister, A., 544–545
Meloen, R.H., 305–309, 306–307t, 311–312, 325
Meng, E.C., 510
Meng, H., 252–254
Meng, H.Y., 562–563
Merchante, R., 544–545
Mergulhão, F.J.M., 234–235
Merino, K.C., 582–583, 596
Merkel, L., 255–257

Merkle, D.L., 562–563
Merrill, N.A., 582–583, 596
Merz, K.M., 511
Mesner, D., 153
Messerschmidt, A., 440
Mesuda, C.K., 170–171
Metz, F., 3–4
Meusinger, R., 475–476
Mewies, M., 380
Meyer, B., 503
Meyer, T.F., 546
Meyer, T.J., 124, 390, 472–473
Migliorini, C., 337–338
Miles, J.A., 283–284
Miley, M.J., 224–225
Miller, A.F., 504–505, 509–510
Miller, J.M., 304
Miller, S., 305, 306–307t
Milles, S., 97–99, 124–125
Mills, J.H., 128
Mills, J.L., 224–225
Mindt, T.L., 475–476
Miner, K.D., 504, 509, 513–514, 517–521, 525t, 527
Minor, Z., 142
Minus, M.B., 2–18
Mirarefi, A.Y., 513–514
Mirts, E.N., 502–529
Mishin, A.S., 122–123
Mishler, D., 207–209, 212–213
Mitchell, R.W., 4
Miteva, M.A., 479, 480f
Mitra, A.K., 304
Mitra, S., 136–138, 568–569
Mitsui, R., 558
Mittl, P.R.E., 410
Miyagawa, S., 304
Miyamoto, S., 542
Mizohata, E., 442–446, 452–453, 540–541
Moal, I.H., 29–30
Mochrie, S.G.J., 206–207
Moehle, K., 341
Moënne-Loccoz, P., 509, 523–526, 525t
Moffat, K., 170–172
Moffet, D.A., 371–372
Molinari, H., 562–563
Molski, M.A., 260
Monera, O.D., 568–569

Monien, B.H., 475
Montclare, J.K., 252–253, 255–257
Monteiro, G.A., 234–235
Montibeller, J.A., 542
Moody, T., 242–243
Moore, A.L., 390
Moore, D.D., 459–460
Moore, J., 192
Moore, J.C., 540
Moore, J.S., 280
Moore, L., 204
Moore, S.J., 29–32, 34–35
Moore, T.A., 390
Morales, G.A., 286–287
Morales, M., 463
Morawski, B., 456
Morell, M., 212–213
Morelli, G., 335–336
Morita, Y., 444–446, 452–453
Moroz, O.V., 192–193, 195, 200–201, 410
Moroz, Y.S., 192–193, 195, 200–201, 410
Mortenson, D.E., 252–253
Mosberg, J.A., 214
Moser, C.C., 366–384, 408, 425
Moss, J., 312–316
Mossin, S., 229–230
Motekaitis, R.J., 230
Mou, Y., 224–225
Moughon, S., 232–233
Movshovitz-Attias, D., 308–309
Moy, V.T., 151, 158, 164
Mueller, R., 97–99
Mui, S., 425, 435–436
Muir, T.W., 150–151, 255–257
Mukherjee, A., 504, 509, 513–514, 518–522, 525t, 527
Mukherjee, M., 521–522
Mukherjee, S., 521–522
Mukherji, M., 90–91
Mulholland, S.E., 366–367, 371–372, 378, 393–395, 394f, 408
Mullaney, E., 150, 159
Muller, J.M., 547
Müller, K., 252
Muller, M., 124–125
Muller, M.M., 192–193
Müller, S., 280–281, 573
Mulligan, V.K., 128

Munro, C., 582–596
Muppidi, A., 308, 310–312
Murakami, Y., 444–446
Murata, D., 441–444, 451–452
Murphy, C.F., 510
Murphy, N.S., 284
Murphy, P., 410
Murray, C.B., 582, 596
Murray, J.F., 347
Murshudov, G.N., 243–244
Musatov, A., 154
Mutter, M., 335–336, 475
Mylvaganam, R., 252–253
Myszka, D.G., 37–41

N

Naarmann, N., 252–254
Nadarajan, S.P., 252–254
Nadassy, K., 235–236
Naganagowda, G.A., 562–563
Nagano, M., 310
Nagaoka, Y., 283–284, 283f, 296
Nagy, Z., 334–335
Naik, R.R., 582–585, 590, 596
Nair, C.M., 334–335
Nair, R.V., 283–284
Nakagawa, T., 558
Nakajima, H., 456–468
Nakajima, T., 252–253, 255–257
Nakamura, T., 151–153, 164
Nakane, D., 379
Nakane, T., 510
Nakashima, R., 114–115
Nakatsu, T., 510
Nakayama, H., 513–514
Nam, K.T., 584
Nam, Y.S., 584, 596
Nanda, V., 192, 378–379, 393–396, 408
Nasertorabi, F., 128
Nash, A.I., 175t, 180, 180–181t
Nastri, F., 136–138, 335–336, 379, 472–477, 484–486, 493–494, 504
Natale, G.D., 334–335
Natale, P., 547
Nathan, A., 203–205, 207, 209, 211–213, 218
Nebel, K., 475
Neil, L.C., 170–171

Nelson, J.C., 280
Nesloney, C.L., 334–335, 339
Neumann, H., 90, 97–99
Neupane, K.P., 136–138, 335–336
Nevin, S.T., 304
New, S.Y., 504–505
Newman, J., 517
Newton, S.L., 565t, 570–575, 575t
Newton-Vinson, P., 380
Ngu-Schwemlein, M., 335–336
Nguyen, D.P., 97–99
Nguyen, P.C., 230–231, 238–239
Nguyen, T.-A.D., 402
Ngyen, H., 335–336
Ni, M., 170
Ni, T., 234
Nichols, R.J., 596
Nicolet, Y., 390
Nicula, S., 335–336
Niedźwiecka, A., 334–335, 573–574
Nielsen, H., 547–548
Nielsen, P.E., 280
Niemz, A., 252–253
Niesen, F.H., 163
Nikic, I., 91, 97–99
Nikiforovich, G.V., 335–336
Nikolovska-Coleska, Z., 176
Nilges, M.J., 504–505, 513–514, 519–521, 525t, 526
Nilsson, B.L., 252–253
Nilsson, L., 511
Niopek, D., 170–173, 173f
Nitz, M., 561, 573–574
Noack, P.L., 408
Noble, R., 312–316
Nocera, D.G., 424–425, 510
Noda, M., 379
Noel, J.P., 91–93
Noguchi, H., 280–281
Nogueira, E.S., 542–545
Nolan, T., 513–514
Nordman, C.E., 136–139, 138f, 142–144
Nossal, N.G., 244–245
Novotny, J., 473–474
Nowicki, M.W., 225, 229–230
Noy, D., 382–383, 393–396
Null, E.L., 509, 513–514, 518–521, 527

Nureki, O., 113
Nussinov, R., 224–225
Nyakatura, E.K., 252–253

O

O'Connor, S.J., 48, 61
O'Halloran, T.V., 136–138, 473–474
O'Neil, K.T., 419–420
O'Sullivan, B., 352
Oas, T.G., 475
Obata, Y., 96
Oberer, L., 280
Obexer, R., 410
O'Connor-McCourt, M.D., 572
Oda, M., 379
Odar, C., 252–253
Oderaotoshi, Y., 254
Odoi, K., 90
O'Donoghue, P., 90
Ofek, G., 305, 308
Ogata, H., 390–391
Ogihara, N.L., 396–397, 408
Ohata, J., 3, 10–12, 15–17
Ohbayashi, J., 444–446, 452–453
Ohki, T., 513–514
Ohlendorf, D.H., 542
Oke, M., 151, 152f
Oktyabrsky, O.N., 544–545
Okuda, J., 540–542
Olafson, B.D., 479
Olds, W., 214
Olsen, A.B., 195, 200–201
Omran, H., 472–473
Onoda, A., 391, 402–403, 404f, 410, 440–441, 540–541
Oohora, K., 440–453
Orlamuender, M., 335–336
Oros, S., 335–336
Orsolya, G., 595–596
Ortega, A., 243
Orthaber, A., 410
Ossetian, R., 459–460
Osyczka, A., 371–372, 378–379, 408
Ősz, K., 334–335
Othon, C.M., 252–253
Ott, S., 410
Ottersen, O.P., 475–476
Otvos, L., 309

Otwinowski, W., 142
Ou, W., 90–91
Overhand, M., 280
Ozaki, S.-i., 390–391, 456–457, 463, 466
Ozawa, T., 162
Ozbas, B., 334–335

P

Paar, A., 456
Pabo, C.O., 48, 475
Pacardo, D.B., 582, 584, 590, 596
Pace, C.J., 252–253
Pace, C.N., 345, 563, 568–569, 572
Paciorek, W., 142–144
Packer, M.S., 150, 178, 549
Packman, K., 305, 308
Page, C.C., 366
Pahwa, S., 308–309
Pai, P.J., 91
Pakstis, L., 334–335
Pal, G., 48, 61
Palafox-Hernandez, J.P., 582–583, 596
Pall, S., 511
Palmer, T., 547–548
Pan, H., 176
Pan, W., 207–209, 212–213
Pan, Y., 91
Pande, U., 440
Pandelia, M.E., 398–399
Pandey, A.K., 562–563
Pandey, R.K., 440
Panke, S., 540–553
Papanikolas, J.M., 390
Papini, A.M., 475–476
Pappalardo, G., 334–336
Park, H., 584, 596
Park, K., 204
Park, M., 546
Parmeggiani, F., 218
Parra, R.D., 280
Parrish, A.R., 90
Parry, D., 563–564
Parslow, R.A., 565t, 566–567, 570–574
Parthasarathy, R., 150–151
Paschke, M., 544
Passard, G., 410
Passioura, T., 308–309
Patching, S.G., 37–41

Patel, A., 304
Patel, S.U., 48
Patelis, L., 15
Paterson, N.G., 472–473
Patgiri, A., 305
Patora, K., 341
Paul, D., 457–458
Paulmurugan, R., 207–209
Pavlov, A.R., 37–38
Pavone, V., 136–138, 379, 472–477, 484–486, 493–494, 504
Peacock, A.F.A., 136–139, 142–144, 472–473, 541, 558–576
Pearce, D.A., 334–335
Pearson, A.D., 128
Pearson, R.G., 337
Pecoraro, C., 509
Pecoraro, V.L., 136–139, 138f, 142–144, 366, 419–420, 473–475, 504, 541, 568–570
Peggion, C., 335–336
Pei, D., 308
Peisach, E., 561, 573–574
Peishoff, C.E., 334–335
Pekka, M., 506–508
Pelletier, J.N., 475–476
Pellizzoni, M., 541–542
Peluso, S., 335–336
Pendley, S.S., 252–253
Penner-Hahn, J.E., 136–138, 419–420
Pentelute, B., 305–308, 306–307t
Peraro, L., 304–327
Pérez Gutiérrez, O.N., 150–151
Perez, A., 234–236
Perez, M., 440–441
Perez-Aguilar, J.M., 408–409, 472–473
Perico, A., 472–473
Perrakis, A., 142–144
Perrett, S., 116
Perrin, M.H., 90, 92, 99–100
Persson, H., 54
Peters, D., 354
Peters, F.B., 91
Peters, J.W., 390
Petersen, M.T.N., 234
Petersen, S.B., 234
Petersen, T.N., 547–548
Peterson, J.C., 15
Peteya, L.A., 170–171
Petillo, P.A., 506–509, 513–514
Petka, W.A., 252–253, 255–257
Petkov, V., 594–595
Petrella, R.J., 511
Petrenko, V., 53–54
Petrey, D., 507
Petrik, I.D., 392–393, 408, 502–529
Petros, A.K., 378, 569–570
Petsko, G.A., 224–225
Pettersen, E.F., 510
Peuser, I., 256t
Peyralans, J.J.P., 540–542
Peyron, S., 465–466
Pfister, T.D., 509–510, 513–514, 519–521, 527–528
Philippart, F., 541–542
Philips, G.N., 440
Phillips, D.M., 596
Phillips, J.C., 511
Phillips-Piro, C.M., 561f, 562, 575t
Pickett, C.J., 503–504
Pidikiti, R., 366–367, 371–372, 408
Pielak, G.J., 265
Piera, J., 255
Pierre, B., 154
Pierre, J.-L., 503–504
Pietzsch, T., 185
Pike, D.H., 393–395
Pinkas, D.M., 410
Pinkas, I., 377
Pires, M.M., 493–494
Pires, S., 334–335
Pirotte, G., 281–282
Pirrung, M.C., 3–4
Planas, A., 150–151
Plante, J.P., 284
Plass, T., 97–99, 124–125
Platsaki, S., 472–473
Platt, G., 562–563
Plegaria, J.S., 398–399, 472–473, 541
Ploegh, H.L., 150–151, 477–478
Plotnikov, V., 163
Plückthun, A., 46, 150–151, 154, 178, 218, 544
Po, J., 305, 306–307t

Pochan, D.J., 334–335
Poeppelmeier, J., 390–391
Poeppelmeier, K., 504
Pogni, R., 462–465
Poisson, J., 511
Pokora, A.R., 456
Polizzi, N.F., 408–409
Pollard, T.D., 23–24, 26
Polyakov, K.M., 116
Polycarpo, C., 113
Pomerantz, W.C., 192–193, 265
Ponamarev, M.V., 373*t*
Pooley, H.M., 544–545
Popp, B.V., 3, 5, 10–12
Popp, M.W.-L., 150–151
Porciatti, E., 337–338
Poreda, R.J., 573
Potel, M., 280
Potterton, E., 142–144
Poulos, T.L., 440, 506–508
Powell, D.R., 280
Powell, H.R., 243–244
Prabhakaran, P., 284
Prabhulkar, S., 366
Prasad, P.N., 582–584, 596
Prat, L., 90
Pratt, R.C., 116, 525*t*
Presta, L.G., 48, 61
Pretzel, D., 335–336
Price, N.C., 265–266
Pritchard, M.P., 459–460
Privalov, P.L., 163
Privett, H.K., 410
Proffen, T., 595–596
Proft, T., 151
Pronk, S., 511
Prosser, R.S., 252
Prusoff, W.H., 36
Prutsman-Pfeiffer, J.J., 573
Puckett, J.W., 284
Puijk, W.C., 305–309, 306–307*t*, 311–312, 325
Pujol, A.M., 335–336
Pullen, S., 410
Punniyamurthy, T., 503–504
Putnam, C.D., 506–509
Pye, C.R., 304
Pyun, J.C., 546

Q

Qayyum, M., 337–338, 357–358
Qi, D.F., 541–542
Qi, H., 53–54
Qi, T., 280–281
Qian, Z., 308
Qiao, J., 561–562
Qiu, G., 516
Qiu, H.-J., 53–54
Qiu, X., 594–595
Quail, P.H., 170
Quan, J., 155
Quan, S., 456
Que, L., 503–504, 525*t*
Quentel, F., 410
Quintanilla, A., 584–585
Qvit, N., 309

R

Rabanal, F., 366–367, 369, 372, 376, 378, 380–382, 393–395, 394*f*, 504
Rachel, N.M., 475–476
Raddatz, R., 22–23, 36–37
Rader, C., 61
Radford, R.J., 230–231, 238–239
Radtke, F., 309–310
Rajagopal, K., 334–335
Rajbongshi, J., 461–462
Raleigh, D.P., 252–253, 268
Ralle, M., 507
Rama, G., 334–335
Ramakrishnan, C., 475
Ramamoorthy, A., 252–253, 256*t*, 265
Ramanathan, K., 459, 461–462
Ramani, A.K., 204
Ramezani-Dakhel, H., 582–584, 596
Ramírez-Alvarado, M., 334–335
Ran, X., 304–305, 306–307*t*, 308
Rao, B.V., 154
Rath, A., 158
Rau, H.K., 475
Raveh, B., 308–309
Ravikumar, Y., 252–254
Ravindran, V., 159
Rawson, J., 408–409
Raymond, E.A., 195, 200–201
Raymond, K.N., 561*f*, 562, 575*t*

Razavi, A.M., 309
Read, R.J., 142–144, 243–244
Reardon, N.J., 504–505
Reback, M.L., 390–391
Rebar, E.J., 48
Reddi, A.R., 418–420, 425–426, 435–436
Reddington, S.C., 150–151
Reddy, K.S., 372
Reece, S.Y., 510
Reed, J., 502–529
Reed, S., 91
Reedijk, J., 504
Reedijk, K., 390–391
Reedy, C.J., 371, 420–425, 430–431, 435–436, 440, 504
Reetz, M.T., 540–544, 549
Regan, L., 203–218
Regula, L.K., 57–58, 61
Reibenspies, J.H., 410
Reichard, P., 472–473
Reichen, C., 218
Reichert, J.M., 304
Reig, A.J., 493–494
Reijerse, E.J., 378–379, 390–391, 395–396
Reinhart, B., 582–583, 596
Reinke, A.W., 473–474
Reinwarth, M., 475–476
Reisner, E., 477
Reissmann, S., 335–336
Remarchuk, T.P., 4–5
Rempel, G.L., 4
Ren, Y., 582–583, 596
Renata, H., 540
Renaudet, O., 335–336, 341–342, 475–476
Renicke, C., 170–171
Renukuntla, J., 304
Reubi, J.C., 475–476
Rey, F.A., 61–62
Reynolds, A.M., 561
Reynolds, M., 559
Rheingold, A.L., 229–230
Richards, F.M., 207–209, 254
Richards, J.H., 234–235
Richardson, D.J., 503–505, 509
Richter, F., 194, 504, 508
Riddles, P.W., 347
Rife, C.L., 396–397
Rinehart, J., 203–205, 207, 213–214, 216

Ringe, D., 224–225
Risse, J.M., 547
Ritter, M., 380
Rizzarelli, E., 334–335
Roach, M.P., 390–391, 456–457
Robb, F., 457–458
Roberts, J.A.S., 391
Roberts, J.D., 66, 70
Robertson, D.E., 366–367, 371–372, 408
Robertson, I.R., 15
Robertson, N.S., 305–308
Robins, K., 192, 540
Robinson, C.V., 151–153, 164
Robinson, H., 504–505, 509–510, 513–514, 517–521, 525t, 527–528
Robinson, J.A., 334–336, 339, 341
Robl, C., 400–401
Robra, K.H., 456
Rochalski, A., 163
Rodriguez-Granillo, A., 393–395
Roelfes, G., 493–494
Roensch, J., 170–173, 173f
Roeske, W.R., 36–37
Rognan, D., 280
Rogulina, S., 214
Rohl, C.A., 232–233
Roisnel, T., 280
Rolland, P.H., 22–23
Romero, A., 463
Rongved, P., 475–476
Rose, G.D., 334–335
Rosen, B.P., 113
Rosen, M.K., 172–173, 175t, 177
Rosenblatt, M.M., 369, 378–379, 473–474
Rosenstrom, P., 507
Rosenthal, H.E., 23
Rosenzweig, A.C., 472–473
Rosi, N.L., 596
Rosner, P.J., 158–159
Ross, M., 504–505, 513–514, 525t, 526
Rossel, T., 542
Rosset, C., 573–574
Rossi, A.M., 217–218
Rossotti, F.J.C., 352
Rossotti, H., 352
Rosu, F., 281–282, 284, 286, 293f, 294, 298f
Röthlisberger, D., 410, 419–420
Rouau, X., 465–466

Rousseau, D.L., 509, 525t
Rousselot-Pailley, P., 335–336, 338–339, 338t, 341
Roux, B., 511
Rovero, P., 475–476
Rovis, T., 542
Rovner, A.J., 130, 214
Roy, A., 283–284, 378–379, 390–393, 396–398, 401–402, 401f, 406, 511
Roy, S., 390–391, 401–402, 401f, 410
Royant, A., 472–473
Rozwarski, D.A., 472–473
Ruan, L., 582–583, 596
Rückle, T., 335–336
Ruckthong, L., 136–145
Ruddick, J.D., 4
Rudick, J.G., 475–476
Rüdiger, O., 390–391
Rudolph, M., 400–401
Ruiz-Duenas, F.J., 463
Rulíšek, L., 225, 229–230
Rumbley, J., 503
Rumpf, J., 316–317
Ruotolo, B.T., 252–253, 256t, 265
Russell, B.S., 507–509, 513–515, 518–522
Russell, W.K., 91
Rutherford, T., 309–310
Ryu, S.-B., 136–138

S

Sabatini, A., 353–354
Sablon, E., 48
Sachdev, P., 93
Sachdeva, A., 97–99
Saddler, J.N., 506–508
Sadeghi, M., 304
Sage, J.T., 525t
Saigo, K., 252–253
Saita, E.-I., 418
Sakamoto, K., 97–99
Sakmar, T.P., 90, 103–104
Salaun, A., 280
Salemme, F.R., 542
Salgado, E.N., 225–230, 232–234
Saltman, P., 521–522
Salwiczek, M., 252–254, 260
Salzmann, S., 382
Sambasivan, R., 5

Samish, I., 472–473
Sancey, L., 475–476
Sanchez, M.E.N., 61–62
Sanders, C.R., 252
Sanders, J.K.M., 503–504
Sandesara, R.G., 48, 57–58
Sandhage, K.H., 583–584
Sani, M., 255
Sanjayan, G.J., 283–284
Sano, Y., 391, 402–403, 404f, 440–441
Santi, D.V., 475
Santini, C.L., 547
Santoro, S.W., 111–112
Saphire, E.O., 48, 57–58
Sarangi, R., 525t
Sarikaya, M., 582–584
Sarkar, M., 209
Sarraju, A., 561, 561f, 575t
Sarrou, I., 378–379, 396–397
Sasa, K., 558
Saschenbrecker, S., 154
Sastry, M., 506–509
Satake, Y., 461–462
Sato, H., 441–444, 451
Sato, M., 180–181t
Satterlee, J.D., 515
Sattler, M., 316–317
Satyshur, K.A., 252–253
Sauer, D.F., 540–541
Savaresi, S., 158
Savchenko, A., 517
Saven, J.G., 280, 408, 472–475
Sawada, A., 452–453
Sawaki, K., 252–253
Sawyer, E.B., 110–111
Sawyer, N., 203–218
Sawyer, T.K., 305, 308
Sazinsky, M.H., 472–473
Scalley, M.L., 268
Scatchard, G., 23
Schaer, T., 311–312
Schaffer, D.V., 170–171
Schafmeister, C.E., 305, 306–307t
Schepartz, A., 260
Schertler, G.F., 90
Schiffrin, D.J., 596
Schindelin, J., 185
Schirf, V., 136–138

Schirrmann, T., 544
Schlapschy, M., 547–548
Schlichting, I., 170–171, 173–174
Schlieker, C., 203
Schlippe, Y.V., 309–310
Schmickler, H., 442–444
Schmidt, F.S., 48
Schmidt, M.J., 91
Schmied, W.H., 91–93, 99
Schmitter, J.-M., 280–281
Schmitz, R.A., 378–379, 392, 396–398
Schneider, G., 472–473
Schneider, J.P., 334–335
Schneider, P., 456
Schnepf, R., 379, 475
Schnölzer, M., 257
Schoene, C., 150–164
Scholl, Z.N., 162–163
Schollmeyer, D., 284
Scholtz, J.M., 563, 568–569, 572
Schomburg, B., 151
Schraidt, O., 97–99
Schreiber, G., 224–225
Schröder, H., 162–163
Schuck, P., 242–243
Schueler-Furman, O., 232–233, 308–309
Schulten, K., 507, 583–584
Schultz, C., 97–99, 124–125
Schultz, P.G., 48, 90–93, 99, 110–114, 119–120, 124–129, 255–257, 380
Schulz, G.E., 546
Schulz, H., 237–238
Schulz, R., 511
Schumann, F.H., 150–151, 153
Schunemann, V., 462–463
Schuster, D., 170–171
Schwalbe, H., 561
Schwartz, D.T., 583–584
Schwartz, J.W., 170–171
Schwarzenbach, G., 347–348
Schwarz-Linek, U., 151, 158, 164
Schwemlein, S., 335–336
Schwochert, J.A., 304
Scopes, R.K., 347
Scott, C.P., 150–151
Scott, M.P., 393–395
Scott, R.A., 504, 513–514, 525t, 526
Scott, T., 159

Scott, W.G., 510
Scrima, M., 475–476
Scrofani, S.D., 236
Scrutton, N.S., 380
Sculimbrene, B.R., 559
Searle, M.S., 562–563
Sedlák, E., 154
Seebach, D., 280
Segal, D.J., 218
Segatori, L., 547
Seidman, J.G., 459–460
Seifert, S., 584, 596
Selle, P.H., 159
Semetey, V., 280
Semple, A., 163
Sénèque, O., 334–336, 562, 575t
Senes, A., 252–253, 383, 408
Senguen, F.T., 252–253
Senn, S., 393–395
Senske, M., 265
Séraphin, B., 513–514
Serfling, R., 90–104
Serrano, L., 334–335
Sessions, R.B., 564
Sessler, J., 442–444
Sethi, M., 582, 584, 590, 596
Seufert, W., 252–254
Seyedsayamdost, M.R., 118–119
Sezer, M., 334–335
Shah, A.S., 15
Shalev, D.E., 309
Shamala, N., 570
Shameem, M., 163
Shaner, S.E., 569–570
Shang, J., 281–282, 284, 286, 293f, 294, 298f
Shanklin, J., 472–473
Shanmugam, M., 525t
Shao-Horn, Y., 584
Sharp, J.T., 15
Sharp, K.A., 136–138
Sharp, R.E., 366–367, 376–377, 380–382
Sharpless, K.B., 124–125, 475–476
Shaw, W.J., 390–391
Sheehan, M.M., 366–384, 408–409
Sheffler, W., 224–225
Shelburne, C.E., 252–253
Shelton, P.T., 314
Sheneman, B.A., 308–309, 311, 316–317

Shepherd, N.E., 304–305, 306–307t, 308
Sherawat, M., 573–574
Shi, C., 308–309
Shi, P., 255–257, 265
Shi, R., 284
Shi, S., 163
Shifman, J.M., 425
Shiga, D., 379
Shimada, A., 114–115
Shimamoto, N., 513–514
Shinde, S., 401, 401f
Shinoda, S., 457–458
Shinzawa-Itoh, K., 114–115
Shiratori, O., 255, 256t
Shiro, M., 252–253
Shoichet, B., 475
Shoji, I., 304
Shoji-Kawata, S., 317
Shook, R.L., 504–505
Shu, J.Y., 475–476
Shults, M.D., 334–335
Si, G., 410
Sia, S.K., 570
Sidhu, S.S., 46, 48, 50, 53, 57–58, 61–62, 68, 70, 72, 204–205
Siegbahn, P.E., 510
Siegert, T.R., 304–327
Sieracki, N., 120, 136–138, 502–504
Sievers, F., 507
Sigel, A., 444–446
Sigel, H., 337–338, 444–446
Sigel, R.K.O., 444–446
Sigman, J.A., 114–115, 504, 506–509, 513–514
Sigoillot, J.C., 456
Sila, U., 335–336
Silchenko, S., 436
Sillitoe, I., 507
Silvage, N.R., 561
Simmons, T.R., 390–391
Simon, B., 316–317
Simon, J., 382
Simon, R.J., 280
Simona, F., 236
Simons, K.T., 268
Simpson, N., 54
Singer, C.P., 136–138, 473–474
Singleton, M.L., 281–282, 410

Sinha, S.C., 90
Sinks, L.E., 408
Sippel, M.L., 436
Sirajuddin, S., 472–473
Sirish, M., 560–561
Sisommay, N., 335–336
Siu, F.Y., 90, 99–100, 101f, 103–104
Sivakumar, K., 13
Skalicky, J.J., 369, 504
Skarina, T., 517
Skerra, A., 48, 544, 547–548
Skrzypczac-Jankun, E., 280
Sligar, S.G., 462–463, 506–508, 513–514
Slocik, J.M., 582–585, 596
Sloof, W.G., 584–585
Slope, L.N., 472–473, 565t, 570–575, 575t
Slutsky, A., 116
Slutsky, M.M., 252–254, 257–259
Smeenk, L.E.J., 311
Smirnova, G.V., 544–545
Smith, A.D., 392–393
Smith, A.E., 265
Smith, A.J., 371–372
Smith, A.J.T., 194
Smith, B.A., 474–475
Smith, C.F., 565t, 570–575, 575t
Smith, G.P., 46, 48
Smith, H.O., 155
Smith, J.A., 334–335, 459–460
Smith, K.M., 371–372, 440, 444, 460–461
Smith, R.M., 230
Smith, S.J., 238–239
Snell, E.H., 517
Snell, J.M., 380
Snyder, R.A., 493–494
Snyder, T.M., 308–309
Soares, C.M., 309, 334–335
Sobolev, A.P., 335–336
Sok, V., 304
Sokolova, L., 503
Soll, D., 130
Sollewijn Gelpke, M.D., 509
Solomon, E.I., 235–236, 337–338, 357–358, 472–473, 507
Solomon, L.A., 366–367, 370–372, 374–377, 380–382, 408–409
Sommer, D.J., 378–379, 390–411
Sommerhalter, M., 475

Son, S., 252–253, 255–257
Song, S., 475–476
Song, W.J., 225, 234–236, 245–247, 541, 543–544
Song, Y.F., 128
Sonnichsen, F.D., 568–569
Sontz, P.A., 225, 234–236, 245–247
Soper, M.T., 252–253, 256t, 265
Sosnick, T.R., 172–173, 175t, 177, 180, 180–181t, 228
Sota, H., 154
Sowdhamini, R., 475
Spadafora, L., 234–235
Spalding, M.H., 218
Späth, J., 334–335
Spelke, D.P., 170–171
Speltz, E.B., 203–218
Spencer, J., 235–236
Spiezia, M.C., 335–336
Spiltoir, J.I., 170–173, 173f
Spirin, A.S., 154
Spiro, S., 504–505, 509
Spokoyny, A., 305–308, 306–307t
Spooner, E., 150–151
Spring, D.R., 475–476
Springer, B.A., 506–508, 513–514
Squire, J., 563–564
Srinivas, K., 281–282
Srinivasan, N., 475
Srivastava, P., 540–541
Sroga, G.E., 544
Stachura, M., 136–138
Stack, T.D.P., 116, 503–504, 525t
Staedel, C., 280–281
Stafford, W.F., 242–243
Starck, M., 335–336
Starling-Windhof, A., 154
Staropoli, I., 61–62
Stayrook, S.E., 368f, 369
Steffes, C., 113
Stein, H.H., 159
Steinem, C., 252–254
Steiner, D., 178
Stenberg, E., 38
Stenkamp, R.E., 472–473
Stephenson, R.P., 23
Sternberg, U., 335–336
Stetter, K.O., 378

Stevens, R.C., 265
Stewart, A., 57–58, 61
Stewart, G., 347
Stewart, M., 367
Stibora, T., 48
Stiefel, E.I., 228–229
Stoilova, I., 311
Stone, M.O., 582, 584, 596
Storch, M., 513–514
Storoni, L.C., 142–144, 243–244
Storr, T., 116
Stout, C.D., 475–476
Strahl, B.D., 170–173, 173f
Strickland, D., 170–173, 173f, 175t, 177
Strupat, K., 513–514
Strzalka, J., 382–383
Stuart, F., 334–335
Stubbe, J.A., 118–119, 366
Stuchebrukhov, A.A., 424–425
Stuckey, J.A., 136–145, 252–253, 263, 541, 569–570
Studier, F.W., 551–552
Stumpp, M.T., 178
Stupák, M., 154
Stupfel, M., 280–281, 286
Subach, F.V., 122–123
Subra, G., 435
Subramanian, R.H., 224–247
Subramanian, S., 150–151
Suchanek, E.G., 475
Sudha, R., 562–563
Suga, H., 304, 308–310
Sugahara, M., 510
Sugimoto, H., 451, 457–458
Suich, D.J., 419–420
Summa, C.M., 136–138, 369, 473–474, 504
Summerer, D., 90–91
Summers, D.K., 234–235
Sumpter, R., 317
Sun, F., 155–156
Sun, J., 304
Sun, L., 309–310
Sun, N., 514–515, 525t
Sun, S.B., 99
Sun, T., 90, 99–100, 101f, 103–104
Sun, X., 335–336
Sun, Y., 91
Sundaramoorthy, M., 506–508

Sundberg, E.J., 204
Sunderland, C.J., 502–503
Sung, M.-K., 209
Sutin, N., 122
Sutter, M., 398–399
Sutton, B.J., 236
Suzuki, H., 180–181t
Suzuki, M., 510
Suzuki, T., 418
Suzuki, Y., 252–253, 256t, 260, 265
Suzumura, A., 457–458
Svendsen, A., 456
Svensson, S.O., 593
Svetlov, M.S., 154
Swartz, T.E., 180–181t
Swers, J.S., 28, 30–32
Swihart, M.T., 582–584, 596
Sykes, B.D., 565t, 566–567, 571–573
Szostak, J.W., 309–310
Szymanski, J., 97–99
Szyperski, T., 224–225

T

Taboy, C.H., 521–522
Tachikawa, H., 513–514
Tachikawa, K., 540–541
Taina, L., 506–508
Tajkhorshid, E., 511
Takafuji, M., 280–281
Takahashi, H., 154
Takatani, N., 461–462
Takenawa, T., 154
Takeuchi, R., 541
Takimoto, J.K., 91–93, 100
Tamerler, C., 583–584
Tamura, N., 96
Tan, C., 475–476
Tanaka, K., 418
Tanatani, A., 283–284, 283f, 296
Tang, Y., 252–257, 256t, 304
Tang, Z., 582–584, 596
Tani, A., 558
Tann, C.M., 541–542
Tanrikulu, I.C., 252–253, 255–257
Tard, C., 503–504
Tarr, G.E., 457–458
Tashiro, S., 558
Taslimi, A., 170–171

Tawfik, D.S., 195, 410
Taxis, C., 170–171
Tayakuniyil, P.P., 150–151, 153
Taylor, C.W., 217–218
Teale, F., 447
Tebo, A.G., 472–473, 541
Tegoni, M., 379, 472–473, 541
Tenne, S.J., 541–542
Teplyakov, A., 142–144, 243–244
Tepperman, J.M., 170
Tereshko, V., 204–205
Terrasse, R., 151–152
Teutloff, C., 462–465
Tezcan, F.A., 224–247, 541, 543–544
Thaler, R.C., 159
Than, M.E., 457–459
Thanh, N.T.K., 596
Thanyakoop, C., 234–235
Theile, C.S., 477–478
Therien, M.J., 383, 408–409, 473–474
Thie, H., 544
Thomas, C.M., 504
Thomas, F., 473–474
Thomas, J.G., 544, 552
Thomas, K.G., 584–585
Thommes, S., 546
Thompson, A.D., 304
Thompson, J., 198–199, 265
Thompson, J.W., 594–595
Thomson, A.R., 564
Thony-Meyer, L., 237–238
Thorn, K.S., 61
Thornton, J.M., 48, 334–335
Thulstrup, P.W., 136–138, 474–475
Thumbs, P., 124–125
Thust, S., 252–254
Tian, H., 366
Tian, J., 155
Tian, S., 392–393, 502–503, 510, 513–514, 517–518, 525t
Tidor, B., 564
Tierney, D.L., 136–138, 419–420, 426, 473–474, 569–570
Tikhonova, T.V., 116
Timmerman, P., 305–309, 306–307t, 311–312, 325
Tinberg, C.E., 541–542
Ting, A.Y., 158

Tirrell, D.A., 155–156, 252–257, 256t
To, K.-H., 305, 308
Tobback, P., 456
Todorova-Balvay, D., 311
Toews, M.L., 26, 35, 37
Tokunaga, K., 444–446
Tokuriki, N., 195
Tolman, W.B., 503–504
Tomatis, P.E., 236
Tomita, Y., 176
Tomlin, J., 406
Tomlinson, D.C., 283–284
Tommos, C., 366–367, 380
Tomson, F.L., 509
Tonikian, R., 50, 53, 68, 70, 72
Toogood, P.L., 255, 256t
Torres, M.E.S., 472–473
Torres-Kolbus, J., 97–99
Toscano, W.A., 440–441
Toulmé, J.-J., 280–281
Touw, D.S., 136–139, 138f, 142–144
Tovar, C., 305, 308
Townsend, C.E., 304
Townsley, F.M., 97–99
Tremmel, D., 218
Trevethick, M.A., 22–24, 26–29, 34
Tronin, A., 408
Trumpower, B.L., 371, 513–514
Tsai, C.-J., 224–225
Tsai, P.C., 32
Tse, B.N., 308–309
Tse, K.M., 510
Tsernoglou, D., 472–473
Tsume, Y., 304
Tsushima, T., 255, 256t
Tuchscherer, G., 335–336, 475
Tucker, C.L., 170–173, 173f
Tucker, M.G., 595–596
Tufféry, P., 479, 480f
Tukalo, M., 113
Tulla-Puche, J., 475–476
Tullman, D., 178
Tung, C.-H., 410
Turkenburg, M., 142–144
Turmo, A., 398–399
Turner, J.M., 284
Tyagi, R., 513–514
Tyka, M., 198–199, 265

U

Ubbink, M., 457–458
Uchida, N., 255, 256t
Uchime, O., 57–58
Uchiyama, S., 379
Ueno, T., 440–441, 456–459, 504, 513–514
Uges, D.R.A., 22–23
Ullah, A.H., 150, 159
Ultsch, M., 472–473
Umezawa, Y., 162
Ungaro, R., 503–504
Uppalapati, M., 57–58
Urbauer, J.L., 366–367, 371–372, 408
Urech-Varenne, C., 309–310
Urvoas, A., 61–62
Uryu, S., 113, 124, 380
Usherenko, S., 170–171
Utschig, L.M., 136–138, 473–474
Uttamapinant, C., 91–93, 99
Uy, R., 114

V

Vacca, A., 353–354
Vaccaro, B., 180–181t
Vadlapudi, A.D., 304
Vagin, A.A., 142–144, 243–244
Vaia, R.A., 582–583
Vaidehi, N., 252–253, 255–257
Vajdos, F., 345
Valderrama, B., 462–466
Valdez, C.A., 503–504
Vale, W.W., 90, 92, 99–100
Valensin, D., 337–338
Valensin, G., 337–338
Valentine, J.S., 228–229
Valeri, P., 595–596
Valverde, I.E., 475–476
Valverde, R., 203, 207, 211
van den Beuken, T., 48
van der Donk, W.A., 366
Van der Eycken, E.V., 291
van der Oost, J., 150–151, 456
van der Post, S., 472–473
Van Deventer, J.A., 252–253
Van Dyke, B.R., 521–522
van Gastel, M., 377
van Lieshout, J.F.T., 150–151

van Maarseveen, J.H., 311, 475–476
van Neer, N., 48
van Pee, K.H., 456
Van Regenmortel, M.H.V., 573
Vanderah, T.A., 595–596
Varadamsetty, G., 218
Varadan, R., 150–151, 153
Varadarajan, R., 504–508
Varani, G., 341
Varedi, M., 170–171
Varshavsky, A., 207–209
Vass, E., 335–336
Vaughn, M.D., 378–379, 396–397, 405–406
Vazquez-Duhalt, R., 456–458, 463, 466
Vedadi, M., 163
Veggiani, G., 151–153, 164
Veitch, N.C., 456
Velusamy, S., 503–504
Venkatachalapathi, Y.V., 334–335
Venkatraman, J., 562–563
Venter, J.C., 155
Ventura, S., 212–213
Verdin, J., 456, 462–465
Verdine, G.L., 305, 306–307t
Verkade, J.M., 97–99
Verkhusha, V.V., 207–209
Verma, C.S., 305
Verma, P., 116
Vernet, T., 151–152
Victor, B.L., 309
Vijayadas, K.N., 283–284
Vijayalakshmi, M.A., 311
Vijayan, M., 334–335
Vila, A.J., 235–236
Vilela, M., 170–173, 173f
Villa, E., 511
Villafranca, J.J., 473–474
Villegas, J.A., 463
Vincent, T.L., 566–567
Vind, J., 456
Vine, W.H., 255
Vita, N., 472–473
Vitello, L.B., 515
Voelz, V.A., 309
Vogel, A.I., 348–349
Vogel, E., 442–444
Vohidov, F., 2–3, 10–12, 14–17

Voigt, C.A., 170
Vojkovsky, T., 341–342
Volbeda, A., 225, 231–232
Volonterio, A., 255
Vomstein, S., 475–476
von Heijne, G., 547–548
von Hippel, P.H., 345
von Stetten, D., 472–473
Vondrášek, J., 225, 229–230
Voss, S., 544
Vrabel, M., 124–125
Vrana, J.D., 170–171
Vroom, W., 150–151
Vulpetti, A., 252

W

Waage, P., 23
Wachtershäuser, G., 378
Wade, J.D., 309
Wagner, E., 170–173, 173f, 175t, 177
Wagner, G.C., 440–441
Wahlgren, W.Y., 472–473
Wahnon, D.C., 150–151
Wakabayashi, C., 418
Wakefield, A.E., 309
Walensky, L.D., 305, 308
Walkinshaw, M.D., 225, 229–230
Wall, M., 150–151
Wallace, S., 97–99
Walters, R.F.S., 382–383
Waluk, J., 442–444
Wan, W., 91
Wand, A.J., 504
Wang, C., 232–233
Wang, C.-I.A., 304
Wang, F., 284, 410
Wang, H., 170–174, 173f, 511
Wang, H.-Y., 410
Wang, J., 91, 110–130, 255–257, 265, 392, 407–408, 509, 527
Wang, J.Y., 111–112, 114–116, 120–122, 127f
Wang, L., 90–92, 99–100, 111–112, 124–127, 255–257, 380, 407–408
Wang, N., 265
Wang, P., 252–253, 255–257, 256t
Wang, Q., 13, 90
Wang, R., 176, 265

Wang, S., 176, 304–305, 306–307t, 308
Wang, T., 255–257, 265
Wang, W., 474–475, 511
Wang, W.-G., 408, 410
Wang, W.Y., 91–93
Wang, X., 150, 218, 366, 504, 506–509, 513–514, 525t
Wang, Y., 305–309, 306–307t
Wang, Y.-P., 207–209
Wang, Y.S., 91
Wang, Z., 91, 308, 310–312, 419–420, 502–503, 513–514
Wang, Z.J., 540
Wang, Z.U., 91
Wanner, R., 163
Ward, T.R., 116–117, 390–391, 504, 540–553
Wardt, T.R., 541–542
Warncke, K., 375–376
Wasbotten, I., 444
Waser, B., 475–476
Wasinger, E.C., 116
Wasowicz, T., 334–335
Wasserman, Z.R., 192–193, 504
Watanabe, C.K., 48, 61–62, 456–468
Watanabe, T., 441–442
Watanabe, Y., 390–391, 440–441, 456–459, 461–462, 504
Waters, M.L., 334–335
Watkins, A.M., 308–309
Watkins, D.W., 370, 372, 408
Watmough, N.J., 503–505, 509, 521–522
Watson, M.D., 475–476
Watters, J.J., 180
Weaver, J.D., 150, 159
Weber, P.C., 472–473, 542
Wehler, P., 170–173, 173f
Wei, H.-P., 207–209
Wei, J., 91
Wei, L., 561–562, 575t
Wei, Y., 408
Weigand, W., 400–401
Weinberg, D.R., 510
Weiner, L., 378–379, 395–396
Weiner, O.D., 170
Weiss, G.A., 48, 61–62
Weissberger, A., 380
Weissensteiner, T., 513–514
Weissig, H., 507

Welch, J.T., 560–561, 575t
Welch, M.J., 560
Wells, J.A., 48, 50, 52–54, 61, 304, 308–309
Wemmer, D., 22–24, 26–27, 29
Wendoloski, J.J., 542
Weng, T.-C., 136–138
Weng, Z.J., 29–30, 456
Wernimont, A., 54
West, K.P., 159
Westbrook, J., 507
Westerlund, K., 366–367, 380
Westlake, B.C., 510
Wetzel, D., 547
Wheadon, M.J., 3, 10–12
White, E.R., 309–310
White, P.D., 141, 289, 341–342
White, S.H., 383
Whitesides, G.M., 390–391, 542
Widdick, D., 547–548
Wiedman, G., 383
Wieghardt, K., 379, 440, 475
Wiessler, M., 97–99
Wijma, H.J., 192
Wikstrom, M., 114–115
Wilcox, S.K., 506–509
Wilkins, M.R., 158–159
Wilkinson, G., 4
Will, S., 444
Willcott, M.R., 5
William, F., 138–139
Williams, J.C., 180
Williams, R.J.P., 228–229, 378
Williams, S.A., 228
Williams, T., 170–173, 173f, 178–180
Williston, S., 163
Wilm, A., 507
Wilson, C.G.M., 207–213
Wilson, I.A., 305, 308, 396–397
Wilson, M.E., 390–391, 542
Wilson, T.D., 504–505
Wilson, Y.M., 542–545
Wiltschi, B., 252–253
Wimley, W.C., 383
Winbush, S.A.M., 113
Windass, J.D., 48
Windsor, M.A., 192–193
Winkler, J.R., 122, 227, 234–235, 240–242, 418, 502–503, 510

Winkler, M., 252–253
Winn, M.D., 142–144, 243–244
Winter, G., 48, 163, 309–310
Winter, M.B., 561f, 562, 575t
Witt, M., 515
Witte, M.D., 477–478
Wittrup, K.D., 25–26, 28–32, 204
Wlodawer, A., 472–473
Wodak, S.J., 235–236
Wöhnert, J., 561
Woicik, J.C., 595–596
Wojcik, J., 57–58, 204–205
Wold, F., 114
Wolf, C.R., 459–460
Wolfson, H.J., 224–225
Woll, M.G., 252–253, 260
Wollacott, A.M., 410, 504
Wolynes, P.G., 280
Wong, S.E., 503–504
Wood, C.W., 564
Wood, S.W., 542
Woolfson, D.N., 473–474, 563–564
Worth, C.L., 509
Wright, D.A., 218
Wright, P.E., 236
Wright, R.D., 305
Wu, B., 91
Wu, J., 113
Wu, L.F., 547
Wu, L.-Z., 410
Wu, N., 90–91, 111–112
Wu, Q., 252, 265
Wu, S.P., 408
Wu, T., 477–478
Wu, Y., 192–193, 200–201, 305–308, 306–307t, 310–312, 323, 410, 475–476, 493–494
Wu, Y.I., 170–171, 173–174
Wuest, W.M., 309
Wuo, M.G., 308–309
Wurtz, N.R., 284
Wuthrich, K., 265

X

Xiang, J., 281–282
Xiang, Z., 90, 99–100, 101f, 103–104
Xiao, H., 90–93, 99
Xiao, W., 308–309
Xie, F., 13
Xie, J., 255–257, 504–505
Xie, J.M., 111–112, 119–120
Xie, Q.H., 154
Xing, X., 252–255, 256t, 259–260
Xiong, J., 377
Xiong, T., 154
Xu, D., 224–225, 511
Xu, F., 378–379, 395–396
Xu, H.B., 225
Xu, M.Q., 150–151
Xu, T., 475–476
Xu, X., 517
Xue, F., 504–505
Xue, G., 525t

Y

Yadav, M.K., 475–476
Yalow, R.S., 22–24, 26
Yamagishi, Y., 304
Yamaguchi, Y., 252–253
Yamamura, H.I., 36–37
Yamanaka, Y., 162
Yamashita, E., 114–115
Yampolsky, I.V., 122–123
Yan, B., 284
Yan, C., 218
Yan, E.C.Y., 93
Yan, J., 151–153, 164
Yan, R., 511
Yan, X., 151, 152f
Yan, Y.L., 114–115
Yanagisawa, T., 97–99
Yang, C.-T., 335–336
Yang, C.-Y., 304–305, 306–307t, 308
Yang, D., 280
Yang, F., 255–257, 265
Yang, G., 410
Yang, H., 150, 170–171, 540–541
Yang, J.J., 511, 561–562, 575t
Yang, P.Y., 91
Yang, S., 150
Yang, W., 162–163, 561–562, 575t
Yang, X., 150
Yang, X.L., 396–397
Yang, Y., 114–115, 407–408
Yang, Y.J., 265
Yano, N., 4
Yao, N., 308–309
Yao, X., 172–173, 175t, 177

Yaono, R., 114–115
Yaremchuk, A., 113
Yariv, J., 472–473
Ye, M., 562–563
Ye, S., 90, 252–253, 382–383
Ye, S.X., 93
Ye, W., 54
Ye, X., 582, 596
Ye, Y., 252, 265, 507
Yee, C.S., 118–119
Yeh, S.-R., 525*t*
Yeung, B.K., 506–509, 513–514
Yeung, N., 120, 136–138, 502–504, 507–509, 513–515, 517–522, 525*t*
Yeung, Y.A., 28, 30–32
Yi, H., 304–305, 306–307*t*, 308
Yi, J.J., 170–173, 173*f*
Yi, Q., 268
Yikilmaz, E., 504–505
Yin, H., 308–309
Yin, J.A., 128–129
Yin, P., 218
Yoder, N.C., 252–253
Yokoyama, S., 97–99
Yoo, J.S., 544–545
Yoo, P.J., 584
Yoo, T.H., 252–254
Yoon, J.H., 192–193, 200–201, 410, 472–473
Yoshida, M., 418
Yoshida, T., 457–458
Yoshikawa, S., 114–115
Young, A.L., 154
Young, L., 155
Young, T.S., 128–129
Yu, F.T., 218, 472–473, 541
Yu, H., 204, 305–309, 306–307*t*
Yu, J.-X., 252
Yu, T., 410
Yu, X., 170–171
Yu, Y., 110–111, 115–116, 123–124, 123*f*, 127*f*, 392–393, 408, 502–504, 509–510, 519–521, 525*t*, 527
Yu, Y.B., 252–253
Yudenfreund Kravitz, J., 474–475
Yüksel, D., 252–253
Yumerefendi, H., 170–188
Yun, D.S., 584, 596
Yun, H., 252–254, 544–545

Z

Zabinski, J.S., 596
Zaccai, N.R., 564, 566–567
Zaitseva, E., 90
Zajicek, J., 584–585
Zak, S.E., 48, 57–58, 60
Zakeri, B., 151, 158, 164
Zakour, R.A., 66, 70
Zamore, P.D., 218
Zampella, G., 136–138
Zanda, M., 255
Zanghellini, A., 419–420, 504
Zarzycki, J., 398–399
Zastrow, M.L., 136–139, 142–144, 366, 472–473, 504, 541, 570
Zayas, G., 347
Zayner, J.P., 170–173, 173*f*, 175*t*, 177, 180, 180–181*t*
Zeng, H., 280
Zeng, J., 516
Zeng, X.C., 280
Zeng, Y., 410
Zerner, B., 347
Zewail, A.H., 252–253
Zewert, T.E., 504–508
Zhang, D., 214
Zhang, H., 91, 310–312
Zhang, J.-L., 3–4, 441–442, 502–504
Zhang, M., 547
Zhang, P., 596
Zhang, Q.C., 507
Zhang, Q.S., 254
Zhang, T., 257–259
Zhang, W., 115–116, 119, 127*f*, 255–257, 265, 407–408, 509
Zhang, W.-B., 155–156
Zhang, X., 254
Zhang, Y., 50, 53, 68, 70, 72, 473–474, 511
Zhang, Z., 90–91, 252, 265, 380
Zhang, Z.-P., 207–209
Zhang, Z.W., 113, 124
Zhao, X., 218, 504–505, 507–509, 513–515, 518–522, 525*t*, 527
Zhao, Y., 5
Zheng, H., 252–253
Zheng, Q., 367
Zhong, A., 48, 61–62
Zhong, G., 204

Zhou, H.X.X., 419–420
Zhou, J.Z., 91, 116
Zhou, L.Z., 265
Zhou, M., 123–124, 123f
Zhou, Q., 116, 119, 124, 407–408, 519–521, 525t
Zhou, Y., 252–253, 256t, 265
Zhou, Y.-F., 207–209
Zhu, J., 280
Zhu, J.-J., 366
Zhu, M., 150
Zhuang, J., 419–420
Zigoneanu, I.G., 265
Zimm, B.H., 304–305
Zimmerman, S.P., 170–188
Zobnina, V., 335–336
Zoldák, G., 154
Zoltowski, B.D., 180–181t
Zondlo, N.J., 562–563
Zou, H., 383, 408
Zou, Y., 305–308, 306–307t
Zou, Z., 317
Zubrik, A., 154
Zuckermann, R.N., 280
Zufferey, R., 237–238
Zuiderweg, E.R.P., 473–474
Zurkiya, O., 561–562, 575t
Zvyagilskaya, R.A., 116
Zwicker, K., 503

SUBJECT INDEX

Note: Page numbers followed by "*f*" indicate figures, "*t*" indicate tables, and "*s*" indicate schemes.

A

α-Amino acid hybrid oligoamides
  aliphatic N-terminal amine acetylation, 296–297
  characterization, 297–298
  via in situ acid chloride formation, 294–296
Abiotic foldamers, 280
Accelrys Discovery Studio 3.0 (DS3), 479
Acid titrant solution, 349
AlleyCatE process
  AutoDockTools, 197
  design process of, 192–193
  docking experiment, 197, 200
  minimalist design of, 196*f*
  mutations on overall fold, 198–200
  overall approach, 194–195
  positions identification to mutate, 198
  predicting feasibility of catalysis, 200–201
  required tools, 193–194
Allosteric regulation
  catalysis, feasibility, 200–201
  mutations impact, 198–200
  possibility, testing, 197–198
  in protein catalysts, 195
  required tools, 193–194
  starting points for design, 195–196
Alpha-helix
  CC designs, 563–564
  stabilization, 304–308
Amino acid
  CC designs, 563
  natural, 366
  nonnatural, 559, 560*f*
  with phosphine side chain, 402
  polar inclusion, 564
Aminoacyl-tRNA synthetase
  bioorthogonal verification, 126, 127*f*
  chloramphenicol resistance assay, 126
  negative screening, 126
  on PBK plasmid, 125
  positive screening, 126
  unnatural amino acids and, 113
7-Amino-8-fluoro-2-quinolinecarboxylic acid-based oligoamides, 293–294, 293*f*
8-Amino-2-quinolinecarboxylic acid-based oligoamides
  characterization, 297
  cleavage from resin, 292–293
  Fmoc deprotection, 289
  low-loading Wang resin, 286–288
  modified DESC test, 291
  N-Fmoc quinoline carboxylic acid coupling, 290–291
  N-Fmoc quinoline carboxylic acid monomer conversion, 289–290
  N-terminal aromatic amine acetylation, 292
  Wang-bromide resin, 288–289
Antenna systems, 390
Apo-(GRAND-CSL12 D LL16C)$_3$, 142–144, 145*f*
Apoproteins, 447–448
ArMs. *See* Artificial metalloenzymes (ArMs)
Aromatic amine acetylation, N-terminal, 292
Aromatic oligoamide foldamers
  α-amino acids, 283–284
  7-amino-8-fluoro-2-quinolinecarboxylic acid-based oligoamides, 293–294, 293*f*
  characterization, 297
  8-amino-2-quinolinecarboxylic acid-based oligoamides
  characterization, 297
  cleavage from resin, 292–293
  Fmoc deprotection, 289
  low-loading Wang resin, 286–288
  modified DESC test, 291
  N-Fmoc quinoline carboxylic acid coupling, 290–291

639

Aromatic oligoamide foldamers (*Continued*)
  N-Fmoc quinoline carboxylic acid monomer conversion, 289–290
  N-terminal aromatic amine acetylation, 292
  Wang-bromide resin, 288–289
  equipment, 285–286
  fluoroquinoline, 281–282, 282f
  production, 284
  pseudodilution effect, 284
  quinoline/α-amino acid hybrid oligoamides
    aliphatic N-terminal amine acetylation, 296–297
    characterization, 297–298
    via in situ acid chloride formation, 294–296
  quinoline-based, 281f
  reagents, 285–286
  single helical architectures, 280–281
  site-site isolation phenomenon, 284
  SPS protocols, 287f
Artificial cofactor, 546, 552
Artificial enzyme, 392, 409
Artificial metalloenzymes (ArMs), 540–541
  anchoring strategies, 541–542, 541t
  computational design algorithms, 541
  directed evolution, 540–541, 543–544, 553
  periplasmic screening, 544
    activity assay and mutant retrieval, 552–553
    advantages, 544–546
    cofactor uptake and, 552
    compartmentalization strategies, 545f
    cultivation and expression, 551–552
    expression strain/translocation strategy, 546–549
    in vitro library generation, 549–551
    workflow, 549–553, 550f
Artificial metalloporphyrinoids
  design and synthesis, 442–446
  dianionic porphyrinoid ligand, 442–444
  hemeproteins reconstitution with, 446, 448f
    apoproteins preparation, 447–448, 447f
    characterization of reconstituted proteins, 449–450
    construction, 441–442
    protocols for, 448–449
    representative characteristics, 451–453
  monoanionic porphyrinoid ligand, 444–446
  trianionic porphyrinoid ligand, 444
Artificial photosynthesis, 390
Association constants, 420
AutoDockTools-1.5.6 program, 196–197
AutoDock Vina, 197–198
Autotransporters, 546
Avidin–biotin technology, 542

B

Bacterial microcompartments (BMCs)
  crystal structure, 398f
  protein-based, 398–399
Base titrant solution, 348–349
Binary patterning, redox proteins, 367–370
Biocatalysis, 540
Bio-combinatorial methods, 583–584, 583f
Biomimetic studies, 503
Biopanning, 46
Biosynthetic models, heteronuclear metalloenzymes, 505
  functional, 508–509
  nitric oxide reductase, 514f
  protein-based, 504
Bis-alkylation reaction
  dithiol, for locking loop epitopes, 316–317
  investigation, 322–325
  optimizing and troubleshooting, 325–327
  robustness of, 322–324
  thiol (*see* Thiol bis-alkylation)
Blue-light photoreceptors, 170–171
BMCs. *See* Bacterial microcompartments (BMCs)
Boc-SPPS methods, 257
β-*t*-Butylalanine, 263, 264f

C

Ca(II)-binding loops, 560–562
Calmodulin (CaM), 192–193, 195, 198–199
Carbon dioxide reduction, 390–392
Carbon-neutral fuels, 390
C-Asp peptide, 357–358
C*c*P. *See* Cytochrome *c* peroxidase (C*c*P)

Subject Index

Cell-binding assays
　binding theory and relevance, of K_d, 23–25
　ligand depletion
　　compensation for, 29
　　concentration of ligand, 27
　　concentration of receptor, 28
　　effects of, 29f
　　experiments, 30t
　　minimum volume, 28–29, 28t
　　ratio for, 27–28
　　volumetric increase, 28
　measurement
　　advantages, 40–41t
　　comparison, 41–42
　　disadvantages, 40–41t
　　kinetic exclusion assay, 37–38
　　in mammalian cells surface, 35–37
　　SPR, 38–41
　　in yeast surface, 30–35
　　time to equilibrium, 26
ChemDraw, 196
Chemical blotting
　fluorescent chemical blot analysis, 17
　SH3 domain, modification, 13
Chemoselective synthetic strategy, 475
Circular dichroism (CD), 310–311, 449, 513–514
Circular polymerase extension cloning (CPEC), 155
Citrobacter freundii, 118–119
Click chemistry, 124–125, 475–477
Co(TDHC), 444–446, 445–446s
Cobalt porphyrins
　artificial, 409
　intrinsic activity, 404, 406
　usage, 391–392
Cofactor binding equilibria, 426–427
Cofactor binding model, 543f
Cofactor self-assembly, 370–371
Coiled coil (CC) design
　amino acid, 473
　gadolinium, MRI efficiency of, 574–575
　heterotrimeric, 259–260
　hydrophobic core, 568–569
　lanthanide, 562–566, 568f
　location within, 570–571
　sequences of peptide, 565t
　stability types, 571–574

CoilSer (CS) peptide, 138–139
Combinatorial alanine scanning, 48, 61
Computational design algorithms, ArMs, 541
CoMyo, 405–406, 406f
Corrole, 444
Covalent anchoring, 541–542
Cry2, 170–171
Cryptochrome, 170–171
Crystallography data collection, 142, 143t
Cu(I)-catalyzed azide–alkyne cycloaddition (CuAAC), 475–477
Cyclic peptides, 562
　structure, 309
　unbiased libraries of, 308–309
Cys residues, 396–399
Cys-rich site, 138–139
Cysteine, 234
Cysteine alkylation
　peptides libraries, 309–310
　rational design using, 310–311
Cytochrome b562, 48
Cytochrome c, 403f, 404–405
Cytochrome c_{552} (Cyt c_{552}), 457–458
　gene constructs, 459–460
　peroxidase activity, 458–459
　pET-cyt c_{552}, 459–460
　proteins, 460–461
　pyridine hemochrome method, 461
　Tyr45 in, 457–459, 457f, 463
　V49D/M69A mutant, 457–458, 462–465
　　active intermediates, Trp45 effect, 465–468
　　replacement of, 458–459
　　thermal stability, 461–462
Cytochrome cb_{562} variant
　crystallization, 243–244, 244t
　expression and purification, 237–238
　iodoacetamide ligand synthesis, 238–240
　periplasmic extraction procedure, 244–245
　protein labeling, 238–240
　SV-AUC
　　oligomeric state determination by, 240–243, 241f
　　protein-protein affinities with, 243
　in vivo screening, 245–247
Cytochrome c oxidase (CcO), 407–408

Cytochrome c peroxidase (CcP), 505
  heteronuclear metal-binding sites, 505–513
  structural alignment of, 508f

D

D-amino acids, 570, 573
De novo protein design, 366, 562–563
  apo-(CSL9C)$_3$, 140f
  coiled-coil structure, 136–138
  crystallization, 141
  D-amino acids in, 570
  data collection and processing, 142
  DF3 protein, 477
    best candidates for synthesis, 484–486
    candidates generation for click reaction, 483–484
    data analysis, 482–483
    geometrical parameter calculations, 479–482
    linker coordinates, 479
    target protein, 479
  helical bundle, 371–372, 379–381
  hydrophobic residues, 136–138
  iron-sulfur cluster, 397
  lanthanide CC design, 565–566, 569–572
  metal–protein interactions, 136–138
  peptide purification, 141
  peptide sequences, 137t
  porphyrin-binding proteins, 404–409
  protein crystallization, 141–142
  PyMOL visualization, 138f
  refinements, 142–145
  structure determination, 142–145
  synthesis and purification, 140–141
  TRI-family peptides, 138–139
DENV domain III (DIII) protein binding, 61–62
DESC test, coupling completion using, 291, 291f
Dianionic porphyrinoid ligand, 442–444
Dichloromethane, 285
Dihydrofolate reductase (DHFR), 154
Diiron proteins, 472–473
N,N-Diisopropylethylamine, 286
Dimethyl sulfoxide (DMSO), 316
Directed evolution, 410

artificial metalloenzymes, 540–541, 543–544, 553
  protein scaffold, 391
Dithiol bis-alkylation, 316–317
Dithiol-containing amino acids, 311–312
D-Leu residue (DLE), 138–139
DNA sequencing, 513–514
dU-ssDNA synthesis, 66–72
Dynamic scanning calorimetry (DSC), 163
Dynamic scanning fluorimetry (DSF), 163

E

Electrochemical equilibrium, 428–430
Electrochemistry, 431–432
Electrolyte stock solution, 348
Electron paramagnetic resonance (EPR)
  cyt c_{552} V49D/M69A mutant, 462–465
  heteronuclear metalloenzymes, 516–517
Electron-transfer reactions, 409
  flavin cofactor binding and, 380–382
  heme maquettes, 374–376
Electron tunneling, redox cofactor, 366, 375–377, 376f
Electrospray ionization (ESI), 516
Electrospray ionization mass spectrometry (ESI-MS), 158–159, 449–451, 450f
Epitope, 308–309
  dithiol bis-alkylation for locking loop, 316–317
  thiol bis-alkylation and, 311–312
EPR. See Electron paramagnetic resonance (EPR)
Equilibrium constants, 418–419
Equilibrium measurements, 435–436
Escherichia coli, 540–541
  artificial cofactor in, 546
  cyt cb_{562}, periplasmic extraction, 244–245
  expressing soluble Fab, 79–81
  periplasmic screening strategy, 544
ESI. See Electrospray ionization (ESI)
Esterase, 195
Ethanedithiol (EDT), 312–316

F

Fab
  as IgG, 81–84
  soluble Fab, 79–81

[FeFe]-hydrogenases
  azadithiolate bridge, 399
  C. pasteurianum, 399f
  design of, 399–403
  H-cluster, 399–400, 399f
  peptide scaffolds, 400–402
  use of natural sequences, 402–403
Ferredoxin, 392
  iron-sulfur proteins, 393
  P. aerogenes, 394f
Fibronectin type III domain, 48
FIJI software package, 185
Firefly luciferase (FLuc), 162–163
FKBP inhibitor, 9–10
Flavin cofactor, 380–382
FLuc, 162–163
9-Fluorenylmethoxycarbonyl (FMOC), 585–586
Fluorescence-based functional assay
  different UaaRS–tRNA pairs, evaluation, 97–99
  dual-luminescence assay, 93
  to evaluate, 94f
  HEK293 cells transfection
    measurement, 96–97
    translational system, 96–97
  in mammalian cells, 93
  materials, 95–96
    growth medium, 95
    plasmids, 95
    reagents for harvesting cells, 96
    Uaa stock solutions, 95
  sensitivity, 93
  UaaRS/tRNA pairs, evaluation, 94f
    for bioorthogonal chemistry, 97, 98f
    dual-fluorescence assay, 99
    protein labeling, 97–99
    substrate selectivity, 97–99
Fluorescence polarization
  to measure binding affinities in lit/dark states, 176
  measurements, 171–172
Fluorescently labeled peptide, 174–176
Fluorescent protein (FP) assays, 207–209
Fluorinated amino acid
  choice of, 255, 256t
  crystallographic analysis, 263
  designing into proteins, 257–260
  GuHCl-induced protein unfolding
  baselines treatment, 267
  determining ΔG° from, 266–267
  MATLAB Code, 269–272
  thermodynamic parameters, 267–269
  thermodynamic stability, measuring, 265–266
  treatment of baseplanes, 269
  protein stabilization by, 254
  synthesis of, 255–257
  thermodynamic analysis, 261–263, 262f
Fluorinated β-peptide foldamer, 265
Fluorine
  electronegativity, 255
  ^{19}F NMR, 252
  NMR activity, 252–253
Fluorous, 254
Fluoview software, 184
Fmoc chemistry, 313f
Fmoc deprotection, 289
Fmoc-protected peptide synthesis, 257
Foldamers, 280
Four-helix bundles, 395–396, 472–473
  antiparallel, 474f
  asymmetric, 475–476
  binary patterning, 367–370, 368f
  de novo design, 366–367, 408–409
  designing functional proteins, 473–477
  DF3 protein, de novo design, 477
    candidates generation for click reaction, 483–484
    data analysis, 482–483
    evaluation of best candidates for synthesis, 484–486
    geometrical parameter calculations, 479–482
    linker coordinates, 479
    target protein, 479
  hotspot and linker selections, 486–492
  Pra/azido-Lys linker, 488, 490–492, 491–492t, 491t
  single-chain designs, 367–369, 382–383
Frog2 web server, 479, 480f
Fusing protein, 173–174

G

Gadolinium
  dissociation constant, 573–574
  MRI and, 574–575
  peptides and proteins, 575t

Genetic code expansion, 111–112, 111f
Gibbs-Helmholtz equation, 261
Glee freeware tool, 352
GRAND-CoilSer (GRAND-CS), 138–139
GRAND-CSL12$_D$LL16C peptide, 141–142
Guanidinium hydrochloride (GuHCl), 261
  baselines treatment, 267
  determining $\Delta G°$ from, 266–267
  MATLAB Code, 269–272
  thermodynamic parameters, 267–269
  thermodynamic stability, measuring, 265–266
  treatment of baseplanes, 269

H

β-Hairpin-forming peptide, 260
HCO. *See* Heme–copper oxidase (HCO)
Heat, thermodynamic parameters, 267–272
HEK293 cells transfection
  measurement, 96–97
  translational system, 96–97
HEK239F cells, 81–84
Helical bundle
  antiparallel, 374
  de novo designs, 371–372, 379–381
  maquette structures, 369–370
  robust frames, 367
  transmembrane design, 383, 384f
Helical interfaces, 369
4-α-Helix bundles, 257–259, 258f
Heme
  affinity, 430–431, 430f
  binding redox cofactors, 371–374
  maquettes, electron-transfer reactions, 374–376
Heme binding four helix bundle
  electrochemistry, 431–432
  heme affinity, 430–431, 430f
  PCET event, 434–435
  proton competition, 432–434
Heme–copper oxidase (HCO), 503, 506–508
  Mb-based, 519–521, 527
  and NORs, 507, 517–518, 522
  Tyr in, 509

Heme electrochemistry
  and affinity, 422–424
  influence of protons, 424–425
  and protein folding, 426
Hemeproteins, 440
  with Fischer nomenclature, 440f
  reconstitution with artificial metalloporphyrinoids, 441f, 446, 448f
  apoproteins preparation, 447–448, 447f
  characterization of reconstituted proteins, 449–450
  construction, 441–442
  protocols for, 448–449
  representative characteristics, 451–453
Heptad repeat, 257–259
Heterodimeric four-helix bundle, 472–473
  antiparallel, 474f
  asymmetric, 475–476
  designing functional proteins, 473–477
  DF3 protein, de novo design, 477
  candidates generation for click reaction, 483–484
  data analysis, 482–483
  evaluation of best candidates for synthesis, 484–486
  geometrical parameter calculations, 479–482
  linker coordinates, 479
  target protein, 479
  hotspot and linker selections, 486–492
  Pra/azido-Lys linker, 488, 490–492, 491–492t, 491t
Heteronuclear metalloenzymes, 502–503
  activity assays, 519–521
  biological reactions in, 510
  biosynthetic models, 504–505, 508–509, 514f
  case studies, 527–528
  computational analysis of design, 510–513
  computational design, 505–513
  EPR, 516–517
  examples of, 502t
  functional characterization of, 518–526
  kinetic and mechanistic studies, 522–524
  mass spectrometry, 516
  rapid freeze-quench, 524–526
  reaction mechanism, 524–526
  redox potential measurement, 521–522

secondary sphere interactions, 504–505, 509
structure-guided design, of primary ligands, 506–508
tuning heme reduction potential, 527
UV–vis spectroscopy, 514–516
X-ray crystallography, 517–518
Hexafluoroleucine (hFLeu), 255, 256t
High-energy X-ray diffraction (HE-XRD), 586–587
  data acquisition
    detector distance, 593
    exposures, 594
    sample preparation, 593
    setting acquisition times, 593–594
  pattern conversion
    calculating PDFs, 594–595
    HE-XRD integration, 594
^{1}H NMR spectrum, 298–299f
Horse radish peroxidase (HRP)., 451–452
Hotspot residues, protein–protein interfaces, 61–66
HP153 plasmid, 54
Human growth hormone, 48, 61
Humanized antibody, 60–61
Hybrid coordination motif (HCM), 230–231
Hydrogenases, 390–391
Hydrogen-bond surrogate approach, 305
Hydrogen peroxide (H_2O_2), 456–458
Hydrogen production
  catalysis of, 409
  CoMP11 catalyzes, 404–405
  light-driven, 403
  microperoxidase-11 and, 404–405
  photocatalytic, 401–402
  porphyrin redox sites, 404–409
Hydrolysis, of pNPP, 195s
Hydrophobic core, lanthanide, 566–567
Hydrophobic effect, 254
Hydroxylase, reconstituted myoglobin, 452
HyperQuad suite, 353–354

I

IgG, Fab as, 81–84
Improved light-induced dimer (iLID) photoswitch, 178–180, 185
Inductively coupled plasma optical emission spectrometry (ICP-OES), 552

Iron corrole complex (FeCor), 444, 445s, 452
Iron porphycene complex (FePc), 442–444, 443s
Iron-sulfur clusters
  cofactors, 378–379
  computational design of model proteins, 395–396
  design of, 392–399
  DSD family, 396–398, 397f
  incorporation into prestructured proteins, 396–399
  use of natural sequences, 393–395
Isopeptide-mediated enzyme cyclization
  characterization
    DSC, 163
    FLuc, 162–163
    industrial relevance, of phytase, 159
    measure aggregation, 159–160
    on thermal resilience, 159–162, 161f
  ESI-MS, 158–159
  gel analysis of, 157f
  IPTG, 155–156
  isopeptide bond formation, 151–152
  with isopeptide bonds, 153
  SDS–PAGE, 158
  selection, 153–154
  Spy0128, 151
  SpyRing cassette, 154–155, 156f, 158
  SpyTag and SpyCatcher for, 151–152, 152f
Isopropyl β-D-thiogalactopyranoside (IPTG), 155–156
Iterative saturation mutagenesis (ISM), 549, 553

J

Jα-helix, 170–176, 173f

K

Kinetic evaluation, 436
Kinetic exclusion assay, 37–38
Kunkel mutagenesis, 66–72, 67f

L

β-Lactamase activity, 245–247
Lanthanide, 558
  Ca(II)-binding loops, 560–562
  CC design, 562–566, 568f

Lanthanide (*Continued*)
  controlling water coordination and access, 568–571
  coordination chemistry, 558
  gadolinium dissociation constant, 573–574
  at helical interface, 565–566
  within hydrophobic core, 566–567
  imaging agents, 558–559
  luminescence imaging agents, 558–559
  MD simulation, 567
  nonnatural amino acids for, 559, 560*f*
  peptide folding, 572–573
  peptides and proteins, 560–562
  proteolytic stability, 573
  selectivity, 571
  sensitization, 558–559, 560*f*
  stability types, 571–574
  steric bulk, removing of, 569–570
Lanthanide-binding tags (LBTs), 561, 561*f*
LED, LOV2-based photoswitches, 187–188
Library-based screening, 177–180, 178*f*
Light-activatable systems, 170
Light-induced dimerization, yeast transcription, 187
LOV2-based photoswitches
  caging affect, 175*t*
  design of, 173*f*
  domain, 170–171
  efficiency, 177
  engineering process, 171–180
  equipment, 182*f*
  functional assays, 187–188
  image analysis, 185–186, 186*f*
  improving initial switches by mutations, 177
  LED, 187–188
  phage display, 178–180, 178*f*
  population-based assays, 181
  rational engineering, 172–174
    embedding peptides, 172–173
    fusing protein domains, 173–174
  single-cell microscopy, 181–185
  timescales of, 180
  tuning half-life, using mutations, 180–181*t*
  validation, 174–176, 175*f*

Luminescence
  imaging agents, 558–559
  and MRI, 559

M

Macrocyclization reaction, 304–305, 308–309
  lactam formation, 324*f*
  scope and versatility, 317–321
Magnetic resonance imaging (MRI)
  gadolinium CCs, 559, 574–575
  lanthanide luminescence and, 559
MALDI mass spectrometry, 14, 16*f*, 322–324
Maltose-binding protein (MBP), 151
Mammalian cell-binding assays
  competition binding, 36–37, 37*f*
  direct binding, 35–36
Manganese peroxidases (MnP), 503–504, 506–508
  crystal structures, 503*f*
  MnCcP, 509–510, 527–528
  structural alignment, 507, 508*f*
Manganese porphycene (MnPc), 442–444
Mass spectrometry, heteronuclear metalloenzymes, 516
Matcher program, 508
MDPSA. *See* Metal-directed protein self-assembly (MDPSA)
MD simulation, lanthanide, 567
Membrane protein design, 383
β-Mercaptoethanol ligands, 393–395
Metal-binding motifs, MDPSA using, 228–230
Metal cation stock solutions, 347–348
Metal-chelating amino acid, 120–122
Metal-coordinating ligands, 230–231
Metal-directed protein self-assembly (MDPSA)
  application, 229*f*
  protein building block for, 226*f*, 227
  using metal chelating motifs, 228–230
  using synthetic metal-coordinating ligands, 230–231
  via hybrid coordination motifs, 231*f*
Metal-ion affinity, 420–422
Metal-ion binding constants, 419

Subject Index 647

Metal ion coordination
  β-turns, 334–335
  cyclic peptides, 335–336
  cyclization, 335–336
  design, short preorganized peptide sequences
    His residue, 337–338
    peptidic scaffolds, 337–339
    systematic modifications, 338–339, 338f
    tolerance, 337
  dPro-Pro unit, 334–335
  peptides, 334–335
  potentiometric titrations
    calibration, of pH electrode, 352
    complexation titration, 353
    complex formation, 350
    concentration of analyte, 352
    data treatment, 353–356, 355t, 356f
    metallopeptide system, 349–350, 356
    peptidic scaffold, 350–351
    pH, 351
    protonation titration, 353
    water leading, self-ionization of, 351
  preorganized peptidic scaffolds design, 340–341
  RAFT prototype, 335–336, 336f
  spectroscopy, characterization, 356–359
Metallopeptide S2E^{Rh}, 10–11
Metalloproteins, 558
  association constants, 420
  essential equilibrium measurements
    cofactor binding equilibria, 426–427
    electrochemical equilibrium, 428–430
    protonation equilibrium, 427–428
  experimental procedures
    equilibrium measurements, 435–436
    kinetic evaluation, 436
    protein synthesis and purification, 435
  heme binding four helix bundle
    electrochemistry, 431–432
    heme affinity, 430–431
    PCET event, 434–435
    proton competition, 432–434
  heme electrochemistry
    and affinity, 422–424
    influence of protons, 424–425
    and protein folding, 426

  metal-ion affinity, 420–422
  protein design, 419–420
  stability constants, 420
Metal-templated interface redesign (MeTIR)
  application, 233f
  ^{C96}RIDC1$_4$, 235–236
  directed evolution, 235f
  disulfide cross-links, enhancing scaffold robustness, 233–234
  noncovalent interfaces redesign with, 231–233
  protein building block by, 226f
  in vivo assembly and functional screening, 234–236
Methionine synthase model, 452–453
MeTIR. *See* Metal-templated interface redesign (MeTIR)
Met69 with alanine (M69A) mutant, Cyt c_{552}, 457–458, 462–465
  active intermediates, Trp45 effect, 465–468
  replacement of, 458–459
  thermal stability, 461–462
  Trp45 effect, 465–468
Microperoxidase-11 (MP11), 404–405, 405f
M13KO7 helper phage, 52–53
MnP. *See* Manganese peroxidases (MnP)
Molecular dynamics simulation, 511
Molecular velcro technology, 542
Monoanionic porphyrinoid ligand, 444–446
Monoclonal phage ELISA, 76–78
MP11. *See* Microperoxidase-11 (MP11)
Murine SUDV antibody, 60–61
Myoglobin, reconstitution of
  with Co(TDHC), 452–453
  hydroxylase, 452
  peroxidase, 451–452

N

Nanoparticle (NP) production
  peptide-based approaches for, 584
  peptides, advantages and limitations, 584–585
NATRO command, 199–200
Nernst equation, 422–423
N-Fmoc quinoline carboxylic acid
  to corresponding acid chloride, 289–290
  coupling to resin bound amine, 290–291

Nitric oxide reductase (NOR)
  biosynthetic models, 514f
  Glu in, 509
  HCO and, 507, 517–518, 522
  Mb-based, 526
  X-ray crystal structure, 507
Nonhelical structures, stabilization, 308–309
Nonnatural amino acids, 559, 560f
Nonpolar cofactors, 370–371
NOR. See Nitric oxide reductase (NOR)
Nuclear magnetic resonance (NMR)
  spectroscopy, 369–370, 561
Nucleophiles, 322–324
  in macrocyclization, 324–325, 324f

O
Olefin metathesis, 305
Oligomerization, 393–396
Oligonucleotides, 72
Optogenetics, 170
Organometallic catalysts, 390–391

P
PA-Rac, 173–174
PDF analysis
  HE-XRD data acquisition, 593–594
  HE-XRD pattern conversion, 594–595
  materials for, 586–587
  sample preparation, 592
  structural modeling, 595–596
Pentafluorophenylalanine (pFPhe), 256t
Peptide-based approaches, 584
Peptide-based therapeutics, 304
Peptide-capped NP synthesis
  catalytic application, 591f
  characterization of, 592
  materials for, 586
  protocols for, 591–592
  solutions required for, 590–591
Peptides
  advantages and limitations, 584–585
  chemical modification of, 304
  drug limitations, 304
  epitope, 308–309, 311–312, 316–317
  folding, 572–573
  gadolinium, 575t
  intramolecular cross-linking, 306–307t
  lanthanide, 560–562

  with material affinity, 583–584
  for NP production, 584
  purification, 141
    materials for, 586
  synthesis protocols, 312–316
  thiol bis-alkylation to constrain, 309–312
Peptide scaffolds. See also Metal ion
    coordination
  design, 340–341
  [FeFe]-hydrogenases, 400–402
  Porphyrin-binding proteins, 408–409
  stock solutions preparation
    Cys-containing scaffolds, 346–347
    scaffolds with No Trp, Tyr, and Cys, 347
    Trp/Tyr/disulfide-containing
      scaffolds, 345–346
  synthesis and characterization
    cyclic synthesis, 344–345
    Fmoc protocols, 341–342
    linear synthesis, 342–344
Peptide synthesis
  cleavage from resin, 589
  materials for, 585–586
  protocols for automated synthesis, 588
  reverse-phase HPLC, purification, 589–590
  solutions required for, 587
Perfluoro-t-butylhomoserine (pFtBSer), 256t
Periplasmic screening
  advantages of, 544–546
  artificial metalloenzymes, 545f
    activity assay and mutant retrieval, 552–553
    cofactor uptake and, 552
    cultivation and expression, 551–552
    expression strain/translocation strategy, 546–549
    in vitro library generation, 549–551
    workflow, 549–553, 550f
  Escherichia coli, 544
Peroxidase
  benefits of, 456
  Cyt c_{552}, 457–459
    gene constructs, 459–460
    proteins, 460–461
    V49D/M69A mutant, 461–468
  mechanistic aspects of, 456–457
  reconstituted myoglobin, 451–452

Phage display, 46, 178–180, 178f
  biopanning, 46
  cell lines, 50–52
  diversity, 46
  energetics of protein, 48
  equipment, 48–49
  library
    expressing soluble Fab, 79–81
    Fab as IgG, 81–84
    monoclonal ELISA, 76–78
    screening, 75–78
    selections, 73–75
    troubleshooting, 84–85
  library design
    amplifying and tittering phage, 64–65
    cloning into HP153, 62–64
    ELISA, 57, 58f, 78f
    humanization of antibodies, 60–61
    mapping of hotspot residues, 61–66
    phage ELISA, 65–66
    protein of interest, 57–58
    protein scaffold, 57–58
    randomization scheme, 57–59
  library production
    dU-ssDNA synthesis, 68–70
    Kunkel mutagenesis, 66–72, 67f
    primer design, 72
    quality control, 72–73
    replication, 66–68
  materials binding affinity, 583–584
  MC1061, 50
  M13KO7 helper phage, 52–53
  phagemid considerations, 53–54
  reagents, 54–57
  selection, 47f
  SS320 cells, 50–51
  stages of, 78–79
  utility of, 48
  XL1-Blue cells, 51–52
Phagemids, 53–54
Photoreceptors, 170–171
Photoswitches, 170
  drawbacks, 170
  efficiency of, 177
  LOV2-based (see LOV2-based photoswitches)

Phytochrome B (PhyB), 170
Phytochrome-interacting family (PIF) proteins, 170
PIKAA command, 199–200
Pilin-C, 151
Plasmid-encoding TPL, 118
Plasmid preparation, 128–129
pNPP model, 194, 196–198
  hydrolysis of, 195s
  relative poses, 200–201
Polar hydrophobic effect, 254
Polymerase chain reaction (PCR)
  amplification, 459–460
  library construction using, 208t
  semirational protein gene libraries using, 208f
Porphycene, 442–444
Porphyrin-binding proteins
  design of, 404–409
  peptide scaffolds, 408–409
  redesign of natural scaffolds, 404–408
Porphyrinoid ligand, 442f
  dianionic, 442–444
  monoanionic, 444–446
  trianionic, 444
Potentiometric titrations
  calibration, of pH electrode, 352
  complexation titration, 353
  complex formation, 350
  concentration of analyte, 352
  data treatment, 353–356, 355t, 356f
  metallopeptide system, 349–350, 356
  peptidic scaffold, 350–351
  pH, 351
  protonation titration, 353
  water leading, self-ionization of, 351
PPIs. See Protein–protein interactions (PPIs)
Pra/azido-Lys linker, 488, 490–492, 491–492t, 491t
Protein building block, for MDPSA, 226f, 227
Protein cofactors, 366
Protein complementation assays, 207–209, 216
Protein crystallization, 141–142
Protein cyclization, 150–151
Protein data bank (PDB), 507

Protein design
 de novo (see De novo protein design)
 minimalist approach to, 192–193
 unnatural amino acid design, 127–128
Protein expression, 129
Protein folding, 426
Protein interactions, 22–23
Protein of interest (POI), 90, 92
Protein–protein interactions (PPIs), 224–225, 308–309
 charge–charge interactions, 207
 cloning, 216
 design approaches, 204, 205f
 expression, 216
 hydrophobic interactions, 207
 incorporating redesign, 218
 modular design, 218, 219f
 purification, 216
 qualitative measurements, 216–217
 quantifying, split FPs, 212–213
 rational protein design, 204–205
 in Saccharomyces cerevisiae, 204
 screening in E. coli, 210–211f, 212
 selection using split FPs, 213–216
 semirational design, 206f
 semirational protein gene libraries, 208f
 split fluorescent protein assays, 207–216
 structure-guided rational design, 206–207, 206f
 in vitro, 217–218
Protein–protein interfaces, 61–66
Proteins
 gadolinium, 575t
 lanthanide, 560–562
 stabilization, 254
Protein scaffolds, 504–505
 directed evolution, 391
 residues in, 507
 visual inspection of, 507
Protein synthesis and purification, 435
Proteolytic stability, 573
Protonation equilibrium, 427–428
Proton competition, 432–434
Proton-coupled electron transfer (PCET) event, 434–435
Pulse electron-electron double-resonance technique, 397

PyMOL, 195–196
 docking structures, 198
 GUI, 489
 scripts, 479–480, 488
 visualization, 138f
 wizard/measurement tool, 200–201
Pyridine hemochrome method, 461
Pyrrolizine (Pyl), 90–91

Q

Quinoline
 aliphatic N-terminal amine acetylation, 296–297
 characterization, 297–298
 via in situ acid chloride formation, 294–296

R

Radioligand labeling, 22–23
Random mutagenesis, peroxidases, 456
Rapid freeze-quench (RFQ), 524–526
Reagents
 antibiotics, 55
 buffers, 55
 media, 54–55
 solutions, 55
Recombinant proteins, 460–461
Red-light photoreceptor, 170
Redox active metalloenzymes
 [FeFe]-hydrogenases mimics, 399–403
 iron-sulfur clusters, 392–399
 porphyrin redox sites, 404–409
Redox cofactors, 366–367
 electron tunneling, 366, 375–377, 376f
 flavin cofactor binding and electron transfer, 380–382
 heme binding, 371–374
 iron–sulfur cluster, 378–379
 maquette designs into membranes, 382–384
 metal binding sites, 379
 ZN tetrapyrrole, 377
Redox mediators, 122–124, 123f
Resin bound amine, 290–291
Resins, 285
Reverse Monte Carlo (RMC) simulations, 595–596

Reverse-phase high-performance liquid
  chromatography (RP-HPLC),
  312–317, 589–590
Rh_2 $(OAc)_3$ $(tfa)_1$
  preparation, 7–8
  synthesis, 4–5, 6f
Rhodium(II) conjugates
  modification
    amino acid side chain, 3
    chemical blotting, 13
    of diazo, 2–3, 2f
    imaging, 12, 14
    protein, 2, 12
    protein transfer, 13
    SDS-PAGE, 13
    of SH3 domain, 14–15, 16f
  preparation
    aqueous solution, 10–11
    in organic solution, 8–10
    Rh_2 $(OAc)_3$ $(tfa)_1$, 7–8
  synthesis
    heteroleptic complexes, 4–5
    materials, 5
    metalation method, 5
    paddlewheel structure, 4, 4f
    Rh_2 $(OAc)_3$ $(tfa)_1$, 4–5, 6f
    tetracarboxylate complexes, 4
  traditional equilibration methods, 3
  visualization, alkyne-tagged protein
    chemical blotting protocol, 15–17, 16f
    fluorescent chemical blot analysis of,
      17–18
Ring-closing olefin metathesis, 305–308
Rink amide resin, 312–316
RMSD, 478, 484, 486t, 492
RosettaCommons, 193
Rosetta score, 193, 198–200
RP-HPLC chromatogram, 297, 297f
Ru(II)-catalyzed azide–alkyne cycloaddition
  (RuAAC), 475–476
Ruthenium, 546

S

SDS-polyacrylamide gel electrophoresis
  (SDS-PAGE), 13
  fluorescent chemical blot analysis, 17
  SH3 domain, modification, 13

Sedimentation velocity analytical
  ultracentrifugation (SV-AUC)
  oligomeric state determination by,
    240–243, 241f
  protein-protein affinities with, 243
SH3 domain, modification
  biophysical methods, 11–12
  chemical blotting, 13
  imaging, 12, 14
  in mammalian cell lysate, 14–15
  protein, 12
  protein transfer, 13
  SDS-PAGE, 13
Single-cell microscopy, 181–185
Solid-phase peptide synthesis (SPPS),
  255–257, 312–316, 313f, 391, 402
Soluble Fab, 79–81
Sperm whale myoglobin (swMb), 505–508,
  519–521
Split FPs
  expression and quantification, 213
  screening PPIs
    cloning, 209–211
    in E. coli, 212
    expression, 211–212
    quantifying PPIs, 212–213
  selecting for PPIs
    cloning, 214
    selection, 214–216
Split mCherry, 207–209, 213
SpyRing cyclization
  characterization
    DSC, 163
    FLuc, 162–163
    industrial relevance of phytase, 159
    measure aggregation, 159–160, 161f
    on thermal resilience, 159–162, 161f
  cloning strategy to, 154–155, 156f
  gel analysis of, 157f
  with isopeptide bonds, 153
  selection, 153–154
Stability constants, 420
Streptavidin–biotin technology, 544
Structure-guided rational design, 206–207,
  206f
SUDV murine antibody, 60–61
Surface plasmon resonance (SPR),
  38–41

Synthetic metal-coordinating ligands, 230–231
Syringe pump ESI, 516

T

Teale's 2-butanone method, 447
Template assembled synthetic proteins (TASP), 379–380, 475
Tetrahydrofuran (THF), 286
Tetratricopeptide repeat affinity proteins (TRAPs), 206–207
Thermal unfolding, 261
Thermus thermophilus, 457–460
Thioalkalivibrio nitratireducens cytochrome *c* nitrite reductase (TvNiR), 116
Thioether ligation, 305–308
Thiol bis-alkylation, 309
    to constrain peptides, 309–312
    crude linear peptides, 315f
    cysteine alkylation, 309–311
    linkers for, 310f
    monitoring progression of, 317
    peptide synthesis/cross-linking, 312–316
Three-stranded coiled coil (3SCC), 136–139, 138f
Tissue culture, 187–188
Transition metal ions, 225
Trianionic porphyrinoid ligand, 444
Trifluoroethylglycine (tFeG), 256t
Trifluoroisoleucine (tFIle), 256t
Trifluoroleucine (tFLeu), 256t
Trifluoromethylmethionine (tFMet), 256t
Trifluoromethyl-phenylalanine (tFmPhe), 256t
Trifluorovaline (tFVal), 256t
Tris-(bromomethyl)benzene, 309–310
tRNA, 90
Twin-arginine translocation (TAT) pathway, 547–548
Tyrosine, 124

U

Ullman coupling, 442–444
Unnatural amino acids (Uaas) design, 112–116
    aminoacyl-tRNA synthetase, 125–127
    catalytic structures, 110
    cell's membrane, 113
    click chemistry reaction reagents, 124–125
    crystal structure, 115f
    enzyme design with, 112
    genetic code expansion, 111–112, 111f
    incorporation, 91–92, 110
        and cross-linking events, 103–104
        fluorescence-based functional assay, 92–99
        into membrane proteins, 99–104
        transfection, of HEK293T cells and cross-linking procedure, 102–103
    metal-chelating amino acid, 120–122
    plasmid preparation, 128–129
    POI, 92
    protein design, 127–128
    protein expression, 129
    redox mediators, 122–124, 123f
    significance, 110–111
    synthetic chemistry-guided, 116–117
    synthetic methods for, 117–125, 118f
    toolkit, 119–120, 121f
    X-ray diffraction, 114–115
UV–vis spectrophotometry, 315
UV–vis spectroscopy
    cytochrome c_{552}, 461–463
    heteronuclear metalloenzymes, 514–516
    reconstituted hemeproteins, 448–449

V

Val49 with aspartic acid (V49D) mutant, Cyt c_{552}, 457–458, 462–465
    peroxidase reaction of, 462–465
    replacement of, 458–459
    thermal stability, 461–462
    Trp45 effect, 465–468
Vilsmeier–Haack reaction, 442–444

W

Wang resin
    bromination of low-loading, 286–288
    loading, 288–289

X

X-ray crystallography
  heteronuclear metalloenzymes, 517–518
  structure analysis, 449–451

Y

YADS1, 60–61
Yeast surface display (YSD)
  materials, 32–33
  method, 32
  performing characterization of, 30–31
  schematic of, 31f

Z

Zimm–Bragg model, 304–305
ZN tetrapyrrole, 377

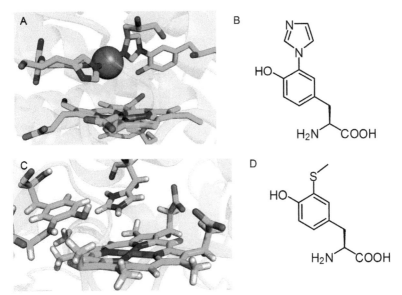

C. Hu and J. Wang, Fig. 2 A scheme for natural inspired unnatural amino acids design. (A) Crystal structure of the Cu_B site of bovine cytochrome c oxidase. (B) The unnatural amino acid imiTyr, with a 3-imidazole phenol group inspired by the natural tyrosine–histidine cross-link. (C) Structure of the catalytic center of *T. nitratireducens* cytochrome c nitrite reductase. (D) The unnatural amino acid 3-methylthio tyrosine inspired by the natural cross-link between tyrosine and cysteine.

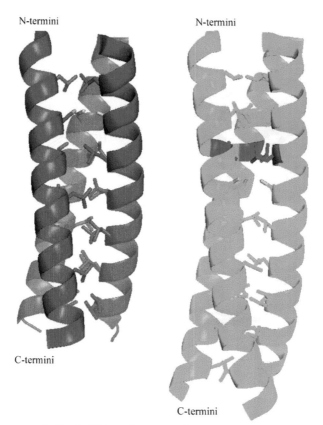

L. Ruckthong et al., Fig. 2 Ribbon diagrams of apo-(CSL9C)$_3$ (*left*) and apo-(GRAND-CSL12$_D$LL16C)$_3$ (*right*) representing the difference in length of the two peptides. The apo-(CSL9C)$_3$ contains four heptad repeats while an additional heptad is added in the apo-(GRAND-CSL12$_D$LL16C)$_3$ subsequently resulting in a longer trimer. Helical core residues are shown as *sticks*. D-Leu in the 12th position of apo-(GRAND-CSL12$_D$LL16C)$_3$ is *red*. Thiols of Cys residues are *yellow*.

L. Ruckthong et al., Fig. 3 Side view of an overlay between the trimeric apo-(GRAND-CSL12$_D$LL16C)$_3$ and apo-(CSL9C)$_3$ structures indicating that there is no change in the overall secondary structures of GRAND-CS and CS derivatives. Moreover, the alignment demonstrates that the incorporation of a nonnatural amino acid, D-Leu, does not perturb the secondary structure of the three-stranded coiled coil. Shown in *green* is the apo-(GRAND-CSL12$_D$LL16C)$_3$ and in *pink* is apo-(CSL9C)$_3$. Side chain residues are omitted for clarity.

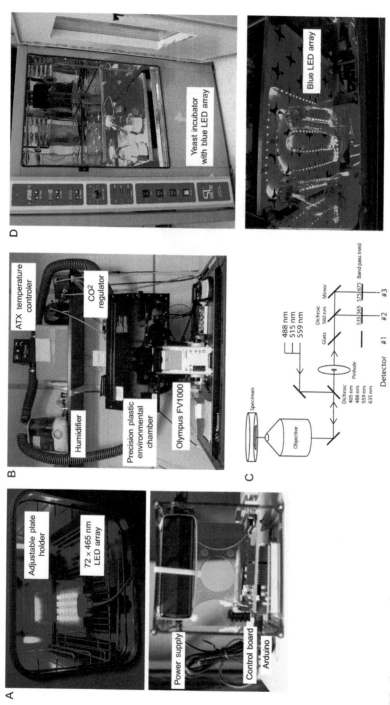

S.P. Zimmerman et al., Fig. 4 Equipment used in the validation and application of LOV2-based photoswitches. (A) Adjustable and programmable tissue culture dish illuminator. *Top*—image of 72 × 465 nm LED array placed in incubator for cell culture experiments. The plate holder can be adjusted to control the distance from the LED array. *Bottom*—image of Arduino-based controller, which allows for programming of LED array brightness and timing. (B) Image of Olympus FV1000 with equipment necessary for imaging and activation of LOV2-based photoswitches in live cells. (C) Diagram of the light path used to image and activate LOV2-based photoswitches on a confocal microscope. (D) Image of yeast incubator outfitted with a blue LED array for growth under activating conditions.

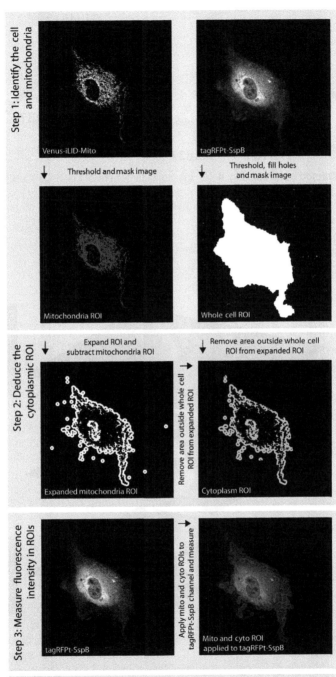

S.P. Zimmerman et al., Fig. 5 Image analysis schematic for quantification of mitochondrial/cytoplasmic fluorescence intensity.

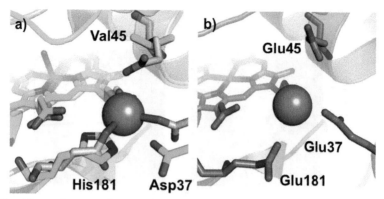

A. Bhagi-Damodaran et al., Fig. 4 (A) Structural alignment of CcP (PDB ID: 2CYP, *green*) and MnP (PDB ID: 1MNP, *gray*) indicates the possible position of primary ligands to be designed in CcP for Mn-binding site. (B) Final MnCcP.1 model binds Mn(II) (PDB ID: 5D6M, *orange*).

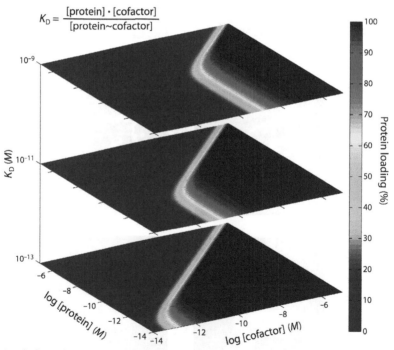

M. Jeschek et al., Fig. 1 Model for cofactor binding. Protein loading (ie, percentage of binding sites occupied by the cofactor) is derived from the law of mass action and displayed as a function of the concentration of the protein and the cofactor as well as the dissociation constant K_D. For high affinities (low K_D) between the cofactor and the protein as for instance for the streptavidin–biotin interaction ($K_D \sim 10^{-13}$ M) quantitative occupation (>99%) of the binding sites occurs even at subnanomolar concentrations for both binding partners.

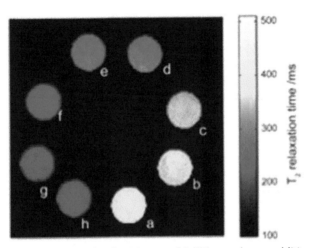

A.F.A. Peacock, Fig. 6 T_2 weighted MR image of Gd(III) samples on addition of increasing concentrations of MB1–2, and formation of Gd(MB1–2)$_3$. *Reprinted from Berwick, M. R., Lewis, D. J., Jones, A. W., Parslow, R. A., Dafforn, T. R., Cooper, H. J., et al. (2014). De novo design of Ln(III) coiled coils for imaging applications.* Journal of the American Chemical Society, 136, 1166–1169.

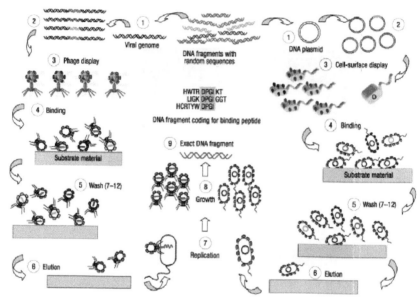

N.M. Bedford et al., Fig. 1 Scheme of biocombinatorial selection approaches for the isolation of peptides with affinity to inorganic materials. The method for phage display is shown on the *left*, while the process for cell surface display is presented on the *right*. Adapted with permission from Sarikaya, M., Tamerler, C., Jen, A. K. Y., Schulten, K., & Baneyx, F. (2003). Molecular biomimetics: Nanotechnology through biology. Nature Materials, 2, 577–585; copyright 2003, Nature Publishing Group.

Edwards Brothers Malloy
Ann Arbor MI. USA
September 6, 2016